WITHDRAWN BY THE
UNIVERSITY OF MICHIGAN

Natural resources research XX

Other titles in the series
Natural resources research

I. A review of the natural resources of the African continent
II. Bibliography of African hydrology
III. Geological map of Africa with explanatory note
IV. Review of research on laterites
V. Functioning of terrestrial ecosystems at the primary production level. Proceedings of the Copenhagen Symposium
VI. Aerial surveys and integrated studies. Proceedings of the Toulouse Conference
VII. Agroclimatological methods. Proceedings of the Reading Symposium
VIII. Proceedings of the Symposium on the Granites of West Africa: Ivory Coast, Nigeria, Cameroon, March 1965
IX. Soil biology. Reviews of research
X. Use and conservation of the biosphere
XI. Soils and tropical weathering. Proceedings of the Bandung Symposium, 16 to 23 November 1969
XII. Natural resources of humid tropical Asia
XIII. Computer handling of geographical data
XIV. Tropical forest ecosystems. A state-of-knowledge report
XV. Review of research on salt-affected soils
XVI. Tropical grazing land ecosystems. A state-of-knowledge report
XVII. Vegetation map of South America: explanatory notes
XVIII. Case studies on desertification
XIX. Solar electricity

The vegetation of Africa

A descriptive memoir
to accompany the Unesco/AETFAT/UNSO
vegetation map of Africa

by F. White

Unesco

Published in 1983 by the United Nations
Educational, Scientific and Cultural Organization,
7 Place de Fontenoy, 75700 Paris
Printed by Courvoisier S.A., 2300 La Chaux-de-Fonds

The designations employed and the presentation of the material
in this publication do not imply the expression of any opinion
whatsoever on the part of Unesco, AETFAT or UNSO
concerning the legal status of any country or territory, or of its
authorities, or concerning the delimitations of the frontiers of
any country or territory.

ISBN 92-3-101955-4

© Unesco 1983
Printed in Switzerland

Preface

The new vegetation map of Africa has been compiled by the Vegetation Map Committee of the Association pour l'Étude Taxonomique de la Flore de l'Afrique Tropicale (AETFAT) in collaboration with Unesco and the United Nations Sudano-Sahelian Office (UNSO). It comprises three map sheets at a scale of 1:5 000 000, a legend, and the present accompanying memoir.

An earlier AETFAT *Vegetation Map of Africa South of the Sahara* was published in 1958 with financial support from Unesco. It soon went out of print, and at the fifth plenary meeting of AETFAT held in Genoa and Florence in 1963 a small committee was asked to look into the possibility of preparing a new edition. At about this time, Unesco had convened a Standing Committee on Classification and Mapping of Vegetation on a World Basis, and had initiated a programme of mapping the world's vegetation at a scale of 1:5 000 000. In this connection, AETFAT was invited to participate in the preparation of a new and more detailed vegetation map of the whole of Africa as part of the world series.

It was originally intended that all the maps in this series should use a uniform legend and colour scheme, but, because of the complexity of the subject-matter and the diversity of approaches, this objective has not been fully achieved. Hence, the map of Africa differs in some important respects from the Unesco–FAO *Vegetation Map of the Mediterranean Basin* and the Unesco *Vegetation Map of South America*. The classification used for the African map also diverges in some respects from that recommended in the Unesco publication *International Classification and Mapping of Vegetation* (Ecology and Conservation Series No. 6, 1973). It is based almost entirely on physiognomy and floristic composition of the vegetation, and not on climate, although a few comparative climatic terms such as moist and dry are occasionally used in the designation of the mapping units. Otherwise, vegetation and climate are dealt with separately, and separate climatic maps are given in the text for each of the major phytogeographic regions.

A feature of the Unesco/AETFAT/UNSO *Vegetation Map of Africa* is that, in the legend, the mapping units are grouped in the traditional manner according to physiognomy, whereas in the text, here, they are grouped according to the floristic regions in which they occur. There are thus two interconnected classifications, which can be used independently but are fully cross-referenced. The legend permits easy comparison of African vegetation with that of other continents, whereas in the text it is possible to deal effectively with complicated spatial and dynamic relationships.

While the legend of the map is composite (English and French), this accompanying memoir, because of its length, has been prepared in separate English and French versions. The memoir aims to provide a succinct though comprehensive account of the vegetation of the African mainland, Madagascar and the other offshore islands. Brief introductory chapters deal with geology, climate, soils, animals, fire, land use and conservation. Their purpose is merely to provide an entrée to the specialist literature and to introduce important topics which recur elsewhere in the main text.

The vegetation of the main floristic regions is described individually in twenty-two chapters, which comprise the greater part of the text. For each region the salient features of the flora, geology and climate are also described, and a black-and-white map is provided. The latter illustrates topographic features mentioned in the text and summarizes the regional climate by means of climatic diagrams. For each of the main vegetation types, references to source materials and other important publications, published photographs and profile diagrams (if available), as well as major synonymy, are given.

Publication of the *Vegetation Map of Africa* forms part of Unesco's long-term programme for the synthesis and diffusion of information on natural resources. The map thus complements other maps such as the vegetation maps of the Mediterranean Basin and of South America, the FAO–Unesco *Soil Map of the World* or the small-scale map showing the world distribution of arid regions. It is also related to a number of other Unesco initiatives for synthesis of information at regional and international levels in order to promote the integrated management of natural resources. Mention might thus be made of *A Review of the Natural Resources of the African Continent* (1963), the more recent Unesco–UNEP–FAO state-of-knowledge reports on tropical forest ecosystems (1978) and tropical grazing land ecosystems (1979), and a series of national case studies on desertification (1980). Several issues in the Man and the Biosphere (MAB) Technical Notes Series also deal with problems of natural resources and their management in the African region, including a review of ecological approaches to land use in the Sahel (MAB Technical

Notes 1, 1975), a study on traditional strategies and modern decision-making in the management of natural resources in Africa (MAB Technical Notes 9, 1978) and an analysis of trends in research and in the application of science and technology for arid-zone development (MAB Technical Notes 10, 1979).

The United Nations Sudano-Sahelian Office (UNSO) was established by the Secretary-General of the United Nations in 1973, in the aftermath of the severe 1968–73 drought which devastated the economy and social life of the Sahelian region, to initiate and assist in the implementation of the medium- and long-term recovery and rehabilitation programme in the eight countries of the area, namely, Cape Verde, Chad, Gambia, Mali, Mauritania, Niger, Senegal and Upper Volta. Since that time, UNSO has developed into the principal body and central co-ordinating point of the United Nations system mandated by the General Assembly and other United Nations organs (a) to assist the eight drought-stricken Sahelian countries—members of the Permanent Inter-State Committee for Drought Control in the Sahel (CILSS)—in the implementation of their medium and long-term recovery and rehabilitation programmes; and (b) to act as the arm of the United Nations responsible for assisting, on behalf of the United Nations Environment Programme (UNEP), eighteen countries of the Sudano-Sahelian region in the implementation of the Plan of Action to Combat Desertification, as a joint UNDP/UNEP venture (Djibouti, Ethiopia, Guinea, Guinea-Bissau, Kenya, Nigeria, Somalia, Sudan, Uganda and the United Republic of Cameroon, in addition to the eight countries mentioned above).

UNSO's work, undertaken in close collaboration with the Sudano-Sahelian countries, CILSS and the respective United Nations agencies, is focused primarily on: (a) assisting the countries and CILSS in the planning and programming of priority projects and programmes in the field of drought-related medium- and long-term recovery and rehabilitation activities and desertification control; (b) providing assistance in the mobilisation of the necessary resources for implementing such projects and programmes, either on bilateral or multilateral bases, or by contributions to the United Nations Trust Fund for Sudano-Sahelian Activities established by the Secretary-General for that purpose; (c) managing the Trust Fund and implementing, from the resources of the Fund, in accordance with the relevant rules and regulations of the United Nations, projects not undertaken bilaterally or multilaterally; and (d) monitoring, reporting and disseminating knowledge on drought-related and desertification-control programmes.

As is shown in Figure 1 (see page 12), the Sudano-Sahelian region extends over a large portion of Africa. Consequently, it is hoped that the information contained in the *Vegetation Map of Africa* and in the accompanying memoir will provide a synthesis of knowledge on African vegetation that will be useful as a source of reference in land-use planning as well as in training for the purposes of drought recovery and rehabilitation and of desertification control.

The task of preparing the map and memoir has been a very long and complex one, and Unesco expresses its sincere thanks to the AETFAT Vegetation Map Committee in seeing the work through to fruition. The committee was made up of the following specialists: A. Aubréville, L.A.G. Barbosa, L.E. Codd, P. Duvigneaud, H. Gaussen, R.E.G. Pichi-Sermolli, H. Wild and F. White (Secretary). In publishing the new *Vegetation Map of Africa*, Unesco is especially grateful to Mr Frank White of Oxford University (United Kingdom), who compiled the vegetation map on behalf of the AETFAT Vegetation Map Committee and who is the author of the present memoir. The views expressed herein are those of the author and are not necessarily shared by Unesco and by UNSO.

Unesco also thanks the Oxford University Press for the preparation of successive proofs of the map.

In finalizing the map and memoir, an attempt has been made to use up-to-date forms of geographic names. However, the designations employed and the delimitations of frontiers on the map and in the accompanying text do not imply the expression of any opinion whatsoever on the part of Unesco or UNSO concerning the legal or constitutional status of any country, territory, city or area or of its authorities, or concerning the delimitation of its frontiers or boundaries.

Contents

Introduction 9

Acknowledgements 13

List of former names of countries 15

Part One *Environment, land use and conservation*

    1. Geology and physiography 19
    2. Climate and plant growth 23
    3. Soils 26
    4. Animals 29
    5. Fire, land use and conservation 32

Part Two *Regional framework, classification and mapping units*

    Introduction 37
    6. Regional framework 39
    7. Classification 44
    8. Mapping units 56

Part Three *Vegetation of the floristic regions*

    Introduction 69

THE AFRICAN MAINLAND

I.	The Guineo–Congolian regional centre of endemism	71
II.	The Zambezian regional centre of endemism	86
III.	The Sudanian regional centre of endemism	102
IV.	The Somalia–Masai regional centre of endemism	110
V.	The Cape regional centre of endemism	131
VI.	The Karoo-Namib regional centre of endemism	136
VII.	The Mediterranean regional centre of endemism	146
VIII/IX.	The Afromontane archipelago-like regional centre of endemism and the Afroalpine archipelago-like region of extreme floristic impoverishment	161
X.	The Guinea–Congolia/Zambezia regional transition zone	170
XI.	The Guinea–Congolia/Sudania regional transition zone	175
XII.	The Lake Victoria regional mosaic	179
XIII.	The Zanzibar–Inhambane regional mosaic	184
XIV.	The Kalahari–Highveld regional transition zone	190
XV.	The Tongaland–Pondoland regional mosaic	197
XVI.	The Sahel regional transition zone	203
XVII.	The Sahara regional transition zone	216
XVIII.	The Mediterranean/Sahara regional transition zone	225

MADAGASCAR AND OTHER OFFSHORE ISLANDS

XIX.	The East Malagasy regional centre of endemism	234
XX.	The West Malagasy regional centre of endemism	240
XXI.	Other offshore islands	244

AZONAL VEGETATION
XXII.	Mangrove, halophytic and fresh-water swamp vegetation 260

Glossary and index of vernacular names of vegetation types and habitat   269

Geographical bibliography   271

Alphabetical bibliography   275

Index of plant names   325

Introduction

The new *Vegetation Map of Africa* and its accompanying text is the fruit of some fifteen years of co-operation between Unesco and AETFAT (Association pour l'Étude Taxonomique de la Flore de l'Afrique Tropicale).

In 1965 the AETFAT Vegetation Map Committee, consisting of A. Aubréville, L. A. G. Barbosa, L. E. Codd, P. Duvigneaud, R. E. G. Pichi-Sermolli, H. Wild and F. White (Secretary), to which the late H. Gaussen was subsequently co-opted, was asked to collaborate with Unesco in the preparation of a new vegetation map of Africa as part of the latter's programme of mapping the world's vegetation at a scale of 1:5 000 000.

The materials used in compiling this map turned out to be exceedingly diverse. Those used for the first draft included:

1. Original contributions by Duvigneaud, Pichi-Sermolli and Gaussen for Zaire, the Ethiopian region, and the Maghreb and Madagascar respectively.
2. Large-scale maps which were being prepared for independent publication elsewhere by Wild & Barbosa for the Flora Zambesiaca region (1:2 500 000) and by Barbosa for Angola (1:2 500 000).
3. The remarkably detailed and accurate map of the Veld types of South Africa (1:1 500 000) by Acocks. Codd advised on the adaptation and simplification of this work for the present purpose.
4. Several published and unpublished maps of parts of francophone Africa communicated by Aubréville.
5. For much of the remainder of Africa, a large number of vegetation maps at various scales, which had been prepared for a wide variety of purposes.
6. For the few parts of Africa without vegetation maps of any description, correspondence with a host of local specialists, many of whom are members of AETFAT, has supplied the missing information.

I have been responsible for attempting to standardize the source materials and weave them into a coherent whole. In doing this I have continued to receive the unstinted help of the committee members and many others, but I must bear full responsibility for the final presentation and the imperfections which remain.

The map and the descriptive memoir cover not only the whole of Africa and the large island of Madagascar, but also all ecologically important islands in the eastern South Atlantic and western Indian Oceans, though space could only be found for a very brief treatment.

The purpose of the map is not to provide *detailed* information on any particular area for the benefit of residents in that area, since that information is usually available locally in published or unpublished form and is also inappropriate on maps of this scale. Rather, the purpose is to indicate to local residents in broad terms the manner in which the main features of their local vegetation can be related to the main features of African vegetation as a whole. Another major objective is to provide a framework on a continental scale within which more detailed local studies can be conducted and compared. Inevitably, in simplifying larger-scale maps arbitrary decisions cannot be avoided and deliberate omissions must be made. This should be borne in mind by users of the map, especially those seeking local detail. It is for considerations such as these that some features of altitudinal zonation have been deliberately suppressed. It would have been cartographically possible to have shown a more complete and accurate zonation but the small gain in factual content would not have been commensurate with the effort and costs involved.

A few years elapsed between the completion of the first draft of the map and its publication. This time was put to good use since it provided many opportunities for testing the accuracy of the map. This was done in different ways, as follows:

1. Several ecologists, travelling extensively in Africa, have been able to check the map against vegetation on the ground. I, myself, in the course of three journeys in East, Central and Southern Africa, undertaken for other purposes, have been able to check the accuracy of several pattern lines. The Maghreb was also specifically visited to check the map and collect information for the accompanying text.
2. Earlier drafts of the map were exhibited at the plenary congresses of AETFAT held in Munich in 1970 and Geneva in 1974, when members of AETFAT, whose collective experience, which embraces the whole of Africa, were invited to comment on the map. By this means several inaccuracies were avoided.
3. For those parts of Africa I have not visited a collection of vegetation photographs, published and unpublished, was assembled in order to check whether the physiognomy they portray agrees with that one would expect from the map.

All this checking has led to some significant corrections, but for much of Africa the accuracy of the map, within the objectives set out above, was confirmed. For a few (relatively) parts of Africa, there is a total dearth of information. For other parts, even where the original documentation was sparse, subsequent work has usually confirmed the validity of the general features shown on the map, though inevitably some local detail was shown to be inaccurate. This reflects, at least in part, inaccuracies in the base maps used at different stages in the compilation.

The relationship between the broad framework attempted in the Unesco/AETFAT/UNSO *Vegetation Map* and more detailed studies can be illustrated by reference to the vegetation of Marsabit District in Kenya, which I visited in 1980, long after the map was completed. The area under consideration occupies approximately 3×3.5 cm of the Unesco map, which shows a pattern of islands of undifferentiated Afromontane communities and degraded evergreen bushland arising from a matrix of lowland deciduous bushland and semi-desert shrubland and grassland. The much more detailed map of Herlocker (1979*a*) at a scale ten times greater (area of map 100 times larger) shows nine primary vegetation types which can readily be interpreted in terms of the Unesco framework, even though the criteria for distinguishing the major types do not always coincide.

The classification of vegetation used in the present work differs in some respects from those in general use. This, in part, is due to the fact that the flora and vegetation of Africa are now better known than those of most other parts of the tropics. For Africa a new synthesis now seems appropriate. The principles underlying the present approach are discussed in Chapters 6 and 7. They can be briefly summarized as follows:

1. Vegetation, in the first instance, should be classified without reference to the physical environment, including climate, or to animals. The extent to which environmental factors and the associated fauna can be used to diagnose vegetation types should be evaluated independently. In the classification adopted here climatic terms are occasionally used, but only as a convenient nomenclatural shorthand for important physiognomic and floristic differences which it would be impossible to designate concisely in purely physiognomic or floristic terms.
2. The physiognomic features which figure prominently in conventional classifications were found to be inadequate.
3. A chorological system based on the patterns of geographical distribution shown by entire floras, as revealed by samples, was found to provide both the basis of an objective framework within which the vegetation of Africa can be described and compared, and an indirect shorthand method of expressing the *entire* physiognomy of regional vegetation types, not merely a few selected features.
4. As to nomenclature, it was found that sufficient English terms were available to cover the main physiognomic types, and no difficulty was experienced in finding French equivalents. It was thus possible to avoid the use of imported vernacular names of equivocal application such as savanna and steppe for classificatory units of high rank. By contrast, indigenous, African vernacular names for local variants of the major physiognomic types have proved most valuable.

For some parts of Africa a rich vocabulary is available and several terms such as 'muhulu', 'mopane', 'miombo' have been taken up and precisely applied by local botanists. Relatively few such terms have been used in a formal sense in the present work, although the potential application of others is indicated in the text. The fifty or so terms mentioned are included in a combined glossary/index at the end of the work.

Because the classification is based solely on the plants themselves, sometimes supplemented by features of the environment which can be seen, such as standing water or outcrops of rock, the colour scheme adopted for the map does not consciously embody information on climate. The colours chosen were determined primarily by economy and clarity of presentation of information on the vegetation.

In order to maintain continuity with the first AETFAT map, the *Vegetation Map of Africa South of the Tropic of Cancer* (Keay, 1959*b*), which is well known, similar colours have been used for the main regional vegetation types. Since the colours of the earlier map were chosen to reflect the principles of Gaussen's (1955) system of climato-ecological cartography, this new map inevitably embodies some climatic information. But this is accidental. In my opinion, in our present state of knowledge, it is preferable to map climate and vegetation separately.

With this in mind, in the present work an attempt has been made to characterize the twenty regional phytochoria by means of selected climatic diagrams. The results are displayed on seventeen pictorialized climatic maps (Figs. 5–8 and 11–23) (not twenty, since some figures include more than one phytochorion). These maps not only summarize a great deal of climatic information, but also allow *visual* comparison of the main climatic features of the different phytochoria. They also show all the main topographical features mentioned in the text.

Within the constraints imposed by economy and continuity with the past, the colours of the mapping units have been chosen to bring out the relationships of the latter. Much of the vegetation of Africa is transitional and this is shown by using 'pyjama' stripes. A white background to stripes of colour indicates landscapes which are largely anthropic. The map was designed so that, when viewed from a distance, it shows features of regional extent. Increasingly closer inspection should reveal detail of increasingly local significance, but for ultimate detail the user should refer to the text.

For reasons given in Part Two, the one hundred cartographic units shown on the map are described in the text under the twenty regional phytochoria (eighteen on the African mainland, two on Madagascar) in which they occur. The outlines of these phytochoria are also boldly indicated on the map. To a considerable extent they coincide with the regional vegetation types, although they were discovered independently.

The main aim of the text is to describe salient features of the vegetation. The original draft was twice as long as this published version, so, for reasons of economy, much detail has been omitted. Similarly little room could be found for detailed discussion of the influence of climate, geology, soils, fire, animals and man on vegetation. Where such influences are particularly striking they are mentioned in the text. Otherwise, detail must be sought in the works referred to in the five introductory chapters dealing with these topics.

The vegetation types are described as concisely as is compatible with their complexity. For each type the principal sources of information, both published and unpublished, are cited, as are references to characteristic photographs and profile diagrams. It is hoped that this information will, at least in part, compensate for the fact that, diagrams and maps excepted, this work lacks illustrations. In order to avoid imprecise generalizations, I have tried to characterize the individual vegetation types by describing one or more concrete examples. Wherever possible, I have chosen examples I am familiar with in the field, or those described by authors who were available for discussion of their work, either orally or by correspondence. Inevitably, because of the great area covered and the diversity of vegetation types, it has not normally been possible to describe detailed patterns of vegetation in relation to environmental factors and human interference. There are, however, a few important exceptions.

This is chiefly because UNSO expressed a wish that such information should be given for the drier parts of Africa, which suffered from the recent Great Sahelian Drought. This provided an opportunity to include detailed accounts of selected areas of outstanding ecological importance in relation to human needs. The examples chosen are the Jebel Marra area and part of Kordofan Province, Sudan Republic in the Sahel Region, and the Serengeti ecosystem, and part of Marsabit District, Kenya, in the Somalia–Masai Region. They clearly illustrate the ways in which the results of such local studies can be accommodated in a more general framework, such as that which is the subject of the present work, and also demonstrate the fundamental importance of the detailed study and mapping of vegetation in land-use planning.

In addition to the detailed studies referred to above, other investigations inspired by the Great Sahelian Drought have resulted in an extensive, more general, literature on the ecology of the Arid Zone (Anon, 1977; Bartha, 1970; Breman & Cissé, 1977; Brown, 1971; Cloudsley-Thompson, 1974; Curry-Lindahl, 1974; Dalby & Harrison-Church, 1973; Dalby *et al.*, 1977; De Leeuw, 1965; FAO, 1977; Gallais, 1975; Konczacki, 1978; Lamprey, 1975, 1978; Lewis, 1975; Monod, 1975; Petrides, 1974; Swift, 1973; and Unesco, 1975). It has not been possible to summarize this in detail, though it has been taken into account in expanding the chapters on the Somalia–Masai and Sahelian regions.

The position of the Sahel and the four study areas mentioned above are shown on Figure 1.

In attempting to characterize the different vegetation types floristically, economy has been exercised in the selection of representative species. In other words equivalents of the 'zone fossils' of geologists have been sought. Even so, it has been necessary to mention some 3000 species in the text. They are included in an index to botanical names together with those major synonyms that feature prominently in the ecological literature. Hence the index provides a means of obtaining both new and previously published information on the autecology of the majority of important plant species in Africa.

A serious attempt has been made to ensure that the names used are those which satisfy the requirements of the International Code of Botanical Nomenclature and that the citation of authors is correct. In this connection I have received much help from the Director of the Royal Botanic Gardens, Kew, and his staff, but for some parts of Africa, particularly North Africa, the relevant information is not readily available and some further checking is required.

The names of countries are based upon United Nations official practice. There are a few exceptions. Thus, for the United Republic of Cameroon, the simple French title of 'Cameroun' is used throughout the publication, in order to retain consistency with the names of geographic locations such as Mount Cameroun and the Cameroun highlands. For reasons of brevity, the Democratic Republic of the Sudan is referred to either as Sudan or as Sudan Republic when the use of term 'Sudan' could be taken to mean either the country or the centre of endemism of the same name. Similarly, 'Guinea' and 'Guinea Republic' are used as synonyms for the Revolutionary People's Republic of Guinea. The Socialist People's Libyan Arab Jamahiriya is referred to as Libya or the Libyan Arab Jamahiriya. Generally, the English-language version of a country's title is used in the text of the memoir (e.g. Ivory Coast), though the French title (Côte d'Ivoire) may be used on the map and climate diagrams, given that a single version of the map and figures has been prepared to accompany the two language versions of this descriptive memoir. Finally, some older administrative names have been maintained because they feature prominently in the botanical literature.

The bibliography of some 2400 items aims to be comprehensive rather than complete, though it is unlikely that many important works have been left out. Apart from a few recent publications, most of the works referred to were used in writing this book, and all but a small minority of the cited references were personally checked by the author in the course of completing the manuscript. For ease of reference a second bibliography is provided in which the publications are arranged geographically.

F. White

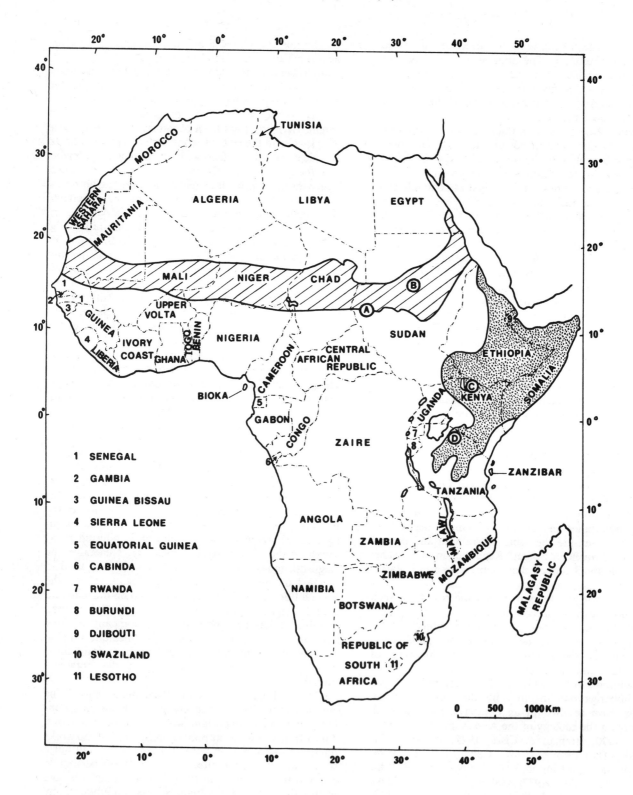

FIG. 1. Map of Africa showing names of countries and position of the Sahel regional transition zone, the Somalia–Masai regional centre of endemism and four Sahel-type study areas described in the text: (A) Jebel Marra (Chapter XVI); (B) Kordofan (Chapter XVI); (C) South-Western Marsabit District (Chapter IV); (D) The greater Serengeti region (Chapter IV). The following countries are members of the Plan of Action to Combat Desertification: Cape Verde, Chad, Djibouti, Ethiopia, Gambia, Guinea, Guinea-Bissau, Kenya, Mali, Mauritania, Niger, Nigeria, Senegal, Somalia, Sudan, Uganda, United Republic of Cameroon and Upper Volta

Acknowledgements

Many people have contributed to the successful completion of this work. It is a pleasure to record their kind help.

In addition to the members of the AETFAT Vegetation Map Committee and others mentioned in the Introduction, the following either read early drafts or provided information: J. P. H. Acocks (South Africa), E. J. Adjanohoun (Benin), L. Aké Assi (Ivory Coast), G. Aymonin (Madagascar), P. Bamps (Zaire and general), J. P. M. Brenan (general), J. F. M. Cannon (general), J. D. Chapman (Malawi, Nigeria), W. D. Clayton (grasslands), M. J. Coe (animals), K. G. Cox (geology), D. Edwards (South Africa), C. Évrard (Zaire), D. B. Fanshawe (Zambia), I. Friis (Ethiopia), M. G. Gilbert (Ethiopia, Kenya), J. B. Gillett (Ethiopia, Kenya), the late P. E. Glover (Kenya), the late P. J. Greenway (Kenya), J. B. Hall (Ghana), A. J. Hall-Martin (Malawi), O. Hedberg (Afroalpine vegetation), C. F. Hemming (arid zones), C. J. Humphries (Canaries), P. James (Ascension), C. Jeffrey (general), E. W. Jones (Nigeria), D. J. B. Killick (South Africa), F. J. Kruger (Cape), H. F. Lamprey (Serengeti), R. M. Lawton (Nigeria, Zambia), J. P. Lebrun (Sahel), O. Leistner (Kalahari), J. Léonard (Sahara), R. Letouzey (Cameroun, general), A. Le Thomas (Gabon), J. Lewalle (Burundi, Maghreb), L. Leyton (*Welwitschia*), G. L. Lucas (general), D. J. Mabberley (general), W. S. McKerrow (geology), F. Malaisse (Zaire), W. Marais (Mascarenes), E. J. Mendes (general), H. Merxmüller (Namibia), E. J. Moll (South Africa), T. Monod (Sahel), J. K. Morton (Ghana, Sierra Leone), R. M. Polhill (Kenya), D. J. Pratt (general), the late J. Procter (Tanzania), P. Quézel (Sahara), A. Radcliffe-Smith (Socotra), the late J. Raynal (general), A. Raynal-Roques (general), S. A. Renvoize (Aldabra), E. R. C. Reynolds (climate and plant growth), W. A. Rodgers (Tanzania), R. Rose Innes (Ghana), J. H. Ross (South Africa), R. Schnell (West Africa), P. J. Stewart (Maghreb), P. Sunding (Mascarenes), M. D. Swaine (Ghana), J. J. Symoens (Zaire), T. J. Synnott (general), H. C. Taylor (South Africa), B. Verdcourt (East Africa), the late D. F. Vesey-FitzGerald (Tanzania), R. Webster (soils), G. E. Wickens (Sudan, general), M. J. A. Werger (South Africa). In the early stages of this work, R. W. J. Keay kindly made his considerable experience of African vegetation available and gave generously of his time.

Assistance in the field was provided by: W. R. Bainbridge (Natal), the Chief Conservator of Forests, Kenya, L. E. Codd and B. de Winter (South Africa), R. B. Drummond (Zimbabwe), J. B. Gillett (Kenya), D. Herlocker (Kenya), Christine Kabuye (East Africa), J.O. Kokwaro (Kenya), J. Kornaś (Zambia), F.J. Kruger (Cape), H. F. Lamprey (Kenya), J. Lewalle (Morocco), E. J. Moll (Natal), T. Müller (Zimbabwe), J. C. Scheepers (Transvaal), John and Lucie Tanner (Tanzania) and H. C. Taylor (Cape).

Three successive Professors of Forest Science in the University of Oxford, M. V. Laurie, J. L. Harley and M. E. D. Poore, have freely made the facilities of their Department available for the execution of this work. A special debt of gratitude is owed to Ernie Hemmings, Librarian in the Department of Forestry, and his staff for the trouble they have taken in obtaining photocopies of rare literature and for other bibliographic assistance. Much help was provided by the Assistants in the Forest Herbarium, notably Rosemary Wise, who prepared the text figures, Frances Bennett and Helen Hopkins, who assisted in the compilation of the map, Serena Marner and Michael Wilkinson, who have helped with the bibliography and index respectively, and Cynthia Styles, who typed the text.

I am also deeply indebted to John Callow and his colleagues at the Oxford University Press for their advice on cartographic matters and their patience in handling a project which extended over several years.

F. W.

List of former names of countries

List of former names of countries which appear in the cited botanical literature.

Old name	New name	Old name	New name
Abyssinia	Ethiopia	Italian Somaliland	Somalia (part)
Anglo-Egyptian Sudan	Sudan	Madagascar	Malagasy Republic
Basutoland	Lesotho	Middle Congo	Congo
Bechuanaland Protectorate	Botswana	Northern Rhodesia	Zambia
Belgian Congo	Zaire	Nyasaland	Malawi
British Somaliland	Somalia (part)	Rhodesia	Zimbabwe
Cameroons	United Republic of Cameroon or Cameroun	Rio Muni	Equatorial Guinea
		Ruanda	Rwanda
Dahomey	Benin	Southern Rhodesia	Zimbabwe
Eritrea	Ethiopia (part)	South-West Africa	Namibia
Fernando Po	Bioko	Tanganyika Territory	Tanzania
French Guinea	Guinea	Ubangi-Shari	Central African Republic
French Somaliland	Djibouti	Union of South Africa	Republic of South Africa
French Sudan	Mali	Urundi	Burundi
Gold Coast	Ghana	Zululand	KwaZulu (part)

Part One   Environment, land use and conservation

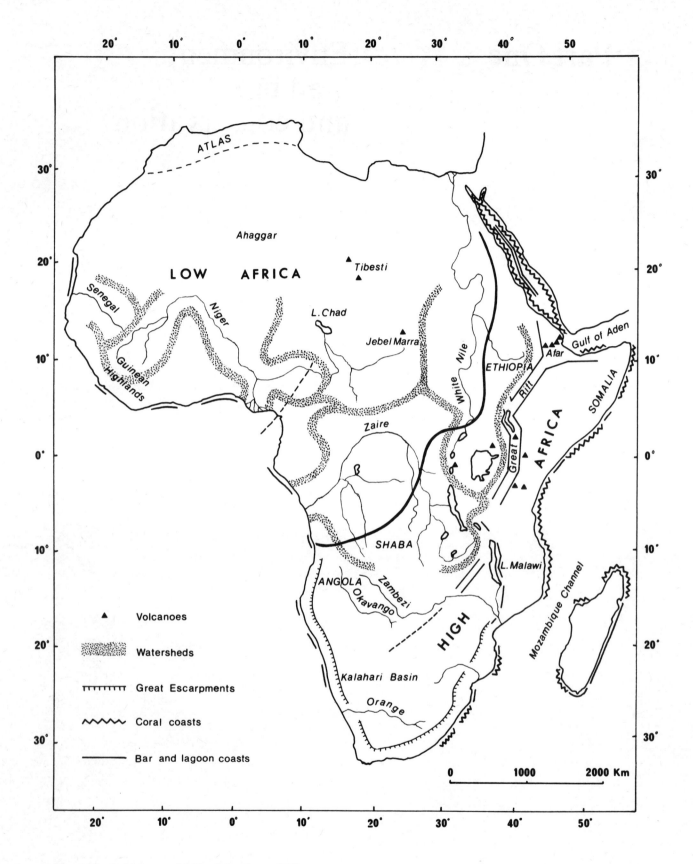

FIG. 2. Physical features of Africa (after Grove, 1978)

1 Geology and physiography

Africa is the second largest continent. It is unique in that, with the exception of the Atlas mountain system in the north-west and the Cape ranges in the south, it consists of a continuous crystalline shield which is exposed over extensive areas. In places, unaltered sedimentary rocks deposited on a metamorphic basement complex are largely of Precambrian age. In the Sahara and the Kalahari region aeolian sands conceal the older rocks over large surfaces.

A general account of the geology of Africa has been published by Furon (1963, 1968), who also wrote the explanatory note to the Unesco/Asga geological map of Africa (Furon & Lombard, 1964). The latter is extremely brief. A somewhat longer treatment accompanies the FAO–Unesco soil map of the world (FAO–Unesco, 1977). Clear summaries of the main features of African geology are few. The following pages rely heavily on that of Grove (1978).

Important regional works include those by Gray (ed., 1971) on the Libyan Arab Jamahiriya, by Whiteman (1971) on the Sudan Republic, by Cahen (1954) on Zaire, by Haughton (1963, 1969) on the stratigraphy of Africa south of the Sahara and on the geological history of southern Africa, by Du Toit (1954) on South African geology, and by King (1967a, 1967b, 1978) on South African scenery and geomorphology. The physiography and geology of East Africa are briefly summarized by Saggerson (1962a, 1962b, 1972). Mäckel (1974) has described the geomorphology of the shallow streamless depressions (dambos) at the heads of drainage systems on the Great African Plateau in south-central Africa.

A line drawn across the map of Africa from Angola to western Ethiopia divides the continent into high and low parts (Fig. 2). Low Africa, in the north-west, is made up of sedimentary basins and upland plains mostly between 150 and 600 m above sea level, comprising the Sahara and the catchments of the lower Nile, Senegal, Niger, Chad and Zaire rivers. Land rising above 1 000 m is confined mainly to: the Atlas mountains in the Maghreb; the Saharan massifs Ahaggar and Tibesti; Jebel Marra in the Sudan Republic; the headwaters of the Niger; the Jos Plateau in Nigeria; and the Cameroun highlands. Nearly all of High Africa, to the south and east, rises above 1 000 m, with the exceptions of Somalia, the broad lowlands of Mozambique and relatively narrow coastal plains and valley strips elsewhere. Even the Kalahari basin is about 1 000 m above the sea, and in east Africa the surface of Lake Victoria stands 1 130 m above sea-level.

The rocks of Africa span a time period of 3 500 million years and the geological structure of the continent is complicated. There are three great areas where the rocks were affected by mountain building more than 1 500 million years ago that have not been disturbed by folding since. They occupy the western lobe of Africa, the Zaire–Angola region and the Zimbabwe–Transvaal–Orange Free State area, and are called *older cratons*. The areas between the older cratons have been affected by mountain building within the last 1 200 million years and are known as the *younger orogens*. Among those minerals which are associated with vegetation anomalies, the principal deposits of chromium and asbestos are found in the older cratons, whereas most of Africa's copper, lead, zinc, and cobalt are contained in the younger orogens.

The ancient rocks of the older cratons and younger orogens, which are mostly of Precambrian age, underlie the whole area, but they are widely buried by younger sedimentary rocks and wind-blown sand. They outcrop in watershed areas and escarpments flanking the Niger basin, and form the rugged massifs of Aïr and Ahaggar as well as the dissected plateaux of the Guinea highlands, and Jos in Nigeria. They build the Cameroun highlands and all the western rim of the continent from the Crystal Mountains to the Orange River in the south. Granites, gneisses and schists, exposed in hill ranges or masked by the products of long-continued weathering, form long watersheds between the Nile, Zaire and Chad basins. In East Africa they appear more extensively than elsewhere, building much of the high country between the Transvaal in the south and the Red Sea Hills of Egypt in the north.

The oldest of the crystalline rocks are intensely folded schists and banded gneisses which are resistant to erosion but generally less so than the granites intruded into them. Over wide areas these old rocks are characterized by extensive, gently sloping surfaces, covered with a deeply weathered layer or comparatively recently transported material which conceals the solid rocks. The granites and some of the gneisses form rugged hills or dome-shaped inselbergs which rise sharply from the surrounding plains, whereas quartz dykes stand up as long narrow ridges. Commonly it is the rocks in the vicinity of igneous intrusions that have been mineralized.

The youngest of the Precambrian rocks have not been strongly folded in most areas. They include the gold-bearing quartzites of the Rand, much-hardened sandstones, of which the older cratonic rocks were the source.

From the Lower Palaeozoic period until the Jurassic, Africa, together with South America, Antarctica, Madagascar, India and Australia, was part of a southern land mass, Gondwanaland. Before its break up it was surrounded by an ocean in which great layers of sediments accumulated. They now form sandstones, shales, limestone and dolomites in the Maghreb, the western Sahara and the Cape. Some of these rocks are very resistant to erosion, notably the tough sandstones of the Cape series which form Table Mountain behind Cape Town.

The sands and clays accumulating on the continental margins and in the oceans around Gondwanaland were derived from the erosion of the surface of the supercontinent. In Ordovician times the seas locally invaded the interior and laid down shales and sandstones which now form escarpments rising from the desert plains at the margins of Tibesti and Tassili des Ajjers. Resting on these marine rocks are tillites deposited by the continental glaciers that, 450 million years ago, occupied much of what is now the north-western Sahara.

From 450 to 250 million years ago Africa was never very far from the South Pole. At that time violent earth movements caused the rocks of the Cape System to be strongly folded. Subsequent erosion has moulded the ranges which run parallel to the coast south of the Great Karoo.

The Karoo gives its name to a system of rocks which accumulated over much of southern and central Africa during the Upper Carboniferous, Permian, Triassic and Lower Jurassic. The system includes glacial tillites, marine clays and coal-bearing continental deposits, debris eroded from the Hercynian fold mountains, deltaic and lacustrine sandstones, and enormous sheets of volcanic lavas. In Lesotho the lavas are as much as 1 800 m thick.

In north Africa, the *Continental Intercalaire*, a system of rocks varying to a similar degree, but much thinner than the Karoo, accumulated at a somewhat later stage. It includes the Nubian sandstone and other permeable water-bearing beds that underlie much of the Sahara and are tapped by deep boreholes.

During the late Mesozoic, Africa became separated from the rest of Gondwanaland. As the oceans around Africa widened, freshwater and then marine deposits accumulated on the continental margins. Great fractures appeared in the east, letting down blocks of Karoo sediments, which subsequently guided the evolution of the relief and drainage. Similarly in West Africa, the Benue fault trough opened from the Gulf of Guinea, penetrating far to the north-east into the Saharan region.

From the Cretaceous (100 million years ago) onwards the geological history of the south-eastern part of the continent is quite different from that of the north-west. The former was left high and dry, and only its margins were covered with marine sediments. The north and west were flooded by seas advancing south from Tethys in which thick layers of sandstone and limestone were laid down. During the late Cretaceous and Tertiary periods Africa moved north, slowly deforming the sediments in the Tethys ocean to produce the European Alps and related structures, including the Atlas. This Alpine deformation was associated with several faults with large horizontal movements. One such series of faults separates the folded rocks of the Maghreb from the rigid Saharan block.

South of the Atlas, the rock strata were not greatly contorted by the late Mesozoic and Tertiary earth movements, except in the Benue trough, where Cretaceous sandstones and clays were quite strongly folded and have since been dissected to give scarpland topography. Elsewhere, the basin form of certain depressions was accentuated, and stresses in the crust caused large-scale faulting.

Evidence exists of successive uplifts of High Africa and of watershed areas throughout the continent during late Mesozoic and Tertiary times. The uplifted areas in High Africa form the Great African Plateau, which is the largest plateau in the world. Most of the plateau is more than 900 m above sea-level.

Generally, the edge of the plateau forms the highest ground and is in the nature of a watershed between the headwaters of plateau rivers and those of the coastal drainage.

In southern Africa the highest parts of the plateau rim are on the Drakensberg (Thabana–Ntlenyana, 3485 m, and many points above 3190 m) and the volcanic cone of Mt Rungwe, 2961 m, in southern Tanzania. In between, the syenite mass of Mlanje in southern Malawi rises to 3 000 m, the granite Namuli Peak in Mozambique to 2 419 m and Inyangani in Zimbabwe to 2 515 m. On the western side of the subcontinent the highest points on the plateau are in Angola (Mt Moco, 2 620 m, Serra da Chela, 2 300 m) and in the Auas highlands south of Windhoek (Molkteblick, 2 485 m), but the highest point in the west lies outside the plateau where the Brandberg rises steeply from the Namib desert to an altitude of 2 695 m.

The original edge of the plateau probably at one time formed the coastal margin of the continent, but as streams cut back into the plateau edge and as the sea floor became exposed during uplifts of the land mass, the edge of the plateau no longer formed the coastline, but became a physical feature separating the plateau from the coastal region. In the course of time, with the continued retreat of the plateau edge, and with further exposures of the sea floor, the area between the plateau and the coast became in places so extensive that it can now no longer be considered coastal in character, but rather as a region marginal to the plateau, its inner areas removed from the coast in some places hundreds of kilometres.

In southern Africa the boundary between the plateau and areas marginal to it is generally called the 'Great Escarpment'. The latter is a variable feature, particularly in height, abruptness, and steepness of slope, depending chiefly on the configuration of the plateau itself, its rock formations, and on climate. In general it is abrupt, cliff-like and linear where hard, resistant formations overlie soft ones. Where the rock is homogeneous and easily decomposed the escarpment is an irregular feature with more gentle slopes.

The Great Escarpment is most abrupt and prominent along the border of Natal and Lesotho, where it is capped by Stormberg lavas and is known as the Drakensberg. Elsewhere in southern Africa it is usually a well-defined feature, but in places it vanishes or is not very pronounced. Thus there is a 95 km wide gap between the Koudeveld and Nieuweld Mts in Cape Province. North of the Swakop River in Namibia there is no distinct edge to the plateau for 480 km except for the western face of the Erongo Mts. In Angola north of the Huilla plateau the escarpment forms the watershed between the Cunene basin and the coastal drainage but does not form a very well-defined feature.

North of the Natal Drakensberg the Great Escarpment is only intermittently conspicuous and no plateau edge is discernible in the Limpopo valley, but in Zimbabwe the feature is again recognizable in the Melsetter–Chimanimani highlands and the Inyanga scarp. Further north, owing to advanced dissection and to complex trough faulting the position of the plateau edge is difficult to locate. To the east of Lake Malawi it is possible that the Njombe highlands represent the position of the plateau edge in pre-rift valley times, but this is conjectural.

In southern Africa two primary divisions of the great plateau can be recognized: the central or Kalahari basin, and the peripheral highlands, which are widest in the east and narrowest in the west.

The Kalahari basin extends some 1930 km from the Orange River to the southern Zaire watershed. Its greatest width is 1300 km and it has an area of about 1640000 km^2. It is almost entirely covered with a mantle of sand, which extends across the watershed into the Zaire basin, and is probably the largest continuous surface of sand in the world.

It is widely believed that each uplift of High Africa was followed by the cutting of an erosion surface graded to a lower base level than the one preceding it. Some geomorphologists have looked upon the continent's relief as consisting essentially of broad erosion levels and intervening erosional escarpments, of which the most striking of all is the Drakensberg in the Republic of South Africa.

Dating of the erosion surfaces and the stages in the evolution of the relief is difficult, and the way in which the erosion surfaces have been cut is still uncertain. King (1967, 1978) groups the surfaces together into a few major 'cycles' and attributes them to pediplanation involving scarp retreat over long distances, but more evidence is required before this interpretation of African landscape can be unreservedly accepted.

The most striking feature of the relief of High Africa is the rift-valley system, which extends from Turkey to Zimbabwe. In east Africa two fault systems can be distinguished. The eastern rift bisects Tanzania, cuts across the Kenya highlands to Lake Turkana (Rudolph), then turns north-east, splitting the Ethiopian plateaux and diverging into the much wider trenches of the Afar depression, the Red Sea and the Gulf of Aden. The western rift can be traced from the upper Nile and Lake Edward, through Lakes Kivu, Tanganyika and Malawi to the coast near Beira, with a branch extending along the Luangwa valley, the middle Zambezi and the southern margins of the Okavango swamps in Botswana.

The general pattern of the rift valleys seems to correspond with the lines of ancient structures in the crystalline floor. In the southern part of the system, some of the troughs such as the Luangwa valley seem to have been subsequently eroded in relatively soft sediments that filled the troughs after they had been let down between parallel faults at the end of Karoo times. They are essentially erosional features. In others, like the Malawi and Tanganyika troughs, the form of the valleys has been sharpened in late Tertiary and Quaternary times by dislocation along earlier fault planes, sometimes involving vertical movements of many hundreds of metres. The separation of Arabia from Africa to form the Red Sea and Gulf of Aden has taken place within the last 15 million years, and locally, as in the Afar and Danakil depressions of Ethiopia, faulting has continued into present times.

Many of the Indian Ocean and Atlantic islands are the fragments of old volcanoes that originated on the crests of mid-ocean ridges and were carried away from the crest as the sea floor continued to spread, but the Seychelles and Madagascar are isolated bits of continent. The oldest volcanic islands are the furthest from the ridge crests. A volcano on Réunion is still active and so are Mount Cameroun and a number of the volcanoes in the vicinity of the rift valley, notably the Virungas near Lake Kivu.

The coastline of Africa, like that of Gondwanaland generally, is remarkably free from indentations. This is probably to be explained by its faulted character, by the lack of folding in late geological times, by continental uplift being dominant and by the deposition of river-borne sediments along the coast as sandbars and deltas during and since the rise in sea-level at the end of the last glaciation in high latitudes.

Climatic conditions have varied all over Africa during the last million or more years of the Quaternary period. Nearly everywhere there is evidence that the climate has been both wetter and drier and warmer and colder in the past. There is, however, much uncertainty concerning the precise pattern of change, even during the last 20000 years for which most evidence is available.

The climatic sequence in the drier parts of north Africa subsequent to the melting of the ice-caps in Europe and North America about 15 000 years ago appears to be fairly well documented. By contrast, some conclusions about other parts of Africa, e.g. those of Livingstone (1967, 1975) on East Africa are not fully supported by the evidence cited, which is partly palaeobotanical (White, 1981). It is clear that the subject is much more complicated than is commonly assumed, especially concerning the effects of climatic change on the distributions of animal and plant species, and hence on the floristic composition of plant communities. Although this subject is important in the understanding of African vegetation it lies outside the scope of the present work.

Quaternary climatic fluctuation has also had profound indirect effects on vegetation through changes in soil, drainage systems and land-forms. Frequent mention is made in the text of Quaternary aeolian sands and lacustrine clays. The influence of Quaternary climatic change on land-form has been described in some detail by Grove (1969) for the Kalahari region and Grove & Warren (1968) for the southern Sahara. Kassas (1953a, 1956a) has described the desert vegetation of Egypt and the Sudan in relation to land-form.

Although our understanding of African vegetation is greatly enhanced if it can be related to parent material and physiography, the influence of the former is usually indirect through its control of physiography rather than through the chemical nature of the soils to which it gives rise. There are, of course, exceptions. The extensive literature on heavy metal and other toxic soils has been summarized by Wild (1978). Such vegetation is comparatively localized and is only briefly mentioned in the present account.

The distribution of saline soils is partly determined by geology in that they can occur in relatively wet regions around springs provided evaporation is sufficiently high, as in parts of the Mediterranean Region. In parts of east Africa, where the mean annual rainfall is 250–1 000 mm, salts derived from volcanic deposits rich in sodium are deposited in lake basins and river valleys (see page 266). In the Zambezian region rocks containing the soda feldspar perthite locally give rise to sodium soils, which, though not sufficiently saline to support halophytic communities, nevertheless carry distinctive vegetation (see page 94).

The colonizing vegetation of recent lava flows has been described by Keay (1959d), J. Lebrun (1959, 1960d), A. Léonard (1959) and Robyns (1932).

2 Climate and plant growth

Important general works include those by Aubréville (1949), S. P. Jackson (1962), B.W. Thompson (1965, 1966), Griffiths (ed., 1972), and I.J. Jackson (1977). The climate of the Zaire basin is summarized by Bernard (1945) and Bultot (1971–77), and of East Africa by Griffiths (1962). The rainfall map of the drier parts of East Africa and Arabia compiled by Griffiths & Hemming (1963) is based partly on ecological information to compensate for the paucity of meteorological data. Swami (1973) has described moisture conditions in the savanna region of West Africa.

Agroclimatological studies have been made for the highlands of eastern Africa by Brown & Cochemé (1969), and for the drier parts of West Africa by Cochemé & Franquin (1967). Woodhead (1970) has used water balance in East Africa as a guide to site potential. Trochain (1952) has mapped the phytogeographic units of West Africa using bioclimatic data, and Papadakis (1966, 1970) has divided the world into climatic regions based on critical temperatures of certain cultivated plants and the water balance of the soil. This is the classification used in conjunction with the FAO–Unesco *Soil Map of the World* (1977). A more detailed survey of climate in relation to world vegetation has been edited by Monteith (1976).

Among more specialized studies, Trapnell & Griffiths (1960) have described rainfall in relation to altitude in Kenya. The importance of light rainfall in semi-arid areas is emphasized by Glover (P.E. Glover *et al.*, 1962, J. Glover & Gwynne, 1962). Rainfall variability is discussed by Nieuwolt (1972) for Zambia, and by Pennycuick & Norton-Griffiths (1976) for the Serengeti ecosystem, Tanzania. Tyson (1978) adduces evidence for a quasi-20-year oscillation in rainfall for the summer rainfall region for South Africa, and a 10-year oscillation for the all-seasons rainfall region of the southern Cape. The literature on mist precipitation and vegetation has been briefly reviewed by Kerfoot (1968). Walter (1936) has described the effects of fog on the vegetation of the Namib desert, and the mist oases of the eastern Sudan are the subject of publications by Troll (1935*a*) and Kassas (1956*b*).

Walter (1939; Walter & Walter, 1953) has also emphasized the importance of soil texture in relation to rainfall in the drier parts of southern Africa, and more generally (1955*a*), as has Smith (1949) for the Sudan. Included among the relatively few other works on water relations of African plants are publications by Okali (1971) on some woody species of the Accra Plains, Ghana, by Ernst & Walker (1973) on the hydrature of trees in miombo woodland, by Vieweg & Ziegler (1969) on the 'Resurrection Plant', *Myrothamnus flabellifolius*, and by Gaff (1977) on the poikilohydrous plants of southern Africa generally.

The recent severe drought in the Sahel zone and the famine associated with it have stimulated several publications on desertification (e.g. Depierre & Gillet, 1971; Boudet, 1972; Delwaulle, 1973; Michon, 1973; Wade, 1974). The general consensus is that destruction of vegetation by domestic animals and man has had far more effect than recent climatic deterioration. Boudet suggests that the *brousse tigrée* patterns of vegetation are probably brought about by wind action and sheet flow following degradation of the original vegetation. L.P. White (1971), however, believes that they are highly stable and have been in existence throughout the Quaternary period, but have migrated in response to successive climatic changes (but see page 27).

The effect of wind on vegetation is described by Pitot (1950*b*, coastal vegetation in Senegal), Jeník (1968, tropical West Africa), Jeník & Hall (1966, effect of the harmattan in the Togo Mountains, Ghana), and Marloth (1907, South Africa). Tyson (1964) describes the berg winds of South Africa, which are usually characterized by a spectacular rise in temperature. Saboureau (1958) provides a graphic account of the destruction wrought by cyclones and floods on the vegetation of Madagascar, and Sauer (1962) deals with the influence of cyclones on coastal vegetation in Mauritius.

Several workers have attempted to classify climate by means of indices based on selected factors which are assumed to be important in influencing plant growth, especially that of cultivated crops. The most widely used classifications of this kind are those of Köppen and of Thornthwaite. More recently Holdridge has tried to define the world's major vegetation units in terms of logarithmic increases of temperature and rainfall units. The use in such systems of mean values which take no account of extremes was criticized by Moreau as long ago as 1938. The systems of Köppen, Thornthwaite, and Holdridge are discussed in some detail for southern Africa by Schulze & McGree (1978, q.v. for references). The application of Thornthwaite's system with various refinements to rangeland management in East Africa is described by Pratt & Gwynne (1977). Emberger's climatic index is referred to in Chapter VII.

As Walter (1963) points out, it is not possible to express climate satisfactorily by figures or formulae, however complex, because of the seasonal rhythm of the most important factors, and variation from one year to another. By contrast, diagrams, although far from perfect, can be used to summarize an enormous amount of relevant information, and permit rapid visual comparison of different stations and different vegetation types or different chorological and climatic regions.

Walter (1955b, 1959, 1963) has adapted a model first proposed by Gaussen (1955), and has published c. 10 000 climatic diagrams in an atlas covering the whole world (Walter & Lieth, 1960–67; Walter, Harnickell & Mueller-Dombois, 1975).

Walter's diagrams (see Fig. 3), besides summarizing up to eleven temperature parameters, show the annual march of mean monthly temperature and mean monthly rainfall, plotted on the same scale such that 20 mm of rainfall is equivalent to each increase of 10°C above zero.

It has been empirically established that a relatively arid period prevails when the rainfall curve falls below the temperature curve, and that a relatively humid period occurs when the rainfall curve rises above the temperature curve.

The vertical extent of the hatched and dotted fields in the diagrams provides an indication of the intensity of the humid or arid periods respectively. It must be emphasized, however, that these are relative values which apply only to the climatic type represented by the diagram. This is because it was necessary to indicate drought by the use of a temperature curve instead of an evaporation curve, since potential evaporation is measured only at very few stations, and measurements of radiation, wind, and air humidity are rarely available for estimating evaporation.

The curves for temperature and potential evaporation often run parallel to each other, but these are not identical. In different climatic types the relation of temperature to potential evaporation is different and the difference increases with aridity. A drought period when shown on a climatic diagram is more severe the more arid the climate. This shortcoming of Walter's method is unavoidable where broad comparison on a continental scale is necessary, and is relatively unimportant if the objective is primarily to *describe* the climates of different vegetation types. The *interpretation* of plant growth, however, both indigenous and exotic, in terms of climate is another matter, and for this purpose all indices and diagrams are of limited value, and must be supplemented by *direct* physiological investigation.

Unfavourable seasons caused by cold are also shown on Walter's diagrams. Months with a mean daily minimum below 0°C are indicated by black blocks, and months with an absolute minimum below 0°C are hatched but frost may occur only in exceptionally cold years.

Additional climatic data are shown by numbers. In the tropics the monthly rainfall is often extremely high, and in order to simplify the diagram, the scale for rainfall above 100 mm per month is reduced 10 times. This area is shown in black and represents a perhumid season.

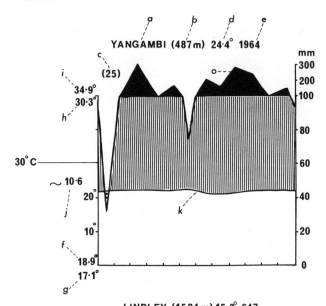

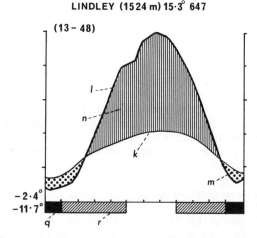

a Station
b Altitude
c Number of years of observation (the first stands for temperature, the second for precipitation)
d Mean annual temperature in °C
e Mean annual precipitation in mm
f Mean daily minimum of coldest month
g Absolute minimum
h Mean daily maximum of hottest month
i Absolute maximum
j Mean daily range of temperature
k Monthly means of temperature (thin line) in °C
l Monthly means of precipitation (thick line) in mm
m Arid period (dotted area)
n Humid period (hatched area)
o Perhumid area (black area), mean monthly rainfall over 100 mm (scale reduced ten times)
q Months with a mean daily minimum below 0°C
r Months with an absolute minimum below 0°C

FIG. 3. Climatic diagrams for Yangambi and Lindley

In the equatorial region the daily variation in temperature is much larger than the seasonal variation. Therefore additional figures are necessary: the mean daily maximum of the hottest month; the absolute maximum temperature; and the amplitude of the mean daily variation of the temperature.

In order to facilitate comparison, the sequence of months for stations in the northern hemisphere starts with January, whereas for the southern hemisphere it begins with July. This means that the warm season is always in the middle of the diagram.

Climatic extremes are often more important for plants than the means, and for agriculture and forestry, in particular, it is important to know how often such extremes may occur. For this purpose climatograms are used. Their construction is nearly the same as that of ordinary climatic diagrams, except for the curves, which show the values for single years (preferably for a period of at least twenty years) instead of the means (see Walter, 1971, p. 56–57). Climatograms have been prepared for very few stations. Although they are indispensable for certain studies they are less essential for the purpose of broad regional comparison.

Walter & Lieth have published more than 1 000 climatic diagrams for Africa in their *Weltatlas*. They distinguish ten principal world climatic types, of which five occur in Africa, namely:

I. *Equatorial* type, humid or with two rainy seasons.
II. *Tropical* type with summer rain.
III. *Subtropical* type, hot and arid.
IV. *Mediterranean* type, with arid summer, and winter rains, frost rare.
X. *Mountain* types.

Each of these is subdivided and the boundaries of 51 subdivisions including transitional types are shown in the *Weltatlas*.

On map 3 of the publication of Walter, Harnickell and Mueller-Dombois (1975), 392 climatic diagrams for the African mainland and 14 for Madagascar and the Comores are superimposed in the approximate location of the meteorological station. The boundaries of the climatic types they represent are not shown on this map. The four main climatic regions in Africa and the transitions between them are, however, shown at a scale of 1 : 30 000 000 on their map 9.

An enormous amount of climatic information is included in the *Weltatlas* and the *Climate Diagram Map*, and certain broad correlations between climate and vegetations are revealed by the latter. Nevertheless, it is still difficult to characterize the main phytochoria and vegetation types of Africa using these publications. For this reason, in the present work separate maps (Figs. 5–8 and 11–23) summarizing their climates have been prepared for all the major African phytochoria. For each phytochorion the climate is also briefly described. Elsewhere in the text climatic features of special significance are mentioned under individual vegetation types.

Although it is relatively easy, using Walter's data, to characterize the climates of the major phytochoria, no attempt is made here to produce a new bioclimatic synthesis, since it is apparent that the relationships between climate and vegetation are much more complex than has generally been supposed. Thus, White (1978b), who has mapped the distributions of the Guineo–Congolian species of *Diospyros*, concluded that historic factors have played an important part. Also some features, such as dry-season atmospheric humidity, which do not contribute to the climatic diagrams, have a considerable influence on the vegetation. Similarly heat advection, as by the harmattan, is largely ignored by Walter's method.

Relatively few attempts have been made to relate growth and phenology to climate factors, other than crudely. A notable exception is the introductory text on tropical forest by Longman & Jeník (1974). The distribution of miombo in relation to temperature and frost has been studied by Ernst (1971).

Various aspects of phenology have been studied by Koriba (1958, tropics generally), Menault (1974, Ivory Coast), Yanney-Ewusie (1968, Ghana), Madge (1965, Nigeria), Njoku (1963, 1964, Nigeria), Huxley & van Eck (1974, Uganda), Kornaś (1977, pteridophytes, Zambia) and Malaisse (1974, Malaisse *et al.*, 1970, 1975, Zaire).

The growth of two rapidly growing pioneer species of secondary forest, *Trema orientalis* (*guineensis*) (Coombe, 1960) and *Musanga cecropioides* (Coombe & Hadfield, 1962), has been analysed in great detail. In both species it was found that the rapid growth appears to be related to the prolonged and efficient development of new leaf-area, rather than to a high rate of dry weight increase per unit leaf-area.

3 Soils

This section is intended to introduce botanists to literature they might find useful and to mention briefly some topics which recur at various places in Part Three, in the context of their associated vegetation.

The soils of Africa have recently been mapped and classified twice for the whole continent. One system, the CCTA classification (D'Hoore, 1964), is local to Africa, created by scientists working in Africa, whereas the other (FAO–Unesco, 1974, 1977) tries to encompass the whole world. The CCTA system is summarized by D'Hoore (1968), Ahn (1970, p. 213–19) and Young (1976, p. 236–40), and the FAO–Unesco system by Young (1976, p. 240–48). Information on the soil resources of tropical Africa is included in a symposium volume edited by Moss (1968). D'Hoore (1959) has briefly compared the soils of South America and Africa.

Regional accounts include those by Ahn (1970) on West Africa, Jones and Wild (1975) on West African savanna soils, and Scott (1962) on East Africa.

Many references to individual countries can be found in the general works cited above but the following publications deserve special mention, in most cases because they attempt to relate vegetation to soils: Ahn (1961, Ghana), Audry & Rossetti (1962, Mauritania), Ballantyne (1968, Zambia), Bawden & Carroll (1968, Lesotho), Bawden & Stobbs (1963, Botswana), Blair Rains & McKay (1968, Botswana), Brown & Young (1964, Malawi), Diniz (1973, Angola), Hemming (1966, Somalia), Hemming & Trapnell (1957, Kenya), Hopkins (1966, Nigeria), Latham & Dugerdil (1970, Ivory Coast), Milne (1947, Tanzania), Morison et al. (1948, Sudan), Perraud (1971, Ivory Coast), Pias (1970, Chad), Streel (1963, Zaire, Upper Shaba, Lufira), A. S. Thomas (1941, Uganda), Thompson (1965, Zimbabwe), Trapnell (1953, Zambia), Trapnell & Clothier (1937, Zambia), Trapnell et al. (1950, Zambia), Webster (in Chapman & White, 1970, Malawi), Wilson (1956, Zambia, Copperbelt) and Young & Brown (1962, Malawi).

The soils of Zaire and Rwanda and Burundi have been mapped in relation to vegetation in greater detail than those of other African countries. Twenty-six publications in the series *Carte des sols et de la végétation du Congo belge et du Ruanda–Urundi* (continued as *Carte… du Congo, du Rwanda et du Burundi*) deal with the following areas:

— Zaire, Rwanda, Burundi (Sys, 1960, general soils map, 1:5 000 000)
— Rwanda, Burundi (Van Wambeke, 1963, soils)
— Kaniama, Upper Lomami (Focan & Mullenders, 1955, soils, vegetation)
— Mvuazi, Lower Zaire (Denisoff & Devred, 1954, soils, vegetation)
— Ruzizi Valley (Germain et al., 1955, soils, vegetation)
— Nioka, Ituri (Holowaychuk et al., 1954, soils, vegetation)
— Mosso, Burundi (Bourbeau et al., 1955, soils, vegetation)
— Yangambi: Weko (Van Wambeke & Évrard, 1954, soils, vegetation)
— Yangambi: Yangambi (Gilson et al., 1956, soils, vegetation)
— Yangambi: Lilanda (Gilson et al., 1957, soils, vegetation)
— Yangambi: Yambaw (Van Wambeke & Liben, 1957, soils, vegetation)
— Bugesera–Mayaga, Rwanda (Frankart & Liben, 1956, soils, vegetation)
— Lufira Valley, Upper Shaba (Van Wambeke & Van Oosten, 1956, soils)
— Lubumbashi, Upper Shaba (Sys & Schmitz, 1956, soils, vegetation)
— Kwango (Devred et al., 1958, soils, vegetation)
— Ubangi (Jongen et al., soils, vegetation)
— Bengamisa (Van Wambeke, 1958, soils)
— Lake Albert (Van Wambeke, 1959, soils)
— Uele (Frankart, 1960, soils)
— Kasai (Gilson & Liben, 1960, soils, vegetation)
— 'Dorsale du Kivu' (Pécrot & A. Léonard, 1960, soils, vegetation)
— Yanonge–Yatolema (Van Wambeke, 1960, soils)
— Karuzi basin (Pahaut & Van der Ben, 1962, soils, vegetation)
— Maniema (Jamagne, 1965, soils)
— Tshuapa–Équateur (Jongen & Jamagne, 1966, soils)
— Paysannat Babua (Frankart, 1967, soils)
— Ubangi (Jongen, 1968, soils)
— Upper Lulua (Gilson & François, 1969, soils)
— Mahagi (Sys & Hubert, 1969, soils)
— Lower Zaire (Compère, 1970, vegetation)
— North Kivu and Lake Edward (Jongen et al., 1970, soils)

Most of the maps are at scales of 1:50 000 or 1:100 000, but others range from 1:10 000 to 1:1 000 000. All are in colour.

General works on weathering and soil formation include those by Thomas (1974) and Nye (1954/5). The origin of laterite is reviewed by McFarlane (1976) and of duricrusts generally by Goudie (1973). Some aspects of soil genesis in Central Africa are discussed by Ellis (1958), Webster (1960) and Paton (1961).

When soils and vegetation are mapped independently the results frequently do not correspond. There are several possible reasons for this. One is that the scale of variation of soil and vegetation is different. Another is that the vegetation is much more attuned to the present-day climate than is the soil, some of which has its origin in a quite different climate from that now. Nevertheless, when soils and vegetation are studied jointly, especially in relation to the evolution of landscape and drainage patterns and the activities of soil-forming animals, good, though sometimes unexpected, correlations are revealed. Studies along these lines are fewer than the purely descriptive and classificatory. Those mentioned below are of particular interest to the botanist.

Trapnell (Trapnell, 1943; Trapnell & Clothier, 1938; Trapnell, Martin & Allan, 1950; see also Astle et al., 1969, and Webster, 1960) was one of the first to combine soil and vegetation in ecological survey. Although he published relatively little his influence has been considerable. Trapnell realized that, at least in Central Africa, climatic effects are often masked by or subsidiary to those of geomorphology and age, and that the processes of intermittent continental uplift, peneplanation and faulting have greatly influenced soil formation and the differentiation of vegetation. He used such topographic terms as 'Plateau', 'Upper Valley' and 'Lake Basin' to denote major soil groups in Zambia. Trapnell's units were based on the dominant, that is, the most extensive, vegetation–soil association in an area. They did not purport to describe fine detail nor to list all the minority vegetation–soil types within each mapping unit. Where such considerations are important the recognition of catenas and land-forms has proved valuable.

The concept of the catena (Milne, 1935, 1936, 1947), which was originally defined 'as a certain sequence of soil profiles in association with a certain topography', is particularly valuable in explaining the relationships between soils and vegetation. Following Milne's example, several investigators in describing soil catenas have also included vegetation. Of special interest to the botanist are the catenas described by Bourguignon et al. (1960), Duvigneaud (1953), Lawson et al. (1968, 1970, Ghana), Morison et al. (1948, Sudan Republic), Radwanski & Ollier (1959, Uganda), Watson (1964, 1965, Zimbabwe), Webster (1965, Zambia) and Williams (1968, stabilized sand dunes, Gezira, Sudan Republic).

A repeating pattern of vegetation change analogous to that associated with soil catenas occurs at many localities in arid and semi-arid regions. Vegetation stripes running parallel to the contours occur on virtually flat to gently sloping surfaces. They are very conspicuous on aerial photographs, and have variously been described as vegetation arcs, bands or ripples and as *'brousse tigrée'* (Boaler & Hodge, 1964; Clos-Arceduc, 1956; Hemming, 1965; Macfadyen, 1950; L.P. White, 1970, 1971; Wickens & Collier, 1971; Worrall, 1959). The vegetation of the bands is denser, taller and physiognomically more complex than that of the intervening lanes, which are sometimes virtually devoid of vegetation. According to L.P. White, in most of the examples described, the stripes are not due to gross soil differences. What differences there are between the soils of the two phases can be accounted for by the influence of the vegetation itself.

For certain purposes of land classification for management planning the catena is too restrictive, and in recent years several studies, inspired by the work of Bourne (1931), have used a framework of land classes based on physiography, within which the different soil and vegetation units are subordinated. This approach has been widely used in Australia. In Africa it provides the bases of the land-resource studies of the British Ministry of Overseas Development. For a general discussion see Astle et al. (1969).

This method recognizes that in any one landscape there are only a few kinds of terrain, each with its particular combination of land-forms, rocks, soil and vegetation. These few terrain types recur in association with one another in the landscape to give a more or less regular pattern always with the same interrelations. A new landscape is recognized where there is a change either in the terrain types or in the relations between them. This approach, which is of wider application than that of the catena, is more appropriate to rapid survey of extensive little-known areas. Astle et al. found that the patterns of landscape are easily distinguished on air photographs and their component parts consistently recognizable under the stereoscope. They recognize composite landscape patterns called Land Systems, the component parts of which are Land Facets. In their survey of the Luangwa Valley, Astle et al. describe the soils and vegetation of forty-six Land Facets which are grouped in nine Land Systems. Similar methods have been used in Botswana (Bawden & Stobbs, 1963; Bawden, 1965), western Kenya (Scott et al., 1971), Lesotho (Bawden & Carroll, 1968), Nigeria (Bawden & Tuley, 1966), Swaziland (Murdoch et al., 1972), and Uganda (Ollier et al., 1969).

The part played by termites in tropical soil formation is certainly considerable but its extent is still controversial. Termites dominate the macrofauna of tropical and subtropical soils in the way that earthworms dominate temperate soils, but their influence descends to considerably greater depths. Termites transport large volumes of mineral and organic matter, both vertically and laterally, and in many places the top metre or so of soil has been radically modified by termite action. The nests of some species of termite are contained in mounds of earth built up above the ground surface. The largest mounds, which may be up to 9 m in height, represent the work of species of *Macrotermes*. They are particularly conspicuous on the unrejuvenated plateau surface in the heart of the Zambezian Region, and most of the recent literature on the soil-forming activities of termites refers to them (Hesse, 1955; Meikeljohn, 1965; Sys, 1955; Watson, 1962a, 1967, 1969, 1974a, 1974b).

The soil within a termite mound is very different from that surrounding it. Usually there is more clay and less coarse sand. The pH is nearly always higher and the content of carbon, nitrogen and exchangeable bases, especially calcium, is greater. The microflora is also distinctive. There are more cellulose decomposers, denitrifiers, ammonifiers and nitrifiers, but fewer nitrogen-fixing bacteria of the genera *Beijerinckia* and *Clostridium*. Termite mounds are subjected to less leaching than the surrounding soils. This should favour the retention of bases within the mounds, but it does not explain how they enter. The work of Trapnell *et al.* (1976) indicates that bases removed from the soil by the woodland canopy become concentrated in termitaria by the termites themselves which feed largely on dead wood and litter. The vegetation of large termite mounds, which is very different from that of the surrounding soil, is briefly described at appropriate places in Part Three.

The soils of the Zambezian Region are more diversified than those of some other parts of Africa and their influence on vegetation has been studied in greater detail. The extensive literature on the vegetation of heavy metal and other toxic soils, which in Africa have been chiefly studied in the Zambezian Region, has been reviewed by Wild (1978). The distinctive soils of mopane woodland and other types of Zambezian vegetation are discussed in Chapter II.

Savory (1963) has shown that the distribution and height of the dominants of miombo woodland in Zambia are closely correlated with effective soil depth. Elsewhere, root systems have been studied by Huttel (1969, 1975, Ivory Coast), Okali *et al.* (1973, Ghana) and Glover (1950–51, Somalia). Kerfoot (1963) has published a short review.

Experiments (Grant, n.d.) have shown that the very acid infertile Kalahari sands of Zimbabwe are deficient in boron and sulphur. Some soils in the Cape Region are also deficient in trace elements and even indigenous species show symptoms of nutritional imbalance (Schütte, 1960). There is evidence that sclerophylly is sometimes associated with nutrient deficiency, especially of phosphorus (Loveless, 1961, 1962; Beadle, 1966, 1968; Grubb & Tanner, 1976; Grubb, 1977), but little information is available from Africa.

The effects on soil of shifting cultivation have been described by Nye & Greenland (1960) and Vine (1968). The former publication remains the most important work in English on soil fertility in the tropics. Nitrogen supply in the forests and savannas of West Africa has been studied by de Rham (1974). Milne & Calton (1944) discuss the effect of clearing vegetation on soil salinity in a semi-arid part of Tanzania. The influence of annual burning on soil structure and fertility is dealt with by Moore (1960, derived savanna zone, Nigeria) and Trapnell *et al.* (1976, Zambia).

Soil factors affecting the distribution of vegetation types and their utilization by wild mammals in East Africa are described by Anderson & Talbot (1965, Serengeti Plains) and Anderson & Herlocker (1973, Ngorongoro crater).

Some leguminous trees are believed to increase soil fertility. In the Sudan, according to Radwanski and Wickens (1967), the yields of sorghum and other crops planted under *Acacia albida* are considerably greater than elsewhere. The pod and leaf-fall and perhaps the dung and urine of cattle which eat the pods and seek the shade of the trees increase the supply of nutrients and improve the physical condition of the soil. Dancette & Poulain (1968) describe a similar study of *Acacia albida* in Senegal.

4 Animals

As explained in the general Introduction, the aim of this book is primarily descriptive, namely to provide a classificatory framework within which more detailed and localized studies, both of plants and animals, can be conducted and compared. The *classification*, however, is based on plants alone. In this context animals have been deliberately left out. This is because their habitats are too imperfectly correlated with vegetation types to provide useful diagnostic features. Their ranges either greatly transgress the boundaries of vegetation types or, alternatively, when they are confined to an individual type, they usually occupy only part of it.

Theoretical concepts such as the *biocoenose*, which attach equal importance to animals and plants, are useful, indeed essential, for the *understanding* of vegetation, but they are of little classificatory value.

The *interpretation* of vegetation is another matter. Until about twenty years ago, the importance of the interaction of plants and animals in shaping African vegetation had, with few exceptions, been gravely neglected by botanists and zoologists alike. Recently, however, there has been much activity in this field, especially by zoologists, but progress has been patchy, with most effort concentrated on large mammals, chiefly in East Africa.

Detailed consideration of the effects of animals on vegetation lies outside the terms of reference of this book, even if sufficient information for a general synthesis were available, which is not the case. Nevertheless, where animals greatly influence vegetation this is briefly mentioned in the text.

In an attempt to compensate for the omission just mentioned, this chapter on 'Animals' is mainly intended to introduce botanists to zoological literature of botanical interest. It is meant to be no more than an entrée to the subject, not a complete review.

General works are few, but Cloudsley-Thompson (1969) and Owen (1976) have respectively written introductions to the zoology and animal ecology of tropical Africa. Curry-Lindahl (1968) deals with zoological aspects of the conservation of vegetation in tropical Africa. Petersen & Casebeer (1971) have compiled a bibliography relating to the ecology of East African large mammals.

Most other general works are devoted to single taxonomic groups of animals, especially mammals, more rarely birds. The books by Delany & Happold (1978) on mammals, by Leuthold (1977) on ungulates, and by Moreau (1966) on the birds of Africa include much of interest to the botanist, as does Kingdon's (1971–77) encyclopaedic work on East African mammals. Bigalke's treatment (1978) of the biogeography and ecology of the mammals of southern Africa is particularly valuable for its references to the literature, as is the review by Bourlière & Hadley (1970) of the ecology of tropical savannas. On a more local scale, Rosevear's (1953) check-list provides much information on the distribution of Nigerian mammals in relation to vegetation. Bourlière & Verschuren (1960) have published a monograph on the ecology of ungulates in the Albert National Park.

Most of the publications mentioned below deal with a single animal species or a few related species. Relatively few deal with animal communities in relation to plant communities. Important works among the latter are those by:

— Chapin (1932, 1939) on the birds of Zaire.
— Moreau (1935*a*) on the birds of the Usambara Mts, Tanzania.
— Fraser Darling (1960) on the ecology of the Mara Plains in Kenya.
— Lamprey (1963, 1964) on ecological separation and population dynamics of the large mammal species in the Tarangire Game Reserve, Tanzania.
— Coe (1967) on the vertebrate fauna of the Afroalpine zone in Kenya.
— Anderson & Herlocker (1973) on soil factors which affect the distribution of vegetation types and the animals which use them in Ngorongoro Crater, Tanzania.
— Sinclair & Norton-Griffiths (1979) on the dynamics of the Serengeti ecosystem in Tanzania.

Acocks (1979) has attempted a reconstruction of the vegetation of the whole of the drier half of South Africa in relation to its indigenous fauna, as it was before the advent of the European.

In recent years much work has been done on the larger mammals, chiefly in relation to their feeding behaviour, population dynamics and influence on vegetation in national parks and game reserves. The pioneer publications of Eggeling (1939), Mitchell (1961*a*), Walter (1961) and Cornet d'Elzius (1964), which emphasized the potentially far-reaching effects of game animals on vegetation, have been followed by numerous detailed studies of several animal species, and by monographic treatments of the elephant (Wing & Buss, 1970; Laws *et al.*, 1975) and the buffalo (Sinclair, 1977).

Many publications deal with the feeding ecology of herbivores, for example:

— baboon (Lock, 1972b).
— black rhinoceros (Goddard, 1968, 1970).
— buffalo (Vesey-FitzGerald, 1969, 1974a; Leuthold, 1972; Sinclair & Gwynne, 1972; Grimsdell & Field, 1976).
— buffalo, hippopotamus, Uganda kob, topi, warthog, and waterbuck (Field, 1972).
— Chanler's mountain reedbuck (Irby, 1977).
— duiker (Wilson & Clarke, 1962; Wilson, 1966).
— elephant (Napier Bax & Sheldrick, 1963; Field & Ross, 1976).
— fringe-eared oryx (Root, 1972).
— gerenuk (Leuthold, 1970).
— giraffe (Innis, 1958; Foster, 1966; Foster & Dagg, 1972; Leuthold & Leuthold, 1972; Field & Ross, 1976).
— greater kudu (Wilson, 1965).
— hares (*Lepus capensis*, *L. crawshayi*; *Pronolagus crassicaudatus*, Stewart, 1971a–c).
— hippopotamus (Field, 1970; Lock, 1972a).
— impala (Stewart, 1971d; Rodgers, 1976).
— lechwe (Vesey-FitzGerald, 1965b).
— lesser kudu (Leuthold, 1971).
— primates generally (Clutton-Brock, 1977, ed.).
— rock hyrax (Sale, 1965).
— rock hyrax and tree hyrax (Turner & Watson, 1965).
— sitatunga (R. Owen, 1970).
— waterbuck (Kiley, 1966).
— wildebeest (Talbot & Talbot, 1963).
— wildebeest and zebra (Owaga, 1975).
— wildebeest, zebra, and hartebeest (Casebeer & Koss, 1970).
— wildebeest, Thomson's gazelle, Grant's gazelle, topi, and impala (Talbot & Talbot, 1962).
— various ungulates (Pienaar, 1963; Gwynne & Bell, 1968; Stewart & Stewart, 1971; Pratt & Gwynne, 1977).

The grazing succession throughout the year of the eight commonest large herbivores in the Rukwa Valley, Tanzania, namely elephant, buffalo, hippopotamus, puku, topi, zebra, bohor reedbuck and eland, has been described by Vesey-FitzGerald (1960, 1965a), who has also (1973b, 1973c) studied browse production and utilization in Tarangire and Lake Manyara National Parks. His surveys indicated that at the time of the study the animals, principally elephant, rhinoceros and giraffe, were utilizing only half of the browse available.

Where large mammals, especially the elephant, have been protected in recent years, they have often shown a dramatic increase in numbers, sometimes accompanied by wholesale destruction of vegetation with concomitant landscape change (see page 114). The extent to which such population changes can be regarded as 'natural' and the desirability of exerting artificial control are still matters of controversy and have given rise to an extensive literature.

Publications which discuss large mammals as agents of habitat and landscape change include those by:

— Agnew (1968, Tsavo National Park (East), Kenya).
— Buechner & Dawkins (1961, elephant, Murchison Falls National Park, Uganda).
— Douglas-Hamilton (1973, elephant, Lake Manyara, Tanzania).
— Glover (1963, elephant, Tsavo).
— Glover & Wateridge (1968, cattle and wild ungulates as a cause of terrace erosion).
— Harrington & Ross (1974, elephant, Kidepo Valley National Park, Uganda).
— Kortlandt (1976, elephant, Tsavo).
— Lamprey *et al.* (1967, elephant, Serengeti National Park, Tanzania).
— Laws (1970a, 1970b, elephant, East Africa).
— Penzhorn *et al.* (1974, elephant, Addo Elephant National Park, Eastern Cape Province, South Africa).
— Thomson (1975, elephant in *Brachystegia boehmii* woodland, Chizarira Game Reserve, Zimbabwe).
— Van Wyk & Fairall (1969, elephant, Kruger National Park, Transvaal).
— Watson & Bell (1969, elephant, Serengeti).

In an attempt to understand the long-term implications of the recent impact of large mammals on the vegetation of reserves, several studies of the movement, habitat utilization, biomass, density, mortality, age structure and population dynamics, again especially of the elephant, have been made, including those by:

— Bourlière (1965, ungulates generally).
— Coe *et al.* (1976, large African herbivores).
— Corfield (1973, elephant).
— Lamprey (1964, large mammals generally).
— Leuthold (1976, elephant).
— Leuthold & Leuthold (1976, ungulates generally).
— Leuthold & Sale (1973, elephant).
— Olivier & Laurie (1974, hippopotamus).
— Sinclair (1974, buffalo).
— Western & Sindiyo (1972, black rhinoceros).

Not all degradation in reserves is brought about by large mammals, and in some cases their influence is indirect.

That fire is often involved in the destruction of vegetation cannot be denied, though its influence varies locally, and interactions between fire and elephants deserve further study. According to Verdcourt (in litt. 18 December 1978) elephants in Kenya have sometimes been blamed for the extensive depredations of charcoal-burners.

Even in areas such as the Akagera National Park in Rwanda, where elephants are virtually unknown today, a high proportion of trees are felled by wind or damaged by lightning, and so simulate elephant damage (Spinage & Guiness, 1971).

In areas of low rainfall (*c*. 350–400 mm per year), especially in basins of closed drainage, elephants might play a catalytic rather than a primary role in the decline of woody vegetation (Western & Van Praet, 1973). Thus, for the Amboseli Game Reserve in Kenya, Western & Van Praet cite evidence which indicates that cyclical climatic change is responsible for a cyclical alternation of *Acacia xanthophloea* woodland and treeless halophytic communities dominated by *Suaeda monoica*. During the wet phase of the cycle the water-table rises by as much as 3.5 m, and the capillary fringe introduces a high level of soluble salts into the rooting horizon of the *Acacia xanthophloea* trees, which ultimately die and fail to regenerate. Elephants merely hasten death brought on by other causes.

Notwithstanding the fact that trees are destroyed by fire, charcoal-burners, wind, lightning and changes in soil salinity, much degradation must be attributed to elephants and other large mammals. It was formerly assumed that elephant/forest systems possess a stable equilibrium point, and that rapid decline of vegetation only occurs when this equilibrium is displaced by man, leading to local high densities of elephants.

Caughley (1976) has proposed an alternative hypothesis. He believes that the relationship is a stable limit cycle in which elephants increase while thinning the forest, and then decline until they reach a density low enough to allow resurgence of the forest. His hypothesis is incomplete since he does not mention the possible effects of cyclical climatic change on the elephant/vegetation cycle nor does he consider the relative importance of decline in elephant numbers due to natural catastrophic mortality and migration. He does, however, discuss modifications brought about by man. Caughley's evidence comes from the Zambezi Valley in Zambia, where the uneven distribution of age classes of elephant-damaged baobab (*Adansonia digitata*), and mopane (*Colophospermum mopane*) trees, suggests a cycle of the order of 200 years.

A cyclical relationship between animal numbers and vegetation, but in this case controlled by rainfall, has also been postulated by Phillipson (1975), who related elephant mortality in Tsavo National Park (East) in Kenya to estimated primary production. He concluded that 'carrying capacity' declines markedly approximately once every 10 years for large mammals other than elephant, and only once every 43–50 years is there a change sufficiently marked to result in a crash in elephant numbers. There is no single level of population at which a steady state is possible.

Nearly all the literature on elephant ecology refers to East and southern Africa. Little has been published on the Guineo–Congolian Region, but elephants are known to retard succession of vegetation at places where they bathe and drink. In Zaire, for instance, the vegetation near elephant baths dominated by *Rhynchospora corymbosa* might represent an elephant-maintained subclimax (Léonard, 1951).

The possibility of domesticating wild mammals for meat production is discussed by Dasmann (1964) and Parker & Graham (1971). Despite promising beginnings game ranching in most places has failed as an economically viable agricultural practice. Alternative ways of cropping wild herbivores are considered by Pratt & Gwynne (1977) for East Africa and by Huntley (1978) for southern Africa.

Notwithstanding the considerable importance of fruit and seed dispersal by birds and mammals the subject has been little studied. The few publications include those by:

— Burtt (1929, 28 plant species by 9 mammal and 4 bird species).
— Clutton-Brock (ed., 1977, primates).
— Gwynne (1969, *Acacia* by ungulates).
— Hladik & Hladik (1967, primates, Gabon).
— Jeník & Hall (1969, *Detarium microcarpum* by elephants).
— Kingdon (1971–77, mammals generally).
— Lamprey (1967, *Acacia* by ungulates; *Commiphora* by birds, baboons, and monkeys).
— Lamprey *et al.* (1974, *Acacia tortilis* by elephant, impala, dikdik, and Thomson's gazelle).
— Leistner (1961*b*, *Acacia erioloba* by elephant, giraffe, black rhinoceros, gemsbok, and eland).
— Phillips (1926*b*, forest trees in the Knysna region, Cape Province, South Africa by wild pig).
— Van der Pijl (1957, bats generally).
— Wilson & Clarke (1962, *Pseudolachnostylis maprouneifolia* by the common duiker).

The effects on vegetation brought about by some groups of insect are no less important than those of mammals, both directly (locusts; armyworm, Edroma, 1977) and indirectly through soil formation (termites), or through population control of mammals including man (tsetse fly). For each group there is an extensive specialized literature. Only a few of the more general works can be mentioned here.

The habitats of the Desert Locust are described by Guichard (1955) and by Hemming & Symmons (1969) and those of the Red Locust by Backlund (1956), Vesey-FitzGerald (1955*a*, 1964) and Rainey *et al.* (1957).

The interrelationships of vegetation and termites have been described by Murray (1938) for South Africa, by Wild (1952*a*, 1975) for Zimbabwe, by Fries (1921) and Fanshawe (1968) for Zambia, and by Malaisse (1976*a*) for Zaire (Upper Shaba). The last author (Malaisse, 1978*b*) has also published a well-documented review of the termite mound as an ecosystem for the whole of southern Africa. References to the soil-forming activity of termites are given in Chapter 3.

Important publications on the tsetse fly include those of Goodier (1968), Nash (1969), Ford (1971) and Ormerod (1976).

Despite a recent renewed interest in symbiotic relationships between ants and plants few publications deal with Africa. The association between ants and gall Acacias has been studied by Brown (1960), Monod & Schmitt (1968), Hocking (1970, 1975), and Foster & Dagg (1972). The protection of the rain-forest tree *Barteria fistulosa* by *Pachysima* ants is described by Janzen (1972).

5 Fire, land use and conservation

Fire

The effects of fire on vegetation are often referred to in publications mainly concerned with other matters. Those mentioned below deal exclusively or principally with the subject, mostly with fires started deliberately or accidentally by man, but Komarek (1964, 1972) reviews the occurrence of fires caused by lightning.

It is generally agreed that frequent uncontrolled fires are harmful both to vegetation and soil, and that for some purposes controlled burning is beneficial. Some controversy, however, still surrounds the precise regime to be followed, and the long-term effects are often uncertain. In this account no attempt is made to resolve these problems. The principal literature is merely referred to. Other references and information are given in Part Three in relation to individual vegetation types.

General works dealing with the ecological effects of fire and its use in land management include those by Humbert (1938), Bartlett (1956), Ahlgren (1960), West (1965), Daubenmire (1968), and Glover (1968, 1972), and for Africa only, Guilloteau (1957) and Phillips (1965, 1968, 1972, 1974). West Africa is covered by Scaëtta (1941), Viguier (1946), Pitot (1953) and Rose Innes (1972), southern Central and eastern Africa by Van Rensburg (1972), and the Mediterranean Region by Naveh (1974).

The role of fire in individual territories has been described by Lamotte (1975b) and Monnier (1968) for the Ivory Coast, by Hopkins (1963, 1965d) for Nigeria, by Robyns (1938) for Zaire, by Spinage & Guiness (1972) for Rwanda, by Masefield (1948), Ross (1968, 1969), Spence & Angus (1971), Wheather (1972), Harrington (1974), and Harrington & Ross (1974) for Uganda, by Edwards (1942), Thomas & Pratt (1967), and Olindo (1972) for Kenya, by Vesey-FitzGerald (1972) for Tanzania, by Lemon (1968) and Chapman and White (1970, p. 31–4) for Malawi, by Austen (1972), Kennan (1972) and West (1972) for Zimbabwe, by Brynard (1964) and Van Wyk (1972) for the Kruger National Park, Transvaal, by Nänni (1969) and Scott (1972) for Natal, by Michell (1922), Martin (1966), and Trollope (1972, 1974) for the Cape Province, and by Humbert (1927b) and Morat (1973) for Madagascar.

Several publications describe the effects on vegetation of controlled burning policies carried out for a number of years. One of the first was Swynnerton's (1917) account of the encroachment of secondary grassland by forest precursor species following 15 years of complete fire protection in Rhodesia (now Zimbabwe). At about this time an experimental study of the effects of controlled burning on vegetation near Pretoria was initiated by E. P. Phillips (Scott, 1972) but was not continued. The results of similar long-term experiments carried out at Frankenwald near Johannesburg are described by Glover & Van Rensburg (1938) and Davidson (1964). An account of burning experiments started in 1927 in secondary upland grassland in the Southern Highlands of Tanzania is given by Van Rensburg (1952).

A short-term experiment described by Levyns (1927) suggests that at least locally in Cape 'fynbos' fire favours the Rhenosterbos, *Elytropappus rhinocerotis*. More recent experiments in this type of vegetation are briefly referred to by Taylor (1978).

Our understanding of the nature of climax and fire-induced vegetation in the Zambezian Region has been enhanced by the fire-control experiments described by Schmitz (1952b), and particularly by Trapnell (1959; Trapnell *et al.*, 1976).

The effects of different types of burning treatment on secondary wooded grassland occupying rain-forest sites in Nigeria have been described for Olokemeji Forest Reserve near Ibadan by MacGregor (1937) and Charter & Keay (1960), and for Anara Forest Reserve near Onitsha by Onochie (1961). The results of similar experiments in northern Ghana have been reported by Ramsay & Rose Innes (1963). Adam & Jaeger (1976) have shown that certain grasses, e.g. *Hyparrhenia subplumosa* and *Rhytachne rottboellioides*, do not flower in the absence of fire. The seed of *Themeda triandra* is buried by hygroscopic movements of the awn to a depth of 1 cm, where it is protected from the heat of fires (Lock & Milburn, 1971).

Land use

General works on land use and conservation relevant to Africa include those by Talbot (1964), and Bourlière & Hadley (1970) on tropical savannas, and by Whyte (1974) on tropical grazing lands. The effect of shifting cultivation on soil fertility is reviewed by Nye & Greenland (1960). *The World Atlas of Agriculture* (Anon, 1976) is a useful work of reference. A review on the use and misuse of shrubs and trees as fodder (United Kingdom, Imperial Agricultural Bureaux, 1947) deals comprehensively with Africa.

The Unesco-UNEP-FAO state-of-knowledge report on tropical grazing land ecosystems published by Unesco (1979) contains detailed case histories of Lamto (Ivory Coast) by Lamotte, and of the Serengeti region (Tanzania) by Lamprey, as well as a more general survey of mid-west Madagascar by Granier.

Among publications exclusively concerned with Africa are studies by Allan (1965) on the African husbandman, and on soil resources and land use (1968), and by Phillips (1959) on agriculture in relation to ecology. The agricultural ecology of the savanna regions of West Africa is dealt with by Kowal & Kassam (1978). The extensive literature on primary production ecology in southern Africa is summarized by Rutherwood (1978).

Harroy (1949) and De Vos (1975) broadly review for the whole continent man's harmful effects on vegetation and soil. Aubréville (1947b, 1949a, 1949b, 1971) has written extensively on the consequences of the destruction of vegetation. The regression of the forest boundary in the Ivory Coast has been described by Lanly (1969). Halwagy (1962a, 1962b) deals with the consequences of overgrazing in the northern Sudan. In a remarkable study, Shantz & Turner (1958) provide photographic evidence from thirty widely scattered localities in Africa of vegetational change over a third of a century. The precise interpretation of some photographs is difficult because of a lack of documentation but the overall impression is that in many places there has been less degradation than might have been expected. For the drier parts of South Africa, Acocks (1979) has emphasized the severe degradation of soils and vegetation following the advent of the European.

Various specialized aspects of land use are dealt with individually below. In addition, for many countries, there are more general accounts which to varying degrees attempt to relate land use in general to natural vegetation. They include works by Knapp (1968b) on Tunisia, by Wills (1962, ed.) on Ghana, by Bawden & Tuley (1966) on Nigeria, by Hawkins & Brunt (1965) on Cameroun, by Gillet (1962b, 1963, 1964) on Chad, by Tothill (1948, ed.) on the Sudan Republic and Uganda (1940, ed.), by Langdale-Brown *et al.* (1964) on Uganda, by Malaisse (1978a) on the miombo ecosystem, by Boaler & Sciwale (1966) on Tanzania, by G. Jackson (1954) on Malawi, by Trapnell & Clothier (1937), Trapnell (1953), Van Rensburg (1968), Astle *et al.* (1969), and Verboom & Brunt (1970) on Zambia, by Vincent & Thomas (1961) on Zimbabwe, by Diniz (1973) on Angola, by Bawden & Stobbs (1963), Blair Rains & McKay (1968) and Blair Rains & Yalala (1972) on Botswana, by Staples & Hudson (1938) and Bawden & Carroll (1968) on Lesotho, and by Pentz (1945) and D. Edwards (1967) on Natal. Two anonymous publications (Huntings Technical Services, 1964; Jonglei Investigation Team, 1954) include important information on the Sudan Republic.

There are many publications on the African rangelands which are defined as 'land carrying natural or semi-natural vegetation which provides habitats suitable for herds of wild or domestic ungulates' (Pratt, Greenway & Gwynne, 1966).

The comprehensive review of the rangelands of East Africa by Pratt & Gwynne (1977) is also relevant elsewhere in Africa. Other general works include the Unesco volumes on arid-zone research, especially volume VI (Unesco, 1955), papers by Grove (1977) and Rainey (1977) and a book on pastoralism in Africa edited by Monod (1975).

Regional and local studies have been published on the rangelands of the Mediterranean Region by Tomaselli (1976), of East Africa by Heady (1960, 1966) and Woodhead (1970), of South and East Africa by Phillips (1956), of southern Africa by Shaw (1875), Acocks (1964) and Pereira (1977), of the Syrte region of the Libyan Arab Jamahiriya by Nègre (1974), and of the Sidi Barani region of Egypt by Migahid *et al.* (1975). The Sahel zone has an extensive literature including a large number of publications from the Institut d'Élevage et de Médecine Vétérinaire des Pays Tropicaux, which are listed by J.-P. Lebrun (1971a). Other publications dealing with the Sahel, many of which describe the effects of drought and overgrazing, include those by Mourgues (1950), Boudet & Duverger (1961), Halwagy (1962a, 1962b), Gillet (1967), Depierre & Gillet (1971), Boudet (1972) and Wade (1974).

Studies made in individual countries have been published by Long (1955, Egypt), Batanouny & Zaki (1973, Egypt), Gillet (1960, 1961a, 1961c, Chad), Dawkins (1954, Uganda), Kelly & Walker (1976, Zimbabwe), Walter & Volk (1954, Namibia) and Volk (1966a, Namibia).

Overgrazing in rangelands sometimes results in an undesirable thickening of the woody plants. This is discussed by Walter (1954), West (1958), Volk (1966a), Lawton (1967b) and Thomas & Pratt (1967).

Secondary grasslands which are used for grazing but occur under wetter climates than those of rangelands are described by Scaëtta (1936, high mountains of East Africa), Trochain & Koechlin (1958, Gabon, Congo), Malato Beliz & Alves Pereira (1965, Guinea-Bissau), Tuley (1966, Obudu Plateau, Nigeria) and Myre (1971, southern Mozambique).

The effect of cultivation on the evolution of landscape in West Africa is described by Portères (1957), Clayton (1963), and Miège *et al.* (1966). The origin of 'derived savanna' is discussed by Keay (1959b) and Clayton (1961).

Jackson & Shawki (1950) deal with shifting cultivation in the Sudan.

The hydrological consequences of changes in land use in East Africa are described by Pereira (1962, ed.).

Wicht (1971) describes the influence of South African montane vegetation on water supplies.

A publication by Wild (1961) deals with harmful aquatic plants in Africa and Madagascar.

Jordan (1964) stresses the importance of a sound knowledge of the natural vegetation and its relation to soil in the successful reclamation of mangrove for rice-growing.

There are few parts of Africa where wild plants are still a principal source of nutrition, but this is true for some tribes of Bushmen whose feeding habits are described by Story (1958), Lee (1966), Heinz & Maguire (1974), and Maguire (1978, q.v. for other references). In some countries rain-forest trees play an important part in the diet of the local people. Okafor (1977) reviews attempts to improve fruit production by selection and propagation in Nigeria.

There is a voluminous literature on forestry practice in relation to the natural environment but there are few regional syntheses. An important exception is Boudy's encyclopaedic work on North Africa (1948, 1950). Afforestation in the Sahel and the Maghreb is described by Fishwick (1970) and Métro (1970) respectively. Martin's (1940) account of forestry in Barotseland in relation to the agricultural practices and the needs for timber of the local people is an important example of an imaginative approach which has been too rarely followed. Leggat (1965) suggests solutions to the conflicting needs of forestry and game preservation in Uganda.

Conservation

Following the publicity given to the need to conserve wild life by Huxley (1958), Darling (1960), Worthington (1961) and others, many game reserves have been established, though sometimes, apparently, with unfortunate consequences for the vegetation (Chapter 4). There has been less deliberate effort to safeguard vegetation, and many plant species and even whole ecosystems are at risk. The symposium volume edited by I. & O. Hedberg (1968) includes summaries of the conservation status of the vegetation in all African countries south of the Sahara, and remains an important working document. More recently, Huntley (1978) has reviewed the situation for southern Africa, and Rodgers and Homewood (1979) have published carefully prepared and realistic proposals for the conservation of the floristically and faunistically rich communities of the East Usambara Mts in Tanzania.

Detailed information on sixty-four species of vascular plants which occur on the African mainland and its offshore islands and are believed to be in danger of extinction is given in the IUCN *Red Data Book* (Lucas & Synge, 1978). This work deals with a world-wide sample of 250 species out of an estimated total of 20000–25000 endangered species. They were selected, not only because they themselves are at risk, but also to emphasize the growing and continuing threats to the ecosystems to which they belong.

Part Two   Regional framework, classification and mapping units

Introduction

One hundred cartographic units are shown on the map, eighty by numbers, the remainder by letters.

The classification used is primarily physiognomic. However, partly for reasons of scale, but also because of the inherent complexity of the vegetation itself, nearly all the units include more than one major physiognomic type. Their relationships with neighbouring units are also complex, and their vegetation has usually, though to differing degrees, been severely degraded by man.

For these reasons, if the vegetation of each unit were to be described separately without any regional grouping there would be much repetition or much would have to be left out. It was therefore decided to arrange the legend of the map in the traditional manner by grouping the units according to the physiognomy of their most extensive or characteristic types, but to group them in the text according to the floristic regions (phytochoria) in which they occur. This permits easy comparison of the broad features with those shown on vegetation maps from other continents, while also doing full justice to the complexity of the situation in Africa. There are thus two interconnected classifications which can be used independently. Care has been taken to see that they are adequately cross-referenced.

The interrelations between vegetation types and floristic regions are discussed in Chapter 6. The mapping units are listed in Table 4 with cross-references to the phytochoria under which they are described, together with other relevant information. The main vegetation types are defined in Chapter 7, with some comments on their distribution and criteria for their subdivision.

The boundaries of the phytochoria are shown in Figure 4 (see overleaf) and by thick lines on the vegetation map. For reasons discussed elsewhere (White, 1979), within fairly narrow limits the boundaries of phytochoria can be somewhat elastic, especially of transition zones. The precise position of those shown on the map has sometimes been determined by cartographic convenience.

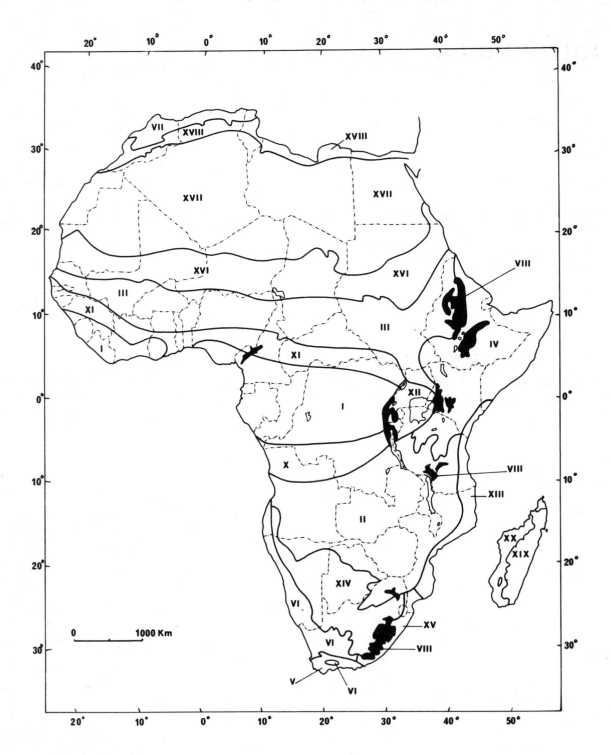

FIG. 4. Main phytochoria of Africa and Madagascar
I. Guineo–Congolian regional centre of endemism. II. Zambezian regional centre of endemism. III. Sudanian regional centre of endemism. IV. Somalia–Masai regional centre of endemism. V. Cape regional centre of endemism. VI. Karoo–Namib regional centre of endemism. VII. Mediterranean regional centre of endemism. VIII. Afromontane archipelago-like regional centre of endemism, including IX, Afroalpine archipelago-like region of extreme floristic impoverishment (not shown separately). X. Guinea–Congolia/Zambezia regional transition zone. XI. Guinea–Congolia/Sudania regional transition zone. XII. Lake Victoria regional mosaic. XIII. Zanzibar–Inhambane regional mosaic. XIV. Kalahari–Highveld regional transition zone. XV. Tongaland–Pondoland regional mosaic. XVI. Sahel regional transition zone. XVII. Sahara regional transition zone. XVIII. Mediterranean/Sahara regional transition zone. XIX. East Malagasy regional centre of endemism. XX. West Malagasy regional centre of endemism

6 Regional framework

Introduction

Main physiognomic types

The need for a regional framework

The main phytochoria

Introduction

The approach to classification described below has developed with little reference to other parts of the world. Africa is the second largest land mass and has a more diversified flora than any equivalent area, though it is not necessarily the richest in species. Its vegetation has also been described in greater detail than that of other parts of the tropics. It has not been possible to classify African vegetation according to such proposed world-wide systems as those of Fosberg (1961), IUCN (1973), and Unesco (1973). This is partly because the units used in these classifications are too briefly defined or are not defined at all and so cannot be recognized, and partly because comparative studies involving Africa and other continents are still too few. The remark made long ago by Richards, Tansley & Watt (1939, 1940) in discussing Burtt Davy's (1938) classification of tropical woody vegetation, namely that existing knowledge is inadequate for the construction of a world-wide natural classification, still remains true.

Main physiognomic types

Vegetation is usually classified on physiognomy but there is wide disagreement concerning the characters to be used (for historical discussion see Du Rietz, 1931), and the extent to which physiognomy should be combined with floristics, environment, and a regional approach (Küchler, 1951).

Physiognomy covers all aspects of the structure of vegetation, but most classifications rely heavily on a few characters such as height, density, thorniness, deciduousness, etc. Comparisons of the vegetation of different regions have all too often in the past been based exclusively on Raunkiaer's system of life-form classes which gives undue prominence to a single character, position of the perennating organs. Although useful for some purposes of superficial comparison, such practice is liable to separate like from like and to bring together things which in many respects are different. This is emphasized by Böcher (1977) who points out that the highest categories in the life-form hierarchy, such as trees, have so few characters in common that it is useless to consider them as biological types.

As long ago as 1913 Drude (for discussion and references see Du Rietz, 1931) rejected Raunkiaer's one-sided emphasis on epharmonic characters believed to be of biological importance. He advocated the use of a wider range of morphological features in the description and classification of vegetation, and suggested that the aim of such studies should be an understanding of the phylogenetic diversification of vegetation. Subsequent progress along these lines has been spasmodic, but some recent studies are revealing an unsuspected structural complexity of tropical and subtropical vegetation. The relationships between plant structure and the environment are also proving to be much more complicated than was previously supposed. It is on the extension of such studies that an improved physiognomic classification will ultimately depend. For Africa, as indeed for the tropics generally, much remains to be done.

Our knowledge of the structure and physiognomy of tropical rain forest is based largely on the methods of analysis and description developed by Richards (Davis & Richards, 1933, 1943; Richards, 1952). More recently, Hallé & Oldeman (1970, 1975; Oldeman, 1974; Hallé, Oldeman & Tomlinson, 1978) have brought a new approach to the study of the architecture and growth of tropical trees which Cremers (1973) has extended to lianes. In a series of papers, Descoings (1971–78) has advocated a more rigorous approach to the classification of African grasslands, but his growth forms are based on only a few characters, so that such structurally and ecologically dissimilar species as the Sudano–Zambezian *Loudetia simplex* and the Mediterranean *Stipa tenacissima* are brought unnaturally together.

Aspects of the origin and adaptive significance of growth forms are discussed by Bews (1925) for South Africa, by Burtt Davy (1922) and White (1976d) for the geoxylic suffrutex flora of south tropical Africa, by Meusel (1952) for certain Mediterranean genera, and by Hedberg (1964) and Mabberley (1973) for Afroalpine vegetation. Several important comparisons have been made in recent years of the physiognomy and ecological physiology of the vegetation of regions with a Mediterranean climate (see Cody & Mooney, 1978 for review and references). The Mediterranean Region itself and the Cape Region, however, have so far contributed less to these studies than the other regions.

In the present work the classification aims to be as simple, informal and non-hierarchical as is compatible with the diversity of its subject-matter. Sixteen major vegetation types are recognized at the highest rank. This is more than is customary, but so large a number is required if unnecessary nomenclatural complexity is to be avoided. Major types such as forest in the past have often been defined so broadly as to include both bamboo and, not merely mangrove forest, but open scrubby communities of mangrove no more than 2 m tall. Such conglomerations are too diverse to be useful and inevitably give rise to unwieldy hierarchies.

For the sixteen main vegetation types the term formation, in its widest sense, namely a kind of plant community characterized by its physiognomy, seems appropriate, although it is not in fact necessary to use it. If formation is defined in this way it can serve as useful and flexible a purpose as other general terms in botany such as taxon, phytochorion and element.

The sixteen formations are very unequal in the size of the areas they occupy and, to a lesser extent, in their degree of physiognomic distinctness, but this feature would remain regardless of the way in which they are classified. This is one of the reasons for classifying them informally. They fall into five main groups (Table 1) and are discussed further and described in Chapter 7.

TABLE 1. The main vegetation types of Africa

1 FORMATIONS OF REGIONAL EXTENT:	2 FORMATIONS INTERMEDIATE BETWEEN THOSE BELONGING TO GROUP 1, AND MOSTLY OF RESTRICTED DISTRIBUTION:	3 EDAPHIC FORMATIONS OF DISTINCT PHYSIOGNOMY:	4 FORMATION OF DISTINCT PHYSIOGNOMY, BUT RESTRICTED DISTRIBUTION:	5 UNNATURAL VEGETATION:
forest *woodland* *bushland and thicket* *shrubland* *grassland* *wooded grassland* *desert* *Afroalpine vegetation* These all occur extensively in at least one major phytogeographic region or transition zone.	*scrub forest* *transition woodland* *scrub woodland* These types are not normally recognized in major classifications, but they are important for the following reasons. First, they enable clearer and less arbitrary distinctions to be made between regional formations. Second, they facilitate the description of transition zones and complex mosaics, and, as in the case of transition woodland, the interpretation of vegetation dynamics.	*mangrove* *herbaceous fresh-water swamp and aquatic vegetation* *halophytic vegetation*	*bamboo*	*anthropic landscapes*

The need for a regional framework

In formal classifications when formations are grouped into higher categories or are subdivided the selection of characters is inevitably arbitrary since there are so many to choose from and many of the facts are often not available. Another drawback of complex classifications is that several descriptive epithets are usually needed to distinguish between the various units, and this both burdens the memory and is an obstacle to verbal communication.

These failings can be avoided if the regional variants of the main formations are designated, not by a few selected physiognomic features but by the phytogeographical areas in which they occur. For Africa at least, the distribution of floras and of major physiognomic types correspond sufficiently closely to make it possible.

In the past the classification of phytogeographical areas or phytochoria has been as arbitrary and subjective as that of vegetation, and there was no general agreement on the criteria to be used. A new chorological map of Africa has recently (White, 1976a, fig. 3) been prepared, the pattern lines of which were first drawn on an early unpublished draft of the Unesco/AETFAT/UNSO vegetation map. They also appear on the published version of the latter, which accompanies this text, in most cases not as the boundaries between single mapping units, but of groupings of related units. For Africa eighteen major phytochoria are recognized and for Madagascar two.

There is no a priori reason why the pattern lines derived from a vegetation map should faithfully delineate floristic regions. Nevertheless, subsequent floristic analysis (Goldblatt, 1978; Moll & White, 1978; White 1976b, 1976c, 1978a, 1978b, 1978c, 1979; White & Werger, 1978), based in part on a large number of detailed published and unpublished distribution maps of individual species, has confirmed the general validity of the phytochoria, though, not unexpectedly, a few minor emendations have been indicated.

It should be emphasized that the distributional data which confirm the boundaries on the chorological map were not used in the compilation of the vegetation map. Most, in fact, were not then available.

This concordance of chorology and physiognomy is doubly significant. It both helps to confirm the objectivity of the mapping units and provides an objective means of designating and differentiating vegetation types of broadly similar physiognomy which occur in different floristic regions. It also can provide a 'shorthand' reference to the entire physiognomy of a vegetation type, since the latter is determined by its entire flora. Reference to sclerophyllous vegetation should make this clear.

Since the time of Schimper (1898, 1903) the vegetation of the various widely separated regions with a Mediterranean climate has been regarded as equivalent both in physiognomy and ecology: 'All districts agreeing with the Mediterranean coast as regards the distribution in time of the rainy and dry seasons repeat in their vegetation essential oecological features of the Mediterranean vegetation.' These regions 'are the home of evergreen xerophilous plants which, owing to the stiffness of their thick leathery leaves, may be termed sclerophyllous woody plants' (Schimper, 1903).

The term sclerophyll has been subsequently extended to similar vegetation in non-Mediterranean climates, such as parts of Australia (see Seddon, 1974) and on high tropical mountains. The leaves of some plants on the latter, however, have some distinctive anatomical features and Grubb (1977) refers to them as 'pachyphylls'. It would be foolish to deny the striking similarities in leaf-form in the different regions with a Mediterranean climate, but no less so to ignore the important differences.

Cody & Mooney (1978), who have made an interesting though limited comparison of the Mediterranean ecosystems of the five main areas in which they occur, found that the structure of the communities, the successional relationships and the growth rhythms all differ substantially among regions, with South Africa being the most dissimilar.

In the Cape Region there is more summer rainfall than elsewhere and the main growth period is not confined to the spring but continues through the summer. The Cape vegetation has a greater diversity of shrubby forms, and is rich in bulbous plants, but poor in annuals and vines. A more detailed comparison would doubtless reveal other differences. Furthermore the successional status of 'Mediterranean' shrubland is different in the different areas. Thus, in the Cape Region 'fynbos' (Chapter V) nearly everywhere represents the climax. True forest, other than some small areas of scrub forest, only occurs as enclaves where summer rainfall or cloud profoundly modifies the Mediterranean climate; floristically it is of Afromontane affinity. By contrast, in the Mediterranean basin evergreen forest dominated by Mediterranean endemic species is the regional climax.

The ecological significance of sclerophylly has long been and still remains a subject of controversy (for recent reviews see Seddon, 1974; Grubb, 1977; Cody & Mooney, 1978). There can be little doubt, however, that in different places different combinations of environmental factors are responsible.

From the above it is clear that sclerophyllous shrubland in the different regions shows appreciable differences in physiognomy, related in part to different environmental conditions which are still imperfectly understood. This poses considerable problems in classification. On the one hand, it is clearly inappropriate that vegetation as different as that of the Mediterranean and Cape Regions should go under the same name, e.g. the 'Hard-leaved scrub bushes (macchia)' of Schmithüsen (in FAO–Unesco, 1977). On the other hand, in view of our incomplete knowledge of the physiognomic differences and particularly of their adaptive role, the selection of discriminating epithets is difficult. However, if the sclerophyllous formations in the different floristic regions are designated and distinguished by the names of those regions, the arbitrary selection of imperfectly understood characters can be avoided. This applies equally to all other multiregional formations. Hence, the name of the phytochorion can serve to embody and provide a key to all the diagnostic features of its vegetation types.

In the main text (Part Three) the descriptions of individual vegetation types include as much physiognomic information as was readily available, but published data are often sparse. Sometimes I have had to rely almost exclusively on my own field records.

An additional advantage of grouping the main vegetation types under the regional phytochoria is that it permits a more effective treatment of mosaics, continua and transitions, and also of the dynamic relationships of important regional types which have suffered degradation at the hands of man. This can be exemplified by the Sudanian and Zambezian Regions, which have broadly similar climates. In both regions, woodland, broadly comparable in physiognomy and floristic composition, is the most widespread type of vegetation. The Sudanian Region, however, belongs to 'Low Africa' (Chapter 1), whereas the Zambezian Region belongs to 'High Africa' and hence has a more diversified physiography and climate. This is reflected in a wider range of vegetation types.

In the Zambezian Region distinctive types of dry evergreen forest, dry deciduous forest, thicket and edaphic grassland are much more widespread than their rather different analogues in the Sudanian Region, and this is only partly due to the effects of man. In the Zambezian Region the subordinate types form complicated mosaics with the regional woodlands and have complex dynamic relationships with them. The pattern itself is more important than the sum of its parts and is best described in a regional context.

The main phytochoria

The chorological map mentioned above differs from earlier maps in two main respects. First, the phytochoria are based on richness (or otherwise) of their endemic floras at the *species* level. Second, no attempt is made to carve up Africa into mutually exclusive areas which are themselves subdivided in a hierarchical manner (Regions, Sectors, Domains, Districts, etc.). Different parts of Africa differ greatly in their richness in endemic species and in the distribution patterns shown by the latter. For this reason a flexible, non-hierarchical system is proposed (for a fuller account see White, 1979). It recognizes at the rank of Region four fundamentally different types of phytochorion, the salient features of which are embodied in the terminology employed (Table 2).

As provisionally defined (White, 1979), a Regional Centre of Endemism is a phytochorion which has both more than 50 per cent of its species confined to it and a total of more than 1000 endemic species. All the phytochoria designated above as Regional Centres of Endemism appear to fulfil these criteria with the exception of the Sudanian Region, the status of which is still uncertain.

The Regional Centres of Endemism are separated by transition zones. If the latter are comparable in size to the former they should be named and given comparable rank. The transition zones between the Cape and Karoo–Namib Regions and between most islands of the Afromontane Region and their surrounding lowland phytochoria are too narrow to justify this. By contrast

TABLE 2. The main phytochoria of Africa and Madagascar

A *Africa*

1 REGIONAL CENTRES OF ENDEMISM:	2 ARCHIPELAGO-LIKE CENTRE OF ENDEMISM:	3 ARCHIPELAGO-LIKE CENTRE OF EXTREME FLORISTIC IMPOVERISHMENT:	4 REGIONAL TRANSITION ZONES AND MOSAICS:
I *Guineo–Congolian*	VIII *Afromontane*	IX *Afroalpine*	X *Guinea–Congolia/Zambezia*
II *Zambezian*			XI *Guinea–Congolia/Sudania*
III *Sudanian*			XII *Lake Victoria*
IV *Somalia–Masai*			XIII *Zanzibar–Inhambane*
V *Cape*			XIV *Kahalari–Highveld*
VI *Karoo–Namib*			XV *Tongaland–Pondoland*
VII *Mediterranean*			XVI *Sahel*
			XVII *Sahara*
			XVII *Mediterranean/Sahara*

B *Madagascar*

1 REGIONAL CENTRES OF ENDEMISM:

   XIX *East Malagasy*
   XX *West Malagasy*

the Guinea–Congolia/Sudania, Guinea–Congolia/Zambezia and the Kalahari–Highveld transition zones are larger than some Regional Centres of Endemism and need to be named. These three transition zones have few endemic species and the majority of their species also occur in adjacent phytochoria.

The transition from the Mediterranean Region to the tropical flora is more complex, and can conveniently be divided into three Regional transition zones, the Sahara, the Sahel and the Mediterranean/Sahara transition zones. The last two have impoverished floras and few endemic species. Endemism is appreciably higher in the Sahara, but both the total and percentage of endemic species are too low for the Sahara to qualify as a Regional Centre of Endemism. Furthermore the Sahara forms a transition between two great floristic kingdoms, the Holarctic and the Palaeotropic, whose *generic* floras, subcosmopolites apart, are almost totally different.

The transition zones mentioned above show a simple gradual replacement of one flora by another which is little complicated by endemism. The three Regional Mosaics, however, are transitional in a more complex manner. In all three the vegetation forms a mosaic of different physiognomic types with different floristic relationships. The Lake Victoria Regional Mosaic has few endemic species, whereas endemism in the Zanzibar–Inhambane and Tongaland–Pondoland Regions is relatively high.

The vegetation of the regional phytochoria is described in Part Three, where the main floristic features of each phytochorion are briefly summarized. The figures relating to the size of the flora and the degree of endemism are estimates based on available information, much of it unpublished (for discussion see White, 1979). Further work will certainly modify the details considerably, though the overall pattern seems to be well established.

7 Classification

Introduction

Description of the main vegetation types
    Forest
    Woodland
    Bushland and thicket
    Shrubland
    Grassland
    Wooded grassland
    Desert
    Afroalpine vegetation
    Scrub forest
    Transition woodland
    Scrub woodland
    Mangrove
    Herbaceous fresh-water swamp and aquatic vegetation
    Saline and brackish swamp
    Bamboo
    Anthropic landscapes

Introduction

The broad principles underlying the present classification are explained in the Introduction and in Chapter 6. Because these principles are somewhat different from those of most other classifications it was originally intended to include a longer discussion on classification in general. But this was precluded for lack of space. Nevertheless it might be useful to mention in this context the following publications, most of which are not discussed elsewhere: Aubréville (1951), Bamps (1975), Beard (1978), Cain (1950), Chevalier (1953*a*, 1953*b*), Dansereau (1951), Dasmann (1972, 1973*a*, 1973*b*), Drude (1913), Du Rietz (1931), Emberger, Mangenot & Miège (1950*a*, 1950*b*), Gaussen (1955, 1958), Kinloch (1939), Küchler (1947, 1949, 1950, 1960, 1967, 1973), Plaisance (1959), Poore (1962, 1963), Redinha (1961), Robyns (1942), Schnell (1970–71, 1977), Udvardy (1975, 1976), Walter (1976*b*), Walter & Box (1976) and Webb (1954).

In this chapter the classification is merely briefly considered in relation to earlier classifications of African vegetation. The distinguishing features of the sixteen major formations are also described, and an outline of their distribution in Africa is given.

The present classification has evolved from earlier classifications, including the so-called 'Yangambi' classification (CCTA/CSA, 1956; Trochain, 1957; Boughey, 1957*b*, 1961; Monod, 1963; Aubréville, 1965; Beard, 1967; Guillaumet & Koechlin, 1971; Descoings, 1973), and more particularly that of Greenway (1943, 1973; Pratt *et al.*, 1966), but differs in several respects from both.

The earliest classifications starting with Schimper (1898, 1903) had too few major categories, and failed to make a clear distinction between forest and other woody vegetation types. They also misinterpreted much lowland grassland now known to be edaphic or secondary.

The Yangambi classification represents an improvement in some respects, especially in its treatment of forest, woodland and thicket, but there are several reasons why it cannot be accepted without considerable modification. The main categories are too few to do justice to the variety of African vegetation and they are heavily biased in favour of West African types. The most serious criticism, however, concerns the use of the terms 'savanna' and 'steppe' and their definition (for fuller discussion see White, in Chapman & White, 1970, and Descoings, 1973, 1978).

There is little justification for using the term steppe in tropical Africa. The fact that the only ecologist (Walter, 1939, 1943, 1962, 1964) who has made a thorough study of the grasslands of eastern Europe and subtropical Africa repudiates the use of the word steppe for the latter should carry great weight.

The term savanna has been defined in so many different ways that it is no longer possible to use it in a precise classificatory sense. In more general contexts, however, both popular and scientific, the antithesis 'forest and savanna' is a commonplace and will doubtless continue to be usefully employed for certain tropical landscapes. But its value is greatly diminished if it is made to include, for instance, wooded *Sphagnum* bogs of temperate regions as is done by Fosberg (1961).

Greenway's system is avowedly simple and intended to be understood by laymen. It drew heavily on the experience of many people, and is so well documented that it is possible to attribute actual stands in the field to their correct position. Greenway, who avoided the use of imported terms like savanna and steppe, adopted seven main types, namely (1) forest, (2) woodland, (3) wooded grassland, (4) grassland, (5) permanent swamp vegetation, (6) bushland, thicket and scrub, and (7) semi-desert vegetation.

Because Greenway was only concerned with East Africa it is not surprising that the system adopted here differs from Greenway's in several respects. The chief modifications to Greenway's system are as follows:
1. Mangrove is separated from forest as a major physiognomic type.
2. Bamboo is treated as a major physiognomic type and not as a type of thicket.
3. Giant-grass thicket is treated as grassland.
4. The term scrub is used in a general sense to designate all woody vegetation other than forest, woodland, mangrove and bamboo, though in most contexts more precise terms such as bushland or shrubland are preferred.
5. Shrubland is recognized as a major physiognomic type.
6. The physiognomically mixed and distinctive Afroalpine vegetation is treated as a major type.
7. Desert is recognized as a major classificatory unit, but semi-desert vegetation is classified as shrubland, grassland, etc., wherever the physiognomy justifies this.
8. The physiognomically diverse vegetation of saline and brackish swamp is treated collectively as a major classificatory unit.
9. In addition to wooded grassland, three other transitional types, namely scrub forest, transition woodland and scrub woodland are recognized.

The sixteen major physiognomic divisions of the present classification have been selected and defined so as to correspond as closely as possible to the vegetation on the ground. Thus, forest, woodland, bushland, shrubland, grassland and wooded grassland are not defined arbitrarily, but in such a way as to accommodate the great regional formations of Africa.

For the most part the regional vegetation types differ from each other in the height and density of their principal growth forms, but there are often atypical variants in these respects which, however, are typical in most other features. This means that although height and density often have considerable diagnostic value, their application must be flexible, and sometimes they need to be subordinated to other physiognomic features. Many applications of this principle are mentioned below where the individual types are described.

Description of the main vegetation types

Forest

Forest is a continuous stand of trees. The canopy varies in height from 10 to 50 m or more, and usually consists of several layers or storeys. The crowns of individual trees interdigitate or overlap each other and are often interlaced with lianes. A shrub layer is normally present. It is usually densest in those types of forest with a more open canopy. The ground layer is often sparse and may be absent or consist only of bryophytes. In tropical and subtropical types grasses, if present, are comparatively localized and inconspicuous, though lianes are usually well represented. Epiphytes, including ferns, orchids and large mosses are characteristic of the moister tropical and subtropical types, but vascular epiphytes are virtually absent from more temperate types, if the word temperate is used in a latitudinal rather than a strictly climatic sense. Large epiphytic lichens, especially *Usnea*, are often conspicuous, especially in upland types.

In forest, woody plants, especially trees, contribute most to the physiognomy and the phytomass, and in number of species often greatly exceed herbs.

Nearly all the forests in Africa are evergreen or semi-evergreen, though deciduous forests occur locally. In Madagascar, by contrast, deciduous forest is widely distributed on the drier western side of the island.

In many types of forest, especially rain forest, different tree species reach different sizes at maturity, and the overall structure is complex and difficult to analyse. To overcome this, Davis & Richards (1933–34) adopted the profile diagram, which has been widely used ever since. It is often claimed, and equally often denied, that forests show a well-defined stratification.

In recent years a new approach to the architecture and growth patterns of rain-forest trees (Hallé & Oldeman, 1970, 1975; Oldeman, 1974; Hallé et al., 1978) and lianes (Cremers, 1973) is greatly enhancing our understanding of structure and stratification, but it has not yet been extensively applied in Africa. In the

TABLE 3. Synopsis of the main vegetation types

FORMATIONS OF REGIONAL EXTENT:		TRANSITIONAL FORMATIONS OF LOCAL EXTENT:	EDAPHIC FORMATIONS:	FORMATION OF DISTINCT PHYSIOGNOMY BUT RESTRICTED DISTRIBUTION:
1. *Forest*. A continuous stand of trees at least 10 m tall, their crowns interlocking.	5. *Grassland*. Land covered with grasses and other herbs, either without woody plants or the latter not covering more than 10 per cent of the ground.	9. *Scrub forest*. Intermediate between forest and bushland or thicket.	12. *Mangrove*. Open or closed stands of trees or bushes occurring on shores between high- and low-water mark. Most mangrove species have pneumatophores or are viviparous.	15. *Bamboo*
2. *Woodland*. An open stand of trees at least 8 m tall with a canopy cover of 40 per cent or more. The field layer is usually dominated by grasses.	6. *Wooded grassland*. Land covered with grasses and other herbs, with woody plants covering between 10 and 40 per cent of the ground.	10. *Transition woodland*. Intermediate between forest and woodland.		UNNATURAL VEGETATION:
3a. *Bushland*. An open stand of bushes usually between 3 and 7 m tall with a canopy cover of 40 per cent or more.	7. *Desert*. Arid landscapes with a sparse plant cover, except in depressions where water accumulates. The sandy, stony or rocky substrate contributes more to the appearance of the landscape than does the vegetation.	11. *Scrub woodland*. Stunted woodland less than 8 m tall or vegetation intermediate between woodland and bushland.		16. *Anthropic landscapes*.
3b. *Thicket*. A closed stand of bushes and climbers usually between 3 and 7 m tall.			13. *Herbaceous fresh-water swamp and aquatic vegetation*.	
4. *Shrubland*. An open or closed stand of shrubs up to 2 m tall.	8. *Afroalpine vegetation*. Physiognomically mixed vegetation occurring on high mountains where night frosts are liable to occur throughout the year.		13. *Halophytic vegetation* (saline and brackish swamp).	

present work details of stratification follow the accounts of the original authors.

Since some of the terms frequently used to classify forest are applied imprecisely or in conflicting senses the following explanation of the usage adopted here is given:

Rain forest. This term is unsatisfactory, both in its connotation and in its application, but is provisionally retained for want of a better. Richards (1952) and others, e.g. Whitmore (1975) and Grubb & Tanner (1976), apply it loosely to include low vegetation, which in the present classification is referred to as scrub forest or thicket.

The term rain forest seems to be appropriate for the most widespread type of forest in tropical Africa, namely Guineo–Congolian forest on well-drained soils (page 75). The upper canopy is nearly everywhere more than 30 m tall. Forest shorter than this occurs very locally in the Guineo–Congolian Region, especially on rocky hills and in upland areas. It is not classified as rain forest, but as short forest, scrub forest or thicket. Guineo–Congolian rain forest includes both evergreen and semi-evergreen variants.

Forest which in structure is almost indistinguishable from Guineo–Congolian rain forest and floristically is closely related to it also occurs outside the Guineo–Congolian Region on the slopes of certain mountains (Chapters VIII and XIII). The forests occurring in the lowlands of the wetter eastern half of Madagascar (Chapter XIX) are somewhat lower in stature than the rain forests on the African mainland but resemble them sufficiently for the term rain forest to be usefully employed.

Dry forest. This term is restricted to forests which experience a dry season lasting several months during which atmospheric humidity is low. They are shorter than rain forest, simpler in structure and poorer floristically. Dry forest is very localized in the Zambezian Region (page 89) and even more so in the Sudanian Region (page 103). It is, however, the most widespread climax type in the West Malagasy Region (Chapter XX) and occurs on the dry coastal plain of Ghana (page 176).

Semi-evergreen forest. Some canopy species are briefly deciduous but not necessarily at the same time. Most members of the understorey are evergreen. The drier types of Guineo–Congolian rain forest are semi-evergreen. It is misleading to call them deciduous.

Deciduous forest. The majority of individuals, both in the upper and lower canopy, lose their leaves simultaneously and usually remain bare for several weeks or months. In some types the largest trees, as in Zambezian dry deciduous forest (page 90), may, on favourable sites or in favourable years, remain evergreen over an almost completely deciduous lower canopy. The West Malagasy forests (Chapter XX) are deciduous.

Undifferentiated forest. This term is applied to forests which undergo rapid and kaleidoscopic change in structure and composition over short distances. The precise pattern is chiefly of local interest, and in general works, such as this, little subdivision is necessary. Most of the forests in the Afromontane, Zanzibar–Inhambane and Tongaland–Pondoland Regions are treated as undifferentiated forest.

With the exception of the sclerophyllous forests of the Mediterranean basin (Chapter VII) most of the important African forests have been mentioned above. The forests of the Guinea–Congolia/Zambezia transition zone form a series linking typical Guineo–Congolian forest, through transition woodland, to Zambezian woodland. A similar transition probably occurred in the Guinea–Congolia/Sudania transition zone, though less evidence survives.

Swamp forest and riparian forest are widespread in the Guineo–Congolian, Zambezian, and Sudanian Regions, but in drier regions are rare or are replaced there by riparian scrub forest and bushland. In the Sahara (Chapter XVII) *Tamarix* very locally forms riparian forest 10 m high, and *Cupressus dupreziana* may formerly have formed forest on the mountains. The Cape Region is bereft of forest other than enclaves of Afromontane forest and small patches of scrub forest.

Woodland

Woodland is land with an open stand of trees the crowns of which form a canopy from 8 to 20 m or more in height and cover at least 40 per cent of the surface.

The crowns of adjacent trees are often in contact but are not densely interlocking. Frequently the trees are more widely spaced and may be up to 1 crown diameter apart. In Africa woodlands of regional extent are confined to tropical and subtropical regions. Most African woodlands are deciduous or semi-deciduous but nearly all types contain a few evergreen species. No examples of completely evergreen woodland have been described from tropical Africa. In woodland the trees (except in transitional types) usually have more or less straight boles which are not branched for at least 2 m. If the boles do branch below this height the branches are usually ascending and progress is unimpeded. The canopy casts little shade and there is usually a ground cover consisting principally of herbaceous tussock grasses, the culms of which are up to 2 m high. On stony or rapidly eroding soils the grass layer is often poorly developed and then other herbs and dwarf shrubs are sometimes prominent. It is the dominance of trees combined with the light open canopy and almost universal presence of heliophilous grasses that distinguish woodland from other vegetation types. The grasses are usually perennial, but annual grasses are predominant in certain drier transitional types especially under the influence of heavy grazing. In most types there is an incomplete understorey of small trees or large bushes, the density of which is very variable. Smaller shrubs also vary greatly in their size and density. Lianes are mostly rare or absent. The scarcity of smaller woody plants is sometimes due to burning of the grassy ground layer, but in some types where the grass cover is sparse, smaller woody plants may show an insignificant increase after many years of protection from fire. Although the grasses usually dominate the ground layer there is often a wealth of herbs and suffrutices, the attractive flowers of which are conspicuous at the end of the dry season and early in the rainy season before the grass is tall enough to conceal them. Vascular epiphytes are often present, though relatively rare, except in secondary woodland occurring on forest sites.

Occasionally, stands of woodland have a closed canopy and, as a consequence, also a poorly developed grass layer. These are not true forests since they differ from forest in almost all respects other than in density of canopy and field layer. They are best referred to as 'closed woodland'. However, when such woodlands contain forest elements, either because they form part of the forest/woodland ecotone or represent a seral stage to forest, they are referred to as 'transition woodland' (see below).

Woodland as defined above is the most widespread vegetation type in tropical Africa. It is especially characteristic of the Sudanian and Zambezian Regions with their continental climate and moderate precipitation, which falls in summer. There is little doubt that much Sudanian and Zambezian woodland is natural, especially when it occurs on shallow stony soils. However, on deeper soils which have been cultivated, similar woodland may be secondary, having replaced dry forest or transition woodland.

Several of the dominant species of woodland under limiting conditions are less than 8 m tall. The communities they form are then scrub woodland. Where scrub woodland occurs in intimate mosaic with normal woodland or is part of a gradual continuum it is not given separate treatment. Only a few types which are more sharply defined are individually described.

In the Somalia–Masai Region, the Sahel transition zone, and in the Kalahari part of the Kalahari–Highveld transition zone only a few trees exceed 8 m in height. The prevalent vegetation is bushland and thicket or various types of wooded grassland. Woodland in these phytochoria is very local and not very typical; it is not described separately from the more characteristic types.

Woodland does not occur in the Afromontane Region, except when *Hagenia abyssinica* forms almost pure semi-open stands with a grassy field layer.

Although forest is the natural vegetation of the Mediterranean Region, managed forests, such as some of the Cork Oak (*Quercus suber*) forests, have the appearance of woodland. Elsewhere, for example in *Tetraclinis articulata* communities on steep rapidly eroding soils, the trees do not always form a closed canopy. The resemblance to woodland, however, is superficial and the community is best regarded as an open forest.

Bushland and thicket

Bushland is land of which 40 per cent or more is covered by bushes. In this work a bush is defined as a woody plant intermediate in habit between a shrub and a tree. Bushes are usually between 3 and 7 m tall, but can be smaller or larger. They are usually multiple-stemmed and the main axes are frequently 10 cm or more in diameter at the base. Bushy trees are also frequently present in bushland. They too are normally less than 7 m tall and, although there is a main bole, it branches low down so that progress is impeded. Taller trees sometimes occur as emergents but they are either localized in groups or are widely scattered. When they are sufficiently numerous to form a distinct but open canopy we are dealing with scrub forest. When they are less frequent the term 'wooded bushland' could be used.

Grasses are present in most types of bushland but are physiognomically subordinate. Where bushes occur scattered in a continuous sward of grasses and cover less than 40 per cent of the surface we can talk of 'bushed grassland', but the proportion of bushes is usually much less than this. For reasons given elsewhere (p. 52) bushed grassland is not recognized separately from wooded grassland in this work. Bushland often occurs in rocky or stony places which are unfavourable to grasses, and elsewhere the grasses have been eliminated or much reduced by grazing. Because of the insignificance of grass, it is inappropriate to use the term savanna in such cases although they have frequently been designated as bush- or shrub-savanna in the past. In places where grasses are sparse, the vegetation remains physiognomically bushland even when the cover is much below 40 per cent. Such vegetation can be referred to as open bushland.

In thicket, the bushes are so densely interlaced as to form an impenetrable community except along tracks made by animals. In most types of bushland larger or smaller patches of thicket also occur without significant change in floristic composition. Some types of thicket, however, e.g. Itigi thicket (page 97), are dominated by species which do not normally occur in more open communities.

Bushland and thicket occur under a wide range of climatic and edaphic conditions which are unfavourable for the growth of taller woody plants.

Bushy plants are most widely distributed in Africa in regions where rainfall is between 250 and 500 mm per year, but they only form communities of regional extent where there are two rainy seasons or rainfall is irregular throughout the year, or where dry-season humidity is very high. Thus, deciduous bushland and thicket is extensively developed in the lowlands of the Somalia–Masai Region (Chapter IV) where in most parts there are two peaks in the annual distribution of rainfall. Elsewhere in Africa, where the annual rainfall is between 250 and 500 mm but falls entirely in summer, grasses are favoured on sandy soils. Consequently, the most widespread vegetation in the Sahel zone (Chapter XVI) and the Kalahari part of the Kalahari–Highveld transition zone (Chapter XIV) is wooded grassland, though the woody plants are possibly less dense now than formerly because of human activity. In both areas bushland is largely confined to rocky and stony places, which are relatively infrequent. Deciduous bushland and thicket is also the regional climax vegetation of southwest Madagascar (Chapter XX). Here rainfall is between 300 and 500 mm per year and the dry season is up to 10 months long. Although droughts sometimes last as long as 18 months, rain may fall in any month, and relative humidity is high throughout the year.

Evergreen and semi-evergreen bushland and thicket occur extensively on the slopes of mountains and other uplands which arise from the lowlands of the Somalia–Masai Region (Chapter IV), and form an ecotone between deciduous bushland and thicket and the drier types of Afromontane forest. Mean annual rainfall is mostly between 500 and 850 mm and is irregularly distributed throughout the year but with two main peaks. Similar but floristically poorer vegetation also occurs in parts of the Lake Victoria basin (Chapter XII), where the rainfall is somewhat higher (850–1 000 mm per year). Evergreen and semi-evergreen bushland is also the characteristic vegetation of river basins in the Tongaland–Pondoland Region (Chapter XV). The rainfall there, which is too low to support forest, is concentrated in a single rainy season but there is also significant precipitation in the dry season. In the Zanzibar–Inhambane Region (Chapter XIII) climatically determined bushland and thicket (other than littoral thicket) is much more localized and occupies only a few pockets of relatively low rainfall.

In the Cape Region (Chapter V) the prevalent vegetation is fynbos. Most fynbos is shrubland, but the tallest examples would be referred to as bushland or thicket if major physiognomic types are defined on artificial criteria such as height alone. Within the Cape Region, however, there are considerable enclaves in which the climax vegetation was probably true evergreen bushland similar to the drier types of valley bushland of the Tongaland-Pondoland Region (Chapter XV).

On most African and Malagasy mountains which are high enough there is a zone of bushland and thicket (Chapters VIII and XIX respectively) above the forest zone. It is usually dominated by Ericaceae. The summits of some mountains which are not high enough to support an Ericaceous belt are covered with elfin thicket. Elfin thicket also crowns a few peaks in the Guineo–Congolian Region (Chapter I).

Evergreen littoral thicket is found on rocky or sandy shores in higher-rainfall areas wherever salt-laden offshore winds are sufficiently severe to prevent the development of forest. It is not included in this work.

In tropical Africa the remaining types of bushland and thicket (mostly the latter) are edaphically determined. The most widespread occupy old termite mounds or rocky outcrops. Bushland and thicket, physiognomically similar to that of the Ericaceous belt on the African mountains, occurs locally on badly drained coastal sands in the Zanzibar–Inhambane Region (Chapter XIII).

In the drier parts of the Zambezian Region (Chapter II) dense deciduous thicket occurs on certain soils which favour the intensive root systems of thicket species. There is an abundant supply of water in the rainy season but the soils dry out, at least in their upper layers, in the dry season.

In the Maghreb (Chapter VII), forest nearly everywhere represents the climax. It is possible, however, that in certain drier areas, especially on clay soils, bushland and thicket were the original vegetation (page 157).

Secondary thicket is also widespread, especially as an early seral stage in the reversion to forest, both lowland and upland. Some examples are described by Clayton (1958a, 1961).

Just as some communities which are more than 3 m tall, phylogenetically must be regarded as shrubland, other communities less than 3 m in height phylogenetically are bushland, and are treated as such in the following pages.

Shrubland

Shrubland is land dominated by shrubs which vary in height from 10 cm to 2 m or more. It occurs where taller woody plants are excluded by low rainfall, summer drought, low temperatures, exposure to wind, or shallowness, salinity, toxicity or extreme oligotrophy of the soil, operating singly or in various combinations. In tropical and subtropical semi-desert areas with summer rainfall the climate is equally favourable to grasses and woody plants. Grasses are dominant on deep sandy soils, and woody plants in stony and rocky places. The latter, however, are often bushes or stunted trees, and shrubs play a relatively minor role.

The most extensive and distinctive shrublands in Africa are those of the Karoo–Namib Region. Almost the whole of this area, with the exception of the Namib desert, is covered with various types of Karoo shrubland consisting almost entirely of shrubs less than 2 m tall. Taller bushes and pachycaul trees also occur locally as scattered emergents but, with the exception of certain types intermediate between Karoo shrubland and Tongaland–Pondoland bushland, they are rarely numerous.

The Karoo shrublands are markedly different in appearance from the bushlands and thickets of the Somalia–Masai Region, most of which receives only a slightly higher rainfall but lies wholly within the tropics.

The dwarf shrublands of the Upper Karoo were formerly separated from the climatic grasslands of the Highveld by a transition zone in which grassy shrubland occurred in mosaic with patches of pure grassland. Overgrazing has largely converted this region into predominantly shrubby communities, and the drier grasslands of the Highveld have suffered a similar fate. The original vegetation of the Karoo mountains is believed to have been more grassy than it is today. Overgrazing has favoured the extension of Karoo shrublets, which originally occurred chiefly on shallow soils. The grassiness of these secondary Karoo shrublands varies greatly, and some could more correctly be referred to as shrub grassland, but that distinction is not made here.

The very distinctive type of sclerophyllous shrubland known as fynbos is the prevalent vegetation of the Cape Region. Today most fynbos is less than 2 m tall and only exceptionally more than 3 m, though some types can reach a height of 6 m. The scarcity of tall fynbos can partly be explained by the effects of man-made fires. Fires, however, also occur naturally, and, in the absence of fire, fynbos becomes moribund and dies, so that tall fynbos was probably never the predominant type in the Cape Region. For this reason and because tall fynbos and the shorter types are part of the same floristic and physiognomic continuum, fynbos is dealt with collectively here—as shrubland.

A striking difference between the Cape Region and the Mediterranean Region is that in the former, forest is of restricted occurrence and shrubland (fynbos) is the most widespread climax, whereas in the latter, evergreen sclerophyllous and coniferous forests represent the climax except on the summits of the highest mountains, where dwarf spinous cushion-shaped shrubs prevail. The well-known Mediterranean shrublands, maquis and garrigue, are almost entirely secondary. The tallest types of maquis are somewhat transitional towards bushland, but since they, and less tall variants, often represent part of the same continuous degradation series they are dealt with together as shrubland.

Afromontane shrubland is usually grassy and often occurs in mosaic with purer grassland which occurs on shallow soil or along drainage lines. Fire has often favoured the spread of grasses at the expense of the shrubs, so that it is difficult to reconstruct their former relative extent. For this reason they are treated collectively.

At higher altitudes, in the Afroalpine belt, dwarf shrubland is a characteristic community, but it is only one component of an extremely diversified assemblage of communities, characterized collectively by the abundance of pachycaul Senecios and Lobelias.

Typical shrublands are absent from the Guineo–Congolian Region and, with the exception of suffrutex grassland mentioned below, are virtually absent from the Sudanian and Zambezian Regions.

In the Zambezian Region mixed communities of grasses and geoxylic suffrutices occur extensively on seasonally waterlogged Kalahari Sand and less extensively on similar soils elsewhere. The suffrutices come into leaf and flower before the onset of the rains while the ground is still bare. At this time of year the community is an open dwarf shrubland. The grasses emerge from dormancy two or three months later and eventually, because of their taller growth, completely conceal the subshrubs. Physiognomically the community is now a grassland. In some types the phytomass of the subshrubs (especially if their frequently massive underground parts are taken into account) may greatly exceed that of the grasses. Since, however, they are physiognomically grassland for a considerable part of the year, and often grade imperceptibly into pure grassland, they are treated, in this work, as grassland.

In the lowlands most primary shrubland, with the exception of Cape fynbos, occurs under a semi-desert climate, where edaphic conditions profoundly influence the vegetation, but are themselves at least in part a consequence of the dry climate. This is true for succulent sub-Mediterranean *Euphorbia* shrubland, which is largely confined to stony soils, and for Somalia–Masai shrubland occurring on gypseous soils.

Halophytic shrubland is described in the chapters on Desert (XVII) and Azonal Vegetation (XXII).

The leaves of the shrubland dominants are usually small and are deciduous or evergreen or sometimes reduced to scales. In some types they are sclerophyllous: in others malacophyllous. Shrubs with succulent stems or leaves may be present or absent. In some types they are dominant or even almost exclusively present.

Grassland

Grassland is land covered with grasses and other herbs, with the former physiognomically dominant.

In Africa grassland is sometimes completely devoid of woody plants, but such pure grassland is often intimately associated in mosaic or zonally with lightly wooded communities. On a continental scale, it would be unsatisfactory to attempt to separate them. In this work communities with up to a 10 per cent cover of woody plants are treated as grassland without further qualification. If the cover of woody plants is between 10 and 40 per cent we are dealing with wooded grassland which is intermediate between grassland and woodland. Climatically determined wooded grassland is extensively developed in the drier parts of tropical Africa and is dealt with below. Edaphic and secondary wooded grasslands, however, are much more restricted in occurrence, and are often difficult to delimit from the more open grasslands with which they are usually associated. Hence in this work they are treated together.

Cyperaceae are present in many types of grassland, particularly edaphic grassland, and locally are more abundant than the grasses themselves, especially in the wettest places. Other herbs, e.g. Acanthaceae, are occasionally dominant. Because of their purely local significance, such non-graminaceous communities are not dealt with separately but are mentioned at the appropriate places in the text.

The dominant grasses may be up to 3 m or more in height, but are usually shorter. Communities dominated by giant grasses such as *Pennisetum purpureum* and species of *Cymbopogon* and *Hyparrhenia* are regarded as grassland since both in physiognomy and ecology they resemble other grasslands much more than they resemble thicket, to which they are sometimes assigned. Vegetation dominated by bamboos, however, is excluded, and is treated as a separate physiognomic type. Some permanent swamps are dominated by grasses. They are described along with other swamp communities in Chapter XXII.

Grasses show a remarkable diversity of growth form notwithstanding their basic vegetative uniformity, which is characterized by the almost universal occurrence of a tubular leaf-sheath, an elongate, usually ribbon-like lamina, and long sustained growth of stems and leaves from intercalary meristems.

In the wetter parts of tropical and substropical Africa the grasses are generally perennial. Annuals are most frequent in the drier parts and are sometimes dominant. There is, however, no simple correlation between the distribution of annuals and rainfall. Their extensive dominance is often due to overgrazing.

Some grasses are procumbent and mat-like and form a dense thin carpet like *Cynodon dactylon*. Many species are tufted, when the culms may be virtually leafless (*Loudetia simplex*) or densely leafy (*Hyparrhenia*). The leaves of some species, e.g. *Schismus barbatus*, form spreading basal rosettes. Other species, e.g. *Imperata cylindrica*, have extensive rhizomes which send up solitary or tufted flowering stems. Some non-rhizomatous species, such as *Schizachyrium platyphyllum*, have single culms which are prostrate below and root at the nodes. The most conspicuous grasses in the Mediterranean Region and on the high mountains of Africa and Madagascar are tufted species with filiform, sclerophyllous leaves, which can function intermittently throughout the year. Some desert and semi-desert grasses are chamaephytes or even nano-phanerophytes.

The grass flora of Africa is rich and diversified. There are more than a thousand species and they are found in all parts of the continent, but communities dominated by grasses have a very irregular and scattered distribution.

Some earlier workers such as Schimper (1898, 1903) believed that grassland occurs in tropical Africa as a distinct zonal formation with a distinct type of climate. Schimper distinguished grassland with trees as 'savanna' and grassland without trees as 'steppe'. It is now known, however, that much of the grassland which Schimper and others, e.g. Robyns (1936), thought represented the climatic climax is in fact secondary and is due to fires, mostly man-made, or is an edaphic climax due to soil conditions unfavourable to trees. Nevertheless, the statement by Richards (1952, p. 316) that 'it is extremely doubtful whether any tropical grassland is a true climatic climax' is possibly too sweeping.

According to Walter (1962, 1971) pure grassland without an admixture of woody plants represents the zonal climax vegetation on level areas and average soils in tropical and subtropical regions where the rain falls in summer and amounts to 100–250 mm per year. Where the rainfall is between 250 and 500 mm per year wooded grassland, which Walter call 'savanna', is the climax. Such grassland and wooded grassland is widespread in the Sahel and Kalahari transition zones. The argument that these 'zonal' wooded grasslands represent edaphic rather than climatic climaxes (see 'Wooded Grassland' below) applies equally to pure grassland.

The areas occupied by edaphic and secondary grassland vary greatly in different parts of Africa.

The most widespread edaphic grasslands are those associated with seasonally or permanently waterlogged soils. They are of limited occurrence in the Guineo–Congolian Region, which has a short dry season or none at all. By contrast, they are widespread in the Sudanian, Zambezian and Somalia–Masai Regions and the Indian Ocean Coastal Belt, which all experience strongly seasonal rainfall. Waterlogged soils usually occur in depressions which receive more water than that supplied by the incident rainfall, but the extent of such water-receiving sites varies greatly from place to place. It depends largely on the stage reached by the landscape in the cycle of geological erosion and on recent geomorphological history. The impeding and reversal of drainage due to warping or tilting of the earth's crust has had a profound effect on the distribution of waterlogged grassland. In some places, parent material has an overriding effect in that waterlogged grassland can occur on eluvial sites such as certain recent volcanic soils (Serengeti Plains, Somalia–Masai Region, Chapter IV), and on soils derived from Karoo mudstone (Luangwa Valley, Zambezian Region). Heavy metal and serpentine soils, which are inimical to the growth of trees in Africa, appear to be almost confined to the Zambezian Region.

The great regional vegetation types vary greatly in the ease with which they can be replaced by secondary grassland following human interference. In the wetter parts of the Guineo–Congolian Region the regeneration of forest after cultivation is so rapid that grasses have difficulty in becoming established. By contrast, the drier peripheral forests are much more vulnerable, and have been replaced by secondary grassland and wooded grassland over extensive areas. Even more vulnerable has been the climax vegetation of the transition zones to the north and south of the Guineo–Congolian Region. Nearly everywhere the original vegetation has gone and the landscape is dominated by fire-maintained grassland.

There is evidence that in the Sudanian and Zambezian Regions various types of dry forest were formerly more extensive and that they have almost entirely been replaced by secondary grassland or wooded grassland following cultivation and fire. The woodlands in these regions, however, are comprised of fire-tolerant trees and grasses, and have probably always been subjected to at least occasional natural fires. They have withstood the effects of fire and cultivation better than the forests mentioned above. Indeed over large areas agricultural practice is based on the ability of the trees to coppice after felling or lopping of their branches. Experimental evidence shows that many Zambezian woodland trees can survive as annual coppice even when subjected to fierce annual fires for more than forty years. Nevertheless, despite the fire-hardiness of some species, large areas of woodland have been converted to secondary grassland by too intensive farming.

In the drier parts of tropical Africa fire is less important. The grass growth is not sufficiently luxuriant to support fierce burns and much is removed during the growing season by wild or domesticated animals. Indeed, grazing animals often favour woody plants vis-à-vis grass by reducing the competitive vigour of the latter. This is also true beyond the tropics. It is believed that the Karoo shrublands were formerly much more grassy than they are today. Within historic times the climax grassland of the drier parts of the Highveld Region in South Africa has been degraded by grazing sheep to secondary Karoo shrubland.

Secondary grassland is unimportant in the Cape Region and the wetter parts of the Mediterranean Region, except where deliberately induced for pasture. In some of the drier parts of the Mediterranean Region, however, and in the transition zone between it and the Sahara the landscape is dominated by sclerophyllous grass species, especially *Stipa tenacissima*. There is increasing evidence that this grassland is secondary and has replaced *Pinus halepensis* forest.

Secondary grassland is now the most extensive community on the African mountains. As in the lowlands, the vulnerability of the original communities varies greatly. The wettest forests on the wettest mountains are difficult to replace by grassland. At the other extreme the drier forests, especially those dominated by the conifers *Juniperus procera* and *Widdringtonia cupressoides*, are very susceptible to fire and can be ignited without being felled.

Except for the wettest and driest types, the original vegetation of Madagascar has been devastated by fire and much of the island is now covered with secondary grassland.

In the present account a clear distinction is made between climatic, edaphic and secondary grassland, but in practice all three factors may operate together and it is not always easy to decide to which category a particular type should belong. The difficulty over climatic grassland was mentioned above. Elsewhere, grassland may occur on soils which are incapable of supporting trees, but the soils themselves may have developed under unusual climatic conditions. This may be true of the edaphic grasslands of the Accra plains (page 178).

The description and classification of African grasslands are in their infancy. For many types even floristic composition has only been sketchily recorded. For the great majority the growth-form spectra and the rhythms of growth in relation to environmental factors and potential competitors remain uninvestigated. Improved classification will depend on a knowledge of the former, and ecological understanding on the latter. The pioneer studies of Walter (1939) on the competitive behaviour of grasses in arid and semi-arid regions, and the publications of Descoings (1971–76) on growth-form analysis, point in the right direction. Descoings, however, recognizes only five main growth-forms, based on the form of the tufts, the manner of branching and the number of culms. This is insufficient since it places together species as different in their ecology (and in other features of growth not used in his classification) as *Loudetia simplex* and *Stipa tenacissima*. A more detailed analysis of growth-form would provide a more powerful analytical tool.

Wooded grassland

Wooded grassland is land covered with grasses and other herbs, with scattered or, more rarely, grouped woody plants, which are often, but not necessarily, trees. The woody plants cover between 10 per cent and 40 per cent of the surface. These figures are somewhat arbitrary, since wooded grassland grades into woodland on the one hand and pure grassland on the other. But by defining wooded grassland in this way, it includes most of the zonal vegetation of the wetter half on the Sahelian Region and of the wetter half of the Kalahari section of the Kalahari–Highveld transition zone.

The woody plants in wooded grassland, which may be trees, bushes, dwarf trees (e.g. *Acacia drepanolobium*), palm trees or shrubs, are nearly always scattered. When the woody plants are grouped it is better to regard the vegetation as a mosaic. Pure edaphic grassland on seasonally waterlogged soils often forms a mosaic with very sharply defined islands of thicket which occur on the better drained soil of old eroded termite mounds. Such 'termite savanna' is not treated here as a single vegetation type but as a patchwork of edaphic grassland and termite-mound thicket.

Greenway (1973) recognizes, in addition to grouped tree grassland, four types of wooded grassland, namely scattered tree grassland, palm-stand grassland, shrub grassland, and dwarf-tree grassland. These categories are certainly useful for local purposes, but extensive areas of wooded grassland often contain trees, bushes and shrubs, and sometimes palm trees, in ever varying proportions, or consist of complicated mosaics of the different variants. For this reason only a single category which embraces all the variants is used in the present work.

Climatically determined wooded grassland is described below. Edaphic wooded grassland is also important, especially in the Sudanian and Zambezian Regions, but since it intergrades with edaphic grassland it is not treated separately from the latter. Secondary wooded grassland is merely an arbitrarily defined phase in a degradation series replacing other vegetation, or in the succession leading to its restitution. It too is not given separate treatment here.

Climatic wooded grassland

According to Walter (1971) wooded grassland is the zonal vegetation on deep *sandy* soils on level sites in tropical and subtropical regions where the rain falls in summer and amounts to 250 to 500 mm per year. He restricts the term savanna to such vegetation, but, for reasons given elsewhere (Chapman & White, 1970), it is not used in this book except in a more general sense. Where the rainfall exceeds 500 mm per year woodland replaces wooded grassland, and where it is between 250 and 100 mm pure grassland occurs on deep sandy soils.

The requirements of wooded grassland are found extensively in the southern half of the Sahel zone and in the wetter parts of the Kalahari section of the Kalahari–Highveld transition zone. It should be pointed out, however, that the sandy soils in both areas are provided by stabilized sand dunes which were formed during the arid phases of the Quaternary. It is doubtful whether similar sandy soils are being actively formed in these transition zones under the present climatic regime. Hence it is somewhat misleading to refer to the wooded grassland as a climatic climax. In the wooded grassland belts, loamy and clayey soils are formed only in periodically wet hollows which are often covered with edaphic grassland. Elsewhere, where there is no sand cover, a true soil is lacking and the surface is covered with weathered rock. Grasses are not dominant in such places and the vegetation is scrub woodland, bushland or shrubland depending chiefly on the rainfall. It would seem that if the Quaternary sand dunes had not been formed, wooded grassland would be relatively unimportant in the transition zones.

The vegetation of the Sahel zone has been greatly modified by man and his domestic animals, and is certainly less dense than formerly. It is therefore difficult to reconstruct the original vegetation, but comparison with sparsely populated parts of the Kalahari region with a similar rainfall suggests that on sandy soils wooded grassland represents the climax.

Desert

Deserts and semi-deserts occur in arid regions where plants suffer from lack of water as a result of low rainfall and high evaporation throughout the greater part of the year. The plant cover is sparse and shows various adaptations to unfavourable water conditions. The transition towards arid regions is always gradual unless a range of mountains provides an abrupt climatic division. An objective criterion for separating arid regions from wet regions does not exist and any separation is somewhat arbitrary. Different arid regions have their own climatic peculiarities and hence different criteria for their delimitation.

In semi-deserts the soil is often more conspicuous than the vegetation, so that the aspect in general view is dominated by the colour and character of the soil rather than by the plants themselves. The latter, however, are still sufficiently evenly distributed and sufficiently numerous for it to be meaningful to refer the communities they form to general physiognomic categories such as semi-desert grassland, semi-desert shrubland, etc. Thus, the grasslands and shrublands of the wetter parts of the Karoo–Namib Region, and the drier parts of the Somalia–Masai Region and the Sahel and Mediterranean–Sahara transition zones comprise semi-desert vegetation.

In Africa semi-desert usually begins to appear where the mean annual rainfall drops to below about 250 mm, but the figure may be higher or lower than this, depending on the distribution of rainfall throughout the year, and its incidence in relation to other factors such as seasonally low temperature, and on soil texture.

Desert, like semi-desert, cannot be defined precisely. Very few deserts are so dry that they can be considered 'absolute deserts'. This term applies only to regions where vegetation is completely absent except for oases, or where only ephemeral vegetation develops after a rare rainfall.

The African deserts comprise the enormous wastes of the Sahara north of the equator, the much smaller but floristically diversified Namib desert south of the equator, and some little-known small areas of coastal desert in Somalia.

The driest parts of northern Kenya, where the rainfall may be as low as 150 mm per year, as at Lodwar, west of Lake Turkana, are sometimes regarded as desert. Extensive areas are covered with stone mantles and are almost devoid of vegetation. Because, however, these stone pavements occur in mosaic with more luxuriant vegetation, mostly semi-desert shrubland and dwarf bushland, and the rainfall in most places is higher than at Lodwar, they are considered to be edaphic desert and are not given separate treatment.

Following Quézel (1965a), the northern limit of the Sahara desert is shown on the map to correspond, more or less, with the 100 mm isohyet, and the southern limit to correspond with the 150 mm isohyet. These criteria, more than any others, are closely correlated with changes in floristic composition which differentiate the flora of the Sahara from those of adjacent areas. Although the vegetation of the moister outer fringes of the Sahara is 'diffuse', in that it is not confined to water-receiving sites but also occurs on sites with pronounced run-off, the vegetation of the latter is usually sparse, and the contrast between the two types of site is very much more pronounced than in semi-desert areas. With increasing aridity, perennial vegetation becomes more and more contracted until it is eventually confined to water-receiving sites which are so situated that they collect most of the incident rainfall from relatively wide catchment areas.

The Namib desert is not nearly so well differentiated from its surrounding semi-desert vegetation as is the Sahara. Following Walter (1971) the 100 mm isohyet is arbitrarily chosen for its delimitation. The small areas of coastal desert in Somalia are similarly defined.

For a number of reasons, most desert vegetation cannot be conveniently accommodated in simple physiognomic classifications. First, some desert plants, e.g. *Welwitschia bainesii* (*mirabilis*), have unusual growth forms so that the communities in which they occur do not fit orthodox categories. Second, vegetation is often too mixed physiognomically, and varies too much and too rapidly from place to place in the relative proportions of its chief physiognomic elements. Such changes are largely dependent on local variation in the distribution of soil moisture, but are greatly modified by anthropic influences. Third, even when the vegetation is physiognomically uniform, the individual plants, except very locally, and then chiefly in the high mountains, are too few and the vegetation is too exiguous for such terms as 'grassland' or 'shrubland' to be appropriate. Fourth, in some habitats much of the vegetation consists of micro-mosaics of psammophytic, chasmophytic and halophytic elements, the complexity of which renders a simple classification impossible. Since, however, a large proportion of desert species and the communities they form are characteristically associated with distinct physiographic or edaphic features, it is most convenient to classify the vegetation they comprise according to those features.

Afroalpine vegetation

Afroalpine vegetation, which is confined to the highest mountains of tropical Africa, is physiognomically very mixed. Nearly half the species of flowering plants belong to the following specialized growth forms: giant rosette plants, mostly species of *Lobelia* and *Senecio*, up to 6 m tall; tussock grasses and sedges with filiform xeromorphic leaves; aculescent rosette plants; cushion plants and sclerophyllous shrubs and dwarf shrubs.

Owing to the frequency of nocturnal frosts extensive areas of bare soil are subjected to pronounced solifluction giving rise to *mobilideserta*. In such places unattached colonies of two mosses, *Grimmia campestris* and *G. ovalis* (*ovata*), the fruticose lichen, *Parmelia* sp. near *vagans*, and one alga, *Nostoc commune*, are conspicuous.

Scrub forest

Scrub forest is intermediate in structure between true forest and bushland and thicket. It is normally 10–15 m high. Trees with well-defined and upright boles are usually present but they do not form a closed canopy; smaller woody plants, principally bushes and shrubs, contribute at least as much as the trees to the appearance of the vegetation and its phytomass.

In most types of scrub forest in the lowlands of tropical and subtropical Africa, cactiform tree Euphorbias, or more rarely species of *Elaeophorbia* or arborescent Aloes, are present, but their density varies greatly.

In some *Euphorbia*-dominated scrub forests the tree Euphorbias are said to provide 70–80 per cent cover. Their crowns, however, are very open and the true figure must be much less than this. In types where broad-leaved trees are emergent their crown cover is rarely more than 50 per cent and often much less.

Also included in scrub forest are a few types which are dominated by bushy trees that rarely have a well-defined bole (and if they do it branches low down). The two most important are characterized by *Argania spinosa* (Mediterranean–Sahara transition zone) and *Olea laperrinei* (Sahel zone). Both species are often multiple-stemmed but their main axes are of massive proportions.

Scrub forest often forms an ecotone between true forest and bushland.

On most of the higher mountains in tropical Africa and Madagascar the stature of the vegetation steadily diminishes with increasing altitude, and scrub forest is sometimes well developed in the transition from forest to montane bushland and thicket.

In the lowlands, scrub forest occurs where the rainfall is intermediate between that required by forest and bushland, but it is only well developed where the rainfall is irregular or bimodal in distribution or where dry-season humidity is high. Where there is a single severe dry season and all the rain falls in summer, as in most parts of the Zambezian and Sudanian Regions, scrub forest is very poorly developed, and is not described separately from other vegetation types with which it is associated.

Extreme edaphic conditions such as those provided by granite inselbergs and acid peat are also responsible for the occurrence of scrub forest in parts of the Guineo–Congolian Region, though only on a very local scale.

Various types of scrub forest are described in the sections dealing with the following phytochoria: Guineo–Congolian, Somalia–Masai, Lake Victoria Basin, Zanzibar–Inhambane, Kalahari–Highveld, Tongaland–Pondoland, Sahel and Mediterranean–Sahara transition.

Transition woodland

The term transition woodland was first used in Nigeria by Jones & Keay (see Forest Department Nigeria, 1948) to designate a type of woodland in which fire-tolerant and fire-sensitive trees occur together. It was originally applied either to secondary, fire-tolerant vegetation occurring on forest sites, which, following a period of fire protection, was reverting to forest, or to degraded forest which had been invaded by fire-tolerant savanna species. Both types are unstable and in the absence of interference are likely to be transient. This type is described in Chapter I.

Keay (1959c) subsequently suggested that the original vegetation of the Guinea–Congolia/Sudania transition zone was a mosaic, with forest in the valley bottom and on the deeper soils of the lower slopes, with *Isoberlinia* woodland on the shallow soils of the crests, and transition woodland occupying an ecotone between. There is evidence that a similar situation also occurred in the Guinea–Congolia/Zambezia transition zone (Chapter X).

In the Zanzibar–Inhambane Region (Chapter XIII) *Brachystegia spiciformis* dominates transition woodland. Some stands are clearly seral, but others appear to be stable.

There is some evidence that in the Zambezian Region transition woodland represents the climax in an ecotone between dry evergreen forest on the deeper soils and miombo woodland on shallower soils (page 91).

Scrub woodland

Scrub woodland is intermediate between true woodland and bushland. It is dominated by stunted trees, sometimes no more than 3 m high, which belong to typical woodland species. Scrub woodland sometimes consists almost exclusively of stunted trees but bushy and shrubby species are frequently also abundant.

In the Zambezian Region (Chapter II) each of the three main woodland types occurs as scrub woodland under limiting conditions. Thus, scrub miombo dominated by dwarf *Brachystegia spiciformis* is found towards the upper altitudinal limits of miombo. Other miombo species including *B. boehmii* dominate scrub woodland on the plateau on unfavourable soils. Throughout the range of mopane (*Colophospermum mopane*) scrubby communities are found on unfavourable soils, and more locally as a consequence of fire or frost. They also predominate towards the drier climatic limits of the species range. South of the Limpopo River the vegetation of mapping unit 29d, southern undifferentiated Zambezian woodland, is intermediate between Zambezian woodland and Tongaland–Pondoland evergreen and semi-evergreen bushland and in places could be regarded as scrub woodland.

The ecotone between Zambezian woodland and the edaphic grassland of waterlogged depressions (dambos) is also dominated by scrub woodland, as are some communities on heavy metal and other toxic soils.

In this work scrub woodland is not described separately from the other communities with which it is associated.

Mangrove

Mangrove is dominated by trees or bushes occurring on shores periodically flooded by sea-water. It is sometimes classified as forest but this is misleading since many mangrove species, especially towards their climatic and edaphic limits of tolerance, form communities which must be referred to as 'thicket' or 'bushland'. All mangrove communities, regardless of stature and density, in *overall* physiognomy resemble each other more than any particular stand of mangrove resembles any other type of vegetation. In a natural classification they should be regarded as a major and quite remarkable physiognomic type.

In Africa the tallest mangrove, at the mouth of the Niger River, attains a height of 45 m, whereas at the limits of its geographical range and on unfavourable soils it is no more than 2 m tall. All true mangrove species have either pneumatophores which are exposed at low tide or are viviparous or almost so. Most African species show both these features. *Rhizophora* has stilt roots which function as pneumatophores. The leaves of mangrove species are thick and leathery.

Herbaceous fresh-water swamp and aquatic vegetation

Permanent swamp, or reed-swamp, occurs in depressions where water accumulates and permanently floods the surface to a shallow depth. In the Guineo–Congolian Region most swampy areas are covered with swamp forest, which is described alongside other Guineo–Congolian forest. Outside the Guineo–Congolian Region most of the shallower lakes have a wide belt of reed-swamp. Truly aquatic vegetation occurs in deeper water. Many swamp and aquatic species are distributed through several chorological regions and the vegetation they occur in is classified as azonal (Chapter XXII).

The dominants of reed-swamp are usually rooted in the soil and their stems rise out of the water. The most abundant is the giant sedge, *Cyperus papyrus*, but other sedges, grasses such as *Miscanthus* and *Phragmites*, the bullrush (*Typha*) and, locally, ferns are also dominant.

True aquatics are either completely submerged or have floating leaves. Of the latter, some are rooted in the mud, whereas others are free-floating. Reed-swamp and aquatic vegetation are often separated by a belt of floating grasses, principally *Vossia cuspidata*, *Paspalidium geminatum* and *Panicum repens*, which is often invaded by the rhizomes of papyrus.

The vegetation of seasonal swamps is often grassland, which is described in conjunction with other types of grassland.

Saline and brackish swamp

The vegetation on saline soils, which is dominated by halophytes, is physiognomically varied and includes grassland, wooded grassland, shrubland, and bushland. It is described in Chapter XXII. Most halophytes have fleshy leaves, which in some species are very reduced. The ground between the plants is often covered with a white saline efflorescence.

Bamboo

Bamboos are giant grasses 2–20 m or more in height, with erect woody stems, which persist for several years. Many species flower gregariously over large areas and then die back to the underground rhizomes or die completely. Sometimes they form almost pure, virtually impenetrable communities, but sometimes are scattered in other vegetation.

Only four species of bamboo (belonging to three genera), namely *Arundinaria alpina*, *A. tesselata*, *Oreobambos buchwaldii*, and *Oxytenanthera abyssinica*, are indigenous to Africa, though the introduced *Bambusa vulgaris* is locally naturalized. The bamboo flora of Madagascar is slightly more diversified but less is known about it.

Arundinaria forms Afromontane communities. *Oreobambos* usually occurs on mountains but is not strictly Afromontane. *Oxytenanthera* is widespread in the Sudanian and Zambezian Regions.

Anthropic landscapes

In most parts of Africa the vegetation has been profoundly altered by human activity and few natural stands remain. The only places, however, where the natural vegetation has been totally eliminated over sufficiently extensive areas to show on the map are in the Mediterranean Region and the Mediterranean/Sahara Transition Zone, where cultivation has been continuous for more than 2000 years. In other parts of Africa sufficient relics remain to permit plausible reconstruction of the original vegetation.

In some of the more densely settled parts of tropical Africa, such as the northern Sudanian Region, cultivation around the big cities is almost continuous. Often the only trees are self-sown individuals of species of economic importance, such as *Parkia biglobosa* (*clappertoniana*), *Diospyros mespiliformis*, *Hyphaene thebaica*, *Acacia albida*, *Anogeissus leiocarpus* and *Butyrospermum paradoxum* (*parkii*), which have been permitted to remain, giving the landscape a park-like appearance (wooded farmland).

8 Mapping units

For reasons given in the Introduction to Part Two the vegetation of the mapping units is described within a chorological framework rather than under the names of the units themselves. In many instances a single section of text deals only with a single mapping unit, but for other units, especially complex mosaics, the relevant information is given in more than one place.

The mapping units are listed in Table 4 (see opposite), with cross-references to the phytochoria under which they are described.

TABLE 4. Mapping units and related information

Mapping units	Phytochoria	Main vegetation types and page references
1. Lowland rain forest: wetter types		
(a) Guineo–Congolian *Note*: Included in this unit, but not shown separately, are large areas of cultivation, and of secondary forest (page 80) and smaller areas of short forest and scrub forest (page 81), elfin thicket (page 83), edaphic grassland (page 83), and secondary grassland and wooded grassland (page 84). The largest areas of Guineo–Congolian swamp forest are mapped separately (units 8 and 9) but smaller areas are scattered throughout 1a.	I. Guineo–Congolian Region	Hygrophilous coastal evergreen Guineo–Congolian rain forest (page 76). Mixed moist semi-evergreen Guineo–Congolian rain forest (page 77). Single-dominant moist evergreen and semi-evergreen Guineo–Congolian rain forest (page 78).
(b) Malagasy *Note*: Also included are large areas of cultivation, secondary forest (page 235) and secondary grassland (page 239).	XIX. East Malagasy Region	See page 235.
2. Guineo–Congolian rain forest: drier types *Note*: Also included but not shown separately are large areas of cultivation and secondary grassland and wooded grassland (page 84) and secondary forest (page 80), and smaller areas of short forest and scrub forest (page 81), swamp forest (page 82), transition woodland (page 83) and edaphic grassland (page 83).	I. Guineo–Congolian Region X. Guinea–Congolia/Zambezia Regional Transition Zone XI. Guinea–Congolia/Sudania Regional Transition Zone XII. Lake Victoria Regional Mosaic	Drier peripheral semi-evergreen Guineo–Congolian rain forest and similar forest in the transition zones. {See page 79. / See page 172. / See page 79. / See page 181.}
3. Mosaic of 1a and 2 *Note*: In this type large areas of hygrophilous and moist rain forest (1a) have been replaced by old secondary forest dominated by species more characteristic of drier peripheral Guineo–Congolian rain forest (page 79) such as *Triplochiton scleroxylon* and *Terminalia superba*.	I. Guineo–Congolian Region	See note in column 1.
4. Transitional rain forest *Note*: The largest occurrence of this unit forms the transition between the Guineo–Congolian Region and a large 'island' of the Afromontane Region to the east. For convenience it is included in the former. There is little published information.	I. Guineo–Congolian Region XII. Lake Victoria Regional Mosaic	See page 85. See page 181.
5. Malagasy moist montane forest	XIX. East Malagasy Region	See page 236.
6. Zambezian dry evergreen forest *Note*: The largest areas, which are dominated by *Cryptosepalum pseudotaxus*, are shown on the map, but in many places they have been degraded or replaced by secondary communities. They also occur in mosaic with smaller areas of edaphic grassland (page 99) and miombo woodland (page 92). Elsewhere this type is shown symbolically as part of a mosaic (see units 14 and 21).	II. Zambezian Region	See page 89.
7. Malagasy dry deciduous forest	XX. West Malagasy Region	See page 241.

Continued overleaf

TABLE 4 contd

Mapping units	Phytochoria	Main vegetation types and page references
8. Swamp forest Notes: (1) This unit also includes small areas of herbaceous swamp and aquatic vegetation (page 264), and edaphic grassland (page 83). (2) Guineo–Congolian swamp and riparian forest extends into the adjacent transition zones, and related forest occurs in the wetter part of the Zambezian Region (page 91).	I. Guineo–Congolian Region XII. Lake Victoria Regional Mosaic	See page 82. See page 181.
9. Mosaic of 8 and 1a	I. Guineo–Congolian Region	
10. Mediterranean sclerophyllous forest Note: Much of the area shown as forest has been destroyed or severely degraded and only scattered fragments remain. Several types of forest can be recognized but their distribution is too complicated for their boundaries to be shown accurately on the map. The occurrence of three major variants is indicated by letters. Among subordinate types, deciduous forest is described on page 156, bushland and thicket are described on page 157, and secondary shrubland (maquis and garrigue) on page 159.	VII. Mediterranean Region XVIII. Mediterranean/Sahara Regional Transition Zone	Mediterranean broad-leaved sclerophyllous forest (page 150). Mediterranean coniferous forest, p.p. (page 152). Sub-Mediterranean forest (page 226).
11. Mosaic of lowland rain forest and secondary grassland (a) Guineo–Congolian Note: Most of the grassland is secondary but there are small areas of edaphic grassland (page 83 and page 173).	I. Guineo–Congolian Region X. Guinea–Congolia/Zambezia Regional Transition Zone XI. Guinea–Congolia/Sudania Regional Transition Zone XII. Lake Victoria Regional Mosaic	Drier peripheral semi-evergreen Guineo–Congolian rain forest and similar forest in the transition zones (page 79). Guineo–Congolian secondary grassland and wooded grassland (page 84). Drier peripheral semi-evergreen Guineo–Congolian rain forest. — See pages 84, 172 and 173. Guineo–Congolian secondary grassland and wooded grassland. — See pages 79 and 84. See pages 84 and 181.
(b) Malagasy	XIX. East Malagasy Region	East Malagasy primary lowland rain forest (page 235). East Malagasy secondary lowland rain forest (page 235). East Malagasy secondary grassland, p.p. (page 239).
12. Mosaic of lowland rain forest, Isoberlinia woodland and secondary grassland Note: Similar to 11a but with patches of Isoberlinia woodland on the poorer soils, principally stony hills and ironstone plateaux. Their distribution in Nigeria is accurately known (J.A.D. Jackson, pers. comm.). Elsewhere information is incomplete.	XI. Guinea–Congolia/Sudania Regional Transition Zone	Drier peripheral semi-evergreen Guineo–Congolian rain forest and similar types in the transition zones (page 79). Guineo–Congolian secondary grassland and wooded grassland (page 84). Sudanian Isoberlinia and related woodlands (page 106).

13. Mosaic of lowland rain forest, secondary grassland and montane elements	XI. Guinea–Congolia/Sudania Regional Transition Zone	See page 176.
14. Mosaic of lowland (Guineo–Congolian) rain forest, Zambezian dry evergreen forest and secondary grassland	X. Guinea–Congolia/Zambezia Regional Transition Zone	Drier peripheral semi-evergreen Guineo–Congolian rain forest (page 172). Zambezian dry evergreen forest and transition woodland (page 172). Grassland and wooded grassland (page 173).
15. West African coastal mosaic	X. Guinea–Congolia/Zambezia Regional Transition Zone	See page 174.
	XI. Guinea–Congolia/Sudania Regional Transition Zone	See pages 176-8.
16. East African coastal mosaic		
(a) Zanzibar–Inhambane *Note*: Except for some small islands of forest (16b below) the vegetation has been so extensively modified by man that it is impossible to map the different physiognomic types separately. Unit 16a almost exactly coincides with the Zanzibar–Inhambane Region but there are also some exclaves in the eastern parts of the Zambezian Region.	XIII. Zanzibar–Inhambane Regional Mosaic	See pages 186–9.
(b) Forest patches (Zanzibar–Inhambane)	XIII. Zanzibar–Inhambane Regional Mosaic	Zanzibar–Inhambane lowland rain forest (page 186). Transitional rain forest (page 186). Zanzibar–Inhambane undifferentiated forest (page 187). Zanzibar–Inhambane lowland rain forest (page 186). Zanzibar–Inhambane lowland rain forest (page 186).
	II. Zambezian Region (as enclaves)	
	IV. Somalia–Masai Region (as enclaves)	
(c) Tongaland–Pondoland *Note*: This unit is a mosaic of relict forest patches in a matrix of secondary grassland and wooded grassland. There are also patches of scrub forest and semi-evergreen thicket on shallow soils and in rain-shadow areas, and, on the coastal plain, of swamp forest and edaphic grassland.	XV. Tongaland–Pondoland Regional Mosaic	See pages 199–202.
17. Cultivation and secondary grassland replacing upland and montane forest in Africa *Note*: The natural vegetation, which is mostly destroyed, probably originally contained a mixture of Afromontane and lowland species. There is virtually no published information.	II. Zambezian Region VIII. Afromontane Region	See note in column 1.
18. Cultivation and secondary grassland replacing upland and montane forest in Madagascar.	XIX. East Malagasy Region	East Malagasy sclerophyllous montane forest (page 236). East Malagasy tapia forest (page 237). East Malagasy secondary grassland, p.p. (page 239).

Continued overleaf

TABLE 4 contd

Mapping units	Phytochoria	Main vegetation types and page references
19. Undifferentiated montane vegetation		
(a) Afromontane	VIII. Afromontane Region	See pages 162–9.
(b) Sahelomontane	XVI. Sahel Regional Transition Zone	Sahelomontane scrub forest (page 207). Sahelomontane secondary grassland (page 207).
(c) Malagasy	XIX. East Malagasy Region	East Malagasy montane bushland and thicket (page 237). East Malagasy rupicolous shrubland (page 238). East Malagasy secondary montane grassland (page 239).
20. Transition from Afromontane scrub forest to Highveld grassland	XIV. Kalahari–Highveld Transition Zone	See page 195.
21. Mosaic of Zambezian dry evergreen forest and wetter miombo woodland *Note*: The largest areas of dry evergreen forest are shown separately (mapping unit 6). Elsewhere, this type has a fragmentary distribution and always occurs in mosaic with transition woodland, miombo woodland and various types of secondary, fire-maintained woodland and grassland (chipya).	II. Zambezian Region X. Guinea–Congolia/Zambezia Transition Zone	Zambezian dry evergreen forest (page 89). Zambezian transition woodland (page 91). Zambezian wetter miombo woodland (page 93). Zambezian chipya woodland and wooded grassland (page 96). Zambezian dry evergreen forest and transition woodland (page 172). Grassland and wooded grassland (page 173).
22. Mosaic of dry deciduous forest and secondary grassland and wooded grassland		
(a) Zambezian *Note*: Forest and secondary communities replacing it occur on deep well-drained Kalahari sand. Other types, principally Kalahari woodland (page 97), Kalahari thicket (page 98) and edaphic grassland (page 100), occupy shallower and less well-drained soils.	II. Zambezian Region	Zambezian dry deciduous forest and scrub forest (page 90). Zambezian secondary grassland and wooded grassland, p.p. (page 101).
(b) Malagasy	XX. West Malagasy Region	West Malagasy dry deciduous forest (page 241). West Malagasy grassland, p.p. (page 242).
23. Mosaic of Mediterranean montane forest and altimontane shrubland	VII. Mediterranean Region	*Cedrus atlantica* forest (page 155). *Abies pinsapo* and *A. numidica* forest (page 156). *Juniperus thurifera* forest (page 156). Mediterranean deciduous forest, p.p. (page 156). Altimontane Mediterranean shrubland (page 158).
24. Mosaic of Afromontane scrub forest, Zambezian scrub woodland, and secondary grassland	XIV. Kalahari–Highveld Transition Zone	See page 196.
25. Wetter Zambezian miombo woodland *Note*: In addition to the islands of other vegetation (units 6, 19a, 21, 37, 40, 47, 60, 64 and 75) shown on the map there are also smaller areas of evergreen swamp and riparian forest (page 91),	II. Zambezian Region X. Guinea–Congolia/Zambezia Transition Zone	See page 93. See page 172.

	transition woodland (page 91), chipya woodland and wooded grassland (page 96) and wet dambo (page 99).	XII. Lake Victoria Regional Mosaic	See page 181.
26.	Drier Zambezian miombo woodland *Note*: This unit also includes smaller areas of dry deciduous thicket (page 97), mopane woodland (page 93), deciduous riparian forest (page 91) and dry dambo (page 99).	II. Zambezian Region IV. Somalia–Masai Region (as enclaves) XIII. Zanzibar–Inhambane Regional Mosaic (as enclaves)	See page 93.
27.	Sudanian woodland with abundant *Isoberlinia* *Note*: This unit also includes small areas of semi-evergreen swamp and riparian forest (page 103), transition woodland (page 105), edaphic (page 107) and secondary grassland (page 108), and rupicolous communities (page 108).	III. Sudanian Region	See page 106.
28.	*Colophospermum mopane* woodland and scrub woodland *Note*: Included in unit 26 are smaller areas of miombo woodland (page 92), dry deciduous forest (page 90), deciduous riparian forest and woodland (pages 91, 95), undifferentiated woodland (page 95) and edaphic grassland (page 99).	II. Zambezian Region	See page 93.
29.	Undifferentiated woodland		
	(a) Sudanian *Note*: Much of the area is now semi-permanent cultivation and bush fallow. In places the original vegetation may have been dry forest. Included in this type are areas, too small to be shown on the map, of dry evergreen forest (page 103), semi-deciduous riparian forest (page 103), transition woodland (page 105), edaphic grassland (page 107) and rupicolous communities (page 108).	III. Sudanian Region	See page 106.
	(b) Ethiopian	III. Sudanian Region	See page 107.
	(c) North Zambezian	II. Zambezian Region	See page 95.
	(d) South Zambezian	II. Zambezian Region	See page 96.
	(e) The transition from undifferentiated Zambezian woodland to Tongaland–Pondoland bushland *Note*: This unit is at the northern end of a rather complex floristic and physiognomic continuum. Where the rainfall is higher, as on the seaward slopes of the Lebombo Mts, there are patches of forest (page 199).	XV. Tongaland–Pondoland Regional Mosaic	Tongaland–Pondoland evergreen and semi-evergreen bushland and thicket, p.p. (page 200).
30.	Sudanian undifferentiated woodland with islands of *Isoberlinia* *Note*: Small islands of *Isoberlinia* occur, especially on rocky hills, in wetter undifferentiated Sudanian woodland and are shown schematically on the map. Their distribution in Nigeria is accurately known (J.A.D. Jackson, pers. comm.). Elsewhere information is incomplete.	III. Sudanian Region	See pages 105–7.
31.	Mosaic of wetter Zambezian woodland and secondary grassland *Note*: This corresponds in part to Devred's (1958) mapping unit 16. There is virtually no published information.	X. Guinea–Congolia/Zambezia Regional Transition Zone	See note in column 1.

Continued overleaf

TABLE 4 contd

Mapping units	Phytochoria	Main vegetation types and page references
32. The Jos Plateau mosaic *Note*: This is the largest upland area in Nigeria above 1000 m. Much of the original vegetation has been destroyed for agriculture or to provide fuel for the tin-mines. Some communities are related to vegetation in the East African Highlands and other upland areas in West Africa (see Keay, 1959). The best-preserved vegetation is in rocky places.	III. Sudanian Region XI. Guinea–Congolia/Sudania Regional Transition Zone	Sudanian transition woodland (page 105). Sudanian rupicolous scrub forest, bushland and thicket (page 108).
33. The Mandara Plateau mosaic *Note*: On the upper slopes (1 300–1 442 m) Sudanian elements are mixed with Afromontane species such as *Olea capensis* and *Pittosporum viridiflorum*, and the succulent tree *Euphorbia*, *E. desmondii*, which also occur on the Jos Plateau (see Letouzey, 1968).	III. Sudanian Region	See note in column 1.
34. Transition from South African scrub woodland to Highveld grassland	XIV. Kalahari–Highveld Regional Transition Zone	See page 196.
35. Transition from undifferentiated woodland to *Acacia* deciduous bushland and wooded grassland (also including mosaics of communities dominated by *Acacia* and broad-leaved trees)		
(a) Zambezian *Notes*: (1) Woodland and wooded grassland, dominated by species of *Acacia* and broad-leaved trees, greatly modified by fire, is extensively distributed on alluvium in the Rukwa Valley, Tanzania (Pielou, 1952; Vesey-FitzGerald, 1970). (2) Wooded grassland dominated by species of *Acacia* and broad-leaved trees occurs in the Zambezian Region on Kalahari Sand between the Okavango basin and the Makarikari depression, but is much more extensive in the transition zone further south.	II. Zambezian Region XIV. Kalahari–Highveld Regional Transition Zone	See notes in column 1. See page 193.
(b) Ethiopian *Note*: This unit forms the transition from the edaphic grasslands of the Flood Region of the Nile to the *Anogeissus*–*Combretum hartmannianum* woodland (page 107) flanking the Ethiopian Highlands to the east. *Acacia seyal* and *Balanites aegyptiaca* occur throughout except for patches of thornless woodland dominated by *Combretum hartmannianum*, *Sterculia setigera*, *Stereospermum kunthianum* and *Adansonia digitata*.	III. Sudanian Region	See note in column 1.
(c) The Windhoek Mountains	XIV. Kalahari–Highveld Regional Transition Zone	See page 193.
36. Transition from *Colophospermum mopane* scrub woodland to Karoo–Namib shrubland	XIV. Kalahari–Highveld Regional Transition Zone	See page 191.

37. *Acacia polyacantha* secondary wooded grassland		See note in column 1.
Note: *A. polyacantha* is widely distributed in the wetter parts of the Zambezian Region and further north as a pioneer species in colonizing forest and as a fire-hardy constituent of secondary wooded grassland. Some occurrences are sufficiently extensive to show on the map, but there is little published information.	II. Zambezian Region X. Guinea–Congolia/Zambezia Regional Transition Zone	
38. East African evergreen and semi-evergreen bushland and thicket	IV. Somalia–Masai Region VIII. Afromontane Region XVI. Sahel Regional Transition Zone XVII. Sahara Regional Transition Zone	See page 115. See note in column 1. See page 206. See page 224.
Note: This unit often forms the transition between the Somalia–Masai and Afromontane Regions. For cartographic reasons it has been placed partly in the former and partly in the latter. In parts of East Africa and the Lake Victoria basin it has been extensively replaced by mapping unit 45 and the surviving remnants are too small to show separately.		
39. South African evergreen and semi-evergreen bushland and thicket	XV. Tongaland–Pondoland Regional Mosaic	See page 200.
40. Itigi deciduous thicket	II. Zambezian Region	See page 97.
41. Malagasy deciduous thicket	XX. West Malagasy Region	See page 242.
42. Somalia–Masai *Acacia*–*Commiphora* deciduous bushland and thicket	IV. Somalia–Masai Region	See page 113.
43. Sahel *Acacia* wooded grassland and deciduous bushland	XVI. Sahel Regional Transition Zone	Sahel wooded grassland (page 206). Sahel deciduous bushland (page 207).
44. Kalahari deciduous *Acacia* wooded grassland and bushland	II. Zambezian Region (as enclaves) XIV. Kalahari–Highveld Regional Transition Zone	See page 193.
45. Mosaic of East African evergreen bushland and secondary *Acacia* wooded grassland	IV. Somalia–Masai Region	Somalia–Masai secondary grassland and wooded grassland, p.p. (page 114). East African evergreen and semi-evergreen bushland and thicket, p.p. (page 115).
46. Mosaic of Malagasy deciduous thicket and secondary grassland	XII. Lake Victoria Regional Mosaic	See page 182.
47. Mosaic of *Brachystegia bakerana* thicket and edaphic grassland	XX. West Malagasy Region	West Malagasy deciduous thicket (page 242). West Malagasy grassland, p.p. (page 242).
Note: Apart from Barbosa's brief account (1970) nothing is known. Patches of *Cryptosepalum* forest (page 89) and Kalahari woodland (page 97) also contribute to the mosaic.	II. Zambezian Region	See pages 98, 100.
48. Tugela basin wooded bushland	XV. Tongaland–Pondoland Regional Mosaic	Tongaland–Pondoland evergreen and semi-evergreen bushland and thicket, p.p. (page 200).
Note: The vegetation is intermediate between that of mapping units 29e and 39. It is described in considerable detail by D. Edwards (1967).		

Continued overleaf

TABLE 4 contd.

Mapping units	Phytochoria	Main vegetation types and page references
49. Transition from Mediterranean *Argania* scrubland to succulent semi-desert shrubland	XVIII. Mediterranean/Sahara Regional Transition Zone	*Argania spinosa* scrub forest and bushland (page 227). Succulent sub-Mediterranean shrubland, p.p. (page 228).
50. Cape shrubland (fynbos)	V. Cape Region	See page 132.
51. Bushy Karoo–Namib shrubland	VI. Karoo–Namib Region	See pages 137, 140.
52. Succulent Karoo shrubland	VI. Karoo–Namib Region	See pages 137–142.
53. Dwarf Karoo shrubland	VI. Karoo–Namib Region	See pages 139–140.
54. Semi-desert grassland and shrubland		
(a) Northern Sahel	XVI. Sahel Regional Transition Zone	See page 206.
(b) Somalia–Masai	IV. Somalia–Masai Region	See page 115.
55. Sub-Mediterranean semi-desert grassland and shrubland	XVIII. Mediterranean/Sahara Regional Transition Zone	Succulent sub-Mediterranean shrubland, p.p. (page 228). Sub-Mediterranean grassland (page 229).
56. The Kalahari/Karoo–Namib transition	XIV. Kalahari–Highveld Regional Transition Zone	See page 193.
57. Grassy shrubland		
(a) Montane Karoo	VI. Karoo–Namib Region	See page 139.
(b) Transition from Karoo shrubland to Highveld	XIV. Kalahari–Highveld Regional Transition Zone	See page 195.
58. Highveld grassland	XIV. Kalahari–Highveld Regional Transition Zone	See page 194.
59. Edaphic grassland on volcanic soils	IV. Somalia–Masai Region	Somalia–Masai edaphic grassland, p.p. (pages 116 and 125).
60. Edaphic and secondary grassland on Kalahari Sand *Note*: In the Upper Zambezi basin most of the grasslands are edaphic and are often fringed with scrub woodland dominated by *Diplorhynchus condylocarpon* (page 99). In Zaire floristically similar grassland occurs on Kalahari Sand on the unrejuvenated plateau. Some of it is edaphic but much is secondary.	II. Zambezian Region X. Guinea-Congolia/Zambezia Regional Transition Zone	See page 100. Grassland and wooded grassland, p.p. (page 173).
61. Edaphic grassland in the Upper Nile basin	III. Sudanian Region	Grassland and wooded grassland on Pleistocene clays, p.p. (page 108).
62. Mosaic of edaphic grassland and *Acacia* wooded grassland	III. Sudanian Region	Grassland and wooded grassland on Pleistocene clays, p.p. (page 108).

Note: Vegetation similar to units 62 and 63 is too restricted in the Zambezian Region to map separately and is included in mapping unit 64.

63. Mosaic of edaphic grassland and communities of *Acacia* and broad-leaved trees	XVI. Sahel Regional Transition Zone	See page 108.
	III. Sudanian Region	Grassland and wooded grassland on Pleistocene clays, p.p. (page 108).

Note: The distribution of this and the preceding unit in the Chad basin are shown on the map schematically. A more detailed representation is given by Pias (1970).

64. Mosaic of edaphic grassland and semi-aquatic vegetation	II. Zambezian Region	See pages 99, 264	[XXI]
	III. Sudanian Region	See pages 107, 264.	
	XVI. Sahel Regional Transition Zone	See pages 107, 264.	
65. Altimontane vegetation in tropical Africa	VIII and IX. Afromontane and Afroalpine Regions	Afromontane evergreen bushland and thicket, p.p. (page 167). Afromontane shrubland, p.p. (page 168). Afromontane and Afroalpine grassland, p.p. (page 168). Mixed Afroalpine communities in tropical Africa (page 169).	
66. Altimontane vegetation in South Africa	VIII and IX. Afromontane and Afroalpine Regions	Afromontane evergreen bushland and thicket, p.p. (page 167). Afromontane shrubland, p.p. (page 168). Afromontane and Afroalpine grassland, p.p. (page 168). Mixed Afroalpine communities in South Africa (page 169).	
67–73. The Sahara Desert			
67. Absolute desert	XVII. Sahara Regional Transition Zone	See page 223.	
68a. Atlantic coastal desert	XVII. Sahara Regional Transition Zone	See page 224.	
68b. Red Sea coastal desert	XVII. Sahara Regional Transition Zone	See page 224.	
69. Desert dunes without perennial vegetation	XVII. Sahara Regional Transition Zone	Psammophilous vegetation, p.p. (page 220).	
70. Desert dunes with perennial vegetation	XVII. Sahara Regional Transition Zone	Psammophilous vegetation, p.p. (page 220).	
71. Regs, hamadas, wadis	XVII. Sahara Regional Transition Zone	Wadis (page 219). Hamadas (page 220). Regs (page 221).	

Continued overleaf

TABLE 4 contd

Mapping units	Phytochoria	Main vegetation types and page references
72. Saharomontane vegetation	XVII. Sahara Regional Transition Zone	See page 221.
73. Oasis	XVII. Sahara Regional Transition Zone	See page 218.
	XVIII. Mediterranean/Sahara Regional Transition Zone	See page 226.
74. The Namib Desert	VI. Karoo–Namib Region	See page 141.
	See note in column 1.	See page 264.
75. Herbaceous swamp and aquatic vegetation Note: Widespread in all but the driest parts, but only sufficiently extensive to be mapped separately in the Zambezian and Sahel Regions. Elsewhere mostly in mosaic (mapping unit 64) or too small to show. Swamp forest is mapped separately (mapping units 8 and 9).		
76. Halophytic vegetation Note: Areas too small to map also occur in the Karoo–Namib (pages 144, 266), Mediterranean and Lake Victoria Regions.	Azonal (page 266), but most extensive in:	
	II. Zambezian Region	See page 267.
	IV. Somalia–Masai Region	See page 266.
	XVII. Sahara Regional Transition Zone	See page 222.
	XVII. Mediterranean/Sahara Regional Transition Zone	See page 230.
77. Mangrove	Azonal	See page 260.
78. Mediterranean anthropic landscapes	VII. Mediterranean Region	See page 159.
79. Western sub-Mediterranean anthropic landscapes	XVIII. Mediterranean/Sahara Regional Transition Zone	Sub-Mediterranean anthropic landscapes, p.p. (page 230).
80. Eastern sub-Mediterranean anthropic landscapes	XVIII. Mediterranean/Sahara Regional Transition Zone	Sub-Mediterranean anthropic landscapes, p.p. (page 230).

Part Three Vegetation of the floristic regions

Introduction

The vegetation of each of the major phytochoria is described in turn. In addition, three vegetation types, namely mangrove, herbaceous swamp and aquatic vegetation, and saline and brackish swamp, are included in Chapter XXII, which deals with azonal vegetation. They are not strictly azonal, since most of their species are confined to the tropics and subtropics. Nevertheless, their ranges transgress the limits of the major phytochoria and it is convenient to give them separate treatment.

The African mainland

1 The Guineo–Congolian regional centre of endemism

Geographical position and area

Geology and physiography

Climate

Flora

Mapping units

Vegetation
    Guineo–Congolian rain forest
        Hygrophilous coastal evergreen Guineo–Congolian rain forest
        Mixed moist semi-evergreen Guineo–Congolian rain forest
        Single-dominant moist evergreen and semi-evergreen Guineo–Congolian rain forest
        Drier peripheral semi-evergreen Guineo–Congolian rain forest and similar forest in the transition zones
        Secondary Guineo–Congolian rain forest
            Pioneer secondary forest
            Young secondary forest
            Old secondary forest
    Guineo–Congolian short forest and scrub forest
        On granite inselbergs
        Upland *Parinari excelsa* forest in West Africa
    Guineo–Congolian swamp forest and riparian forest
    Guineo–Congolian transition woodland
    Guineo–Congolian elfin thicket
    Guineo–Congolian edaphic grassland
        On hydromorphic soils
        On rocky outcrops
    Guineo–Congolian secondary grassland and wooded grassland
    Transitional rain forest

Geographical position and area

The main area of the Guineo–Congolian Region extends as a broad band north and south of the equator from the Atlantic seaboard eastwards through the Zaire basin to the western slopes of the 'dorsale du Kivu'. A smaller western satellite occurs in Upper Guinea from Guinea Republic to Ghana. The dry 'Dahomey interval' separates the two areas. (Area: 2 800 000 km².)

Geology and physiography

Nearly everywhere the altitude is less than 1 000 m but at the eastern end of the Zaire basin the land rises steeply and Guineo–Congolian vegetation gives way to Afromontane communities. Between Guinea Republic and Gabon there are a few limited areas above 1 000 m where Guineo–Congolian vegetation is either diluted by Afromontane species or local upland endemic species, or entirely replaced by Afromontane communities.

The Zaire basin has an average altitude of 400 m and gradually rises to the uplands and plateaux which form its rim. In the basin the Precambrian rocks are covered by continental sediments ranging from Palaeozoic to recent. Quaternary sediments cover a 150 000 km² alluvial plain in the centre of the basin where there are remnants of former lakes and extensive swampy areas.

Surrounding the Zaire basin are the equatorial uplands, a region of dissected plateaux which merge with it. Near the basin the plateaux are mostly composed of slightly metamorphic Upper Precambrian sandstone, quartzite and schist.

To the east the Zaire basin rises to the Kivu ridge, which consists principally of strongly metamorphic Precambrian gneiss, amphibolite, quartzite and micaschist as well as granitic intrusives. The upper slopes lie outside the Guineo–Congolian Region and carry Afromontane vegetation.

Towards the north-west the undulating plateau of east Cameroun, which has an elevation from 600 to 800 m, rises gently to the Bamenda–Adamawa highlands which reach 1 500 m or more, and consist of basement rocks partly overlain by volcanic deposits. Mount Cameroun, a still active volcano (4 095 m), stands separate from the main range.

At the western rim of the Zaire basin the contact with the rather narrow Atlantic coastal plain is formed by the Cameroun–Gabon plateaux with altitudes of 600–1000 m, which extend south as the Crystal Mts and the Mayombe Mts. These ranges consist of gneiss, granite, migmatite, quartzite, greenstone, diorite, mica-schist and amphibolite.

The coastal plain itself, between Angola and Cameroun, varies greatly in width and is especially wide near Libreville where the Ogooue River enters the plain, and also around Douala. The plain is crossed by numerous rivers with mangrove bordering their estuaries. There are also lagoons, lakes and swamps.

In West Africa almost the whole of the Guineo–Congolian Region is underlain by Precambrian rocks. The landscape is formed of relatively low plateaux and plains interrupted by residual inselbergs and small higher plateaux. The most important of the latter are Fouta Djalon, the Upper Guinea Highlands (Loma-Man 'dorsale') and the Togo–Atacora range, all of which also extend into the Guinea–Congolia/Sudania transition zone.

Fouta Djalon is a remarkably level plateau with an average altitude of 1000 m, but ascending locally to 1500 m. It is extensively dissected by the angular drainage pattern.

The Guinea Highlands attain 1752 m in Mt Nimba and 1947 m in the Loma Mts. In contrast to Fouta Djalon there are few level surfaces, and the hills are rounded. The hardness of the quartzite of the Nimba chain has helped it to withstand erosion. The Man massif consists of a continuous granite-norite series.

The Togo–Atacora range rises to over 1000 m in Mount Agou.

In places along the West African coastal plain between Nigeria and Guinea–Bissau, under the influence of west–east longshore currents, huge sand-bars have formed between the ocean and the lagoons. North of the lagoon stretch the shore is rocky. In Guinea Republic south-west of Fouta Djalon the coastal plain penetrates far inland along river valleys. There are no sand-bars or lagoons but muddy creeks and estuaries ('rias') covered with mangrove and flanked by marshes. In Guinea–Bissau mangrove covers relatively large areas.

Climate

Compared with rain-forest areas in other continents, most of the Guineo–Congolian Region is relatively dry and receives between 1600 and 2000 mm of rainfall per year. Areas receiving more rain than this are largely confined to the coastal parts of Upper and Lower Guinea. Only a small part of the Zaire basin receives more than 2000 mm per year. Rainfall in excess of 3000 mm per year falls only in two relatively restricted areas, namely a coastal belt from Guinea Republic to Liberia, but here there is a very pronounced dry season, and a narrow coastal region of Cameroun adjacent to the Gulf of Biafra. Very locally in the latter at the foot of Mt Cameroun annual precipitation exceeds 10000 mm. (See Fig. 5.)

The rainfall is, in general, not only lower than in some rain forest regions elsewhere, but its distribution throughout the year is less even. Virtually nowhere in the Guineo–Congolian Region is mean monthly rainfall higher than 100 mm throughout the year. In the equatorial parts of the Zaire basin one or two months usually have a rainfall under 100 mm but more than 50 mm. Further away from the equator, but also towards the Atlantic coast at equatorial latitudes, the length and severity of the dry period increases.

In the main eastern block of the Guineo–Congolian Region, nearly everywhere the rainfall shows two peaks, separated by one relatively severe and one less severe dry interval. There is a single peak only in the very high-rainfall belt of the Gulf of Biafra, which unlike most of tropical Africa lies within the tropical rain belt throughout the year.

Individual dry periods are both more frequent and more severe than the climatic diagrams, which are based on mean values, indicate. Thus, near the equator in the heart of the Zaire basin periods of several successive rainy days are rather rare and dry periods are frequent throughout the year even during the wettest seasons. At Yangambi, for instance (Bultot, 1954, summarized by Évrard, 1968), dry periods of 6–10 days' duration occur on average 1.6 times per year in the driest month and 0.6 times in the wettest month. Dry periods lasting 30 days or more probably occur once every 12 years.

In Nigeria the dry season lasts for three months from December to February, and each receives less than 50 mm. In January and February the harmattan, a north-easterly desiccating wind from the Sahara, at times reaches the rain-forest zone. In Ghana, when the harmattan blows, relative humidity at 15.00 GMT falls to 53 per cent. Further east in Zaire a wind from the same direction also influences the rain-forest climate, but since it emanates from the Ethiopian highlands and the Nile Valley its influence is probably less pronounced than that of the harmattan proper.

In Ghana rainfall varies appreciably over short distances but it is uniformly more than 1750 mm per year in the south-west. Elsewhere this figure is reached only on hills above 600 m elevation. In general there are four to five months with less than 100 mm rainfall, even in the wettest south-west corner.

Further west, in Liberia, Sierra Leone and Guinea Republic the rainfall becomes increasingly concentrated in a single season. In Guinea Republic rainfall in places exceeds 4000 mm per year, but for four months there is virtually no rain. According to Aubréville (1938) the climax vegetation under this type of climate is not rain forest, because of the severity of the dry season, but there is little published information, and forest of Guineo–Congolian affinity certainly occurs on favourable soils and even extends further to the north-west into a yet more strongly seasonal climate (page 178).

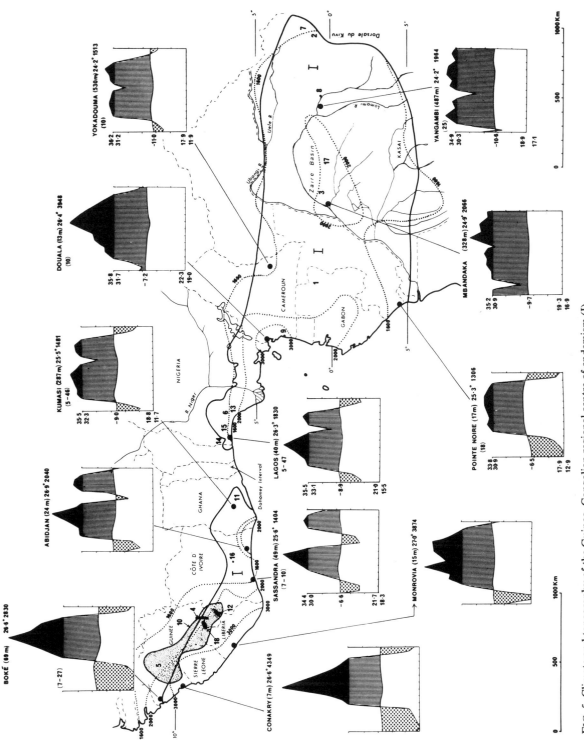

FIG. 5. Climate and topography of the Guineo–Congolian regional centre of endemism (I)
1. Belinga Mts. 2. Beni. 3. Équateur. 4. Fon. 5. Fouta Djalon. 6. Idanre Hills. 7. Irumu. 8. Kisangani. 9. Kribi. 10. Loma–Man Dorsale. 11. Mpraeso. 12. Nimba. 13. Okomu Forest Reserve. 14. Olokemeji Forest Reserve. 15. Shasha Forest Reserve. 16. Singrobo. 17. Tshuapa. 18. Ziama
The 1600, 2000 and 3000 mm isohyets are also shown

At the other extreme a drier type of Guineo–Congolian rain forest can occur where annual rainfall is as low as 1230 mm, as at Ibadan in Nigeria, provided the dry season is no longer than three months and atmospheric humidity is high throughout the year.

The distribution of Guineo–Congolian species in relation to moisture conditions is complex, and the relative importance of rainfall, relative humidity and soil moisture is imperfectly understood. For many species distribution is poorly correlated with rainfall (White, 1979).

Throughout the Guineo–Congolian Region mean monthly temperature is almost constant throughout the year.

Tornadoes are less important than in some other parts of the tropics, but in the more seasonal parts violent squalls of wind are frequent at the beginning of the wet season.

Flora

There are about 8000 species, of which more than 80 per cent are endemic.

Endemic families. The endemic families are Dioncophyllaceae, Hoplestigmataceae, Huaceae, Lepidobotryaceae, Medusandraceae, Octoknemaceae, Pandaceae, Pentadiplandraceae and Scytopetalaceae. Some of these, e.g. Octoknemaceae and Pandaceae, are not universally accepted as families.

Other characteristic families. Large trees are represented by several species in Leguminosae: Caesalpinioideae and Mimosoideae, Chrysobalanaceae, Guttiferae, Irvingiaceae, Meliaceae, Moraceae, Myristicaceae, Sapotaceae, Sterculiaceae and Ulmaceae, and by fewer species in Combretaceae and Lecythidaceae. Smaller woody plants are abundantly represented by Anacardiaceae (*Sorindeia, Trichoscypha*), Annonaceae, Apocynaceae, Celastraceae (mostly lianes), Dichapetalaceae, Ebenaceae (*Diospyros*), Euphorbiaceae, Flacourtiaceae, Guttiferae, Icacinaceae, Ochnaceae (*Ouratea*), Olacaceae, Rubiaceae, Tiliaceae, Sapindaceae and Violaceae (*Rinorea*).

Endemic genera. About one-quarter of the genera are endemic, but since most of them are small the great majority of species belong to non-endemic genera. Leguminosae: Caesalpinioideae is particularly rich in endemic and near-endemic genera which include, among others, *Amphimas, Anthonotha, Aphanocalyx, Chidlowia, Didelotia, Distemonanthus, Eurypetalum, Hylodendron, Hymenostegia, Gilbertiodendron, Gossweilerodendron, Librevillea, Loesenera, Monopetalanthus, Oxystigma, Pachyelasma, Sindoropsis, Stemonocoleus, Talbotiella, Tessmannia* and *Tetraberlinia.* Endemic genera in other families include *Afrobrunnichia, Aneulophus, Antrocaryon, Aubrevillea, Aucoumea, Anopyxis, Baillonella, Brenania, Buchholzia, Calpocalyx, Coelocaryon, Coula, Crotonogyne, Cylicodiscus, Decorsella (Gymnorinorea) Desbordesia, Discoglypremna, Duboscia, Endodesmia, Erismadelphus, Fegimanra, Grossera, Heckeldora, Hypodaphnis, Monocyclanthus, Ophiobotrys, Poga, Turraeanthus* and *Tieghemella.*

Linking elements. About 3 per cent of Guineo–Congolian species, including *Diospyros ferrea, D. hoyleana, D. pseudomespilus, Ekebergia capensis (senegalensis), Magnistipula butayei, Parinari excelsa* and *Polyscias fulva,* are chorological and ecological transgressors.

About 2 per cent, including *Albizia adianthifolia, Chlorophora excelsa, Croton sylvaticus, Erythrophleum suaveolens (guineense), Parkia filicoidea* and *Sapium ellipticum,* are widespread linking species which are too uniform in their ecology to be regarded as transgressors.

About 5 per cent are near endemics which extend beyond the adjacent transition zones to other phytochoria, where they occur as marginal intruders or form small distant satellite populations. Among them *Afzelia bipindensis, Aningeria altissima, Diospyros gabunensis, Garcinia punctata, Sterculia tragacantha, Syzygium owariense, Trichilia prieuriana* and *Xylopia aethiopica* occur in the Zambezian Region. The affinity between the Indian Ocean Coastal Belt and the Guineo–Congolian Region is much less than has been previously supposed (White, 1979). Nevertheless, several species are common to both, and some, such as *Balanites wilsoniana, Chrysophyllum perpulchrum, Funtumia africana, Greenwayodendron suaveolens, Paramacrolobium coeruleum, Pterocarpus mildbraedii, Ricinodendron heudelotii, Schefflerodendron usambarense* and *Tetrapleura tetraptera,* show wide disjunctions.

The Guineo–Congolian flora is remarkably pure. Species belonging to linking elements more characteristic of other phytochoria are localized. Afromontane species are found in the transition zone which separates islands of Afromontane vegetation from lowland rain forest. A few Afromontane species, including *Blaeria mannii, Hypericum roeperanum, Ilex mitis, Nuxia congesta, Peddiea fischeri, Piper capense* and *Pittosporum viridiflorum,* occur in upland areas west of the 'Dahomey interval' but do not form extensive Afromontane communities.

Sudanian and Zambezian species occur towards the periphery of the Guineo–Congolian Region as marginal intruders on edaphically specialized sites (see for instance page 84). Apart from ruderal species, very few penetrate far into the interior.

Mapping units

1a. Wetter Guineo–Congolian rain forest.
2 (p.p.). Drier Guineo–Congolian rain forest.
3. Mosaic of 1a and 2.
4. Transitional rain forest.

8 (p.p.). Swamp forest.
9. Mosaic of 8 and 1a.
11a (p.p.). Mosaic of Guineo–Congolian rain forest and secondary grassland.

Vegetation

The greater part of the Guineo–Congolian Region was formerly covered with rain forest on well-drained sites and swamp forest on hydromorphic soils. Today, little undisturbed rain forest remains and secondary grassland and various stages of forest regrowth are extensive. There are also small patches of edaphic grassland on certain hydromorphic and other soils not suited to the growth of trees. Stunted forest and various types of bushland and thicket occur in upland areas, above about 1000 m, especially in rocky places. Numerous Afromontane species are found in upland areas but it is only on the highest peaks such as Mt Cameroun that they form distinct Afromontane communities which have to be excluded from the Guineo–Congolian Region.

Guineo–Congolian rain forest
(mapping units 1a, 2, 3, 9, 11a, 12, 13 and 14)

There is no comprehensive review. Much useful information has been obtained from published and unpublished work on the forests of Ghana by Hall & Swaine (1974, 1976, 1981). Of the seven types of forest recognized by these authors, two (southern marginal and south-east outliers) lie wholly, and one (semi-deciduous) lies partly, in the Guinea–Congolia/Sudania transition zone.

By African standards individual stands of Guineo–Congolian rain forest are floristically diverse in that up to 200 species of vascular plants have been recorded from 0.06 ha plots. Rain forest on other continents, however, especially in Malesia, is often much richer.

Guineo–Congolian rain forest is usually at least 30 m tall and is often much taller. In mixed moist semi-evergreen rain forest in Ghana the tallest trees are commonly 55–60 m high. Most rain-forest species are woody. Thirty-seven per cent are non-climbing phanerophytes, mostly trees. Smaller woody plants, even those less than 2 m tall, are often pygmy trees (treelets). Most tree species have slender trunks covered with thin smooth bark; they often branch only near the top and often have buttresses at the base. Their crowns are frequently narrow, except those of emergent species which are commonly 30–40 m in diameter, as in *Entandrophragma utile* and *Piptadeniastrum africanum*. Some tree species are cauliflorous, bearing their flowers on the trunks or branches. The leaves (or leaflets of compound leaves) of most tree species are dark green, lanceolate or elliptic in shape, with an entire margin; they are frequently acuminate at the apex (drip tip). In size most are mesophylls (area 20–180 cm^2).

In Ghana, 31 per cent of rain-forest species are climbers, which commonly constitute up to 40 per cent of the flora of small sample plots. Most climbers are woody. Among them, giant lianes in genera such as *Agelaea*, *Combretum*, *Salacia* and *Strychnos* reach the canopy and have stems up to 30 cm in diameter.

Epiphytic herbs are usually present, but are abundant only in the wetter, especially upland, variants. In Ghana 10 per cent of the forest flora belongs to this category, which is mainly represented by orchids and ferns. Among woody epiphytes, species of *Ficus* are frequent. Some, e.g. *F. sagittifolia*, remain epiphytic but others send roots to the ground and eventually strangle the host. Semi-parasitic Loranthaceae seem to be more abundant on isolated trees left by farming than in the forest itself.

In Ghana terrestrial herbs comprise 22 per cent of the rain-forest flora though most are confined to paths and farmland. In undisturbed forest they are normally poorly represented and contribute little to the appearance of the vegetation. The most abundant erect herbs in Ghana are ferns such as *Adiantum vogelii* and species of *Asplenium*, *Bolbitis* and *Pteris*, and various Marantaceae such as *Ataenidia*, *Marantochloa* and *Sarcophrynium*. Other members of this synusia include broad-leaved grasses such as *Leptaspis*, broad-leaved sedges such as *Hypolytrum* and *Mapania*, various Acanthaceae, and creeping species of *Costus*, *Geophila* and *Hymenocoleus*. There is a single holoparasite, *Thonningia sanguinea*, and a few rare holosaprophytes including species of *Auxopus*, *Gymnosiphon* and *Burmannia*.

Although some Guineo–Congolian species are deciduous, the forests themselves are evergreen or semi-evergreen. Some types have frequently been referred to as 'semi-deciduous' or even 'deciduous', but in a pan-African classification this is untenable. In Ghana 9 per cent of non-climbing phanerophytes are said to be 'deciduous'. The great majority are emergent or canopy trees. The understorey is virtually completely evergreen although one rare species, *Schumanniophyton problematicum*, is deciduous. A few canopy species, such as *Terminalia superba*, have a well-marked leafless period, but most 'deciduous' species either produce their new leaves as the old ones are shed (*Lophira alata*) or some branches retain their leaves when others are bare.

Whether well-defined strata can be usefully recognized in rain forest remains a matter of controversy. In the following pages details of stratification follow the accounts of the original authors.

The classification of Guineo–Congolian rain forest is difficult. This is partly because variation in floristic composition, physiognomy, and phenology is largely gradual and continuous (Aubréville, 1951; Keay, 1959a; Hall & Swaine, 1974, 1976), and partly because the distribution of many species is very imperfectly correlated with obvious environmental factors (Hall & Swaine, 1981; White, 1978b).

The majority of Guineo–Congolian rain-forest species are widespread. A sizeable minority, however, are confined to the wetter parts of a relatively narrow coastal band. The forests of the relatively dry peripheral parts of the Guineo-Congolian Region are more deciduous than the others, and also lack many characteristic species but include only a relatively small number of endemic species. These considerations permit the recognition of three main types of forest. By far the most extensive, which separates the coastal forests from those of the periphery, shows mixed dominance, but includes within it small islands of single-dominant forest, which collectively constitute a fourth variant. The salient feature of these four types are summarized in Table 5.

Although the boundaries between these types are somewhat arbitrary, and there is appreciable floristic overlap, each has its own distinct endemic flora. As one proceeds from the wettest to the driest types there is almost complete floristic replacement. In West Africa, for instance, only about 20 per cent of the larger woody species occurring in the wettest types also occur in the driest.

Hygrophilous coastal evergreen Guineo–Congolian rain forest

Refs.: Aubréville (1957–58); Guillaumet (1967); Guillaumet & Adjanohoun (1971, p. 168–76); Hall & Swaine (1974); Letouzey (1957; 1960; 1968a, p. 124–53); Taylor (1952, p. 3–4; 1960, p. 39–40); Voorhoeve (1965).

Phots.: Letouzey (1968a: 1); Voorhoeve (1965: 7).
Syn.: forêt biafréenne (Letouzey, 1968a); forêt littorale (Letouzey, 1968); forêt dense humide sempervirente à légumineuses (Aubréville, 1957–58, p.p.); rain forest (Taylor); wet evergreen forest (Hall & Swaine, 1974, p. 16).

This type occurs in the form of three blocks of varying width along the Atlantic coast of Africa from Sierra Leone to western Gabon. It has a very rich and distinctive endemic flora. The interval between the eastern and the two western blocks, which is an extension of the 'Dahomey gap', is approximately 600 km wide.

Sacoglottis gabonensis (Distr. Pl. Afr., 3, map 80, 1971) is one of the few species which is both endemic to this type and also occurs throughout, although, as is often the case with 'constant' species, it also slightly transgresses the boundaries of the type. Several other species which occur in both the eastern and western blocks, e.g. *Gluema ivorensis*, *Tarrietia* (*Heritiera*) *utilis* and *Crudia gabonensis*, show a wider interval than *Sacoglottis*, and some are completely absent from Nigeria or occur there only in the extreme south-east. Other species are confined to either the eastern or the western centres of endemism, but are replaced by a closely related species in the other, e.g. *Didelotia idae* (east) and

TABLE 5. Salient features of the four main variants of Guineo–Congolian rain forest

1. *Hygrophilous coastal evergreen rain forest*	Mean annual rainfall is often more than 3000 mm, but elsewhere is between 2000 and 3000 mm; the atmospheric humidity then is very high throughout the year. Most individuals of most tree species are evergreen and shed their leaves intermittently. Where rainfall is very high, but there is also a pronounced, though short, dry season, as in the coastal parts of Liberia, many species shed all their leaves simultaneously, followed immediately by the new flush.
2. *Mixed moist semi-evergreen rain forest*	Mean annual rainfall is mostly between 1600 and 2000 mm and is well distributed (Zaire basin) or the dry season is mitigated by moist air from the sea (West Africa). These climatic conditions prevail over the greater part of the Guineo–Congolian Region and moist semi-evergreen rain forest covers most of this area. It is nearly always mixed forest. Some species are evergreen but many are briefly 'deciduous'.
3. *Single dominant moist evergreen and semi-evergreen rain forest*	This type occurs as small islands in a matrix of mixed moist semi-evergreen rain forest and is found scattered throughout the latter. Its total area is small and it is always dominated by one (sometimes two) of a small number of caesalpiniaceous trees. The dominants shed and renew their leaves almost continuously throughout the year, except sometimes towards the limits of their range.
4. *Drier peripheral semi-evergreen rain forest*	Rainfall is between 1200 and 1600 mm per year but dry-season relative humidity is very high. Most individuals of the commoner larger tree species are deciduous and lose their leaves during the well-defined dry season, but any individual is deciduous for only a short period, usually a few weeks. Relatively few individuals are leafless at the same time and different species and even different individuals of the same species may lose their leaves at different times. Many individuals are never completely bare, since some branches acquire their new leaves before other branches have shed their old crop. This type is usually referred to as 'semi-deciduous' or even 'deciduous' forest but these terms detract from its essentially evergreen nature.

D. unifoliolata (west), and *Tieghemella* (*Dumoria*) *heckelii* (east) and *T. africana* (west).

One of the most abundant of the widespread taller trees is *Lophira alata* (Distr. Pl. Afr., 2, map 44, 1970), although it is not confined to this type. When it is present in quantity it usually indicates former cultivation. It also occurs as a secondary species in moist semi-evergreen rain forest in West Africa and in the western part of the Zaire basin.

At its most typical, hygrophilous coastal evergreen rain forest is very rich in Caesalpinioideae, many of which are gregarious. For Cameroun, Letouzey lists the following gregarious species: *Brachystegia cynometroides*, *B. laurentii*, *B. mildbraedii*, *Cryptosepalum staudtii*, *Cynometra hankei*, *Didelotia brevipaniculata*, *Gilbertiodendron brachystegioides*, *Hymenostegia afzelii*, *Julbernardia pellegriniana*, *J. seretii*, *Microberlinia bisulcata*, *Monopetalanthus hedinii*, *Schotia africana*, *Tetraberlinia bifoliolata* and *T. polyphylla*. These species often form almost pure stands in which there is abundant regeneration and all size-classes are well represented. Among them, *Brachystegia laurentii* and *Julbernardia seretii* are more characteristic of the single-dominant moist evergreen rain forests of the Zaire basin (see below), but most of the others are endemic to hygrophilous coastal evergreen rain forest.

Relatively gregarious Caesalpinioideae are also prominent in hygrophilous coastal evergreen and semi-evergreen rain forest west of the Dahomey gap but are represented there by relatively few species, and of them probably only *Tetraberlinia* is truly gregarious. *Brachystegia leonensis*, *Cynometra ananta*, *C. leonensis*, *Gilbertiodendron preussii*, *Monopetalanthus compactus* and *Tetraberlinia tubmaniana* occur in the westernmost block centred on Liberia, but only *Cynometra ananta* and *Gilbertiodendron preussii* also occur in the small island of hygrophilous coastal evergreen rain forest in south-east Ivory Coast and adjacent Ghana. In Liberia *Tetraberlinia tubmaniana* occurs in almost pure stands and regenerates freely in its own shade. Other important members of this western community in addition to those mentioned earlier include: *Berlinia occidentalis*, *Gilbertiodendron bilineatum*, *G. splendidum* and *Kaoue stapfiana* among Caesalpinioideae, *Coula edulis*, *Gluema ivorensis*, *Oldfieldia africana* and *Soyauxia grandifolia* among other trees, the treelet *Diospyros chevalieri*, and several sedges of the genus *Mapania*.

Forests similar to those described above from Cameroun occur in Gabon (Letouzey, 1968a, summarizing de Saint Aubin, 1963), though with some floristic differences. One of the most abundant species, *Aucoumea klaineana*, originally was probably a natural component of forest in swampy depressions, but now occurs abundantly in old secondary forest on well-drained sites.

In Ghana (Hall & Swaine, 1974, 1976), and probably elsewhere, hygrophilous coastal evergreen rain forest is appreciably shorter than semi-evergreen rain forest. Its mean maximum height is 30 m and few trees are more than 40 m tall. A higher proportion of species have leaves or leaflets with 'drip tips'.

Mixed moist semi-evergreen Guineo–Congolian rain forest

Refs.: Évrard (1968, p. 86–96); Jones (1955, 1956); Lebrun & Gilbert (1954, p. 19–20); Letouzey (1968a, p. 154–80); Louis (1947a, p. 904–6).
Phots.: Jones (1955: 1–2); Lebrun & Gilbert (1954: 3).
Profile: Jones (1955: 2); Louis (1947a: 4).
Syn.: forêt congolaise (Letouzey, 1968a); forêts semi-caducifoliées subéquatoriales et guinéennes: alliance *Oxystigmo–Scorodophloeion* (Lebrun & Gilbert, 1954); forêt dense humide sempervirente à légumineuses (Aubréville, 1957–58, p.p.).

Most Guineo–Congolian rain forest belongs to this type. It occurs on well-drained soils throughout the whole of the Guineo–Congolian Region except for the wettest and driest extremities. In West Africa it is relatively poorly developed because of the rapid increase in dry-season severity away from the coast. By contrast, it covers an enormous area in the heart of the eastern block of rain forest, comprising north-east Gabon, south-east Cameroun, south-west Central African Republic, northern Congo Republic and most of the Zaire basin and its periphery. Here, the mean annual rainfall only very locally exceeds 2 000 mm or falls below 1 600 mm. The prevalent vegetation is moist semi-evergreen rain forest of mixed composition, though small islands of single-dominant forest, which are described in the next section, are also found scattered throughout.

Mixed moist semi-evergreen rain forest is relatively rich floristically. In Okomu Forest Reserve, near Benin in Nigeria, Jones recorded 170 species more than 30 cm in girth in a plot of 18.4 ha, of which 52 belonged to the upper, emergent, storey. Most species in this type of forest are widely distributed. The following large trees, among many others, occur west of the Dahomey gap, and also in southern Nigeria and throughout the greater part of the Zaire basin: *Entandrophragma angolense*, *E. candollei*, *E. cylindricum*, *E. utile*, *Guarea cedrata*, *G. thompsonii* and *Lovoa trichilioides* (all Meliaceae), *Maranthes* (*Parinari*) *glabra* (Chrysobalanaceae), *Nauclea diderrichii* (Rubiaceae), *Parkia bicolor* (Leguminosae: Mimosoideae), *Pericopsis* (*Afrormosia*) *elata* (Leguminosae: Papilionoideae), and *Petersianthus macrocarpus* (*Combretodendron africanum*, *C. macrocarpum*, Lecythidaceae).

No detailed descriptions of this type of forest have been published for Zaire, though Évrard lists the principal trees for the Tshuapa–Equateur region. Of the thirty commonest large species all but six, i.e. 80 per cent, extend as far west as Nigeria, and in many cases far beyond. Species which are particularly significant in the Zaire basin but are absent or only sporadic further west include *Oxystigma oxyphyllum* and *Scorodophloeus zenkeri*, which lend their names to the alliance

Oxystigmo–Scorodophloeion, to which phytosociologists in Zaire assign this type of forest (Lebrun & Gilbert, 1954).

Few of the species mentioned above are absolutely confined to mixed moist semi-evergreen lowland rain forest, but all achieve their maximum development there. Most of them also occur both in the drier peripheral semi-evergreen lowland rain forest, and in hygrophilous coastal evergreen lowland rain forest, but mostly only in the wetter types of the former and in the drier types of the latter. In both they tend to be rare and localized.

Some of the more abundant emergent species of mixed moist semi-evergreen lowland rain forest, e.g. *Canarium schweinfurthii*, *Piptadeniastrum africanum*, *Ricinodendron heudelotii*, *Sterculia oblonga* (*Eriobroma oblongum*) and *Terminalia superba*, are also found in dry peripheral semi-evergreen rain forest. In the former they usually occur in secondary forest. *Lophira alata*, another secondary forest species in mixed moist semi-evergreen rain forest, is also an abundant component of secondary forest in the hygrophilous coastal evergreen rain forest region, of which it is more characteristic.

Single-dominant moist evergreen and semi-evergreen Guineo–Congolian rain forest

Refs.: Évrard (1968, p. 81–6); Gérard (1960); Germain & Évrard (1956); Lebrun (1936, p. 88–95); Lebrun & Gilbert (1954, p. 13–14); Louis (1947a, p. 906–8); Pecrot & Léonard (1960, p. 67–8); Peeters (1964); Robyns (1948, p. xlix).

Phots.: Évrard (1968: 3–5); Gérard (1960: 5–8, 14, 15, 23); Germain & Évrard (1956: 1–7); Lebrun (1936: 39–40); Lebrun & Gilbert (1954: 1).

Profiles: Germain & Évrard (1956: 2); Louis (1947a: 6).

In addition to the single-dominant forests described above from the coastal parts of the Guineo–Congolian Region, similar small islands of single-dominant forest, often no more than a few hectares in extent, are also widely distributed further inland, where they are usually surrounded by mixed moist semi-evergreen lowland rain forest. They occur extensively in a broad aureole surrounding the Zaire basin, but are much more localized in the basin itself because of the extensive development of swamp forest there. They are usually dominated by one or more of the following five species of Leguminosae: Caesalpinioideae–*Brachystegia laurentii*, *Cynometra alexandri*, *Gilbertiodendron dewevrei*, *Julbernardia seretii* and *Michelsonia microphylla*. *Cynometra* forests also occur in Uganda (Chapter XII), *Gilbertiodendron* forests in south-east Cameroun, and *Julbernardia* forests in Gabon. Similar islands of single–dominant moist evergreen forest appear to be virtually absent from the moist semi-evergreen forests of West Africa. Of the five gregarious dominants, only *Julbernardia* and *Gilbertiodendron* extend as far as Nigeria, where the former is confined to hygrophilous coastal evergreen forest, and the latter occurs only on river banks and in swamp forest.

The upper stratum in single-dominant forest is uniform and dense, usually 35–45 m high, and is composed of a single or at most very few species. The dominant species are shade-bearers and, apparently, cannot tolerate high radiation intensities in their early stages of development. All size-classes are well represented and the dominants appear to be able to maintain themselves indefinitely.

Brachystegia laurentii and *Gilbertiodendron dewevrei* are normally completely evergreen: *Julbernardia seretii* and *Cynometra alexandri* are less completely so. The leaves of *Brachystegia* and *Gilbertiodendron* are shed and renewed almost continuously throughout the year. Their young shoots are reddish-purple in colour and confer a characteristic appearance. In Uganda, at least, individual *Cynometra* trees shed their leaves simultaneously, though they are leafless only for a few days, and most trees in a stand are never completely leafless at the same time. *Julbernardia* is also said to behave in a similar fashion. Except for the attack of old individuals by *Fomes annosus*, and occasional defoliation by caterpillars, the dominants are not known to have any serious pests or diseases.

The lower tree strata are less dense and are formed principally of recruitment of the dominants. The herb layer is poorly developed and much of the soil is covered by a dense leaf litter which decomposes slowly. Heliophilous trees are rare, and lianes and giant monocotyledonous herbs are poorly represented.

The seeds of the dominants are explosively discharged but they are large and heavy and dispersal is slow. Évrard (1968) estimates the rate of migration of *Brachystegia* and *Gilbertiodendron* at 100 metres every two or three centuries. The dominants, besides being able to regenerate in their own shade, can also invade mixed moist semi-evergreen rain forest, which is usually deficient in regeneration of its own upper-canopy species. For this reason, Évrard regards single-dominant forest as the climax type, but suggests that because of its slow penetration of mixed forest the latter might persist for long periods, possibly by means of mosaic regeneration, as postulated by Aubréville and confirmed by Jones for Nigeria. In view of the slow migration rate of single-dominant forests it would be unwise to assume that they owe their present-day restricted distribution entirely to man's destructive activity. It is possible that recent climatic change has been too rapid to allow this type of forest to achieve its maximum potential range during those climatic phases of the Pleistocene most favourable for its expansion.

Brachystegia laurentii is widely distributed in the Zaire basin but it has been closely studied only in the neighbourhood of Yangambi, where it forms many, more or less pure, stands, mostly only a few hectares in extent. The latter are found on well-drained soil of the interfluves of the plateau at about 500 m.

Cynometra alexandri is widely distributed in the Zaire basin (Distr. Pl. Afr., 2, map 46, 1970) especially above 700–800 m towards the eastern periphery. Below

1200 m in the Beni–Irumu region, it constitutes 50–70 per cent of the forests on dry land. Small populations are also found throughout the Zaire basin and in the drier peripheral forests of Kasai. On the 'dorsale du Kivu' it is co-dominant with *Julbernardia seretii* and *Staudtia stipitata* between 1000 and 1350 m, and is abundant in transitional rain forest above 1350 m with *Pentadesma lebrunii*, *Lebrunia bushaie* and *Staudtia stipitata*.

Gilbertiodendron dewevrei occurs throughout the Zaire basin and its peripheral regions, and extends westwards to Gabon, Cameroun and southern Nigeria (Distr. Pl. Afr., 2, map 47, 1970). Towards the northern and southern limits of its range it is confined to certain river valleys where it occurs in fringing forest or swamp forest on sandy soils. It is most abundant in a wide aureole on the plateau surrounding the Zaire basin, but only forms extensive forests on well-drained but water-retaining red-clay soils in the regions of Ubangi, Uele, and east of Kisangani. In the centre of the Zaire basin *G. dewevrei* is much more restricted. Together with *G. ogoouense* it occurs on more or less leached colluvial sands bordering swamp forest in valley bottoms. The distribution of *Julbernardia seretii* is similar to that of *Gilbertiodendron dewevrei* but not quite as extensive. The relative abundance of the two species varies greatly from place to place. In the Uele region *Gilbertiodendron* is much commoner than *Julbernardia*, but in Lomami the latter is dominant over extensive areas and *Gilbertiodendron* occurs only in small patches (P. Bamps, pers. comm.).

Michelsonia microphylla forms extensive almost pure stands 30–35 m tall, at the eastern rim of the Zaire basin, where, between 650 and 1200 m, the undulating landscape indicates the approach to the 'dorsale du Kivu'. It is usually associated with *Julbernardia seretii* and *Staudtia stipitata*. *Michelsonia* forest often interdigitates with *Gilbertiodendron dewevrei* forest. The latter occupies the valley bottoms and the former the summits of the hills.

Drier peripheral semi-evergreen Guineo–Congolian rain forest and similar forest in the transition zones

Refs.: Aubréville (1957–58); Clayton (1961, p. 596–7); Guillaumet & Adjanohoun (1971, p. 192–7); Hall & Swaine (1974, 1976); Hambler (1964); Jones (1963a, 1963b); Lebrun & Gilbert (1954, p. 20–1); Mullenders (1954, p. 389–449); Letouzey (1968a, p. 181–237); White (MS, 1963).
Phots.: Guillaumet & Adjanohoun (1971: 14–15); Letouzey (1968a: 5).
Syn.: forêt dense humide semi-décidue de moyenne altitude (Letouzey, 1968a); forêts semi-caducifoliées subéquatoriales et périguinéennes (Lebrun & Gilbert, 1954); forêt semi-décidue (Guillaumet & Adjanohoun, 1971); forêt semi-décidue à malvales et ulmacées (Aubréville, 1957–58).

This type occurs in the form of two bands running transversely accross Africa, to the north and south of the moister forests described above. In addition to forming a fringe to the Guineo–Congolian region, it was also formerly widespread in the adjoining transition zones and had a patchy distribution in the Lake Victoria basin. Only generalized descriptions have been published. Some tree species are more or less confined to this type and others are most abundant there, but some also extend their distributions beyond the rain-forest region into drier regions, especially along watercourses.

Species which are frequent in peripheral semi-evergreen lowland rain forest in West Africa, but are absent, or virtually absent, from wetter types include: *Afzelia africana*, *Aningeria altissima*, *A. robusta*, *Aubrevillea kerstingii*, *Chrysophyllum perpulchrum*, *Cola gigantea*, *Hildegardia barteri* (especially on shallow soils and rocky outcrops), *Khaya grandifoliola*, *Mansonia altissima*, *Morus mesozygia*, *Nesogordonia papaverifera*, and *Pterygota macrocarpa*. Some of these, e.g. *Aningeria altissima*, *Chrysophyllum perpulchrum*, *Cola gigantea*, *Khaya grandifoliola* and *Morus mesozygia*, extend eastwards as far as Uganda or beyond. Others such as *Hildegardia* and *Mansonia* are confined to West Africa.

Some other species which are important components of peripheral semi-evergreen lowland rain forest also occur in mixed moist semi-evergreen lowland rain forest, e.g. in the forests of Benin in Nigeria, for instance *Celtis mildbraedii*, *C. zenkeri*, *Holoptelea grandis*, *Sterculia oblonga* and *S. rhinopetala*.

Several other species which figure prominently in peripheral semi-evergreen lowland rain forest are very extensively distributed in wetter types of rain forest but are usually found there only in secondary forest. To this group belong: *Trilepisium madagascariense* (*Bosqueia angolensis*), *Canarium schweinfurthii*, *Chlorophora excelsa*, *Piptadeniastrum africanum* and *Ricinodendron heudelotii*. All of them are very widely distributed in the Zaire basin, but (except for *Piptadeniastrum*) are absent, at least locally, from the hygrophilous evergreen lowland rain forest of the Atlantic seaboard.

Triplochiton scleroxylon and *Terminalia superba*, two rapidly growing, light-demanding, valuable timber trees, are of special interest. They often occur gregariously, and can regenerate abundantly on abandoned farmland, in contrast to the behaviour of species like *Khaya grandifoliola*, which, although it is a light-demander, can only regenerate in gaps in the forest. Both these species have greatly extended their range in recent times following the destruction of forest for agricultural purposes. In Cameroun, *Terminalia superba* has penetrated deeply into mixed moist semi-evergreen lowland rain forest and hygrophilous coastal evergreen lowland rain forest and has even reached the coast at Kribi (Letouzey, 1960). The status of these two species, even in dry peripheral semi-evergreen rain forest, remains equivocal. According to Letouzey (1968a, p. 183), gaps in the canopy of mature forest caused by the deaths of old trees do not provide adequate opportunities for their regeneration. In the forests north-west of Yokadouma seedlings of *Triplochiton* are scarcely ever found in such situations.

Where peripheral Guineo–Congolian rain forest is in contact with savanna it is liable to damage by ground fires which burn the litter and kill shrubs and young trees. In Ghana (Hall & Swaine, 1976) such fires, which occur at intervals of up to fifteen years, give rise to a distinct 'fire-zone' variant. Structurally these forests are exceptional in the paucity of trees in the lower size-classes, but it is conceivable that if fires are sufficiently infrequent the forest may nevertheless be able to maintain itself. Floristically they are distinguished by the presence of facultative savanna tree species such as *Anogeissus leiocarpus* and *Afzelia africana*, by abundant oil palms (*Elaeis guineensis*), and by the absence of some thin-barked trees such as *Hymenostegia afzelii*, which are abundant in forest of similar dryness in more southerly parts of the forest area. The relative openness of the forest canopy allows the development of a profuse growth of marantaceous forbs (though savanna grasses are absent), and a high proportion of secondary forest species.

Less is known of the peripheral semi-evergreen lowland rain forests south of the equator. Floristically they seem to be less well-characterized than the northern variants. Some of their characteristic species also occur in the latter, e.g. *Celtis zenkeri*, *C. brownii* (*C. philippensis*), *Trilepisium madagascariense*, *Canarium schweinfurthii*, *Chlorophora excelsa* and *Klainedoxa gabonensis*, but all of these also occur in secondary forest more or less throughout the Zaire basin. A few species, e.g. *Pteleopsis diptera*, are endemic, while others are transgressors from other forest types. The main distribution of *Newtonia buchananii*, for instance, which is absent from peripheral semi-evergreen rain forest in West Africa, is on the lower slopes of the East African Mountains, but it also occurs in fringing forest in the Zambezian Region and extends into the peripheral rain forests south of the equator. Other transgressors include *Prunus africana* (*Pygeum africanum*) and *Celtis africana*.

Secondary Guineo–Congolian rain forest

Refs.: Aubréville (1974*a*); Charter & Keay (1960); Keay (1957); Lebrun & Gilbert (1954, p. 45–62); Léonard (1953, p. 58–9); Richards (1952, p. 377–92); Ross (1954); White (MS, 1963).
Phots.: Charter & Keay (1960: 7–8); Lebrun & Gilbert (1954: 14–16); Richards (1952: 14*b*).
Profiles: Keay (1957: 1); Léonard (1953: 1).

Outside the forest reserves much of the remaining Guineo–Congolian rain forest on well-drained soils occurs on land which has been formerly cultivated; it is therefore secondary. Some of the forest inside forest reserves is also secondary, but very old secondary forest is often difficult or impossible to distinguish from primary forest. This section deals only with the earlier stages of succession. Their dominants are all light-demanding and are intolerant of shade; usually they are unable to regenerate beneath their own canopy. Many occur in primary forest but are rare there and are confined to small opening caused by the death of canopy trees, or to larger clearances resulting from natural catastrophes such as hurricanes or landslides. Once such openings occur they may be maintained for a long time by the browsing of large mammals, especially elephants. Other pioneer species have probably been recruited from the more open forests of river valleys. Many pioneer species are gregarious. They all grow rapidly and are short-lived; *Musanga cecropioides*, for instance, reaches its maximum height of 24 m in 15–20 years but dies soon after, sometimes earlier. They have efficient means of seed dispersal. Either their fruits, or rarely (*Pycnanthus angolensis*) their seeds, are dispersed by animals, or their winged or plumose seeds are dispersed by wind. In general, pioneer species have wide distributions. Many occur, not only throughout the Guineo–Congolian Region, but extend far beyond its limits to Madagascar in the case of *Harungana madagascariensis*, and to eastern Asia in the case of *Trema orientalis* (*guineensis*), as their specific names suggest. Other species, such as *Musanga cecropioides*, however, are strictly confined to the Guineo–Congolian Region.

Epiphytes are rare in secondary forest. According to Lebrun and Gilbert large lianes in Zaire, although heliophilous, grow too slowly to enable them to gain a substantial foothold.

Much of our knowledge of forest succession is based on inference. Historical evidence, both circumstantial and the result of planned experiments, is available from only a few places. Ross (1954) has described the changes which took place on formerly cultivated land in moist semi-evergreen rain forest in Shasha Forest, Southern Nigeria, which had been left uncultivated for 5½, 14½, and 17½ years respectively.

Forest succession in a plot of secondary fire-maintained wooded grassland in Olokemeji Forest Reserve in the peripheral semi-evergreen rain forest region in Nigeria has been described by Charter & Keay (1960) and is summarized on page 83.

In the wetter Guineo–Congolian forests the following stages in the forest succession can usually be recognized:

Pioneer secondary forest. This stage varies in height from 4–6 to 8–12 m. The dominant bushes and small trees are mixed with many coarse herbs, softly woody shrubs and small climbers. The dominants include *Anthocleista* spp., *Caloncoba welwitschii*, *Chaetocarpus africanus*, *Harungana madagascariensis*, *Rauvolfia vomitoria*, *Tetrorchidium didymostemon*, *Trema orientalis* and *Vernonia conferta*.

Young secondary forest. Characteristically this stage is dominated by the parasol tree, *Musanga cecropioides*, which is the most abundant and characteristic secondary forest tree in tropical Africa, though not the most widespread. It can become dominant after 3 years and normally reaches its optimum after 8–10 years. Other less common dominants of this stage are similar to *Musanga* in their ecology. *Musanga* overtops and shades out the dominants of the previous stage but provides conditions suitable for the seedlings and saplings of the next. Characteristic species of this stage include: *Buchnerodendron speciosum, Caloncoba glauca, Croton mubango, Lindackeria dentata, Macaranga monandra, M. spinosa, Maesopsis eminii* and *Myrianthus arboreus*. The most abundant species (*Macaranga, Musanga* and *Myrianthus*) have stilt roots.

Old secondary forest. This stage is dominated by semi-heliophilous species of moderately rapid growth which reach a height of 35 m. Characteristic species occurring in the canopy in Zaire are: *Alstonia boonei, Antrocaryon micraster, Trilepisium madagascariense, Canarium schweinfurthii, Ceiba pentandra, Chlorophora excelsa, Discoglypremna caloneura, Zanthoxylum gillettii (Fagara macrophylla), Funtumia africana, Holoptelea grandis, Khaya anthotheca, Morus mesozygia, Pentaclethra macrophylla, Petersianthus macrocarpus, Pterygota macrocarpa, Pycnanthus angolensis, Ricinodendron heudelotii, Terminalia superba, Triplochiton scleroxylon* and *Xylopia aethiopica*. Some of these, e.g. *Canarium, Chlorophora, Morus, Ricinodendron, Terminalia,* and *Triplochiton,* are also characteristic species of dry semi-evergreen Guineo–Congolian forest and it is not always easy to determine the status of the forests in which they occur. *Chlorophora, Terminalia,* and *Triplochiton* can regenerate abundantly on abandoned farmland without the necessity of an intervening *Musanga* phase.

The succession outlined above is characteristic on relatively fertile soil in the moister parts of the Guineo–Congolian Region. On inhospitable soils towards the drier limits of the Region the reversion to forest proceeds much more slowly and the species are different. As an example the succession at Olokemeji Forest Reserve in Nigeria from fire-maintained wooded grassland, on a seasonally waterlogged soil, towards forest with abundant *Manilkara obovata* and *Diospyros mespiliformis,* is described on page 83.

Guineo–Congolian short forest and scrub forest
(mapping units 1a, 2, 3 and 11a)

Refs.: Jaeger & Adam (1968, 1971, 1975); Richards (1957); Schnell (1952a, 1961); White (1976; MS, 1963).
Photos.: Schnell (1952a: p. 37, 40, 42, 45, 46, 49).

Forest similar in composition to Guineo–Congolian rain forest, but floristically poorer and of shorter stature and simpler structure, occurs on rocky hills and other upland, though not Afromontane, areas inside the Guineo–Congolian Region. Stunted rupicolous forest is usually associated with various types of bushland and thicket which occupy the shallower soils.

On granite inselbergs

The granite inselbergs within the rain-forest zone in Nigeria and Cameroun support some vegetation types and species which are absent from the surrounding lowlands. The highest (945 m) and best known are the Idanre Hills in Ondo Province in Nigeria.

The vegetation ranges from almost bare rock to dry semi-evergreen lowland rain forest in the larger and more sheltered gullies. The latter is very similar to the surrounding lowland forest. Of the various types of forest of low stature the most distinct is what Richards refers to as 'semi-montane forest'. This occurs above 800 m, where atmospheric humidity due to cloud is significantly higher than elsewhere. At the forest limit the maximum height of the trees, among which *Anthonotha obanensis* is abundant, is no more than 15 m. Bryophytes and ferns growing on boulders and as epiphytes on trees are much more abundant and luxuriant than on the lower slopes and there are more lianes. The fern *Asplenium dregeanum* thickly covers nearly every boulder and shaded tree trunk. Other abundant epiphytes are species of *Plagiochila,* various Lejeuneaceae and other foliose hepatics, and the mosses *Lepidopilum callochlorum* and species of *Ectropothecium* and *Pilotrichella*.

In the Idanre Hills short forest in ravines is usually fringed with narrow bands of less luxuriant scrub forest, bushland, and thicket. The tallest trees, which reach a height of 15 m, are *Hildegardia barteri, Alstonia boonei, Albizia ferruginea, Diospyros monbuttensis,* and *Holarrhena floribunda*. Some pioneer forest species such as *Clausena anisata, Harungana madagascariensis* and *Newbouldia laevis* appear to be natural members of the community. Climbers, which are abundant, are represented by *Acacia kamerunensis, Acridocarpus smeathmannii, Bowringia mildbraedii, Cissus quadrangularis, Combretum paniculatum, C. racemosum, C. mucronatum, Entada mannii, E. pursaetha* and *Uvaria chamae*. The 3–5 m tall bushland is dominated by the deciduous lithophyte *Hymenodictyon floribundum,* which becomes established in mats formed by the arborescent sedge *Afrotrilepis pilosa*.

Upland Parinari excelsa *forest in West Africa*

The uplands of the Loma–Man 'dorsale', which include the plateau of Foula Djalon and the massifs of Nimba, Ziama, and Fon, are not high enough to support distinct Afromontane communities, though a number of Afromontane species occur, mixed with 'lowland' elements. Above 1000 m the forests are dominated by *Parinari excelsa,* which is often the only tree present in the main canopy. *Parinari excelsa* is one of the most widespread trees in tropical Africa, both in the lowlands and on the mountains (White 1976a, 1976b). It is often an upper-canopy or emergent species 30 or more m high, but in the Upper Guinea highlands it progressively diminishes in stature with increasing altitude and ultimately dominates dwarf forest only 10 m tall. These highlands are surrounded by lowland rain forest or secondary grassland derived from it. *P. excelsa* is rare in lowland rain forest in this region, but at about 800–900 m it becomes locally abundant in forest which both floristically and structurally is still lowland rain forest.

Schnell has described in considerable detail the *Parinari* forests of Nimba (1952) and Fon (1961). For Nimba he recognizes the following three types:

1. Tall forest (20–30 m) rich in lowland species between 1000 and 1300 m.
2. Tall and semi-tall forest in the upper ravines between 1300 and 1600 m, where *P. excelsa* is often the only large tree, and lowland species are rare or absent.
3. Short forest, 8–12 m tall, on shallow soils of the upper slopes and crests.

Above 800 m, and especially above 1000 m, mists are frequent and ferns and epiphytes become plentiful. Mean annual rainfall in the surrounding lowlands is 1750 mm or more, and is well distributed throughout the year. Only three or four months have less than 100 mm and only one has less than 20 mm. Nearer the crest rainfall is probably higher.

These upland *Parinari* forests in West Africa contain very few endemic species, at least among larger woody plants. Their tree flora is made up almost entirely of species which also occur in the lowlands. Many are species at their upper limit of tolerance and are no more (or not much more) abundant there than at lower altitudes. Such are: *Alstonia congensis, Antiaris toxicaria (africana), Canarium schweinfurthii, Chrysophyllum perpulchrum, Entandrophragma utile, Guarea cedrata, Khaya grandifoliola, Morus mesozygia, Newtonia aubrevillei, Parkia bicolor, Piptadeniastrum africanum, Stereospermum acuminatissimum, Sterculia tragacantha* and *Tetrapleura tetraptera*. Other species, like *Parinari* itself, though present in lowland forest, are much more abundant above 1000 m. They include *Carapa procera, Cryptosepalum tetraphyllum, Drypetes leonensis, Garcinia smeathmannii, (polyantha)* and *Ochna membranacea*. A few species in this community, at least in West Africa, are usually found in upland areas, though they are not strictly Afromontane. Among them are: *Dracaena arborea, Syzygium guineense* subsp. *occidentale, Lycopodium mildbraedii, Marattia* sp., *Peperomia fernandopoana (staudtii)*, and *Trichomanes mannii*.

Schnell (1952a) regards the short forest as an impoverished, edaphic variant of the taller *Parinari* forest. It occurs on very shallow soils between 1200 and 1600 m. The main canopy is at 8 m and consists of small trees of several species with slender boles, surmounted here and there by the rounded crowns of *Parinari*, which reach a height of 12 m. The understorey is not dense and lianes are rare. The principal associates include *Craterispermum laurinum* s.l., *Cryptosepalum tetraphyllum, Drypetes leonensis, Eugenia leonensis, Gaertnera* sp., *Hymenodictyon floribundum, Schefflera barteri* and *Syzygium guineense* subsp. *occidentale*.

The upland *Parinari* forests in West Africa have been greatly reduced by fire. Remnants survive in ravines and where rock outcrops provide natural fire-breaks. On the south-western slopes of the Nimba range in Guinea Republic and on the western slopes of Fon, patches of short forest survive, but are virtually absent from the opposite slopes. Fires burn less fiercely on the seaward-facing slopes because of protection from the desiccating harmattan.

The south-western part of the Nimba massif is situated in Liberia and progressively decreases in altitude down to 1040 m. Rainfall is higher (up to 3500 mm) than in Guinea Republic and the dry season is less severe. The *Parinari excelsa* forests, except near the frontier with Guinea Republic, have not been damaged by fire (Jaeger & Adam, 1975). They have, however, been largely destroyed by mining operations. They are similar to those described by Schnell, but there are some floristic differences. On shallow soils the *Parinari* forests are 12–15 m tall; on deeper soils up to 25 m. Principal associates include: *Syzygium guineense* subsp. *occidentale, Santiria trimera, Uapaca chevalieri* and *Amanoa bracteosa*, with *Cyathea manniana* in the understorey. Among the trees, *P. excelsa* is exceptional in the exuberance and richness of its vascular epiphytes. The latter extend from the base of the bole to the periphery of the crown. As many as fifteen species may be found on a single individual (Johansson, 1974). Vascular epiphytes are virtually absent from the understorey, but bryophytes festoon the stems of climbers which link one tree to another.

Guineo–Congolian swamp forest and riparian forest
(mapping units 1a, 2, 3, 8, 9, and 11a)

Refs.: Bouillenne *et al.* (1955); Chipp (1927, p. 62–4); Évrard (1968); Jeník (1970); Keay (1959a, p. 13); Lebrun & Gilbert (1954, p. 33–43); Léonard (1953, p. 62–5); Letouzey (1975); Richards (1939, p. 42–7; 1952, p. 288–90); White & Werger (1978).
Phots.: Bouillenne *et al.* (1955: 5, 6, 10); Chipp (1927: 29); Évrard (1968: 7–33); Lebrun & Gilbert (1954: 9–13); Léonard (1953: 6); Richards (1952: 10a, 10b).
Profiles: Bouillenne *et al.* (1955: 1–3); Keay (1959a: 4).

Certain types of swamp forest and riparian forest are very different floristically, but they are connected by a complex series of intermediates and are treated collectively here.

Swamp forest (including riparian forest) occurs throughout the Guineo–Congolian Region wherever conditions are suitable, but is most extensively developed in the Zaire basin and the Niger delta. Floristically impoverished variants extend beyond the Guineo–Congolian Region into other regions. At its most luxuriant, Guineo–Congolian swamp forest is similar in appearance to rain forest, and the tallest trees attain a height of 45 m. The main canopy, however, is irregular and rather open and superficially resembles broken or secondary rain forest caused by man's disturbance. Until recently swamp forest was usually more or less virgin as it was considered unsuitable for farming. Nowadays, however, it is cleared on a large scale for rice-farming. Dense tangles of shrubs and lianes fill the gaps, in which climbing palms (*Ancistrophyllum*,

Eremospatha and *Calamus*) with their evil hooked spines are particularly characteristic, as are clumps of the large aroid *Cyrtosperma senegalense*.

Guineo–Congolian swamp forest has a diversified endemic flora though it is somewhat poor in species. The most characteristic trees include *Berlinia auriculata*, *Carapa procera, *Coelocaryon botryoides*, *Diospyros longiflora*, *Entandrophragma palustre*, *Guibourtia demeusii*, *Irvingia smithii*, **Mitragyna ciliata*, **M. stipulosa*, **Nauclea pobeguinii*, *Oubanguia africana*, *Oxystigma mannii*, **Pandanus candelabrum*, *Parinari congensis*, *P. congolana*, **Phoenix reclinata*, species of *Raphia, *Scytopetalum pierreanum*, **Spondianthus preussii*, **Symphonia globulifera*, **Uapaca guineensis*, **U. heudelotii* and **Voacanga thouarsii*. Species marked with an asterisk are widespread. In the lowlands *Carapa* and *Symphonia* are largely confined to swamp forest, but they also are abundant on freely drained soils above 1000 m. Many swamp forest trees have pneumatophores and some have stilt roots.

Guineo–Congolian transition woodland
(mapping units 2, 11a, 12, 13, and 14)

Refs.: Adjanohoun (1964, p. 130–1); Charter & Keay (1960); Clayton (1958a); Keay (1951, p. 63–4); Letouzey (1968a); MacGregor (1937); White (MS, 1962–63).
Phots.: Charter & Keay (1960: 7, 8); Letouzey (1968a: 11–15, 35).
Profile: Charter & Keay (1960: 6).

The secondary grasslands and wooded grassland which replace both Guineo–Congolian rain forest and related but floristically poorer forests in the transition zones to the north and south are described on pages 84 and 173. There is abundant evidence that when fire is excluded or its intensity reduced such grassland will revert to forest provided propagules of forest species are available. This succession has been studied in detail at Olokemeji in Nigeria (Charter & Keay, 1960; Clayton, 1958a; MacGregor, 1937).

The experiments at Olokemeji have shown that secondary wooded grassland can be directly invaded by forest species. After six years of protection from fire the following had invaded: *Holarrhena floribunda*, *Hildegardia barteri*, *Zanthoxylum xanthoxyloides*, *Malacantha alnifolia*, *Ceiba pentandra*, *Diospyros mespiliformis*, *Manilkara obovata*, *Celtis brownii* and *Antiaris toxicaria*. The succession to forest on this site, however, is very slow, probably owing to the unfavourable seasonally waterlogged soil. After thirty-one years of fire-protection the 8–11 m tall canopy was composed principally of the forest species *Manilkara obovata*, *Hildegardia barteri*, *Afzelia africana* and *Diospyros mespiliformis*. More than a dozen savanna tree species still persisted, however, but as suppressed and dying individuals or otherwise only near the edges of the plot. They included *Annona senegalensis*, *Butyrospermum paradoxum*, *Crossopteryx febrifuga*, *Daniellia oliveri*, *Maranthes* (*Parinari*) *polyandra*, *Parkia biglobosa* (*clappertoniana*), *Piliostigma thonningii*, *Pseudocedrela kotschyi*, *Pterocarpus erinaceus* and *Stereospermum kunthianum*. There were also a few 15 m tall individuals of *Anogeissus leiocarpus*.

Adjanohoun (1964) has described transition woodland in the Ivory Coast. Inside the forest but not far from its northern margin in the region of Singrobo there are populations of savanna species, especially *Borassus aethiopum*. They appear to represent the vestiges of former islands of secondary wooded grassland in the process of replacement by forest. Savanna trees are rare and some individuals, e.g. of *Terminalia glaucescens*, *Borassus aethiopum*, *Crossopteryx febrifuga* and *Cussonia arborea* (*barteri*) are dead. *Borassus* is sometimes killed by the strangling fig, *Ficus vogelii*. Other individuals of *Terminalia glaucescens* and *Vitex doniana* are etiolated, and *Nauclea latifolia* assumes a sarmentose habit. There are a few scattered tufts of the grasses *Andropogon tectorum* and *Imperata cylindrica*. The forest species include *Paullinia pinnata*, *Elaeis guineensis*, *Albizia adianthifolia*, *Harungana madagascariensis*, *Trema orientalis*, *Ceiba pentandra*, *Rauvolfia vomitoria*, *Ficus exasperata*, *Albizia zygia*, *Anthocleista nobilis*, *Alchornea cordifolia*, *Setaria chevalieri*, *Musanga cecropioides*, and the naturalized species *Psidium guajava*.

In Cameroun aerial photographs show that hundreds of thousands of hectares of secondary wooded grassland have recently been invaded by forest species (Letouzey, 1968a).

Guineo–Congolian elfin thicket
(mapping unit 1a)

Ref.: Hallé, Le Thomas & Gazel (1967).
Profile: Hallé *et al.* (1967: 3).

The crests of the Belinga Mts in Gabon, 400 km from the sea, are covered between 950 and 1000 m with a very dense 4–8 m tall thicket in which epiphytic orchids, bryophytes, and lichens clothe the stems down to ground level. It consists mainly of erect treelets with well-defined though narrow boles and relatively narrow crowns. Lianes, which include *Asparagus warneckei*, but mostly belong to Apocynaceae, Annonaceae, Celastraceae, Rubiaceae, and Loganiaceae, are extremely abundant. Trees and shrubs are represented by only eighteen species including: *Cassipourea* cf. *congoensis*, *Garcinia chromocarpa* (*echirensis*), *G. punctata*, *Homalium* sp., *Hymenocardia ulmoides*, *Hymenodictyon floribundum*, *Ocotea gabonensis*, *Picralima nitida*, *Santiria trimera*, *Schefflera barteri* and species of *Canthium*, *Ochna*, and *Ouratea*.

Guineo–Congolian edaphic grassland
(mapping units 1a, 2, and 3)

Refs.: Adjanohoun (1962, 1965); Ahn (1959); Bellier *et al.* (1969); Bouillenne *et al.* (1955); Deuse (1960); Germain (1965); Lebrun (1936a, p. 182–5).

Phots.: Ahn (1959: 1–2); Bellier *et al.* (1969: 9–10); Bouillenne *et al.* (1955: 12–15); Léonard (1950, on p. 373).
Profile: Bouillenne *et al.* (1955, fig. 3).

Secondary Guineo–Congolian grassland, which has replaced forest on well-drained soils, is described in the next section. Small patches of grassland surrounded by forest also occur on hydromorphic soils. Their status has long been and still remains a matter of controversy. It would appear that some of this grassland represents a transient stage in the succession from aquatic vegetation to forest, and, in the absence of fire, would soon disappear. Since, however, fire is frequent today, especially fires lit by man for hunting purposes, such grassland may persist indefinitely. There is also evidence that some hydromorphic soils are incapable of supporting forest, and the grassland is truly edaphic, even though it may also be subjected to annual or more frequent fires. There has been a tendency for authors to seek a single explanation to account for the occurrence of all types of grassland on hydromorphic soils, and the literature is not always easy to interpret. The fact that many of the species mentioned below (or their close relatives) are both fire-sensitive and also occur outside the Guineo–Congolian Region on hydromorphic soils which have been shown by experiment to be incapable of supporting trees suggests that the communities they form inside the Guineo–Congolian Region also represent an edaphic climax.

Grassland also occurs very locally on rocky outcrops in the Guineo–Congolian Region on very shallow soils which experience alternate waterlogging and drought.

On hydromorphic soils

This type has been described from the Ivory Coast by Bellier *et al.* (1969), from the Ivory Coast and Benin Republic by Adjanohoun (1962, 1965), from Ghana by Ahn (1959), and from Zaire by Bouillenne *et al.* (1955), Deuse (1960), Germain (1965), and Lebrun (1936*a*). The principal grasses are *Anadelphia afzeliana*, *A. leptocoma*, *A. trispiculata*, *Hyparrhenia mutica*, *Jardinea congoensis*, *J. gabonensis*, *Panicum parvifolium* and *Rhytachne rottboellioides*. Cyperaceae are well represented by *Bulbostylis abortiva*, *B. laniceps*, *Fuirena umbellata*, *Rhynchospora candida*, *R. corymbosa*, *R. holoschoenoides*, *R. rubra*, *R. rugosa* and *Scleria aterrima*. Other herbs include *Lycopodium affine*, *L. carolinianum*, *L. cernuum*, *Mesanthemum radicans*, *Neurotheca congolana*, *Selaginella scandens* and species of *Burmannia*, *Drosera* and *Xyris*. *Sphagnum* is also often present.

On rocky outcrops

Detailed information is only available from three sites in Ghana (R. Rose Innes, in litt. 23 March 1977), namely Nyinahin, about 64 km SW of Kumasi; Krobo Hill on the Mampong scarp in the Mampong Ashanti area, NE of Kumasi; and Kwahu Tafo, on the mountains near Mpraeso.

The Nyinahin grassland is in a forest reserve at about 610 m altitude and is surrounded by tall forest. It lies on a broad crest, and the soil which overlies impermeable bauxite is no more than 7.5 cm deep in the centre. *Andropogon perligulatus*, *Loudetia kagerensis* and *Panicum lindleyanum* occur in the central area, with *Andropogon tectorum* nearer the surrounding forest. There is also a fair amount of the shrub *Dichrostachys cinerea* near the forest edge.

The Krobo Hill grassland occurs perched on a small shelf at the edge of a sheer sandstone cliff with tall forest behind and dense forest and cocoa plantations at the foot of the scarp. There has been no disturbance except for rare fires started by man. The humic soil is only a few centimetres deep over solid rock. The grasses include: *Andropogon curvifolius*, *A. perligulatus*, *Loudetiopsis ambiens*, *Monocymbium ceresiiforme*, *Sporobolus sanguineus*, *S. infirmus*, *Eragrostis scotelliana*, *Panicum griffonii* and *Pennisetum polystachion*. The shrub *Dichrostachys cinerea* also occurs. Both Nyinahin and Krobo Hill are a long way from any other grassland.

The Kwahu Tafo grassland is only 12–16 km from the northern edge of the present forest zone. There is much farming near by but the grassland apparently is not due to fire or any influence other than edaphic. The very shallow humic sandy soil occurs on exposed sandstone slabs on the broad crest of the mountain. The grasses include: *Andropogon perligulatus*, *Monocymbium ceresiiforme*, *Rhytachne rottboellioides*, *Sporobolus infirmus*, *S. sanguineus*, *Loudetiopsis glabrata*, *Loudetia simplex*, *Elymandra androphila*, *Setaria sphacelata*, *Panicum griffonii* and *P. pilgeri*.

Guineo–Congolian secondary grassland and wooded grassland
(mapping units 1a, 2, 3, and 11a)

Refs.: Adjanohoun (1964); Aubréville (1948*a*, p. 29-44; 1949*a*); Charter & Keay (1960); Clayton (1958*a*, 1961); Descoings (1973); Devred (1956); Devred *et al.* (1958); Duvigneaud (1949*b*, 1950, 1952, 1953); Keay (1951, 1959*a*, 1959*c*); Keay & Onochie (1947); Koechlin (1961); Lebrun (1936*a*); Léonard (1950); Letouzey (1968*a*, p. 265–73); Makany (1976, p. 40–72); Mullenders (1954); Sillans (1958, p. 94–6); White (MS, 1962–3); White & Werger (1978).
Phots.: Adjanohoun (1964: 21, 22, 46, 50, 52, 55–7); Charter & Keay (1960: 1–3); Clayton (1958*a*: 4); Léonard (1950: 388); Letouzey (1968*a*: 29–32); Mullenders (1954: 1–6, 9–13, 15–17); Sillans (1958: 42, 43, 94).
Profiles: Descoings (1973: 1, 2, 4, 8); Duvigneaud (1949*b*: 2–8).

Much of the rain forest at the northern and southern limits of the Guineo–Congolian Region has been destroyed by cultivation and fire and replaced by secondary grassland, which often occurs in mosaic with small, usually severely degraded, patches of the original forest, and small patches of secondary thicket and secondary forest. The grassland is often 2 m or more tall and usually contains an admixture of fire-hardy, often fire-trimmed trees, the density of which varies greatly depending mainly on the precise history of the site. These grasslands are usually burned at least once a year. In the absence of fire, provided seed is available, they would revert to forest.

Similar secondary grassland, which is even more widely distributed in the Guinea–Congolia/Sudania and Guinea–Congolia/Zambezia transition zones and in the Lake Victoria basin, is included in the present account.

There are also small patches of secondary grassland far inside the Guineo–Congolian Region. They are usually treeless. Among them the grasslands of the Sobo plains in south Benin, Nigeria, are dominated by *Loudetia arundinacea* (Keay & Onochie, 1947). Three types of secondary grassland dominated by *Panicum maximum*, *Pennisetum purpureum* and *Imperata cylindrica* respectively occur in the heart of the Zaire basin and have been described by Léonard (1950).

The secondary grasslands and wooded grasslands fringing the Guineo–Congolian Region and in the transition zones beyond show considerable local variation in floristic composition, but most of their constituent species are widespread and occur both north and south of the equator. Exceptions are indicated by (N) or (S) below.

The principal grasses are: *Andropogon gayanus*, *A. schirensis*, *A. tectorum*, *Pennisetum unisetum* (*Beckeropsis uniseta*), *Brachiaria brizantha*, *Ctenium newtonii*, *Hyparrhenia diplandra*, *H. familiaris*, *H. nyassae*, *H. rufa*, *H. subplumosa*, *Imperata cylindrica*, *Loudetia arundinacea*, *L. phragmitoides*, *L. simplex*, *Monocymbium ceresiiforme*, *Panicum phragmitoides*, *Pennisetum purpureum* and *Schizachyrium sanguineum* (*semiberbe*).

Principal trees are: *Annona senegalensis*, *Afzelia africana* (N), *Borassus aethiopum*, *Bridelia ferruginea*, *Burkea africana*, *Butyrospermum paradoxum* (N), *Combretum collinum*, *Crossopteryx febrifuga*, *Cussonia arborea*, *Daniellia oliveri* (N), *Detarium senegalense* (N), *Dialium engleranum* (S), *Dichrostachys cinerea*, *Entada abyssinica*, *Gardenia ternifolia* (N), *Hymenocardia acida*, *Lophira lanceolata* (N), *Maranthes polyandra* (N), *Maytenus senegalensis*, *Nauclea latifolia*, *Parinari curatellifolia*, *Parkia biglobosa* (N), *Pericopsis* (*Afrormosia*) *laxiflora* (N), *Piliostigma thonningii*, *Pseudocedrela kotschyi* (N), *Psorospermum febrifugum*, *Pterocarpus erinaceus* (N), *Securidaca longepedunculata*, *Stereospermum kunthianum*, *Strychnos madagascariensis* (*innocua*), *S. pungens* (S), *S. spinosa*, *Syzygium guineense*, *Terminalia glaucescens* (N), *T. laxiflora* (N), *Uapaca togoensis* (N), *Vitex doniana* and *V. madiensis*.

Transitional rain forest
(mapping unit 4)

Refs.: Lebrun (1935, p. 12 and map opposite p. 36; 1936a, p. 33, 85, 137, 140–2, 174–6); Pécrot & Léonard (1960, p. 68–9).

Syn.: forêt de basse montagne (Pécrot & Léonard, 1960); forêt de transition (Lebrun, 1935, 1936a).

Although this type occupies the transition zone at the eastern end of the Zaire basin, between the Guineo–Congolian and Afromontane Regions, it is convenient to describe it here. It occurs between 1 100 and 1 750 m on the lower slopes of the chain of mountains which form the western rim of the Great Rift Valley. It formerly provided a connection between lowland rain forest and various types of Afromontane forest, but, owing to its wholesale destruction, direct contact remains in only a few places today. It is a mixture of Guineo–Congolian, Afromontane and endemic species. The pattern of altitudinal replacement is far from simple. Within the transitional forest zone isolated summits more than 1 750 m tall are not necessarily covered with montane forest, although the latter often descends below 1 750 m in valleys and ravines probably because of edaphic and especially local climatic conditions. Elsewhere the continuity of rain forest is broken by a band of more xerophilous forest which owes its presence to the influence of 'föhn' winds.

Of canopy trees, lowland species include *Cynometra alexandri*, *Julbernardia seretii*, *Maranthes glabra*, *Pycnanthus angolensis*, *Staudtia stipitata*, *Strombosia grandifolia*, *Symphonia globulifera* and *Uapaca guineensis*; ecological and chorological transgressors include *Newtonia buchananii* and *Parinari excelsa*; and species centred on transitional forest include *Carapa grandiflora*, *Lebrunia bushaie*, *Musanga leo-errerae*, *Pentadesma lebrunii* and *Ocotea michelsonii*. Afromontane species include: *Aningeria adolfi-friedericii*, *Entandrophragma excelsum*, *Mitragyna rubrostipulata* and *Ocotea usambarensis*.

II The Zambezian regional centre of endemism

Geographical position and area

Geology and physiography

Climate

Flora

Mapping units

Vegetation
    Zambezian dry forest
        Zambezian dry evergreen forest
        Zambezian dry deciduous forest and scrub forest
    Zambezian swamp forest and riparian forest
    Zambezian transition woodland
    Zambezian woodland
        Zambezian miombo woodland
        Zambezian mopane woodland and scrub woodland
        North Zambezian undifferentiated woodland and wooded grassland
            Riparian woodland
            On Upper Valley Soils in Zambia
        South Zambezian undifferentiated woodland and scrub woodland
        Zambezian 'chipya' woodland and wooded grassland
        Zambezian Kalahari woodland
    Zambezian thicket
        Itigi thicket and related types
        Kalahari thicket
        Zambezian termite-mound thicket
        Zambezian rupicolous bushland and thicket
    Zambezian scrub woodland
    Zambezian grassland
        Zambezian edaphic grassland
            Dambo grassland
            Flood-plain grassland
            Kalahari and dambo-edge suffrutex grassland
        Zambezian secondary grassland and wooded grassland
    Zambezian vegetation on heavy metal and other toxic soils

Geographical position and area

The Zambezian Region extends from 3°S to 26°S and from the Atlantic Ocean almost to the Indian Ocean. It includes the whole of Zambia, Malawi and Zimbabwe, large parts of Angola, Tanzania and Mozambique, and smaller parts of Zaire (Shaba), Namibia (previously South-West Africa), Botswana, and South Africa (Transvaal). (Area: 3 770 000 km².)

Geology and physiography

The larger part of the Zambezian Region is occupied by the Great African Plateau and lies more than 900 m above sea-level, rising in places near the rim to over 2 500 m. This higher ground supports Afromontane communities.

Locally the Great Escarpment, which delimits the plateau, is a conspicuous feature, but north of the Zambezi it is difficult to locate owing to advanced dissection and complex trough faulting associated with the southern extension of the Great Rift Valley.

To the east of the Great Escarpment the marginal regions gradually merge into the coastal plain of Mozambique, which belongs however to the Zanzibar–Inhambane Region (Chapter XIII). In parts of southern Angola the transition from the Zambezian vegetation of the escarpment to the coastal desert of Mossamedes (page 144) is much more abrupt.

Most of the Zambezian Region is drained by the Zambezi River but the northern fringes lie within the catchment area of the Zaire.

In the Zambezian Region the plateaux take the form of peripheral uplands surrounding the northern part of the Kalahari basin. The latter, which is covered by an almost continuous mantle of Kalahari Sand, lies mostly between 1 000 and 1 250 m and is drained by the upper Zambezi.

The peripheral uplands, which have extremely flat surfaces, lie mostly between 1 200 and 1 500 m. Locally, as in the Muchinga Mountains west of the Luangwa valley, there is a more rugged relief. The floors of the trough-faulted Zambezi and Luangwa valleys may be as much as 1 000 m below the adjacent plateaux from which they are separated by steep, stony escarpments.

Gentle warping of the plateau surface has caused ponding or seasonal flooding of the main rivers, and extensive areas of hydromorphic soils occur in the Lake Bangweulu and Upper Kafue basins and elsewhere.

The Kalahari Sands of the Upper Zambezi basin were originally laid down under desert conditions on a late Cretaceous erosion surface. Locally they have been redistributed by water. Their thickness varies greatly and in places they are up to 150 metres deep. Basement rocks protrude through them in the north on the Zambezi–Zaire watershed.

The peripheral uplands are formed of a wide variety of Precambrian rocks which are extensively granitized, but the Katanga formation consists principally of limestone, dolomite, shale, schist, quartzite, sandstone, and conglomerate. The great dyke of Zimbabwe is an extraordinary feature which extends over 480 km in a NNE–SSW. direction and is 5–6 km wide, cutting through the Precambrian Basement granites. It consists of ultrabasic and basic intrusives, principally diorite, gabbro, peridotite, and serpentinite, and is an important source of nickel and chrome. In many places its vegetation is distinctive because of the toxic soils (see Wild, 1978, for summary and references). There are many other small outcrops of ultrabasic rocks in Zimbabwe, and some extensive areas in the northern Transvaal.

The floors of the down-faulted valleys are formed of Karoo strata, in places overlain by recent alluvium. Basalts of late Triassic age occur at the top of the Karoo system near the Victoria Falls, south of the Makarikari depression and elsewhere.

Climate

Almost the whole of the Zambezian Region lies within Walter's tropical summer-rainfall zone. Except towards the coast the climate is continental in character with an appreciably larger seasonal variation in temperature than that of the Guineo–Congolian Region. (See Fig. 6.)

There is a single rainy season, chiefly from November to April. In some places it may be interrupted by a dry spell lasting for two or three weeks.

Rainfall is between 500 and 1 400 mm per year and in general decreases from north to south, but there are pronounced regional variations. Mean annual temperature, which varies from 18° to 24° C, is correlated more with altitude than latitude.

There are three main seasons, one wet and two dry. The following details refer to Zambia and Malawi. Elsewhere they may be slightly different:
1. *Wet season, November to April.* The rain falls mostly as thunderstorms and heavy showers, with only rare periods of continuous rain and over several days. There is a fair amount of sunshine.
2. *Cool season, May to August.* The day temperatures are moderately high with continuous sunshine but night temperatures are low and ground frost occurs occasionally in sheltered valleys.
3. *Hot season, September to November.* Temperatures and atmospheric humidity progressively increase until the oppressive feeling in the air is relieved by the advent of the rains.

In the heart of the Zambezian Region dry-season precipitation is exceptional, and there is usually no measurable rain for six months or more. Relative humidity during most of this period is low. Towards the periphery of the region the dry season, though still severe, is somewhat less pronounced. In Malawi and Mozambique, for instance, there are dry-season invasions of moist air from the Mozambique channel which bring periodic spells of mist or drizzle and occasionally even rain.

Frost is more widespread and frequent than the climatic diagrams indicate, although except in the extreme south-west it is nevertheless still localized. On the plateaux above 1 200 m light frosts may occur very locally nearly every year, but only in depressions where cold air accumulates, e.g. at Lilongwe in Malawi. Much more severe and widespread frosts occur at intervals of ten to twenty years (Ernst, 1971; Willan, 1957).

Flora

There are at least 8 500 species, of which *c.* 54 per cent are endemic.

Endemic families. None.

Endemic genera. Few. Woody endemic genera include *Bolusanthus, Cleistochlamys, Colophospermum, Diplorhynchus, Pseudolachnostylis* and *Viridivia*, which are all monotypic. *Androstachys* and *Xanthocercis* otherwise occur only in Madagascar. The centre of variation of *Brachystegia* and *Monotes* is in the Zambezian Region.

Linking elements. About 24 per cent of Zambezian tree species also occur in the Sudanian Region. Some 'dry country' species, such as *Balanites aegyptiaca, Boscia angustifolia, B. salicifolia, Commiphora africana* and *Maerua angolensis*, also occur both in the Somalia–Masai and the Sudanian Regions and have a quasi-continuous distribution. Other species, more characteristic of higher rainfall areas, such as *Amblygonocarpus andongensis, Burkea africana, Erythrophleum africanum, Isoberlinia angolensis* (including *I. tomentosa*) and *Swartzia madagascariensis*, do not occur in the Somalia–Masai Region and show an appreciable interval between their Zambezian and Sudanian areas. Other 'wet country' species, including *Acacia hockii, Combretum collinum (mechowianum), C. molle, Parinari curatellifolia* and *Piliostigma thonningii*, though absent from typical Somalia–Masai vegetation occur there in small enclaves of pronounced Zambezian affinity.

In general, the herbaceous and small woody Zambezian species show a similar relationship to the Sudanian Region as do the trees, but it is less well

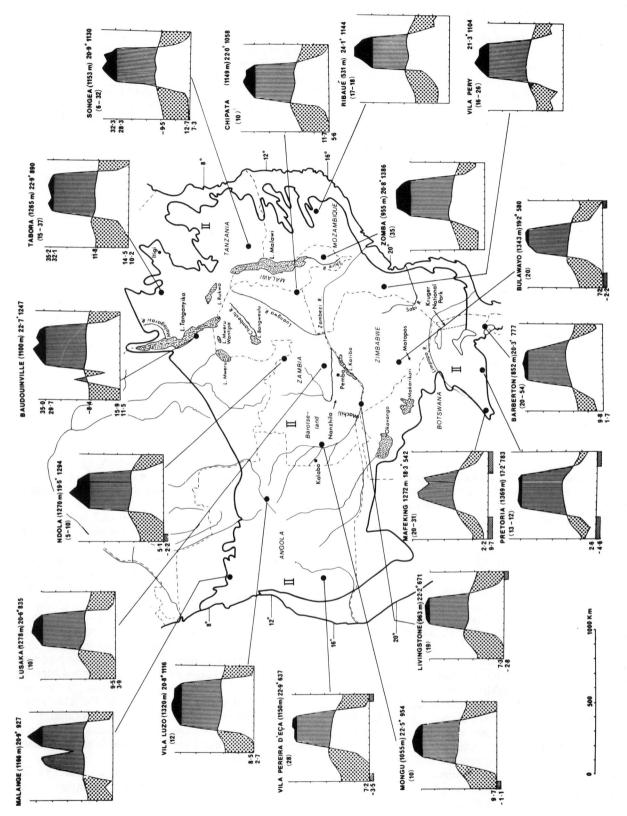

FIG. 6. Climate and topography of the Zambezian regional centre of endemism (II)

defined since a higher proportion of herbaceous species also extend to more distant phytochoria.

Although Afromontane linking species do not feature prominently in Zambezian vegetation, about fifty species occur in small distant satellite populations, the collective range of which extends through most of the wetter parts of the Region (White, 1978a, fig. 4).

Mapping units

6. Zambezian dry evergreen forest.
16a. Enclaves of Zanzibar–Inhambane coastal mosaic (see Chapter XIII).
16b. Enclaves of Zanzibar–Inhambane forest (see Chapter XIII).
17 (p.p.). Cultivation and secondary grassland replacing upland and montane forest (see page 59).
21 (p.p.). Mosaic of Zambezian dry evergreen forest and wetter miombo woodland.
22a. Mosaic of Zambezian dry deciduous forest and secondary grassland.
25. Wetter Zambezian miombo woodland.
26. Drier Zambezian miombo woodland.
28. *Colophospermum mopane* woodland and scrub woodland.
29c. North Zambezian undifferentiated woodland and wooded grassland.
29d. South Zambezian undifferentiated woodland and scrub woodland.
35a (p.p.). Transition from undifferentiated Zambezian woodland to *Acacia* deciduous bushland and wooded grassland (see Chapter XIV).
37 (p.p.). *Acacia polyacantha* secondary wooded grassland (see page 63).
40. Itigi deciduous thicket.
44. Enclaves of Kalahari deciduous *Acacia* bushland and wooded grassland (see Chapter XIV).
47. Mosaic of *Brachystegia bakerana* thicket and edaphic grassland.
60. Edaphic and secondary grassland on Kalahari Sand.
64 (p.p.). Mosaic of edaphic grassland and semi-aquatic vegetation (see below and Chapter XXII).
75 (p.p.). Swamp and aquatic vegetation (see below and Chapter XXII).
76 (p.p.). Halophytic vegetation (see Chapter XXII).

Vegetation

The Zambezian Region, after the Sahara, is the largest major phytochorion in Africa. It probably has the richest and most diversified flora, and certainly shows the widest range of vegetation types, as the following pages testify. The literature dealing with its vegetation has recently been reviewed by Werger & Coetzee (1978).

Zambezian dry forest

In the Zambezian Region forest occurs, or formerly occurred, on deep, freely drained soils with an adequate supply of moisture in their lower horizons during the dry season. Its area has been greatly diminished by fire and cultivation, but it has probably always been relatively restricted during the present climatic era. The evergreen and deciduous types are almost completely different in floristic composition. Only in a few places do their floras intermingle.

Zambezian dry evergreen forest
(mapping units 6, 14, and 21)

Refs.: Cottrell & Loveridge (1966); Fanshawe (1961; 1969, p. 11–18); Lawton (1963, p. 60–2; 1964; 1978b); Schmitz (1962; 1971, p. 268–87); White (MS, 1952, 1960, 1973); White & Werger (1978).
Phots.: Fanshawe (1969: 1); Lawton (1964: 1–4); Schmitz (1971: 31); Trapnell & Clothier (1937: 4).
Profile: Cottrell & Loveridge (1966: 2).

Zambezian dry evergreen forest, which rarely exceeds 25 m in height, except for a few emergents, represents part of a transition, both floristic and physiognomic, from Guineo–Congolian rain forest to Zambezian woodland. That part of the transition which occurs in the Guinea–Congolian transition zone is described in Chapter X. In the Zambezian Region dry evergreen forest is confined to the wetter northern parts where the mean annual rainfall is more than 1200 mm, except on Kalahari Sand. There it extends into regions with a rainfall as low as 900 mm.

Dry evergreen forest is simpler in structure than rain forest and is relatively poor floristically. The leaves of the dominant trees are more coriaceous than those of rain-forest species and few have 'drip-tips'.

Most dry evergreen forest has been destroyed by cultivation and fire and only tiny, mostly disturbed, fragments remain, usually in a matrix of secondary grassland and wooded grassland, and, where there has been some degree of protection from fire, by various stages of forest regrowth. The catenary and successional relationships between dry evergreen forest and other vegetation types are complex and are probably different in different parts of its range. They are still controversial. Schmitz (1962), for instance, believes that dry evergreen forest represents climax vegetation in the northern parts of the Zambezian Region, whereas according to others (Duvigneaud, 1958; Fanshawe 1960) it is confined to edaphically favourable sites. Evidence from the burning plots at Ndola (Trapnell, 1959; White, in prep.) supports the latter view. At Ndola it appears that forest was formerly confined to the deepest soils, and miombo woodland to the shallowest soils, and that they were separated by transition woodland (page 91).

Zambezian dry evergreen forest varies greatly in floristic composition from place to place. There are eight dominant and emergent tree species, namely, *Berlinia giorgii, Cryptosepalum pseudotaxus, Daniellia alsteeniana, Entandrophragma delevoyi, Marquesia acuminata, M. macroura, Parinari excelsa* and *Syzygium guineense* subsp. *afromontanum*. Although none of the dominants occurs throughout, each overlaps considerably with most of the others. *Cryptosepalum pseudotaxus* dominates the most distinctive variant of dry evergreen forest, which is known locally as 'mavunda' and occurs on Kalahari Sand. Many of the species occurring in dry evergreen forest are either Guineo–Congolian linking species or Afromontane linking species. Most are readily killed by fire and are normally absent from woodland.

In Mbala District in Zambia there are small patches of semi-evergreen forest of more markedly Guineo–Congolian affinity, which are characterized by *Celtis gomphophylla (durandii), Aningeria altissima* and *Trichilia prieuriana*.

Zambezian dry deciduous forest and scrub forest
(mapping unit 22a)

Refs.: W. R. Bainbridge (pers. comm.); Barbosa (1970, p. 21–5; 207–14); Fanshawe (1969, p. 21–7; MS); Fanshawe & Savory (1964); Hall-Martin (1975); Martin (1940); Miller (1939); White (MS, 1952, 1960, 1973).
Phots.: Barbosa (1970: 23.1, 23.3); Fanshawe (1969: 2); Hall-Martin (1975: 2–8).
Profiles: Hall-Martin (1975: 1a, 1b).

Dry deciduous forest occurs in those parts of the Zambezian Region where rainfall is between 600 and 900 mm per year, and characteristically is found on certain deep, usually sandy soils which absorb all the incident rainfall or receive lateral seepage water and remain moist at depth throughout the greater part of the dry season.

The canopy, which is usually about 20 m high, varies from 12 to 25 m, and is not always continuous. The understorey is often dense and thicket-like. On unfavourable sites the thicket-component sometimes occurs, usually in mosaic with forest, without its emergent trees. It is not treated separately here. Nearly all the species are deciduous, but there is considerable variation from species to species and year to year. The thicket understorey of *Baikiaea* forest is always deciduous, in the driest types for up to 5–6 months. In the wettest types the canopy in some years is semi-evergreen.

The most extensive deciduous forests are the *Baikiaea* forests on Kalahari Sand in the southern part of the upper Zambezi basin. Related forests without *Baikiaea* occur in the valleys of the middle and lower Zambezi and its tributaries, but there is progressive floristic change towards the east and those in the Shire valley have little in common with the *Baikiaea* forests of Barotseland.

In *Baikiaea* forest, *B. plurijuga* forms an almost pure canopy which is usually about 20 m high and almost closed. *Pterocarpus antunesii* is abundant as a subdominant. *Entandrophragma caudatum* is a local emergent. The invasive *Acacia erioloba (giraffae)* and *Combretum collinum* are widespread, and *Ricinodendron rautanenii* is locally abundant in degraded types near the Zambezi River. There is no well-defined lower canopy, but several trees are subdominant, especially *Boscia albitrunca, Commiphora angolensis* (local), *Croton gratissimus, Excoecaria (Sapium) bussei, Lonchocarpus nelsii, Strychnos madagascariensis* and *S. potatorum (stuhlmannii)*. A large strangling fig, *Ficus fischeri*, is sometimes conspicuous. *Acacia fleckii, Croton pseudopulchellus* and *Markhamia obtusifolia* are common in old fireholes.

The shrub layer of *Baikiaea* forest, the mutemwa, forms a well-defined deciduous thicket of tall coppicing shrubs 5–8 m high. The commonest species are *Acacia ataxacantha, Baphia massaiensis (obovata), Bauhinia petersiana* (including *macrantha*), *Combretum celastroides, C. elaeagnoides, Dalbergia martinii* and *Popowia (Friesodielsia) obovata*. Of less general occurrence are *Acalypha chirindica, Alchornea occidentalis, Byrsocarpus orientalis, Canthium frangula, C. martinii, Citropsis daweana, Zanthoxylum trijugum (Fagara trijuga), Grewia flavescens, Markhamia acuminata, Tarenna luteola* and *Tricalysia allenii*. Beneath the thicket there are scattered smaller shrubs, including *Croton scheffleri, Erythrococca menyharthii* and *Grewia avellana*, and shrubby herbs such as *Achyranthes aspera, Blepharis maderaspatensis, Hypoestes verticillaris, Plumbago zeylanica, Pupalia lappacea* and *Triumfetta annua*.

The commonest woody climbers in *Baikiaea* forest are *Baissea wulfhorstii, Combretum mossambicense*, and *Hippocratea parviflora*, but several shrubs, especially *Acacia ataxacantha, Bauhinia petersiana, Combretum celastroides* and *Dalbergia martinii*, are in fact scramblers and occasionally climb into the canopy. The herb layer is only conspicuous during the rainy season. Grasses vary from sparse to dense, and include *Leptochloa uniflora, Oplismenus hirtellus, Panicum heterostachyum* and *Setaria homonyma*. Other conspicuous herbs are *Aneilema johnstonii* and *Kaempferia rosea*. Epiphytes and bryophytes are virtually absent.

Baikiaea plurijuga is almost confined to Kalahari Sand, but *Pterocarpus antunesii* occurs on suitable sites throughout the Zambezi Valley. In the Lower Shire Valley (Hall-Martin, 1975), it is co-dominant with *Newtonia hildebrandtii*; other associates, which do not occur in *Baikiaea* forest, include *Adansonia digitata, Balanites maughamii, Cordyla africana* and *Diospyros quiloensis*.

Dry scrub forest occurs locally in western Angola west of the escarpment which delimits the great interior plateau. In it *Adansonia digitata, Sterculia setigera* and *Euphorbia conspicua* emerge from a dense thicket of *Strychnos henningsii* and *Combretum camporum*. When this type grows on alluvium the trees, which then include *Acacia welwitschii, Berchemia discolor* and *Diospyros mespiliformis*, as well as *Adansonia* and *Euphorbia conspicua*, grow up to 25 m tall. The thicket understorey comprises *Capparis erythrocarpos*, *Balanites angolensis*, *Grewia carpinifolia*, *Ximenia americana*, *Maytenus senegalensis*, *Garcinia livingstonei*, *Cassine aethiopica* (*Mystroxylum aethiopicum*) and *Bauhinia tomentosa*.

In the Matopos Hills and a few places elsewhere the rupicolous communities are locally sufficiently luxuriant to be classified as scrub forest.

Zambezian swamp forest and riparian forest
(not shown on map)

Refs.: Fanshawe (1969, p. 32–8; MS); Lawton (1967a); Lebrun & Gilbert (1954, p. 43–4); Simpson (1975); White (MS, 1952, 1959–60, 1973); White & Werger (1978).
Phot.: Fanshawe (1969: 5).

In the wetter parts of the Zambezian Region, where annual rainfall is above 1000 mm, permanent swamp forest occurs around springs at the sources of tributary streams and locally along watercourses where the water movement is sluggish. Swamp forest in the latter situation merges into other types of riparian forest in which the water-table is some distance below the surface for at least part of the year. Riparian forest varies greatly in relation to substrate, climate, and the depth and duration of flooding. Only a few variants are briefly described below.

The most abundant dominant trees of swamp forest, *Mitragyna stipulosa*, *Syzygium owariense*, *Xylopia aethiopica*, *X. rubescens* and *Uapaca guineensis*, are also widespread in the Guineo–Congolian Region, as are the most characteristic members of the understorey, *Aporrhiza nitida*, *Garcinia smeathmannii* and *Gardenia imperialis*, and of the shrub layer, *Psychotria* (*Cephaelis*) *peduncularis*, *Craterispermum laurinum* s.l., and *Dracaena camerooniana*. Other important trees are *Ficus congensis*, *Raphia* sp., *Syzygium cordatum* (usually on the fringes) and the Afromontane species *Ilex mitis*. Climbers are rare, but epiphytes, especially ferns, are frequent.

Where rainfall exceeds 1000 mm per year perennial streams are usually fringed with evergreen or semi-evergreen forest, 20 m or more in height. The most abundant species are *Adina* (*Breonadia*) *microcephala*, *Khaya nyasica* and *Newtonia buchananii*. Of less frequent occurrence are, among many others, *Anthocleista schweinfurthii*, *Canarium schweinfurthii*, *Dacryodes edulis*, *Erythrophleum suaveolens*, *Monopetalanthus richardsiae*, *Nauclea pobeguinii*, *Parkia filicoidea*, *Treculia africana*, and the endemic species *Monopetalanthus trapnellii* and *Tessmannia burttii*.

Where rainfall is less than 800 mm per year the streams and smaller rivers are seasonal, and at best support a poor scrubby growth. Well-developed forest, which may reach a height of 18–24 m, is confined to the banks of the larger watercourses such as the Zambezi and its major tributaries. Most of its tree species are deciduous for at least two months. In the most luxuriant types the trees and climbers form a continuous canopy, but it is rare to find undisturbed examples and, in any case, this vegetation has probably always been kept open by the movements and browsing of large mammals. This may explain the almost universal presence of several heliophilous species of *Acacia* and other genera. The most characteristic trees are: *Acacia albida*, *A. galpinii*, *A. polyacantha* subsp. *campylacantha*, *A. robusta* subsp. *clavigera*, *A. tortilis*, *A. xanthophloea*, *Albizia versicolor*, *Combretum imberbe*, *Cordyla africana*, *Croton megalobotrys*, *Diospyros mespiliformis*, *Ficus capensis*, *F. sycomorus*, *Kigelia africana*, *Lecaniodiscus fraxinifolius*, *Manilkara mochisia*, *Mimusops zeyheri*, *Newtonia hildebrandtii*, *Strychnos potatorum*, *Trichilia emetica* and *Xanthocercis zambesiaca*.

The flood plain of the upper Zambezi in Barotseland is flooded each year to a depth of 2 m or more from mid-February to mid-June. Few Zambezian tree species can withstand such a pronounced fluctuation in the water-level and the outer fringe of riparian forest, which is 9–12 m tall, is dominated almost exclusively by *Syzygium guineense* subsp. *barotsense* with an understorey of *Rhus quartiniana*.

Zambezian transition woodland
(not mapped separately, occurring in mapping units 6, 21, and 25)

Refs.: Endean (MS); Lawton (1978b); Schmitz (1962); Trapnell (1959); White (MS, 1952, 1959–60, 1973).
Phot.: Trapnell (1959: 3).

According to Schmitz (1962), in Upper Shaba dry evergreen forest (page 89) or 'muhulu', following cultivation and fire, has been extensively replaced by secondary evergreen miombo woodland dominated by *Marquesia macroura* and *Brachystegia taxifolia*.

In places the destruction of muhulu was not complete and a few large forest trees and the rootstocks of smaller trees and climbers have survived in the secondary woodland to act as potential foci for the re-establishment of forest. There are extensive areas, where, following human depopulation, *Marquesia* and *Brachystegia taxifolia* woodlands have been invaded by forest species. Schmitz concludes that muhulu represents the climatic climax in the wetter part of the Zambezian Region and that most miombo woodland and hence also the transition woodland just described is secondary. The burning and fire-protection experiments at Ndola in Zambia, however, suggest that some transition woodland is secondary and some represents an ecotone between dry evergreen forest and climax miombo woodland.

Near Ndola, *Marquesia* and *Brachystegia taxifolia* are rare but miombo woodland dominated by *Julbernardia paniculata*, *Isoberlinia angolensis*, *Brachystegia spiciformis*, *B. longifolia* and *Erythrophleum africanum* has been protected from fire for more than forty years. When the experiment began there was evidence that at least the deeper soils were capable of supporting forest. On the deepest soils there has been an extensive invasion of forest shrubs, climbers and canopy dominants. The latter, *Parinari excelsa* and *Syzygium guineense* subsp. *afromontanum*, have locally formed closed stands. There is no regeneration of the woodland species and there is every appearance of a reversion to forest.

On the shallowest soils there has been virtually no invasion of forest elements and the miombo elements appear to be in permanent possession.

On soils of intermediate depth evergreen shrubs and climbers are now abundant and locally form thickets, but the dominants of evergreen forest have failed to establish themselves. There has been a dramatic decline in the regeneration of the woodland dominants associated with the invasion of forest shrubs and climbers, and many of the original trees of the woodland canopy have died, presumably because of competition from the forest element. The miombo element, however, still persists and seems capable of coexisting indefinitely in a state of dynamic equilibrium with the forest shrubs and climbers. A sequence of dry years or natural fires and the activity of large mammals would favour miombo species. Wetter periods and protection from fire would favour forest species.

Zambezian woodland

Woodland is the most widespead and most characteristic vegetation of the Zambezian Region. There is little doubt that in many places it represents the climax, but it is equally certain that much woodland elsewhere is secondary or has been profoundly modified by cultivation and fire. Three main types of woodland, namely miombo woodland (mapping units 25 and 26), mopane woodland (28) and undifferentiated Zambezian woodland (29c and 29d), are sufficiently distinct to show on the map.

Zambezian miombo woodland
(mapping units 24, 25, and 26)

Refs.: Astle (1969, p. 74–6); Astle, Webster & Lawrance (1969); Barbosa (1970, p. 133–85); Burtt (1942, p. 73–86); Duvigneaud (1958, p. 201–7); Fanshawe (1969, p. 38–44); Lewalle (1972, p. 79–87); Malaisse (1977); Schmitz (1963a, p. 308–52; 1971, p. 79–87); Trapnell (1953, p. 15–17); Trapnell & Clothier (1937, p. 10–12); White (MS, 1951–52, 1959–60, 1973, 1975); Wild & Barbosa (1968, p. 18–34).
Phots.: Astle (1969: 6); Astle *et al.* (1969: 8); Barbosa (1970: 15.1, 16.1, 16.2, 18.2).

Throughout the greater part of the Zambezian Region, especially on the main plateau surface and its flanking escarpments, miombo is the prevalent vegetation where the soils are freely drained, but the rooting environment is restricted. It is, however, absent from the southern and extreme western fringes of the region, and, although it occurs in the low-lying Zambezi and Luangwa valleys, it is very localized there. It is also absent from large parts of the Kalahari Sand of western Zambia and eastern Angola.

Floristically and physiognomically miombo is very different from other types of woodland. It is nearly always dominated by species of *Brachystegia*, either alone or with *Julbernardia* or *Isoberlinia*. The dominants are extremely gregarious and only rarely occur in other vegetation types.

The most characteristic miombo soils are leached and acid. They are often shallow and stony, or the rooting environment is otherwise restricted by the occurrence of laterite or a gley horizon near the surface. On rocky outcrops and on stony slopes which are too steep to cultivate the surviving miombo is probably little modified, but the status of miombo on the deeper soils of the plateau is much more uncertain. Nearly all the miombo on the plateau has been subjected to agricultural practices of the 'citemene' or similar types. As a consequence its structure and floristic composition have been altered and probably greatly simplified. The trees in such secondary miombo are often uniform in age and size, and are 'kneed' or forked at breast height or show other evidence of former mutilation dating from the previous cycle of cultivation.

On the drier and shallower plateau soils the original vegetation was almost certainly miombo but on the moister and deeper soils in the higher-rainfall areas miombo has probably replaced dry evergreen forest or transition woodland following cultivation and fire.

In appearance miombo is distinctive because of the shape of the dominant trees. Their boles are mostly short but relatively slender and the branches are at first markedly ascending before spreading out to support the light, shallow, flat-topped crown which bears pinnate leaves.

Miombo is mostly 10–20 m tall but scrub miombo can be as short as 3 m. By contrast, on certain deep soils *Brachystegia spiciformis*, *B. longifolia* and *B. utilis* reach a height of 30 m (Savory, 1963). The climax in such situations, however, is possibly dry evergreen forest or transition woodland rather than miombo.

Nineteen species of *Brachystegia* and three species in related genera, namely *Julbernardia globiflora*, *J. paniculata* and *Isoberlinia angolensis*, occur in miombo as dominants.

The *Brachystegia* dominants are unevenly distributed between the eastern and western halves of the Zambezian Region. All but two, *B. puberula* and *B. tamarindoides*, occur in the eastern part, east of the Kalahari Sand, and nine, namely *B. allenii*, *B. angustistipulata*, *B. bussei*, *B. manga*, *B. microphylla*, *B. stipulata*, *B. taxifolia*, *B. torrei* and *B. utilis*, are confined to it. The Kalahari Sand is poor in *Brachystegia* species; in addition to the bushy endemic species, *B. bakerana* (page 98) and the geoxylic suffrutex, *B. russelliae*, only *B. spiciformis* is widespread. *B. boehmii* is abundant on transitional Kalahari Sand to the east, and *B. glaberrima*, *B. longifolia*, *B. puberula* and *B. wangermeeana* are almost confined to the higher-rainfall areas. All these species extend further west into Angola, where they are joined by *B. floribunda* and the endemic *B. tamarindoides*. The *Brachystegia* flora of Angola is thus much poorer than that of the eastern half of the Zambezian Region and consists predominantly of taxonomically critical species.

Of the commoner species, *B. allenii*, *B. bussei*, and *B. microphylla* (including *B. glaucescens*) are virtually confined to rocky hills and escarpments; *B. boehmii*, *B. utilis* and *B. taxifolia* occur both on escarpments and ridges and on certain plateau soils, whilst *B. floribunda*, *B. glaberrima*, *B. longifolia*, *B. manga*, *B. spiciformis* and *B. wangermeeana* are more characteristic of deeper plateau soils and, in some cases, Kalahari Sand.

Most miombo woodlands are semi-deciduous, but some are completely deciduous and some are almost evergreen. In the higher-rainfall areas, with the exception of the evergreen *Brachystegia taxifolia*, all Zambezian *Brachystegia* species are semi-deciduous or briefly deciduous. The old leaves are shed as the new leaves unfold some weeks or even months before the end of the dry season. Of associated species, a few, e.g. *Diospyros batocana*, are completely evergreen, and some, e.g. *Pterocarpus angolensis*, are completely deciduous, but most behave similarly to *Brachystegia*. At low altitudes, where temperatures are high and rainfall somewhat marginal, as on the escarpment of the Zambezi and Sabi valleys, miombo is completely deciduous for up to two months.

Most miombo species are both semi-heliophilous and show some degree of fire-resistance, but the dominants cannot survive repeated fierce fires (Trapnell, 1959; White, MS).

Because the dominants of miombo are extremely gregarious, few other species enter the canopy, except in the more stunted variants. The principal canopy associates are *Afzelia quanzensis*, *Anisophyllea pomifera*, *Erythrophleum africanum*, *Faurea saligna*, *Marquesia macroura*, *Parinari curatellifolia*, *Pericopsis* (*Afrormosia*) *angolensis* and *Pterocarpus angolensis*.

Several species of *Uapaca* and *Monotes* occur scattered in miombo as small trees less than 10 m tall. They are frequently dominant on shallow soils and in secondary miombo, and are also abundant in the scrub woodland which represents the ecotone between miombo and the edaphic grassland of waterlogged depressions (dambos). Similarly, the half-dozen small tree species of *Protea* which occur in miombo are more characteristic of the shorter open types and dambo edges than of well-developed woodland. Other small trees and large shrubs in miombo are numerous in species but are rarely abundant.

Lianes, pteridophytes, and bryophytes are normally absent from miombo except on fire-protected sites, rocky places, termite mounds, and in secondary miombo which is developing towards forest or transition woodland, though a few pteridophytes are adapted to fire (Kornaś, 1978). Vascular epiphytes and epiphytic bryophytes are plentiful only in the moister types of miombo, though epiphytic lichens and the orchid *Ansellia gigantea* (*nilotica*) are more widespread. Hemi-parasitic Loranthaceae belonging to many species are conspicuous. *Pilostyles aethiopica* is an internal parasite of *Brachystegia* and *Julbernardia*. The root parasite *Thonningia sanguinea* is locally plentiful in higher-rainfall areas.

The field layer is usually rather sparse and locally is replaced by patches of leaf-litter or stony, eroding soil. The grasses are mostly 0.6–1.2 m tall with scattered culms of taller species up to 2 m. There is usually little foliage between the field layer and the lower canopy, and visibility is unimpeded for 100 m or more.

Most of the miombo dominants are widely distributed and have wide ecological amplitudes. Since they combine kaleidoscopically a detailed classification would be of limited value. Nevertheless some well-defined trends are apparent. On the vegetation map a distinction is made between wetter miombo (mapping unit 25) and drier miombo (mapping unit 26). This might be difficult to justify on the distribution of the dominants alone, but the associated vegetation types are quite different. They may be characterized, local exceptions apart, as follows:

1. *Wetter miombo*. Rainfall usually more than 1000 mm per year but less on Kalahari Sand. Canopy height often more than 15 m. Floristically rich. Includes nearly all the miombo dominants. *Brachystegia floribunda*, *B. glaberrima*, *B. taxifolia*, *B. wangermeeana* and *Marquesia macroura* are widespread. Associates in rocky places include many species which otherwise occur in evergreen forest and thicket. Associated vegetation includes dry evergreen forest and thicket, swamp forest, evergreen riparian forest, and wet dambos.

2. *Drier miombo*. Rainfall less than 1000 mm. Canopy height usually less than 15 m. Floristically poor. *Brachystegia floribunda* etc. absent or very local. *Brachystegia spiciformis*, *B. boehmii* and *Julbernardia globiflora* are often the only dominants present. Associates in rocky places include many species which otherwise occur in deciduous forest and thicket or other dry types. Associated vegetation includes dry deciduous forest and thicket, deciduous riparian forest, and dry dambos.

Where climate changes rapidly as on the escarpments flanking the Lake Tanganyika and Lake Malawi troughs and in the Eastern Highlands of Zimbabwe it has not always been possible to map wetter and drier miombo separately. Various types of scrub miombo are described on page 99.

Zambezian mopane woodland and scrub woodland (mapping units 28 and 36)

Refs.: Astle, Webster & Lawrance (1969); Barbosa (1970, p. 185–200); Brynard (1964); Ellis (1950); Fanshawe (1969,

p. 48–50, 57–8; MS, 1969); Giess (1971, p. 10); Seagrief & Drummond (1958); Tinley (1966, p. 67–72); Volk (1966b); Whellan (1965); White (MS, 1952, 1960, 1973); Wild & Barbosa (1968, p. 34–5; 50–2).

Phots.: Astle et al. (1969: 3); Barbosa (1970: 20.1–4); Brynard (1964: 2b); Fanshawe (1969: 8); Giess (1971: 28–31); Seagrief & Drummond (1958: 8); Tinley (1966: 10, 25, 30–7).

Profiles: Seagrief & Drumond (1958: 1–3).

Communities dominated by *Colophospermum mopane* are widespread in the drier half of the Zambezian Region. The tallest are woodlands 10–20 (25) m tall. There is a broad correlation between rainfall and height but vigour also varies greatly according to local site factors, and nearly everywhere mopane woodland and scrub mopane occur in mosaic. They are treated collectively in this account.

Mopane is the most extensive vegetation type in parts of the Zambezi, Luangwa, Limpopo, Shashi and Sabi valleys, but further east it is absent from the Indian Ocean Coastal Belt. It is also widespread in the Nanzhila and Machili basins in Zambia, and in the Makarikari and Okavango depressions in Botswana but is otherwise almost absent from the region of the Kalahari Sand. Further west it covers extensive areas in Namibia and south-west Angola, and scattered shrubby individuals penetrate deep into the Namib Desert along the beds of normally dry watercourses to within 20 km of the coast. In Angola and Namibia, towards the limits of its range, mopane is associated with *Welwitschia bainesii* (pages 143 and 191).

Both in Zambia and Zimbabwe, mopane also occurs very locally on the Plateau between 1280 and 1400 m but is generally absent from the escarpment country which separates the plateau from the floors of the great river valleys.

Despite the diversity in height and density of mopane communities, they have a remarkable physiognomic uniformity. This is due to the almost complete dominance of mopane itself and its very characteristic appearance. Regardless of size, except when damaged, it is usually a single-stemmed tree or treelet. Its sparse crown is distinctive because of the rigid, irregular, markedly ascending branches which give off slender, tortuous, more spreading laterals. The butterfly-shaped leaves which consist of a single pair of large leaflets are unmistakable.

Mopane is capable of growing under a wide range of climatic and edaphic conditions but its actual distribution is severely restricted because of fire and competition from other species. Over most of its range, mopane experiences low rainfall and high temperatures, but the tallest stands occur towards the upper limits of its rainfall tolerance (*c.* 800 mm per year). The fact that its upper altitudinal limit is at 1400 m shows that low temperatures if not accompanied by frost are not inimical. Like most Zambezian tree species, however, it is killed back by frost.

Where rainfall is 500 mm or less, mopane grows on most types of soil, but towards the drier limit of its range it is not found on heavy clay. Where annual rainfall is higher than 500 mm, mopane only flourishes on soils which are shallow, or have a heavy or deflocculated subsoil. These conditions are often brought about by a high concentration of sodium. Hence in the higher-rainfall part of its range mopane is associated with rocks containing sodium, such as Karoo sediments in the down-faulted valleys, and granite or Karoo rocks on the plateau. The sodium is derived from the soda feldspar perthite which occurs as large crystals in the granite and as small transported particles in the Karoo sediments. The sodium disperses the clay particles in the soil which then accumulate in the lower horizons to form an impervious layer. Such mopane soils have a low water-storage capacity and poor depth penetration. They are frequently unable to absorb all the water they receive. The experiments of Thompson (1960) have shown that in the absence of competition mopane grows better on ordinary soils than on sodium soils even on the Plateau. In Zambia the more alkaline the soil the poorer the growth of mopane. Mopane does not occur on true saline soils in which water-soluble salts exceed 0.2–0.3 per cent. Such soils are usually bare pans with scattered palms, *Acacia tortilis* and *Adenium obesum* (*multiflorum*), all of which are more salt-tolerant than mopane.

Mopane has a shallow root system with a dense concentration of fine roots in the top 25 cm of soil. Grass is sparse or absent in well-developed mopane, hence fire-damage is minimal. This favours the regeneration of mopane which is unable to regenerate in a dense sward.

In the Zambezi Valley and further south, in most situations, mopane is deciduous for about five months, but near the shore of Lake Kariba it is almost evergreen. Further north, in the Luangwa Valley, it loses its leaves for only about three months.

The mopane plant is very resinous and the tree itself is flammable once the bark is burnt, or if there is a crown fire. Normally, because of the sparse ground layer, this is a rare event, but if the canopy is opened up by browsing elephants, and coarse grasses invade, the fire hazard is greatly increased. Browsing elephants frequently snap off quite large trees at a height of 0.6–2 m or fell them completely, in which case coppice shoots are formed at the base of the felled tree. As a direct result of fire, mopane woodland may be converted to shrubby mopane grassland, in which multiple-stemmed coppice varying in height from 0.3 to 1.6 m is produced from the base of the charred original stems. There is usually an equally tall cover of grass between individual coppice clumps. Fire-maintained mopane of this kind covers extensive areas in Botswana and in the Kruger National Park in the Transvaal.

Colophospermum mopane and the miombo dominants scarcely ever occur together and their associated floras are almost totally dissimilar.

In the tall mopane in the Luangwa Valley the most conspicuous associates are *Acacia nigrescens*, *Adansonia digitata, Combretum imberbe, Sclerocarya caffra* and *Kirkia acuminata.*

In Namibia and Angola where annual rainfall is 400–600 mm, mopane occurs as a 7–10 m high tree and forms light stunted woodland with a shrubby understorey on stony or sandy permeable soils derived from a wide range of parent material. In Angola the principal associates are: *Acacia erubescens, A. kirkii, Balanites angolensis, Boscia microphylla, B. rehmanniana, Catophractes alexandri, Combretum apiculatum, C. oxystachyum, Commiphora anacardiifolia, C. angolensis, C. pyracanthoides, Grewia villosa, Rhigozum brevispinosum, R. virgatum, Spirostachys africana, Terminalia prunioides, T. sericea, Ximenia americana* and *X. caffra.*

North Zambezian undifferentiated woodland and wooded grassland
(mapping unit 29c)

Refs.: Astle, Webster & Lawrance (1969); Barbosa (1970, p. 106–10; 200–6); Fanshawe (1969, p. 50–5); Simpson (1975, p. 192–3); Tinley (1966, p. 43-55); Trapnell (1953, p. 18–19); Trapnell & Clothier (1937, p. 13); White (MS, 1952, 1959–60, 1973); Wild & Barbosa (1968, p. 39–40, 43–50, 52–4, 59).

Phots.: Astle *et al.* (1969: 7); Fanshawe (1969: 9, 9a); Simpson (1975: 6); Tinley (1966: 14, 16–21); Trapnell & Clothier (1937: 7 and 8).

This type occurs north of the Limpopo. It is floristically rich and is more easily defined by the absence of the miombo and mopane dominants than by its own floristic composition. Despite its small area it is composed of many more tree species than either miombo or mopane. It occupies a wide range of soils and is very variable floristically, but there is considerable floristic overlap between different examples and variation is more or less continuous. Although the dominants of miombo are normally absent from undifferentiated woodland some of their associates are frequently present, notably *Afzelia quanzensis, Burkea africana, Dombeya rotundifolia, Pericopsis angolensis, Pseudolachnostylis maprouneifolia, Pterocarpus angolensis* and *Terminalia sericea.*

Towards or slightly beyond the drier climatic limits of miombo small patches of undifferentiated woodland occur on certain soils unsuitable for mopane, e.g. on freely drained granite soils in the Matopos Hills, Zimbabwe, and on Karoo sandstone ridges in the Zambezi Valley. The most extensive occurrences of this vegetation type, however, appear to be largely secondary. The two main variants are described below.

Riparian woodland

Riparian woodland and wooded grassland are extensively developed on alluvium fringing the larger permanent watercourses in the drier half on the Zambezian Region. The most characteristic trees which are frequently more than 20 m tall are *Acacia albida, A. robusta* subsp. *clavigera, A. erioloba, A. nigrescens, A. polyacantha* subsp. *campylacantha, A. sieberana, A. tortilis, Adansonia digitata, Albizia harveyi, Berchemia discolor, Borassus aethiopum, Combretum imberbe, Cordyla africana, Croton megalobotrys, Diospyros mespiliformis, Ficus sycomorus, Hyphaene ventricosa, Kigelia africana, Lannea stuhlmannii, Lonchocarpus capassa, Sclerocarya caffra, Tamarindus indica, Trichilia emetica, Xanthocercis zambesiaca, Xeroderris (Ostryoderris) stuhlmannii* and *Ziziphus mucronata.*

Most riparian woodland probably represents degraded riparian forest, or rather transition woodland. It is unlikely that the climax was true forest, since the habitat is a favourite resort of elephants, which do much damage in obtaining their food. Thus, *Acacia nigrescens* trees are often pushed over for the sole purpose of obtaining the tufts of a parasitic *Loranthus* in their crown (Tinley, 1966). Some riparian woodland, however, is clearly seral to riparian forest, and other stands, especially on seasonally flooded heavy clay soils, may represent an edaphic climax.

On Upper Valley Soils in Zambia

Trapnell & Clothier (1937, p. 7) designate as 'Upper Valley Soils' those that have formed at lower altitudes than the surrounding plateau in regions of more modified topography. They commonly occupy broken or gently rolling country, mostly above 760 m, with free drainage, and are especially associated with limestones and mica schists. They lack the ironstone formations which frequently occur in the Plateau soils and the subsoil tends to show a basic reaction. Their fertility, higher than that of the Plateau soils, is due to a higher degree of base saturation and a higher phosphate and nitrogen content. According to R. Webster (pers. comm.) they favour intensively rooting species and even if shallow provide a good rooting environment.

The original vegetation was probably thicket with numerous emergent trees, or even dry forest or transition woodland. Only fragmentary and degraded relics survive and nearly everywhere, where the land is not cultivated, it has been replaced by secondary woodland and wooded grassland. The principal woodland trees, some of which reach a height of 20 m, are: *Acacia polyacantha* subsp. *campylacantha, A. sieberana, Albizia amara, A. harveyi, A. versicolor, Azanza garckeana, Cassia abbreviata, Combretum collinum, C. fragrans (ghasalense), C. molle, Dalbergia boehmii, Ficus sycomorus, Kigelia africana, Lonchocarpus capassa, Markhamia obtusifolia, Peltophorum africanum, Pericopsis angolensis, Piliostigma thonningii, Pterocarpus rotundifolius, Terminalia mollis, Trichilia emetica, Xeroderris stuhlmannii* and *Ziziphus abyssinica.* The grass layer is up to 4 m tall.

*South Zambezian undifferentiated woodland
and scrub woodland*
(mapping units 29d and 29e)

Refs.: Acocks (1975, p. 27–30, 33–7, 44–50); White (MS, 1973).
Phots.: Acocks (1975: 14, 16, 21, 22, 25, 33–7, 39, 40–2).
Syn.: lowveld sour bushveld, lowveld, arid lowveld, arid sweet bushveld, mixed bushveld, sourish mixed bushveld, sour bushveld (all of Acocks, 1975).

This type occupies the south-eastern extremity of the Zambezian Region, between the floor of the Limpopo Valley and the northern limits of the Highveld, and extends southwards through Swaziland as a narrow tongue between the northern extension of the Drakensberg and the coastal plain. Altitude varies from 150 to 1 525 m, and mean annual rainfall from 325 to 1 000 mm.

In structure and floristic composition it is intermediate between north Zambezian undifferentiated woodland and Tongaland–Pondoland semi-evergreen bushland and thicket (Chapter XV). Well-developed woodland more than 9 m tall is localized. Elsewhere it is mostly scrub woodland. Most of the larger woody plants are *c.* 7 m high with taller emergent trees. In a natural state the vegetation is usually rather dense or even closed but is scarcely ever impenetrable.

Of the larger woody species about half are widespread in the Zambezian Region. The remainder are more or less confined to the southern fringes of the Zambezian Region or also extend southwards into other phytochoria.

The former category includes *Acacia gerrardii, A. nigrescens, A. nilotica, A. rehmanniana, A. sieberana, A. tortilis, Adansonia digitata, Albizia antunesiana, A. tanganyicensis, Balanites maughamii, Bolusanthus speciosus, Burkea africana, Cassia abbreviata, Combretum apiculatum, C. collinum, C. hereroense, C. imberbe, C. molle, C. zeyheri, Commiphora mollis, C. pyracanthoides, Dalbergia melanoxylon, Dichrostachys cinerea, Diospyros mespiliformis, Diplorhynchus condylocarpon, Dombeya rotundifolia, Euphorbia ingens, Faurea saligna, Ficus ingens, F. sycomorus, Heeria (Ozoroa) reticulata, Kirkia acuminata, Lannea discolor, Lonchocarpus capassa, Ochna pulchra, Parinari curatellifolia, Peltophorum africanum, Piliostigma thonningii, Pseudolachnostylis maprouneifolia, Pterocarpus angolensis, P. rotundifolius, Sclerocarya caffra, Steganotaenia araliacea, Strychnos pungens, Terminalia prunioides, T. sericea, Trichilia emetica* and *Ziziphus mucronata.* The majority of these are trees, but south of the Limpopo their stature is less than elsewhere.

The more important members of the southern element are: *Acacia caffra, A. davyi, A. luederitzii (gillettiae), A. permixta, A. robusta* subsp. *robusta, A. tenuispina, Aloe arborescens, A. marlothii, Androstachys johnsonii, Berchemia zeyheri, Cadaba aphylla, Carissa bispinosa, Dalbergia armata, Diospyros villosa, Ekebergia pterophylla, Grewia flava, Kirkia wilmsii, Manilkara concolor, Protea caffra, Ptaeroxylon obliquum, Pterocelastrus* spp., *Rhigozum obovatum, Rhus* spp. including *R. chirindensis* and *R. leptodictya, Schotia brachypetala, Sesamothamnus lugardii, Spirostachys africana, Sterculia rogersii* and *Tarchonanthus galpinii.* Many of these are bushes or small bushy trees. Some are deciduous, others evergreen.

*Zambezian 'chipya' woodland
and wooded grassland*
(not mapped separately, but occurring in mapping units 6, 21, and 25)

Refs.: Cottrell & Loveridge (1966, p. 93–5); Fanshawe (1969, p. 15–20); Lawton (1963, p. 51–4, 70–1; 1964, p. 472–3; 1972; 1978*b*); Trapnell (1943, p. 11–18; 1950, p. 14; 1959); White (MS, 1952, 1960, 1973).
Phots.: Fanshawe (1969: 1a, 1b); Lawton (1963: 1–5; 1964: 5–7); Trapnell (1943: 8, 9).
Syn.: chipya (high grass) woodland (Trapnell, 1943); *Brachystegia spiciformis* ('*hockii*') woodland on the transition to Lake Basin soils (Trapnell, 1943); chipya forest (Trapnell, 1959).

Trapnell (1943) applied the term chipya (from Bemba 'cipya') to vegetation in which various trees other than *Brachystegia, Julbernardia* and *Isoberlinia* grow mixed in very tall grass. Such vegetation burns fiercely and the trees are remarkably fire-resistant. It occurs locally on suitable soils on the Central Africa Plateau in parts of Zambia, Shaba, and Malawi where rainfall exceeds 1 000 mm per year, but is most extensively developed on the alluvial soils of lake basins, especially Lake Bangweulu, and their associated river systems. Small patches of evergreen thicket known as 'mateshi' are often locally included in chipya, as well as scattered individuals of *Entandrophragma delevoyi.* The latter is one of the most characteristic species of Zambezian dry evergreen forest. It is now well established that chipya occurs on sites formerly occupied by forest or transition woodland and owes its existence to cultivation and fire. Three herbaceous species, namely *Aframomum biauriculatum, Pteridium aquilinum* and *Smilax kraussiana,* which are absent from most types of miombo woodland, are almost universally present in chipya.

Chipya usually consists of a complex mosaic representing different stages of degradation and re-establishment of the original vegetation, though in most places it is the more degraded phases that predominate. At one extreme, it consists of tall, almost pure grassland, though coppice of many fire-hardy trees may persist after the death of the original trunks by fire. At the other extreme the canopy is virtually closed, evergreen species are plentiful, and the community is well on its way to reverting to forest. The whole spectrum is treated as a single dynamic continuum and is classified here as woodland only for convenience.

The fire-hardy trees in chipya, which are sometimes 20 m or more high, include *Afzelia quanzensis* (local), *Albizia antunesiana, Amblygonocarpus andongensis, Burkea africana, Erythrophleum africanum, Parinari curatellifolia, Pericopsis angolensis* and *Pterocarpus angolensis*. Smaller trees are *Anisophyllea boehmii, Combretum collinum, C. celastroides, C. zeyheri, Diospyros batocana, Diplorhynchus condylocarpon, Heeria reticulata, Hymenocardia acida, Maprounea africana, Ochthocosmus lemaireanus, Oldfieldia dactylophylla, Pseudolachnostylis maprouneifolia, Swartzia madagascariensis, Syzygium guineense* subsp. *guineense, Terminalia sericea, Xylopia odoratissima* and *Zanha africana*.

In typical chipya the field layer is dense and usually between 2 and 3 m high. The dominant grasses are species of *Hyparrhenia* and *Andropogon gayanus*.

Zambezian Kalahari woodland
(not mapped separately, but occurring in mapping units 6, 21, 21a, and 25)

Refs.: Barbosa (1970, p. 167–72; 219–36); Duvigneaud (1952), p. 104–5); Fanshawe (1969, p. 44–7); Trapnell & Clothier (1937, p. 11–13); White (MS, 1952, 1960).
Phots.: Barbosa (1970: 17.1); Fanshawe (1969: 7–8); Trapnell & Clothier (1937: 5, 6).

On the Kalahari Sands of the upper Zambezi basin, forest was the original climax, at least on the most favourable soils: evergreen *Cryptosepalum pseudotaxus* forest in the north and deciduous *Baikiaea plurijuga* forest in the south. Most of the vegetation today has been modified by fire and cultivation, and various types of woodland and wooded grassland are widespread. Much of the woodland is clearly secondary but there can be little doubt that primary woodland and wooded grassland originally formed part of the catenary sequence from forest on the ridges to edaphic grassland in the seasonally waterlogged depressions between. Today it is often impossible to distinguish these two types of woodland.

The following trees are widespread in Kalahari woodland: *Afzelia quanzensis, Albizia antunesiana, Amblygonocarpus obtusangulus, Brachystegia spiciformis, Burkea africana, Combretum psidioides, C. zeyheri, Dialium engleranum, Diospyros batocana, Diplorhynchus condylocarpon, Erythrophleum africanum, Hymenocardia acida, Lannea discolor, Maprounea africana, Ochna pulchra, Parinari curatellifolia, Pseudolachnostylis maprouneifolia, Pterocarpus angolensis, Strychnos pungens, Swartzia madagascariensis, Terminalia sericea,* and *Vangueriopsis lanciflora*.

In addition, *Brachystegia longifolia, B. puberula, B. wangermeeana, Cryptosepalum pseudotaxus* and *Julbernardia paniculata* are characteristic of the northern sands, and *Acacia erioloba, Baikiaea plurijuga* and *Ricinodendron rautanenii* of the southern.

Zambezian thicket

Various types of thicket occur scattered throughout the Zambezian Region but only Itigi thicket is sufficiently extensive to show on the map.

Itigi thicket and related types
(mapping unit 40)

Refs.: Burtt (1942, p. 104); Fanshawe (1969, p. 25); Jacobsen (1973); Schmitz (1971, p. 287–8); Trapnell & Clothier (1937, p. 13); White (MS, 1952, 1959–60); Wild 1968e).
Phots.: Burtt (1942: 41–6); Trapnell & Clothier (1937: 7).
Syn.: Entandrophragmeto–Diospyretum hoyleanae, sous-assoc. Pseudoprosopitietosum (Schmitz, 1963a).

Dense deciduous thicket occurs on specialized soils in various places in the drier parts of the Zambezian Region and towards its periphery. Although the geological nature of the soils is variable they appear to have certain features in common. The soil is well aerated and well supplied with water in the rainy season but dries out, at least in its upper layers, during the dry season. It is less stony than many miombo soils and favours the intensive root systems of thicket species. The most extensive thicket of this type takes its name from the village of Itigi in Tanzania. Related communities occur in Zambia and Zimbabwe.

Itigi thicket has a markedly discontinuous distribution. In the Central Province of Tanzania it covers 620 km². It also occurs in Zambia in the depressions between Lake Mweru Wantipa and the southern end of Lake Tanganyika and in a few localities in adjacent parts of Zaire.

In Tanzania it is composed almost entirely of much-branched coppice-like shrubs which are deciduous for about four months each year and form a canopy 3–5 m in height. The shrubs are interlaced overhead to form a thick continuous cover which is very dense when in leaf. *Baphia burttii, B. massaiensis, Burttia prunoides, Combretum celastroides* subsp. *orientale (C. trothae), Grewia burttii, Pseudoprosopis fischeri* and *Tapiphyllum floribundum* are the principal canopy species with scattered 8 m high emergents of *Albizia petersiana (brachycalyx)*, or somewhat taller, semi-evergreen trees of *Craibia brevicaudata* subsp. *burttii*, or smaller trees of the semi-evergreen *Bussea massaiensis*. In places the underwood is almost impenetrable to man, but when elephants pass through the dense shrubs spring back behind them.

The canopy is so dense that light is excluded and the ground layer is virtually absent except for the slender grass, *Panicum heterostachyum*, and a few small herbs. Succulents are poorly represented. A large candelabra *Euphorbia, E. bilocularis*, occurs, but only on termite mounds, together with other species alien to the community. Itigi thicket is thornless and climbers are insignificant. It is sharply demarcated from the surrounding *Brachystegia–Julbernardia* woodland and there is no transition zone, though the *Brachystegia* trees near the thicket are often stunted. Fierce fires kill the thicket but they are unable to penetrate far.

The soil under Itigi thicket varies in depth from 0.6 to 3 m and is sandy. It is soft in the rainy season but hardens considerably on drying. It overlies a cement-like duricrust which in turn overlies the granite floor. In Zambia the shallow, stony soils have impeded drainage and are well supplied with water in the rainy season, but dry out during the dry season.

A related type of thicket, 'Pemba' thicket occurs on shallow transitional soils on the border of the unrejuvenated miombo-dominated plateau of the Southern and Central Provinces of Zambia. It takes its name from the village of Pemba and is known locally as 'kasaka'. The soils are rather clayey and usually contain abundant undecomposed minerals, especially feldspar. They are commonly 30–90 cm deep above the weathering zone. The upper layers are intensively occupied by the roots of the thicket species. During the rainy season the soil is saturated with non-stagnant water.

The thicket is normally 6–7 m tall and is almost impenetrable except locally where it has been kept open by fires or wild pigs and buffaloes. Most species are deciduous but a few are evergreen.

The following are the more abundant thicket-forming species: *Acalypha chirindica, Aeschynomene trigonocarpa, Byrsocarpus orientalis* (sometimes scandent), *Canthium burttii, Cassipourea gossweileri, Combretum celastroides* (sometimes scandent), *Haplocoelum foliolosum, Indigofera rhynchocarpa, I. subcorymbosa, Popowia obovata* (sometimes scandent), *Rytigynia umbellulata* and *Tarenna neurophylla.*

Emergent trees are normally present but they are rare and usually quite small. Many, such as *Brachystegia spiciformis* (15 m), *Combretum collinum* (9 m), *Lannea discolor* (9 m), *Parinari curatellifolia* (9 m), *Pericopsis angolensis* (12 m), *Peltophorum africanum* (8 m), *Pterocarpus angolensis* (12 m) and *P. rotundifolius* (12 m) are heliophilous species which do not regenerate in the shade of the thicket and they often occur as slender, weakly individuals. It seems that their natural occurrence in this community is dependent on the activities of large mammals and possibly natural fires. Other tree species including *Pteleopsis anisoptera, Phyllanthus (Margaritaria) discoideus* and *Strychnos potatorum* are normal constituents but they rarely emerge very far above the thicket canopy. The 20 m tall strangling fig, *Ficus fischeri,* is a striking feature, though it is rare. Cactoid tree Euphorbias are virtually absent from this community.

In Lomagundi District in Zimbabwe thicket occurs on the flanks of the Zambezi Valley on certain localized water-retaining soils. Jacobsen (1973) has published a long list of species occurring in thicket on dolomites and cherts, and Wild (1968e) has described thicket on graphitic slates.

Kalahari thicket
(mapping units 22a and 47)

Refs.: Barbosa (1970, p. 219–26); Fanshawe (MS, 1969); Fanshawe & Savory (1964); White (MS, 1952).

In a few places in Sesheke District in Zambia thicket which resembles the mutemwa understorey of *Baikiaea* forest (page 90) but contains dwarf individuals of *Baikiaea* less than 2 m tall occurs at the edges or heads of certain dambos. The dwarfing of *Baikiaea* is believed to be a response to imperfect drainage but is not yet fully understood.

Thicket dominated by *Brachystegia bakerana* occurs in similar situations further north in Kalabo District, where *B. bakerana* is sometimes no more than 1.3 m high. Fanshawe (MS) has suggested that its low stature is due to frost. This species, however, is usually appreciably taller than this, and the fact that the thickets it forms frequently occur in the ecotone between hydromorphic grassland and woodland on well-drained Kalahari Sand suggests that dwarfing is at least partly related to unfavourable soil conditions. In eastern Angola *Brachystegia bakerana* thickets are an important feature of mapping unit 47, but little is known about them.

Zambezian termite-mound thicket

Refs.: Astle (1965a); Fanshawe (1968; 1969, p. 55–60); Malaisse (1978b); Vesey-FitzGerald (1963); White (MS, 1960, 1975–6); Wild (1952a).
Phots.: Astle (1965a: 2, 8); Malaisse (1978b: 1a, 1b); Vesey-FitzGerald (1963: 2); Wild (1952a: 1, 2).

Termites pervade the soil of Africa, and indeed contribute greatly to its formation. Their termitaria are usually conspicuous, but are absent at high altitudes and from some swampy areas and pure sand. When termite mounds are more than a metre or so in diameter, unless they have been newly built or are in the final stages of erosion, they are usually covered with dense thicket, often with one or more emergent trees.

The largest mounds in the Zambezian Region are built by species of *Macrotermes*, and may be up to 8 or 9 m high. Their vegetation is completely different from that on surrounding normal soil. This is particularly true for the Central African Plateau, but at lower altitudes the contrast is less striking.

In toto, the flora of Zambezian termitaria is astonishingly rich. According to Fanshawe no fewer than 700 woody species occur in this habitat in Zambia alone. There is so much variation from one mound to the next that it is impossible to generalize, except in the broadest terms. Genera which are particularly characteristic include: *Acacia, Albizia, Asparagus, Canthium, Cassia, Cassine, Combretum, Commiphora, Euclea, Ficus, Grewia, Popowia, Pterocarpus, Sansevieria, Ximenia* and *Ziziphus.* Among widespread species the following are noteworthy: *Carissa edulis, Diospyros lycioides, Euphorbia candelabrum, Pappea capensis, Peltophorum africanum, Rhoicissus tridentata, Securinega virosa, Steganotaenia araliacea* and *Strychnos potatorum.*

Zambezian rupicolous bushland and thicket

Rocky outcrops, especially granite kopjes, support a distinctive vegetation, which is usually different from that occurring on nearby normal soils and has much in common with the vegetation on termite mounds. The most luxuriant vegetation, which is rooted in joints in the rocks where water accumulates, is often bushland or thicket or locally even scrub forest. Many species occur in this habitat.

Widely distributed bushes, small trees, and climbers include: *Bauhinia petersiana, Canthium burttii, C. lactescens, Cassia abbreviata, Commiphora mossambicensis, Erythroxylum emarginatum, Euclea natalensis, Euphorbia candelabrum, Feretia aeruginescens, Ficus ingens, F. sonderi, Haplocoelum foliolosum, Hippocratea indica, Lannea discolor, Landolphia parvifolia, Steganotaenia araliacea, Strychnos potatorum, Tarenna neurophylla* and *Thunbergia crispa*.

Several tree species occasionally occur as 10 to 20 m tall emergents. They include: *Afzelia quanzensis, Diospyros mespiliformis, Entandrophragma caudatum, Kirkia acuminata, Mimusops zeyheri, Pterocarpus rotundifolius* and *Sclerocarya caffra*.

On very shallow soil and in small crevices, succulents and poikilohydrous species are chiefly found. Among the latter *Myrothamnus flabellifolius*, the 'Resurrection Plant', is the most striking.

Zambezian scrub woodland

Refs.: Astle (1969, p. 76); Astle, Webster & Lawrance (1969, p. 146, 152); Barbosa (1970, p. 151–4); Boughey (1961, p. 69–70); Chapman (1962); Fanshawe (1969, p. 41); Phipps & Goodier (1962); White (1978a, p. 484–5; MS, 1952, 1959–60, 1973); Wild & Barbosa (1967, p. 61).
Phots.: Astle *et al.* (1969: 8); Boughey (1961: 6); Phipps & Goodier (1962: 3).
Profile: Boughey (1962: 8).

Scrub mopane (page 93) and scrub woodland associated with the southern variant of undifferentiated Zambezian woodland (page 96) have already been mentioned. Miombo frequently occurs as scrub woodland at high altitudes and on certain shallow soils. Scrub woodland also forms the ecotone between miombo woodland and dambo grassland, and between Kalahari woodland and Kalahari suffrutex grassland.

Towards its upper altitudinal limit, which is usually between 1 600 and 2 100 m, miombo occurs as stunted, floristically impoverished scrub woodland no more than 6 m tall, usually dominated by *Brachystegia spiciformis*, more rarely by *B. floribunda, B. taxifolia, B. microphylla* (*glaucescens*), or *Uapaca kirkiana*. The trees are often festooned with *Usnea*; epiphytic orchids also occur. The grass cover is sparse, and ferns, e.g. *Pellaea* and *Arthropteris orientalis*, are often present. *Philippia benguelensis*, a characteristic member of Ericaceous montane shrubland, is also normally found.

On shallow soils overlying laterite on the Zambezi–Zaire watershed, *Brachystegia boehmii*, associated with *Ochna schweinfurthiana, Parinari curatellifolia* and *Uapaca pilosa* forms scrub woodland only 3 m tall.

In the Luangwa Valley, *Brachystegia stipulata* and *Julbernardia globiflora* form 3–5 m tall scrub woodland on certain shallow soils derived from siltstone. A mantle of stones up to 30 cm thick usually covers the surface.

The scrub woodland at the edges of dambos is usually between 4 and 7 m tall. It is rather open and often shows signs of damage by fire, more rarely by frost. Species of *Monotes, Uapaca* and *Protea* are usually present. Other frequent woody associates include *Burkea africana, Faurea speciosa, Hymenocardia acida, Ochna schweinfurthiana, Parinari curatellifolia, Swartzia madagascariensis, Syzygium guineense* subsp. *guineense, Terminalia brachystemma* and *Vangueriopsis lanciflora*. The miombo dominants except for *Brachystegia boehmii* are normally absent. Analogous vegetation is more extensive on the Kalahari Sand of the Upper Zambezi basin. It is characterized by *Burkea africana, Hymenocardia acida, Parinari curatellifolia, Terminalia brachystemma* and, above all, *Diplorhynchus condylocarpon*.

Zambezian grassland

The most characteristic Zambezian grasslands occur on seasonally waterlogged soil. There are also extensive areas of secondary grassland, but the latter usually contain trees. Grasslands are also associated with heavy metal and other toxic soils (page 101).

Zambezian edaphic grassland
(mapping units 22a, 25, 26, 47, 60, and 64)

Refs.: Astle (1965a, p. 1969); Astle, Webster & Lawrance (1969); Barbosa (1970, p. 110–15, 225–6, 231); Fanshawe (1969, p. 45; MS); Vesey-FitzGerald (1955a; 1963; 1970); White (MS, 1952, 1959–60); White & Werger (1978).
Phots.: Astle (1965a: 1, 2, 7–11; 1969: 1, 5, 7); Barbosa (1970: 24.1); Fanshawe (1969: 45); Vesey-FitzGerald (1955a: 4–12; 1963: 1–3).

Grassland on seasonally waterlogged soils is widespread in the Zambezian Region and occurs principally in the following habitats:
1. In shallow depressions (dambos) which form the headwater reaches of drainage lines on the unrejuvenated surface of the Central African Plateau.
2. On the flood plains in river valleys and in basins with internal drainage.
3. On Kalahari Sand of low relief in the Upper Zambezi basin.

Dambo grassland

This type occurs above 1 200 m and occupies up to 20 per cent of the plateau surface. Drainage is sluggish but ultimately connected with the through drainage of the country. Locally there is seasonal flooding and parts may remain boggy throughout the year, though elsewhere the surface layers dry out and become very compact during the dry season. The soil is normally acid.

The vegetation is usually a medium-dense grass mat of rather uniform appearance and height. It shows considerable local variation in floristic composition. The mat is composed principally of fine-leaved perennial bunch grasses, and in the wetter types usually with abundant Cyperaceae and Xyridaceae and frequent flowering herbs. The foliage more or less completely conceals the ground, but beneath it about 25–30 per cent of the surface may be bare. The leaf-table is of variable height, from 50 to 100 cm, and the flowering culms extend above this to about 1 m to 2 m.

Loudetia simplex is the most characteristic grass and is dominant over extensive areas. Other common grass species are: *Andropogon schirensis, Hyparrhenia bracteata, H. diplandra (pachystachya), H. newtonii, Miscanthus teretifolius, Monocymbium ceresiiforme, Themeda triandra* and *Trachypogon spicatus*.

Cyperaceae are always present in dambos and in the wetter parts are often dominant. Commonly occurring species are *Ascolepis anthemiflora, A. elata, Bulbostylis cinnamomea, Cyperus esculentus, C. margaritaceus, C. platycaulis, Fuirena pubescens, Kyllinga erecta, Mariscus deciduus, Pycreus aethiops, Scirpus microcephalus* and *Scleria bulbifera*.

Conspicuous forbs include: *Acrocephalus sericeus, Dipcadi thollonianum, Eriospermum abyssinicum, Hypoxis angustifolia, Icomum lineare, Moraea natalensis, Pachycarpus lineolatus, Pandiaka carsonii, Saxymolobium holubii, Stathmostelma pauciflorum* and *S. welwitschii*.

Dambo grasslands are frequently burnt. When protected from fire there is a marked increase in the occurrence of bryophytes and orchids, especially in the moister parts. Most dambos are fringed by a narrow zone of sparse wiry grassland with abundant geoxylic suffrutices, similar to Kalahari suffrutex grassland.

Flood-plain grassland

Where the valleys of the larger rivers have reached their base level of erosion their floors are covered with alluvium, mostly heavy clay. In regions of seasonal rainfall these almost flat valley bottoms are usually flooded annually, or are at least seasonally waterlogged, and are covered with a complex and unstable, constantly changing mosaic of different types of edaphic grassland intermingled with permanent swamp vegetation. The levees fringing the main channels of the river are usually covered with woody plants occurring as thicket, fringing forest, woodland or wooded grassland. Permanent swamp vegetation fills former river channels and occurs as patches elsewhere where the water is impounded by rank herbage. Where the flood water is shallow, 'bush group' grassland, which is a mosaic of pure grassland and termite-mound thicket (page 98), often occurs extensively. The complicated patterns formed by valley grassland, permanent swamp vegetation, and 'bush-group' grassland make it impossible for them to be mapped separately. They are shown collectively as mapping unit 64. Valley grassland is also usually associated with permanent swamp and aquatic vegetation, mapping unit 75.

The most extensive areas of flood-plain grassland are in the Malagarasi and Rukwa valleys in Tanzania, in the Upper Zambezi, Kafue, and Chambeshi valleys and in the Mweru Wantipa and Bangweulu basins in Zambia, and in the Lake Chilwa basin in Malawi. The Rukwa and Mweru Wantipa grasslands are important breeding places of the Red Locust (*Nomadacris septemfasciata* Serv.). In both basins alkaline grassland (Chapter XXII) also occurs.

The principal grasses of the wetter types of flood-plain grassland are *Acroceras macrum, Echinochloa pyramidalis, E. scabra, Leersia hexandra, Oryza longistaminata, Panicum repens, Paspalum scrobiculatum, Sacciolepis africana* and *Vossia cuspidata*. In better-drained places other species, principally *Andropogon brazzae, Entolasia imbricata, Loudetia simplex, Monocymbium ceresiiforme, Setaria sphacelata* and *Themeda triandra*, prevail.

Kalahari and dambo-edge suffrutex grassland

This type is a short sparse wiry grassland. It occurs on oligotrophic Kalahari Sand, which has often been redistributed by wind or water, and is seasonally waterlogged. Trees are virtually absent and are replaced by rhizomatous geoxylic suffrutices most of which are closely related to forest or woodland trees or lianes, and are usually less than 0.6 m tall. At least under present-day conditions, their stems are normally burnt back to ground level every year. Flowering occurs precociously before the end of the dry season, either from the axils of fallen leaves at the base of the burnt-back shoots or on the new shoots before the latter have completed their development. At this time the grasses are still dormant.

In the absence of fire, obligate suffrutices (the majority) are capable of a limited amount of upward growth, but eventually their stems become moribund and die. The underground parts are usually of massive proportions and the phytomass of the suffrutices greatly exceeds that of the grasses. The communities they form are really 'underground forests' (White, 1976d), but for most of the year they look like grasslands and are treated as such here.

Geoxylic suffrutex grassland is most extensively developed on the Kalahari Sand of the Upper Zambezi basin west of the Zambezi River in Barotseland and adjacent parts of Angola, where it is known as 'chanas da borracha'. It occupies the wide, virtually flat, interfluves between the sluggish, meandering tributary rivers. It also occurs, though less extensively, on seasonally waterlogged depressions on the northern extension of Kalahari Sand on the Kwango plateau far into Zaire, and in similar situations on Kalahari Sand covering the remnant high-plateaux surfaces of Manica, Kibara, Kundelungu, Biano, and Marungu in Shaba. Beyond the limits of Kalahari Sand, similar sparse grassland with suffrutices is found at the sandy edges of seasonally waterlogged depressions or dambos (page 99).

The most widespread dominant grasses in suffrutex grassland are *Loudetia simplex* and *Monocymbium ceresiiforme*. They are often associated with other wiry grasses such as *Andropogon schirensis, Aristida stipitata (graciliflora), Elionurus argenteus, Eragrostis ciliaris, Rhynchelytrum amethysteum, R. repens, Schizachyrium sanguineum, Sporobolus subtilis (barbigerus), Trachypogon thollonii*, and many *Cyperaceae*. The most abundant suffrutex is *Parinari capensis*. Others are listed by White (1976d).

Zambezian secondary grassland and wooded grassland

Refs.: Cottrell & Loveridge (1966); Lawton (1964); Trapnell (1953; 1959); Vesey-FitzGerald (1963); White (MS, 1952, 1959, 1973).
Phots.: Lawton (1964: 5, 7); Trapnell (1959: 2); Vesey-FitzGerald (1963: 8).

Most woody Zambezian vegetation, following destruction by fire and cultivation, is replaced by secondary grassland, which is maintained as such if fires are sufficiently frequent. It is rare, however, to find pure secondary grassland. Fire-resistant trees are nearly always present, and often survive from the original vegetation. Chipya grassland replacing dry evergreen forest is included in the general account of chipya vegetation (page 96).

The principal grasses replacing *Baikiaea plurijuga* dry deciduous forest are *Brachiaria brizantha, Dactyloctenium giganteum, Digitaria milanjiana, Panicum maximum* and *Sporobolus pyramidalis*.

Miombo woodland subjected to the citemene agricultural system is rapidly converted to open woodland and wooded grassland if the resting period between successive cycles of cultivation is too short. The dominant grasses are *Andropogoneae*, especially *Hyparrhenia dichroa, H. newtonii* and *Hyperthelia dissoluta*.

Zambezian vegetation on heavy metal and other toxic soils

Refs.: Duvigneaud (1958; 1959); Duvigneaud & Denaeyer de Smet (1963); Wild (1965; 1968d; 1970; 1974a; 1974c; 1974e; 1978).
Phots.: Duvigneaud (1958: 1–4); Duvigneaud & Danaeyer de Smet (1963: 1–7, 11, 20, 22–4, 27, 28); Wild (1965: 1–8; 1968d: 1–7; 1970: 3–5; 1978: 1–7).
Profiles: Duvigneaud (1958: 2); Duvigneaud & Denaeyer de Smet (1963: 1–2, 12, 14).

Locally in the Zambezian Region the uniformity of the prevailing woodland is broken by not only the edaphic grassland of the dambos but also certain hills and other areas without or with a very sparse cover of woody plants. The treelessness of these hills is due to the presence in the soil of abnormally large and more or less toxic amounts of heavy metals. By far the most widespread and abundant is copper, which is usually accompanied by important quantities of cobalt and often of nickel and sometimes of uranium. Movement of water down the slopes results in the accumulation of metals in an aureole of contamination on lower-lying slopes and in the dambos into which the toxic water drains. The less heavily contaminated soils support an open bushland or wooded grassland. Similar treeless or sparsely wooded grassland is also found on serpentine which, however, at least in Zimbabwe, is usually contaminated with nickel. The vegetation of toxic soils in southern Africa has been comprehensively reviewed by Wild (1978).

III The Sudanian regional centre of endemism

Geographical position and area

Geology and physiography

Climate

Flora

Mapping units

Vegetation
    Sudanian dry forest
    Sudanian swamp forest and riparian forest
    Sudanian transition woodland
    Sudanian woodland
        Sudanian *Isoberlinia* and related woodlands
        Sudanian undifferentiated woodland
            Drier types in Nigeria
            Wetter types in Nigeria
            Undifferentiated woodland in the Sudan Republic
            Ethiopian undifferentiated woodland
    Sudanian grassland
        Sudanian edaphic grassland and wooded grassland
            Valley and flood-plain grassland
            Grassland and wooded grassland on Pleistocene clays
            Grassland and other herbaceous communities on shallow soil over ironstone
        Sudanian secondary grassland
    Sudanian rupicolous scrub forest, bushland and thicket

Geographical position and area

The Sudanian Region extends as a relatively narrow band across Africa from the coast of Senegal to the foothills of the Ethiopian Highlands. It is mostly between 500 and 700 km wide but becomes narrower in the west and broader in the east. (Area: 3 731 000 km².)

Geology and physiography

Nearly everywhere the mostly level or gently undulating surfaces are below 750 m altitude. In West Africa land above 1 000 m is principally confined to two small areas, the Jos Plateau, which extends south into the Guinea–Congolia/Sudania transition zone (page 175) and the Mandara Plateau (mapping unit 33, page 62) which is the northernmost outlier of the Cameroun Highlands. Further east, the Nile–Chad watershed is mostly below 750 m but it rises steadily north to Jebel Marra (3 057 m) which is an inactive volcano of late Tertiary age. It stands on the boundary of the Sudanian and Sahelian Regions and its upper slopes support distinct communities (Chapter XVI). Towards the south-eastern limit of the Sudan Region the crystalline mountains of the Imatong (3 187 m), Dongotona (2 623 m), and Didinga (2 963 m) ranges, which support Afromontane vegetation on their upper slopes, rise steeply from the surrounding plain at about 600 m.

Large areas are covered with superficial deposits of Pleistocene age. Consolidated dunes of wind-blown sand occur towards the northern fringes of the zone, especially in Nigeria, and in the Sudan Republic where they are known as 'qoz'. The Pleistocene clays deposited in broad down-warped basins, especially the Chad basin and the Upper Nile, are of great extent.

The soils covering most of the remainder of the Sudanian Region are formed from Precambrian rocks, but in Mali there are extensive outcrops of 'Infra Cambrian' sandstone which give rise to a distinctive landscape. Cretaceous sediments occur in the Niger and Benue valleys and also in the Sudan Republic (Nubian Series).

Climate

Like the Zambezian Region, the Sudanian Region lies inside Walter's tropical summer rainfall zone and the climates of the two are broadly similar, especially with regard to rainfall. Temperatures, however, in the Sudanian Region are appreciably higher (mean annual temperature 24–28° C), and, because of the harmattan wind, the dry season is more severe. Frost is unknown, (See Fig. 7.)

Flora

There are possibly no more than 2750 species, of which about one-third are endemic.

There are no endemic families. The very few endemic genera include *Butyrospermum*, *Haematostaphis* and *Pseudocedrela*, all of which are monotypic. A high proportion of Sudanian linking species are very widespread in the moderately dry parts of Africa and many exend to other parts of the tropics.

Mapping units

27. Sudanian woodland with abundant *Isoberlinia*.
29a. Undifferentiated Sudanian woodland.
29b. Undifferentiated Ethiopian woodland.
30. Undifferentiated Sudanian woodland with islands of *Isoberlinia* (see page 61).
32 (p.p.). The Jos Plateau mosaic (see page 62).
33. The Mandara Plateau mosaic (see page 62).
35b. The transition from undifferentiated Ethiopian woodland to *Acacia* deciduous bushland and wooded grassland (see page 62).
61. Edaphic grassland in the Upper Nile basin.
62 (p.p.). Mosaic of edaphic grassland and *Acacia* wooded grassland.
63. Mosaic of edaphic grassland and communities of *Acacia* and broad-leaved trees.
64 (p.p.). Mosaic of edaphic grassland and semi-aquatic vegetation (see below and Chapter XXII).

Vegetation

Of the surviving stands of natural and semi-natural vegetation the most numerous and characteristic belong to various types of woodland. Apart from a small amount of swamp forest and riparian forest, and some outliers of forest of Guineo–Congolian affinity in the extreme south, there is virtually no true forest. Nevertheless, some authors, e.g. Keay (1949), Aubréville (1950) and Chevalier (1951), have suggested that dry forest was the original climax over extensive areas before the Region became densely settled. According to Chevalier, undisturbed dense dry forest still exists between Bria and Ndélé in the Central African Republic where there are fewer than 0.5 inhabitants per square kilometre. Some interesting relics of dry evergreen forest also persist on the sandstone plateaux of western Mali.

In most places where cultivation is possible, the natural vegetation has been profoundly modified. In the less densely populated areas most of the land is bush fallow, that is, woodland in various stages of regeneration following a period of cultivation. Where the fallow period is short and fires are frequent, the trees are often represented only by coppice shoots and mature trees of specially preserved species of economic importance; sometimes trees are eliminated completely. Around the large towns cultivation is permanent or semi-permanent up to a distance of tens of kilometres, but valuable trees are protected and the landscape is one of wooded farmland.

Most of the Sudanian Region lies below 1000 m but some small areas such as the Jos and Mandara plateaux are sufficiently elevated to support distinctive communities (mapping units 32 and 33).

In some parts of the Sudanian Region, especially in the valleys of the larger rivers and on the sites of Pleistocene lakes, the prevalent vegetation is edaphic grassland and wooded grassland. The latter mostly occurs on hydromorphic soils, or vertisols, and is often associated with aquatic and semi-aquatic vegetation. Small patches of grassland occur on seasonally waterlogged soils at the heads of some tributary streams, and on shallow soils on ironstone and other rocky outcrops. In the Sudanian Region, bushland and thicket are very poorly represented compared with the Zambezian Region, and typical suffrutex grassland does not occur.

Sudanian dry forest

Ref.: Jaeger (1956).

Only a few examples remain. The best known and best preserved occur in deep ravines and other fire-protected situations on the sandstone plateaux of western Mali. They are dominated by two evergreen species of caesalpiniaceous tree, *Gilletiodendron glandulosum* and *Guibourtia copallifera*, which rarely occur together.

Forests dominated by *Gilletiodendron glandulosum* occur in the triangle delimited by Kita, Kayes and Kéniéba, but only in rocky places where there is protection from fire and the roots have access to subterranean water. The canopy is formed almost exclusively of *Gilletiodendron* and is usually 12–15 m high, but sometimes attains 20 m.

Guibourtia copallifera is not confined to the Sudanian Region but also occurs further south with a scattered distribution from Guinea–Bissau to the Ivory Coast. On the Bamako plateau it forms almost pure forests 15–20 m high. Unlike *Gilletiodendron*, *Guibourtia* is not confined to forest but also occurs in woodland and bushland.

Sudanian swamp forest and riparian forest

Refs.: Keay (1949, p. 358; 1959a); Letouzey (1968a, p. 304–6); Tothill (1948, p. 50); Wickens (1977a, p. 25); White (MS, 1963).

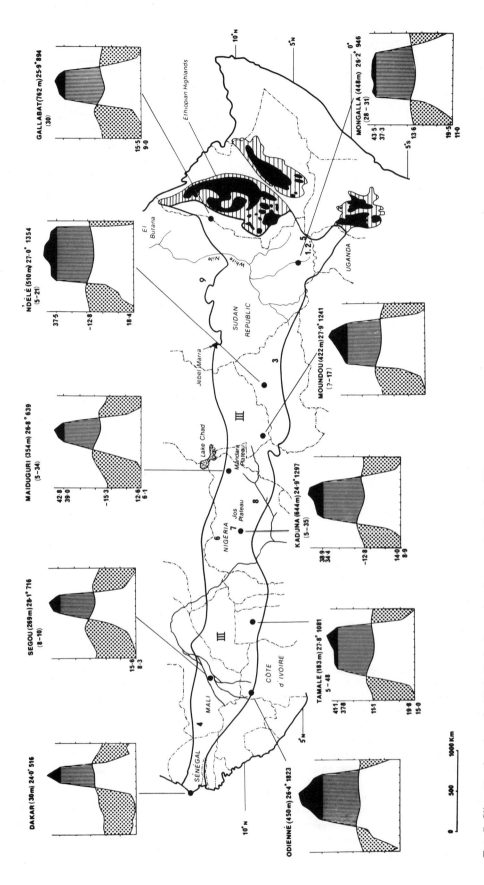

FIG. 7. Climate and topography of the Sudanian regional centre of endemism (III)
1. Imatong Mts. 2. Dongotona Mts. 3. Bria. 4. Sandstone cliffs of western Mali. 5. Imatong Mts. 6. Zamfara Forest Reserve. 7. Zaria. 8. Yankari Game Reserve. 9. El Obeid

The flora of riparian forest in the Sudanian Region, like the Sudanian flora in general, is extremely poor. Semi-evergreen riparian forest in the southern half of the Sudanian Region (the Northern Guinea zone of Keay, 1959a) is often dominated by *Syzygium guineense* subsp. *guineense*. Other characteristic trees are *Khaya senegalensis*, *Terminalia glaucescens* and *Vitex doniana*.

In the northern half of the Sudanian Region (the Sudan zone of Keay, 1959a) the riparian forest is semi-deciduous but has frequently been degraded to riparian woodland. The most frequent large trees are *Khaya senegalensis*, *Diospyros mespiliformis*, *Tamarindus indica*, *Ficus glumosa*, *F. sycomorus*, *Acacia sieberana* and, in rocky places, *Adina microcephala*.

Sudanian transition woodland

Refs.: Harrison & Jackson (1958, p. 23); Schnell (1977, p. 273, 292); White (MS, 1963).

Transition woodland was probably formerly extensive, but only tiny remnants remain, especially in the ecotone between riparian forest and various types of savanna.

In the Yankari Game Reserve in Nigeria (White, MS, 1963) well-developed transition woodland occurs on the outer fringe of riparian forest. It is dominated by *Khaya senegalensis* and *Diospyros mespiliformis*, both of which are regenerating freely. The more conspicuous savanna trees are *Anogeissus leiocarpus*, *Borassus aethiopum*, *Combretum molle*, *Kigelia africana*, *Piliostigma thonningii*, *Stereospermum kunthianum*, *Terminalia laxiflora*, *Vitex doniana* and *Ziziphus mucronata*.

On the Jos Plateau in Nigeria at c. 1400 m (White, MS, 1963) rocky inselbergs are crowned with transition woodland dominated by *Olea capensis* (including *hochstetteri*) and *Anogeissus leiocarpus*. Most of the associates are forest species and include *Albizia zygia*, *Dialium guineense*, *Ficus capensis*, *Harungana madagascariensis* and *Pachystela brevipes*.

Near the western limit of the Sudanian Region, the forest of Bandia near Thiès (Schnell, 1977) is comprised of savanna and forest species. Several woodland species, including *Parkia biglobosa*, *Adansonia digitata*, *Celtis integrifolia* and *Anogeissus leiocarpus* contribute to the 20–30 m tall canopy and assume the habit of forest trees. Their forest associates are *Khaya senegalensis*, *Morus mesozygia*, *Ceiba pentandra* and *Antiaris toxicaria*. Transition woodland in the Sudan Republic near the eastern limit of the Sudanian Region has been described by Harrison & Jackson (1958). Important species include *Terminalia glaucescens*, *Albizia zygia*, *Vitex doniana* and *Anogeissus leiocarpus*.

Sudanian woodland

Most Sudanian trees have very wide geographical ranges, both longitudinal and latitudinal, and wide ecological tolerances. Well-defined centres of endemism of relatively small extent within the Region do not exist.

Too few species occur in sufficiently constant association with others over sufficiently extensive areas to permit the recognition of well-defined ecological elements, though, more locally, poorly defined 'associations' can sometimes be recognized (Kershaw, 1968). Furthermore, any well-defined floristic pattern which may formerly have existed has been greatly obscured because of the degradation most Sudanian vegetation has suffered.

Most of the Sudanian Region is relatively low-lying and without pronounced relief. The climate changes gradually, and this, combined with the wide tolerances of the species, renders difficult the recognition of distinct zones and distinct vegetation types. Some species, however, have a distinctly northern or southern distribution, though the majority are more widespread.

Species characteristic of the drier northern parts of the Sudanian Region include (S = Sudanian endemic): *Acacia albida*, *A. macrostachya* (S), *A. nilotica* subsp. *adstringens* (S), *A. senegal*, *A. seyal*, *Albizia chevalieri* (S), *Balanites aegyptiaca*, *Bauhinia rufescens* (S), *Boscia salicifolia*, *Commiphora africana*, *C. pedunculata*, *Dalbergia melanoxylon*, *Ficus sycomorus*, *Lannea humilis*, *Lonchocarpus laxiflorus* (S), *Maerua angolensis*, *Piliostigma reticulatum* (S), *Sclerocarya birrea* (S), *Tamarindus indica* and *Ziziphus mauritiana*. Some of them extend northwards into the Sahel zone.

Species characteristic of the wetter southern parts of the Sudanian Region include: *Acacia dudgeonii* (S), *A. gourmaensis* (S), *Antidesma venosum*, *Faurea saligna*, *Lophira lanceolata* (S), *Maprounea africana*, *Maranthes polyandra* (S), *Monotes kerstingii* (S), *Ochna afzelii*, *O. schweinfurthiana*, *Protea madiensis*, *Terminalia glaucescens* (S) and *Uapaca togoensis* (S).

Species occurring both in the drier and wetter parts of the Sudanian Region include: *Acacia hockii*, *A. macrothyrsa*, *A. polyacantha* subsp. *campylacantha*, *A. sieberana*, *Afzelia africana* (S), *Amblygonocarpus andongensis*, *Annona senegalensis*, *Anogeissus leiocarpus* (S), *Bombax costatum* (S), *Boswellia dalzielii* (S), *Bridelia ferruginea*, *Burkea africana*, *Butyrospermum paradoxum* (S), *Cassia sieberana* (S), *Combretum collinum*, *C. fragrans*, *C. glutinosum* (S), *C. molle*, *C. nigricans* (S), *Crossopteryx febrifuga*, *Cussonia arborea*, *Daniellia oliveri* (S), *Detarium microcarpum* (S), *Dichrostachys cinerea*, *Diospyros mespiliformis*, *Ekebergia capensis*, *Erythrophleum africanum*, *Ficus glumosa* (S), *Haematostaphis barteri* (S), *Hymenocardia acida*, *Khaya senegalensis* (S), *Isoberlinia doka* (S), *I. angolensis*, *Lannea schimperi*, *Mitragyna inermis* (S), *Nauclea latifolia*, *Parinari curatellifolia*, *Parkia biglobosa* (S), *Pericopsis laxiflora* (S), *Piliostigma thonningii*, *Prosopis africana* (S), *Pseudocedrela kotschyi* (S), *Pterocarpus erinaceus* (S), *Steganotaenia araliacea*, *Sterculia setigera* (S), *Stereospermum kunthianum*, *Strychnos madagascariensis*, *Swartzia madagascariensis*, *Syzygium guineense* subsp. *guineense*, *Terminalia avicennioides* (S), *T. laxiflora* (S), *T. macroptera* (S), *Trichilia emetica*, *Vitex doniana*,

Xeroderris stuhlmannii, Ziziphus abyssinica and *Z. mucronata*.

It is convenient to divide the Sudanian woodlands into two types, though their floristic differences should not be over-emphasized. The wetter woodlands in the south are often dominated by *Isoberlinia doka*. The drier northern woodlands characteristically lack *Isoberlinia*, except for small pockets which nearly always occur on rocky hills.

Isoberlinia doka does not cross the Nile Valley, except in the extreme south, where it extends a short distance into Uganda. The woodlands occurring between the valley of the White Nile and the lower slopes of the Ethiopian Highlands appear to be related to Sudanian undifferentiated woodland and are briefly described below (page 107).

Sudanian Isoberlinia
and related woodlands
(mapping units 12, 27, 30, 32, and 33)

Refs.: Adjanohoun & Aké-Assi (1967); Guillaumet & Adjanohoun (1971, p. 222–4); Chevalier (1951); Grondard (1964, p. 20–1); Harrison & Jackson (1958, p. 21–3); Keay (1959a, p. 22–5; 1960); Kershaw (1968a, p. 244–68; 1968b, p. 467–82); Lawson, Jeník & Armstrong-Mensah (1968); Morison, Hoyle & Hope-Simpson (1948); Rosevear (1953, p. 13); Sillans (1958, p. 101–11, 183–5); White (MS, 1963).
Phots.: Guillaumet & Adjanohoun (1971, p. 29, 30); Lawson *et al.* (1968: 1–4); Morison *et al.* (1948: 14, 15, 17).
Profile: Keay (1959a: 9).
Syn.: Northern Guinea zone (Keay, 1959a).

This type (mapping unit 27) is shown on the map as a broad band extending almost without interruption from Mali in the west to north-west Uganda in the east. In places, as on the Kaduna–Zaria plateau in Nigeria, *Isoberlinia doka* woodland is widespread and in uncultivated places dominates the landscape. *Isoberlinia*, however, is not so extensively dominant everywhere. In Ghana, for instance, *Isoberlinia* shows the same kind of distribution as in Nigeria, but is relatively less abundant (J. B. Hall, pers. comm.). *Daniellia oliveri* and *Burkea africana* are more plentiful, and *Erythrophleum africanum* is probably equally common.

Floristically, Sudanian *Isoberlinia* woodland can be regarded as an impoverished variant of miombo woodland, the most characteristic vegetation of the Zambezian Region, and has been referred to as 'miombo' by some authors (e.g. Keay, 1951), but the two most characteristic genera of miombo, *Brachystegia* and *Julbernardia*, are entirely absent, and two others, *Monotes* and *Uapaca*, are represented only by single species each. Sudanian 'miombo' also differs in its lower stature—it is scarcely ever more than 15 m tall—and in the greater relative importance of non-miombo species. It also has a less compact distribution than Zambezian miombo, in that *Isoberlinia doka* also has a very extensive scattered distribution south of the main *Isoberlinia* belt, and, along with *I. angolensis* also to the north.

Sudanian undifferentiated woodland
(mapping units 29a, 30, 32, and 33)

Refs.: Clayton (1963); Grondard (1964, p. 22–5); Harrison & Jackson (1958, p. 15–16, 17–19); Jaeger (1956, 1959); Keay (1949; 1959a, p. 25–7); Pichi-Sermolli (1957, p. 70–2); Ramsay (1964); Ramsay & De Leeuw (1964, 1965a, 1965b); White (MS, 1973); Wickens (1977a, p. 24).
Phots.: Rosevear (1953: 16, 17).
Profiles: Keay (1959a: 5, 6).
Syn.: forêts claires et savanes boisées soudaniennes à Combrétacées dominantes (Grondard); *Combretum glutinosum* (*cordofanum*), *Dalbergia, Albizia amara* ('*sericocephala*') savanna woodland (Harrison & Jackson); *Terminalia, Sclerocarya, Anogeissus, Prosopis* savanna woodland (Harrison & Jackson).

To the north of the belt of *Isoberlinia* woodlands described above there is a belt of somewhat drier country, extending from Senegal almost to the Red Sea, where the rainfall is sufficient for agriculture, but does not support a vegetation cover so dense that it is difficult to clear with primitive tools. This area is to a great extent free from tsetse fly and has for long been a highway of civilization across the continent.

The original vegetation was probably a floristically rich woodland in which *Isoberlinia* was either absent, or, in the wetter parts, was of localized occurrence and confined there to rocky hills.

Nearly everywhere, however, the land is heavily cultivated or has been in the past. Formerly the farmers allowed the vegetation to recover to restore fertility between short periods of cultivation, but now in the more densely populated regions the period of fallow has been progressively shortened, and, over extensive areas, especially surrounding the larger centres of population, cultivation is almost continuous and the fallow is taken under grass. Self-sown trees of economic importance have, however, been permitted to remain and the landscape has a park-like appearance. On uncultivated soils the woodland has been heavily degraded and locally replaced by secondary thicket and shrubland. Some areas, which were formerly cultivated, became depopulated during periods of warfare in the last century and now support secondary woodland. Relatively undisturbed undifferentiated Sudanian woodland only survives on rocky hills and in places where there is no available water.

There is little published information on undifferentiated Sudanian woodland. Two representative variants in Nigeria, and similar vegetation in the Sudan Republic and Ethiopia, are briefly described below.

Drier types in Nigeria

The most extensive vegetation in Zamfara Forest Reserve west of Katsina, where rainfall is approximately 700 mm per year, is *Anogeissus* woodland, which is kept rather open by frequent fires and domestic animals (Keay, 1949). The area was cultivated during the first half of last century but became uninhabited, probably between 1850 and 1870.

The principal trees associated with *Anogeissus* are: *Acacia seyal, A. senegal, Adansonia digitata, Albizia chevalieri, Annona senegalensis, Balanites aegyptiaca, Boswellia dalziellii, Butyrospermum paradoxum, Combretum glutinosum, Commiphora africana, Diospyros mespiliformis, Entada africana, Gardenia sokotensis, Hyphaene thebaica, Lannea microcarpa, L. schimperi, Lonchocarpus laxiflorus, Piliostigma reticulatum, Prosopis africana, Sclerocarya birrea, Strychnos spinosa, Tamarindus indica, Terminalia avicennioides, Ximenia americana* and *Ziziphus mucronata*. Few are more than 10 m high.

Wetter types in Nigeria

Woodland somewhat intermediate in floristic composition between *Isoberlinia* woodland and the drier types of undifferentiated Sudanian woodland described above is found to the north of the main belt of *Isoberlinia* woodland. Small stands of *Isoberlinia*, especially *I. tomentosa*, occur scattered throughout, especially on rocky hills. Relatively undisturbed woodland of this type occurs in Yankari Game Reserve in Bauchi State and has been described by Keay (1961) and White (MS, 1963). Similar vegetation occurs in Cameroun (Mildbraed, 1932).

The prevalent vegetation on the level sandy plateau in Yankari Game Reserve is a rather open woodland about 8 m high with the tallest species up to 12 m. The most conspicuous trees are: *Afzelia africana, Burkea africana, Anogeissus leiocarpus, Pteleopsis suberosa, Combretum glutinosum, C. nigricans, Pericopsis laxiflora, Lonchocarpus laxiflorus, Terminalia avicennioides, T. laxiflora, Lannea schimperi* and, in rocky places, *Detarium microcarpum*. Nearly all the above belong to two families, namely, Combretaceae and Leguminosae. The genus *Acacia*, which is well represented in the drier undifferentiated Sudanian woodland, plays a very subordinate role in the wetter types. Rocky hills of hard metamorphosed sandstone have a much more diversified flora.

Woodland and wooded grassland at the edge of grassy flood plains ('fadamas') of the rivers is of a somewhat different composition and is chiefly composed of: *Acacia sieberana, Adansonia digitata, Balanites aegyptiaca, Borassus aethiopum, Daniellia oliveri, Nauclea latifolia, Piliostigma thonningii, Pseudocedrela kotschyi, Tamarindus indica* and *Vitex doniana*.

Undifferentiated woodland in the Sudan Republic
(mapping unit 29a)

In Darfur Province *Anogeissus leiocarpus* occurs on the deeper Basement Complex soils, either in almost pure stands or accompanied by *Combretum glutinosum, Terminalia laxiflora, Sclerocarya birrea*, and *Dichrostachys cinerea*. Stony ridges are dominated by *Boswellia papyrifera* (Wickens, 1977a).

Ethiopian undifferentiated woodland
(mapping unit 29b)

In the extreme east of the Sudan Republic, against the frontier with Ethiopia, a narrow strip of dark cracking clays on sloping ground is dominated by *Anogeissus leiocarpus* and *Combretum hartmannianum* with sporadic *Sterculia setigera*. In western Ethiopia the woodland consists chiefly of *Anogeissus leiocarpus, Balanites aegyptiaca, Boswellia papyrifera, Combretum collinum, C. hartmannianum, Commiphora africana, Dalbergia melanoxylon, Erythrina abyssinica, Gardenia ternifolia (lutea), Lannea schimperi, Lonchocarpus laxiflorus, Piliostigma thonningii, Stereospermum kunthianum* and *Terminalia brownii*.

Sudanian grassland

There is little published information. Pure grassland is rare. In most grasslands, both edaphic and secondary, there is an admixture of woody plants.

Sudanian edaphic grassland and wooded grassland
(mapping units 62, 63, and 64)

Refs.: Harrison & Jackson (1958); Keay (1959a; 1960); Letouzey (1968a, p. 320–3); White (MS, 1963).
Phot.: Letouzey (1968a: 47).

The recent geomorphological history of the Sudanian Region has been very different from that of the Zambezian Region. Hence the nature and extent of their edaphic grasslands are also dissimilar. In the Sudanian Region grasslands on hydromorphic soils associated with the drainage lines are relatively restricted. By contrast, grassland and wooded grassland on vertisols formed from Pleistocene alluvium are very extensive. The latter occupies former shallow lake basins which often extend into the Sahel zone. Their vegetation is treated collectively here.

Valley and flood-plain grassland

Wide river valleys usually have a seasonally inundated flood plain known as 'fadama' in the Hausa language, which are covered with a dense grassland up to 3 m or more in height with widely spaced trees, notably *Terminalia macroptera, T. glaucescens, Mitragyna inermis*, and the fan palm *Borassus aethiopum*. The principal grasses are *Hyparrhenia cyanescens, Pennisetum unisetum* and *P. polystachion*.

Terminalia macroptera also occurs in lightly wooded grassland in seasonally flooded depressions with impeded drainage on the plateau at the source of tributary streams. *Hyperthelia dissoluta* (1.5 m high) is dominant and *Brachiaria jubata (fulva)* is frequent.

The drainage lines connecting the headwater valleys to the lower-lying flood plains are occupied by seasonally flooded grassland dominated by *Setaria sphacelata* or *Andropogon gayanus* var. *gayanus* with a few scattered trees of *Mitragyna inermis*.

Grassland and wooded grassland on Pleistocene clays

Refs.: Harrison & Jackson (1958); Letouzey (1968a, p. 320–3); Pias (1970); White (MS, 1963).
Phot.: Letouzey (1968a: 47).

Lake Chad and the Upper Nile basin are the two best-known of several extensive areas of dark, cracking Pleistocene clays in the Sudanian and Sahelian Regions. They are almost flat with gradients often no more than 1 in 5000. In both areas grassland occurs in mosaic with woody communities of varying density (scattered tree or bushed grassland, woodland and thicket) mostly dominated by species of *Acacia*.

In the Chad basin the grassland, known as 'yaéré', occurs where the flood is 1–2 m deep, and the flooding prolonged. *Echinochloa pyramidalis* is the most characteristic grass, followed by *Vetiveria nigritana*, *Oryza longistaminata* and *Hyparrhenia rufa*. During the dry season the vegetation dries up completely, even if it has escaped destruction by grazing or fire. In places where flooding is shallow and of short duration wooded grassland known as 'karal' or 'firki' is normally present. It is dominated by *Acacia seyal*, but in depressions the latter is replaced by *A. nilotica* subsp. *nilotica*. The field layer, which is 2–3 m high, consists of tall herbs and coarse grasses, principally *Caperonia palustris*, *Echinochloa colona*, *Hibiscus asper*, *Hygrophila auriculata*, *Sorghum arundinaceum* and *Thalia welwitschii*.

The Pleistocene clays of the Nile Valley extend from the southern fringe of the Sahara desert to regions of more than 1000 mm annual rainfall. In the driest semi-desert areas, in the Butana, where the rainfall is less than 400 mm, the cracking clays are virtually treeless and the Acanthaceous herb *Blepharis ciliaris* (*edulis*) locally forms pure stands. Elsewhere grasses are dominant, principally *Cymbopogon nervatus*, *Sorghum purpureo-sericeum* and *Schoenefeldia gracilis*.

Where rainfall is between 400 and 570 mm, *Acacia mellifera* often forms dense, impenetrable, almost pure thickets, which alternate with grassy areas of *Schoenefeldia gracilis*, *Cymbopogon nervatus*, *Sorghum purpureo-sericeum*, *Hyparrhenia anthistirioides* and *Sehima ischaemoides*. Grass and thicket probably have a cyclical relationship.

Acacia seyal replaces *A. mellifera* where the rainfall is more than 570 mm per year, and is said to have a similar cyclical relationship with grassland, which is dominated by *Sorghum purpureo-sericeum*, *Hyparrhenia anthistirioides*, and *Cymbopogon nervatus*.

On level sites, the dark, cracking clays cannot absorb much more than 700 mm of rainfall without flooding occurring. As the flooding increases the trees disappear, to be replaced by open grass plains of *Setaria incrassata*, and the conditions of the Flood Region of the Nile are reached. Within the Flood Region, however, areas of slightly higher ground, which are only flooded to a shallow depth, carry typical *Acacia seyal* communities and associated grassland under rainfall of up to 1000 mm per year. The transition zone between swamp grassland and the better-drained areas with *Acacia seyal* is sometimes dominated by *Hyphaene thebaica* and *Borassus aethiopum*, singly or together.

The Flood Region comprises a mosaic of various types of wooded grassland, grassland, and swamp communities. By far the largest constituent is seasonally flooded grassland dominated by *Hyparrhenia rufa* or *Setaria incrassata*.

Grassland and other herbaceous communities on shallow soil over ironstone

Typical dambos are very localized in the Sudanian Region, but similar seasonally waterlogged grassland, often not more than a few hectares in extent, occurs where drainage is impeded by the occurrence of impervious ironstone near the surface, especially on flat mesa-like residual hills, e.g. in *Isoberlini doka* woodland in Ghana (J. B. Hall pers. comm.; Lawson et al., 1968). The sparse grassland, which may be peaty in places, is often dominated by *Rhytachne rottboellioides*, associated with *Lycopodium affine*, and species of *Xyris*, *Utricularia* and *Drosera*. Outcrops of ironstone are frequent in other parts of West Africa where they are known locally as 'bowal' (pl. bowe). In Central Africa they are known as 'pengbele'. They support not pure grassland but a very open seasonal marsh vegetation.

Towards the margins of the Pleistocene basins, and in better-drained areas generally, the Acacias are mixed with broad-leaved trees (mapping unit 63) which locally form pure stands. The principal species include *Balanites aegyptiaca*, *Combretum glutinosum*, *Diospyros mespiliformis*, *Gardenia ternifolia*, *Mitragyna inermis*, *Nauclea latifolia*, *Piliostigma reticulatum*, *Pseudocedrela kotschyi* and *Terminalia macroptera*.

Sudanian secondary grassland

In the more densely populated parts of the *Isoberlinia* belt well-developed woodland is rarely seen. Around the larger towns and villages, beyond the permanent farmland, there is a zone of grassland with abundant regrowth shoots of *Isoberlinia* and other trees. This is alternately cultivated for short periods and then grazed. It never gets a chance to revert to woodland, because of shifting cultivation and grass fires. Beyond this zone there is open and very irregular woodland or wooded grassland which provides fuel, and poles for building, as well as land for sporadic shifting cultivation and grazing. When *Isoberlinia* dies out because of frequent cultivation it is often replaced by *Terminalia avicennioides*, *T. laxiflora* and *Butyrospermum paradoxum*.

Sudanian rupicolous scrub forest, bushland and thicket

Little information is available and the rupicolous flora is poor compared with that of the Zambezian Region. On inselbergs on the Jos Plateau the following species occur in bushland and scrub forest: *Carissa edulis, Dalbergia hostilis, Diospyros abyssinica, D. ferrea, Dodonaea viscosa, Euphorbia desmondii, E. kamerunica, E. poissonii, Ficus glumosa, Kleinia cliffordiana, Rhus longipes, R. natalensis, Ochna schweinfurthiana, Olea capensis, Opilia celtidifolia* and *Pachystela brevipes*.

On the sandstones of western Mali, *Guibourtia copallifera*, which is more characteristic as a dominant of dry evergreen forest (page 103) also forms a 5–6 m tall bushland or thicket on the less protected upper cliff slopes. Its woody associates include: *Bombax costatum, Boscia salicifolia, Combretum collinum, C. micranthum, Erythroxylum emarginatum, Euphorbia sudanica, Zanthoxylum xanthoxyloides, Ficus lecardii, Gardenia sokotensis, Gyrocarpus americanus, Hexalobus monopetalus, Spondias mombin* and *Zanha golungensis*.

IV The Somalia–Masai regional centre of endemism

Introduction

Geographical position and area

Geology and physiography

Climate

Flora

Mapping units

Vegetation
    Somalia–Masai *Acacia–Commiphora* deciduous bushland and thicket
    Somalia–Masai secondary grassland and wooded grassland
    East African evergreen and semi-evergreen bushland and thicket
    Somalia–Masai semi-desert grassland and shrubland
    Somalia–Masai edaphic grassland
    Somalia–Masai scrub forest
    Somalia–Masai riparian forest

Vegetation pattern in Marsabit District, Kenya
    Introduction
    Main vegetation types
      1. Barrenland
      2. Semi-desert annual grassland
      3. Semi-desert dwarf shrubland
      4. Stunted deciduous bushland
      5. Deciduous bushland
      6. Woodland
      7. Perennial grassland
      8. Evergreen and semi-evergreen bushland
      9. Afromontane evergreen forest, scrub forest, and related types
     10. Palm stand

Vegetation pattern in the greater Serengeti region
    Introduction
    Main vegetation types
      1. Edaphic grassland of the Serengeti Plains
      2. Secondary grassland of the Loita Plains
      3. *Acacia–Commiphora* deciduous bushland and thicket
      4. *Acacia–Commiphora* deciduous wooded grassland (and related types)
      5. *Combretum–Terminalia* secondary wooded grassland
      6. Evergreen and semi-evergreen bushland and thicket
      7. Evergreen forest
      8. Afromontane communities

Introduction

This chapter, in common with Chapter XVI, which deals with the Sahel transition zone, includes additional information on the detailed pattern of vegetation in relation to environmental factors, and also on the effects of the recent Great Drought on the natural and semi-natural vegetation.

The length and severity of the Great Drought were not uniform throughout the Somalia–Masai Region, and its effects on the natural vegetation and indigenous herbivores, and on the pastoral populations and their livestock, for various reasons have varied greatly from place to place. This is shown by reference to the following areas, two of which are occupied by pastoral man, and two by game animals:

1. In Somalia, where 60 per cent of the people are nomads or semi-nomads engaged in livestock-raising, the drought was severe and its effects were exacerbated because of political instability. A disastrous drought struck in 1974 after nearly three years of failing rainfall. Although relatively few people died of starvation, nearly 1 million required assistance. Livestock losses were estimated to be 80 per cent of cattle, 40 per cent of camels and 60 per cent of sheep and goats in the affected regions. For several decades previously the rangelands had been subjected to progressive deterioration owing to overpopulation and overgrazing. The sinking of boreholes and the extension of veterinary services, in particular, had served to widen the gap between the supply of forage and the demand for it. In northern Somalia population pressure was increased by an influx of thousands of nomads from the Ogaden, where they had been affected earlier by the Ethiopian famine. It is generally agreed that the vegetation of Somalia has suffered extreme degradation, but detailed botanical studies have not yet been made, and plans for controlled grazing have still to be implemented (Konczacki, 1978).

2. In Marsabit District in Kenya, where the drought was also severe, its influence on livestock was partly mitigated because of the availability of dry-season pasture on the upper slopes of certain mountains.

3. In the Greater Serengeti region, where rainfall is mostly higher than in other parts of the Somalia–Masai

Region, the drought was less severe than elsewhere, and the indigenous herbivore population continued to increase steadily throughout this period. It does not yet appear to have reached equilibrium.

4. This is in marked contrast to the Tsavo East National Park in Kenya where rainfall is lower and the game animals are more circumscribed in their movements. In Tsavo the vegetation was severely degraded followed by a crash in animal numbers which was particularly severe in 1973 (see pages 31 and 114).

The vegetation of Marsabit District and the Serengeti have recently been studied in considerable detail as part of long-term investigations of arid-zone ecology. In Marsabit District the objectives were to identify and describe the causes of ecological degradation and desert encroachment in the arid zone and to investigate and demonstrate suitable rehabilitation management procedures. Man and his stock have had relatively little influence on a large part of the Serengeti ecosystem and in recent years have been excluded completely. More than twenty years ago it was realized that the Serengeti could provide an incomparable opportunity to study an ecosystem still relatively unaffected by human influence. The vegetation of these two contrasting grazing ecosystems, Marsabit and Serengeti, is described in some detail at the end of this chapter.

Geographical position and area

The Somalia–Masai Region occupies a large part of the African mainland between 16° N. and 9° S. and 34° E. and 51° E., and also the island of Socotra. It includes eastern and southern Ethiopia (except the mountains), SE. Sudan, NE. Uganda (Karamoja), most of Kenya between the Highlands and the coastal belt, and the dry lowlands of north and central Tanzania south to the Great Ruaha River valley. It also extends across the Red Sea into southern Arabia, though its area there is uncertain. (Area, African mainland and Socotra only: 1 873 000 km².)

Geology and physiography

Nearly everywhere the land is below 900 m and in the north-east descends to sea-level. In places it rises considerably higher, especially where it abuts on islands of the Afromontane archipelago, though the altitude at which the transition to Afromontane vegetation takes place is variable. The underlying lithology is extremely diverse and includes extensive areas of marine sediments of Jurassic, Cretaceous and lower Tertiary age, and less extensive areas of Tertiary and Pleistocene lava flows. Quaternary continental deposits and Precambrian outcrops are more localized.

Climate

The climate is arid or semi-arid. Rainfall nearly everywhere is less than 500 mm per year and in places as low as 20 mm. Temperatures are high (mean monthly temperature mostly between 25° and 30° C).

In most places there are two rainy seasons separated by drought periods. This is related to the influence of the SW. monsoon in summer and the NE. monsoon in winter. For the most part, however, these monsoons do not bring rain. Instead, rainfall occurs during the intervening calm periods. Where there are not two well-defined peaks, rainfall is irregular. Throughout the Somalia–Masai Region there are great fluctuations in rainfall from year to year. (See Fig. 8.)

Flora

There are about 2 500 species, of which possibly half are endemic.

Endemic family. Dirachmaceae (one sp., *Dirachma socotrana*), confined to Socotra and Somalia.

Endemic genera. About 50. Those occurring on the African mainland, but sometimes also in Arabia or Socotra include: *Allmaniopsis* (1), *Arthrocarpum* (2), *Bottegoa* (1), *Calyptrotheca* (2), *Capitanya* (1), *Cephalopentandra* (1), *Chionothrix* (3), *Cladostigma* (2), *Cordeauxia* (1), *Dasysphaera* (2), *Dicraeopetalum* (1), *Drakebrockmania* (1), *Erythrochlamys* (5), *Gyroptera* (1), *Harmsia* (3), *Harpachne* (2), *Hildebrandtia* (9), *Kanahia* (4), *Kelleronia* (9), *Loewia* (3), *Myrmecosicyos* (1), *Neocentema* (2), *Pentanopsis* (1), *Platycelyphium* (1), *Pleuropterantha* (2), *Poskea* (2), *Psilonema* (1), *Puccionia* (1), *Sericocomopsis* (4), *Socotora* (1), *Spathionema* (1), *Volkensinia* (2), *Wissmannia* (1), *Xylocalyx* (4). The number of species is indicated in brackets. There are also four or five endemic genera of Asclepiadiaceae/Ceropegieae.

Several endemic genera, namely *Angkalanthus* (1), *Ballochia* (3), *Dendrosicyos* (1), *Haya* (1), *Lachnocapsa* (1), *Lochia* (1), *Mitolepis* (1), *Nirarathamnos* (1), *Placopoda* (1), *Socotranthus* (1) and *Trichocalyx* (2), are only known from Socotra.

Endemic species. The following non-endemic genera have important concentrations of endemic species (approximate numbers are given in brackets): *Acacia* (30), *Aloe, Boscia* (7), *Boswellia* (6), *Cadaba* (10), *Ceropegia, Commicarpus* (7), *Commiphora* (60), *Crotalaria* (30), *Euphorbia, Farsetia* (8), *Indigofera* (20), *Ipomoea* (20), *Jatropha* (6), *Maerua* (10), *Moringa* (9), *Neuracanthus* (8), *Otostegia* (5), *Psilotrichum* (7) and *Terminalia* (5). Of the 120 or so Stapelieae known from the Region, all but about eight are endemic.

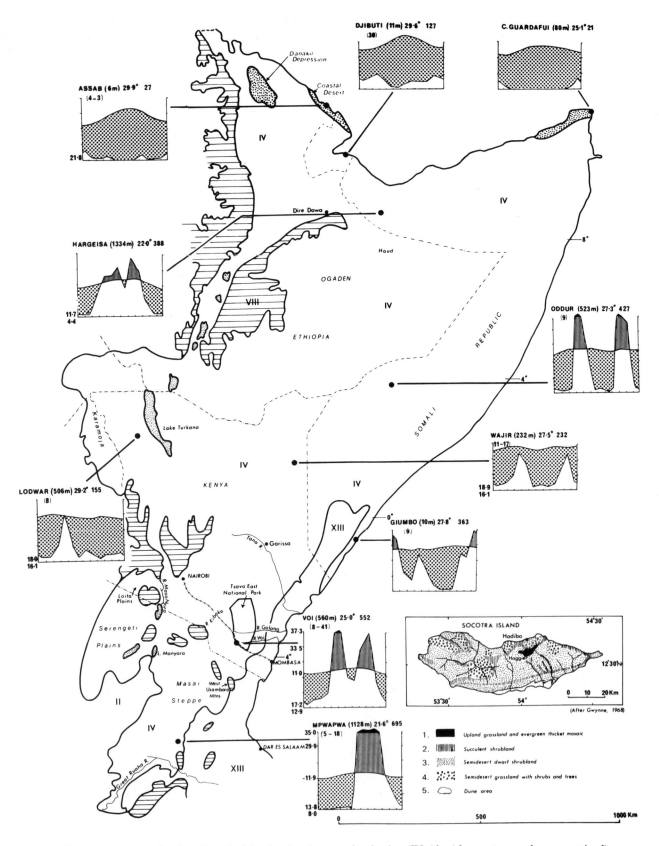

FIG. 8. Climate and topography of the Somalia–Masai regional centre of endemism (IV) (the Afromontane enclaves are striped)

Endemic species which extend to Arabia include *Adenia venenata, Socotora visciformis, Dorstenia foetida, Euphorbia phillipsiae, Kissenia capensis* and *Wissmannia carinensis*.

Linking elements. Above the species level there are interesting links with Madagascar and tropical America, though detailed comparisons have not yet been made. Thus, *Cadia* is otherwise only known from Madagascar, and some geophytic Somalia–Masai Euphorbias are more closely related to Malagasy than to African species. Similarly, the arid-disjunct genus *Kissenia* (see below) belongs to a family (Loasaceae) otherwise confined to tropical and subtropical America.

Most of the non-endemic species are also widespread in other dry parts of Africa, and several extend into Asia. *Acacia tortilis, Kohautia aspera, Stipagrostis hirtigluma* and *S. uniplumis* occur both north and south of the Somalia–Masai Region.

Species only extending to the north include *Aristida mutabilis* (Mauretania to India), *Cadaba glandulosa* (Mauretania to Arabia), *Combretum aculeatum* (westwards to Senegal), *Dobera glabra* (Sudan Republic to India), *Seddera latifolia* (Senegal to Pakistan), *Tamarix aphylla* (Morocco to India) and *T. nilotica* (Egypt to Israel and Arabia).

Species which are confined to the Somalia–Masai Region and the drier parts of South Africa, and hence show a pronounced disjunction, include *Asthenatherum glaucum, Tribulocarpus dimorphanthus*, and *Xerophyta humilis*. At the generic level this disjunction is shown by *Duvalia, Kissenia, Megalochlamys, Orbea, Sesamothamnus* and *Wellstedia*. A small antelope, the dikdik (*Madoqua kirki*), shows a similar distribution (Tinley, 1969). Further information on the floristic relationships between the Somalia–Masai Region and South Africa is given by Verdcourt (1969), de Winter (1966, 1971), J. P. Lebrun (1971b, 1975, 1977) and Monod (1971).

Mapping units

16b. Enclave of Zanzibar–Inhambane forest (see Chapter XIII).
26. Enclaves of drier Zambezian miombo woodland (see Chapter II).
38 (p.p.). East African evergreen and semi-evergreen bushland and thicket.
42. Somalia–Masai *Acacia–Commiphora* deciduous bushland and thicket.
45 (p.p.). Mosaic of East African evergreen bushland and secondary *Acacia* wooded grassland.
54b. Somalia–Masai semi-desert grassland and shrubland.
68b. Red Sea coastal desert.
71. Stony desert.
76 (p.p.). Halophytic vegetation (see Chapter XXII).

Vegetation

Most of the Region is covered with deciduous bushland and thicket which grade into and are replaced by semi-evergreen and evergreen bushland and thicket on the lower slopes of the mountains. There are smaller areas of scrub forest, riparian forest, secondary grassland and wooded grassland, seasonally waterlogged grassland, semi-desert grassland and shrubland, and desert. Virtually nothing has been published on some of these types. Upland evergreen bushland and lowland deciduous bushland have few species in common but the ecotone between them has not been studied in detail. Similarly, there is no published information on the transition from Zanzibar–Inhambane forest through scrub forest to the deciduous bushland of the interior.

Somalia–Masai
Acacia–Commiphora **deciduous bushland and thicket**
(mapping unit 42)

Refs.: Burtt (1942, p. 109–11); Greenway (1969, p. 172–6); Hemming (1966, p. 221–5); Pichi-Sermolli (1957, p. 39–49); White (MS, 1973, 1975–6).
Phots.: Burtt (1942: 17, 19, 25-9); Hemming (1966: 22); Pichi-Sermolli (1957: 4–5).
Syn.: deciduous *Acacia* and *Commiphora* thorn savanna (Burtt, 1942); haud-type mixed bush (Hemming, 1966).

Deciduous bushland and thicket is the climax over the greater part of the Somalia–Masai Region. Characteristically it is a dense bushland, 3–5 m tall with scattered emergent trees up to 9 m. Locally it is impenetrable and then forms thickets. The dominant Acacias and some of the Commiphoras are spinous and so impede progress even in the more open types except along game and cattle tracks. In higher-rainfall areas, especially on rocky hills, the emergent trees occur closer together and are a little taller, though scarcely ever more than 10 m. Greenway (1969) refers to this variant as woodland. Most species are deciduous. Evergreens, however, occur throughout, but they contribute no more than 2.5 to 10 per cent to the phytomass. Succulents also are usually present, but are much less abundant than in Malagasy deciduous thicket, and in most types of evergreen and semi-evergreen bushland. In many places grasses contribute little to the phytomass and are represented by a few annual and short-lived perennial species. When the grasses are inconspicuous it is misleading to use the word 'savanna', or to refer to the more open types as 'wooded grassland'. Even when the bush cover is less than 40 per cent the bushes remain physiognomically dominant and contribute most to the phytomass.

Although there is appreciable variation in floristic composition, species of *Acacia, Commiphora*, Capparidaceae, and *Grewia* are nearly always present. The bushland occurring between Garissa and Voi in southern Kenya which is described below may be regarded as typical.

Characteristic species in the main canopy are: *Acacia bussei, A. mellifera, A. nilotica* subsp. *subalata, A. reficiens* subsp. *misera, A. thomasii, Commiphora africana, C. boiviniana, C. campestris, C. erythraea, C. mollis* (*C. riparia*), *C. schimperi* (*C. trothae*), *Balanites orbicularis, Boscia coriacea* (evergreen), *Boswellia neglecta* (*hildebrandtii*), *Cadaba farinosa, C. heterotricha, Cassia abbreviata* subsp. *kassneri, Cordia ovalis, C. sinensis* (*gharaf, rothii*), *Dobera glabra* and *D. loranthifolia* (evergreen), *Euphorbia scheffleri, Givotia gosai, Hymenodictyon parvifolium, Lannea alata, L. triphylla, Sesamothamnus rivae, Platycelyphium voense, Premna hildebrandtii, Salvadora persica* (evergreen), *Sterculia africana, S. rhynchocarpa, S. stenocarpa, Terminalia orbicularis*, where drainage is impeded (J. P. M. Brenan, pers. comm.), *T. parvula* and *Thylachium thomasii*.

The majority of these are multiple-stemmed bushes or small bushy trees which are branched near the base. In some *Commiphora* species several massive more or less prostrate branches radiate from a common base. *Terminalia orbicularis* has a similar habit and forms impenetrable thickets up to 12 m across and 5 m tall.

Only a few species have well-defined trunks which carry the crown well above the main canopy. They include *Acacia tortilis*, the baobab (*Adansonia digitata*), *Delonix elata, Melia volkensii, Terminalia spinosa*, and the candelabra *Euphorbia, E. robecchii*. Such emergents are virtually absent from the driest areas; elsewhere they are mostly widely scattered. They only exceptionally attain a height of 9–10 m. Even the baobab, which in other parts of Africa is a huge tree, in this community is often only 8 m tall with a short but massive trunk 3–4 m long.

Smaller bushes and shrubs include: *Bauhinia taitensis, Bridelia taitensis, Caesalpinia trothae, Caucanthus albidus, Combretum aculeatum, Carphalea* (*Dirichletia*) *glaucescens, Ecbolium amplexicaule, E. revolutum, Ehretia teitensis, Erythrochlamys spectabilis, Grewia fallax, G. tembensis, G. tenax, G. villosa, Maerua denhardtiorum, M. subcordata, Premna resinosa, Sericocomopsis hildebrandtii* and *S. pallida*.

Succulents occur more or less scattered throughout but are rarely plentiful. The most conspicuous is the large candelabra-like *Euphorbia robecchii*, which is widely scattered but only plentiful in higher-rainfall areas in the transition towards semi-evergreen bushland. Other candelabra Euphorbias (*E. nyikae, E. quinquecostata*) are much more restricted. The metre-high cactiform *E. grandicornis* locally forms low thickets. Other stem-succulents are *Adenium obesum, Calyptrotheca somalensis, C. taitensis*, and *Monadenium invenustum*. A few climbers, namely *Cissus quadrangularis, Sarcostemma viminale* and related species, and *Vanilla roscheri*, have succulent photosynthetic stems; *Cissus rotundifolia* has succulent leaves. Two climbers, *Adenia globosa* and *Pyrenacantha malvifolia*, have enormous half-submerged water-storing tubers up to 1 m or more in diameter and almost as high. There are half a dozen species of *Sansevieria*. The 1–2 m high *S. arborescens*, in places, is a conspicuous feature of the landscape. Succulent stapeliads are represented by a few species of *Caralluma* and *Echidnopsis* and one of *Edithcolea*. There are four acaulescent or short-stemmed species of *Aloe* but arborescent species are lacking. Species of *Kalanchoe* represent the Crassulaceae.

Climbers, additional to those mentioned above, are *Gerrardanthus lobatus, Ipomoea* spp., *Kedrostis gijef, Pergularia daemia* and *Thunbergia guerkeana*.

Grasses are inconspicuous and are represented by a few ephemeral species such as *Aristida adscensionis, A. barbicollis, Brachiaria eruciformis*, and *B. leersioides*, and short-lived perennials which include *Cenchrus ciliaris, Chloris roxburghiana* and *Schmidtia pappophoroides*.

Deciduous bushland similar to the above continues northwards through the Haud region of Somalia and the Ogaden to the coastal plain. There is appreciable floristic overlap but there are more species of *Commiphora* in the north, though collectively they are no more abundant in the vegetation. There are also extensive areas of *Acacia bussei* scrub woodland in northern Somalia.

In those parts of Kenya and Somalia where rainfall is somewhat less than 250 mm per year, the vegetation is intermediate between bushland and shrubland, and consists of low 2–3 m high bushes and stunted trees, principally of *Acacia reficiens* subsp. *misera*, which form a thin cover over a ground layer consisting largely of small shrubs. Ephemeral grasses appear after rain but perennial grasses are insignifiant.

Somalia–Masai secondary grassland and wooded grassland
(mapping units 42 and 45)

In many places bushland has been destroyed by man and his domestic animals or by elephants and other large mammals.

In Tsavo East National Park in Kenya deciduous bushland has been extensively degraded and in places destroyed, chiefly by elephants. The shallow-rooted *Commiphora*, which is easily pushed over, is the first to go, but eventually nearly every woody species succumbs. The tall-growing, sturdy *Melia volkensii* persists longer than most but its foliage shows a well-defined browse line, produced by giraffe. In places its truncated crowns are the most striking feature of the landscape. The baobab also sometimes shows a browse-line but more frequently it is ring-barked and is soon demolished. Some species, like *Platycelyphium voense* and *Ehretia teitensis*, which are not eaten if other food is available, dominate the intermediate stages of decline. Ultimately, however, all woody plants are eliminated except for scattered browsed bushes of *Boscia coriacea* and a few other species. The landscape is now a sparse secondary grassland littered with the bleached skeletons of the former dominants. The monotony is broken only by the large rock-like tubers of *Pyrenacantha malvifolia* which break the surface at intervals of 5 to 20 m. There is some evidence (Chapter 4) which suggests that the conversion of deciduous bushland to secondary grassland may be part of a cyclical succession. The heavy rainfall of the late 1970s has been accompanied by abundant regeneration of *Commiphora* (M. J. Coe, pers. comm.).

In parts of Kenya where domestic animals are numerous, evergreen thicket has been severely degraded and invaded by species of *Acacia* (mapping unit 45). It is not uncommon to find evergreen species, such as *Carissa edulis, Dodonaea viscosa, Euclea divinorum, E. racemosa* subsp. *schimperi*, and *Tarchonanthus camphoratus*, occurring with *Acacia drepanolobium, A. hockii, A. kirkii* and *A. seyal*. Similar communities occur in the Lake Victoria basin (page 182).

Charcoal-burners have also been responsible for the conversion of bushland to grassland over hundreds of square kilometres (B. Verdcourt, pers. comm.).

East African evergreen and semi-evergreen bushland and thicket
(mapping units 38 and 45)

Refs.: Hemming (1966, p. 216–18); Pichi-Sermolli (1957, p. 53–61); Popov (1957); White (MS, 1973, 1975–76).

Phots.: Hemming (1966: 18); Kassas (1956b: 5, 7); Pichi-Sermolli (1957: 7); Popov (1957: 12, 16); Wettstein (1906: 25–30).

This type occurs on the drier slopes of mountains and upland areas in East Africa from central Tanzania to Eritrea and beyond. It often forms an ecotone between montane forest, especially *Juniperus* forest, and deciduous *Acacia–Commiphora* bushland and thicket. It also occurs on Socotra. It reaches its greatest development on the steep slopes of the Ethiopian highlands, but it appears that not all of mapping unit 38 belongs to this type, especially in the south-west (Ib Friis, pers. comm.). Because of this uncertainty and for cartographic reasons mapping unit 38 in Ethiopia has been included within the boundary of the Afromontane Region on the map. Its transitional nature must, however, be borne in mind.

It varies greatly in composition and richness, but certain genera and species are nearly always present, such as *Carissa edulis, Dodonaea viscosa, Olea africana, Tarchonanthus camphoratus*, species of *Acokanthera, Euclea, Sansevieria* and *Teclea*, and succulent species of *Aloe* and *Euphorbia*.

On the Kedong escarpment near Nairobi, dense, relatively undisturbed bushland occurs between 1875 and 2080 m and is locally almost impenetrable. The height of the canopy varies from 3 to 7 m and consists mostly of the crowns of *Olea africana, Gnidia subcordata, Teclea simplicifolia, Euclea divinorum, Acokanthera schimperi*, and especially in disturbed areas, *Tarchonanthus camphoratus*. Other large bushes include *Canthium keniense, Croton dichogamus, Dodonaea viscosa, Dombeya burgessiae, Grewia similis, G. tembensis, Maytenus heterophylla* and *Rhus natalensis*. *Olea* is particularly abundant but occurs as a bush rather than as a tree. The cactoid stem-succulent *Euphorbia candelabrum* occurs throughout as a scattered emergent up to 9 m tall and with a mean spacing of about 20 m. The sparsely branched rosette tree, *Dracaena ellenbeckiana*, is also prominent, especially on the more open rocky slopes where it forms colonies about 6 m tall, but elsewhere in Kenya it is very local. Shrubs are few in species but *Aspilia mossambicensis, Psiadia arabica (punctulata), Tinnaea aethiopica* and *Turraea mombassana* are abundant. The few climbers include *Capparis fascicularis (elaeagnoides), Pterolobium stellatum, Senecio petitianus* and *Scutia myrtina*. Non-arborescent succulents are the climber *Sarcostemma viminale*, and species of *Sansevieria, Kalanchoe*, and *Crassula*, which are conspicuous in the ground layer. The 2 m tall *Aloe kedongensis* occurs throughout in rocky places. Apart from the succulents, the ground layer is sparse, but includes shade-tolerant grasses such as *Ehrharta erecta*. A few epiphytic orchids occur, especially on *Acokanthera*. With increasing altitude scattered stunted individuals of the trees *Schrebera alata, Cassine (Elaeodendron) buchananii, Calodendrum capense, Cussonia holstii, Drypetes gerrardii* and *Juniperus procera* appear, and herald the transition to *Juniperus* forest from which *Dracaena, Euphorbia candelabrum* and the other succulents are absent.

In Somalia the principal dominants of evergreen bushland are *Acokanthera schimperi, Buxus hildebrandtii, Cadia purpurea* and *Dodonaea viscosa*. Locally *Buxus* forms scrub forest 9 m tall. Other important species are *Aloe eminens* (15 m), *Barbeya oleoides, Cussonia holstii, Dracaena schizantha, Euclea racemosa* subsp. *schimperi, Euphorbia grandis, Pistacia lentiscus, Rhus somalensis* and *Sideroxylon (Monotheca) buxifolium*, several of which also occur in *Juniperus* forest.

On Socotra evergreen bushland and thicket occurs on the granite massif, the Hagghier, above 750 m, but is absent from the steeper peaks. The commonest plants in order of abundance are *Cephalocroton socotranus, Carissa edulis, Buxus hildebrandtii, Dodonaea viscosa, Ficus socotrana, Indigofera sokotrana, Ruellia insignis, Boswellia ameero* and *Euphorbia socotrana*.

One of the most singular communities on Socotra is a semi-evergreen bushland dominated by *Dracaena cinnabari*, which occurs on limestone slopes. *D. cinnabari* has a short stout trunk and a very dense umbrella-shaped crown. Its principal associates are *Boswellia ameero, B. elongata, B. socotrana, Aloe perryi, Adenium socotranum* and *Mitolepis intricata*.

Somalia–Masai semi-desert grassland and shrubland
(mapping unit 54b)

Where rainfall is between 100 and 200 mm per year, semi-desert grassland dominated by *Eragrostis hararensis, Panicum turgidum* or *Asthenatherum glaucum* occurs on deep sand. Shrubland occurs on stony soils.

On the coastal plain of Somalia the principal shrubby species are *Aerva javanica, Jatropha pelargoniifolia (glandulosa)* and *Farsetia longisiliqua* (Gillett, 1941).

Further inland, dwarf shrublands on gypseous soils are composed of *Aloe breviscapa, A. rigens, A. scobinifolia, Euphorbia cuneata, E. multiclava, Ipomoea sultani, Kelleronia quadricornuta, Lasiocorys argyrophylla, Lycium europaeum, Ochradenus baccatus* and *Zygophyllum hildebrandtii*. The shrubby species are most abundant in overgrazed and eroded areas and it is possible that grasses including *Chrysopogon plumulosus* and *Dactyloctenium robecchii* were formerly dominant (Hemming, 1966). Gypseous soils support a considerable variety of endemic succulents including several Euphorbias such as *E. columnaris, E. sepulta,* and *E. mosaica,* and *Dorstenia gypsophila* and *Pelargonium cristophoranum*. Stapeliads, however, appear to occur on limestone rather than gypsum (M. G. Gilbert, in litt. 15 March 1979).

Near Lake Turkana in Kenya, on sandy alluvial plains *Indigofera spinosa* forms dwarf shrublands usually associated with the perennial grass *Sporobolus spicatus* and the annual grass *Aristida mutabilis*. On rocky plateau surfaces *Helichrysum glumaceum* is dominant. In some rocky places of limited extent most of the plants are small succulents, especially species of *Aloe, Euphorbia, Sansevieria, Caralluma, Kleinia* and *Sarcostemma* (Hemming, 1972). Communities of dwarf succulents are also important in similar situations elsewhere, e.g. near Dire Dawa, where eleven species of *Caralluma* and *Echidnopsis* occur in a small area with *C. penicillata* and *C. edithae* forming much of the phytomass (M. G. Gilbert, in litt. 15 March 1979).

The semi-desert communities of Socotra have been described by Popov (1957).

Somalia–Masai edaphic grassland
(mapping units 42, 45, and 59)

Refs.: Anderson & Talbot (1965); Burtt (1942, p. 87, 94–7); Hemming (1966, p. 208–9, 215–16, 223–5).
Phots.: Burtt (1942: 17–21); Gillman (1949: 24, 30); Hemming (1966: 13); Pratt *et al.* (1966: 2).
Profile: Anderson & Talbot (1965: 2).

Seasonally waterlogged grassland has a very uneven distribution in the Somalia–Masai Region. It covers large areas in Tanzania but is less well developed further north. There is little information for Kenya. The treelessness of the Serengeti Plains is due, at least in part, to unfavourable edaphic conditions, but the seasonal waterlogging of the soil is caused by parent material rather than physiographic position.

In Somalia seasonally waterlogged grassland is of limited occurrence. Treeless plains or 'bans' dominated by *Chrysopogon plumulosus* occur in *Acacia bussei* scrub woodland and in *Acacia–Commiphora* bushland. Small temporary pools are dominated by *Andropogon kelleri*, and larger, ill-defined areas of inland drainage are covered with *Chloris roxburghiana* and *Cynodon dactylon* with scattered *Acacia tortilis*. Clay plains in the zone of *Acacia etbaica* scrub woodland are dominated mainly by *Andropogon, Chrysopogon plumulosus, Panicum coloratum, Cenchrus ciliaris, Aristida adscencionis* and *Eragrostis* sp.

In Central Tanzania water-receiving depressions are extensively developed because run-off is insufficient to carve distinct stream beds and merely collects in hollows from which it quickly evaporates in the dry season. The principal grasses covering the black, cracking clays of these 'mbugas' are *Setaria incrassata (holstii)* and *Themeda triandra*. Characteristically, mbugas are treeless, but they are usually separated from the bushland and thicket by an ecotone of wooded grassland dominated by gall Acacias, especially *A. drepanolobium, A. seyal, A. malacocephala* and *A. pseudofistula*.

Seasonally waterlogged grassland also occurs on non-cracking calcimorphic 'hard pan' soils inside the *Acacia–Commiphora* bushland, where it forms ill-defined glades which are often not very obviously associated with the drainage lines. The dominants are dwarf grasses, principally species of *Sporobolus*, and *Microchloa indica*, among which are scattered several conspicuous herbs. In the wettest parts *Blepharis 'acanthoides'* and the sedge, *Kyllinga alba*, are abundant.

The glades are devoid of trees or have scattered individuals of *Acacia drepanolobium, A. mellifera, A. tanganyikensis, A. tortilis, Albizia amara, A. harveyi, Commiphora schimperi, Dalbergia melanoxylon, Lannea humilis, Sclerocarya birrea* and *Terminalia stuhlmannii*.

In the ecotone between forest and grassland in the Nairobi National Park a combination of grazing, browsing and fire can create grassland in areas capable of supporting evergreen bushland or forest. In dry periods animals from the plains, including zebra and wildebeest, congregate near the forest where there is more grass. In the plains they move away and the grass is often enough to support fires. Browsing animals, especially giraffe and impala, are there all the time keeping the bushes down, and in the past elephants broke down trees (J. B. Gillett, in litt. 8 March 1979).

Somalia–Masai scrub forest
(mapping unit 42 p.p.)

Refs.: Greenway (1973, p. 56–7); Greenway & Vesey-FitzGerald (1969, p. 133–4); White (MS, 1975–76).
Phots.: Gillman (1949: 8); Pratt *et al.* (1966: 9).
Syn.: *Euphorbia* bushland and thicket (Greenway, 1973); woodland thicket (Pratt *et al.*, 1966).

In a few places in East Africa, at relatively low altitudes where the rainfall is higher than that of deciduous bushland and thicket but too low to support true forest, a 7–10 m tall scrub forest dominated by species of *Commiphora* and candelabra Euphorbias is found.

On the escarpment above Lake Manyara (945 m) in Tanzania the characteristic trees are *Commiphora baluensis, C. campestris, C. engleri, C. merkeri* and *Sterculia stenocarpa*. They are rather widely and irregularly spaced but the 3–5 m tall understorey is quite dense. Apart from succulent species, this community is almost completely deciduous. Large baobabs (*Adansonia digitata*) occur throughout and candelabra Euphorbias are locally abundant. Succulents in the understorey include thickets of *Sansevieria ehrenbergii*, which is very abundant, groups of tall *Aloe ballyi* and tangles of *Cissus quadrangularis* and *Sarcostemma viminale*.

A similar community occurs between 700 and 960 m on the steep northern slopes of the Western Usambara Mts. In addition to *Commiphora* and *Euphorbia* the following species occur in the canopy: *Acacia tortilis, Afzelia quanzensis, Brachylaena huillensis (hutchinsii), Cussonia zimmermannii, Manilkara sulcata, Newtonia hildebrandtii, Pappea capensis* and *Scorodophloeos fischeri*.

Somalia–Masai riparian forest

Refs.: Bogdan (1958); Burtt (1942, p. 118–25); Greenway (1969, p. 171–2); White (MS, 1975–76).
Phot.: Burtt (1942: 52).

Riparian forest occurs only on the banks of the larger rivers such as the Tana and Galana. *Acacia elatior* and the interesting endemic *Populus ilicifolia* are very important on the Tana R. and the latter also occurs on the Uaso Nyiro and Galana. *Garcinia livingstonei* is also important on the Tana.

On the banks of the Voi River in Tsavo East National Park the principal trees in the 18 m tall forest are *Acacia robusta* subsp. *usambarensis, Albizia glaberrima, A. zimmermannii, Dobera glabra, Ficus ingens, F. sycomorus, Kigelia africana, Lecaniodiscus fraxinifolius, Newtonia hildebrandtii, Tamarindus indica* and *Terminalia sambesiaca*. Forest fringing the Kiboko River 100 km further inland is much poorer floristically and consists principally of *Acacia robusta* subsp. *usambarensis* and *Newtonia hildebrandtii*.

The riparian forests of Tanzania include many species which are widespread in Africa, such as *Albizia glaberrima, Diospyros mespiliformis, Ficus sycomorus, Khaya nyasica, Kigelia africana, Parkia filicoidea, Tamarindus indica* and *Trichilia emetica*. As the coast is approached endemic species of East African coastal forest such as *Fernandoa magnifica* become frequent.

Vegetation pattern in Marsabit District, Kenya

Introduction

Refs.: Edwards et al. (1979); FAO (1971); Herlocker (1979a, 1979b); Lamprey (1978); Lewis (1977); Sobania (1979); Synnott (1979a, 1979b); Unesco (1977); White (MS, 1979).

This section deals with the vegetation of the study area (see Fig. 9) of the Integrated Project on Arid Lands (IPAL) in the Mt Kulal area of northern Kenya.[1] The objectives of the project are basically to identify and describe the causes of ecological degradation and desert encroachment in the arid zone and to investigate and demonstrate suitable rehabilitation management procedures. The Mt Kulal study area of 22 500 km^2, which is situated in Marsabit District, northern Kenya, lies entirely in the Somalia–Masai floristic region, except for small enclaves of Afromontane vegetation on the highest peaks, mostly above 2 000 m.

Although the working area was selected as far as possible as a self-contained unit containing a large proportion of the total ranges of the two principal nomadic tribes, there is an almost continuous exchange of people, livestock and wildlife across the borders of the area as well as continuous nomadic and transhumance movement within it. Hence, the IPAL study area is not a self-contained complete grazing ecosystem. A resident population of cattle stays within the area throughout the year, but its numbers are small, probably fewer than 15 000. By contrast, it is estimated that 200 000 domestic animals use the area at least occasionally. These animals form part of a larger ecosystem, which also includes the montane forests, both inside and adjacent to the study area.

Two major geological formations occur within the study area: Pleistocene basaltic lavas and Quaternary sediments. The lavas occupy 57 per cent of the surface and are primarily centred on Mt Kulal, Marsabit Mountain, and the Hurri Hills.

Quaternary sediments of various kinds occupy 39 per cent of the area. The most extensive are those derived from Precambrian gneissic rocks of Nyiru, Ol Doinyo Mara, and the Ndoto Mountains. The other sediments, which are principally associated with the old bed of Lake Chalbi, include saline and alkaline alluvium, which constitutes the Chalbi Desert, and stabilized sand dunes at the edges of the desert. A third geological formation, Precambrian gneisses, although restricted in extent (3 per cent), is extremely important because it was the source of most of the Quaternary sediments within the study area.

The study area is made up of a large central plain mostly between 530 and 760 m elevation. Around this plain are situated three volcanic hill masses; the Hurri Hills (1 310 m) to the north; Mt Marsabit (1 836 m) to the east; and Mt Kulal (2 295 m) to the west. The mountains to the south-west, Mt Nyiru (2 752 m), Ol Doinyo Mara (2 067 m) and the Ndotos (2 637 m), are mostly formed from the Basement Complex. The Chalbi Desert, a former lake bed, forms a large depression between 435 and 500 m elevation in the north of the study area, whereas Lake Turkana lies at 410 m in the extreme west. Most of the study area lies within an interior drainage system leading to the Chalbi Desert, but it is probable that much of the surface run-off is lost by evaporation or transpiration and never reaches the Chalbi.

1. This project forms part of the Kenyan contribution to Unesco's Man and the Biosphere (MAB) Programme. From 1976 to mid-1980, the project was a co-operative activity between Unesco and UNEP. From July 1980, the project is being funded for a three-year period by the Federal Republic of Germany, through funds-in-trust arrangements with Unesco.

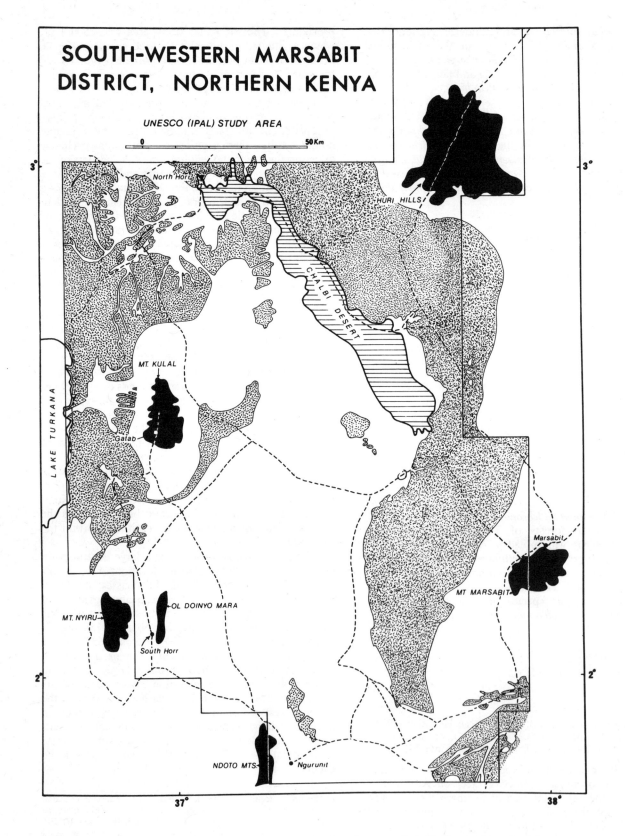

FIG. 9. Map of South-Western Marsabit District, Kenya, showing places mentioned in the text (the stippled area represents annual grassland)

The study area is situated in the most arid climatic region in East Africa, embracing large parts of northern Kenya, Somalia and eastern Ethiopia. Rainfall is low, particularly at lower elevations. For instance, North Horr, the only station below 1 333 m with long-term rainfall records, has a mean annual rainfall of 150 mm. Potential evaporation is high, exceeding 2 600 mm per year over most of the area. With increasing altitude rainfall increases in amount and duration and on the higher mountains approaches or slightly exceeds 1 000 mm per year. In general, the main rainfall occurs in two seasons, from March to May and October to December, but there is great variation from year to year. Rainfall in lowland areas has a coefficient of variation of more than 50 per cent. In addition to fluctuations in rainfall from year to year, there appear to be longer cycles. During the drought years between 1968 and 1976, mean annual rainfall in the lowlands was less than 50 mm, whereas in 1977 over the same area it was approximately 400 mm.

The study area, in common with most of the arid lowlands of the Somalia–Masai Region, is rangeland occupied by pastoralists who are nomadic to varying degrees. They are similar in many respects to pastoralists of other tribes living in the Sahel zone (*sensu stricto*), who depend for their subsistence almost entirely upon their herds. However, unlike many areas in the Sahel, there is very little arid-zone agriculture in the Somalia–Masai Region. The principal ethnic groups in the study area are the Gabra, Rendille and Samburu. The latter, who are basically cattle keepers, live mostly in the mountains and do not make extensive use of lowland areas. The Gabra and Rendille are basically lowland camel and small stock keepers, though some families also have cattle and in times of drought may seek pasture and browse in the mountains. Culturally camels are the more important to these people, but sheep and goats contribute more to their diet and make more impact on the environment.

The main species of indigenous grazing herbivores in the study area are Beisa oryx and Grevy's zebra. Other wild herbivores include dik dik, generuk, Grant's gazelle, Reticulated giraffe, rhinoceros and elephant. Lion, leopard and cheetah also occur. Unlike cattle, oryx avoid the highland areas and are widely distributed throughout the lowlands. Even in the dry season they are found far from water. Lewis (1977) believes that the oryx could be domesticated in this area, as it has been on Galana Ranch (Field, 1975), and that it would make better use of the available vegetation and water than cattle. Synnott (1979) suggests that large mammals, especially elephant, buffalo, and kudu, might have been important in the past in maintaining grassy glades in the forests, though, at present, their numbers are so reduced that any such influence must be negligible.

In the study area, as in most other arid and semi-arid parts of Africa, expanding human populations in recent decades have inevitably been accompanied by increased livestock numbers and concomitant degradation of the habitat. This has been most accentuated around the few springs and wells and especially around boreholes. In very favourable years there may be a temporary abundance of animal fodder, resulting in an increase in the size of animal herds and subsequent overstocking of the grazing lands during the drier periods which follow. The drought years from 1968 to 1976 clearly showed that the long-term carrying capacity of the rangeland is being exceeded. Although there was no massive human starvation as in earlier droughts, large numbers of livestock died and the upland forest areas were subjected to intense grazing and browsing not only by the local Samburu, but also by other pastoralists who travelled great distances to make use of the few remaining patches of herbage.

It was with the object of finding direct solutions to the most urgent environmental problems associated with desert encroachment and ecological degradation of arid lands that the Integrated Project on Arid Lands (IPAL) was launched in the Mt Kulal region. The far-reaching objectives of the research programme are summarized in general by Lamprey (1978), and for forestry in particular by Synnott (1979a, 1979b) and Herlocker (1979b).

Those parts of the work programme which are particularly concerned with vegetation include:
— Measurements of plant productivity using fenced enclosures over a wide range of ecological conditions, and involving assessment of overall biomass and productivity, and also of those species which form the bulk of the diets of the various livestock species.
— The impact of livestock, especially camels, sheep, and goats on subdesert herbaceous and dwarf-shrub vegetation, using exclosure plots and measuring consumption under controlled conditions, with the objective of calculating optimum stocking densities.
— The collection of data, using aerial surveys, on the numbers and distribution of livestock and wild ungulates in relation to seasonal variations in water availability and vegetation condition.
— A study of the food habits of camels, involving chemical analysis of six of the most important plant species in their diet, namely *Boscia coriacea, Duosperma eremophilum, Indigofera spinosa, Leptothrium senegalense, Maerua crassifolia* and *Salvadora persica*.
— An ecological study of the montane forests, including preliminary ecological surveys and an account of botanical composition, and with particular emphasis on successional trends and the interactions of the forests with the human populations.
— A study of the feasibility of arid-zone forestry using drought-adapted tree species, both indigenous and exotic, to provide fodder, firewood, building materials, and other commodities.

The results of studies such as those outlined above can only be effectively applied if the vegetation of the study area has been satisfactorily classified, described, and mapped. The preparation of a vegetation map was regarded as one of the most important initial steps in the research programme and was undertaken by Herlocker

(1979a). His accompanying account of the vegetation is briefly summarized below, with supplementary information on the montane communities from Synnott (1979) and my own observations.

Main vegetation types

1. Barrenland

This largely coincides with the Chalbi Desert and covers 4.1 per cent of the area. The Chalbi Desert is edaphic desert and is part of a closed drainage basin. In addition there are numerous springs at the edge of the desert which apparently originate from subsurface flow from the surrounding mountains (Kulal, Marsabit, and the Hurri Hills). The accumulation of salts after evaporation of seasonal floodwaters inhibits plant growth except very locally, as when annual grasses and herbs, such as *Drakebrockmania somalensis*, a halophytic grass, grow near outlets of the major tributary streams following discharge after seasonal rainfall.

2. Semi-desert annual grassland

This is the most extensive vegetation type and covers 33.3 per cent of the total area. Together with barrenland, it occupies the driest parts and occurs almost entirely below 1 000 m, even on the west and north-west aspects of the major mountains where dry conditions occur at the highest elevations. It is principally found on shallow, poorly developed, stony, loam to clay-loam soils, often on water-shedding sites overlying lava. The grasses *Aristida adscensionis* and *A. mutabilis* dominate, but during drought periods they may be absent for years on end, and the 'pure' grasslands temporarily become desert. Woody plants, however, are nearly always present and provide a 2–20 per cent cover, sometimes in the form of shrubs, e.g. *Duosperma eremophilum*, sometimes as bushes or bushy trees, e.g. *Acacia reficiens, A. seyal, A. tortilis, A. horrida, A. senegal* and *Commiphora* spp.

3. Semi-desert dwarf shrubland

This is the second-most-extensive vegetation type, occupying 27.6 per cent of the area. It is dominated by shrubs less than 1 m high, notably *Duosperma eremophilum* and *Indigofera spinosa*, but the following are also sometimes present or locally dominant: *Euphorbia schimperi, Kleinia kleinioides, Plectranthus ignarius, Sericocomopsis hildebrandtii, Suaeda monoica, Lagenantha nogalensis* and *Dasysphaera prostrata*. Annual grasses, principally *Aristida adscensionis* and *A. mutabilis*, and to a lesser extent forbs, such as *Blepharis linariifolia*, make up the herbaceous layer. Extensive areas are without large woody plants, but bushes and small trees frequently have a scattered occurrence with a 2–20 per cent cover. They include: *Acacia reficiens, Commiphora* spp., *Acacia mellifera, A. tortilis, A. senegal, A. seyal, Boswellia neglecta* and *Acacia etbaica*.

Dwarf shrubland is most extensively developed on the drier western and north-western slopes of the larger mountains. *Duosperma eremophilum* and *Indigofera spinosa* dominate or co-dominate 71 per cent and 63.6 per cent of all dwarf shrubland respectively. They also occur in the understorey of bushland and woodland, and in terms of area covered are the most important woody plants within the study area.

Indigofera spinosa occupies the drier sites, whereas the more moisture-demanding *Duosperma eremophilum* is found on somewhat heavier, wetter soils. Thus, *Indigofera* dominates old, stabilized sand dunes, whereas *Duosperma* dominates sedimentary plains soils, and also occurs at higher elevations on the mountains, and is the major dwarf shrub in the understorey of bushland and woodland. When *Indigofera* and *Duosperma* occur together, they often show a catenary relationship with *Indigofera* dominant on the compact soils of low broad ridge tops, and *Duosperma* dominant in alternating shallow depressions.

Among the less important shrubs, *Lagenantha nogalensis*, a gypsum-tolerant succulent species, similar in appearance to, but smaller than, *Suaeda monoica*, forms almost pure stands, though with only a 20 per cent cover, on white calcareous soils of the old Chalbi lake bed. *Dasysphaera prostrata* occurs on saline or alkaline soils at the margins of Lake Turkana and at the edges of the Chalbi Desert.

4. Stunted deciduous bushland

This type is intermediate between bushland and shrubland (see Chapter 7) and is referred to by Herlocker as shrubland. It covers 20.2 per cent of the area. The dominant deciduous bushes are mostly less than 4 m tall and usually have multiple, gnarled stems. Trees are absent. The dominants are *Acacia reficiens* subsp. *misera, A. mellifera* and several species of *Commiphora*. The grasses in the herb layer are mostly annual, though *Stipagrostis uniplumis* is a short-lived perennial. About half of this type has an understorey of dwarf shrubs, especially *Duosperma eromophilum*, though succulent species are also important.

Stunted bushland occurs on loam to clay-loam, lava-derived soils at higher elevations and on gneiss-derived sandy-silty soils at lower elevations. The latter apparently have a greater soil-moisture potential. It also extends into drier regions along drainage lines.

Acacia reficiens is the most extensive dominant of stunted bushland and is the third-most-abundant woody species in the area. *A. mellifera* is much more restricted but is abundant on the lower southern slopes of Marsabit Mt. Nearly everywhere it occupies wetter sites than *A. reficiens*. Where the two species occur together, *A. mellifera* suffered higher mortality during the recent eight-year drought. The *Commiphora* species characteristically occur on shallow and extremely well-drained soils.

Some other non-dominant bushy species are either of potential economic importance or indicate environmental degradation. Thus, *Acacia senegal* var. *kerensis* may prove to be a valuable source of gum arabic. *Acacia nubica* is normally widely scattered but becomes abundant in degraded areas. *Calotropis procera* is a widespread pioneer on newly deposited alluvium, but since it is inedible and unusable, it also becomes abundant in degraded areas near settlements. *Balanites* (probably *orbicularis*) is not a pioneer, but it may be an indicator of environmental degradation through over-cutting, because it is retained for its edible fruit when the surrounding vegetation is destroyed.

There are also small areas of stunted evergreen bushland, dominated by *Salvadora persica*, which occur on saline soils associated with the Chalbi drainage system.

5. Deciduous bushland

This type is taller than the last and also has scattered short trees. It occupies only 6 per cent of the area. It occurs mainly on the rocky slopes and pediments of the Precambrian gneissic mountains Nyiru, Ol Doinyo Mara and Ndoto and their outliers, between 665 and 1335 m. Elsewhere, it occurs only on Mt Kulal on lava-derived soils. The gneiss-derived soils seem to be more water-retaining than adjacent but differently derived soils which support principally dwarf deciduous bushland, dwarf shrubland and annual grassland.

The dominant species of deciduous bushland are *Acacia mellifera*, *A. reficiens*, *A. senegal*, and *Commiphora* spp. Emergent trees include *Acacia tortilis*, *Balanites aegyptiaca* and single-stemmed species of *Commiphora*.

6. Woodland

Woodland occupies only 3.5 per cent of the study area. The trees are from 5 to 15 m tall. An understorey of dwarf shrubs is usually present. Woodland occurs either at higher elevations on the major mountains where climatic conditions are suitable, or at low elevations on sandy alluvial soils along the larger seasonal streams. In the latter situation increased moisture storage-capacity of the sandy soils and periodic stream flow compensate for low rainfall and high potential evaporation.

Upland woodland is dominated by *Combretum molle*, *Acacia etbaica*, *A. nilotica* subsp. *subalata*, *A. drepanolobium* and *A. tortilis*, with *A. seyal* locally important. The principal perennial grasses are *Chrysopogon plumulosus*, *Themeda triandra* and *Dichanthium insculptum*.

Combretum molle woodland occupies the wetter sites. It is possibly a fire-induced type which has replaced evergreen bushland or scrub forest. In places it has been further degraded to wooded grassland. *Acacia*-dominated upland woodland usually occurs immediately below *Combretum molle* woodland.

Below 1000 m *Acacia tortilis* is the dominant of deciduous woodlands, beneath which *Duosperma eremophilum* and/or *Indigofera spinosa* form a dwarf-shrub understorey. *Leptothrium senegalense* is an important perennial grass. These woodlands occur along seasonal streams and also on alluvial and colluvial soils at the base of Mt Nyiru, Ol Doinyo Mara and the Ndoto Mts.

7. Perennial grassland

This type occupies 3.4 per cent of the study area. It is mostly pure grassland, but about 20 per cent has a very open cover of scattered woody plants. It occurs mostly in the Hurri Hills. It is of much more limited extent on Mt Kulal.

On the Hurri Hills the principal perennial grasses are *Themeda triandra* and *Chrysopogon plumulosus*, and the grassland occupies a zone which topographically and climatically appears to be analogous to that supporting evergreen bushland elsewhere in the study area. The status of the Hurri Hills grasslands is controversial. Herlocker suggests that they are in part fire-induced and in part edaphically controlled. By contrast, there is convincing evidence that most of the grasslands on Mt Kulal are secondary and, in the absence of human interference or grazing by wild ungulates, would revert to evergreen forest and bushland (see 9 below).

8. Evergreen and semi-evergreen bushland

This type covers only 1.4 per cent of the area. It is restricted entirely to the upper elevations of the major mountains and is most extensive on their wetter southern and eastern faces, but in the Hurri Hills it is confined to steep-walled, narrow, rocky canyons on the eastern slope. On the other mountains it forms a transition zone between evergreen forest and deciduous bushland. In places it has ascended the mountains following the degradation of forest.

On Mt Kulal the principal evergreen species are *Carissa edulis*, *Dovyalis abyssinica*, *Euclea racemosa* subsp. *schimperi*, *Grewia similis*, *Olea africana*, *Pappea capensis*, *Pistacia lentiscus*, *Rhamnus staddo*, *Rhus vulgaris*, *Scutia myrtina*, *Teclea simplicifolia* and *Turraea mombassana*. At lower elevations the community is mostly open with succulent Aloes and Euphorbias and scattered individuals of deciduous species such as *Acacia etbaica*. At higher elevations, if there has been little disturbance, the community is often closed, and *Juniperus procera* becomes increasingly numerous. On wind-swept slopes towards the lower limit of its altitudinal range, *Juniperus* occurs as widely scattered, severely wind-pruned individuals no more than 4–6 m high.

When evergreen bushland is degraded by stock and subjected to annual fires it is readily transformed into *Dichanthium insculptum*, *Themeda triandra* grassland.

Further over-use leads to the occurrence of unpalatable species such as *Eragrostis tenuifolia*, *Chenopodium* spp. and *Solanum incanum*. When protected from fire the grasslands are invaded by *Ocimum* spp., *Pavonia urens*, *Lippia ukambensis* etc., and eventually revert to bushland.

9. Afromontane evergreen forest, scrub forest, and related types

These types occupy only 0.5 per cent of the area. Evergreen forests cap the major mountains where rainfall is higher, cloud cover most extensive, and potential evaporation and temperatures lower than in the plains. Precipitation is probably substantially augmented from mist condensation.

On Mt Kulal evergreen forest and scrub forest extend from 1 835 m to the summit (2 335 m). It is convenient to recognize two main types: coniferous forest dominated by *Juniperus procera*; and broad-leaved forest (but see page 164). *Juniperus procera* is confined to the lower, drier part of the forest belt and has been greatly reduced by fire. Today, apart from a few small relict stands, it occurs chiefly as a scattered relict in fire-induced grassland and as a pioneer in fire-protected places at the forest edge. It is also a scattered emergent inside the broad-leaved forest, especially on shallow soils and in rocky places where the canopy is more open. There are also small groves of relatively young *Juniperus* forest in secondary grassland below the existing forest edge.

Towards its lower altitudinal limit broad-leaved forest, especially on shallow soils, occurs chiefly as scrub forest with the main canopy at 9–10 m. The principal species are *Olea capensis*, *O. africana*, *Diospyros abyssinica*, *Teclea simplicifolia* and *Strychnos mitis*.

As one ascends the mountain the forest, at least in sheltered places, increases in stature to 15–20 m. The principal species are *Cassipourea congoensis*, *Diospyros abyssinica*, and *Olea capensis*, associated with *Allophylus abyssinicus*, *Apodytes dimidiata*, *Casearia battiscombei*, *Ilex mitis*, *Lepidotrichilia volkensii*, *Nuxia congesta*, *Ocotea kenyensis*, *Prunus africana* and *Xymalos monospora*. On exposed ridges 8–10 m high scrub forest occurs with a more open canopy, composed of *Cassipourea*, *Brucea antidysenterica*, *Clausena anisata*, *Olea capensis*, *Rapanea melanophloeos* and *Teclea nobilis*. On some steep slopes in gorges the forest is kept open by frequent landslips and there are small communities dominated by *Dombeya goetzenii* and *Phoenix reclinata*.

On Mt Marsabit (1 865 m) the forest is said to have the same basic composition as on Mt Kulal, though *Juniperus* has not been recorded. The forest descends to 1 165 m, but has not been thoroughly studied. Most of the species so far recorded are typical members of the Afromontane forest flora. Lowland elements, however, are likely to occur at lower altitudes, but, so far, few have been definitely recorded.

The origin of open grassy glades, partly or completely surrounded by forest, is uncertain, and their history is probably complex. On Kulal they were probably initiated by fires (started by lightning or by hunters, pastoralists or honey-gatherers) and do not appear to be edaphically controlled. Below about 1 500 m grassland on Mt Kulal has been derived from bushland and usually includes many bushy plants. Above 1 700 m, by contrast, the glades contain mostly grasses. It is probable that, in the past, resident populations of wildlife, particularly buffaloes and various antelopes such as the greater kudu, have had an important effect in maintaining these grasslands by intermittent grazing all the year round. At present, however, the influence of large wild mammals on Mt Kulal is much reduced, and the periodic grazing of livestock, which may be extremely heavy during times of drought, plus occasional fires in the glades, is much more important. If the influence of fires and grazing were withdrawn it seems that many of the glades would develop into closed forest, some very quickly, although some areas of bushland on steep eroded slopes and of swampy herbaceous vegetation near springs might remain.

During the eight-year drought the grassy glades were heavily grazed and any tendency to revert to forest was held in check, but during the exceptionally wet years following the drought there was very little grazing inside the forest because of the abundance of pasture elsewhere and the fringe of colonizing shrubs at the edge of the forest rapidly invaded the grasslands. On Mt Kulal a strip of secondary thicket up to 10 m wide, and composed principally of *Leonotis mollissima*, *Solanum indicum* subsp. *grandifrons*, *Acanthus eminens*, *Ocimum suave* and *Aspilia mossambicensis*, has extended into the *Setaria sphacelata* grasslands.

The factors controlling the advance and retreat of the forests are complex and it seems that variations in the intensity of land use by humans are as important as variations in rainfall. The two seem to interact in unsuspected ways. During dry periods forest is often favoured because there is little grass growth and fires may be reduced for lack of fuel. Grass fires occurring in the dry season during the wetter places of the climatic cycle may be much more damaging. During the severe 1968–76 drought many large and well-established *Juniperus* trees died, but this did not necessarily result in forest destruction and retreat. Many of the trees occurred in relatively non-flammable evergreen bushland, within which the forest is now regenerating, usually with some young junipers developing. A more damaging factor during that drought was browsing of the forest with some lopping of branches for fodder, which, if sustained, could have gradually destroyed the ability of the forest to maintain itself.

Synnott (1979a) points out that the grassy areas inside the forest are a valuable grazing resource in times of drought and are of great importance in sustaining the local pastoral economy. If all the grassland areas above 1 600 m were allowed to revert to forest, there might be a small improvement in water catchment, but this would be seriously offset by the loss of a vital component of the grazing ecosystem. Synnott concludes that the main importance of the forests is for water conservation. He recommends that the forests should be scientifically managed in the interests of the human population, and that controlled grazing during periods of extreme drought should be permitted, not as a privilege, but as an accepted part of a land-use plan.

10. Palm stand

Small stands dominated by *Hyphaene coriacea* occur on sites with permanent ground water at the edge of the Chalbi Desert and at the base of Mt Kulal.

Vegetation pattern in the greater Serengeti region

Introduction

Refs.: Anderson & Talbot (1965); Darling (1960); Herlocker (1975); Herlocker & Dirschl (1972); Glover & Trump (1970); Glover, Trump & Wateridge (1964); Glover & Wateridge (1968); Glover & Williams (1966); Lamprey (1979); Pearsall (1957); Sinclair & Norton-Griffiths (1979).

The Serengeti Region (see Fig. 10) of the East African Plateau comprises some 35 000 km² of grassland and wooded grassland in northern Tanzania extending into southern Kenya. It has a unique status as the habitat of the greatest concentration of large wild mammals in the world. Some 2 million wild ungulates occupy the region, the great majority having a migratory regime which enables them to utilize the highly seasonable availability of grass and water. The region is also unique in that during the last twenty years or so its fauna and flora and their interactions have been subjected to more detailed study than those of any comparable part of the world. Most of this research has been undertaken by the Serengeti Research Institute which has taken the limits of the migratory range of the ungulate population as the boundaries of the so-called Serengeti ecosystem. Except for the Afromontane vegetation of certain high mountains, the Serengeti ecosystem lies entirely within the Somalia–Masai Regional Centre of Endemism.

The Serengeti National Park (13 000 km²) lies wholly within the Serengeti ecosystem, as do the whole or parts of the following administrative areas: Ngorongoro Conservation Area (7 000 km²), Loliondo Area (5 000 km²), Maswa Area (2 000 km²), Musoma Area (3 000 km²), all in Tanzania, and the Masai Mara Game Reserve (2 000 km²) and the Loita Plains of Narok District in Kenya (3 000 km²).

Since the establishment of its present boundaries, the Serengeti National Park has been uninhabited, but formerly the grasslands in the east were used by the pastoral Masai. The latter also occasionally entered the wooded area which constitutes the remainder of the Park to visit water-holes and salt-licks or to raid agricultural peoples living near Lake Victoria, but, because of the high incidence of the tsetse fly, the carrier of sleeping sickness, the wooded area was largely uninhabited and principally used for hunting. Man-made fires have probably been a feature of the ecosystem for centuries or longer.

This history of human activity, in particular the impact of pastoralism and frequent grass fires, precludes the assessment of the region as an entirely natural ecosystem. Outside the Park the intensity of anthropic influences varies greatly from place to place. In some areas it is slight, but parts of Kenya Masailand on the northern fringes of the ecosystem are being turned into a desert because of overgrazing by domestic cattle (Glover & Gwynne, 1961).

Despite these human influences, much of the Serengeti ecosystem, in comparison with surrounding areas, is relatively little affected, and the grasslands inside the Park show little evidence of the widespread degradation and erosion to be seen over the greater part of the Masai steppe in Tanzania and in parts of Kenya Masailand.

More than twenty years ago it was realized that the Serengeti could provide an incomparable opportunity to study an ecosystem still relatively unaffected by human influence. In view of the widespread ecological degradation of arid and semi-arid vegetation in many parts of Africa during recent decades, it was thought that studies of near-natural communities might provide indications of the level of productivity such habitats are capable of sustaining under varying climatic conditions.

The principal effort of the Serengeti Research Institute has so far been concentrated on the behaviour and ecology of the large mammals, and especially the interactions of the grazing and browsing ungulates with their habitats. The results so far achieved have been summarized by Lamprey (1979) on whose review this brief account heavily depends. Although the research programme described by Lamprey is far from complete, it has revealed an ecosystem of unsuspected complexity, and clearly demonstrates the necessity of attaching equal importance to the plant and animal components.

The greater part of the Serengeti region lies between 1 500 and 1 800 m above sea-level, but descends to 1 200 m at Lake Victoria. Along the eastern border it rises to 3 350 m in the Crater Highlands and to 2 500 m in the Loita–Loliondo Hills. Most of the area is covered by various types of woody vegetation, especially wooded grassland, except for the principally edaphic grassland of the Serengeti Plains in the south-east and the secondary grassland of the Loita Plains in the north.

The landscape of the northern and western areas has principally evolved from the ancient Precambrian peneplain, whereas in the east the surface is covered with a thick mantle of volcanic ash mostly derived from the extinct volcano Kerimasi.

The main seasonal rains begin in November and reach their maximum in March. Rainfall is 380 to 660 mm per year in the Serengeti Plains and increases in

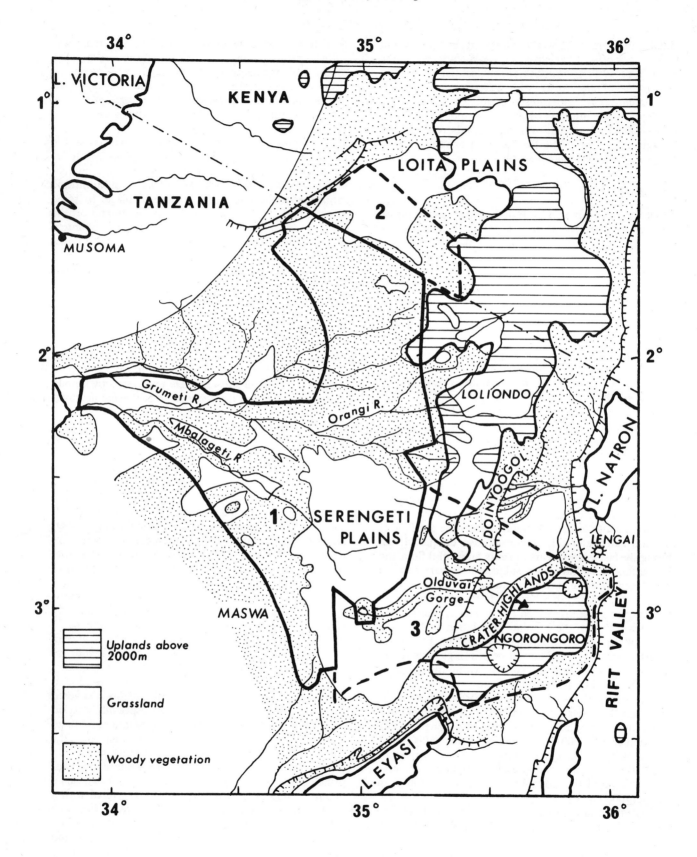

FIG. 10. Map of the Serengeti Region, showing features mentioned in the text

the wooded areas to the west and north, reaching more than 1000 mm per year in the extreme north-west. The 750 mm isohyet separates a dry south-east region with 4–6 wet months (> 50 mm) from a wet north-west region with 7–9 wet months. The north-western area has exceptionally high dry-season rainfall, which is an important factor in dry years in the maintenance of the large wildebeest and zebra populations during September and October when the rest of the region is very dry, grass growth has ceased, and much of the grass elsewhere has been burnt. The region dries out from the south-east in May to the north-west in July and becomes wetter again from the north-west in August–September to the south-east in December. The annual migrations of wildebeest, zebra, and Thomson's gazelle are closely correlated with this sequence of wetting and drying, as is the timing of grass fires, with the south-western areas burning first in early July. The driest south-eastern areas normally do not burn at all owing to the shortness of the heavily grazed grass. This zone is grazed by the very numerous migratory animals only during the short growing season, and by a few drought-tolerant animals, principally the oryx and Grant's gazelle during the dry season.

The commonest large mammal in the Serengeti region is the wildebeest. Its numbers have increased from 263 000 in 1961 to 1 400 000 in 1978. The wildebeest is followed by Thomson's gazelle (c. 400 000) and zebra (c. 200 000). These three species are predominantly migratory and together constitute c. 60 per cent of the numbers of large mammals in the region. The remaining large herbivores include buffalo, elephant, Grant's gazelle, topi, hartebeest, and giraffe. In all, twenty-three large herbivores occur in the Serengeti region. Although no integrated account has yet been published of habitat preference and ecological separation, there are indications of the complex relationships which exist. Of the three principal migratory species, the zebra (grazer) selects grass of intermediate length and takes a high proportion of stem and sheath as well as leaf. By contrast, the wildebeest (grazer) is more dependent on green grass and predominantly selects the leaf component of short grass. Thomson's gazelle is a mixed feeder and uses very short grass, frequently after burning or close grazing by wildebeest.

An outstanding feature of the Serengeti ecosystem is the seasonal migrations of the wildebeest, zebra and Thomson's gazelle populations answering the need to seek seasonally available water and grazing. The wildebeest population performs an annual migration, spending the wet season on the Serengeti Plains, then moving westwards towards Lake Victoria at the beginning of the dry season (normally early June), and later northwards to the northern part of the Serengeti Park or into the Masai Mara Game Reserve in Kenya (normally in August or September). When the rain begins again (most frequently in early November) the wildebeest return to the Serengeti Plains, but if the rain is delayed they may move south-west first. Variations in the annual pattern of migration can be correlated with differences in rainfall distribution.

Like the wildebeest, the zebra and Thomson's gazelle occupy the Serengeti Plains in the rains, and the woody communities to the west and north during the dry season. The zebra usually follows the whole range of movement of the wildebeest, but tends to occupy areas of taller grassland, and the two species only occasionally occupy the same areas at the same time. The Thomson's gazelle population also migrates over the same range as the wildebeest, but does not move as far north.

It has been assumed that the animal populations are subject to natural regulation so that they remain within the long-term carrying capacity of their habitats. The precise means by which this is attained, however, is far from clear. Despite the incredible increase in size of the wildebeest population during the last twenty years there is no indication that it is approaching the equilibrium level determined by prevailing climatic conditions. However, a reduction in grass productivity through drought would be expected to result in increased mortality and, depending on the severity of the drought and the population size at the time, this might amount to a population 'crash'.

Since this section was written an important synthesis has been published (Sinclair & Norton-Griffiths, eds., 1979).

Main vegetation types

The grasslands of the Serengeti Plains have been described by Anderson & Talbot (1965) in relation to the distinctive plains soils. Herlocker (1975) has written a synopsis of the woody vegetation of the remainder of the Serengeti National Park, and Glover and Trump (1970) have prepared a detailed account of the vegetation of the Narok District of Kenya Masailand which includes the northern fringe of the Serengeti ecosystem. The vegetation of the Ngorongoro Conservation Area is dealt with by Herlocker & Dirschl (1972) in a richly illustrated account. The main vegetation types recorded by these authors are summarized below.

The Serengeti ecosystem lies almost entirely within the Somalia–Masai Region but has some distinctive features. The grasslands of the Serengeti Plain are unique, and the wooded grasslands to the north and west are different from most other Somalia–Masai *Acacia*-dominated communities, principally in the insignificance of bushy plants other than *Acacia* and *Commiphora* and the relative abundance of grasses, especially perennial species. The extent to which these features might be due to the prevalent grass fires and a large ungulate population is uncertain.

1. Edaphic grassland of the Serengeti Plains

The grasslands of the Serengeti Plains grow on soils derived from volcanic ash and are briefly mentioned on page 116 in relation to other edaphic grasslands of the

Somalia–Masai Region. Grasslands occurring on volcanic ash are of very restricted occurrence in Africa and are almost confined to the Serengeti region, where the ash was supplied by two volcanoes, one extinct, the other active. This unique pedological feature may provide the explanation for the unique concentration of large mammals in this part of Africa (page 127).

The extinct volcano, Kerimasi, at the northern end of the Crater Highlands, in the last stages of eruption, *c.* 150 000 B.P., produced huge quantities of fine whitish-grey ash which included a high proportion of calcium carbonate. The ash fell over a wide area, including the south-eastern Serengeti, where it settled in successive layers, filling the hollows and providing a relatively flat surface to the formerly undulating peneplain. It hardened to form grey and light-brown calcareous tuffs, and, with the accumulation of lime through downward leaching, almost continuous layers of calcitic hard-pan formed at several successive horizons.

The present surface of the eastern plains of the Serengeti consists of grey, dusty, easily erodible soil derived from the upper layers of the tuff and is precariously protected from the erosive effects of wind, rain, and animals by the short plant cover.

As the activity of Kerimasi ended, a new volcano, Oldoinyo Lengai, arose 11 km to the north, and is now the only active volcano in the region. The last eruption was in 1967. Its blackish-brown ash has locally formed tuffs and agglomerates in the eastern Serengeti but much has remained unconsolidated as loose black sand, which forms elongated dunes moving in a south-westerly direction.

The Serengeti Plains, which lie partly inside partly outside the Serengeti National Park, cover some 6250 km^2 of gently rolling grassland between 1350 and 1650 m above sea-level. More than fifty species of grass are common and a further fifty species also occur. At certain times of year the grasslands provide pasture for the world's greatest concentration of large mammals, estimated at more than 1 million individuals and belonging to thirty species of ungulate, and eight species of large predator. Rainfall gradually increases from 380 mm per year in the east to 780 mm in the west. Thus, the gradient from the juvenile ash soils in the east to the more mature brown calcareous soils of the west is paralleled by a climatic gradient. Anderson & Talbot (1965), who have described zonation of soils and vegetation, and indirectly the distribution of the large ungulates, in relation to these gradients, recognize six types of grassland, although the change is gradual. It seems that the soils derived from volcanic ash, especially vertisols, at least under the present climatic regime, favour grass rather than woody vegetation, whereas on the granite soils in the higher-rainfall areas to the west, trees are excluded, at least in part, by fire, though the matter is by no means fully elucidated. The main features of Anderson & Talbot's grassland types are as follows:

1. *Sparse grassland on juvenile soils on volcanic ash.* The ash forms both mobile and stable dunes. Rainfall is 380–500 mm per year, but, because of high porosity and low moisture retentivity of the soils, the area is edaphic desert. The soils are fully saturated with bases but salt accumulation is very local; pH is 7.4 near the surface and 7.9 at 110 cm; at 200 cm, in the subsoil of the previous soil surface, it reaches 9.7. The vegetation is sparse with about 15–20 per cent basal cover. Among the early colonizers *Chloris gayana*, *Dactyloctenium* sp., *Digitaria macroblephara*, *Sporobolus ioclados* and *S. kentrophyllus* are common. On more stable areas between the dunes scattered bushes of *Acacia mellifera* sometimes occur.

2. *Short grassland on calcimorphic soils with hardpans.* This type covers a large part of the Eastern Serengeti Plains. The parent material is fine volcanic dust overlying calcareous tuff. A very hard calcareous pan occurs at a depth of about 95 cm. The soils are base-saturated and have a high exchangeable sodium content; pH increases from 8.1 at 15 cm to 9.8 at 100 cm. The vegetation is short and sparse (basal cover ± 20 per cent). The most characteristic species are a sedge (*Kyllinga*) and a dozen species of *Sporobolus*. Inside the Park this area is heavily grazed by wild ungulates during the wet season, and outside by domestic animals. The soils are very friable and vulnerable to wind erosion when the vegetative cover has been damaged, as it has been in parts of the Ngorongoro Conservation Area by excessive livestock.

3. *Intermediate grassland on calcimorphic soils with soft pans.* This type is intermediate between grassland type 2 and taller grasslands to the west. The soils are not quite fully base-saturated above, but are completely so below. The upper layers are salt-free and pH increases from 6.2 near the surface to 9.1 at 100 cm. The taller species are *Pennisetum mezianum*, *Eragrostis tenuifolia* and *Sporobolus* spp. with lower patches of *Andropogon greenwayi*, *Panicum coloratum*, *Cynodon dactylon*, etc. Basal cover is *c.* 30 per cent. This area is heavily grazed but occasionally accumulates sufficient fuel to burn.

4. *Tall grassland on vertisols of lithomorphic origin.* These soils are derived from fine ash overlying tuff; pH is 6.8 near the surface and 7.4 at 100 cm. They have a much deeper rooting zone than those described above. There is a high degree of base saturation but an absence of soluble salts to a depth of 170 cm. There are concretions of calcium carbonate but there is no pan. The most abundant species are *Andropogon greenwayi*, *Digitaria macroblephara*, *Cynodon dactylon*, *Eustachys paspaloides*, with *Themeda triandra*, *Pennisetum stramineum*, *P. mezianum* and *Michrochloa kunthii* also present. Basal cover averages 50 per cent. These grasslands are only grazed sporadically by the herds of animals. The latter avoid heavy-textured soils when they are wet, and move eastwards where many of the same plants are available

in shorter and apparently more palatable forms. Most of this grassland is burned at least once annually.

5. *Intermediate grassland on southern vertisols of lithomorphic origin.* The soils are somewhat similar to the northern vertisols but appear to be derived solely from calcareous tuff, and are not fully saturated with bases until a depth of 100 cm. The dominant grasses are *Pennisetum mezianum, P. stramineum, Cynodon dactylon* and *Andropogon greenwayi.*

6. *Tall grassland on brown calcareous soils.* These lighter-textured soils occur in the extreme west and along the northern edge of the plains. The parent material is a calcareous conglomerate with some quartzitic gravel, originating from the decomposition of granite at and near the surface. There is negligible ash deposition. The soils are much better drained than the vertisols and have lower base status and pH (6.2 near the surface but 8.6 at 130 cm). Soluble salts are present in the less-permeable lower layers. The dominant grass is *Themeda triandra*, closely followed by *Pennisetum mezianum* and *P. straminium*. Basal cover averages 45 per cent. The grasslands are heavily grazed in relatively dry years, but are only lightly grazed in wet years. They are normally burned annually.

In the wooded areas of the Serengeti ecosystem very gentle slopes towards the bottom of the catena have alkaline soils with impeded drainage which support grassland, but their area is very much less than that of the woody communities.

Subsequent to the publication of Anderson & Talbot's paper much research, mostly unpublished, has been done on the Serengeti grasslands and is summarized by Lamprey (1979). Fire-control experiments indicate that the *Themeda* grassland of the western Serengeti Plains is maintained mainly by the fires, which occur at least once every three years. They also show that *Acacia tortilis* will only regenerate if fire does not occur for five consecutive years. In the absence of fire most *Acacia* species grow at *c.* 1 m/year and become tolerant to moderate fires after 3–4 years. Where giraffe are numerous their browsing offtake may nearly equal the growth for many years and render the regenerating trees vulnerable to burning for an extended time.

When the short grass plains are fenced to exclude animals they grow a dense grass cover 40–50 cm high, while the surrounding grassland remains at less than 10 cm high owing to ungulate grazing. This taller sward is easy to burn. After burning there is an increase in density of the fire-tolerant species *Themeda triandra*, which is normally scarce or absent on the eastern plains.

The different types of grassland vary greatly in the quality of their grazing. The short grassland provides the best herbage, but as the rainfall is low and irregular, this good forage is available for only a short time. The intermediate grassland has a longer growing period and since the growth of stem in the predominant species, *Andropogon greenwayi*, begins late, this vegetation is acceptable for a longer period. The tall grassland offers good forage at the beginning of the wet season, but as it is not normally grazed frequently during the growing season, its quality rapidly declines.

During dry periods in the wet season the short grasslands are the first to become unproductive. The intermediate and tall grasslands continue to produce for a longer time and the grazing animals use them as a temporary food supply until the short grasslands receive rain again. During the dry season, the Serengeti Plains as a whole receive very little rainfall, forage production virtually ceases, and water and shade are virtually absent. The animals consequently move to the woodlands where conditions are more favourable.

Various reasons have been offered to explain why the ungulates prefer the shorter grasslands. Anderson & Talbot (1965) believed that the animals avoid the sticky soils of the taller grasslands and prefer certain growth forms and developmental stages of the shorter grasses. Bell suggested that grass height is an important factor in relation to feeding adaptations and also that the hazard of predation is lower in short grassland.

Kreulen (pers. comm. in Lamprey, 1979) considers that the main reason is the availability in the short grasslands of essential nutrients to meet the high requirements of lactation during the months February–May, the first four months of growth of wildebeest calves. He observed that the areas of highest occupance during lactation are those in which the grasses and the water have a relatively high calcium content, which is necessary for milk production without loss of condition in the mother. He also found that *Themeda triandra*, the predominant grass in the tall western grasslands, contains too little calcium to enable the lactating females to maintain a positive calcium balance.

Hence it appears that: 'The unique concentration of ungulate animals (primarily wildebeest) which use the eastern Serengeti Plains may depend upon the equally unique occurrence of the highly calcitic ash soils produced by the eruption of the now extinct volcano, Kerimasi' (Lamprey, 1979).

2. Secondary grassland of the Loita Plains

The Loita and similar contiguous plains occur in the Narok district of Kenya Masailand 90 km north of the northern fringe of the Serengeti Plains. In some ways the Loita grasslands are similar to those of the Serengeti Plains and are part of the same ecosystem, but their origin is different, since they occur on truncated soils and have replaced evergreen bushland, following its degradation by fire and browsing.

The Loita plains lie roughly between 1° S. and 2° S. and 35° E. and 36° E., between 1700 and 1900 m, and cover an area of 4500 km^2. In the north and west they lie on Tertiary-Recent volcanic rocks and in the south on the Basement Complex. The volcanic rocks are phonolites with thin intercalated tuffs. The soils to the north and east are composed of lava dust and sediments forming brown calcareous loams and grey, compacted, loamy sands. To the south they are shallow and stony with rock outcrops and large areas of black clay. All these soils are severely truncated.

The rainy season lasts from November to June and mean annual rainfall ranges from 1 000 mm in the north and west to 500 mm in the extreme east towards the edge of the Rift Valley.

The plains are grazed by Thomson's gazelle, Grant's gazelle and kongoni, and by large migratory herds of wildebeest, zebra, and topi. Except for small fringe areas to the south and west, inhabited by tsetse flies, the whole of the plains are heavily grazed throughout the year by Masai cattle, sheep and goats.

Glover & Trump (1970) recognize two kinds of grassland, tall and short. Tall grassland is 45 cm to 2 m tall and is composed principally of *Pennisetum mezianum, P. schimperi, Hyparrhenia cymbaria, H. filipendula, H. hirta, Hyperthelia dissoluta, Themeda triandra* and *Dichanthium insculptum*.

Short grassland is closely cropped vegetation often no more than 7 cm tall, but there may be some flowering culms up to 45 cm or more. It usually occurs on shallow or compacted soils in severely overgrazed and trampled areas. The principal species are *Microchloa kunthii, Sporobolus festivus, Cynodonon dactylon* and some constituents of the tall grass zones such as *Themeda* and *Dichanthium insculptum*. The short grasslands are favoured by many wild ungulates and Masai sheep and goats because they produce fresh green shoots throughout the year.

The distribution of tall and short grassland is often controlled by soil depth and micro-relief associated with termite mounds (Glover *et al.*, 1964) and erosion terraces (Glover & Wateridge, 1968).

3. Acacia–Commiphora *deciduous bushland and thicket*

This type, the most extensive and characteristic of the Somalia–Masai Region, is very poorly represented inside the Serengeti National Park. In most places the rainfall is too high; elsewhere it is precluded by the volcanic ash soils of the Serengeti Plains. It is, however, well developed outside the Park along the drier eastern fringe of the Serengeti ecosystem.

Inside the Park it is dominated by 2–6 m high *Acacia mellifera* and occurs on termite mounds, and in disturbed places where bare soil is associated with sheet and gully erosion or salt-licks.

Outside the Park it is widespread in the Ngorongoro Conservation Area, e.g. in the Doinyoogol Hills (Herlocker & Dirschl, phot. p. 16), on the Lake Eyasi escarpment and flats (Herlocker & Dirschl, phots. p. 27) and in Oldupai Gorge (Herlocker & Dirschl, phots. p. 18). Characteristic species include: *Acacia drepanolobium, A. mellifera, A. seyal, A. tortilis, Adansonia digitata, Cissus cactiformis, C. quadrangularis, Commiphora madagascariensis, C. merkeri, Cordia sinensis, Croton dichogamus, Euphorbia candelabrum, E. nyikae, E. tirucalli, Salvadora persica* and *Sansevieria ehrenbergii*.

4. Acacia–Commiphora *deciduous wooded grassland and related types*

This type is classified by Herlocker as woodland. In most places, however, the canopy cover is less than 40 per cent and the trees are only 4–7 m high, so that the vegetation fits the category of wooded grassland as defined in this work (page 52).

Acacia–Commiphora wooded grassland is the most extensive woody vegetation type in the Serengeti National Park and covers 7 260 km^2 or 88 per cent of all woody vegetation. It consists of a single open stratum of *Acacia* or *Commiphora* thorn-trees mostly from 3 to 7 m high but in a few species from 9 to 20 m. Shrubs and bushes are poorly represented, but scattered single bushes or small groups of *Grewia fallax* and *Cordia ovalis* sometimes form a very open understorey. There is a characteristic grass stratum 0.5–1.5 m high. It is dominated by such species as *Digitaria macroblephara, Themeda triandra* and *Eustachys paspaloides* on relatively well-drained soils, and by *Pennisetum mezianum* on poorly drained soils.

Herlocker, for mapping purposes, divides this vegetation type into 39 species types, 38 of which have one or more of 11 species of *Acacia* (*A. nilotica, A. hockii, A. senegal, A. gerrardii, A. robusta* subsp. *usambarensis, A. drepanolobium, A. seyal, A. xanthophloea, A. sieberana, A. tortilis, A. polyacantha*) as dominants or co-dominants. *Commiphora schimperi* is the sole dominant of the remaining species type. Most of these species combine and recombine in a kaleidoscopic fashion and, as little is known of their ecology, interpretation of the species types is sometimes difficult. There are, however, some distinctive variants.

Acacia gerrardii occupies large areas of poorly drained clay soils. It also dominates secondary wooded grassland replacing evergreen bushland.

Acacia drepanolobium often forms almost pure 1–8 m high open stands on poorly drained soils of valley bottoms. It is common to find stands of dead trees, suggesting recent extreme variation in edaphic conditions.

Commiphora schimperi is dominant on well-drained soils derived from granite or granitic gneiss on ridge tops and slopes. Its most important associates are *Acacia tortilis, A. robusta* subsp. *usambarensis, A. senegal* and *A. hockii*.

Acacia tortilis is most frequently dominant in the drier eastern part of the Park, where it forms 9–14 m high stands on the fringes of the Serengeti Plains. There is little regeneration and the stands are degenerating because of the death of old trees.

Acacia robusta subsp. *usambarensis* is most frequently dominant in the wetter, western half of the park where it often forms almost pure, even-aged stands 8–12 m high with dense canopies and little regeneration.

Acacia xanthophloea, which reaches a height of 22 m, is a riparian species, or otherwise dependent on ground water.

Acacia sieberana and *A. polyacantha* are also riparian species in this part of their geographical range. They are not typical Somalia–Masai species but are more characteristic of regions with higher rainfall.

5. Combretum–Terminalia *secondary wooded grassland*

This type is a fire-climax which has replaced dry evergreen forest on ridge tops and upper slopes in the northern part of the Park. It occupies *c.* 500 km². The open overstorey is dominated by *Combretum molle* and *Terminalia mollis*, which grow to 10–13 and 15–17 m in height, respectively. The open understorey is dominated by *Heeria reticulata*, *Acacia nilotica* subsp. *subalata*, and *A. hockii*. The 1–2 m high grass stratum, which burns fiercely every dry season, is dominated by species of *Diheteropogon*, *Hyparrhenia*, *Loudetia* and *Themeda*. The principal species on termite mounds are *Rhus natalensis* and *Grewia trichocarpa* with emergent *Lannea stuhlmannii* and *Sclerocarya birrea* which reach the main canopy. *Parinari curatellifolia* is characteristic of termite mounds along seepage lines. It is believed that a significant decrease in the *Combretum* and *Terminalia* populations has taken place in recent years because of elephants, but that large trees of *Terminalia* have suffered least, because they can withstand pushing over. Both *Combretum molle* and *Acacia hockii* regenerate freely from their roots and persist despite repeated burning, which, however, retards their further development.

6. *Evergreen and semi-evergreen bushland and thicket*

This type has a scattered distribution, chiefly on stream banks, rocky hills, along seepage lines and on termite mounds, throughout most of the Park, but is best developed in the north, where, however, is has been extensively destroyed by fire (see 7 below). On rocky hills near the Kenya border it is dominated by *Euclea racemosa* subsp. *schimperi*, *Haplocoelum foliolosum*, *Tarenna graveolens*, *Teclea nobilis*, and *T. trichocarpa*, associated with *Aloe*, sp., *Cordia ovalis*, *Euphorbia candelabrum*, *Grewia trichocarpa*, *Pappea capensis*, *Rhus natalensis* and *Strychnos henningsii*.

Evergreen bushland is the climax throughout much of south Narok District in Kenya Masailand immediately north of the Park, but very little primary or undisturbed vegetation remains, and nearly everywhere it has been replaced by secondary communities, including the severely degraded secondary grasslands of the Loita Plains (see above). The principal dominants are *Acacia brevispica* (on shallow, stony soils), *Carissa edulis*, *Croton dichogamus*, *Grewia similis*, *Osyris* sp., *Rhus natalensis*, *Tarenna graveolens* and *Teclea simplicifolia*. On deeper soils and along dry watercourses the following occur as stunted emergent trees: *Albizia harveyi*, *Cassine buchananii*, *Euclea divinorum*, *Lannea stuhlmannii*, *Olea africana*, *Pappea capensis* and *Ziziphus mucronata*. Succulents include *Euphorbia candelabrum*, *Aloe volkensii* and species of *Sansevieria* and *Kalanchoe*.

7. *Evergreen forest*

The total area occupied by the surviving remnants of evergreen forest and evergreen bushland with which it is usually associated is small, amounting to no more than 240 km².

Evergreen forest occurs on alluvial soils as narrow and often discontinuous riverine communities in the Mara River basin and along the lower Grumeti, Orangi and Mbalageti rivers. Towards the north of the Park there are also small relict patches of dry evergreen forest and scrub forest on the deep sandy loams of broad ridge tops. In this situation they are usually associated with evergreen bushland on shallower soils. These two types were probably dominant north of the Grumeti–Mara River divide. Their area has been greatly reduced by fire, and they have been extensively replaced by *Combretum molle*, *Terminalia mollis* wooded grassland and *Acacia robusta* subsp. *clavigera*, *A. gerrardii* wooded grassland respectively.

The canopy of riparian forest is composed principally of *Aphania senegalensis*, *Ekebergia capensis*, *Ficus* spp., *Garcinia livingstonei*, *Lecaniodiscus fraxinifolius*, *Tamarindus indica* and *Ziziphus pubescens*.

In rain-fed dry evergreen forest the main canopy constituents are *Diospyros abyssinica*, *Drypetes gerrardii*, *Cassine buchananii*, *Lecaniodiscus fraxinifolius*, *Suregada procera* and *Teclea nobilis*, with *Chaetacme aristata*, *Euclea divinorum*, *Olea africana*, and *Schrebera alata* occurring less frequently. *Capparis erythrocarpos*, *Croton dichogamus* and *Teclea trichocarpa* are the most abundant members of the understorey. The broad-leaved grass *Setaria chevalieri* occurs in the field layer.

8. *Afromontane communities*

The Crater Highlands are outside the Serengeti ecosystem but within the greater Serengeti region. They rise from Lake Eyasi at 1000 m to an extensive high plateau with an average elevation of 2150–2450 m. Several extinct volcanoes, of which the highest is Lolmalasin Mountain (3350 m), rise above the plateau and several calderas, including Ngorongoro, are sunk into it. The Crater Highlands are drier than most massifs in Africa of comparable size and altitude, and their Afromontane communities show several distinctive features. Also, because of the complex physiography and heavy grazing of the vegetation by wild ungulates and domestic stock, Afromontane vegetation is sometimes less sharply differentiated from 'lowland' vegetation than is often the case. Herlocker & Dirschl (1972) recognize the following main types:

— *Artemisia afra*, *Erica arborea* montane heath above 2450 m.

— *Croton macrostachyus, Calodendrum capense, Olea* spp., *Albizia gummifera* montane forest above 2450 m.
— *Vernonia auriculifera, Crotalaria agatiflora* subsp. *imperialis* secondary thicket above 2450 m.
— *Hagenia abyssinica, Gnidia glauca* upper montane forest, above 2700 m.
— *Juniperus procera* dry evergreen forest in steep canyons between 2450 and 2900 m.
— *Eleusine jaegeri, Pennisetum schimperi* montane grassland, c. 2300 m.
— *Acacia lahai* woodland, which is probably secondary, 2100–2450 m.
— *Arundinaria alpina* bamboo, above 2300 m.

v The Cape regional centre of endemism

Geographical position and area

Geology and physiography

Climate

Flora

Mapping unit

Vegetation
    Cape shrubland (fynbos)
    Secondary Cape shrubland (Rhenosterbosveld)
    Coastal bushland and thicket
    Riparian bushland and thicket
    The transition to Karoo

Geographical position and area

This region covers the south-western and southern part of the Cape Province, South Africa, between 32° and 35° S. and 18° and 27° E. Typical Cape vegetation does not occupy the whole of this area. There are large enclaves of Karoo and Afromontane vegetation and small patches of bushland of Tongaland–Pondoland affinity. The easternmost extremity of Cape vegetation in the Suurberge is separated from the main block by the bushlands of the Sundays River valley. Outliers of Cape vegetation to the north occupy the highlands above Van Rhynsdorp and the summit of the Kamiesberg. (Area: 71 000 km².)

Geology and physiography

The landscape is dominated by subparallel folded mountain ranges with an average altitude of 1 000–1 500 m and individual peaks exceeding 2 000 m. The major ranges are constructed of Table Mountain sandstone and the minor ones of smaller sandstone folds or of Witteberg quartzite. In the western part the foothills and lower slopes are commonly formed of Cape granite. The valleys and parts of the coast belt are formed from Bokkeveld shales and sandstones of the Cape System and Malmesbury shales of the late Precambrian. The coastal fringe itself consists of Tertiary to Recent sands, conglomerate, and limestone.

Climate

Rainfall exceeds 250 mm per year, and is mostly from 300 to 2 500 mm, but locally in the mountains reaches 5 000 mm. The western part receives 60–80 per cent of its rain in winter but from Swellendam eastwards the rainfall is more evenly distributed throughout the year. On the higher mountains summer drought is alleviated by moisture-bearing clouds from the south-east, and on the west coast sea mists are fairly frequent. During winter, snow falls regularly on the higher mountains, especially in the west, but persists only on the southern slopes. In general the winters are mild. Frosts are unknown at the coast, but occur inland and in the mountains where they are infrequent or not severe. Strong desiccating winds blow at certain seasons. (See Fig. 11.)

Flora

There are about 7000 species, of which more than half are endemic.

Endemic families. Bruniaceae (12 genera, 75 species). Geissolomataceae (1 species). Grubbiaceae (2 genera, 5 species). Penaeaceae (5 genera, 25 species). Retziaceae (1 species). Roridulaceae (1 genus, 2 species). Stilbaceae (5 genera, 12 species).

Other characteristic families. Ericaceae (18 endemic genera and *c.* 650 endemic species). Proteaceae (11 endemic genera and *c.* 320 endemic species). Restionaceae (*c.* 10 endemic genera and *c.* 180 endemic species). Rutaceae–Diosmeae (10 endemic genera and *c.* 150 endemic species).

Endemic genera. About 210 genera are confined to the Cape Region and a further 70 have their greatest concentration of species there. Among the latter are *Agathosma* (130 endemic species), *Aspalathus* (240), *Cliffortia* (70), *Crassula* (145), *Erica* (520), *Ficinia* (50), *Metalasia* (30), *Muraltia* (100), *Phylica* (140), *Protea* (85), and *Restio* (40). Most of these genera are virtually confined to the Cape Region. *Erica*, however, has about 35 species on the mountains of tropical Africa and in the Holarctic Kingdom, and *Protea* has about 40 species in tropical Africa.

Mapping unit

50. Cape shrubland (fynbos).
Only a single unit appears on the map, but as is shown below, the typical Cape vegetation is modified in places by the occurrence of species of karroid and tropical affinity.

Vegetation

The prevalent vegetation of the Cape Region is fynbos, which most characteristically occurs in the form of 1–3 m tall sclerophyllous shrubland. Large parts of the Cape lowlands, however, where they are not cultivated, are today occupied by secondary shrubland dominated by the 'rhenosterbos', *Elytropappus rhinocerotis*. There is some evidence that the original vegetation, which has been replaced by *Elytropappus*, included many species of tropical and karroid affinity. On the coastal plain itself coastal fynbos is the prevalent vegetation, but locally there are patches of bushland and thicket dominated mainly by tropical species. Mountain streams in many parts of the Cape Region are fringed with riparian thicket and scrub forest dominated by a mixture of Cape endemics and Afromontane species.

The term fynbos is applied to virtually all the terrestrial vegetation of the Cape Region other than the enclaves mentioned above. Despite its wide range of variation in floristics and structure, most fynbos fits the definition of shrubland or bushy shrubland adopted in this work. Only a few communities of specialized or otherwise localized habitats belong to other physiognomic types.

The *Protea* bushes and other tall plants that often occur scattered in bushy fynbos may thicken up and form dense impenetrable thickets 4–6 m high if they receive sufficiently long protection from fire. Experiments have shown, however (F. J. Kruger, pers. comm.), that most fynbos species cannot regenerate under these conditions and become moribund and die. Accordingly, such thickets, even in the absence of fire, represent an impermanent phenomenon, and are not described separately below.

Cape shrubland (fynbos)
(mapping unit 50)

Refs.: Acocks (1975, p. 104–7); Adamson (1927; 1938*a*, p. 86–95); Duthie (1929); Marloth (1908); H. C. Taylor (1963*b*; 1972*a*; 1972*b*; 1978); Werger *et al.* (1972).

Phots.: Acocks (1975: 102, 103, 104); Adamson (1927: 3–8; 1938*a*; 1, 2); Marloth (1908: 21, 49, 50, 55, IV, V, VI, X, XI); H. C. Taylor (1978: 1–19); Werger *et al.* (1972: 1–5).

Syn.: false macchia (Acocks, 1975); macchia (Acocks, 1975).

The first botanist to use the Afrikaans word fynbos in print appears to have been Bews (1916). It aptly conveys the small leaves and bushy habit of the dominant plants.

Most stands of fynbos contain many species, and single-species dominance does not occur other than locally. Taylor (1972) has recorded 121 species of flowering plants from a single 100 m^2 quadrat in a homogeneous stand.

Floristics. In most types of fynbos other than extreme variants the following genera and families are usually well represented: *Protea, Leucadendron, Leucospermum* and *Serruria* (Proteaceae), *Erica, Simocheilus, Philippia* and *Blaeria* (Ericaceae), all genera of Restionaceae, Bruniaceae and Penaeaceae, *Aspalathus, Podalyria* and *Cyclopia* (Leguminosae), *Phylica* (Rhamnaceae), *Tetraria, Ficinia* and *Chrysithrix* (Cyperaceae), *Diosma* and *Agathosma* (Rutaceae), *Cliffortia* (Rosaceae), *Metalasia, Helichrysum, Stoebe, Elytropappus* and many others (Compositae), *Lobostemon* (Boraginaceae), *Polygala* and *Muraltia* (Polygalaceae), *Grubbia* (Grubbiaceae), and many genera of Liliaceae, Amaryllidaceae and Iridaceae.

Physiognomy. Fynbos often contains scattered taller bushes and, less often, widely spaced trees. There is always a conspicuous admixture of monocotyledonous 'switch' plants belonging to Restionaceae, which in some extreme habitats become physiognomically dominant. It

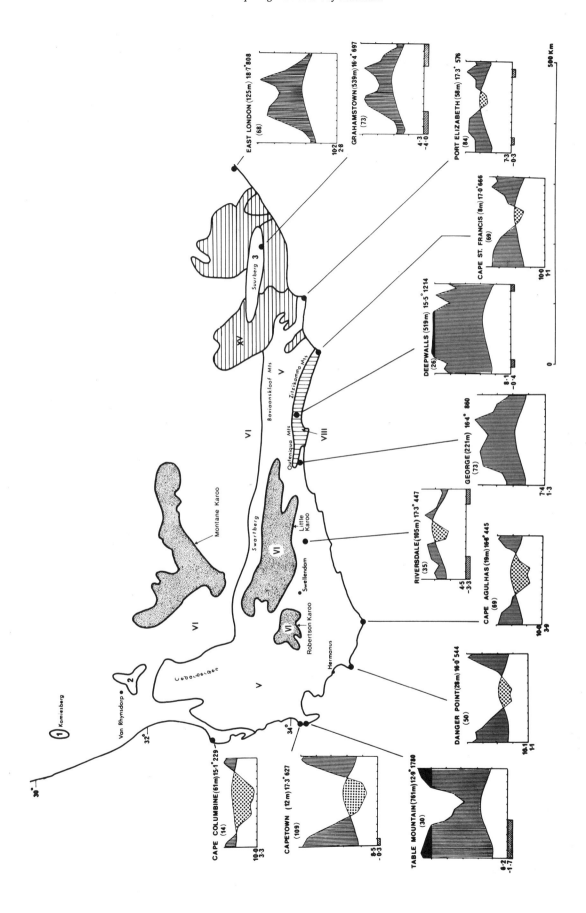

FIG. 11. Climate and topography of the Cape regional centre of endemism (V) (1–3 indicate exclaves of Cape vegetation)

is the constant presence of Restionaceae, above all else, that distinguishes fynbos from the vegetation of other regions with a broadly similar climate.

The growth-form of Restionaceae, which is also found in some Cyperaceae, is referred to by Taylor as 'restioid'. They are tufted or rhizomatous plants ranging in height from 20 cm to 2 m or more. Their green, tubular or wiry stems, which bear reduced, non-photosynthetic scale leaves, always persist for more than one year, but usually for less than four. Geophytes are normally plentiful in fynbos but annuals are conspicuous only in the drier types. Grasses are uncommon today, but according to Acocks (1949) were much more abundant before the advent of the European. Except in certain disturbed areas, they now contribute little to the appearance of the vegetation. Apart from 'weedy' species, they mostly belong to the 'southern' genera *Merxmuellera* (*Danthonia*), *Pentaschistis*, *Ehrharta*, *Plagiochloa* and *Lasiochloa*. Cyperaceae on the other hand are often abundant.

The shrubs and bushes of fynbos vary greatly in height and density. They are mostly richly branched and have twisted boles. In typical fynbos true trees are virtually absent. The only species with well-defined boles are *Leucadendron argenteum*, the 'Silver Tree', and *Widdringtonia cedarbergensis* and *W. schwarzii*. In a few places on Table Mountain, *Leucadendron* occurs as an emergent up to 10 m high scattered in typical fynbos shrubland. Under present conditions, *Widdringtonia cedarbergensis* is normally a small tree 5–7 m high, but can attain a height of 20 m, and formerly frequently did so. It occurs scattered chiefly in rocky places in the Cedarberg between 915 and 1525 m, but never forms closed stands (H. C. Taylor, pers. comm.). *W. schwarzii* is confined to rocky ravines in the Baviaanskloof and Kouga Mountains. It can reach a height of 30 m or more but is usually smaller. Like *W. cedarbergensis* it does not form a closed canopy. Two other species, *Widdringtonia cupressoides* and *Olea capensis*, which often occur as trees elsewhere, in fynbos are usually branched low down, of bushy habit, and less than 7 m tall.

Bushy species. Only about 50 true fynbos species normally exceed 3.5 m in height, but few of them grow taller than 6 m, and many are frequently much smaller than this. It is quite exceptional, today, to find them occurring at all plentifully as tall plants other than very locally. The more important among them are: *Cassine parvifolia*, *Cliffortia arborea*, *C. grandifolia*, *Cryptocarya angustifolia*, *Diospyros glabra*, *Erica caffra*, *E. caterviflora*, *E. inconstans*, *Heeria argentea*, *Hyaenanche globosa*, *Laurophyllus capensis*, *Leucadendron procerum* (*concinnum*), *L. eucalyptifolium*, *L. nobile*, *L. sabulosum*, *Leucospermum conocarpodendron*, *Maytenus oleoides*, *Metalasia muricata*, *Mimetes fimbriifolius*, *Oldenburgia arbuscula*, *Passerina filiformis*, *Philippia chamissonis*, *Phylica buxifolia*, *P. oleifolia*, *P. paniculata*, *P. villosa*, *Polygala myrtifolia*, *Protea arborea*, *P. glabra*, *P. laurifolia*, *P. longiflora*, *P. lorifolia*, *P. nereifolia*, *P. obtusifolia*, *P. repens*, *P. susannae*, *Psoralea pinnata* and *Wiborgia sericea*. Only one arborescent *Aloe* is a member of typical fynbos, namely the 5 m tall, repeatedly branched *Aloe plicatilis* which occurs in the mountains between French Hoek and the Elandskloof Mts, where the rainfall is about 2000 mm per year.

Leaves. The foliage of the woody plants is mostly brownish-green or greyish. The leaves are sclerophyllous —small, stiff, thick, coriaceous and entire. They are strongly cuticularized and rich in sclerenchyma with reduced intercellular spaces and often have struts which prevent leaf collapse. The lower surface is commonly hairy. Two leaf-shapes, the ericoid and proteoid, are particularly prevalent. Ericoid leaves are small and narrow with rolled margins. They are found in all types of fynbos and occur in a wide range of unrelated families, e.g. Ericaceae (*Erica*), Bruniaceae (*Brunia*), Polygalaceae (*Muraltia*), Leguminosae (*Aspalathus*), Thymelaeaceae (*Passerina*), Rosaceae (*Cliffortia*), Rhamnaceae (*Phylica*) and Compositae (*Metalasia*). Proteoid leaves are associated with less branched, usually taller, bushy plants, especially members of the Proteaceae (*Leucadendron*, *Leucospermum*, *Mimetes*, *Protea*). They are mostly elliptic or oblanceolate, up to 15 cm long, phyllodic in appearance and very sclerophyllous. Whereas restioids and ericoids are always present in fynbos, proteoids are sometimes absent, especially from drier types and at high altitudes.

Major variants. The structural complexity of fynbos decreases with increasing altitude. On slopes below 900 m, fynbos is dense and usually has three layers: a discontinuous, bushy, proteoid upper layer 1.5–3(4) m tall; an ericoid layer of shrubs up to 1 m, and a ground layer of smaller woody plants, herbs, geophytes, and, especially, Restionaceae. At higher altitudes the fynbos becomes progressively shorter and less stratified and the proteoid element disappears. In the eastern extension of this mountain fynbos, rainfall is more evenly distributed and grasses increase.

Dense fynbos thickets up to 5 m tall, dominated by a single species, e.g. *Leucadendron salicifolium* or *Berzelia lanuginosa*, sometimes occur on the banks of streams. Some Restionaceae also become virtually exclusively dominant in hollows and by streams where permanent ground water is present. For instance, *Chondropetalum mucronatum* (*Dovea mucronata*) often forms pure communities in poorly drained places on the horizontal sandstone plateau of Table Mountain at about 900 m. Occurrences such as these, however, are unusual. In general fynbos is an exceptionally mixed type.

Fire. Fynbos is very combustible, especially during the hottest, driest, windiest time of year, when large areas are frequently devastated. It is now widely believed that fynbos evolved in relation to recurrent fires from natural causes and that fire is necessary for its healthy maintenance. Most fynbos species (e.g. *Protea arborea*,

Euclea lancea, most Restionaceae) can sprout again after the fiercest fires. Otherwise their seeds are effectively protected from fire (e.g. *Leucadendron salicifolium, Widdringtonia*). In the absence of fire many species become moribund and die, even in open communities (Restionaceae). Some rarities (e.g. *Orothamnus zeyheri*) have almost become extinct because of over-protection from fire.

Aliens. Large areas of fynbos have been invaded and, locally, completely replaced by aliens, originally introduced for reclamation or forestry purposes from other regions with a similar climate. Chief among them are *Hakea acicularis* and various wattles (*Acacia cyclops, melanoxylon* and *cyanophylla*) introduced from Australia, and *Pinus pinaster* from the Mediterranean region.

Secondary Cape shrubland (Rhenosterbosveld)
(mapping unit 50)

Refs.: Acocks (1975, p. 86–7); Marloth (1908, p. 98–106); Muir (1929, p. 14–21, 37–49); H. C. Taylor (1978, p. 215–18).
Phots.: Acocks (1975: 79); Marloth (1908: 21); H. C. Taylor (1978: 18).
Syn.: Coastal Renosterveld (Taylor); Coastal Rhenosterbosveld (Acocks).

There are two main blocks, one in the south, the other in the west. They occur below 300 m between the foot of the mountains and the coastal plain. Rainfall is between 300 and 500 mm per year. The soils which are derived from shales are more fertile than those of the mountains and the coast, and have been intensively farmed for centuries. Rhenosterbosveld is usually 1 m tall or less, rarely up to 2 m. Its ability to invade agricultural land has been known for more than 200 years. In 1775 Sparrman described its encroachment and predicted that it would transform the landscape. Although Rhenosterbosveld is rich in species, in the southern block the typical fynbos families, Ericaceae, Proteaceae and Restionaceae are lacking; the original vegetation was probably evergreen scrub dominated by *Olea africana* and *Sideroxylon inerme* with *Cussonia spicata, Diospyros dichrophylla, Pterocelastrus tricuspidatus* etc.

The southern Rhenosterbosveld is much more grassy than the western and many of its grass species, such as *Hyparrhenia hirta*, are widespread in the tropics. The western block has a greater admixture of fynbos species and the characteristic grasses, such as *Lasiochloa echinata* and *Pentaschistis patula*, are non-tropical annuals. Bushy species are fewer but include *Olea africana*.

Coastal bushland and thicket
(mapping unit 50)

On the south coast certain scrubs and trees, which are not true fynbos species, locally form thicket or scrub forest up to 10 m tall. They include: *Cassine peragua, *Euclea racemosa, E. tomentosa, E. undulata, Maytenus heterophylla, Myrsine africana, *Olea africana, Chionanthus foveolatus, Pterocelastrus tricuspidatus, Putterlickia pyracantha, Rhus crenata, R. glauca, R. laevigata, *R. lucida, *R. tomentosa, *Sideroxylon inerme, Tarchonanthus camphoratus* and *Zygophyllum morgsana*. They are often associated with big restioids, especially *Willdenowia stricta*, and tall fynbos species of karroid appearance, e.g. *Eriocephalus racemosus*.

Tall scrub (3–6 m high) in Hermanus District has been described by Taylor (1961). It is partly seral to forest of Afromontane affinity but is also an edaphic subclimax on the dry northern slopes of limestone outcrops where the soil is shallow and well drained. It is made up of those species marked with an asterisk in the list above together with *Carissa bispinosa, Chrysanthemoides monilifera, Chionanthus (Linociera) foveolatus, Maytenus heterophylla* and *Osyris* sp.

In the western coastal fynbos, bushy species of tropical affinity are fewer than in the southern belt, principally *Diospyros austro-africana* subsp. *rugosa, Euclea natalensis* subsp. *capensis, E. racemosa, Maytenus heterophylla, Osyris* sp., *Pterocelastrus tricuspidatus, Putterlickia pyracantha, Rhus glauca* and *R. mucronata*. They occur as scattered individuals 2–3 m high or locally form small thickets. The more conspicuous true fynbos species are *Leucadendron salignum, Metalasia muricata, Protea repens, Thamnochortus erectus, T. spicigerus* and *Willdenowia striata*.

Riparian bushland and thicket
(mapping unit 50)

The lower and less steep watercourses are fringed with 5–7 m tall, dense thickets of *Brabeium stellatifolium, Freylinia oppositifolia* and *Metrosideros angustifolia*. At higher altitudes *Rapanea melanophloeos, Kiggelaria africana, Maytenus acuminata, Olea africana, Olinia* and *Podocarpus elongatus* are the characteristic species. *Cunonia capensis, Hartogia capensis, Ilex mitis* and *Maytenus oleoides* are common to both types.

The transition to Karoo
(mapping unit 50)

Along the inner margin of the Cape Region there is a narrow band of Arid Fynbos which forms the transition from typical Cape to typical Karoo vegetation. Ericaceae are absent and Proteaceae and Restionaceae though conspicuous, especially the latter, are few in species. Typical shrubby Karoo genera such as *Chrysocoma, Hermannia, Euryops, Pteronia, Eriocephalus, Selago, Walafrida,* and *Lightfootia* are well represented. Succulents, including *Euphorbia mauritanica* and *Aloe ferox* (in the east), are often present.

VI The Karoo–Namib regional centre of endemism

Geographical position and area

Geology and physiography

Climate

Flora

Mapping units
    Bushy Karoo shrubland
    Succulent Karoo shrubland
    Dwarf Karoo shrubland
    Montane grassy Karoo shrubland

Vegetation
    Semi-desert vegetation of the Karoo
        Karoo shrubland
            Dwarf succulents and succulent shrubs
            Arborescent succulents
            Non-succulent bushes, bushy trees and tall shrubs
            Dwarf non-succulent shrubs
            Grasses
            Geophytes and annuals
        Karoo riparian scrub forest
        The transition to Tongaland–Pondoland evergreen bushland
    The Namib desert
        The Outer Namib fog desert
            Sand dunes
            Gravel desert
            Rocky outcrops
        The Inner Namib desert
        The *Welwitschia bainesii* transition zone
        River-bed communities
    The desert of Mossamedes

Geographical position and area

This region occupies the central, northern and north-western parts of the Cape Province immediately to the north of the Cape floristic Region (but also has important exclaves within the latter), mostly north of 33° S. and between 17° and 25° E. It extends northwards as an increasingly narrow band along the entire length of Namibia into south-west Angola to about 11° S. This northern extension of the Karoo–Namib Region not only includes the coastal plain but also the escarpment of the interior plateau and locally the fringes of the plateau itself. (Area: 661 000 km^2.)

Geology and physiography

Both the geology and the physiography are very diverse. Altitude ranges from sea-level to 2 695 m. The region includes four of the Geomorphic Provinces of King (1951), namely Cape Middle Veld, Karoo, Namib, and Kaokoveld, as well as parts of 'Highveld' and Damaraland.

In the interior of the Cape Province the surface is formed of the Karoo System and is extremely even except where broken by dolerite dykes, sills and other intrusions. The soils, mostly derived from Dwyka tillite and dolerite, are clayey and tend to accumulate salts. Brackish seasonal swamps or 'vloere' are extensive.

In the North-West Cape the Karoo beds have been removed to expose granite and other primitive rocks with numerous later igneous intrusions, which provide abundant sand for distribution by wind. This area is mostly of subdued relief but rugged mountains occur in western Namaqualand and in the gorge tract of the Orange River.

The Namib desert occupies a coastal peneplain, extensive areas of which are covered with moving sand of recent origin. Elsewhere, granite, gneiss, or Stormberg lava outcrop at the surface. Further inland in Namibia the rocks are very varied and give rise to a variable relief.

Climate

Rainfall in the Namib Desert is less than 100 mm per year. Elsewhere it rarely exceeds 250 mm. Seasonality of rainfall varies greatly. West of a line running from Spencer Bay through Calvinia to Sutherland more than 60 per cent of the rain falls in winter. East of a line from Swakopmund through Pofadder and Fraserburg to Willimore more than 60 per cent falls in summer. Nevertheless, in most of the 'summer rainfall' parts of the Region there is more dry-season precipitation than in most parts of the Zambezian Region, or, as in the coastal belt, dry-season mists are frequent (see page 141). There is considerable variation in the amount and distribution of rainfall from year to year, especially in the driest parts. Even in the wetter parts of the summer-rainfall belt, winter influences are dominant about one year in twelve.

The coastal belt is frost-free except in the south, where occasional light frosts occur in July as at Port Nolloth. Further inland in southern Namibia and throughout the interior of the Cape the frost period lasts for 5–6 months, though the mean minimum temperature of no month is below zero except in the extreme east towards the contact with the climatic limit of Highveld grassveld. (See Fig. 12.)

Flora

There are about 3 500 species, of which more than half are endemic.

Endemic family. Welwitschiaceae (1 species, *Welwitschia bainesii*).

Other characteristic families. Asclepiadaceae: Stapelieae (6 endemic genera and *c.* 160 endemic species). Aizoaceae (Mesembryanthemaceae) (95 endemic genera and *c.* 1 500 endemic species).

Endemic genera (in addition to above). About 60 including *Adenolobus* (2 species), *Arthraerua* (1), *Augea* (1), *Ceraria* (5), *Didelta* (2), *Grielum* (6), *Kaokochloa* (1), *Leucosphaera* (2), *Monelytrum* (1), *Nymania* (1), *Phaeoptilum* (1), *Phymaspermum* (9), *Sisyndite* (1), *Xerocladia* (1).

Endemic species. The following genera have important concentrations of endemic species: *Aloe, Anacampseros, Babiana, Chrysocoma, Cotyledon, Crassula, Eriocephalus, Euphorbia, Gasteria, Haworthia, Hermannia, Pentzia, Pteronia, Sarcocaulon, Stipagrostis, Tetragonia, Zygophyllum*. Some of these genera, e.g. *Pteronia*, are almost confined to the Karoo–Namib Region. At the other extreme, some, e.g. *Euphorbia*, are cosmopolitan.

Linking elements. At the species level there is little intermingling with the Cape flora. Most linking species extend to the east or the north or both.

Species common to the Karoo–Namib and Tongaland–Pondoland Regions, and in some cases extending slightly beyond their combined area, include *Aloe speciosa, Carissa haematocarpa, Crassula portulacea, Euclea undulata, Euphorbia grandidens, Montinia caryophyllacea, Portulacaria afra, Schotia afra* and *S. latifolia*.

Tree species which extend into the Karoo–Namib Region from the Zambezian Region, and in some cases from much further north, include *Acacia mellifera* subsp. *detinens, A. erioloba, A. karroo, Boscia albitrunca, Diospyros lycioides, Dodonaea viscosa, Euclea crispa, Pappea capensis* and *Ziziphus mucronata*.

Grasses which extend at least as far as the Zambezian Region include *Cymbopogon plurinodis* (*pospischilii*), *Eustachys paspaloides, Fingerhuthia africana, Hyparrhenia hirta, Schmidtia pappophoroides* and *Themeda triandra*.

Mapping units

51. Bushy Karoo shrubland.
52. Succulent Karoo shrubland.
53. Dwarf Karoo shrubland.
57a. Montane Grassy Karoo shrubland.
74. The Namib desert (see below).

The four mapping units into which the Karoo has been divided have been adapted from those of Acocks (1975). There is a fair amount of floristic information, but for large areas ecological information is virtually non-existent. For this reason, after the mapping units have been briefly characterized, the Karoo, apart from its riparian and transitional vegetation, is treated as a single continuum.

Bushy Karoo shrubland
(mapping unit 51)

Refs.: Acocks (1975, p. 59–63, 71–5); Barbosa (1970, p. 245–51); Giess (1971, p. 9–12); de Matos & de Sousa (1970); White (MS, 1973).
Phots.: Acocks (1975: 54, 55, 64–7); Cannon (1924: 8b & c, 11a, 19b, 21b); Giess (1971: 21, 22, 24–7); Marloth (1908: 105, 107, 108, XVI, XVIII); de Matos & de Sousa (1970: 4); Shantz & Turner (1958: 11, 12, 13, 15).
Syn.: karroid broken veld; Namaqualand broken veld; Orange River broken veld (all of Acocks, 1975).

This is shrubland dotted with small bushy trees and large shrubs. It occurs in the Great Karoo, the Little Karoo, Robertson Karoo, and on the rocky hills of Namaqualand and the lower valley of the Orange River and northward to south-west Angola. Succulents are usually abundant, especially in the south. Non-succulent dwarf shrubs are always present but are usually subordinate to the succulents. Grasses though inconspicuous are represented by many species.

Succulent Karoo shrubland
(mapping unit 52)

Refs.: Acocks (1975, p. 69–71).
Phots.: Acocks (1975: 62); White, Dyer & Sloane (1941: 190d, 276).

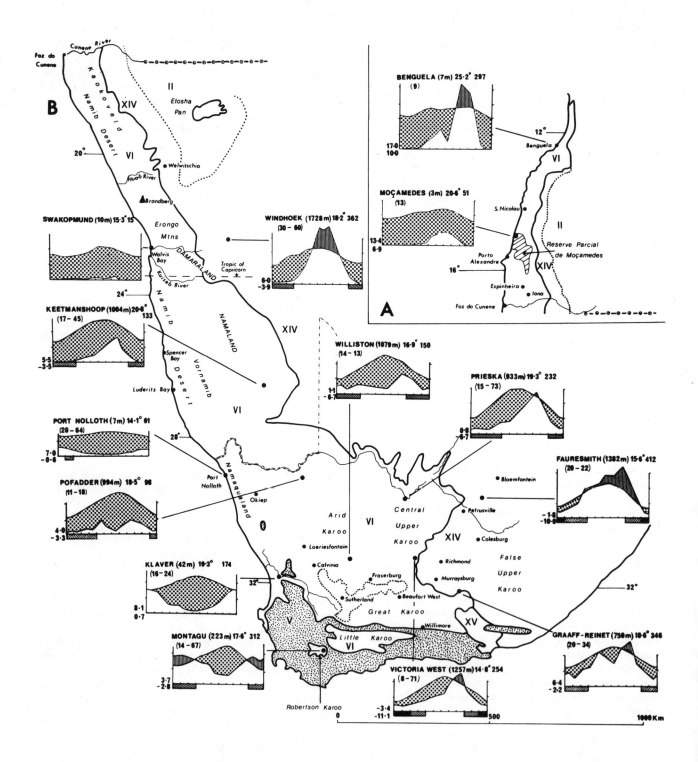

FIG. 12. Climate and topography of the Karoo–Namib regional centre of endemism (VI)
A. North of Cunene River. B. South of Cunene River

Succulent Karoo shrubland, except for secondary types, is largely confined to the sandy coastal plain of Namaqualand and the heavier stony soils of the foothills of the Namaqualand escarpment. There is an important outlier, enclosed by high mountains in the rain shadow valleys of the Tanqua and Doorn Rivers. These areas occur mostly below 610 m and are frost-free. The annual rainfall is less than 200 mm and falls in winter. In the coastal areas the effects of aridity are reduced by sea mists.

Succulents are dominant throughout. They range in height from almost subterranean species to shrubs 2 m or more tall, but the average height is between 0.3 and 1 m. In the driest and most degraded areas ground cover is very low but elsewhere may be as high as 50 per cent. Large shrubs and bushes are virtually absent except along the rivers. Dwarf non-succulent shrubs are represented by numerous species but they contribute little to the physiognomy. Grasses are few in species and usually inconspicuous.

Dwarf Karoo shrubland
(mapping unit 53)

Refs.: Acocks (1975, p. 63–9, 76–7); Adamson (1938*a*, p. 179–88); Marloth (1908, p. 280–90).
Phots.: Acocks (1975: 56–61), Adamson (1938*a*: 14); Marloth (1908: 17a); Shantz & Turner (1958: 19 & 20).
Syn.: Arid Karoo; False Arid Karoo; Central Lower Karoo; Central Upper Karoo; Western Mountain Karoo (all of Acocks, 1975).

This is the most extensive type of Karoo shrubland. It is dominated by dwarf shrubs, most of which belong to Compositae. It occupies the Arid and Central Karoo and the lower slopes of the Karoo mountains. Bushes and trees are absent. Large shrubs are few in species and local, and are represented chiefly by *Rhigozum trichotomum*. Succulents are always present but are mostly inconspicuous; there are relatively few species. Grasses are more abundant than in other types and increase towards the east. The soils are often slightly saline, and halophytes, particularly *Salsola tuberculata*, are widespread, and locally dominant. In the Arid Karoo there are enormous brackish flats or 'vloere', in some cases covered with *Salsola aphylla* and other halophytes, but elsewhere practically bare.

Montane grassy Karoo shrubland
(mapping unit 57a)

Published information is almost non-existent. Acocks (1975, p. 81) suggests that the original vegetation was formerly more grassy. Today *Merxmuellera disticha* and *M. stricta* are the chief relict grasses; Karoo shrublets belonging to *Chrysocoma, Eriocephalus, Pentzia, Ruschia* etc., as well as *Elytropappus rhinocerotis*, are abundant.

Vegetation

The desert vegetation of the Namib, and the semi-desert vegetation of the Karoo, including its northern extension into Angola, are described separately below. A richly illustrated review of the literature on the Karoo–Namib Region has been published by Werger (Werger, 1978*b*, in Werger (ed.), 1978*c*).

Semi-desert vegetation of the Karoo

Except along the larger watercourses, which support a fringe of scrub forest, bushland, or thicket, almost the whole area is covered with shrubland. Only locally in the transition to wetter regions is the vegetation sufficiently luxuriant to be classified as bushland. Although Karoo shrubland includes a wide range of physiognomic types there is insufficient published information to justify their separate description in a general work such as this.

Karoo shrubland

Except very locally the shrubs are less than, and frequently much less than, 2 m tall. Over extensive areas, however, the landscape is dotted with larger woody plants, either arborescent succulents, or non-succulent bushes or bushy trees. These taller plants rarely exceed 5 m in height. They are chiefly confined to places, usually rocky, where the water supply is increased by run-off from surrounding slopes. Hence they are not uniformly distributed and the landscape is a mosaic of bushed and unbushed areas. In the former the bushes vary from 5 to 100 m apart. They reach their greatest luxuriance in places where there is mist condensation, such as in the Richtersveld.

In Karoo shrubland there is as wide a diversity of growth forms as in Cape fynbos, but the subject has been little studied. For convenience, in addition to grasses and forbs, the following principal types may be recognized: (a) dwarf succulents and succulent shrubs; (b) arborescent succulents; (c) dwarf non-succulent shrubs; (d) taller non-succulent shrubs, bushes and bushy trees.

Their relative abundance varies greatly. Dwarf succulents and succulent and non-succulent shrubs occur throughout. Arborescent succulents and non-succulent bushes and bushy trees, however, are more or less confined to 'broken veld' (see mapping unit 51 above). Fewer bushy species occur in the southern broken veld of the Little and Great Karoo than elsewhere.

Dwarf succulents and succulent shrubs

Stem succulents are represented by species of *Euphorbia*, Asclepiadaceae (especially *Stapelia, Caralluma, Hoodia, Huernia* and *Trichocaulon*), *Senecio* ('*Kleinia*'), and by species of *Sarcocaulon* and *Pelargonium* which shed their small mesomorphic leaves during drought. Most species are less than 1 m tall, but the largest Euphorbias are up to 2 m. The cactoid *E. avasmontana* is conspicuous in the mountainous country on both sides of the Orange River. Of the non-spinous species *E. mauritanica* is the most widespread. *E. gummifera* and *E. gregaria* are abundant in parts of Namaqualand and Namibia.

Leaf-succulents are pre-eminently represented by Mesembryanthemaceae. The most prevalent are small shrubs from 30 cm to 1 (2) m tall, especially species of *Ruschia*, but the annuals and stone plants (*Lithops*, *Titanopsis*, *Argyroderma* etc.) are locally important. Other widespread genera with many succulent species include *Anacampseros* (Portulacaceae), *Cotyledon* and *Crassula* (Crassulaceae), *Aloe*, *Haworthia* and *Gasteria* (Liliaceae), *Lycium* (Solanaceae) and *Zygophyllum* (Zygophyllaceae).

Arborescent succulents

Aloe dichotoma (up to 5 m high), which extends from Namaqualand almost as far north as Windhoek, is the most abundant species. The more sparsely branched *A. pillansii* (10 m) is almost confined to the Richtersveld. *Cotyledon paniculata* (3 m), *Cerararia namaquensis* (5 m), and the unbranched *Pachypodium namaquanum* (5 m) extend from Namaqualand into southern Namibia. In northern Namibia the last two are replaced by *C. longepedunculata* and *Pachypodium lealii* respectively. Towards the northern end of the Karoo–Namib Region two succulent Euphorbias, *E. currorii* (10 m) and *E. eduardoi* occur. There are also several species with enlarged water-storing stems, including *Cyphostemma currorii* (7 m), *Moringa ovalifolia* (7 m) and *Sesamothamnus guerichii* and *S. benguellensis* (5 m). South of Namaqualand and the Orange River Valley arborescent succulents are much less plentiful and are represented only by *Crassula arborescens* (3 m, Little Karoo eastwards) and *Portulacaria afra* (4 m, Little Karoo eastwards).

Non-succulent bushes, bushy trees and tall shrubs

There are about 100 species in this group. *Dodonaea viscosa*, *Euclea undulata*, *Nymania capensis*, *Pappea capensis*, *Rhigozum obovatum*, *Rhus undulata* and *Schotia afra* are widespread south of the Orange River and extend at least a short way further north into Namibia.

Ehretia rigida, *Boscia albitrunca*, *B. foetida*, *Acacia erioloba* and *A. mellifera* subsp. *detinens* and *Grewia flava*, which are widespread in the drier parts of South tropical Africa, extend south to beyond the Orange River. In northern Namibia and south-west Angola many more Zambezian linking species enter the Karoo–Namib Region and some, e.g. *Colophospermum mopane*, almost reach the sea.

Acacia redacta, *Adenolobus garipensis*, *Commiphora capensis*, *C. gracilifrondosa*, *C. namaensis*, *C. oblanceolata*, *Diospyros ramulosa*, *Ficus cordata*, *F. guerichiana*, *Heeria concolor*, *H. crassinervia*, *Parkinsonia africana* and *Rhigozum trichotomum* occur in Namaqualand and the Orange River Valley. Some, e.g. *Acacia redacta*, are endemic to this region. Others such as *Parkinsonia* extend as far north as Angola.

Species confined to the northern part of the Karoo–Namib Region include *Acacia montis-usti*, *A. robynsiana*, *Adenolobus pechuelii*, *Euphorbia guerichiana*, *Rhigozum virgatum* and several species of *Commiphora*.

Most of the above are from 2.5 to 4 m tall and are branched from the base or have short irregular boles. *Acacia erioloba*, however, is sometimes taller and has a straight bole.

Dwarf non-succulent shrubs

South of the Orange River nearly all of the most abundant dwarf shrubs belong to non-tropical genera, most of which are virtually confined to South Africa or have their main or subsidiary centres of endemism there. They include *Aster*, *Berkheya*, *Chrysocoma*, *Didelta*, *Eriocephalus*, *Euryops*, *Garuleum*, *Helichrysum*, *Lightfootia*, *Osteospermum*, *Pentzia* and *Pteronia* in Compositae, and *Galenia*, *Hermannia*, *Lebeckia*, *Nestlera*, *Plinthus*, *Selago*, *Sutera*, *Wahlenbergia* and *Walafrida* in other families. The relative importance of these genera steadily diminishes towards the tropics and only a few species, e.g. *Pteronia glauca*, extend into northern Namibia. In the northern part of the Karoo–Namib Region small shrubs and suffrutices (and also herbs) belonging to tropical genera such as *Barleria*, *Blepharis*, *Crotalaria*, *Hibiscus*, *Indigofera*, *Monechma*, *Petalidium*, *Pterodiscus*, *Ruellia* and *Tephrosia* become more important.

The dwarf non-succulent Karoo shrubs are rarely more than 1 m tall and commonly are no more than 25 cm. In heavily grazed areas they may be no more than 10 cm.

Grasses

About 130 species of grass occur in the Karoo–Namib Region, and 80 or so are confined or almost confined to it. The endemic species belong to 29 genera, of which *Aristida* and *Stipagrostis* are the most important. Grasses occur throughout the Karoo but are only locally physiognomically dominant.

There can be little doubt that the importance of grasses in relation to woody plants has declined within recent times owing to overgrazing, and there is some photographic evidence in support of this (Shantz & Turner, 1958), but it is extremely unlikely that pure grassland anywhere represents the climax vegetation, except possibly very locally on deep sandy soils and in a few other edaphically favourable places.

Grasses are much less conspicuous in succulent Karoo and bushy Karoo than in dwarf non-succulent Karoo. In the latter the most conspicuous species are the silvery-white desert grasses, so called because of the bleached appearance of their dead persistent inflorescences. The principal species are *Aristida diffusa*, *Eragrostis lehmanniana*, *Stipagrostis brevifolia*, *S. ciliata*, *S. namaquensis*, *S. obtusa* and *S. uniplumis*. Normally they are perennial, but they sometimes complete their growth-cycle within a single season. In times of drought *Stipagrostis brevifolia*, which is the hardiest of all the plants of the Arid Karoo, sheds its leaves and is transformed into a small woody shrublet. After good rains grasses sometimes temporarily overtop and conceal the shrublets they are growing with. Elsewhere, grasses may besome physiognomically dominant following extreme overgrazing because they regenerate more rapidly from seed than their associates.

Geophytes and annuals

Geophytes and annuals are well represented in the Karoo flora, especially in the winter-rainfall area of Namaqualand. They are conspicuous only after periods of good rainfall when, for a few weeks, their attractive flowers transform the appearance of the veld. Important genera of geophytes include *Babiana*, *Bulbine*, *Homeria*, *Lachenalia*, *Lapeirousia* and *Oxalis*. The most abundant annuals belong to species of *Arctotis*, *Cotula*, *Dimorphotheca*, *Felicia*, *Osteospermum*, *Senecio*, *Ursinia*, *Venidium* (all Compositae) and *Heliophila*, *Hermannia* and *Grielum*.

In overgrazed areas, especially on sandy soils towards the northern limits of the Karoo, the noxious weed *Tribulus zeyheri* forms similar extensive colourful carpets.

Karoo riparian scrub forest

Acacia karroo is gregarious and widespread in the Karoo and is often the only tree present, especially in the interior. Nearer the Atlantic coast, as in the Pofadder–Augrabies sector of the Orange River, the riparian bushland flora is more diversified, and *Acacia karroo* is sometimes rarer there than *Pappea capensis* (6 m), *Euclea pseudebenus* (8 m), *Tamarix usneoides* (7 m), *Diospyros lycioides* (7 m), *Rhus undulata* (3 m) and *Euclea undulata* (6 m). Other associates include *Diospyros acocksii*, *Rhus lancea*, *Combretum erythrophyllum* and *Ziziphus mucronata*.

Towards the northern end of the Karoo–Namib Region several species which are widespread in the Zambezian Region (and sometimes also elsewhere) as quite large trees penetrate far into the desert along watercourses, usually in the form of bushy trees. They include *Acacia albida*, which is sometimes kept small by browsing zebra, *A. erioloba*, *Colophospermum mopane*, *Combretum apiculatum*, *C. imberbe*, *Ficus sycomorus*, *Sterculia africana* and *Ziziphus mucronata*.

The transition to Tongaland–Pondoland evergreen bushland

Communities dominated by *Portulacaria afra* (Acocks, 1975, p. 58–9), and known as 'Spekboomveld' (see page 201), occur in the Southern and Eastern Cape on steep mountain slopes where annual rainfall is 250–300 mm. They are intermediate in structure and floristic composition between the drier types of Tongaland–Pondoland evergreen bushland and the more luxuriant types of bushy Karoo–Namib shrubland. The Noorsveld of the Eastern Cape (Acocks, 1975, p. 58; Van der Walt, 1968), which occurs at lower altitudes and has a slightly lower rainfall, is similarly transitional.

The Namib desert
(mapping unit 74)

Refs.: Giess (1968a, 1971); Marloth (1909); Walter (1971, p. 338–74); Werger (1978b).

Phots.: Adamson (1938: 16); Coetzee & Werger (1975: 13–17); Giess (1968a: 1–6); Marloth (1909: 4); Walter (1971: 209, 225); Werger (1978b: 3, 4, 6, 10, 11).

The Namib desert runs the whole length of Namibia and continues a short distance further north into Angola as the desert of Mossamedes, and a short distance to the south along the coast of Namaqualand. For most of its length it is about 100 km wide and extends from the coastline to the foot of the scarp which delimits the interior highlands of southern Africa. The boundary of the Namib is arbitrarily defined by the 100 mm isohyet, but the greater part of it receives very much less rainfall than this.

Giess (1971) divides the Namib into three. The Northern Namib extends as far south as the Huab River. The Central Namib lies between the Huab and Kuiseb Rivers. The Southern Namib, which extends into northern Namaqualand, experiences summer rainfall in the northern part and winter rainfall in the south. Walter (1971) distinguishes between the Outer Namib, which receives frequent fogs, and the fog-free Inner Namib.

The drier parts of the Namib receive rainfall only rarely, but the outer coastal belt is characterized by many days of fog. In the southern part as far north as Luderitz Bay such rain as there is falls in winter; further north it falls in summer.

In the coastal part of the Central Namib near Swakopmund the mean annual rainfall is only 10 mm, and on average rain occurs only once in two years, but at remote intervals high rainfall is experienced. In 1934, for instance, there was a rainfall of nearly 150 mm; by contrast in certain years rainfall is hardly measurable.

At Swakopmund there are 94 to 215 fog days per annum and the frequency is no less in the Southern Namib. Their effects rarely penetrate as far as 50 km inland. The fog belt which is carried constantly above the cold offshore Benguela current is brought inland by south-westerly winds during the night and only disperses during the day when the desert soil becomes heated. The precipitation from individual fogs is mostly less than 0.1 mm and is never more than 0.7 mm. The total annual precipitation from fog is between 40 and 50 mm but in most places this is of no significance to vascular plants since the water only penetrates to 3.5 cm and evaporates immediately the fog dissipates. The situation is different, however, where fog is condensed on rock faces. Then water from a large catchment area runs into crevices and vascular plants can become established, even though the true rainfall is too low to measure. The plants grow mainly during the cool season, when fogs are frequent, and not during the summer when there may be occasional falls of rain. The fog-water contains sodium chloride and the soils of the Outer Namib are brackish as far as the inland limit of coastal fog, while the Inner Namib has no saline soils.

The Outer Namib fog desert

Sand dunes

Phots.: Giess (1971: 1, 2, 14, 15).

There are two main areas, one in the Northern Namib and the other in the northern half of the Southern Namib.

The former is up to 40 km wide and extends to the north of the Cunene River far into Angola. The extremely sparse vegetation consists of isolated plants of *Barleria solitaria*, *Ectadium virgatum*, *Indigofera cunenensis*, *Merremia multisecta*, *Petalidium angustitubum*, *P. giessii*, and the grasses *Stipagrostis ramulosa* and *Eragrostis cyperoides*.

South of Swakopmund and extending as far south as Luderitzbucht there is a large belt of shifting dunes about 320 km long and 120 km wide. Although botanically unexplored it is believed to be virtually without vegetation. The few species known to occur include *Monsonia ignorata*, *Trianthema hereroensis*, *Stipagrostis gonatostachys* and *S. sabulicola*.

In the Central Namib there is frequently a narrow strip of small dunes up to 200 m wide with scattered cushions of *Psilocaulon salicornioides*, *Salsola aphylla* and *S. nollothensis*.

Gravel desert

Phots.: Giess (1968a: 1; 1971: 7, 16); Marloth (1909: 4).

North of Swakopmund gravel desert occupies most of the outer parts of the Central Namib. Fifty per cent of the surface is covered with a stony pavement. The soil is cemented into a rock-hard layer by deposition of lime and gypsum forming a hardpan at a depth of 1–5 cm. Soil wetting occurs to this depth. All the stones are covered with colourful foliose and crustose lichens, such as species of *Parmelia* and *Usnea* and *Teloschistes capensis*. Otherwise the gravel desert is normally devoid of vegetation. After the heavy rains of 1934 numerous plants appeared. Their mean cover was 20 per cent and, in small depressions, as much as 50–90 per cent. Typical plants were *Psilocaulon* (*Mesembryanthemum*) *salicornioides*, *Mesembryanthemum cryptanthum* (*Hydrodea bossiana*), *Drosanthemum luederitzii* (*paxianum*), *Aizoanthemum* (*Aizoon*) *dinteri* and *Zygophyllum simplex*. These are all ephemeral halophilous succulents with extremely low rates of transpiration. The exceptionally high rainfall of 1934 enabled some of them to remain alive for more than a year but eventually all succumbed to drought. A few individuals of two perennial species, *Arthraerua leubnitziae* and *Zygophyllum stapfii*, which are normally confined to drainage lines, colonized the plains but did not persist. Annual grasses, including *Stipagrostis hermannii*, *S. namibensis* and *S. subacaulis*, also appear plentifully after adequate rain has fallen.

The Southern Namib has a higher rainfall than the Central Namib, and perennial species such as the stem succulent, *Euphorbia gummifera*, and small shrubs of *Zygophyllum retrofractum* and *Sarcocaulon spinosum* occur in places on gravelly flats.

Rocky outcrops

In the outer part of the Central Namib, apart from ground-water areas, rocks provide the only habitats where perennials can survive. The rocks are usually covered with lichens. The following dwarf succulents predominate among the vascular plants rooted in crevices:

Stem succulents: *Trichocaulon clavatum* (*dinteri*), *T. pedicellatum*, *Hoodia currorii*.

Leaf succulents: *Lithops* spp., *Anacampseros albissima*, *Aloe asperifolia*, *Cotyledon orbiculata*.

Stem succulents with deciduous leaves: *Pelargonium otaviense* (*rössingense*), *Sarcocaulon mossamedense* (*marlothii*), *Othonna protecta*, *Senecio longiflorus*, *Adenia pechuelii*.

The Inner Namib desert

Phots.: Giess (1971: 5, 8); Walter (1971: 209).

Information on the region is sparse because most accounts of the Namib desert are generalized and do not draw a clear distinction between the Outer and Inner Namib or between the latter and adjoining semi-desert regions. According to Walter (1971) the plains of the Inner Namib are devoid of halophytes, and in their place occur grasses, chiefly species of *Stipagrostis*, associated with herbaceous Compositae and Acanthaceae. Succulents are represented by a single species, *Sesuvium sesuvioides* (*digynum*) (Aizoaceae). *Stipagrostis obtusa* is probably the most generally occurring dominant. It is often mixed with *S. ciliata* in the sandier parts and with *Eragrostis nindensis* on more stony or gravelly soils. *Kaokochloa nigrirostis* sometimes forms pure stands in the Northern Namib.

South of the Orange River similar desert grassland occurs on Kalahari Sand between Okiep and Pofadder. The rainfall is less than 40 mm. *Stipagrostis brevifolia* is the dominant species. It is a low shrubby plant with the perennating shoots branched above ground. During the dry season it sheds its leaves. At this time, when no inflorescences are present, the plants have the appearance more of dicotyledonous shrubs than of typical grasses. Although the tufts are widely spaced (mean distance c. 1 m) and rather small, being only 20–30 cm high and wide, the landscape from a distance looks as if it is densely shrubby. There are few associated species. The large shrub *Parkinsonia africana* may occur as widely scattered individuals. Typical Karoo dwarf shrubs are virtually absent, being represented by a few individuals of *Lycium* and *Hermannia*. After a good fall of rain, annuals, especially annual grasses and *Tribulus zeyheri*, are plentiful.

The Welwitschia bainesii *transition zone*

Refs.: Bornman *et al.* (1972–73); Giess (1969); Walter (1936; 1971, p. 369–73).
Photos.: Bornman *et al.* (1972: 2–3); Giess (1969: 12–18; 1971: 6, 11).

The distribution of *Welwitschia bainesii*, one of the most remarkable plants in the world, has been described in detail by Kers (1967), Giess (1969), and Barbosa (1970). It extends from the Kuiseb River just south of the Tropic of Capricorn to San Nicolau (14° 20′ S.) in southern Angola. In contrast to earlier beliefs its range is now known to be almost continuous.

It has been studied most in the Central Namib towards the southern limit of its range where it is the characteristic species of the transition zone between the Outer and Inner Namib, and occupies a narrow strip about 50 km from the coast between the Kuiseb and Swakop Rivers (Walter, 1971; Bornman *et al.*, 1972–73). According to Walter, fog is rare here (but see below) and is of little significance to the plant, although light summer rains are frequent.

In the Kuiseb–Swakop area the distribution of *Welwitschia* is not related to ground water and its taproots only go down to a depth of 1–1.5 m. It does not occur on the plains themselves, which are covered, after rains, with annual grasses such as *Stipagrostis subacaulis* and annual forms of perennial species such as *S. hochstetterana*. *Welwitschia* is found in broad flat channels in the plains which are so shallow as to be barely discernible. These channels receive floodsheet waters from higher areas, and the soil becomes moist to a depth of 1.5 m. This subsurface moisture can be retained for years. The *Welwitschia* plants are usually more than 20 m apart and associated plants other than ephemerals are sparse. *Welwitschia* can store a certain amount of water but has no specialized water-storing tissue. With lack of water the leaves die except for their meristematic bases. *Welwitschia* also occurs on slopes covered with coarse scree and in crevices in weathered rock but the plants are isolated and not very vigorous (Walter, 1971).

The ecology of *Welwitschia* has been studied by Bornman and his collaborators (Bornman *et al.*, 1972–73), both in the field and in the laboratory. Where mean annual rainfall is only 25 mm, precipitation from coastal fog is equivalent to a further 50 mm. Bornman has suggested that, contrary to the statements of Walter, some of this fog condensation is absorbed by the leaves, probably through the stomata, but the evidence is inconclusive (L. Leyton, pers. comm.).

The plant body of *Welwitschia* resembles a large fibrous carrot. The stem is up to 1.5 m tall and up to 10.8 m in circumference. Bornman estimates that the largest individuals are 2 500 years old.

There are only two leaves, which grow from a terminal groove in the photosynthetic tissue of the stem. They persist throughout the entire life of the plant and under favourable conditions grow from the basal meristems at the rate of 13.8 cm per year, so that the oldest plants would be capable of producing leaves well over 100 m long if it were not for their intermittent growth and continuous dieback from the extremities. When the leaf tip comes into contact with the ground it withers following the death of protoplasts caused by the high surface temperatures, and it becomes frayed from being scoured on the gravel surface by winds. However, even under gale force conditions, the living part of the leaf, which is rarely more than 3 m long, remains remarkably rigid and stable.

Welwitschia frequently forms groups of individuals of equal size, which are presumably of similar age. Bornman suggests that germination follows a freak downpour or series of downpours of approximately 25 mm. Germination will not take place until an inhibitor has been leached from the seed. The equivalent of 6.25 mm of rain is needed for this.

Walter (1936, 1971) implies that the transition zone between the Outer and Inner Namib is the most characteristic habitat of *Welwitschia*, but this is now known to be true only for the southernmost localities and some of the northern ones. It is not valid for the general distribution of the species (Kers, 1967). *Welwitschia* can occur within 8 km of the coast and extends up to 144 km inland. Its altitudinal range is from 100 to 900 m and it tolerates a wide range of precipitation and soil salinity and grows in a variety of vegetation types.

Welwitschia is not confined to the Namib–Mossamedes desert nor to the Karoo–Namib Region, but extends some way into the Karoo–Namib/Zambezia transition zone. At its eastern limit in Namibia near the small town of Welwitschia, where the rainfall is 200 mm per year, it occurs on gravel soils of deltaic deposits. Here, it is most abundant in the shade of riparian bushland dominated by *Colophospermum mopane* and *Terminalia prunioides*. Individuals are relatively large but are much smaller than the grotesque giants of the desert. On the stony flats between the watercourses, *Welwitschia* is the only 'woody' plant. The associated vegetation is composed of annual grasses, principally *Stipagrostis hirtigluma* and *Anthephora schinzii*, which sometimes completely conceal it. *Welwitschia* also occurs in this area on rocky sandstone hills covered with a relatively dense bushland dominated by *Colophospermum mopane*, *Terminalia prunioides*, *Acacia* spp. and *Commiphora* spp. At its eastern limits in Namibia *Welwitschia* is confined to compact gravel soils or rocky places. It cannot establish itself on less compact soils, since young plants are uprooted and washed away by run-off from violent rainfall (Kers, 1967).

River-bed communities

Photos.: Giess (1971: 3, 9, 10, 11, 12); Walter (1971: 225).

As in all deserts, erosion channels and dry valleys are the most favourable habitats for plants. When the rivers flow, the sand in their otherwise dry channels becomes wetted to an appreciable depth and then remains damp for several years. Most river beds are at least slightly saline but the degree of salinity is variable and depends on the geochemistry of the catchment area and certain geomorphological features. The vegetation varies greatly in relation to the amount of water available and its salinity.

In small erosion channels with non-brackish water one finds *Citrullus ecirrhosus* and a few annual herbs (*Cleome*, *Tribulus*).

When the amount of storm-water increases certain shrubs such as *Adenolobus pechuelii*, *Parkinsonia africana* and *Commiphora saxicola* (*dulcis*) appear.

On the margins of larger rivers taller shrubs and small trees, such as *Rhus lancea*, *Salvadora persica*, *Ficus sycomorus*, *Euclea pseudebenus*, *Acacia erioloba* and *A. albida* occur in riparian woodland.

In the bed of the lower Kuiseb on sand dunes with subterranean water, *Acanthosicyos horridus*, a leafless gourd with green thorns, forms dense impenetrable thickets.

Where the water is brackish the salt-secreting *Tamarix usneoides* occurs with *Lycium tetrandrum*.

In river estuaries, flood plains with accessible ground water support the following halophytes: *Zygophyllum stapfii*, *Arthraerua leubnitziae*, *Salsola* spp., *Suaeda plumosa* and *Arthrocnemum dunense*.

Springs with slightly brackish water are surrounded by communities of *Phragmites*, *Odyssea* (*Diplachne*) *paucinervis* and *Cyperus laevigatus*.

The desert of Mossamedes

Refs.: Barbosa (1970, p. 251–61); de Matos & de Sousa (1970); Diniz (1973, p. 269–90); Whellan (1965).
Photos.: Barbosa (1970: 28.3, 29.1–3); de Matos & de Sousa (1970: 1–3, 6).

The Namib desert continues into the south-western corner of Angola to a short distance north of Mossamedes. Rainfall is less than 100 mm per year, but atmospheric humidity is high. The flora is a continuation of that of the Namib proper, and the vegetation types are similar, but species of tropical affinity are much more numerous than further south.

Welwitschia is absent from the coastal belt of mobile dunes south of Porto Alexandre, but otherwise is scattered throughout. The vegetation of the Reserve Parcial de Mossamedes towards the northern limit of the desert has been described by de Matos and de Sousa (1970), who recognize the following main types:

1. Near the coast on saline soils there are halophytic communities characterized by *Salsola zeyheri*, *Sesuvium* spp., *Suaeda fruticosa*, *Scirpus littoralis* and *Asthenatherum* (*Danthonia*) *forskalii*.

2. A little further inland, on mesa-like terraces and in ravines on less saline, calcareous and gypsaceous soils, the following are found: *Aizoon virgatum*, *A. mossamedense*, *Euphorbia bellica*, *Zygophyllum orbiculatum*, *Z. simplex*, *Rhynchosia candida*, *Indigofera daleoides*, *Geigeria spinosa* and *Berkheyopsis angolensis*.

3. Riparian bushland fringing the Bero and Flamingos rivers is characterized by *Tamarix usneoides*, *Cordia sinensis* and *Euclea pseudebenus*. Associates include: *Sporobolus robustus*, *Atriplex halimus*, *Lotus arabicus* (*mossamedensis*), *Arthrocnemum indicum*, *Psoralea obtusifolia* and *Asthenatherum* (*Danthonia*) *mossamedense*.

4. On well-drained gravel soils near the city of Mossamedes there are communities of *Euphorbia virosa* (*dinteri*) associated with species of *Aristida*, *Stipagrostis*, and *Eragrostis*.

5. *Sarcocaulon mossamedense* is dominant in the dry rocky zone at the north of the Reserve. Other species are virtually absent owing to the strong prevailing winds. *Salvadora persica* occurs in saline depressions.

6. By far the most widespread type is a grassland dominated by desert annuals with scattered individuals of *Welwitschia bainesii*. The largest plants of *Welwitschia* have crowns 1 m or more in diameter and leaves up to 2 m long. They are found near the coast, where they are the only conspicuous feature of the vegetation, and are scattered about the arid plain at intervals of 50 to 100 metres. Lichens are locally plentiful. Except after falls of rain, other species are not apparent, except for the dead remains of annual species of *Stipagrostis* and other plants. Further inland the *Welwitschia* plants are smaller and the associated vegetation becomes more luxuriant. Within the desert area, the latter consists largely of grass which forms quite a sward and is probably the main food of the Springbok and other antelopes, which are fairly numerous (Whellan, 1965).

Bushes and dwarf trees are rare and local, and occur as scattered individuals in water-receiving depressions in the more humid eastern parts. In the western part of the Reserve, *Welwitschia* occurs on gravel soils and is most abundant along drainage lines. Its principal associates are: *Stipagrostis subacaulis*, *S. hirtigluma*, *Eragrostis porosa*, *Enneapogon cenchroides*, *Tricholaena monachne*,

Dicoma foliosa, Indigofera teixeirae, Geigeria spinosa, Hibiscus micranthus, Aloe littoralis, Sarcocaulon mossamedense, Sesuvium portulacastrum, Lophiocarpus polystachyus and *Lotononis tenuis*. Further to the east the following grasses join the assemblage listed above: *Aristida hordeacea, Stipagrostis hochstetterana, Danthoniopsis dinteri, Tetrapogon tenellus* and *Rhynchelytrum repens* (*villosum*). Further east still, towards the edge of the desert, *Welwitschia* occurs in depressions with the microphanerophytes, *Acacia tortilis, A. reficiens* subsp. *reficiens*, and *Maerua angolensis*. Other associates include *Lycium decumbens, Hoodia currorii, Monsonia senegalensis, Aristida hordeacea, A. rhiniochloa, Stipagrostis uniplumis, S. hochstetterana, Schmidtia pappophoroides* and *S. kalahariensis*. Rocky outcrops in this part of the Reserve support *Euphorbia subsalsa, Commiphora* sp., *Sterculia setigera* and *Sansevieria cylindrica*.

Less information is available for other parts of the Mossamedes desert. North of Porto Alexandre, *Welwitschia* almost reaches the coast, but to the south and extending far into Namibia, a belt of mobile sand dunes 30–50 km wide creates conditions approaching absolute desert (P. Bamps, pers. comm.). Thus, between Espinheira and Foz do Cunene, *Welwitschia* drops out at km 50 (its southern limit in Angola), and, apart from a few cactiform euphorbias and rare individuals of *Zygophyllum orbiculatum* sheltered by rocks, there are no plants until the halophytic communities of the coast are reached (P. Bamps, pers. comm.). Further north between Porto Alexandre and Iona, *Welwitschia* does not appear until km 49 where it occurs in sand dunes with *Acanthosicyos*. At the mouth of the Cunene, halophytic communities include *Cotula coronopifolia, Heliotropium curassavicum, Samolus valerandi, Chenopodium ambrosioides, Tetragonia reduplicata* and *Cyperus laevigatus*. The sand dunes near by are devoid of perennial plants except for a few individuals of *Rhigozum angolense* and *Phyla* cf. *nodiflora* (P. Bamps, pers. comm.).

VII The Mediterranean regional centre of endemism

Area, geographical position, geology and physiography

Climate

Flora

Mapping units

Vegetation
    Mediterranean forest
        Mediterranean broad-leaved sclerophyllous forest
            Quercus ilex sclerophyllous forest
            Quercus suber sclerophyllous forest
            Quercus coccifera sclerophyllous forest
        Mediterranean coniferous forest
            Juniperus phoenicea forest
            Cupressus sempervirens and *C. atlantica* forest
            Tetraclinis articulata forest
            Pinus halepensis forest
            Pinus pinaster forest
            Cedrus atlantica forest
            Abies pinsapo and *A. numidica* forest
            Juniperus thurifera forest
        Mediterranean deciduous forest
            Quercus faginea forest
            Quercus pyrenaica forest
            Quercus afares forest
    Mediterranean bushland and thicket
    Mediterranean shrubland
        Altimontane Mediterranean shrubland
        Secondary Mediterranean shrubland
        (maquis and garrigue)
    Mediterranean anthropic landscapes

Area, geographical position, geology and physiography

This section deals only with the African part of the Mediterranean Region, the Maghreb. Essentially it is the region of folded mountains at the north-western extremity of the continent. (Area: 330 000 km^2.)

The landscape is dominated by the Atlas mountains, which are largely the product of Tertiary folding and uplift of sediments deposited over a long period in the ocean which lay between the African and Tyrrhenian shields. The various ranges of the Atlas are separated by plateaux and basins. Coastal lowlands occupy a relatively small area.

The Atlas mountains extend for over 3 000 km from north Morocco to Tunisia. They trend from WSW. to ENE. and run roughly parallel to the Mediterranean coast. They are best developed in Morocco, where Mount Toubkal reaches an altitude of 4 165 m in the High (Great) Atlas, a mountain range with many snow-capped peaks. In Algeria altitudes do not exceed 2 500 m and in Tunisia 1 500 m.

The oldest of the fold mountains, the Rif Atlas, forms a coastal range which extends from south-east of Tangier to the Molouya river valley and continues in Algeria as the Tell Atlas. In some places there is a separate coastal range, the Maritime Atlas, between the Tell Atlas and the sea.

The High Atlas extends from the Atlantic coast near Agadir, and in Algeria becomes the Saharan Atlas, which rarely exceeds 2 000 m and mostly lies in the Sahara/Mediterranean transition zone. In eastern Algeria the Saharan Atlas and the Tell Atlas approach each other. Here occurs Algeria's highest peak, the 2 328 m high Djebel Chélia in the Aurès Mts, which are structurally part of the Saharan Atlas.

In Morocco the Middle Atlas, which diverges from the High Atlas in a north-easterly direction, consists mainly of a plateau bordered by mountain chains on the south and east.

The Anti-Atlas, which for most of its length is in the Mediterranean/Sahara transition zone, is an elevated part of the African shield. It has a tabular surface at about 1 500 metres, but its highest peak is nearly 3 900 m above sea-level. It is joined to the High Atlas by the volcanic formations of Djebel Siroua (3 304 m), but further west the two ranges are separated by the Souss alluvial plain, which occupies a structural depression.

The Atlas ranges extend into Tunisia as the Northern Tell, the High Tell and the Low Tell.

The lithology of the Mediterranean Region is diverse. The prevalent rocks are sediments, sometimes metamorphosed, of Triassic, Jurassic, and Cretaceous age, especially limestones. More recent Upper Tertiary and Quaternary deposits are relatively restricted. There are also small exposures of the Precambrian basement, and outcrops of volcanic rocks.

Climate

Most rain falls in winter, and nearly everywhere is between 250 and 1000 mm per year. The summer is hot and dry and is more extreme than that of the Cape Region. Frost is widespread but some parts of the coastal lowlands are frost-free, whereas parts of the interior may experience frost for up to seven months each year. In the high mountains snow frequently lies for long periods. (See Fig. 13.)

The relationships between climate and vegetation in the Mediterranean Region have been studied by several authors who have defined various climatic indices. Those of Emberger (Emberger, 1955a; Sauvage, 1961, 1963) and Bagnouls and Gaussen (1957), which are the best known, are summarized by Quézel (1976). The method of Bagnouls and Gaussen provided the basis of a bioclimatic map of the Mediterranean zone (Unesco–FAO, 1963). The classification of Emberger, however, has been more widely used in North Africa and his[1] *étages bioclimatiques* are frequently referred to in the following pages.

Emberger defines four main 'étages' using a climatic index based on mean annual rainfall, mean minimum temperature of the coldest month, and mean maximum temperature of the warmest month. They roughly correspond, however, to annual precipitation as follows:
— arid 'étage': 300–500 mm per year;
— semi-arid 'étage': 500–700 mm per year;
— subhumid 'étage': 700–1000 mm per year;
— humid 'étage' > 1000 mm per year.

For each 'étage' there are five variants based on the mean minimum temperature of the coldest month, as follows:
— hot > $7°C$;
— temperate: $3°–7°C$;
— fresh: $0°–3°C$;
— cold: minus $5°–0°C$;
— very cold: < minus $5°C$.

---

1. In French ecological literature the word *étage* is used in two senses, either to designate strictly altitudinal vegetation belts, or to designate areas which can be characterized in terms of climate and vegetation, but of which the spatial relationships are more often of the checkerboard type than strictly altitudinal. It is the latter sense which has been adopted in North Africa by Emberger and others.

Emberger also recognized a Mediterrano-Saharan 'étage' and a Mediterranean high-mountain 'étage'. The latter largely corresponds to the very cold variants of the 'étages' listed above.

The Saharan 'étage' is represented in the Mediterranean Region *sensu stricto* only by very small enclaves.

Flora

About 4000 species occur in the north African part of the Mediterranean Region (excluding the Mediterranean–Sahara transition zone). Of these c. 72.5 per cent are Mediterranean endemics, though only c. 20 per cent of them are confined to North Africa.

Endemic families. None. *Aphyllanthes* (monotypic, Liliaceae) is sometimes given family rank. Globulariaceae, which has most species in the Mediterranean Region, is also represented by two species of *Poskea* in Somalia and Socotra. The Cneoraceae is confined to the Mediterranean Region, the Canaries and Cuba.

Endemic genera. About 250 genera have their greatest concentration of species in the Mediterranean Region, but many are not strictly endemic. Thus, among genera occurring in North Africa, *Cyclamen* reaches Persia, and *Cistus* extends from the Canary Islands to Trans-Caucasia. Other near-endemic genera include *Ceratonia, Helianthemum, Genista, Lavandula* and *Ononis*. More strictly endemic are *Anagyris, Chamaerops, Coriandrum, Halimium, Spartium, Tetraclinis* and many genera of Cruciferae.

Endemic species. Non-endemic genera with more than 50 endemic species in Mediterranean North Africa include: *Astragalus, Centaurea, Euphorbia, Linaria, Silene, Teucrium, Trifolium* and *Vicia*. Many of the most characteristic Mediterranean species, including *Arbutus unedo, Cedrus atlantica, Laurus nobilis, Myrtus communis, Nerium oleander, Quercus coccifera, Q. ilex, Q. suber* and *Vitex agnus-castus*, are endemic, but belong to non-endemic genera.

Linking elements. Approximately 1.5 per cent of the species occurring in Mediterranean North Africa *sensu stricto* are cosmopolites; 20 per cent, including *Acer campestre, Alnus glutinosa, Betula pendula (B. alba* auct.), *Calluna vulgaris, Carex capillaris, Digitalis purpurea* and *Prunus padus*, are Boreal linking species. Only 3 per cent, including *Lupinus varius (pilosus). Retama retam, Stipagrostis (Aristida) pungens* and *Ziziphus lotus* are Saharan linking species, and 2.2 per cent are Irano-Turanian linking species. Other linking species include *Erica arborea*, which also occurs on the high mountains of the Sahara and East Africa, and *Pistacia atlantica*, which extends from the Canary Islands to Afghanistan.

Vegetation of the floristic regions

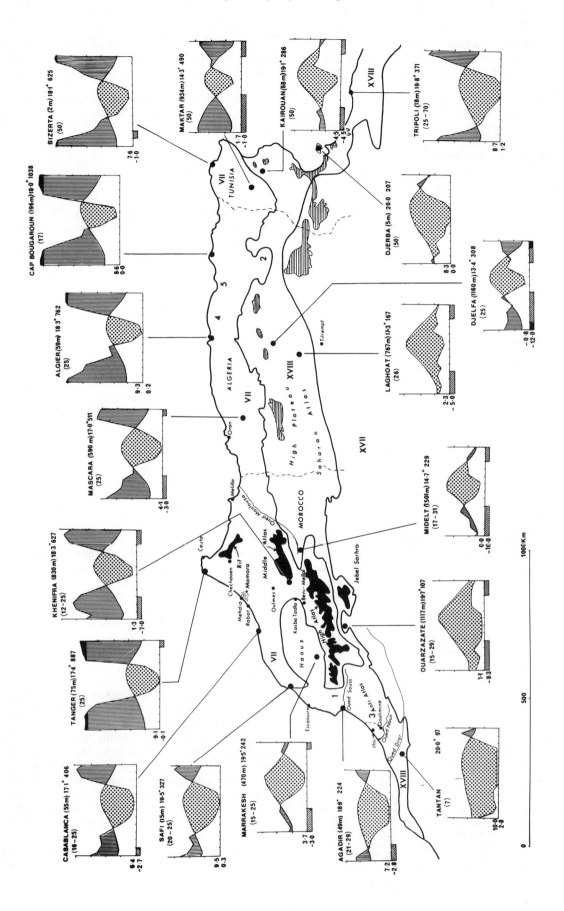

FIG. 13. Climate and topography of the Mediterranean regional centre of endemism (VII) and the western half of the Mediterranean/Sahara regional transition zone (XVIII) 1. Argana. 2. Aurès. 3. Bou Izakarn. 4. Djaradjura. 5. Jebel Babor

Floristic relationships with the Cape Region are slight (Burtt, 1971). Only 30–35 Mediterranean species, of which 15 are in *Erica*, belong to characteristic Cape genera.

Mapping units

10. Mediterranean sclerophyllous forest.
23. Mediterranean montane forest and altimontane shrubland.
78. Mediterranean anthropic landscapes.

Vegetation

Most of the Maghreb was formerly covered with forest, but on clay soils in the semi-arid 'étage' scrub forest dominated by *Olea europaea* and various types of bushland or thicket may formerly have represented the climax. Otherwise, non-forest woody vegetation was confined to shallow soils, wind-swept ridges and coastal habitats, and the summits of the higher mountains.

Ionesco and Sauvage (1962) recommend the use of the Spanish word 'matorral' to cover all woody types of non-forest vegetation in the Mediterranean Region. This usage, which corresponds to the English 'scrub' (bushland plus shrubland) has much to recommend it.

Although grasses are well represented in the Mediterranean flora, climax grasslands were formerly of limited extent. Grasslands dominated by *Stipa tenacissima* and *Lygeum spartum*, which are probably largely secondary, are found in the drier parts of the Mediterranean Region. They are more characteristic, however, of the Mediterranean/Sahara transition zone and are described in Chapter XVIII. *Ampelodesma mauretanicum* also locally forms a secondary 'steppe' grassland in higher-rainfall areas where heavy cutting and grazing on steep clay slopes have led to soil removal.

Communities dominated by *Argania spinosa*, *Acacia gummifera* and succulent Euphorbias, which represent the most characteristic vegetation types of the western end of the Mediterranean/Sahara transition zone, also occur very locally in the Mediterranean Region in Morocco, but only as small enclaves or marginal intrusions. They too are described in Chapter XVIII.

Mediterranean forest
(mapping units 10 and 23)

Most of the North African part of the Mediterranean Region *sensu stricto* was formerly covered with forest. Only small relicts survive, but they include at least 60 tree species, of which 16 are major dominants. Three species, *Quercus ilex*, *Q. suber* and *Q. coccifera*, dominate evergreen sclerophyllous forest. Ten species, *Abies numidica*, *A. pinsapo* subsp. *marocana*, *Cedrus atlantica*, *Cupressus atlantica*, *C. sempervirens*, *Juniperus phoenicea*, *J. thurifera*, *Pinus halepensis*, *P. pinaster* and *Tetraclinis articulata*, dominate coniferous forest, and three species, *Quercus faginea*, *Q. pyrenaica*, and *Q. afares*, dominate deciduous oak forest.

Of the remainder, the majority are Mediterranean endemics, or, if they occur elsewhere, their distributions are centred on the Mediterranean basin. They include: *Acer monspessulanum*, *Arbutus unedo*, *Celtis australis*, *Ceratonia siliqua*, *Chamaerops humilis*, *Crataegus azarolus*, *Fraxinus angustifolia*, *F. xanthoxyloides*, *Juniperus oxycedrus*, *Laurus nobilis*, *Olea europaea*, *Phillyrea angustifolia* (including *P. latifolia* and *P. media*), *Pinus pinaster*, *Pistacia atlantica*, *P. lentiscus*, *P. terebinthus*, *Prunus lusitanica*, *Pyrus gharbiana*, *P. cossonii* (*longipes*), *P. mamorensis* and *Rhus pentaphylla*.

Several Eurosiberian linking species also occur in Mediterranean North Africa, notably *Acer campestre*, *Alnus glutinosa*, *Betula pendula*, *Crataegus monogyna*, *Ilex aquifolium*, *Populus tremula*, *Prunus avium*, *P. padus*, *Sorbus aria*, *S. domestica*, *S. torminalis*, *Taxus baccata* and *Ulmus campestris*. All have restricted distributions and are virtually confined to the humid and subhumid 'étages' of the Rif in Morocco, the wetter parts of the coastal plain and coastal ranges in Algeria, and the wetter parts of the seaward slopes of the Atlas Mts.

Most types of Mediterranean forest in North Africa have suffered extreme degradation nearly everywhere, and have disappeared completely from large areas. Some are represented today by no more than tiny relictual stands. It is possible that some types have gone completely except for a few relict individual trees. Three such types are briefly mentioned below.

1. *Celtis australis* is one of the tallest deciduous trees in Mediterranean Africa. It reaches a height of 25 m. Today it occurs as rare individuals in the lowlands and is even more scattered inland, where it ascends to 1 300 m. According to Monjauze (1958) this species has suffered more at the hands of man and his flocks than any other Mediterranean tree. Monjauze presents evidence which suggests that, on deep soils in the subhumid 'étage' and in the lower warmer part of the humid 'étage', *Celtis australis* formerly dominated forests which had an understorey of *Laurus nobilis* and a herb layer which included *Acanthus mollis*.

2. *Pistacia atlantica* is also a fine deciduous tree which reaches a height of 20 m and a diameter of 1 m. It has a scattered distribution from the Canary Islands to Afghanistan. In the Maghreb it is one of the most widely distributed species but occurs as isolated individuals, not in forest stands. It is most plentiful in the warm semi-arid 'étage' and ascends to 2 000 m in the west and 3 000 m in the east. It regenerates freely from seed, especially in the protection of clumps of *Ziziphus lotus*. Its rarity today is due to its susceptibility when young to browsing and fire, and to the limited occurrence of deep soil because of extensive erosion. Monjauze (1968) postulates that on deep soils in the semi-arid 'étage' the

climax forest was formerly dominated by *Pistacia atlantica* mixed with sclerophyllous oaks.

3. In marshy places on the Algerian coast the climax forest was probably composed of *Ulmus campestris* mixed with *Fraxinus angustifolia*, *Populus alba*, *Salix alba* and *Laurus nobilis*.

The forests of Morocco are floristically richer and more diversified than those occurring elsewhere in Mediterranean Africa, and the following account is based on them, but the forests of Algeria and Tunisia are essentially similar and the dominants are mostly the same. *Quercus pyrenaica* and *Abies pinsapo*, subsp. *marocana*, however, are confined to Morocco, and are replaced by *Q. afares* and *A. numidica* in Algeria; neither occurs in Tunisia. *Cedrus atlantica* and *Pinus pinaster* are also absent from Tunisia, and *Cupressus sempervirens*, *sensu lato*, is absent from Mediterranean Algeria.

Five forest dominants, *Quercus ilex*, *Q. coccifera*, *Pinus halepensis*, *Cupressus sempervirens* and *Juniperus phoenicea*, also occur in Cyrenaica. Their former distribution was restricted and only small degraded remnants remain today. *Tetraclinis articulata* also occurs in Cyrenaica but is doubtfully native.

Argania spinosa is extremely localized in the Mediterranean Region *sensu stricto* but forms extensive communities at the western end of the sub-Mediterranean transition zone. They are described in Chapter XVIII.

Because many of the dominants of Mediterranean forest grow within a wide range of climate, classification is difficult. In the following account the distribution of the main types is described in outline in relation to the 'étages' bioclimatiques of Emberger (page 147).

Mediterranean broad-leaved sclerophyllous forest

This type occupies about half of the total forest area. Everywhere it is dominated by *Quercus ilex*, *Q. suber* or *Q. coccifera*, the distributions of which are almost mutually exclusive. By far the largest area (2 100 000 ha, or more than two-thirds of the total) is *Q. ilex* forest. *Q. coccifera* is relatively restricted (44 000 ha).

Quercus ilex *sclerophyllous forest*

Refs.: Boudy (1948, p. 139–40; 1950, p. 299–351); Emberger (1939, p. 107–10, 111–14, 135–6); Métro (1958, p. 68–73); Peyerimhoff (1941, p. 53); Quézel (1976).

Phots.: Boudy (1948: 6–8; 1950: 38–46); Emberger (1939: 9.2); Métro (1958: 4).

Quercus ilex is the most widely distributed and most abundant tree in Mediterranean Africa, and probably in the Mediterranean basin generally. In the Maghreb it is virtually absent from the lowlands below 400 m, but forms forests above that altitude up to 2 400 m in the Atlas. Scattered individuals ascend to the tree line at *c*. 2 900 m.

Quercus ilex is essentially a species of the mountains. In the Rif it ascends to 2 200 m, and is very widespread at higher altitudes above the *Tetraclinis*, *Quercus suber* and *Pinus halepensis* forests of the lower slopes. In the Middle and High Atlas it covers an enormous area between 600 and 2 900 m on the Atlantic slopes, but is much less abundant on the Saharan slopes and on Mediterranean slopes of pronounced relief. In the Anti-Atlas it forms forest only at the western end.

The ecology of *Quercus ilex* is diverse. It occurs principally under temperate and cold climates in the semi-arid, subhumid, and humid 'étages'. In Morocco, despite the fact that the forests have been decimated over large areas, *Quercus ilex* has survived in anthropic landscapes more frequently than other Mediterranean forest trees because of its resistance to fire and ability to sucker from stumps and damaged roots. It occurs on a wide range of soils but is absent from most clayey depressions.

Towards its lower altitudinal limits in the hot semi-arid 'étage' it is in contact with *Tetraclinis articulata*, *Juniperus phoenicea* and *Pinus halepensis* and, very locally, with *Quercus suber*. In Morocco it occurs in the lowlands of the semi-arid 'étage' only south of Rabat where, on siliceous soils, it forms a band between the forests of *Quercus suber* and those of *Tetraclinis*. These forests are low and open, and are formed of stunted twisted trees. They differ from the nearby *Quercus suber* forests only in their dominant species. The shrub and field layers of the two forest types are virtually identical. This is because during the wet season their climates are similar. During the dry season the *Quercus ilex* forests experience hotter and drier conditions. It is summer drought which eliminates *Q. suber*.

Towards the upper limit of its range in the cold semi-arid 'étage', *Quercus ilex* is in contact with *Juniperus phoenicea*, *Juniperus thurifera* and *Cupressus atlantica*.

In the subhumid and humid 'étages', where it attains its maximum development, it occurs in mosaic with, and sometimes intermingled with, *Quercus suber*, *Q. faginea*, *Q. pyrenaica*, *Pinus pinaster* and *Cedrus atlantica*.

Under favourable conditions *Q. ilex* attains a girth of 3 m and a height of 20 m with a magnificent wide-spreading crown. More frequently, however, it is a smaller tree with a short trunk and more compact crown.

Its coppice sometimes forms thickets, in which seedlings of *Cedrus*, *Pinus halepensis* and *P. pinaster* find protection from browsing animals and can regenerate.

The ability of *Quercus ilex* to withstand competition is remarkable. It is capable of persisting as suppressed individuals beneath the thick canopy of *Cedrus* forest. When a cedar dies, *Quercus ilex* grows rapidly to occupy the vacant space, but regenerating cedars ultimately overtop it and cast their shade again.

In the subhumid 'étage'

Nearly all the surviving stands of *Quercus ilex* forest in Morocco occur in this 'étage'. They cover enormous areas on the lower slopes of the Rif and the Atlantic slopes of the Middle Atlas and the High Atlas. In the Rif *Q. ilex* plays an important part only on the Mediterranean slopes, where it occurs both on calcareous and siliceous soils. On the moister Atlantic slopes it is replaced by other species of *Quercus*, especially *Q. suber*, except on calcareous soils, which are of limited extent. All the stands of subhumid *Q. ilex* forest in the Middle Atlas are on calcareous soils. Those in the High Atlas occur both on calcareous and siliceous soils.

At the western extremity of the High Atlas and in the mountains north of Essaouira (Mogador), *Q. ilex* forms islands of forest above 650 m, at which altitude it replaces *Tetraclinis*. The southernmost stand of subhumid *Quercus ilex* forest is in the massif of Kest in the western Anti-Atlas. A little further to the south-west, scattered individuals of *Q. ilex* occur in populations of *Tetraclinis* on the highest peaks of Ifni (1 250 m).

Subhumid *Quercus ilex* forest has a low, dense, almost closed canopy when it is intact; the underwood is sparse. Such stands are rare. Most have been degraded and invaded by heliophilous species. Emberger (1939) recognizes four types of subhumid *Quercus ilex* forest based on temperature and substrate, but their floristic characterization is somewhat weak.

In the humid 'étage'

This type of *Q. ilex* forest differs from drier types in its greater stature and denser canopy, the abundance of bryophytes and lichens, and in the deep, humus-rich soil. Floristically it shares several species with the forests of the Eurosiberian Region. In Morocco it is confined to the Rif and the Middle Atlas.

The best-known examples occur near Azrou in the Middle Atlas. The canopy is almost pure but there are scattered individuals of *Acer monspessulanum*, *Taxus baccata* and *Sorbus torminalis*. In the underwood many species occur including: *Coronilla valentina* (*glauca*), *Cotoneaster fontanesii*, *Crataegus monogyna*, *Cytisus battandieri*, *Daphne laureola*, *D. gnidium*, *Ilex aquifolium*, *Juniperus oxycedrus*, *Lonicera etrusca*, *Rosa* sp., *Rubus ulmifolius*, *Ruscus aculeatus*, *Viburnum lantana* and *V. tinus*. Climbers are represented by *Asparagus acutifolius*, *Clematis cirrhosa*, *Hedera helix*, *Rubus*, *Smilax aspera*, and *Tamus communis*. In ravines one encounters *Euonymus latifolius*, *Ligustrum vulgare*, *Rhamnus catharticus*, *Salix cinerea* and *Sorbus aria*. The diversified field layer includes many geophytes but annuals by contrast are very rare when the forest is not degraded.

Quercus suber *sclerophyllous forest*

Refs.: Boudy (1948, p. 137–9; 1950, p. 29–180); Emberger (1939, p. 101–5, 118–19, 136–7); Métro (1958, p. 51–68); Peyerimhoff (1941, p. 54–5); Quézel (1976); Sauvage (1961).
Phots.: Boudy (1948: 4–5; 1950: 2–25); Métro (1958: 3).

Together with the Atlas Cedar (*Cedrus atlantica*), the Cork Oak is the most valuable tree in North Africa. Unlike *Q. ilex* it is confined to the western half of the Mediterranean basin and does not extend further east than Italy. In the Maghreb it occupies 843 000 ha.

The canopy of *Q. suber* forest is usually low and somewhat open, and occurs at a height of 6–12 m in the drier types, but is closed and taller (15 m) in the wetter types. There is usually a well-developed shrub layer, 2–4 m tall, except when the canopy is dense and continuous; then the underwood is less abundant or disappears.

Q. suber avoids wet, cold conditions. In Morocco it occurs locally almost at sea-level and ascends to 1 600 m in the Rif, 1 550 m in the Middle Atlas, and 2 100 m in the High Atlas. At different altitudes its climatic and edaphic requirements are somewhat different. Like *Q. ilex* it occurs in the semi-arid, subhumid and humid 'étages'. At the lowest altitudes on clay soils it is replaced by *Olea–Pistacia lentiscus* scrub, on calcareous soils by *Pinus halepensis*, and on the littoral by *P. pinaster* or *Juniperus phoenicea*. In the mountains it is in constant competition with *Q. faginea*, and in Algeria with *Q. afares*, which are better adapted to cold, grow taller, form a more continuous canopy, and regenerate more prolifically. Where *Q. suber* and the deciduous oaks occur in mixed stands the latter would be dominant if it were not for the effects of fire, to which they are less well adapted than *Q. suber*. At higher altitudes and on drier soils or where the dry season is severe, *Q. suber* is replaced by *Q. ilex*. *Q. suber* never forms forests on calcareous soils.

Semi-arid *Q. suber* forest covers extensive areas in NW. Morocco but in Algeria it occurs only as relics in the Oran–Mascara region. Nearly all the *Q. suber* forests in Algeria and Tunisia are of the humid and subhumid types. In Morocco, these wetter types occur only in the Rif. There are certainly some floristic differences between the semi-arid, subhumid, and humid types but superimposed on this are floristic differences which do not appear to reflect ecological conditions.

Except in the remotest places, the bark of *Q. suber* is stripped from the boles of all mature trees every nine years to provide the cork of commerce.

In the semi-arid 'étage'

This type is a xerophilous open forest with a floristically poor underwood but with a field layer rich in therophytes. It occurs principally on Pliocene sands or on schist.

The best-known example in Morocco is the Mâmora forest which occurs on deep sands. An endemic Pear, *Pyrus mamorensis*, is scattered in the canopy. In most of the forest the shrub layer is dominated by *Cytisus* (*Teline*) *linifolius*, and in the eastern part by *Halimium halimiifolium*. In the more open parts *Cytisus arboreus*, *Daphne gnidium*, *Lavandula stoechas* and *Ulex* (*Stauracanthus*) *boivinii* are found. *Halimium libanotis* is abundant in the western part of the forest. *Cistus salviifolius* is common only in those parts which are moist in winter. *Chamaerops humilis* is especially abundant in places where there is a clay horizon near the surface.

In large clearings and at the edge of the forest *Thymelaea lythroides* and the conspicuous composite *Ormenis multicaulis* are particularly common. At the centre of large clearings the woody plants have often disappeared and are replaced by open seasonal herbaceous vegetation, in which bulbous plants, especially *Urginea maritima, Dipcadi serotinum, Asphodelus microcarpus* and *A. aestivus*, are abundant. In this extremely degraded community regeneration of *Q. suber* is sporadic or absent.

In the subhumid 'étage'

The shrub layer is better developed than in the humid type. Below 1 200 m in the Rif *Q. suber* forms a canopy at 15 m beneath which there is a dense shrub layer 4–5 m tall of *Arbutus unedo* and *Erica arborea*. Beneath this double canopy there are scattered small shrubs such as *Cistus salviifolius, Cytisus monspessulanus* and *Lavandula stoechas*. In the meagre field layer *Carex distachya* and *Eryngium tricuspidatum* are found. If a tree dies it is eventually replaced but larger gaps in the canopy are filled by a very dense growth of *Arbutus* and *Erica arborea*, which seriously hinders the regeneration of the Cork Oak. Repeated cutting or burning results in the invasion and eventual dominance of heliophilous shrubs such as *Cistus populifolius, C. crispus, Halimium lasiocalycinum, Erica umbellata* and *Calluna vulgaris*. Further degradation, as in the humid zone, is followed by the establishment of even more xerophilous species, which are otherwise confined to the semi-arid 'étage'.

In the humid 'étage'

In Morocco only a few undisturbed stands survive in the Bab-Azhar forest in the Rif between 1 200 and 1 500 m, but this type occurs extensively in eastern Algeria and Tunisia. Beneath the closed canopy the shrub layer is sparse and consists of scattered individuals of *Cytisus villosus (triflorus), C. maurus, Cistus salviifolius* and *Ulex boivinii*. The field layer, which is also exiguous, includes *Dactylis glomerata* and *Pteridium aquilinum*. If a small gap in the canopy is formed by the death or removal of a tree the above-mentioned shrubs increase in abundance and are joined by more heliophilous species. Young plants of *Q. suber* can establish themselves in such gaps but only outside the area occupied by the roots of established trees, which would inhibit their growth.

Humid *Q. suber* forest differs from the subhumid type in the presence of *Alnus glutinosa, Prunus avium, Quercus faginea* and *Taxus baccata* in valley bottoms.

Quercus coccifera *sclerophyllous forest*

Refs.: Boudy (1950, p. 378–81); Emberger (1939, p. 115); Métro (1958, p. 101): Peyerimhoff (1941, p. 53–4).

The Kermes Oak, *Quercus coccifera* (including *Q. calliprinos*), has a patchy circum-Mediterranean distribution. In the Maghreb it occupies 44 000 ha. In Morocco, it occurs only in the Rif, except for a single station further south. Towards the east, in Algeria and Tunisia, it is much more abundant and occurs on coastal sands and extends a short distance inland, where it is replaced by *Q. suber*. *Q. coccifera* grows on a very wide range of soils, both acid and alkaline, and under a mean annual rainfall from 450 to more than 1 000 mm.

Q. coccifera is usually seen as a dense, much-branched shrub, but locally it is a tree, as in the mountains above Ceuta in Morocco, where it occurs with *Taxus baccata*. Here it is possible for a man to walk unimpeded beneath its boughs. It is possible that *Q. coccifera* formerly formed extensive forests. Almost everywhere it has been cut for charcoal, but, because of its capacity to sucker and withstand repeated cutting, it still covers large areas in its familiar shrubby form. In a few places in Algeria it can still be seen in relict patches of relatively undisturbed forest as a tree up to 12 m tall with a bole about half that length (P. J. Stewart, pers. comm.). Here, on maritime sands, it is often associated with 8–9 m tall *Juniperus phoenicea*. At Mostaganem in Algeria, *Pistacia lentiscus, Olea europaea, Phillyrea angustifolia, Ephedra altissima* and *Ceratonia siliqua* (on the landward side only) occur as shrubs or occasionally as small trees in the understorey. Small shrubs are: *Retama monosperma, Calicotome villosa (intermedia), Withania frutescens, Clematis cirrhosa, Asparagus acutifolius, Lycium intricatum, Lavandula dentata* and *Teucrium polium* (P. J. Stewart, pers. comm.).

Mediterranean coniferous forest

Nearly half of the area occupied by Mediterranean forest in the Maghreb is dominated by coniferous forest. There are 10 main dominants: *Abies numidica, A. pinsapo* subsp. *marocana, Cedrus atlantica, Cupressus atlantica, C. sempervirens, Juniperus phoenicea, J. thurifera, Pinus halepensis, P. pinaster* and *Tetraclinis articulata*. The areas they occupy vary greatly from a few hundred hectares (*A. numidica*) to 1 300 000 ha (*Pinus halepensis*).

Four other conifers occur in the Maghreb, but not as dominants. Thus:

— *Taxus baccata* is frequently associated with *Abies numidica, A. pinsapo, Cedrus atlantica* or *Quercus ilex* in the humid 'étage'.

— *Pinus nigra* is only known from two localities, one in Algeria, where it occurs in the middle of *Cedrus* forest, and the other in the Rif in Morocco where it is mixed with *Pinus pinaster*.

— *Juniperus oxycedrus* occurs throughout the Maghreb, either as a tree 10 m tall with a bole 1 m in diameter, or more often as a smaller bushy plant. It ascends from sea-level to 3 000 m and is nearly always associated with *Quercus ilex* or *Juniperus phoenicea*.

— The Holarctic species, *Juniperus communis*, forms dense cushions above the timber-line.

Juniperus phoenicea *forest*

Refs.: Boudy (1948, p. 134–5; 1950, p. 741–53); Emberger (1939, p. 78–86); Métro (1958, p. 77–8); Peyerimhoff (1941, p. 50).
Phot.: Boudy (1950: 116).

Juniperus phoenicea extends from the Canary Islands to Arabia and Jordan. It has two principal habitats, coastal sands and the high plateaux and mountains of the interior. It is almost confined to the semi-arid 'étage' in which it occurs on a variety of soils. In Morocco it often occurs in a zone between *Tetraclinis* forest and *Quercus ilex* forest, but in the coldest parts of the semi-arid 'étage' it replaces *Tetraclinis*. In Algeria it is often mixed with *Pinus halepensis*, but it is in the Saharan Atlas bordering the desert that it achieves its greatest extension.

Today, *Juniperus phoenicea* usually occurs as a bushy tree less than 7 m tall in open stands, which physiognomically are wooded grassland. They probably represent degraded forest. This is suggested by the fact that it can occur as a tree 8–9 m tall (P. J. Stewart, pers. comm.) with a massive trunk up to 2 m in girth. Although *J. phoenicea* locally dominates thicket on coastal sands where it is exposed to strong wind, or bushland in the interior where the soils are too shallow to support forest, it is likely that formerly it also dominated short forest which covered considerable areas. Emberger, for instance, states that the dunes near Essaouira (Mogador) were once covered with *J. phoenicea* forest. The fact that 'tout Mogador est construit avec du bois de *Juniperus phoenicea*' testifies to its former abundance as a relatively large tree.

For Morocco, Emberger describes the following three types of *J. phoenicea* communities.

1. Littoral communities

J. phoenicea only grows well under the protection of the first ridge of dunes. At Mehdia on the Atlantic coast the tiny remaining fragments are in the form of wind-trimmed thicket (White, MS, 1974). They contain *Phillyrea angustifolia, Pistacia lentiscus, Ephedra fragilis, Rhamnus alaternus, R. oleoides, Jasminum fruticans, Smilax aspera, Clematis cirrhosa, Asparagus albus* and *Osyris* sp. Near Essaouira *J. phoenicea* is associated with many of the species which occur at Mehdia, but with southern elements, including *Periploca laevigata* and *Helianthemum canariense*, in addition.

2. Cis-Atlas communities

These occur below 2 200 m in the Great and Middle Atlas. Where *J. phoenicea* forms a horizon between *Tetraclinis* and *Quercus ilex* it has no floristic individuality. In the lower part its associates are the most tolerant members of the former and in the upper part of the latter. *J. phoenicea* only occurs at two places in the Rif.

3. Trans-Atlas communities

These are the driest types and are very degraded, often almost to the point of total disappearance. They occur on the southern slopes of the Great Atlas and Anti-Atlas and on the lower slopes of the Upper Moulouya Valley. They occupy a zone above *Stipa tenacissima* grassland which may be, at least in part, secondary (page 229). On this side of the Atlas, *Q. ilex* forests formerly occupied only a narrow band, and, at least locally, with increasing altitude, *J. phoenicea* is in contact with *J. thurifera*. The more prominent associates of *Juniperus phoenicea* in these degraded communities include: *Fraxinus xanthoxyloides, Buxus balearica, Rhamnus alaternus, R. oleoides, Adenocarpus bacquei, Carthamnus fruticosus, Genista myriantha, Globularia alypum, Lavandula multifida, Artemisia herba-alba* and *Stipa tenacissima*.

Cupressus sempervirens *and* C. atlantica *forest*

Refs.: Boudy (1950, p. 764–70); Destremau (1974, p. 67–76); Emberger (1939, p. 100); Métro (1958, p. 79); Peyerimhoff (1941, p. 49–50).
Phots.: Boudy (1950: 122–3).

Cupressus sempervirens s.l. is a circum-Mediterranean species which extends as far east as Jordan. In the Maghreb it is indigenous only in two principal localities, namely the Mactar region of central Tunisia, and in Morocco, in the basin of Oued N'fis, south of Marrakech, where it occupies 10 000 ha between 1 100 and 1 800 m. There are also a few small scattered populations in the High Atlas between 1 000 and 2 000 m.

C. sempervirens has been so widely planted since classical times that there is some doubt concerning the extent of its natural distribution. The fastigiate form, var. *sempervirens*, which probably does not occur in the wild, is absent from the Tunisian population, which is probably indigenous (P. J. Stewart, pers. comm.). The Moroccan plant is so different from the naturally occurring *C. sempervirens* that in the opinion of most botanists it should be treated as a distinct species, *C. atlantica*.

The main population of *C. atlantica* forms an island in the *Juniperus phoenicea* horizon between those of *Tetraclinis articulata* and *Quercus ilex* in the semi-arid 'étage'. The associates of *Cupressus atlantica* and *Juniperus phoenicea* are virtually identical.

C. atlantica can develop into a fine tree 40 m or more tall but today, among the older trees, it is mostly mutilated wrecks that are seen. It has, however, been protected for more than thirty years and is reverting to its natural condition.

Tetraclinis articulata *forest*

Refs.: Boudy (1948, p. 133–4; 1950, p. 706–39); Emberger (1939, p. 71–8); Métro (1958, p. 79–83); Peyerimhoff (1941, p. 49).
Phots.: Boudy (1948: 3; 1950: 105–12); Emberger (1939: 5.1); Métro (1958: 7).
Syn.: association du Thuja (*Callitricetum*) (Boudy, 1948); forêt de Thuja de Barbarie (Emberger, 1939); la Callitraie (Emberger, 1939).

Tetraclinis articulata is confined to North Africa, except for a small population in Malta and another in the extreme south-eastern corner of Spain. It extends from southern Morocco to Tunisia, with a gap between Algiers and the Tunisian frontier. In Morocco it occurs on the lower Mediterranean slopes of the Rif, and then extends eastwards into Algeria. It also occurs on the lower northern slopes and in deep rain-shadow valleys of the Middle Atlas and High Atlas and extends round the western end of the latter to the northern slopes of the Anti-Atlas. Extensive *Tetraclinis* forests occur in the hinterland between Essaouira and Agadir above the *Argania* scrubland, and in deep valleys of the upper courses of rivers inland from Rabat and Casablanca.

Tetraclinis is generally confined to the oceanic and maritime semi-arid 'étages' between sea-level and 1500 m. Cold, especially humid cold, prevents it from ascending higher. It does not occur on the southern slopes of the High Atlas east of Siroua nor of the Middle Atlas, except at the northern end where the climate is not too continental because of the proximity of the sea. It is found on both calcareous and siliceous soils, but they are nearly always lithosols. It does not tolerate badly drained conditions. At the moister limits of its range it is confined to calcareous soils; on other well-drained soils it is replaced by *Quercus suber*, and on clay soils by *Olea–Pistacia* scrubland. At the moister limits of its range, where it is in contact with *Quercus ilex* or *Q. suber*, it usually occupies the xerocline, but at the drier limits, where it intermingles with *Argania*, it prefers the mesoline.

Well-grown *Tetraclinis* forest is from 12 to 15 m tall, but it is often much shorter than this. The narrow crowns form only a light canopy and most of the associated species are heliophilous. Certain species are constantly present in *Tetraclinis* forest. Others are of more local occurrence.

Species which are always present in *Tetraclinis* forest include *Cistus villosus*, *Ebenus pinnata*, *Lavandula multifida*, *Osyris* sp. and *Teucrium polium*, but they are not exclusive. *Cistus villosus* and *Teucrium polium* are the least faithful.

Other less constant species are *Ampelodesma mauritanicum*, *Anthyllis cytisoides*, *Brachypodium ramosum*, *Cistus clusii*, *Clematis cirrhosa*, *Ephedra fragilis*, *Erica multiflora*, *Genista retamoides*, *Helianthemum lavandulifolium*, *Jasminum fruticans*, *Lavandula dentata*, *Quercus coccifera*, *Rosmarinus officinalis*, *Teucrium fruticans* (a good indicator of calcareous soil), *Viola arborescens* (strictly littoral) and the endemic *Polygala balansae*.

Pinus halepensis *forest*

Refs.: Boudy (1948, p. 132–3; 1950, p. 639–90); Destremau (1974, p. 5–28); Emberger (1939, p. 94–100); Métro (1958, p. 74–7); Peyerimhoff (1941, p. 48–9).
Phots.: Boudy (1950: 90–101); Emberger (1939: 2); Métro (1958: 5).

Pinus halepensis, which reaches a height of 20 m, is distributed throughout the greater part of the Mediterranean basin from the southern shores of the Black Sea to Spain and Morocco. It occurs in Cyrenaica. In natural forests it does not extend to the Atlantic seaboard, but grows well when planted there. In southern Morocco the nearest populations to the Atlantic Ocean are 145 km inland. In North Africa there are immense populations in Tunisia and in South Oran, but they have often been degraded by fire. In Morocco the only extensive forests are in the High Atlas, but there are many scattered smaller stands. The total area occupied by *Pinus halepensis* in North Africa is estimated at 1 250 000 ha.

P. halepensis extends from sea-level to 2000 m, but it is confined to the semi-arid 'etage' and the drier part of the subhumid 'étage'. In the Atlas it does not descend lower than 1200 m. At the centre of its distribution *P. halepensis* occurs on a wide range of soils but towards the edges of its range, where low temperatures or high humidity are limiting, it is confined to calcareous soils. It is often associated with *Tetraclinis articulata*, *Juniperus phoenicea* or *Quercus ilex*. It is less xerophilous than *Tetraclinis* and does not ascend as high as *J. phoenicea*. In eastern Morocco it forms islands in *Tetraclinis* forest, as on the Melilla peninsula. In this part of its range when *P. halepensis* forest is degraded it is invaded by *Tetraclinis*, which, because of its ability to regenerate vigorously from suckers, is better adapted to recurring fires. In western Morocco the islands of *P. halepensis* are surrounded by *Q. ilex* forest and occur on edaphically favourable sites. Here, where it is at the moister limits of its range, it expands when nearby *Q. ilex* forest or *Cedrus atlantica* forest is damaged by fire.

P. halepensis occurs on the Saharan Atlas in Algeria, but not on the Anti-Atlas in Morocco. Nor has it been found on the southern slopes of the High Atlas, but the vegetation there has been terribly degraded by man.

In Morocco *Pinus halepensis* forests have no floristic individuality. *P. halepensis* normally occurs in mixed forest with *Tetraclinis articulata* or *Quercus ilex*, in which the pine is an emergent and its tree associates form the lower canopy. This condition is often brought about by the action of man, but some mixtures are probably natural. Thus, on very steep slopes in deep valleys in the High Atlas (White, MS, 1974) *P. halepensis* occurs as an emergent over an open 'matorral' of *Pistacia lentiscus*, *Quercus ilex*, *Juniperus phoenicea*, *J. oxycedrus*, *Olea europaea* and *Phillyrea angustifolia*. These slopes are very unstable. Doubtless some of the contemporary erosion is due to human influences, but the relief is so accentuated that natural erosion is probably sufficiently active to ensure a permanent habitat for the heliophilous *P. halepensis*.

In Algeria certain species are found more frequently in *Pinus halepensis* forest than in any other forest type, notably *Globularia alypum*, *Leuzea conifera*, *Rosmarinus eriocalyx* (*tournefortii*) and, as a forest grass, *Stipa tenacissima*.

Pinus pinaster *forest*

Refs.: Boudy (1950, p. 691–702); Destremau (1974, p. 29–66); Emberger (1939, p. 115–17, 137–8); Métro (1958, p. 48–51); Peyerimhoff (1941, p. 49); Quézel (1976).
Phots.: Boudy (1950: 102–4).

Pinus pinaster is virtually confined to the western half of the Mediterranean basin. It extends no further east than Italy and Tunisia. In North Africa its distribution is restricted and it occupies only 28 000 ha. It is, however, of considerable, and even more of potential, economic importance. In Morocco *P. pinaster* grows only in the mountains, in Algeria and Tunisia only on the coastal plain where it never exceeds an altitude of 700 m. In Morocco it scarcely descends below 1 000 m and ascends to 1 900 m in the western Rif. On the northern slopes of the Middle Atlas and the High Atlas it occurs between 1 500 and 2 200 m. It is possible that the Moroccan and Algerian variants are taxonomically different (Monjauze, *Bull. Soc. Hist. Nat. Afr. Nord*, vol. 45, 1954, p. 39–54).

P. pinaster is confined to the subhumid and humid 'étages'. On the North African littoral east of Algiers the climate is humid. Rainfall is 1 000–1 200 mm per year and frosts do not occur. In the mountains of Morocco it occurs in the humid and semi-humid 'étages'. Rainfall is believed to be 800–1 000 mm per year and winter temperature may be below zero for long periods.

P. pinaster grows on a wide range of soils but is more selective than *P. halepensis*. In the littoral zone it is found only on Numidian sandstone. In Morocco it grows both on siliceous rocks and on Jurassic and Cretaceous dolomites, though the soils derived from the latter are often free of calcium carbonate. It always grows on well-drained soils. At its upper limit it mingles with *Cedrus atlantica* and *Abies pinsapo*. It is rarely associated with *P. halepensis*.

The *Pinus pinaster* forests in Algeria and Tunisia usually contain *Quercus suber* and *Q. faginea*. None of the species of the shrub layer, which includes the grass *Ampelodesma mauritanicum*, is characteristic. All occur in humid *Quercus suber* forest.

In the subhumid *Pinus pinaster* forests in Morocco, *Quercus suber* or *Q. ilex*, according to the nature of the soil, are present in the Rif, and in the Middle Atlas *Q. ilex* sometimes mixed with *Q. faginea* usually occurs. In the driest forests scattered tufts of *Stipa tenacissima* occur in the field layer.

In the humid *P. pinaster* forests in the Rif, *Abies pinsapo*, *Cedrus atlantica*, *Q. ilex* or *Q. suber* are usually present. *Cedrus* and *Quercus ilex* are scattered in the canopy of humid *P. pinaster* forests in the Middle Atlas, which sometimes occur as magnificent, almost pure stands.

Cedrus atlantica *forest*

Refs.: Boudy (1948, p. 135–6; 1950, p. 529–611); Destremau (1974, p. 77–90); Emberger (1939, p. 123–31); Métro (1958, p. 34–46); Peyerimhoff (1941, p. 48); Quézel (1976).
Phots.: Boudy (1950: 75–87); Métro (1958: 1).

Cedrus atlantica is confined to the mountains of Algeria and Morocco. It often exceeds a height of 60 m, and can live for 750 years or more. Two related species, *C. brevifolia* and *C. libani*, occur at the eastern end of the Mediterranean, and *C. deodara* forms extensive forests in the Himalayas. The surviving forests of *C. atlantica* in North Africa cover more than 200 000 ha, a fragment of their former extent. In Algeria there are extensive forests in the Aurès and smaller forests on other mountains. In Morocco *C. atlantica* is almost confined to the Rif and the Middle and High Atlas. Its upper limit is well defined and occurs at 2 700–2 800 m. At altitudes higher than that the climate is too dry and cold, and *Juniperus thurifera* becomes dominant. The lower limit at which *Cedrus* forest occurs is less precise and has been obscured by man. It is sometimes as low as 1 350 m. Isolated trees locally descend to 900 m.

Cedrus atlantica is characteristic of the cold humid and subhumid 'étages'. The most luxuriant stands occur on mountain slopes which intercept rain-bearing winds from the Atlantic or Mediterranean, but in parts of its range, as in the High Atlas, the rainfall may be as low as 364 mm per year. When, however, it is as low as this it is well distributed and, by contrast to the general situation in North Africa, an appreciable amount of rain regularly falls in September, and the months of June, July and August, although dry, are not devoid of rain. *C. atlantica* grows on a wide range of soils derived from limestone, basalt, schist, marl, granite, and sandstone.

At its lower limits *C. atlantica* is replaced by *Juniperus phoenicea*, *Pinus halepensis*, *P. pinaster* or *Quercus ilex*. In the wettest regions it is sometimes mixed with deciduous oaks and *Abies pinsapo* or *A. numidica*. It is never in contact with *Tetraclinis articulata*.

Although many of the associates of *Cedrus atlantica* are typical Mediterranean species, Eurosiberian linking species are well represented, especially in the wetter types. Thus in the Rif, *Carex leporina*, *Digitalis purpurea*, *Luzula fosteri* and *Solidago virgaurea* are plentiful. In the wettest places in marshes and near springs Eurosiberian species are more numerous and include *Athyrium filix-femina*, *Carum verticillatum*, *Nardus stricta*, *Parnassia palustris*, *Pinguicula vulgaris*, *Primula vulgaris*, *Sieglingia decumbens*, *Triglochin palustris* and *Viola palustris*. In very humid ravines *Cedrus* is replaced by *Betula pendula*, most of whose associates are Eurosiberian species. In addition to those already mentioned are: *Alnus glutinosa, Aquilegia vulgaris, Carex distans, Dryopteris filix-mas, Luzula multiflora, L. sylvatica, Osmunda regalis, Rhamnus frangula (Frangula alnus), Salix cinerea, S. purpurea* and *Sanicula europaea*.

In the wetter *Cedrus* forests in the Atlas Mountains and Algerian coastal ranges *C. atlantica* is usually accompanied by *Quercus ilex, Acer monspessulanum, Ilex aquifolium, Lonicera arborea, Sorbus aria, S. torminalis, Juniperus oxycedrus, Hedera helix, Fraxinus xanthoxyloides* and *Taxus baccata*. Two rarer associates are *Cytisus battandieri* and *Prunus padus*. Eurosiberian species, though less plentiful than in the Rif, are nevertheless prominent. Degradation of these forests gives rise first to thickets of *Quercus ilex* with *Juniperus oxycedrus, Crataegus* and climbers (*Rosa, Rubus, Asparagus*), and then to *Festuca* grassland with scattered bushy *Q. ilex, J. oxycedrus* or *Crataegus*.

Juniperus thurifera is present in the canopy of the drier and more continental *Cedrus* forests, which is always open. This type occurs on the southern slopes of the Middle Atlas, the eastern, Mediterranean parts of the High Atlas and in the Aurès. *Quercus ilex* is rare. Degradation gives rise to a mixed community of grasses such as *Festuca hystrix*, and spinous Altimontane shrubs such as *Bupleurum spinosum* and *Erinacea anthyllis*.

Abies pinsapo *and* A. numidica *forest*

Refs.: Emberger (1939, p. 131–2); Métro (1958, p. 32–4); Peyerimhoff (1941, p. 47); Quézel (1956, p. 18–24).
Phots.: Quézel (1956: 3, 4a).

Abies pinsapo and *A. numidica* belong to a group of ten closely related species which collectively show an interrupted circum-Mediterranean distribution. The group as a whole is closely related to the central European *A. alba*.

A. pinsapo is only found in southern Spain and a small part of the Rif above Chechaouen where it occupies 15 000 ha on calcareous soils. The Moroccan plant is given separate recognition as subsp. *marocana*. *A. numidica* is confined to a few hundred hectares on the twin summits of Babor and Tababor in Algeria. Both species occur in the humid 'étage'.

A. pinsapo subsp. *marocana*, which attains a height of 20 m and a diameter of 1.5 m, becomes plentiful at 1 500–1 600 m but descends in ravines to 1 300 m. It is usually scattered but locally forms dense forests on inaccessible northern slopes. *A. pinsapo* forests always include a few *Cedrus atlantica, Quercus ilex, Q. faginea, Acer granatense* and *Taxus baccata*. The floristic composition of the underwood is similar to that of adjacent *Cedrus* forests.

A. numidica also occurs on calcareous islands inside the main distribution of *Cedrus*. It is usually mixed with *Taxus baccata, Q. faginea, Q. afares, Q. ilex, Acer campestre, A. obtusatum, Sorbus torminalis, S. aria, S. domestica, Populus tremula* and *Ilex aquifolium*.

Juniperus thurifera *forest*

Refs.: Boudy (1948, p. 135; 1950, p. 754–9); Emberger (1939, p. 86–91); Métro (1958, p. 78); Peyerimhoff (1941, p. 50).
Phots.: Boudy (1948: 1; 1950: 118–21); Emberger (1939: 1–2); Métro (1958: 6).

Juniperus thurifera has a disjunct distribution in the western Mediterranean. It occurs in the French Alps, the Pyrenees, central Spain and the Maghreb. In Algeria it is rare and is confined to the Aurès, but in Morocco it is much more extensive and covers 50 000 ha. It is found on nearly all Moroccan mountains but is absent from the Rif and the western Anti-Atlas. It reaches its maximum development on the southern slopes of the Grand Atlas.

J. thurifera, which is indifferent to substrate, is virtually confined to the cold semi-arid 'étage' between 1 800 and 3 150 m. Towards the lower limits of its range below 2 200 m it is often mixed with *Cedrus atlantica*. When unmolested it is a magnificent tree of massive proportions, up to 15 m tall and with a bole 5 m in diameter. More frequently, however, it is mutilated by herdsmen who, in times of snow, lop off the branches for fodder and fuel. *J. thurifera* does not regenerate from suckers, and, because of browsing animals, rarely from seed. It resists fire, however, much better than *Cedrus*, and is often the only remnant of mixed *Cedrus atlantica, Juniperus thurifera* forest. Although it no longer forms forests above the *Cedrus–Juniperus* zone there is little doubt that it would do so if it were not for human influences.

J. thurifera has few faithful associates. Most species also occur at higher or lower altitudes than *Juniperus* itself. Above 2 500 m notable associates include *Bupleurum spinosum, Prunus prostrata* and *Daphne laureola*. Bushy trees often found with it include *Crataegus laciniata, Buxus sempervirens* and *Lonicera arborea*.

Mediterranean deciduous forest

Refs.: Boudy (1948, p. 140–1; 1950, p. 252–98); Emberger (1939, p. 132–4); Métro (1958, p. 46–8); Peyerimhoff (1941, p. 52–3, 55); Quézel (1956).
Phots.: Boudy (1950: 34–7); Métro (1958: 2); Quézel (1956: 1–2, 4b, 5b).

There are three species of deciduous oak in North Africa, namely *Quercus faginea* (*Q. lusitanica*, 'chêne zéen'), *Q. pyrenaica* (*Q. toza*, 'chêne tauzin') and *Q. afares* ('chêne afarès'). *Quercus faginea* extends from Iran to the Iberian peninsula but is absent from Italy, France, and the Balearics. It is widespread in North Africa from Morocco to Tunisia but only in rather small, widely scattered stands. *Q. pyrenaica* is essentially an Atlantic species, occurring in Morocco, Spain and France. *Q. afares* is confined to Algeria. The total area occupied by deciduous oak forests in the Maghreb is small, amounting to no more than 100 000 ha, nearly all in the humid 'étage'. The deciduous oak forests of Algeria have been the subject of a profound study by Quézel (1956).

Quercus faginea *forest*

Q. faginea is a large tree reaching a height of 30 m and a diameter of 1.5 m. It is deciduous for a few weeks in winter but the dead, brown leaves are retained for several months, especially on young trees and are shed only a few weeks before the new ones come, while in summer its dense crown casts a heavy shade, beneath which it regenerates freely. In *Q. faginea* forest, the soil is always moist and the underwood usually consists of saplings of the dominant.

In Morocco it forms rather small scattered populations, principally in the Rif, the Middle Atlas and in deep valleys on the El-Harcha-Oulmes plateau. It is confined to the subhumid and humid 'étages' and extends from sea-level in Tanger to 1 800 m in the High Atlas. It occurs on many types of parent material, but the most impressive stands are on rich volcanic soils. In the Rif there are many islands of *Q. faginea* forest on Quaternary volcanic soils in a matrix of *Quercus ilex* and *Cedrus* forest on the surrounding calcareous plateau.

Quercus faginea usually occurs in pure stands or is diluted by only a few scattered individuals of *Q. ilex* or *Q. suber*. Even at the limits of its range it does not mix freely with other species, except when it has been subjected to fire and subsequently invaded by *Q. suber*.

On calcareous soils, *Q. faginea* forest has a sparse underwood of *Crataegus monogyna*, *Daphne gnidium*, *Lonicera etrusca*, *Rosa* and *Rubus ulmifolius*, but the very dense field layer includes *Bromus erectus*, *Cynosurus echinatus*, *Festuca triflora*, *Geum sylvaticum*, *Milium vernale*, *Pimpinella villosa* and several species of *Lathyrus*, *Trifolium* and *Vicia*.

On basalt the underwood, in which *Cistus laurifolius*, *Cytisus battandieri* and *Viburnum tinus* play in important role, is better developed.

At higher altitudes several species equally characteristic of *Cedrus* forest put in an appearance: *Acer monspessulanum*, *Ilex aquifolium*, *Sorbus torminalis* etc.

The *Q. faginea* forests of the Rif differ from the others only in their richness in northern species such as *Digitalis purpurea* and *Prunus avium*, and the occurrence of endemics including *Linum villarianum* and *Halimium atlanticum*.

Quercus pyrenaica *forest*

Q. pyrenaica forms a series of small populations in the western Rif and Tangier. It occurs pure, or mixed with *Q. faginea*, *Q. ilex* or *Q. suber*. At its upper limits it may be mixed with *Cedrus*. The pure forest occurs in a zone where mist is frequent even in summer. In humid ravines ferns are plentiful, especially *Aspidium aculeatum*, *Asplenium adiantum-nigrum*, *Athyrium filix-femina*, *Blechnum spicant* and *Pteridium aquilinum*.

Quercus afares *forest*

This type is similar to *Q. faginea* forest but occurs on somewhat drier slopes and at higher altitudes. *Q. faginea* is often scattered in the canopy. The underwood is denser than that of *Q. faginea* forest but is still sparse compared with that of typical *Q. ilex* or *Q. suber* forest. Associates include *Ampelodesma mauritanicum*, *Cytisus villosus*, *Erica arborea*, *Genista tricuspidata*, *Paeonia atlantica* and *Ruscus aculeatus*.

Mediterranean bushland and thicket
(mapping units 10, 23, and 78)

Refs.: Boudy (1950, p. 436–41); Emberger (1939, p. 91–4, 117–18).
Syn.: la brousse semi-aride à Olivier–Pistachier–Chamaerops (Emberger); Oleo-lentiscetum.

It is possible that the climax vegetation occurring on clay soils in the semi-arid 'étage' was bushland or thicket, or, at best, scrub forest dominated by *Olea europaea* rather than true forest. However, since the original vegetation has virtually disappeared during centuries of intensive cultivation, the matter must remain largely conjectural. The degraded remnants of such vegetation are briefly described below.

The spontaneous form of *Olea europaea* occurs more or less throughout the Mediterranean Region, but is absent from France. It can grow into a small tree, 10–12 m high, with a short, stout, often twisted, trunk, but is more often seen as 4–5 m high coppice or as small heavily browsed bushes. In the Maghreb it is characteristic of the semi-arid 'étage' and occurs only exceptionally in the subhumid 'étage'. It is virtually absent from the arid and humid 'étages'. Its upper limit in the mountains depends on humidity. It ascends to only 1 200 m in the Algerian Tell, but attains 1 650 m in the drier Grand Atlas. If often occurs in forests in the semi-arid 'étage' dominated by *Argania spinosa*, *Juniperus phoenicea*, *Pinus halepensis*, *Quercus ilex*, *Q. suber* or *Tetraclinis articulata*. All these species avoid heavy compact clay soils. The latter have been intensively cultivated for centuries and it is difficult to reconstruct their original vegetation. Emberger suggests that it was a bushland or scrub forest dominated by *Olea europaea* and *Pistacia lentiscus* with *Pistacia atlantica* and *Chamaerops humilis*.

Within the area of the *Olea–Pistacia* community the smallest outcrops of rock or islands of sand support other species, in particular *Tetraclinis*, *Q. ilex*, and *Q. suber*. The extensive forests of *Q. suber* near Rabat which separate the *Olea–Pistacia* area of the southern Rif from that of Chaouia, illustrate the way in which, within the semi-arid 'étage', the vegetation is determined by the soil. Many of the associates of *Olea* and *Pistacia lentiscus* also occur in *Q. suber* and *Tetraclinis* communities in the semi-arid 'étage'.

Most of the land formerly occupied by the *Olea–Pistacia* community, where it is not cultivated, supports extremely degraded vegetation characterized by the abundance of the dwarf palm *Chamaerops humilis*, occurring mostly in its acaulescent form. Larger woody associates, which are rare, besides *Olea* and *Pistacia lentiscus*, include *Ceratonia siliqua*, *Pistacia atlantica*, *Ziziphus lotus*, *Crataegus monogyna*, *Rhamnus oleoides*, *Rhus pentaphylla* and *Jasminum fruticans*. Among smaller plants the following are conspicuous: *Anagyris foetida*, *Asparagus albus*, *A. stipularis*, *Calicotome villosa* and *Daphne gnidium*.

Geophytes are particularly well represented in the *Chamaerops* community by species of *Aceras*, *Allium*, *Anacamptis*, *Asphodelus*, *Bellevalia*, *Colchicum*, *Crocus*, *Dipcadi*, *Erythrostictus*, *Gagea*, *Gladiolus*, *Iris*, *Leucojum*, *Muscari*, *Narcissus*, *Ophrys*, *Orchis*, *Ornithogalum*, *Anthericum*, *Romulea*, *Scilla* and *Urginea*. These plants begin their growth several weeks before the first rains appear and so announce the end of summer.

In spring the degraded plains and fallow lands of Morocco are like an immense multi-coloured garden or a gigantic Persian or Moroccan carpet because of the profuse flowering of brightly coloured annuals and geophytes. Large patches of orange *Calendula algeriensis* alternate with or intermingle with purple *Fedia*, violet *Linaria*, yellow *Diplotaxis* and *Chrysanthemum*, white *Ormenis* and blue *Convolvulus gharbensis*, *C. tricolor* and *Echium*, among which emerge the spikes of *Gladiolus byzantinus* and *Asphodelus microcarpus*, and large clumps of *Ferula communis* and *Foeniculum vulgare*.

In Morocco the *Olea–Pistacia* community also occurs locally on clay soils in the subhumid 'étage' in Tangier. Here there are thickets of *Pistacia lentiscus* and *Olea*, associated with *Acanthus*, *Clematis flammula*, *Echium boissieri* (*pomponium*), *Myrtus communis*, *Phillyrea angustifolia*, *Quercus coccifera*, *Teucrium fruticans* and the grass *Ampelodesma mauritanicum*.

Mediterranean shrubland

With the exception of halophilous shrubland (page 266) and the shrublands of the sub-Mediterranean transition zone, true Mediterranean shrubland is almost confined to high mountains above the timber-line, or is secondary.

Altimontane Mediterranean shrubland
(mapping unit 23)

Refs.: Emberger (1939, p. 138–45); Quézel (1957a, p. 109–78, 193–211, 418–20); Taton (1966); White (MS, 1974).
Phots.: Emberger (1939: 9.1–4); Quézel (1957a: 5, 6, 8, 14, 20).
Syn.: les garrigues montagnardes à xérophytes épineux (Quézel); l'horizon à xérophytes épineux en coussinets (Emberger).

Dwarf shrubland dominated by dense, cushion-shaped, very spinous shrubs is one of the most widespread and characteristic vegetation types on the North African mountains. It grows under a typical Mediterranean climate, since the rain falls during the cold season as in the plains. It is, however, an extreme climate and snow lies for several months in winter. The soils frequently show polygonal structure due to frost action and Quézel regards them as periglacial. Altimontane shrubland nearly always grows on skeletal soils and during the summer is exposed to very dry conditions.

In Morocco, Altimontane shrubland represents the climatic climax between the tree-line and 3800–3900 m. It also occurs very extensively inside the forest belt down to c. 2000 m on the drier mountain slopes; in this situation it is mostly secondary. There can be little doubt, however, that, before the forests were destroyed, the shrubland species formed small colonies on shallow soil and rocky outcrops where the trees were unable to form a closed canopy. Because of the widespread destruction of forest it is difficult to determine the lower climatic limit of Altimontane shrubland. For Morocco, Emberger believes it to occur at about 2800 m.

Altimontane shrubland is extensively developed in Morocco, especially in the Great Atlas and on the summits of the Middle Atlas. In the Anti-Atlas it is only found on the massif of Siroua. The Rif is too low for its occurrence. In Algeria it is much more restricted than in Morocco, but it occurs on the summits of the Djurdjura and the Aurès. Physiognomically similar vegetation occurs on all the high mountains in the Mediterranean basin, but floristically similar types are found only in southern and eastern Spain.

The flora of Altimontane shrubland is poor and uniform, but rich in endemic species. According to Quézel, the total Altimontane flora in the Maghreb amounts to no more than 650 species. Most of them are confined to the Mediterranean Region. One hundred and sixty species are endemic to the high mountains of the Maghreb. Nearly all belong to Mediterranean endemic genera. Eurosiberian species are relatively poorly represented, and *Carex capillaris* is one of the few northern species which in the Maghreb is confined to the high mountains.

The dominants of Altimontane shrubland are dwarf shrubs. In Morocco there are 18 species, namely *Alyssum spinosum* (3850 m), *Amelanchier ovalis* (3500 m), *Arenaria dyris* (3750 m), *A. pungens* (3790 m), *Berberis hispanica* (3200 m), *Bupleurum spinosum* (3400 m), *Cytisus balansae* (3600 m), *Erinacea anthyllis* (3600 m), *Juniperus communis* (3300 m), *J. oxycedrus* (3150 m), *Lonicera pyrenaica* (3500 m), *Ononis atlantica* (3250 m), *Prunus prostrata* (3200 m), *Rhamus alpinus* (3200 m), *Ribes alpinum* (3300 m), *R. uva-crispa* (3400 m), *Sorbus aria* (3000 m) and *Vella mairei* (3200 m). Upper altitudinal limits are shown in brackets. The spinous cushion plants (*Alyssum*, *Arenaria pungens*, *Bupleurum*, *Cytisus* and *Erinacea*) are normally about 0.5 m tall. Because of heavy grazing the ground between them is often almost bare, and other species in the community can only grow with their protection.

The spinous cushion plants are rare or absent from certain habitats such as escarpments, mobile screes, earthy slopes, and from the highest altitudes, where they are replaced by herbaceous communities.

Secondary Mediterranean shrubland (maquis and garrigue)
(mapping units 10, 23, and 78)

Refs.: Gimmingham & Walton (1954); Ionesco & Sauvage (1962); Tomaseli (1976).
Phots.: Gimmingham & Walton (1954: 1, 2, 4–8).

The climate of Mediterranean Africa is a forest climate, and forest, scrub forest, or, locally, bushland and thicket represent the climax nearly everywhere below the natural tree line. As one approaches the Sahara there is a progressive diminution in the height of the vegetation, and various types of bushland and shrubland occur in the Mediterranean–Sahara transition zone (Chapter XVIII).

Within the Mediterranean Region proper, nearly all communities dominated by shrubs have been derived from forest, scrub forest, or bushland. Some of the taller shrublands are sometimes referred to as 'maquis' or 'macchia', and some of the shorter as 'garrigue', but considerable confusion is associated with the application of these terms.

Classical maquis is dense and tall (up to c. 4 m), difficult to penetrate, and is dominated by *Erica arborea* and *Arbutus unedo*. It occurs on siliceous soils and degrades to a grassland dominated by *Ampelodesma mauritanicum*.

The term garrigue is derived from the Catalan name for *Quercus coccifera* (garric) and was originally applied to low bushy communities dominated by that species which occur on calcareous soil. Its application is sometimes extended to embrace all open shrubby communities of medium height occurring on calcareous soil in the Mediterranean Region.

The value of this distinction is doubtful and in practice many authors have used both terms in senses different from those quoted above. The importance of the lithological origin of the soil in the Maghreb has probably been exaggerated, since many species, including *Arbutus unedo* and *Quercus coccifera*, grow equally well on calcareous and siliceous soil. In North Africa, *Q. coccifera* has a restricted distribution. Its most characteristic habitat is *Olea europaea, Pistacia lentiscus, Ceratonia siliqua* bushland on clay soils. It is also frequent in maquis and is not found on skeletal calcareous soils (P. J. Stewart, pers. comm.).

In practice, the terms maquis and garrigue have usually been defined in the past in such a way as to cover only a part of the wide range of secondary shrublands that occur in the Mediterranean. Some authors, however, e.g. Tomaselli (1976) in a review of maquis covering the whole of the Mediterranean basin, take a more comprehensive view and draw a purely arbitrary distinction based on height. To Tomaselli, maquis is a formation more than 2 m tall; it is usually dense and composed of woody plants without a well-defined trunk. It is indifferent to substrate.

Since the stature and density of secondary shrubland depends as much on the intensity of degradation as on the nature of the parent material, and floristic composition varies in a complex manner from place to place in relation to the floristic composition of the original forest, there seems to be little justification for the use of the terms maquis and garrigue in a pan-African classification, though doubtless endless litigation will continue to be associated with their local use. By contrast, the use of a vernacular term, 'fynbos' (page 132), to designate a somewhat physiognomically similar type of shrubland in the Cape Region, is free from ambiguity, possibly because the vegetation it refers to is a regional climatic climax which has considerable floristic uniformity throughout its range.

The structure and floristic composition of secondary Mediterranean shrubland are so variable that little would be gained from an attempt to describe them. Some examples are briefly described elsewhere in this chapter, along with the forests from which they have been derived.

Gimmingham & Walton (1954) briefly describe three stages in the degradation of *Cupressus sempervirens, Juniperus phoenicea, Olea europea, Quercus coccifera, Ceratonia siliqua* scrub forest on calcareous soil in Cyrenaica.

The first stage is dominated by *Arbutus pavarii* often associated with *Ceratonia siliqua, Phillyrea angustifolia,* and *Pistacia lentiscus*, with an irregular field layer of *Poterium (Sarcopoterium) spinosum, Cistus parviflorus* etc. *Arbutus pavarii*, which forms a thicket of slender stems 1.8–3 m high, is less susceptible to grazing than saplings of trees which have a single trunk, and so persists under moderate grazing. It is, however, liable to perish under conditions of more heavy grazing, when *Pistacia lentiscus* forms dense thickets 1.5–2.4 m high. Under the most intensive grazing even *Pistacia* disappears and its place as dominant is taken by the low shrub *Poterium spinosum*. In the eastern Mediterranean, communities dominated by *Poterium* are referred to as 'batha'.

Mediterranean anthropic landscapes
(mapping unit 78)

The most fertile lowlands have been cultivated since Roman times, and few vestiges of the natural vegetation remain. Wheat is the most widely planted crop, but peas, beans, and onions are plentiful, and there are many groves of olive, *Citrus*, fig, and *Vitis*.

The hedges are mostly of *Agave, Acacia karroo, Arundo donax* and *Opuntia*. The latter is frequently naturalized. *Eucalyptus* and *Pinus halepensis* are planted locally as wind-breaks and for fuel and timber, but in general the landscape is treeless.

The sparse plant cover of shallower soils is grazed by sheep and cattle, but owing to overstocking there is much soil erosion and in places the stony mantle supports only *Ziziphus lotus*, dwarf *Chamaerops*, and unpalatable herbs such as *Ferula communis*, *Asphodelus microcarpus* and *Urginea maritima*. Marshy hollows are often dominated by *Juncus acutus*.

VIII/IX The Afromontane archipelago-like regional centre of endemism and the Afroalpine archipelago-like region of extreme floristic impoverishment

Geographical position and area

Geology and physiography

Climate

Flora

Mapping units

Vegetation
    Afromontane forest
        Afromontane rain forest
        Undifferentiated Afromontane forest
        Single-dominant Afromontane forest
            Juniperus procera forest
            Widdringtonia cupressoides forest
            Hagenia abyssinica forest
        Dry transitional montane forest
    Afromontane bamboo
    Afromontane evergreen bushland and thicket
    Afromontane and Afroalpine shrubland
    Afromontane and Afroalpine grassland
    Mixed Afroalpine communities
        In tropical Africa
        In South Africa

Geographical position and area

The Afromontane Region is an archipelago-like centre of endemism (White, 1978a) which extends from the Loma Mts and the Tingi Hills (11° W.) in Sierra Leone in the west to the Ahl Mescat Mts (49° E.) in Somalia in the east, and from the Red Sea Hills (17° N.) in the Sudan Republic in the north to the Cape Peninsula (34° S.) in the south. A few Afromontane species descend almost to sea-level even in the tropics, but outside the Afromontane Region they are always very rare. In the tropics most Afromontane communities are found only above 2000 m, but where the climate is more oceanic, as in the West Usambara Mts in Tanzania, they can occur as low as 1200 m. Further south, where latitude compensates for altitude, they descend progressively further, and in the Cape Region exclaves of Afromontane forest are found only a few hundred metres above sea-level. The inclusion of the West African mountains west of Cameroun, and the highlands of Angola, in the Afromontane Region is still problematical, since the Afromontane species occurring there appear to be diluted by the presence of many lowland species. (Area: 715 000 km^2.)

Geology and physiography

The Afromontane 'archipelago' is very diverse in lithology and physiography, which have been little studied from a botanical point of view.

Some of the largest 'islands' and many of the smaller ones are largely volcanic in origin, though the lavas are of different ages.

Most of the Ethiopian highlands are formed of basalt, though Precambrian rocks outcrop locally. The oldest volcanic rocks date from the Eocene and have been intermittently augmented by subsequent eruptions which have continued into the Quaternary.

The Kenya highlands are mostly formed of volcanic deposits, including phonolite, nephelinite and basalt, which were erupted in post-Miocene time during the formation of the eastern Rift Valley. By contrast, the Cherangani Hills (3600 m) are composed of Precambrian metamorphic rocks with conspicuous quartzite ridges.

The highest parts of the Natal Drakensberg and adjacent Lesotho are capped by more or less horizontal basaltic lava flows which terminate the Stormberg Series of Triassic age. Adjacent less-elevated parts of the Afromontane Region, however, are underlain by Karoo sediments of the Stormberg and Beaufort series.

The Kivu ridge and the contiguous uplands which extend from the northern end of Lake Tanganyika to Ruwenzori are largely composed of Precambrian rocks, but with local islands of volcanic deposits including those formed by the still active Virunga volcanoes (4507 m).

The Cameroun Highlands are formed partly of volcanic, partly of ancient crystalline rocks. Mount Cameroun (4095 m), a still active volcano, stands separate from the main range.

Of the more isolated mountains some are volcanic in origin, e.g. Mt Elgon (4315 m), Mt Meru (4566 m) and Mt Kilimanjaro (5890 m), whereas others are formed of crystalline basement rocks. Some of the latter, e.g. the Chimanimani Mts in Zimbabwe, represent the resistant remnants of the uplifted rim of the Great African Plateau, while others, notably Ruwenzori (5119 m), have been uplifted by compressional forces associated with rift faulting.

Climate

The climate is extremely varied but reliable published records are sparse. Information on the Afroalpine zone is provided by Hedberg (1964), on the Austro-afroalpine zone by Killick (1978a, 1978b, 1978c), on Mt Cameroun by Richards (1963b), on the forest belt in Malawi by Chapman & White (1970), and on the mountains at the eastern end of the Zaire basin by Bultot (1950, 1971–74) and Scaëtta (1933, 1934).

In the forest belt, mean annual rainfall is usually more than 1000 mm, but is less in drier types transitional to lowland vegetation. Above the forest belt precipitation diminishes and in the Afroalpine belt of some mountains appears to be much less than 1000 mm per year. Cloud is a feature of most mountains but its importance is very uneven and it has been little studied. The incidence of frost varies considerably from complete absence on some lower slopes to a nightly occurrence on the highest summits. (See Fig. 14.)

Flora

There are at least 4000 species of which c. 3000 are endemic or almost so.

Endemic and near-endemic families. Barbeyaceae, Oliniaceae. *Curtisia* (Cornaceae) is sometimes given family rank. *Barbeya*, which also occurs in the Yemen, is not strictly Afromontane, but is more characteristic of the ecotone between dry Afromontane forest and Somalia–Masai evergreen bushland.

Endemic genera. About one-fifth of the tree genera, including *Afrocrania, Balthasaria, Ficalhoa, Hagenia, Kiggelaria, Leucosidea, Platypterocarpus, Trichocladus,* and *Xymalos*, are endemic. For smaller plants the proportion is probably less and includes *Ardisiandra, Cincinnobotrys,* and *Stapfiella*.

Linking elements. See White (1978a, p. 475–80).

Mapping units

19a. Undifferentiated Afromontane vegetation.
65. Tropical altimontane vegetation.
66. South African altimontane vegetation.

In addition to the above, Afromontane species also occur in the following mapping units:

4. Transitional rain forest.
13. The Fouta Djalon mosaic of lowland rain forest and secondary grassland with montane elements.
17. Cultivation and secondary grassland replacing upland montane forest.
19b. Undifferentiated Sahelomontane vegetation.
32. Jos Plateau mosaic.
33. Mandara Plateau mosaic.

Vegetation

On any particular mountain there is usually a very wide range of vegetation. Extreme types may have few species in common, but all types are intimately connected by complex series of intermediates. The floristic differences between extremes on a single mountain are usually greater than the differences between the Afromontane assemblage as a whole on that mountain and the assemblage found on nearby, or indeed on distant, mountains so that the collective flora of the archipelago shows a remarkable continuity and uniformity.

On most mountains the lowermost vegetation is forest, beneath which one would expect to find a transition zone connecting the Afromontane and lowland phytochoria. Nearly everywhere, however, the vegetation of this transition zone has been destroyed by fire and cultivation, but the remnants of a dry type of transition forest are briefly described below. Forests transitional to Guineo–Congolian, Zanzibar–Inhambane and Tongaland–Pondoland communities are described in Chapters I, XIII, and XV respectively.

On nearly all African mountains the vegetation diminishes in stature from the lower slopes to the summit, but this regularity is so often modified by local features of aspect, exposure, incidence of frost, and depth of soil, and by overall patterns of climate dependent on the size and configuration of the mountain in relation to distance from the sea and other sources of moisture, that generalized schemes of zonation, even for relatively restricted regions, are impossible to devise. Nevertheless, the three broad belts,

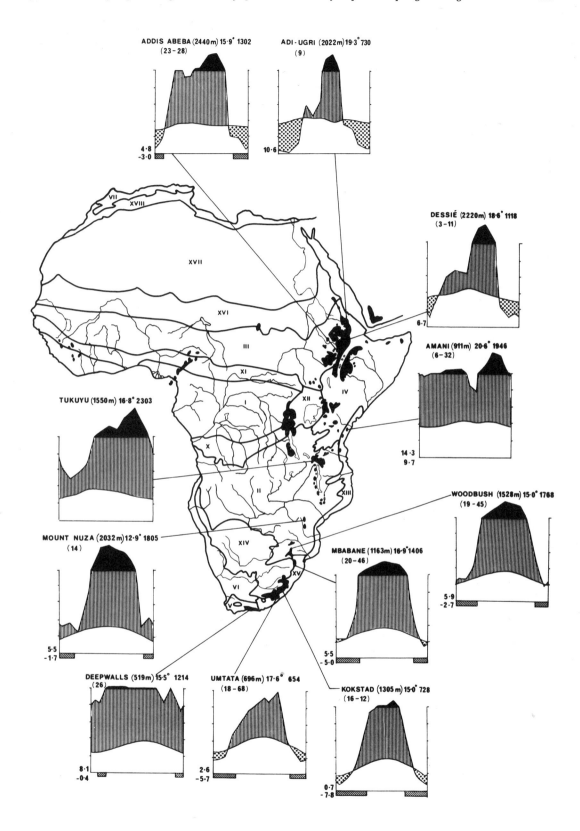

FIG. 14. Climate and topography of the Afromontane archipelago-like regional centre of endemism (VIII) and the Afroalpine archipelago-like region of extreme floristic impoverishment (IX)
(The two phytochoria, shown as solid black areas, are not separately distinguished)

forest, Ericaceous and Afroalpine, recognized by Hedberg (1951) for the high mountains of East Africa, can usually be detected, but the details of zonal replacement vary greatly, even on different slopes of the same mountain. In the Cape Region, Afromontane vegetation, which is represented there only by forest, is no longer associated with the highest parts of mountains but is confined to the lower slopes.

Except on the wettest mountains where the original vegetation is sometimes well preserved the most widespread vegetation is secondary, fire-maintained grassland.

The Ericaceous and Afroalpine belts on the high mountains of tropical East Africa are mapped collectively as Altimontane vegetation (unit 65). Ecologically similar but floristically somewhat different vegetation in South Africa is mapped as unit 66. No attempt is made to map the remaining Afromontane types separately, except that large areas of *Juniperus* forest and mixed Afromontane forest are indicated by letters. The small areas shown as Afromontane in Angola and in the Loma–Man highlands in West Africa harbour many Afromontane species which do not seem, however, to form extensive Afromontane communities.

Afromontane forest

Most Afromontane tree species have wide geographical distributions and wide ecological amplitudes. Many also exhibit a wide range of growth forms. For these reasons classification of the forests they form is difficult, and the latter are probably best regarded as comprising a continuum which is only slightly structured. It is, however, convenient to separate off the most luxuriant types as Afromontane rain forest, and to give individual treatment to certain extreme floristic variants dominated by single species, although it appears that the latter are nearly always secondary in origin.

Rainfall received by Afromontane forest varies from 800 mm to considerably more than 2500 mm per year. Nevertheless, the distinction drawn by earlier workers (e.g. Greenway, 1973) between 'wet' and 'dry' types is difficult to apply, chiefly owing to the wide tolerance of many dominant species.

Little has been published on the transition from Afromontane forest to lowland communities. Dry transitional Afromontane forest in East Africa is described in this chapter, and the bushland which replaces it at lower altitudes is dealt with in Chapter IV. Transitional rain forest is described in Chapters I, XII, and XIII.

Afromontane rain forest
(mapping unit 19a)

Refs.: Chapman & White (1970); Langdale-Brown, Osmaston & Wilson (1964, p. 42–3, 109–10, p.p.); Lewalle (1972, p. 107–14); Mildbraed (1914, p. 623–6); Pichi-Sermolli (1957, p. 82–4); Pitt-Schenkel (1938); Pócs (1976b, p. 486–7); White (1978a, p. 485; MS, 1952, 1975, 1976).

Phots.: Chapman & White (1970: 1–7, 12, 24, 40–4); Langdale-Brown *et al.* (1964: 3).
Profiles: Chapman & White (1970: 1–4, 6); Lewalle (1972: 21, 23).
Syn.: foresta umida sempreverde montane (Pichi-Sermolli, 1957); forêt ombrophile de montagne: horizons inférieur et moyen (Lewalle, 1972, p.p.); *Pygeum* moist montane forest (Langdale-Brown *et al.*, 1964); submontane rain forest (Pócs, 1976b); submontane seasonal rain forest (White, in Chapman & White, 1970).

Afromontane rain forest is very similar in structure and physiognomy to certain types of Guineo–Congolian lowland rain forest. At the species level, however, it is almost completely different, but many of its species are closely related to Guineo–Congolian species, or have their closest relatives elsewhere in the lowland tropics. These are the 'nephews' and 'orphans' of White (in Chapman & White, 1970).

The most characteristic tree species of Afromontane rain forest include *Aningeria adolfi-friedericii, Chrysophyllum gorungosanum, Cola greenwayi, Cylicomorpha parviflora, Diospyros abyssinica, Drypetes gerrardii, Entandrophragma excelsum, Ficalhoa laurifolia, Mitragyna rubrostipulata, Myrianthus holstii, Ochna holstii, Ocotea usambarensis, Olea capensis, Parinari excelsa, Podocarpus latifolius* (including *milanjianus*), *Prunus africana, Strombosia scheffleri, Syzygium guineense* subsp. *afromontanum, Tabernaemontana johnstonii*, and *Xymalos monospora*.

Afromontane rain forest occurs mostly between 1200 and 2500 m, but its precise altitudinal limits vary greatly according to distance from the equator, proximity to the sea, and size and configuration of the massif on which it occurs. It is found on the wetter slopes of most of the higher mountain massifs from southern Ethiopia to Malawi.

The mean annual rainfall of Afromontane rain forest lies mostly between 1250 and 2500 mm, but is sometimes higher. There is usually a dry season lasting from one to five months, but dry-season mists are frequent. This may explain the fact that upland rain forest is often much less deciduous than lowland semi-evergreen rain forest experiencing a similar rainfall. Apart from secondary species, only a few of the larger tree species, e.g. *Aningeria adolfi-friedericii* and *Entandrophragma excelsum*, lose their leaves, and then only for a few days. Frosts occur occasionally but are not severe.

Detailed published information on Afromontane forest is available only for Burundi (Lewalle), the West Usambara Mts (Pitt-Schenkel), and Malawi (Chapman & White).

In physiognomy Afromontane rain forest is similar to certain types of Guineo–Gongolian lowland rain forest. The trees of the upper stratum are 25–45 m tall (average 30–38 m). Their crowns, which are not in lateral contact, are raised well above the middle tree stratum and are heavy-branched and wide-spreading. The middle tree stratum is 14–30 m tall; its crowns are often narrow and conical and may be discontinuous to continuous but do not form a dense canopy. The lower tree stratum is 6–15 m tall and usually forms a dense canopy. The shrub layer, 3–6 m tall, is poorly differentiated from the lower tree layer. The herb layer is usually sparse and consists largely of forest grasses and ferns. Lianes and strangling epiphytes are abundant.

Among vascular epiphytes, ferns and lycopods are more or less abundant throughout, and species of *Begonia*, *Impatiens*, *Streptocarpus* and *Peperomia* are widespread. Orchids, though present, are not abundant, and *Rhipsalis* occurs locally. Epiphytic bryophytes are generally present but are not abundant except in some of the wetter types. Epiphytic lichens are not generally abundant except in the crowns of certain species.

A few of the larger trees are buttressed and some are briefly deciduous, but the general impression throughout the year is that the forests are evergreen.

Physiognomically, Afromontane rain forest differs from most types of Guineo–Congolian rain forest chiefly in the occurrence of tree ferns (*Cyathea*) and conifers (*Podocarpus*). The latter, however, are more characteristic of other types of Afromontane forest. There is also a greater degree of bud protection, and 'drip tips' of the leaves are less well developed.

Undifferentiated Afromontane forest
(mapping unit 19a)

Refs.: Acocks (1975, p. 18–23, 25–7, 82–6); Chapman & White (1970, p. 107–9, 131–2; 137–9, 148–51); Edwards (1967, p. 174–80); Jackson (1956, p. 365–70); Killick (1963, p. 43–57); Letouzey (1968a, p. 325–48); Lewalle (1972, p. 114–23); Lind & Morrison (1974, p. 32–43, p.p.); Pichi-Sermolli (1957, p. 73–84, p.p.); Pócs (1976b): Richards (1963b); von Breitenbach (1972); White (1978a); Wild & Barbosa (1967, p. 10–11).

Phots.: Acocks (1975: 8); Chapman & White (1970: 45, 46); Dyer (1937: 17); Edwards (1967: 111–15); Killick (1963: 16); Moll (1966: 1–3; 1968c: 10); Philipps (1931); Phipps & Goodier (1962: 2); Richards (1963b: 1, 2); von Breitenbach (1972: on pp. 19–26).

Profiles: Boughey (1961: 3, 4); Chapman & White (1970: 5); Jackson (1956: 4); Moll (1968c: 6, 7); Richards (1963b: 1).

Syn.: broad-leaved montane forest (Chapman & White, 1970); Dohne sourveld (Acocks, 1974); forêt ombrophile de montagne: horizon supérieur (Lewalle, 1972); highland sourveld (Acocks, 1975); Knysna forest (Acocks, 1975); mist-belt mixed *Podocarpus* forest (Edwards, 1967); moist broad-leaved montane forest (Wild & Barbosa, 1967); mountain *Podocarpus* forest (Edwards, 1967); Natal mist belt 'ngongoni' veld (Acocks, 1975); north-east mountain sourveld (Acocks, 1975); 'ngongoni' veld (Acocks, 1975); Pondoland coastal plateau sourveld (Acocks, 1975, p.p.).

Undifferentiated Afromontane forest is usually shorter than Afromontane rain forest and despite some floristic overlap is of distinctive composition. It usually replaces rain forest at higher altitudes on the wetter slopes and at comparable altitudes on the drier slopes. On some mountains it also occurs below Afromontane rain forest. It usually but not always receives a lower rainfall.

Most stands of undifferentiated Afromontane forest are floristically mixed, but sometimes after fire they are replaced by almost pure stands of *Juniperus procera*, *Widdringtonia cupressoides* (*whytei*) or *Hagenia abyssinica* (see below).

The majority of tree species in this type are very widespread. Among them are *Apodytes dimidiata*, *Halleria lucida*, *Ilex mitis*, *Kiggelaria africana*, *Nuxia congesta*, *N. floribunda*, *Ocotea bullata* (including *O. kenyensis*), *Podocarpus falcatus* (including *gracilior*), *P. latifolius*, *Prunus africana*, *Rapanea melanophloeos* s.l. and *Xymalos monospora*. Indeed this assemblage of species could almost be used to define the Afromontane Region as a whole. No single species occurs throughout, but the assemblage is represented on virtually every 'island' of Afromontane vegetation, usually by several species. All are widespread in South Africa and all except four (*Kiggelaria*, *Nuxia floribunda*, *Podocarpus latifolius* and *Xymalos*) extend as far north as Ethiopia. Only five (*Halleria*, *Kiggelaria*, *Nuxia floribunda*, *Ocotea* and *Podocarpus falcatus*) are absent from West Africa.

Some species such as *Combretum kraussii*, *Cryptocarya latifolia*, *C. woodii*, *Curtisia dentata* (*faginea*), *Chionanthus foveolatus*, *Ptaeroxylon obliquum*, *Schefflera umbellifera*, *Scolopia mundii* and *Podocarpus henkelii* (including *ensiculus*), which are important in South Africa, either do not cross the Limpopo River or are very localized further north.

Single-dominant Afromontane forest
(mapping unit 19a)

Juniperus procera *forest*

Refs.: Chapman & White (1970, p. 108–9, 130–1); Hemming (1966, p. 218–21); Langdale-Brown, Osmaston & Wilson (1964, p. 43–4, 110–11); Lind & Morrison (1974, p. 41–2); Pichi-Sermolli (1957, p. 75–80); White (MS, 1975, 1979); Wimbush (1937).

Phots.: Chapman & White (1970: 13–16); Herlocker & Dirschl (1972: 22); Pichi-Sermolli (1957: 11); Wimbush (1937: between pp. 50 and 51).

Juniperus procera has a scattered distribution on the eastern side of Africa from the Red Sea Hills in the Sudan, Eritrea, and Arabia to the Nyika Plateau in northern Malawi. A single individual has also been reported from the van Niekerk ruins in Zimbabwe 1280 km further south. It mostly occurs on the drier slopes of mountains between 1800 and 2900 m but it occasionally descends to as low as 1000 m. Rainfall is between 1000 and 1150 mm per year, but well-developed stands of forest more than 30 m tall occasionally occur where the rainfall exceeds 1250 mm per year (Glover & Trump, 1970). It is also present as an emergent in scrub forest and evergreen bushland where the annual rainfall is as low as 650 mm (Hemming, 1966), and this may represent its original habitat.

Juniperus procera most frequently forms forests in which it is by far the most abundant species. It is, however, a strong light-demander and does not regenerate in its own shade. Its seedlings also seem to be intolerant of a deep layer of humus on the surface of the soil (Gardner, 1926). It is clear that its presence as a forest tree is largely dependent on fire, either natural or man-made (Wimbush, 1937).

Widdringtonia cupressoides *forest*

Refs.: Chapman & White (1970, p. 108–9, 162–9); Van der Schijff & Schoonraad (1971, p. 472); White (MS, 1973); Wild & Barbosa (1967, p. 11).
Phots.: Chapman & White (1970: 51, 53–7).
Profile: Chapman & White (1970: 7).
Syn.: dry montane conifer forest (Wild & Barbosa, 1967: 11, p.p.).

Widdringtonia cupressoides extends along the eastern side of Africa from Table Mountain in the south to Mt Mulanje in the north. Over the greater part of its range it occurs as a small, usually bushy, often multiple-stemmed, tree from 4 to 9 m tall, and scarcely ever forms forests. It is only on Mt Mulanje that it occurs as a tall forest tree. Although individual trees attain a height of 40 m the main canopy of mature forest is normally *c*. 27 m high. The forests, which occur between 1525 and 2135 m, are very susceptible to fire and the behaviour of *Widdringtonia cupressoides* in relation to fire is similar to that of *Juniperus procera*.

Hagenia abyssinica *forest*

Refs.: Chapman & White (1970, p. 131–2); Demaret (1958); Jackson (1956, p. 361); Langdale-Brown, Osmaston & Wilson (1964, p. 43, 110); Lebrun (1942, p. 52–6); Lind & Morrison (1974, p. 48–50); Robyns (1948a, p. xli–xlii); Spinage (1972, p. 198); White (MS, 1975, 1978).
Phots.: Lebrun (1942: 23a); Lind & Morrison (1974: 12); Robyns (1937: 3b); Spinage (1972: 7).
Profile: Jackson (1956: 2).
Syn.: *Hagenia* woodland (Lind & Morrison, 1974); *Hagenia–Rapanea* moist montane forest (Langdale-Brown *et al.*, 1964); la forêt–prairie à *Hagenia abyssinica* (Lebrun, 1942).

Hagenia abyssinica is found on most of the higher mountains between Ethiopia and the Nyika Plateau in northern Malawi. It occurs both on the wetter mountains such as Ruwenzori, where it is rare, and the drier mountains such as Mt Meru. Its abundance does not seem to be at all closely related to moisture conditions. Its altitudinal range lies between 1800 and 3400 m but it is normally absent from Afromontane rain forest and the taller types of undifferentiated montane forest.

Characteristically *Hagenia* forms almost pure stands 9–15 m tall in a narrow zone (often interrupted) between taller types of montane forest and the thickets and shrublands of the Ericaceous belt. The biggest trees have boles up to 2 m long and 1.6 m in diameter which support massive spreading branches. The best-developed stands are clearly forest, though of a simpler structure than most African forests. Other stands have more the structure of woodland or scrub forest.

Hagenia is a heliophyte which can withstand at least some burning, though it is killed by repeated fierce fires. At lower altitudes it is always seral. Thus, in the Imatong Mts when secondary grassland derived from lower montane forest is protected from fire, *Hagenia* rapidly invades, followed by other forest species (Jackson, 1956).

At higher altitudes, where *Hagenia* is much more abundant, its status is still uncertain. Lind & Morrison (1974) suggest that *Hagenia* forest may be climax where low night temperatures exclude many other trees, and competition is low. There is little doubt, however, that, even at high altitudes, its abundance is at least partly due to disturbance.

On the Nyika Plateau (Chapman & White, 1970) it occurs in short broad-leaved montane forest 8–15 m tall, in association with about 20 other tree species. Because the canopy is low and is kept open by large mammals, heliophilous species like *Hagenia* can establish themselves. *Hagenia*, however, is more abundant at the forest edge. Elsewhere, small patches of unmodified forest are surrounded by wide aureoles of forest in the process of degradation by fire, in which *Hagenia abyssinica* is usually the dominant species. Repeated burning leads to the replacement of the trees by thicket and ultimately secondary grassland.

On the north-eastern slopes of Mt Kenya (White, MS, 1975) the patches of *Hagenia* forest just below the Ericaceous belt have been much influenced by buffalo. The understorey has been degraded by browsing, and largely replaced by a sward of grass and *Trifolium*. No regeneration of *Hagenia* is to be seen. Lower down the mountain *Hagenia* forms small groves inside *Juniperus* forest.

In the Virunga Mts, at *c.* 3000 m, *Hagenia*, in association with *Hypericum revolutum*, forms a low forest, 10–12 m high, but there is no regeneration (P. Bamps, pers. comm.).

Dry transitional montane forest
(mapping unit 19a)

Refs.: White (MS, 1973, 1975, 1979).

The drier lower slopes of those East African mountains and uplands which rise from the bushland-covered Somalia–Masai plains formerly supported a dry type of forest in which Afromontane and non-Afromontane species occurred together. Only small fragments remain and there is little published information.

There are some well-preserved examples between 1650 and 1800 m near Nairobi where the rainfall is *c.* 800 mm per year. The main canopy is at 15–18 m with emergents up to 25 m high. The larger trees include *Albizia gummifera* (near streams), *Apodytes dimidiata*, *Brachylaena discolor*, *Calodendrum capense*, *Cassipourea congoensis* (including *malosana*), *Chaetacme aristata*,

Chrysophyllum viridifolium, Croton megalocarpus, Diospyros abyssinica, Drypetes gerrardii, Euclea divinorum, Fagaropsis angolensis, Manilkara obovata, Markhamia hildebrandtii, Newtonia buchananii (near streams), *Olea africana, Phyllanthus discoideus, Schrebera alata, Strychnos usambarensis, Suregada procera, Teclea* spp. and other Rutaceae, *Trichocladus ellipticus, Uvariodendron anisatum* and *Warburgia salutaris (ugandensis)*.

Afromontane bamboo
(mapping unit 19a)

Refs.: Acocks (1975, p. 97); Chapman & White (1970, p. 166); Demaret (1958, p. 332); Fries & Fries (1948, p. 31–9); Glover & Trump (1970, p. 17–21); Greenway (1965, p. 98); Hedberg (1951); Hendrickx (1944, p. 5; 1946, p. 39); Jackson (1956, p. 368, 370); Keay (1955, p. 142); Kerfoot (1964a, p. 298); Langdale-Brown et al. (1964, p. 44, 111); Lebrun (1942, p. 47–9; 1960, p. 89); Letouzey (1968a, p. 336, 338); Lewalle (1972, p. 124–31); Lind & Morrison (1974, p. 45–7); Mabberley (1975a, p. 4); Pichi-Sermolli (1957, p. 84–6); Pócs (1976a, p. 489; 1976b, p. 169); Robyns (1937, p. 12–14; 1948, p. xli, xlvii); Snowden (1953, p. 63–4); Spinage (1972, p. 198); Tweedie (1976, p. 240); White (MS, 1949, 1963, 1973, 1975).
Phots.: Langdale-Brown et al. (1964: 7); Lind & Morrison (1974: 10, 11); Robyns (1937: 4a).
Profile: Lewalle (1972: 26).
Syn.: *Arundinaria alpina* forest or thicket (bamboo) (Langdale-Brown et al., 1964); foresta a bambù (*Arundinaria*) (Pichi-Sermolli, 1957); moist bamboo grass thicket (Greenway, 1973, p. 64).

Arundinaria alpina occurs on most of the high mountains in East Africa, from Ethiopia to the Southern Highlands in Tanzania. Further south it is only known from the North Viphya Plateau, Dedza Mt and Mt Mulanje in southern Malawi. In south Africa it is replaced by *A. tessellata*. In West Africa it is found sporadically on some of the mountains of Cameroun, though not on Mt Cameroun itself. The extent and vigour of *A. alpina* varies greatly on different mountains and the reasons for this are not yet understood.

In East Africa *A. alpina* is mostly found between 2380 and 3000 m, but on Mt Kenya it ascends to 3200 m, and in the Uluguru Mts it descends to 1630 m. It appears to grow most vigorously and to form continuous stands on deep volcanic soils on gentle slopes where the rainfall exceeds 1250 mm per year. The largest areas are on the Aberdare Range (65000 hectares, 250^2 miles), Mau Range (51000 hectares, 200^2 miles) and Mt Kenya (39000 hectares, 150^2 miles). On Mt Elgon it has a patchy distribution on the drier eastern side but forms continuous stretches on the moister western side. On Ruwenzori it is poorly developed on very steep slopes but elsewhere is dominant between 2200 and 3200 m. *Arundinaria* is almost absent from the 'dry' mountain Kilimanjaro but forms a belt between 2130 and 2740 m on the adjacent and no less dry Mt Meru (Greenway, 1965). It is mostly of sporadic occurrence in the Cherangani and Uluguru Mts but in the latter forms pure stands at 2400–2650 m on the highest peak, the summit of Kimhandu.

Luxuriance varies from an almost impenetrable thicket of stems as thick as a finger and only 4 m tall on shallow soil as at the foot of Sabinio, to well-spaced stems 8 cm in diameter and 15 m tall between which it is easy to walk, as on the south-west slopes of Mt Elgon. Flowering is gregarious, though rarely simultaneously over large areas. After flowering the whole plant dies and regeneration is by seed. Individual stems live for 5–10 years and the interval between flowering is believed to be at least 30 years. It is possible that the trees which are often found scattered in bamboo become established at these times, when the dead patches are covered with a strong growth of *Rubus, Sambucus africana, Lobelia bambuseti* and *Impatiens*. The extent to which *Arundinaria* has been able to occupy forest sites because of fire is uncertain. On the Mau Range, according to Glover & Trump (1970), it is fire-induced, and charred stumps of *Juniperus* can frequently be seen sticking through it. Elsewhere on the Mau, *Arundinaria* forms an understorey in *Juniperus* forest and here too there is evidence of past burning.

The most frequent trees which occur scattered in *Arundinaria alpina* bamboo are *Afrocrania volkensii, Dombeya goetzenii, Faurea saligna, Hagenia abyssinica, Ilex mitis, Juniperus procera, Lepidotrichilia volkensii, Nuxia congesta, Podocarpus latifolius, Prunus africana, Rapanea melanophloeos* and *Tabernaemontana johnstonii*.

Afromontane evergreen bushland and thicket
(mapping units 19a, 65, and 66)

Refs.: Chapman & White (1970, p. 138, 148, 169–70); Edwards (1967, p. 189–90); Greenway (1955, p. 560; 1965; 1973, p. 55–6); Hedberg (1951); Killick (1963, p. 41–4, 80–4); Langdale-Brown et al. (1964, p. 33, 109); Lebrun (1942, p. 65–8); Lewalle (1972, p. 146–9); Lind & Morrison (1974, p. 145); Phipps & Goodier (1962, p. 306–7); Pócs (1974; 1976b, p. 488–9); Richards (1963b); White (1978a; MS, 1949, 1963, 1973, 1975–76).
Phots.: Chapman (1962: 14); Chapman & White (1970: 52); Edwards (1967: 120); Hedberg (1951: 1b); Langdale-Brown et al. (1964: 2); Lebrun (1942: 27b, 28a, 29a); Pócs (1974: 8).
Profile: Lewalle (1972: 28).
Syn.: Cave Sandstone scrub (Killick, 1963); Ericaceae-*Stoebe* high montane heath (Langdale-Brown et al.); Ericaceous wooded grassland (Lind & Morrison, 1974); *Passerina–Philippia, Widdringtonia* fynbos (Killick, 1963); *Philippia* forest (Hedberg, 1951); subalpine elfin forest (Pócs, 1976b); upland moor p.p. (Greenway, 1973).

Afromontane evergreen bushland and thicket occur on most of the higher African mountains. They are also found on the crests and summits of some smaller mountains, especially those which are situated close to the sea or a large lake. Characteristically they occupy a large part of the 'Ericaceous belt' of Hedberg (1951). They also occur locally on shallow soils inside the montane forest belt, and on the exposed summits of mountains too low to support an Ericaceous belt. They vary greatly in floristic composition but some members of the Ericaceae (species of *Blaeria*, *Erica*, *Philippia* and *Vaccinium*) are nearly always present, and are sometimes exclusively dominant. Ericaceae are virtually absent, however, from the wetter types of elfin thicket. Where the ground is not very rocky and has been protected from fire for several years the dominants form almost impenetrable thickets. These conditions are found on the wetter mountains such as Ruwenzori, where the tallest thickets occur. On drier, rocky slopes the vegetation is often discontinuous and the bushes form an open canopy. Ericaceous bushland and thicket burn readily, and, especially on the drier mountains, they have been extensively replaced by secondary grassland. Unburnt Ericaceous bushland and thicket is normally between 3 and 13 m tall. On shallow soils and exposed slopes, however, they merge into Afromontane shrubland.

Tall elfin thicket (3–7 m high) occurs on the crests of certain low crystalline mountains in East African which rise steeply from the lowlands and are situated relatively close to the sea. They are too low to support an Ericaceous belt but their summits are in the zone of permanent mist. On Bondwa Peak (2 120 m) in the Uluguru Mts, Tanzania, the estimated mean annual rainfall is 3 000 mm (Pócs, 1974). Here *Syzygium cordatum* is the most abundant of the larger woody species. Its gnarled and semi-prostrate habit is totally different from the erect growth it shows lower down the mountain.

Afromontane and Afroalpine shrubland
(mapping units 19a, 65, and 66)

Refs.: Chapman (1962, p. 23); Chapman & White (1970, p. 170); Greenway (1955, p. 560–2); Jackson (1956, p. 370–1); Killick (1963, p. 78, 92–3); Phipps & Goodier (1962, p. 306–8).

Phots.: Chapman (1962: 23); Greenway (1955: 3); Killick (1963: 27, 34, 44–6).

Syn.: *Erica-Helichrysum* heath (Killick, 1963); open upland moorland (Greenway, 1955).

On shallow soils, and especially on exposed rocky ridges, on the high African mountains the Ericaceous bushland and thicket described above give way to much shorter Afromontane shrubland, in which Ericaceae still usually play an important role. These short shrublands are very mixed communities in which, besides shrubs, grasses, Cyperaceae, forbs (especially geophytes), bryophytes and lichens are conspicuous. Afromontane shrubland consists partly of stunted individuals of the dominants of Ericaceous bushland and thicket, and partly of species which are normally absent from the latter communities. Patches of dwarf shrubland also occur at higher altitudes on the highest mountains as part of the Afroalpine mosaic.

Afromontane and Afroalpine grassland
(mapping units 19a, 65, and 66)

Refs.: Chapman (1962); Chapman & White (1970); Fries & Fries (1948, p. 24–7); Greenway (1955, p. 555–8); Hedberg (1964, p. 114–18); Herlocker & Dirschl (1972); G. Jackson (1969); J. D. Jackson (1956, p. 361–3, 368–9); Killick (1963); Lind & Morrison (1974, p. 150–1); Maitland (1932); Phipps & Goodier (1963); Pócs (1976b, p. 494); Richards (1963b, p. 548–53); Van Zinderen Bakker & Werger (1974); White (1978a, p. 495–8, 504; MS, 1973, 1975–76); Wood (1965).

Phots.: Chapman & White (1970: 1–4, 9–10, 13–15, 19–20, 28–30, 32, 38–9); Greenway (1955: 4); Hedberg (1964: 10, 13, 14, 18, 19, 30, 69); Killick (1963: 2, 8, 21); Maitland (1932: 6, 7).

Grassland, today, is the most widespread vegetation type on the African mountains, especially the drier ones. There are undoubtedly some small areas of edaphic grassland and it is equally certain that, in the absence of human interference, grassland could be maintained by naturally occurring fires, which result from lightning, landslides, or volcanic activity. The former extent of such edaphic and natural fire-climax grassland is a matter of controversy. There is now, however, no reasonable doubt that most Afromontane grasslands have originated, or have been greatly extended, relatively recently as a result of man's destructive activity.

On the high mountains of tropical Africa the secondary grasslands of the Ericaceous and Afroalpine belts are quite different from those of the Forest belt in their floristic composition and chorological relationships.

Most of the species which dominate secondary grassland in and above the Ericaceous belt belong to the tribes Festuceae, Aveneae, and Agrosteae, and are Afromontane (including Afroalpine) endemics, or, at least in Africa, are confined to the high mountains. They are normal constituents of Ericaceous and mixed Afroalpine communities or grow on rocky slopes or in boggy hollows and are not dependent on fire for their presence. Indeed several of them are intolerant of fire.

By contrast, nearly all the species which dominate secondary grassland in the Forest belt belong to the tribes Andropogoneae and Paniceae, and are also widespread in the African lowlands. They may have invaded the forest belt from the lowlands following the destruction of forest, or they may formerly have occurred within the forest zone as marginal intruders in a few places where forest was excluded by unfavourable edaphic conditions. On the tropical mountains this distinction between the two chorological elements is well marked. In South Africa, however, it is partly obscured because the 'temperate' genera descend much lower and the most abundant 'tropical' species, *Themeda triandra*, ascends relatively higher.

In general, secondary grassland develops more readily on the drier mountains, but other factors are sometimes important. Steep well-drained slopes such as those formed by the porous lava flows of Mt Cameroun, a wet mountain, are covered with secondary grassland. Readily combustible communities such as those dominated by Ericaceae or conifers are more vulnerable to fire than most types of broad-leaved forest, and have been more extensively replaced than the latter.

Secondary montane grassland is sometimes invaded by small fire-resistant trees. Species of *Protea* are the most characteristic and occur on African mountains from Ethiopia to South Africa. They are usually small bushy trees 3–5 m tall with very short twisted boles and thick deeply fissured bark. Where fires are severe they are often fire-trimmed and there is no regeneration. Persistent severe fire eliminates them completely.

When secondary montane grassland is protected from fire for several years it is eventually invaded by forest-precursor shrubs and climbers, which form a dense thicket. The latter suppresses the grass and is itself invaded by secondary forest trees, which sometimes invade the protected grassland direct. Most examples of this succession are to be found in forest reserves where a deliberate policy of fire-protection has been followed. A similar development towards forest can, however, occur when landslips provide fire-protected niches in otherwise annually burnt grasslands (Chapman & White, 1970, p. 139, phot. 34).

The commonest grasses in secondary grassland in the forest belt on tropical mountains are *Elionurus argenteus*, *Exotheca abyssinica*, *Loudetia simplex*, *Monocymbium ceresiiforme*, *Themeda triandra*, and species of *Andropogon*, *Brachiaria*, *Digitaria*, *Hyparrhenia*, *Pennisetum*, and *Setaria*. In the Ericaceous and Afroalpine belts they are largely replaced by species of *Agrostis*, *Deschampsia*, *Festuca*, *Koeleria*, *Pentaschistis*, and *Poa*. In the Natal Drakensberg above the forest belt the commonest species are *Bromus speciosus*, *Festuca costata*, *Pentaschistis tysonii* and *Themeda triandra*.

Mixed Afroalpine communities
(mapping units 65 and 66)

In tropical Africa

Refs.: Hauman (1933, 1955); Hedberg (1951–69; 1975); Mabberley (1973, 1974, 1976); Salt (1954).
Phots.: Hedberg (1964: 6–109); Salt (1954: 6–9).
Profiles: Hedberg (1964: 84, 96, 102, 104).

The vegetation of the highest mountains of tropical Africa, those which reach altitudes between 3800 and 6000 m, namely Ruwenzori, the Virunga Volcanos, Elgon, Aberdare, Mt Kenya, Kilimanjaro and Mt Meru, is so different from that occurring at lower altitudes that it has attracted the attention of travellers and scientists since the earliest days of botanical exploration. It is characterized by the occurrence of Giant Senecios (*Senecio* subgen. *Dendrosenecio*), Giant Lobelias, shrubby Alchemillas and other plants of remarkable life-form. In recent years the flora and vegetation of these high peaks have been regarded by those with specialist knowledge (Hauman, 1955; Hedberg, 1965) as being sufficiently distinct to justify the recognition of a separate phytogeographical Afroalpine Region, which Hedberg (1961) extended to include the higher peaks of Ethiopia. If the Afroalpine Region is delimited to coincide with the Afroalpine belt (Hedberg, 1951) then its total flora is small (*c.* 280 species on the East African mountains, though there are others in Ethiopia which are still insufficiently known). In the Afroalpine belt there are virtually no endemic genera, and very few species which do not also occur in the Ericaceous and forest belts. For these reasons White (1978a) has suggested that the Afroalpine and Afromontane Regions should be combined, though for certain purposes the former can be recognized as an archipelago-like region of extreme floristic impoverishment.

Afroalpine vegetation is physiognomically very mixed and does not readily fit the major physiognomic categories. Hedberg, on whose scholarly studies our knowledge of Afroalpine vegetation is almost entirely based, recognizes five distinctive Afroalpine life-forms, all of which are closely paralleled in other genera in the 'páramo' vegetation of the northern Andes in South America. These five life-forms (and other less-specialized types) combine kaleidoscopically to give an infinite variety of mixtures. Vegetation consisting exclusively of a single life-form occupies only small areas or extreme habitats.

The tropical African islands of Afroalpine vegetation are too small to show on the map. They are included with the Ericaceous belt as part of mapping unit 65 (Altimontane vegetation).

In South Africa

Refs.: Coetzee (1967); Killick (1978a, 1978b, 1978c); Van Zinderen Bakker & Werger (1974).
Phots.: Killick (1978c: 14–18); Van Zinderen Bakker & Werger (1974: 2–4).

The authors cited above place in the Afroalpine Region all vegetation above the forest belt in the southern Drakensberg. This is at variance with Hedberg's treatment for tropical Africa since he excludes the Ericaceous belt. Although there are some resemblances between East and South Africa there are also important differences, notably the absence of giant Lobelias and Senecios from the latter. Killick has reviewed the Afroalpine Region in southern Africa (1978c) and described its vegetation for one part in great detail (1963). In the present work the vegetation of the crest of the Drakensberg is referred to as 'Altimontane' (mapping unit 66).

x The Guinea–Congolia/Zambezia regional transition zone

Geographical position and area

Geology and physiography

Climate

Flora

Mapping units

Vegetation
    Drier peripheral semi-evergreen Guineo–Congolian rain forest
    Zambezian dry evergreen forest and transition woodland
    Grassland and wooded grassland
    The coastal mosaic

Geographical position and area

The transition zone which separates the Guineo–Congolian and Zambezian Regions extends from the Atlantic Ocean to the high ground flanking the northern end of Lake Tanganyika. Its maximum width is nearly 500 km. (Area: 705 000 km^2.)

Geology and physiography

Most of the transition zone forms part of the dissected plateau which extends south from the Zaire basin to the Zambezi–Zaire watershed. The western limit of the plateau is marked by a well-defined escarpment cut into Precambrian rocks. There is a narrow coastal plain of Cretaceous and more recent sediments which is separated from the escarpment by a somewhat wider belt of undulating but low-lying country underlain by Precambrian rocks.

Much of the plateau in eastern Angola and Kwango is overlain by a thick mantle of Kalahari Sand, but in the deeply entrenched river valleys which mostly run in a north–south direction the underlying Karoo strata are exposed. Further east in Kasai, Karoo beds predominate in the north and Precambrian rocks in the south, but the long narrow ridges between the valleys are mostly covered with Kalahari Sand. Further east still, Precambrian rocks predominate. The altitude of the plateau is mostly between 1 000 and 1 500 m.

Climate

In most places the climate is intermediate between those of the Guineo–Congolian and Zambezian Regions. The dry season is more severe than that of the former but less so than that of the latter. Rainfall diminishes very rapidly near the Atlantic coast to below 800 mm per year, but dry-season relative humidity is high. Frost is unknown in the transition zone. (See Fig. 15.)

Flora

Excluding marginal intruders there are probably no more than 2 000 species, very few of which are endemic.

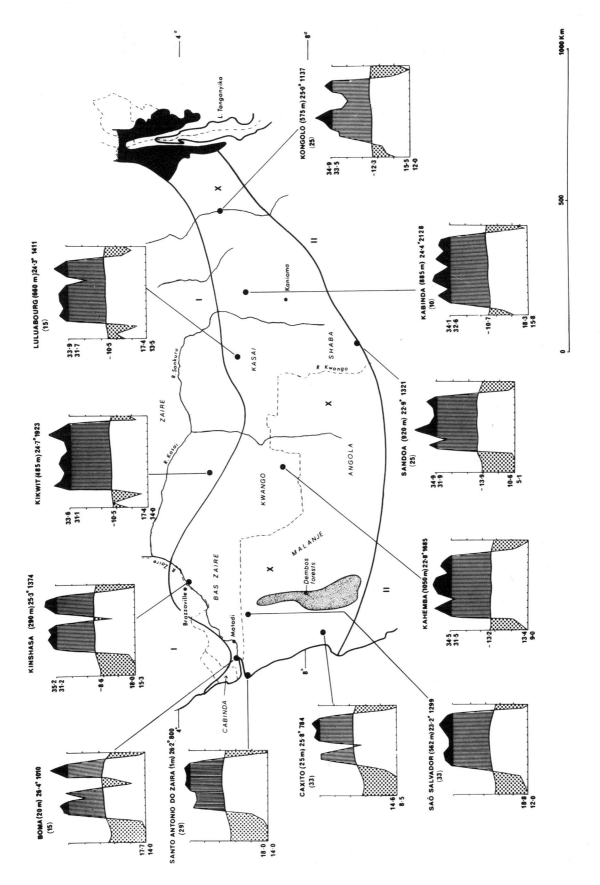

FIG. 15. Climate and topography of the Guinea–Congolia/Zambezia regional transition zone (X)

Of the 841 species recorded from the Kaniama District of Lower Shaba by Mullenders (1954, 1955), 31.3 per cent are Guineo–Congolian species and 31.4 per cent are Zambezian ('Sudano–Zambezian') species. Most of the remainder are linking species with wider distributions. Only 14 species (1.6 per cent) are endemic.

Other endemic species which are confined to the transition zone or only slightly transgress its boundaries include *Combretum camporum*, *Croton dybowskii*, *Diospyros grex*, *D. heterotricha*, *D. wagemansii*, *Hymenostegia laxiflora*, *Pteleopsis diptera* and *Rinorea malembaensis*. Most of these only occur near the coast.

Mapping units

2 (p.p.). Drier peripheral semi-evergreen Guineo–Congolian rain forest.
11a (p.p.). Mosaic of Guineo–Congolian rain forest and secondary grassland.
14. Mosaic of Guineo–Congolian rain forest, Zambezian dry evergreen forest and secondary grassland.
15 (p.p.). West African coastal mosaic.
21 (p.p.). Mosaic of Zambezian dry evergreen forest, wetter miombo woodland and secondary grassland.
31. Mosaic of wetter Zambezian woodland and secondary grassland (see page 61).
37 (p.p.). *Acacia polyacantha* secondary wooded grassland (see page 63).
60 (p.p.). Edaphic and secondary grassland on Kalahari Sand.

Vegetation

Ref.: White & Werger (1978).

The floras of the Guineo–Congolian and Zambezian Regions are almost mutually exclusive. There is, however, a transition zone between them up to 500 km wide and of considerable complexity. In it an impoverished Guineo–Congolian flora and an even more impoverished Zambezian flora interdigitate or occur in mosaic, and locally intermingle.

The greater part of the transition zone is occupied today by secondary grassland and wooded grassland dominated almost exclusively by Zambezian species. The latter, however, have greatly increased in abundance following destruction of the original vegetation, but it seems that formerly they occurred in most parts of the transition zone, though often confined to edaphically specialized sites such as rocky places and seasonally waterlogged grassland. In general, the Zambezian element becomes more abundant towards the south, and, where soil permits, the transition forms a continuum. Elsewhere, edaphic conditions override the climate to produce a mosaic. At the western end of the transition zone climate changes rapidly from the arid coastal plain, where the vegetation is predominantly Zambezian, to the humid, essentially Guineo–Congolian, Dembos 'cloud' forests on the escarpment of the interior plateau.

In addition to the vegetation described below, miombo woodland, similar to that described in Chapter II, occurs in the southern part, but it is floristically poor and little is known about it.

Drier peripheral semi-evergreen Guineo–Congolian rain forest
(mapping units 2, 11a, and 14)

Forest of Guineo–Congolian type has an uneven distribution in the transition zone.

The most extensive of the surviving forests are the Dembos 'cloud' forests in Angola. They occur between 350 and 1000 m about 100 km from the sea on the escarpment leading up to the plateau. Rainfall is between 1100 and 1500 mm per year, but the existence of forest is largely dependent on the constant condensation of water-vapour brought in by moisture-laden sea breezes. The surrounding country is too dry for the occurrence of forest.

Further west in north-west Angola and Zaire vast tongues of Guineo–Congolian forest penetrate towards the south in the wide valleys of the major tributaries of the Zaire River. These forests are not confined to the banks of rivers but also occur on the fertile soils of the rejuvenated land surface and are often several kilometres wide.

Further west still, in the Kaniama region of Lower Shaba, rain forest of Guineo–Congolian affinity was formerly the prevalent vegetation on soils derived from gabbro and tonalite. All that survives today are narrow bands on the slopes of small valleys which have cut into the plateau, and a few isolated remnants on the plateau itself (Mullenders, 1954).

Larger tree species, not all of which are generally distributed, include *Albizia zygia*, *Antiaris toxicaria*, *Trilepisium madagascariense*, *Canarium schweinfurthii*, *Celtis zenkeri*, *Chlorophora excelsa*, *Cynometra alexandri*, *Dacryodes edulis*, *Entandrophragma angolense*, *Khaya anthotheca*, *Klainedoxa gabonensis*, *Lovoa trichilioides*, *Pachystela brevipes*, *Parkia filicoidea*, *Petersianthus macrocarpus*, *Piptadeniastrum africanum*, *Pycnanthus angolensis*, *Ricinodendron heudelotii*, *Staudtia stipitata* and *Treculia africana*.

Zambezian dry evergreen forest and transition woodland
(mapping units 14 and 21)

Dry evergreen forest of pronounced Zambezian affinity is widely distributed on the Kalahari Sand-covered plateau of Kwango, where it is known as 'mabwati', and in adjacent parts of Angola. It is up to 25 m tall. The leaves of the trees are more coriaceous than those of rain-forest species and lack 'drip tips'. The most characteristic taller trees in the mabwati described by Duvigneaud (1950, 1952) are *Marquesia macroura, M. acuminata, Berlinia giorgii, Lannea antiscorbutica, Daniellia alsteeniana, Brachystegia spiciformis, B. wangermeeana* and *Parinari curatellifolia.*

Smaller trees include *Uapaca nitida, U. sansibarica, Memecylon sapinii, Diospyros batocana, Anisophyllea gossweileri, Monotes dasyanthus* and *Diplorhynchus condylocarpon.* The two species of *Marquesia, Berlinia giorgii* and *Daniellia alsteeniana,* only occur in the Guinea–Congolia/Zambezia transition zone and in the wetter northern parts of the Zambezian Region. Most of the other species are very widely distributed in the Zambezian Region. The floristic composition listed above suggests transition woodland (page 91) or a degraded community rather than true forest.

The floristic composition of mabwati changes from north to south. In the north of Kwango, in the Kalahari–Karoo contact zone, there is a strong admixture of Guineo–Congolian species, especially those which also occur on Kalahari Sand in the teke forests of the Brazzaville–Kinshasa region.

Some of the dominants of mabwati, including *Daniellia alsteeniana, Marquesia acuminata* and *M. macroura,* are also widely distributed to the west of Kwango in the Malanje Province of Angola (mapping unit 14), though there is little published information. The miombo dominants seem to be virtually absent from this region.

Tall transition woodland (8–10 m), dominated by *Berlinia giorgii* and *Uapaca nitida* and related to the mabwati of Kwango, occurs in the Kaniama region on granite outcrops where it forms a zone between forest of Guineo–Congolian affinity in the valleys, and a sparsely wooded grassland of Zambezian affinity on the plateau above (Mullenders, 1954). *Berlinia giorgii* is the most abundant species, followed by *Uapaca nitida* and *Combretum psidioides.* Other Zambezian species such as *Albizia versicolor, Cussonia sessilis, Maprounea africana, Monotes dasyanthus, Piliostigma thonningii, Sterculia quinqueloba, Stereospermum kunthianum, Strychnos cocculoides* and *Terminalia mollis* are rarer. Although fire rarely passes through the *Berlinia giorgii* forests they show little tendency to change in the direction of Guineo–Congolian forest. Guineo–Congolian linking species are few, e.g. *Canarium schweinfurthii,* and mostly occur on termite mounds.

Grassland and wooded grassland
(mapping units 11a, 14, 34, 37, and 60)

These communities are mostly secondary but some small patches may be primary. They vary greatly in floristic composition and luxuriance, chiefly in relation to parent material, catenary position, and degree of degradation. Some general information is given on page 50. The grasslands occurring in Lower Zaire have been described in considerable detail by Duvigneaud (1952). Those of Kwango and the Kaniama region of Shaba are briefly considered below.

In Kwango the mabwati forests of the Kalahari Sand-covered plateau have largely been replaced by a wooded grassland ('mikwati') in which the most frequent fire-resistant trees are *Erythrophleum africanum, Dialium engleranum, Burkea africana, Hymenocardia acida, Diplorhynchus condylocarpon, Pterocarpus angolensis, Protea petiolaris, Combretum celastroides* subsp. *laxiflorum* and *Strychnos pungens.* The herb layer consists largely of the grasses *Hyparrhenia diplandra, H. familiaris, Loudetia arundinacea, Digitaria diagonalis* (*uniglumis*), *Brachiaria brizantha* and *Ctenium newtonii.*

On the deepest sands, which may be more than 100 m thick, the structureless soil is without humus and is extremely deficient in nutrients. Although the rainfall is 1 600–1 800 mm per year, it percolates rapidly and the water-table is at a considerable depth. Fires occur each year and have been responsible for degrading the vegetation to a sparse short grassland which includes many geophytes and geoxylic suffrutices, and is referred to by Belgian authors as 'steppe', 'pseudosteppe' or 'savane steppique'. According to Devred *et al.* (1958) these grasslands are secondary and have relatively recently replaced forest and woodland. It is likely, however, that at least locally where the water-table is high for part of the year, a similar community represents an edaphic climax from which the species of secondary grassland have been recruited.

In Kwango the principal grasses of secondary suffrutex grassland are *Aristida vanderystii, Ctenium newtonii, Digitaria brazzae, Diheteropogon grandiflorus* (*emarginatus*), *Elionurus argenteus, Loudetia demeusii, L. simplex, Monocymbium ceresiiforme, Rhynchelytrum amethystinum, Schizachyrium thollonii* and *Tristachya nodiglumis* (*eylesii*). The geoxylic suffrutices include *Anisophyllea quangensis, Brackenridgea arenaria, Erythrina baumii, Gnidia kraussiana, Landolphia camptoloba, Ochna manikensis, Parinari capensis* and *Rauvolfia nana.* These grasslands are sometimes lightly wooded with scattered trees of *Combretum, Dialium engleranum, Erythrophleum africanum,* and *Daniellia alsteeniana,* but often are treeless for considerable distances, or contain only stunted shrubby individuals of *Swartzia madagascariensis, Burkea africana, Oldfieldia dactylophylla* and *Hymenocardia acida.*

The secondary grasslands of the Kaniama region of Shaba have been described by Mullenders (1954). Similar grassland extends for 300 km to the north.

The flora of the secondary grasslands near Kaniama is extremely poor, consisting of a mere 252 species, of which 19 are grasses. A few forest pioneers, such as *Albizia adianthifolia, Clausena anisata, Harungana madagascariensis* and *Phyllanthus muelleranus,* are Guineo–Congolian linking species but the remainder, almost without exception, are Zambezian species some of which also occur in other savanna regions.

The most notable grasses are *Andropogon schirensis* (dominant on the more degraded sites), *Hyparrhenia confinis* (dominant on the less-degraded sites), *Pennisetum unisetum, Brachiaria brizantha, Digitaria diagonalis, Elymandra androphila, Hyparrhenia filipendula, H. newtonii (lecomtei), H. rufa, Hyperthelia dissoluta (Hyparrhenia ruprechtii), Imperata cylindrica, Loudetia arundinacea, Panicum baumannii (fulgens), P. phragmitoides, Schizachyrium brevifolium* and *Urelytrum giganteum*.

The Zambezian woody flora is represented by the following heliophilous, pyrophytic trees: *Acacia hockii, A. polyacantha* subsp. *campylacantha, A. sieberana, Albizia versicolor, Annona senegalensis, Bridelia ferruginea, Dombeya shupangae, Erythrina abyssinica, Gardenia ternifolia, Grewia mollis, Hymenocardia acida, Maprounea africana, Maytenus senegalensis, Monotes caloneurus, M. mutetetwa, Ochna schweinfurthiana, Parinari curatellifolia, Pericopsis angolensis, Piliostigma thonningii, Psorospermum febrifugum, Pterocarpus angolensis, Schrebera trichoclada, Sclerocarya caffra, Securidaca longepedunculata, Sterculia quinqueloba* and *Stereospermum kunthianum*.

There is some evidence that many of the Zambezian species which today dominate secondary wooded grassland in the Kaniama region have been recruited from edaphically specialized refugia in which the Zambezian flora has probably persisted from a drier epoch. Thus, Mullenders describes a *Loudetia arundinacea, Ochna leptoclada* wooded grassland from the upper slopes and ridges of granite outcrops which appears to be an edaphic climax. Many of the trees, including *Stereospermum kunthianum, Entada abyssinica, Parinari curatellifolia, Sterculia quinqueloba, Albizia versicolor, Sclerocarya caffra* and *Terminalia mollis*, which occur among piles of rocks or rooted in the granite pavement, are also typical members of secondary wooded grassland.

The coastal mosaic
(mapping unit 15)

The rainfall of the coastal belt to the west of the cloud forests of Dembos and northwards into Lower Zaire and Cabinda is too low to support Guineo–Congolian vegetation except along watercourses.

The prevalent vegetation is grassland and wooded grassland, most of it probably secondary. *Adansonia digitata*, which is rare or absent further inland, is a conspicuous feature of the landscape. The two introduced trees, *Anacardium occidentale* and *Mangifera indica*, are also plentiful. On sandy soil, on elevated country of the littoral, where there is regular condensation of moisture from the westerly winds, *Strychnos henningsii* forms dense thickets thousands of square kilometres in extent. Granite outcrops near Matadi support a special vegetation rich in lichens and succulents such as *Sansevieria cylindrica, Aloe, Rhipsalis* and *Euphorbia*.

At the mouth of the Zaire River there are 250 km^2 of swamp forest on the landward side of the mangrove. Although the annual rainfall is no more than 700 mm, the constituent species are almost exclusively Guineo–Congolian rain-forest species and some, e.g. *Sacoglottis gabonensis*, are especially characteristic of the wetter types. *Elaeis guineensis*, besides occurring in this community, forms natural societies further south in very hot sheltered situations where it is dependent on phreatic water, since annual rainfall is only 600 mm.

XI The Guinea–Congolia/Sudania regional transition zone

Geographical position and area

Geology and physiography

Climate

Flora

Mapping units

Vegetation
    The Coastal Plain of Ghana
        West African dry coastal forest
        Seasonally waterlogged grassland on the Accra Plains
        Termite-mound thicket
    The Coastal Plain of Basse Casamance

Geographical position and area

This transition zone, which separates the Guineo–Congolian and the Sudanian Regions, extends across Africa from Senegal to western Uganda. Between eastern Ghana and Benin Republic (Dahomey) it reaches the coast, where the well-known 'Dahomey gap' separates the Guineo–Congolian rain forests into two blocks of unequal size. The vegetation of the driest parts of the coastal strip is not truly transitional but is included here for convenience. (Area: 1 165 000 km^2.)

Geology and physiography

Nearly everywhere the altitude is less than 750 m. In places the Cameroun Highlands rise to more than 2000 m but their highest peaks belong to the Afromontane archipelago. Otherwise the land reaches more than 1000 m only in Fouta Djalon, the Guinea Highlands and the Togo–Atacora range, which all extend into the Guineo–Congolian Region, and on the Jos Plateau in Nigeria, which is shared with the Sudanian Region.

The Jos Plateau is the largest area of land in Nigeria over 1200 m altitude. From its surface some rocky hills rise 150 to 300 m above this. It is composed principally of granite and basalt. The former is chiefly responsible for its rugged topography, including the steep escarpment up to 600 m high on the west and south.

The geology of this transition zone is very diverse. Palaeozoic rocks predominate in Guinea–Bissau and the Volta basin. In Nigeria there are extensive areas of Cretaceous sediments, especially in the Benue and lower Niger valleys. Most other places are underlain by the Precambrian.

Climate

Nearly everywhere the climate is transitional between those of the Guineo–Congolian and Sudanian Regions. A narrow strip of coastal plain in West Africa extending from Ghana eastwards to Benin Republic has, however, an anomalously dry climate. In the driest part, near Accra, rainfall is only 733 mm per year. The dryness of the plains is further accentuated by the desiccating action of strong onshore breezes which blow throughout the year. Because of wind action forest is restricted to the protected leeward slopes of hills, thicket clumps are elongated in the direction of the prevailing winds and the crowns of isolated trees are severely pruned (Jeník & Hall, 1976).

The controlling effect of the harmattan wind on vegetation in the Togo Mts in Ghana, which lie astride the northern boundary of the transition zone, has been described by Jeník and Hall (1966). (See Fig. 16.)

Flora

There are probably fewer than 2000 species, nearly all of which are Guineo–Congolian or Sudanian wides, or linking species with even wider distributions. The upland areas of Guinea Republic and adjacent Sierra Leone between 700 and 1000 m, however, support a few endemic species including *Bafodeya benna* and *Fleurydora felicis*, which both belong to monotypic genera, and *Diospyros feliciana*. A few Afromontane species also occur in Fouta Djalon in Guinea Republic.

The Accra Plains, for their size, harbour a remarkable concentration of endemic and disjunct species (Jeník & Hall, 1976). The former include *Commiphora dalzielii*, *Grewia megalocarpa*, *Talbotiella gentii* and *Turraea ghanensis*. Among the disjuncts, *Crossandra nilotica* and *Ochna ovata* elsewhere occur only in East Africa. Other disjuncts, including *Capparis fascicularis*, *Grewia villosa* and the grasses *Aristida sieberana*, *Chloris prieurii* and *Schoenefeldia gracilis*, have their main areas in the Sahel and northern Sudan zones far to the north.

Mapping units

2 (p.p.). Drier peripheral semi-evergreen Guineo–Congolian rain forest (see Chapter I).
11a (p.p.). Mosaic of Guineo–Congolian rain forest and secondary grassland (see Chapter I).
12. Mosaic of Guineo–Congolian rain forest, *Isoberlinia* woodland and secondary grassland.
13. Mosaic of Guineo–Congolian rain forest, secondary grassland and montane elements.
15 (p.p.). West African coastal mosaic.

Vegetation

Refs.: Clayton (1961); Keay (1948, 1959a, 1959c).

Today the greater part of the Guinea–Congolia/Sudania transition zone is covered with secondary grassland and secondary wooded grassland similar to that described in Chapter I. Various types of forest were formerly widespread, but they have been extensively destroyed by fire and cultivation. Of the surviving remnants the most luxuriant are indistinguishable from the drier types of peripheral semi-evergreen rain forest (page 79), and indeed represent a northern extension or outliers of it. In addition, shorter, floristically poorer forest also occurs. At one time it was thought that virtually the whole of the transition zone had been covered with forest. It now seems likely, however, that patches of *Isoberlinia* woodland and *Monotes* woodland similar to some of the most characteristic communities of the Southern Sudan zone (page 106) formerly occurred on shallow soils, and that transition woodland (page 105) formed the ecotone between forest and woodland.

Most of the forest formerly occurring in the uplands of Fouta Djalon (mapping unit 13, alt. *c.* 1000–1500 m) has been replaced by cultivation and secondary grassland. As in other parts of the Upper Guinea highlands, the most abundant tree in the forest is *Parinari excelsa* (page 81). There are also some Afromontane species such as *Nuxia congesta*, but they are less numerous than in the more elevated parts of the highlands further east.

The most characteristic species of swamp forest and riparian forest in the western half of the Guinea–Congolia/Sudania transition zone are *Berlinia grandiflora*, *Cola laurifolia*, *Cynometra vogelii*, *Diospyros elliotii*, *Parinari congensis*, and *Pterocarpus santalinoides*. They all also penetrate some distance into the Guineo–Congolian Region or are extensively distributed within it.

The Guinea–Congolia/Sudania transition zone includes both the Derived Savanna zone and the Southern Guinea zone of Keay (1959a). It corresponds quite closely to the 'zone des savanes subforestières avec galeries' of Chevalier (1938) which separates his 'zone soudanaise proprement dite' from his 'zone nord de la grande forêt dense'. Within the transition zone there is no simple relationship between latitude and the extent to which the original forest has survived. This is partly because rainfall does not always show a regular diminution northwards. Also the original vegetation was profoundly influenced by parent material, and its modification by man has been closely related to population density.

The coastal plain of Ghana and that of Basse Casamance have distinctive local features and are described separately below.

The Coastal Plain of Ghana
(mapping unit 15)

In the driest part of the plain, near Accra, the most extensive soils are unfavourable to tree growth and support a sparse short grassland, only 80 cm tall, dotted with thicket clumps which occur on low flat mounds. Elsewhere various types of forest, less luxuriant than Guineo–Congolian rain forest, represent the climax.

West African dry coastal forest

Refs.: Hall & Swaine (1974; 1976); Jeník & Hall (1976, p. 203–4).
Phot.: Jeník & Hall (1976: 3).
Profiles: Hall & Swaine (1976: 15); Jeník & Hall (1976: 8).

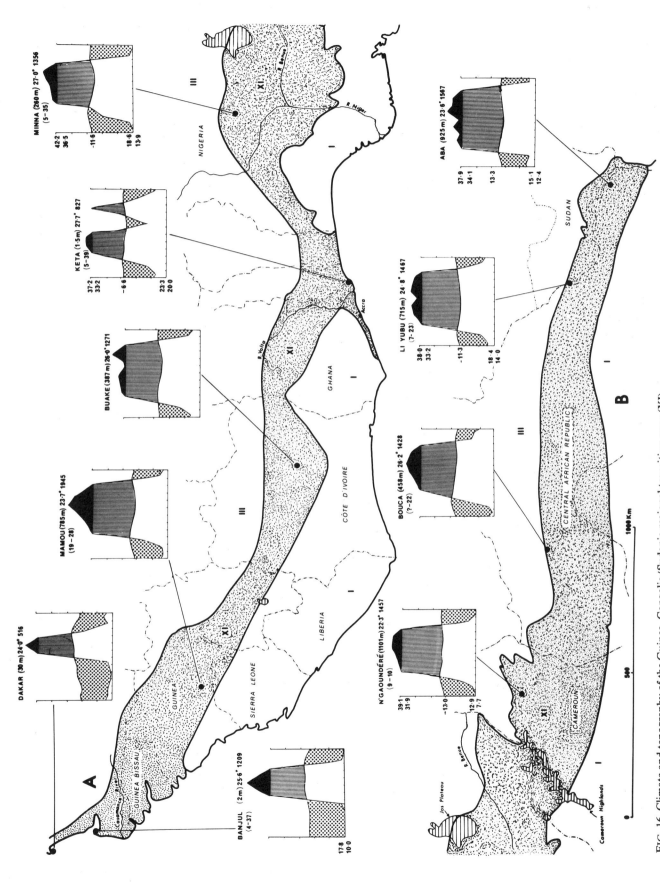

FIG. 16. Climate and topography of the Guinea–Congolia/Sudania regional transition zone (XI). A. West of Nigeria. B. East of Nigeria

There are two types of evergreen or semi-evergreen forest.

In the western type (J. B. Hall, pers. comm.) there are two or three tree layers. The main canopy is at 15–20 m and is characterized by *Cynometra megalophylla* and *Manilkara obovata*, which are much more abundant in this type than in rain forest. Emergents up to 30 m tall of *Nesogordonia papaverifera*, *Celtis mildbraedii*, *Antiaris toxicaria* and *Ceiba pentandra* often occur. Lianes are abundant, but epiphytes are virtually absent and the field layer is poorly developed.

The eastern type, which occurs on inselbergs on the Accra Plains, is always less than 20 m tall. *Diospyros abyssinica* and *Millettia thonningii* dominate the canopy, while the lower storey is composed chiefly of *Drypetes parvifolia*, *D. floribunda* and *Vepris heterophylla*. Most species are capable of coppicing and the undergrowth is extremely sparse. Large lianes are present, especially *Griffonia simplicifolia* and *Premna quadrifolia*, but epiphytes other than lichens are lacking. Grasses are absent from the herb layer but a few succulents, e.g. *Sansevieria liberica*, occur.

Seasonally waterlogged grassland on the Accra Plains

Refs.: Jeník & Hall (1976).
Phots.: Jeník & Hall (1976: 4, 5).

The soils are shallow and parent rock is encountered within two metres of the surface. Drainage is impeded and root development, particularly of woody plants, is inhibited. There are two main types of soil, developed over acidic and basic gneiss respectively. The most extensive soils overlying acidic gneiss are Pallid Sands with an underlying impervious stone/clay pan, which locally reaches the surface. The soils occurring over basic gneiss are Black Earths, which have abundant calcium carbonate concretions and develop deep cracks in the dry season.

The general dominant of the drier grasslands of the Accra plains, both on acidic and basic soils, is *Vetiveria fulvibarbis*. Its most widespread associates are *Brachiaria falcifera*, *Andropogon canaliculatus*, *Cassia mimosoides*, *Fimbristylis pilosa* and *Polygala arenaria*.

Termite-mound thicket

Refs.: Jeník & Hall (1976); Okali et al. (1973).

The mounds are up to 15 m in diameter but less than 0.5 m high. Termite activity is greater on the mounds than in the surrounding grasslands. Within the thicket patches there are small termitaria occupied by *Odontotermes pauperans* and *Amitermes evuncifer*, and larger termitaria, up to 3 m tall and 3 m wide at the base, occupied by *Macrotermes bellicosus*. Although it is likely that the mounds have been built up by termite activity over a long period of time, there is no evidence that new mounds are being formed in the grasslands today. On the contrary, Okali et al. suggest that these clumps may be remnants of formerly continuous thicket.

The patches of thicket have a dense closed canopy at 5 m, composed principally of *Flacourtia indica* (*flavescens*), *Zanthoxylum zanthoxyloides*, *Grewia carpinifolia*, *Securinega virosa*, *Capparis erythrocarpos* and *Uvaria chamae*. Mature emergent trees of *Elaeophorbia drupifera* and *Diospyros mespiliformis* reach a height of 10 m. The succulent-leaved herb, *Sansevieria liberica*, is conspicuous in the field layer.

The Coastal Plain of Basse Casamance
(mapping unit 11a, p.p.)

Refs.: Adam (1961a); Aubréville (1948b).

Forest of Guineo–Congolian affinity extends along the West Coast of Africa into Basse Casamance and Sénégal far beyond the climatic limits of rain forest. In Basse Casamance rainfall is 1 500–1 800 mm per year but most falls in only five months of the year and the dry season is too long and too severe to permit the development of typical rain forest. The country, however, is more or less flooded during the rainy season and the water-table is not far beneath the surface during the dry season. This enables some Guineo–Congolian forest species to extend their range into an inhospitable climate. Most of the natural vegetation has been replaced by rice-fields or ground-nuts (*Arachis*), but until recently sufficient relicts remained to enable a reconstruction of the original vegetation to be made. *Parinari excelsa* was formerly abundant both in permanent swamp forest in depressions and in drier forest on better-drained soils.

Forest on the latter, which is about 18–20 m tall, is dominated by *Parinari excelsa*. Other abundant species are *Erythrophleum suaveolens*, *Detarium senegalense*, *Afzelia africana* and *Khaya senegalensis*. Rarer associates include *Albizia adianthifolia*, *A. ferruginea*, *A. zygia*, *Antiaris toxicaria*, *Chlorophora regia*, *Cola cordifolia*, *Daniellia ogea*, *Dialium guineense*, *Morus mesozygia*, *Schrebera arborea* and *Sterculia tragacantha*.

Most species of the drier *Parinari–Erythrophleum–Detarium* forests of Basse Casamance are also widely distributed in Guineo–Congolian rain forest, especially in the drier semi-evergreen types. Although the Casamance forest is without endemic species, it is a very distinct type, both floristically and in structure (Aubréville, 1949b, p. 41). The canopy is at 18–20 m and is made up of large trees which usually branch close to the ground and have leaning boles and very wide crowns. Trees with straight boles are rare, chiefly *Khaya*. Lianes of all sizes are abundant. The canopy replaces its leaves before the end of the dry season, during which the 3–5 m tall understorey is mostly deciduous. Those leaves which are not shed have a wilted appearance.

XII The Lake Victoria regional mosaic

Geographical position and area

Geology and physiography

Climate

Flora

Mapping units

Vegetation
    Drier peripheral semi-evergreen Guineo–Congolian rain forest
    Transitional rain forest
    Swamp forest
    Scrub forest
    Evergreen and semi-evergreen bushland and thicket and derived communities

Geographical position and area

This Region includes most of Uganda, the whole of eastern Rwanda and Burundi, and small parts of Zaire, Kenya, and Tanzania. A small exclave occupies the Ruzizi Valley north of Lake Tanganyika. (Area: 224 000 km^2.)

Geology and physiography

The Lake Victoria basin was formed during the middle Pleistocene by earth movements associated with the evolution of the western arm of the Great Rift Valley. From the lake itself (altitude 1 134 m) the land surface falls gradually to the north. It rises gently to the south but much more abruptly to the east and west towards important islands of the Afromontane Region. Nearly everywhere the underlying rocks belong to the Precambrian but locally they are covered by recent alluvium.

Climate

Climatic gradients are often steep and are related to the complex physiography and to distance from Lake Victoria, which is an important source of precipitation.

Locally the rainfall is sufficiently high (1 500–2 000 mm per year) and well distributed throughout the year to support rain forest. Elsewhere it is too low for rain forest but is not sufficiently seasonal for woodland. Hence scrub forest and semi-evergreen bushland and thicket represent the climax. A few miombo species reach their northern limits near the southern shore of Lake Victoria. The dry season there is much less severe than in most of the Zambezian Region and the communities in which miombo species occur are not typically Zambezian. (See Fig. 17.)

Flora

There are possibly no more than 3 000 species, of which very few are endemic. There are probably no endemic genera.

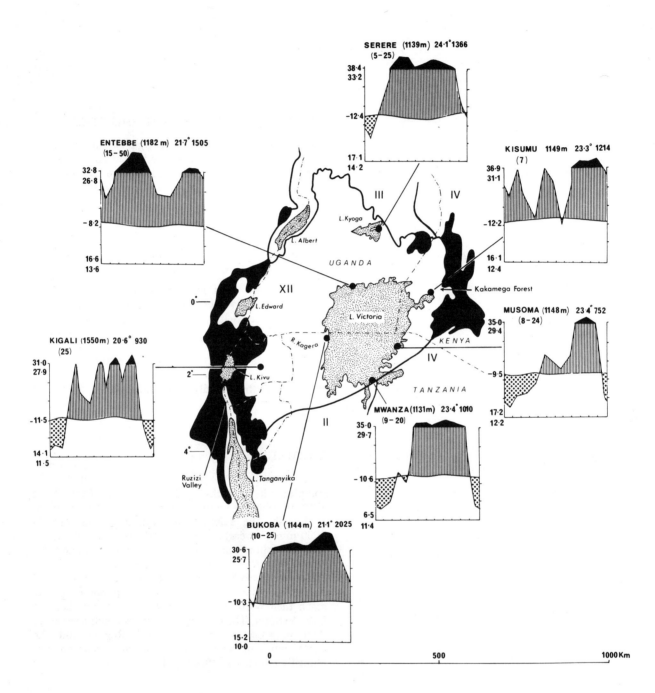

FIG. 17. Climate and topography of the Lake Victoria regional mosaic (XII)

Mapping units

2 (p.p.). Drier peripheral semi-evergreen Guineo–Congolian rain forest.
4 (p.p.). Transitional rain forest.
8 (p.p.). Swamp forest.
11a (p.p.). Mosaic of Guineo–Congolian rain forest and secondary grassland.
25 (p.p.). Wetter Zambezian miombo woodland (see Chapter II).
26 (p.p.). Drier Zambezian miombo woodland (see Chapter II).
42 (p.p.). Somali–Masai *Acacia–Commiphora* deciduous bushland (see Chapter IV).
45 (p.p.). Mosaic of East African evergreen bushland and secondary *Acacia* wooded grassland.

Units 25, 26, and 42 reach their limits at the southern end of Lake Victoria. Their vegetation there is not very typical and little is known about it. It is not considered further here. Most of the species of secondary grassland and wooded grassland in unit 11a also occur in Guineo–Congolian secondary grassland (page 84).

Vegetation

The Lake Victoria Regional Mosaic is the meeting-place of five distinct floras: Guineo–Congolian, Sudanian, Zambezian, Somalia–Masai and Afromontane. Its vegetation consists of a mosaic of floristically impoverished variants of the characteristic vegetation of the first four, in some cases with an admixture of Afromontane species.

Drier peripheral semi-evergreen Guineo–Congolian rain forest
(mapping units 2 and 11a)

Refs.: Eggeling (1947); Langdale-Brown, Osmaston & Wilson (1964, p. 44–51); G. H. S. Wood (1960, p. 26–31).
Phots.: Eggeling (1947: 1–4, 6, 7).
Profiles: Eggeling (1947: 6–9); Langdale-Brown *et al.* (1964: 10).

The majority of species are widespread in the Guineo–Congolian Region. Large trees in this category include *Albizia* spp., *Alstonia boonei*, *Aningeria altissima*, *Antiaris toxicaria*, *Chrysophyllum albidum*, *Celtis* spp., *Chlorophora excelsa*, *Cynometra alexandri*, *Entandrophragma angolense*, *E. cylindricum*, *E. utile*, *Holoptelea grandis*, *Khaya anthotheca*, *K. grandifoliola*, *Maesopsis eminii*, *Mildbraediodendron excelsum*, *Morus mesozygia* (*lactea*), *Piptadeniastrum africanum* and *Pycnanthus angolensis*.

Transitional rain forest
(mapping unit 4)

Refs.: Faden (1970); Lewalle (1972); Lucas (1968).

Little has been published and only small fragments remain.

In western Burundi between 1600 and 1900 m the larger trees include *Alangium chinense*, *Albizia gummifera*, *Anthonotha pynaertii*, *Carapa grandiflora*, *Chrysophyllum gorungosanum*, *Diospyros gabunensis*, *Newtonia buchananii*, *Parinari excelsa*, *Prunus africana*, *Strombosia schefferi*, *Symphonia globulifera*, *Syzygium guineense* and *Xymalos monospora*. These species also occur in western Rwanda and eastern Kivu at the same altitudes (P. Bamps, pers. comm.).

Kakamega forest in Kenya, which has fewer Afromontane elements, is shown on the map as lowland forest. Several Guineo–Congolian lowland rain-forest species, including *Aningeria altissima*, *Cordia millenii*, *Entandrophragma angolense*, *Maesopsis eminii* and *Monodora myristica*, reach their easternmost limit here. This forest, which is situated at 1520 to 1680 m on the Nandi escarpment east of Lake Victoria, contains the following Afromontane species: *Apodytes dimidiata*, *Macaranga kilimandscharica*, *Neoboutonia macrocalyx*, *Prunus africana*, *Strombosia schefferi* and *Turraea holstii*.

Swamp forest
(mapping unit 8)

Refs.: Eggeling (1935, p. 431; 1947, p. 60–1, 86–7); Langdale-Brown *et al.* (1964, p. 74–5); Lind & Morrison (1974, p. 50–8); G. H. S. Wood (1960, p. 30–1).
Phot.: Langdale-Brown *et al.* (1964: 24).
Profile: Lind & Morrison (1974: 1.8).

Swamp forest dominated by species which are widespread in tropical Africa occurs extensively on the shores of Lake Victoria and elsewhere. The more important species are *Anthocleista schweinfurthii*, *Erythrina excelsa*, *Ficus congensis*, *Macaranga monandra*, *M. pynaertii*, *M. schweinfurthii*, *Mitragyna stipulosa*, *Musanga cecropioides*, *Parkia filicoidea*, *Pseudospondias microcarpa*, *Spondianthus preussii*, *Syzygium cordatum*, *Uapaca guineensis*, *Voacanga thouarsii* and the palms *Phoenix reclinata* and *Raphia farinifera*. *Acacia kirkii* subsp. *mildbraedii* occurs sparingly in places.

The forests occurring at *c.* 1200 m on alluvial deposits at the mouth of the Kagera River on the western shore of Lake Victoria are unique in tropical Africa in being composed in almost equal proportions of lowland, predominantly Guineo–Congolian, and Afromontane species. Undisturbed forest is dominated by the Guineo–Congolian *Baikiaea insignis* (*eminii*) and the Afromontane *Podocarpus falcatus* (*P. usambarensis* var. *dawei*). Other Guineo–Congolian species include: *Canarium schweinfurthii*, *Klainedoxa gabonensis*, *Maesopsis eminii*, *Pseudospondias microcarpa*, *Pycnanthus angolensis*, *Symphonia globulifera* and *Tetrapleura tetraptera*. The principal Afromontane tree species are: *Apodytes dimidiata*, *Croton megalocarpus*, *Ilex mitis*, *Podocarpus latifolius*, *Strombosia schefferi*, *Suregada procera*, *Trichocladus ellipticus* and *Warburgia salutaris*. The epiphytic lichen *Usnea* is abundant as in most types of Afromontane forest in East Africa. There has been heavy exploitation for timber, especially *Podocarpus*.

Scrub forest
(mapping unit 45)

Refs.: Germain (1952, p. 255–64); Lebrun (1947, p. 703–21; 1955, p. 59–64); Lewalle (1972, p. 57–69).
Phots.: Germain (1952: 61); Lebrun (1947: 50, 51.1; 1955: 6).
Profiles: Lewalle (1972: 14, 15).
Syn.: la forêt tropophile à *Albizia grandibracteata* et *Strychnos potatorum* ['*stuhlmannii*'] (Germain, 1952); la forêt xérophile des crêtes: groupement à *Croton dichogamus* et *Euphorbia dawei* (Lebrun, 1955); la forêt sclérophylle à *Euphorbia dawei* (Lewalle, 1972); la forêt sclérophylle à *Strychnos potatorum* (Lewalle, 1972).

Vegetation intermediate between rain forest and evergreen and semi-evergreen bushland at one time probably occurred more extensively in the Lake Victoria basin than in any other part of Africa, but few relics remain and published information is fragmentary. In Uganda and Burundi, *Cynometra alexandri*, which elsewhere is a canopy or emergent species of rain forest, also occurs in shorter forest and scrub forest and sometimes is less than 10 m tall. In scrub forest it is usually associated with *Euphorbia dawei*.

In the basin of Lake Edward, *Euphorbia dawei* forms forest at 900–1000 m in bands up to 3 km wide along the banks of rivers and on the lower slopes of escarpments. The 12–15 m tall canopy is composed almost exclusively of this species and has a cover of 70–80 per cent. Since the trunks are weak and easily blown over, gaps in the canopy are numerous, but they are soon filled by recruitment from young individuals, which regenerate freely in the shade of the parent trees. A few other species such as *Euclea racemosa* subsp. *schimperi*, *Spathodea campanulata* and *Dombeya kirkii* (*mukole*) are present in the canopy but only as isolated individuals. Lianes, especially *Cissus quadrangularis*, *Bonamia poranoides*, *Senecio bojeri* and *Cissus petiolata*, are numerous, and after reaching the canopy hang down in draperies. Less vigorous lianes are *Scutia myrtina* and *Cissus rotundifolia*. Epiphytes are virtually absent except for the bracket fern *Platycerium elephantotis*. An 8–10 m tall lower canopy is feebly developed beneath the Euphorbias but becomes dominant in the openings. It consists of *Canthium vulgare*, *Cordia ovalis*, *Euclea racemosa* subsp. *schimperi* and *Olea africana*. Other woody species include *Cassine aethiopica*, *Grewia similis*, *Carissa edulis*, *Erythrococca bongensis*, *Rhus natalensis* and *Teclea nobilis*. The herb layer consists principally of *Asystasia gangetica*, *Achyranthes aspera*, *Panicum deustum* and *Justicia flava*. A few mosses, including *Fissidens sciophyllus*, *Racopilum speluncae* and *Archidium capense*, are confined to the base of the trunks of *Euphorbia dawei*.

In the Ruzizi valley, *Euphorbia dawei* forms scrub forest only in a single locality (Lewalle, 1972). Here it occurs as a 17–18 m tall emergent above a discontinuous 10–12 m tall canopy of *Cynometra alexandri* and *Tamarindus indica*. This particular floristic assemblage is one of the most remarkable in the whole of Africa. *Euphorbia dawei* and *Cynometra alexandri* also occur in scrub forest in Uganda (T. J. Synnott, pers. comm.).

Tall scrub forest (15 m tall) with an upper canopy of *Strychnos potatorum*, *Tamarindus indica*, *Grewia mollis*, *Albizia grandibracteata*, and *Euphorbia candelabrum* is thought to represent the climax vegetation in the Ruzizi valley, though no more than tiny degraded vestiges remain.

Evergreen and semi-evergreen bushland and thicket and derived communities
(mapping unit 45)

Refs.: Germain (1952, p. 233–51); Lebrun (1947, 2, p. 638–61; 1955, p. 52–9); Lebrun & Gilbert (1954, p. 29–31, p.p.); Liben (1961); Troupin (1966).
Phots.: Germain (1952: 55–9); Lebrun (1942: 36.2; 1947: 39–42; 1955: 5); Liben (1961: 1).
Profiles: Lebrun (1947: 95–6); Liben (1961: 1).
Syn.: les forêts sclérophylles montagnardes et submontagnardes: ordre Oleo-Jasminetalia, alliance submontagnarde: Grewia-Carission edulis p.p. (Lebrun & Gilbert, 1954); les bosquets xérophiles à *Maerua mildbraedii* et *Carissa edulis* (Maerueto–Carissetum edulis); les bosquets xérophiles: association à *Jasminum fluminense* ('*mauritianum*') et *Carissa edulis* (Lebrun, 1955).

Evergreen and semi-evergreen bushland and thicket, and related types of scrub forest (see above) probably represent the climax vegetation of large parts of this region. They have, however, been extensively destroyed, and today are represented by small degraded relics which survive chiefly on shallow soils, and by small patches of secondary regrowth. Today the landscape is one of lightly wooded *Acacia* grassland with no more than small islands of secondary evergreen thicket (mapping unit 45).

Lebrun (1947, 1955) and Liben (1961) have suggested how patches of thicket become established in secondary wooded grassland dominated by *Acacia hockii*, *A. gerrardii*, *A. kirkii* subsp. *mildbraedii*, *A. senegal* and *Euphorbia candelabrum*. Lianes germinate in the shade of the Acacias and eventually smother their crowns. The shade they cast provides conditions suitable for the establishment of shrubs and bushes which, after a time, completely suppress the heliophilous Acacias which are unable to regenerate in the shade of the thicket. *Euphorbia candelabrum* (*calycina*) can also initiate the development of thicket. The shade it casts causes a diminution in the vigour of the grass layer which permits the invasion of hemi-heliophilous or even hemi-sciaphilous woody plants. The principal species involved are: *Allophylus africanus*, *Azima tetracantha*, *Canthium schimperanum*, *Carissa edulis*, *Capparis fascicularis*, *C. tomentosa*, *Erythrococca bongensis*, *Grewia bicolor*, *Maerua triphylla* (*mildbraedii*), *Olea africana*, *Rhus natalensis*, *Tarenna graveolens* and *Turraea nilotica* among bushy species, and *Cissus quadrangularis*, *C. rotundifolia*, *Senecio stuhlmannii* and *Vernonia brachycalyx* among climbers. Two mosses, *Bryum argenteum* and *Archidium capense*, sometimes occur on the surface of the soil. Lock (1977b) suggests that *Capparis tomentosa* plays an important part in the establishment of thicket.

The successional status of these thickets is uncertain. Lebrun originally (1947) thought that in the absence of human interference they would be replaced by *Euphorbia dawei* scrub forest, but subsequently (1955) suggested that the latter is edaphically restricted, especially to rocky slopes, thereby implying that the thicket would be climax over extensive areas.

XIII The Zanzibar–Inhambane regional mosaic

Geographical position and area

Geology and physiography

Climate

Flora

Mapping units

Vegetation
    Zanzibar–Inhambane lowland rain forest
    Transitional rain forest
    Zanzibar–Inhambane undifferentiated forest
    Zanzibar–Inhambane scrub forest
    Swamp forest
    Zanzibar–Inhambane transition woodland
    Zanzibar–Inhambane woodland and scrub woodland
    Zanzibar–Inhambane evergreen and semi-evergreen bushland and thicket
    Zanzibar–Inhambane edaphic grassland
    Zanzibar–Inhambane secondary grassland and wooded grassland

Geographical position and area

This region occupies a coastal belt from southern Somalia (1° N.) to the mouth of the Limpopo River (25° S.). It is 50 to 200 km wide except where it penetrates further inland along broad river valleys. Small exclaves also occur to the west on the windward slopes of mountainous massifs below 1 500 m where the local increase in precipitation and dry-season relative humidity favours the development of lowland and transitional rain forest. (Area: 336 000 km².)

Geology and physiography

Most of the land lies below 200 m but in the northern part there are scattered hills and plateaux rising considerably higher. They include the Shimba Hills (*c.* 400 m) and Mrima Hill in Kenya, the Pugu Hills and Rondo (Mwera) Plateau in Tanzania, and the Macondes Plateau (986 m) in northern Mozambique. Only the East Usambara Mts (1 500 m) in Tanzania exceed 1 000 m.

The outer part of the coastal belt, the coastal plain proper, is underlain by marine sediments of various ages from Cretaceous to recent, though in Kenya there are also small areas of Jurassic outcrops. Inland from the coastal plain the more undulating topography is underlain principally by Precambrian rocks, but locally by Triassic sediments. The width of the coastal plain varies considerably and in northern Mozambique is very narrow.

Climate

Rainfall is mostly between 800 and 1 200 mm per year and there is a well-defined dry season. Appreciably higher rainfall is experienced only in a few places such as the East Usambara Mts (Amani, 1 946 mm) and on the islands of Zanzibar and Pemba (Wete, 1 964 mm). In these places the amount and distribution is sufficient to support rain forest.

In most parts of the Zanzibar–Inhambane Region the rainfall is comparable in amount to that of the Zambezian Region, but the dry season is less severe since relative humidity is high and no month is absolutely dry. Mean annual temperature is *c.* 26° C north of the Zambezi, but diminishes steadily southwards. Frosts are unknown. (See Fig. 18.)

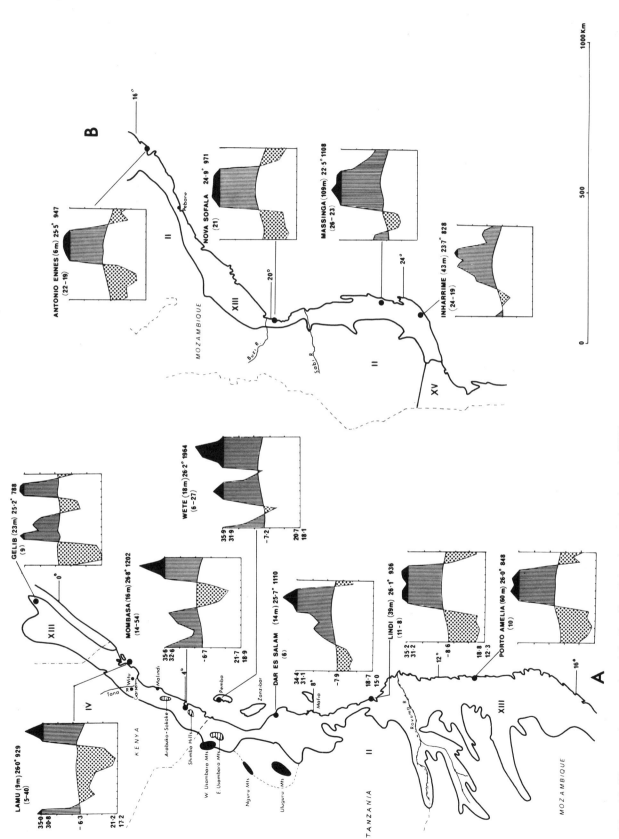

FIG. 18. Climate and topography of the Zanzibar–Inhambane regional mosaic (XIII) A. North of 16° S. B. South of 16° S.

Flora

There are about 3000 species of which at least several hundred are endemic.

Endemic genera. These include *Cephalosphaera, Englerodendron, Grandidiera* and *Stuhlmannia*. A further seven genera, including *Bivinia, Hirtella, Ludia* and *Hymenaea* (*Trachylobium*), which also occur in Madagascar, are confined or almost confined on the African mainland to this region. *Hirtella* and *Hymenaea*, although absent from West and Central Africa, are abundantly represented in the American tropics.

Endemic species. The total number is unknown but of the 190 forest tree species so far recorded, 92 (48.4 per cent) are endemic. The greatest concentration is in Kenya and northern Tanzania, centred on the Shimba Hills and East Usambara Mts.

Linking elements. Of the 190 forest trees species at present known, 15.3 per cent are Tongaland–Pondoland linking species, 4.7 per cent are Malagasy linking species, 25.8 per cent are Guineo–Congolian linking species, 3.7 per cent are Afromontane linking species, and 7.0 per cent are ecological and chorological transgressors. Details of the Zambezian/Zanzibar-Inhambane linking element are given by Moll & White (1978).

Mapping units

16a. Zanzibar–Inhambane coastal mosaic.
16b. Zanzibar–Inhambane forest.

Vegetation

Ref.: Moll & White (1978).

Forest is the most widespread climax vegetation but has been largely replaced by secondary wooded grassland and cultivation. There are also extensive areas of scrub forest and edaphic grassland, and smaller areas of transition woodland, bushland, and thicket. Typical woodland does not occur in Kenya, but from Tanzania southwards a floristically impoverished type of miombo woodland becomes increasingly important.

The forest is rich in species, but owing to the rapidity of change in amount and distribution of rainfall, dry-season atmospheric humidity, and the availability of soil moisture, it changes rapidly in floristic composition and physiognomy over quite short distances and is difficult to classify. Only the most luxuriant types are classified as rain forest here. The latter is lowland rain forest except for that occurring on the highest ground in the Zanzibar–Inhambane Region, namely the summits of the most elevated peaks in the East Usambara Mts, which are clothed with transitional rain forest. All forest other than rain forest is treated as a single continuum, namely undifferentiated Zanzibar–Inhambane forest, much of which has been referred to as rain forest by previous authors.

Zanzibar–Inhambane lowland rain forest
(mapping units 16a and 16b)

Refs.: Chapman & White (1970, p. 87, 97, 173); Pócs (1976*b*, p. 478–9); Polhill (1968); White (MS, 1975–76); Wild & Barbosa (1967, p. 4).
Phots.: Chapman & White (1970: 57).
Profile: Chapman & White (1970: 7).
Syn.: lowland rain forest (Tanzania, Polhill, 1968); lowland seasonal rain forest (Malawi, Chapman & White, 1970); moist evergreen forest at low and medium altitudes (Wild & Barbosa, 1967).

Lowland rain forest was formerly extensively developed in Tanzania along the lower parts of the eastern highland arc, especially the Uluguru, Nguru, and Usambara Mts, and in parts of Ulanga and Iringa Districts. Only small fragments remain. Similar forest occurs on the lower slopes of mountains further inland, such as the Malawi Hills (Chapman & White, 1970), and thus forms small exclaves in the Zambezian Region. Most of the surviving forest in the East Usambara Mts occurs above 900 m and, because of its admixture of Afromontane species, is classified as lower transitional rain forest (see below).

The main canopy of Zanzibar–Inhambane lowland rain forest, which is almost evergreen, is up to 20 m high, with emergents up to 40 m or more. This type differs from Guineo–Congolian rain forest in the greater degree of bud protection, less well-developed 'drip-tips' to the leaves, and the paucity of vascular and non-vascular epiphytes. The larger tree species include *Aningeria pseudoracemosa* (very local), *Antiaris toxicaria, Burttdavya nyasica, Chlorophora excelsa, Khaya nyasica, Lovoa swynnertonii, Maranthes* (*Parinari*) *goetzeniana, Newtonia buchananii, Parkia filicoidea, Ricinodendron heudelotii, Sterculia appendiculata* and *Terminalia sambesiaca*.

A small outlier of similar forest near Taveta in the Somalia–Masai Region is shown on the map. It is dominated by *Chlorophora excelsa, Cordyla africana, Diospyros mespiliformis* and *Newtonia buchananii*, and owes its presence to the high water-table.

Transitional rain forest
(mapping unit 16a)

The rain forest near Amani in the East Usambara Mts in Tanzania is probably the best known in East Africa, at least to taxonomists, although it has never been fully described. The summits of the East Usambara Mts (1254 m) are not high enough for the occurrence of Afromontane rain forest, but several Afromontane species occur on their upper slopes at altitudes appreciably lower than their normal lower limits on other mountains. The forests in which they occur are thus transitional, though lowland elements predominate. The rainfall at Amani is 1946 mm per year. With decreasing altitude it falls off rapidly and rain forest probably does not descend below c. 800 m, although some species are found at lower altitudes in riparian forest.

In relation to their small size the transitional rain forests of the East Usambara Mts have a remarkably rich and diversified flora. More than 40 per cent of the larger woody species are endemic to them or almost so, or, in a few cases, occur as endemic subspecies. Two monotypic tree genera, *Cephalosphaera* (*C. usambarensis*) and *Englerodendron* (*E. usambarense*), are also endemic. Nearly 30 per cent of the larger woody species are Afromontane or, in East Africa at least, show upland but not Afromontane distributions, and are characteristic members of lower transitional rain forest, though none are confined to it. Most of the remaining species are Guineo–Congolian linking species.

The majority of endemic species have their closest relatives in the lowland rain forests of the Guineo–Congolian Region. Some species, e.g. *Anonidium usambarense*, *Enantia kummeriae*, *Isolona heinsenii* and *Polyceratocarpus scheffleri*, belong to genera which are otherwise confined to the Guineo–Congolian Region. A very wide interval separates the Usambara endemics from the other species in their respective genera. A similar wide interval separates the endemic subspecies (*Greenwayodendron suaveolens* subsp. *usambaricum*, *Pterocarpus mildbraedii* subsp. *usambarensis* and *Magnistipula butayei* subsp. *greenwayi*) of those species which otherwise occur exclusively or principally in the Guineo–Congolian Region.

The Afromontane element is represented by *Alangium chinense*, *Allanblackia stuhlmannii*, *Aningeria adolfi-friederici*, *Cylicomorpha parviflora*, *Isoberlinia scheffleri*, *Myrianthus holstii*, *Ocotea usambarensis*, *Strombosia scheffleri*, *Syzygium sclerophyllum*, *Xymalos monospora* and *Zenkerella capparidacea* s.l.

The species characteristic, *inter alia*, of lower transitional rain forest are: *Macaranga capensis*, *Maranthes goetzeniana*, *Morinda asteroscepa*, *Newtonia buchananii*, *Strychnos mitis* (*mellodora*) and *Trichilia dregeana*.

Guineo–Congolian linking species include: *Afrosersalisia cerasifera*, *Antiaris toxicaria*, *Trilepisium madagascariense*, *Chrysophyllum perpulchrum*, *Cleistanthus polystachyus*, *Ficus capensis*, *Funtumia africana*, *Pachystela msolo*, *Parkia filicoidea*, *Rauvolfia caffra*, *Ricinodendron heudelotii*, *Schefflerodendron usambarense* and *Treculia africana*. Some of these, e.g. *Trilepisium*, *Ficus capensis* and *Parkia*, are widespread outside the Guineo–Congolian Region and occur in suitable habitats in the intervening region. Others, such as *Chrysophyllum perpulchrum*, *Pachystela msolo* and *Schefflerodendron usambarense*, outside the Guineo–Congolian Region are virtually confined to the Usambara Mts.

The fact that the East Usambara Mts harbour so many species which are separated from their closest relatives by wide intervals suggests that they have served as a refugium for a formerly widespread flora which has become extinct over much of its former area.

Floristically poorer transitional rain forest also occurs as small exclaves in the Zambezian Region, e.g. in Malawi at about 1370 m in the Misuku Hills, on Nchisi Mt, at Lisau Saddle and Chaone Hill in the Shire Highlands, and on Machemba Hill near Mt Mulanje (Chapman & White, 1970). The Chirinda forest in Zimbabwe (Banks, 1976) also belongs here.

Zanzibar–Inhambane undifferentiated forest
(mapping units 16a and 16b)

Refs.: Dale (1939, p. 8–11, 14-15); Moomaw (1960, p. 22–9, 35–40); White (MS, 1975–76).
Phots.: Dale (1939: 2); Moomaw (1960: 4, 7, 10).
Syn.: lowland evergreen rain forest (Dale): *Sterculia–Chlorophora–Memecylon* lowland rain forest (Moomaw); *Combretum schumannii–Cassipourea* lowland dry forest on coral rag (Moomaw); evergreen dry forest (Dale); *Manilkara–Diospyros* lowland dry forest (Moomaw).

Some species are confined to the wetter types and some to the drier, but many occur throughout. *Julbernardia magnistipulata*, for instance, forms magnificant 30 m tall stands of moist forest but also occurs in scrub forest and evergreen bushland.

In the moister variants the main canopy occurs at 15–20 m and from it emergents rise to a height of 30 or 35 m. Very few individuals are taller than that. Many of the canopy species are briefly deciduous but not concurrently so. Nevertheless this type is appreciably more deciduous than semi-evergreen lowland rain forest. None of the trees is buttressed, though some, e.g. *Terminalia sambesiaca*, are fluted at the base. Lianes are plentiful but vascular epiphytes are usually scarce, as are bryophytes in general. Most tree trunks are without bryophytes except sometimes for a few minute hepatics.

The richest forests are in Kenya and northern Tanzania. Towards the south there is progressive floristic impoverishment and relatively few species reach Mozambique.

The larger trees include: *Afzelia quanzensis* (20 m), *Albizia adianthifolia* (25 m), *Antiaris toxicaria* (35 m), *Apodytes dimidiata*, *Balanites wilsoniana* (30 m), *Trilepisium madagascariense* (20 m), *Celtis wightii* (20 m), *Cola clavata* (20 m), *Combretum schumannii* (25 m), *Cordyla africana* (25 m), *Chlorophora excelsa* (35 m), *Diospyros abyssinica* (very rare), *Diospyros mespiliformis* (30 m), *Erythrina sacleuxii* (20 m), *Erythrophleum suaveolens* (25 m), *Fernandoa magnifica* (20 m), *Ficus vallis-choudae* (20 m), *Inhambanella henriquesii* (25 m), *Julbernardia magnistipulata* (30 m), *Lannea welwitschii* (25 m), *Lovoa swynnertonii* (35 m), *Macaranga capensis* (25 m), *Malacantha alnifolia* (20 m), *Manilkara sansibarensis* (25 m), *Mimusops aedificatoria* (25 m), *Newtonia paucijuga* (25 m), *Nesogordonia parvifolia* (20 m), *Paramacrolobium coeruleum* (25 m), *Parkia filicoidea* (30 m), *Pachystela brevipes* (25 m), *Rhodognaphalon schumannianum* (30 m), *Ricinodendron heudelotii* (35 m), *Sterculia appendiculata* (35 m), *Terminalia sambesiaca* (35 m), *Hymenaea verrucosa* (30 m) and *Xylopia parviflora* (*holtzii*) (25 m).

The drier forests cover a larger area than the moister forests and extend further to the north and south. They are even more diverse floristically than the wetter forests and most of the larger tree species are sometimes gregarious and locally dominant or co-dominant. The principal trees are: *Acacia robusta* subsp. *usambarensis* (20 m), *Afzelia quanzensis* (15 m), *Albizia petersiana* (15 m), *Balanites wilsoniana*, *Trilepisium madagascariense* (15 m), *Brachylaena huillensis* (15 m), *Cassipourea euryoides* (15 m), *Combretum schumannii* (15 m), *Cussonia zimmermannii* (15 m), *Cynometra webberi* (12 m), *Julbernardia magnistipulata* (10–15 m), *Manilkara sansibarensis* (18 m), *M. sulcata* (10 m), *Memecylon sansibaricum* (9 m), *Newtonia paucijuga* (15 m), *Oldfieldia somalensis* (12 m), *Pleurostylia africana* (15 m), *Scorodophloeus fischeri* (15 m), *Tamarindus indica* (12 m) and *Hymenaea verrucosa* (18 m).

In many places 25 m or more tall *Chlorophora excelsa* and *Sterculia appendiculata* are emergent. The cycad, *Encephalartos hildebrandtii*, and two succulent Euphorbias, *E. nyikae* and *E. wakefieldii*, are locally plentiful.

Zanzibar–Inhambane scrub forest
(mapping unit 16a)

Ref.: White (MS, 1975–76).

In Kenya and southern Somalia scrub forest dominated by *Diospyros cornii* and *Manilkara mochisia* forms a quasi-continuous narrow band separating the forests of the coastal region from the bushlands of the interior. The annual rainfall is between 500 and 750 mm. Scrub forest reaches the coast between Malindi and Lamu where the rainfall is lower than elsewhere. Similar vegetation extends to southern Tanzania.

D. cornii forms a discontinuous upper canopy 9–15 m tall. Sometimes the crowns are in contact but usually the cover is no more than 50 per cent. Among other canopy trees: *Manilkara mochisia* is an almost constant associate but is less plentiful; *Dobera glabra* is often abundant, especially where the water-table is near the surface; *Newtonia erlangeri* only occurs in the northern forests; and *Terminalia spinosa* indicates disturbance. Cactiform Euphorbias of tree dimensions, which are rare and often absent, are represented only by scattered individuals of *E. candelabrum*. The 1 m high *E. grandicornis*, however, frequently forms dense communities in the understorey.

The lower canopy, which reaches a height of 7 m, is very rich in species. It is usually dense but is only impenetrable where species of *Sansevieria*, *Euphorbia grandicornis* or the viciously spinous *Adenia globosa* form dense thickets. The more characteristic members of this layer include: *Bivinia jalbertii*, *Carissa* sp., *Croton pseudopulchellus*, *Diospyros consolatae*, *Euclea natalensis*, *E. racemosa* subsp. *schimperi*, *Excoecaria venenifera*, *Grandidiera boivinii*, *Haplocoelum foliosum*, *H. inoploeum*, *Ochna thomasiana*, *Sideroxylon inerme*, *Suregada zanzibarensis*, *Thespesia danis*, *Thylachium africanum* and *Xeromphis nilotica*.

Climbers are rather rare, and epiphytes are virtually absent except between Witu and Garsen. The herb layer is poorly developed except for colonies of *Sansevieria*. Most species in scrub forest are evergreen. Only a few, e.g. species of *Commiphora*, are more than briefly deciduous.

In most places *D. cornii* scrub forest has been degraded and converted into secondary deciduous bushland dominated by *Albizia anthelminthica*, *Acacia bussei*, *A. mellifera*, *A. nilotica*, *Hyphaene compressa*, *Terminalia spinosa* etc.

Swamp forest
(mapping unit 16a)

Fresh-water swamp forest is of restricted occurrence. *Barringtonia racemosa* forest with *Acrostrichum aureum*, *Hibiscus tiliaceus*, *Pandanus* spp. and *Phoenix reclinata* often occurs immediately behind the mangrove zone. In places *Barringtonia* extends upstream for a considerable distance. On Pemba Island *Raphia* swamp forest occurs in shallow valleys with very slow drainage. Associates include *Elaeis guineensis*, *Voacanga thouarsii* and the aroid *Typhonodorum lindleyanum* (Greenway, 1973).

Zanzibar–Inhambane transition woodland
(mapping unit 16a)

Refs.: Dale (1939, p. 15–16); Moomaw (1960, p. 30–5); White (MS, 1975–76); Wild & Barbosa (1967, p. 21, 26, 31).

In places Zanzibar–Inhambane forest species occur in intimate mixture with heliophilous Zambezian woodland species to form communities which are intermediate between forest and woodland. Some such communities are clearly seral but others seem to be stable.

In the Shimba Hills there are patches dominated by the woodland species *Brachystegia spiciformis* with an almost pure understorey of saplings of the forest species *Paramacrolobium coeruleum*. Near Witu in northern Kenya the Doum palm, *Hyphaene compressa*, is a conspicuous feature of the secondary grasslands. Locally the forest is encroaching and dead and dying *Hyphaene* have been overtopped by the forest species *Trichilia emetica*, *Erythrophleum suaveolens* and *Manilkara sansibarensis*.

Elsewhere in Kenya, as in the Arabuko–Sokoke forest, *Brachystegia spiciformis* forms almost pure stands on white sterile sands. The forest species *Manilkara sansibarensis* and *Hymenaea verrucosa* occur scattered in the canopy, and locally there are patches of evergreen thicket formed of forest shrubs and climbers. The forest element, however, is not very vigorous and it appears that the soil is too unfavourable to permit the completion of the succession to forest.

Zanzibar–Inhambane woodland and scrub woodland
(mapping unit 16a)

In a few dry rain-shadow areas, as in the foothills of the West Usambara Mts, the vegetation is a scrub woodland dominated chiefly by Zambezian linking species such as *Cassia singueana*, *Combretum collinum*, *Dichrostachys cinerea*, *Heeria reticulata*, *Lonchocarpus bussei*, *Pappea capensis*, *Sclerocarya caffra*, *Stereospermum kunthianum* and *Ziziphus mucronata*. The tallest species, *Sclerocarya*, is up to 8 m high.

Floristically poor miombo woodland has a scattered distribution south of the Rovuma River, where it interdigitates or occurs in mosaic with patches of forest and other vegetation. It forms parts of Wild & Barbosa's (1967) mapping units: 9, 10, 13, 25, 26, 27, 31, 32, and 33.

Zanzibar–Inhambane evergreen and semi-evergreen bushland and thicket
(mapping unit 16a)

Various types of bushland and thicket are found where unfavourable soil conditions prevent the development of forest.

Dense thicket occurs on termite mounds in seasonally waterlogged grassland which occupies parts of the coastal plain, as between Garsen and Lamu in northern Kenya. The thicket of *Capparis*, *Carissa*, *Commiphora*, *Euclea natalensis*, *Diospyros consolatae*, *Sideroxylon inerme* etc. is usually overtopped by emergent trees of *Diospyros cornii*, *Dobera glabra*, *Manilkara mochisia* or *Tamarindus indica*.

According to Birch (1963) evergreen thicket represents the climax on shallow soils overlying coral limestone in parts of Kenya where the rainfall is between 950 and 1200 mm per year. Characteristic species include *Carpodiptera africana*, *Cussonia zimmermannii*, *Diospyros squarrosa*, *Zanthoxylum* (*Fagara*) *chalybeum*, *Grewia plagiophylla*, *G. truncata*, *Haplocoelum inoploeum*, *Harrisonia abyssinica*, *Lannea stuhlmannii*, *Ludia mauritiana* (*sessiliflora*), *Manilkara sansibarensis*, *Millettia usaramensis*, *Monanthotaxis fornicata*, *Pycnocoma littoralis*, *Sterculia rhyncocarpa*, *Suregada zanzibarensis*, *Tabernaemontana elegans* and *Uvaria leptocladon*.

At a few places on the East African coast Ericaceous bushland dominated by species of *Philippia* occurs on the waterlogged sites of former shallow lagoons or lake basins. Thus on Mafia and Pemba Islands (Greenway, 1973), *P. mafiensis* forms a loose open canopy at a height of 8 m. Associates include *Syzygium cordatum*, *Uapaca sansibarica*, *Parinari curatellifolia*, *Manilkara sansibarensis*, *Euclea natalensis* and scattered *Pandanus goetzei*. When *Philippia* is exclusively dominant the ground is covered with leaf litter or the lichen *Cladonia medusiana*; otherwise the fern *Phymatodes scolopendria* is conspicuous.

Another species of *Philippia*, *P. simii*, forms an open shrubland on poorly drained soils on the coast of Mozambique near Pebane (Wild & Barbosa, 1967).

Zanzibar–Inhambane edaphic grassland
(mapping unit 16a)

In northern Kenya edaphic grassland studded with thicket-covered termite mounds (see above) covers large areas of grey-black cracking clay soil near the mouth of the Tana River. It is largely treeless except for a few widely spaced individuals of *Acacia zanzibarica*, *Hyphaene compressa*, *Terminalia spinosa* and *Thespesia danis*.

On the coastal plain of Mozambique between the R. Sabi and the R. Buzi, the 'tandos' or seasonally flooded clayey depressions occurring on sandy Quaternary or calcareous Cretaceous deposits are covered with *Hyparrhenia*, *Ischaemum*, *Setaria* grassland and are bordered by wooded grassland with *Parinari curatellifolia*, *Uapaca nitida*, *Syzygium guineense* etc. There are also extensive areas of badly drained grassland on the deltas of the larger rivers.

Zanzibar–Inhambane secondary grassland and wooded grassland
(mapping unit 16a)

These communities are extensive but published information is meagre. In Kenya, between Mombasa and the Tanzania frontier, the landscape is a mosaic of agricultural crops, grassy fallows, secondary thicket often dominated by *Lantana*, and orchards of *Cocos*, *Anacardium* and *Mangifera*, which, when the canopy is not too dense, often have a carpet of grass. In places trees from the original forest, especially *Chlorophora excelsa* and *Sterculia appendiculata*, have been left standing. The palms, *Borassus aethiopum* and *Hyphaene compressa*, are locally conspicuous. Other scattered non-forest trees include *Acacia senegal*, *Adansonia digitata*, *Afzelia quanzensis*, *Annona senegalensis*, *Antidesma venosum*, *Crossopteryx febrifuga*, *Dalbergia melanoxylon*, *Dichrostachys cinerea*, *Flacourtia indica*, *Harrisonia abyssinica*, *Lannea stuhlmannii*, *Lonchocarpus bussei*, *Maytenus senegalensis*, *Piliostigma thonningii*, *Sclerocarya caffra*, *Securidaca longependunculata*, *Stereospermum kunthianum*, *Strychnos madagascariensis*, *S. spinosa* and *Vitex mombassae*.

XIV The Kalahari–Highveld regional transition zone

Geographical position and area

Geology and physiography

Climate

Flora

Mapping units

Vegetation
    The Zambezia/Kaokoveld–Mossamedes transition
    Kalahari thornveld and the transition to Zambezian broad-leaved woodland
    The Windhoek Mountains
    The Kalahari/Karoo–Namib transition
    Highveld grassland and associated communities
        Grassland
        Riparian scrub
        Rupicolous bushland and shrubland
        Scrub forest
    The Highveld/Karoo transition
    The transition from Afromontane scrub forest to Highveld grassland
    The Afromontane/Tongaland–Pondoland transition
    The Zambezia/Highveld transition

Geographical position and area

The Kalahari–Highveld Transition Zone separates the Zambezian and Karoo–Namib Regional Centres of Endemism. It runs diagonally across Africa from 13° S. in southern Angola to 33° S. in the Eastern Cape. Its width varies considerably. In the widest parts of the Highveld and Kalahari sectors it is more than 1 800 km across, but north of Windhoek it suddenly narrows. (Area: 1 223 000 km².)

Geology and physiography

Most of the Kalahari–Highveld transition zone occurs on the great Interior Plateau of southern Africa. In only a few places does it extend on to or beyond the Great Escarpment. The Kalahari basin occupies the central part. From this area of extremely low relief, which lies mostly between 850 and 1 000 m, the land rises gradually to the east and west.

The peripheral highlands in the west, which are formed of Precambrian and Palaeozoic rocks, occupy only a narrow strip and reach their maximum elevation of 2 484 m in the Windhoek Mts.

To the east and south-east of the Kalahari basin the land rises gradually to more than 2 000 m towards the plateau rim and the Great Escarpment, though for much of its length the latter lies well within the Afromontane Region. The whole of this eastern part of the transition zone is underlain by Karoo rocks.

The Kalahari basin is filled with sand and only locally are there outcrops of older rocks of Precambrian, Palaeozoic, and Karoo age. Its geological history has not been fully worked out, but it is thought that dune sands began to form at the end of the Cretaceous or in the early Tertiary. Since then there have been several changes of climate and the dunes in the extreme south may have assumed their present shape within the last 10 000 years. Today the area is largely one of internal drainage but there have been periods of strong river flow during the wetter phases of the Pleistocene.

The Kaap Plateau, west of Kimberley, which is formed of Precambrian rocks and lies between 1 220 and 1 830 m, separates the Kalahari basin from the Karoo beds further east.

Climate

Rainfall is intermediate between that of the Zambezian and Karoo–Namib Regions. Nearly everywhere it is between 250 and 500 mm per year, but it increases somewhat in the east as the Drakensberg is approached.

Most rain falls in summer. It is less concentrated than in the Zambezian Region but less evenly distributed than in the Karoo–Namib.

Winter temperatures are low everywhere except in the narrow north–eastern extension into Angola. Frost is widespread and severe, and absolute minimum temperatures are mostly lower than in the Karoo–Namib Region. (See Fig. 19.)

Flora

The total flora is fairly large, possibly c. 3 000 species, but this is because of the large number of marginal intruders which penetrate a short distance from the four contiguous major phytochoria. There are very few endemic species and the greater part of the interior has a very poor flora.

Thus, Mostert (1958) records only 738 species of indigenous flowering plants from his study of the Bloemfontein and Brandford magisterial districts (2 590 km^2). A high proportion of them, including *Celtis africana, Commelina benghalensis, Crotalaria podocarpa, Juncus effusus, Phyllanthus maderaspatensis, Sarcostemma viminale, Tarchonanthus camphoratus, Themeda triandra* and *Typha australis*, are pluriregional species. Some 112 species (15.2 per cent) listed by Mostert are Compositae and 111 (15.1 per cent) are Gramineae.

The flora of the southern Kalahari, which almost exactly corresponds to mapping unit 56, is even poorer. Leistner (1967) records 438 species of flowering plants from the South African part (58 000 km^2), and estimates that the total flora of the Southern Kalahari (124 320 km^2), which is appreciably larger than the entire Cape floristic region, amounts to no more than 550 species.

Fewer than half the species listed by Leistner can be regarded as typical Karoo–Namib species. They include *Leucosphaera bainesii, Nymania capensis, Parkinsonia africana, Phaeoptilum spinosum, Rhigozum trichotomum, Stipagrostis amabilis* and *Tamarix usneoides*. Most of the remainder are widespread in southern Africa and many extend to the Zambezian Region or further, e.g. *Acacia erioloba, A. hebeclada, A. mellifera, Albizia anthelmintica, Boscia albitrunca, Diospyros lycioides, Terminalia sericea, Echinochloa colona, Pogonarthria squarrosa* and *Sporobolus pyramidalis*. One of the most characteristic grasses of the Southern Kalahari, *Asthenatherum glaucum*, is otherwise confined in South Africa to the Karoo–Namib Region, but it also occurs near Lake Turkana in Kenya.

The relatively few species which are more or less confined to the Kalahari–Highveld Region include *Acacia haematoxylon, Anthephora argentea* and *Schmidtia kalahariensis*.

Mapping units

20. Transition from Afromontane scrub forest to Highveld grassland.
24. Mosaic of Afromontane scrub forest, Zambezian scrub woodland, and secondary grassland.
34. Transition from South African scrub woodland to Highveld grassland.
35a (p.p.). Transition from Zambezian undifferentiated woodland to Kalahari *Acacia* deciduous bushland and wooded grassland.
35c. The Windhoek Mountains.
36. Transition from *Colophospermum mopane* scrub woodland to Karoo–Namib shrubland
44. Kalahari deciduous *Acacia* bushland and wooded grassland.
56. The Kalahari/Karoo–Namib transition.
57b. The Highveld/Karoo transition.
58. Highveld grassland.

Vegetation

The Kalahari–Highveld Region abuts on and provides transitions connecting four major phytochoria. For this reason its vegetation pattern is complex. In the following account the vegetation is not described primarily on its physiognomy as for other phytochoria, but the region is divided into nine subordinate areas, which largely coincide with the units shown on the map and are described separately below.

The greater part of the Kalahari–Highveld Region is included in the Zambezian Domain of the Sudano–Zambezian Region by Werger (1978a), who does not recognize transition zones. He places the remainder in the Karoo–Namib Region.

The Zambezia/Kaokoveld–Mossamedes transition (mapping unit 36)

Refs.: de Matos & de Sousa (1970); Giess (1971, p. 10); Tinley (1971); Volk (1966b); Whellan (1965).

Zambezian and Karoo-Namib species occur in intimate mixture. Even *Colophospermum mopane* and *Welwitschia bainesii* grow in the same community. At its western limits in Angola at the edge of the Mossamedes desert, *Colophospermum* occurs as a scattered stunted tree 3 m tall associated with the Zambezian species *Acacia mellifera, Albizia anthelmintica, Commiphora* sp., and *Terminalia prunioides* and abundant *Welwitschia*.

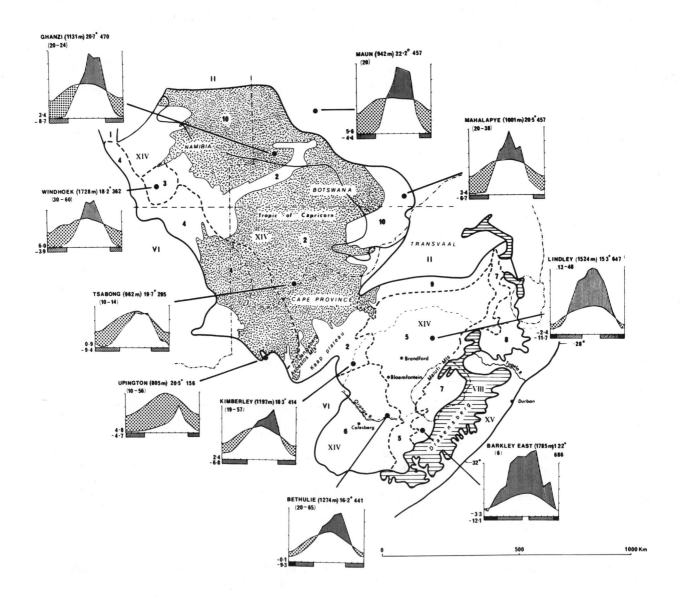

FIG. 19. Climate and topography of the Kalahari/Highveld regional transition zone (XIV)
Figures 1–10 indicate the mapping units shown on the accompanying *Vegetation Map of Africa* as follows: (1) mapping unit 36, transition from *Colophospermum mopane* scrub woodland to Karoo–Namib shrubland; (2) mapping unit 44, Kalahari *Acacia* bushland and wooded grassland; (3) mapping unit 35c, the Windhoek Mts; (4) mapping unit 56, the Kalahari/Karoo–Namib transition; (5) mapping unit 58, Highveld grassland; (6) mapping unit 57b, the Highveld/Karoo transition; (7) mapping unit 20, transition from Afromontane scrub forest to Highveld grassland; (8) mapping unit 24, mosaic of Afromontane scrub forest, Zambezian scrub woodland and secondary grassland; (9) mapping unit 34, transition from South African scrub woodland to Highveld grassland; (10) mapping unit 35a, transition from Zambezian undifferentiated woodland to Kalahari *Acacia* deciduous bushland and wooded grassland.
The stippled area shows the distribution of Kalahari Sand

Kalahari thornveld and the transition to Zambezian broad-leaved woodland
(mapping units 35a, p.p., and 44)

Refs.: Acocks (1975, p. 39–43); Cole & Brown (1976); Wild & Barbosa (1967, p. 45, 46, 60).
Phots.: Acocks (1975: 31, 32); Cole & Brown (1976: 7, 10); Volk (1966a: 3, 11); Walter (1971: 136, 137, 140, 141, 144–6).

Wooded grassland is the characteristic vegetation of the thick mantle of Kalahari Sand. In Botswana the more-or-less continuous grass sward is less than 1 m high and consists principally of *Anthephora argentea*, *A. pubescens*, *Digitaria pentzii*, *Eragrostis biflora*, *E. ciliaris*, *E. lehmanniana*, *E. pallens*, *Panicum kalaharense*, *P. lanipes*, *Pogonarthria squarrosa*, *Schmidtia kalahariensis*, *S. pappophoroides* and *Stipagrostis uniplumis*. Of these, *Anthephora argentea* in almost endemic, and a few species, e.g. *Panicum lanipes* and *Schmidtia kalahariensis*, are shared with the Karoo–Namib Region, but the majority extend from the Karoo–Namib at least as far as the southern part of the Zambezian Region.

The principal trees and bushes in the southern variant (mapping unit 44) are all Zambezian species, namely *Acacia erioloba*, *A. fleckii*, *A. hebeclada*, *A. luederitzii*, *A. mellifera*, *A. tortilis*, *Boscia albitrunca*, *Dichrostachys cinerea* and *Terminalia sericea*. In the northern variant (mapping unit 35a) broad-leaved trees are more abundant, and include *Combretum collinum*, *Commiphora africana*, *C. angolensis*, *Ochna pulchra* and *Ziziphus mucronata*, but *Acacia* is still dominant.

The trees are always less than 7 m tall and usually are much smaller. Normally they are widely spaced. On the shallow soils of the quartzite and limestone outcrops which form the Ghanzi Ridge the woody plants are denser, grasses are relatively less important and shrubs are plentiful. The latter include the Karoo species *Rhigozum brevispinosum*, *Leucosphaera bainesii*, *Phaeoptilum spinosum* and *Montinia caryophyllacea*.

In the Cape Province north of the Orange River, Kalahari Sand alternates with extensive areas of stony soil on the Kaap Plateau, Langeberg, and Asbestos mountains.

On Kalahari Sand in this area the vegetation has been severely degraded. *Themeda triandra* was formerly the dominant grass but it has been extensively replaced, because of overgrazing, by 'white' desert grasses, chiefly species of *Aristida*, *Eragrostis* and *Stipagrostis*. Further degradation leads to a uniform sward of *Schmidtia pappophoroides*, and ultimately to invasion by the Karoo shrublets *Pentzia incana* and *Chrysocoma tenuifolia*. The most abundant and characteristic tree, *Acacia erioloba*, has been removed from large areas to provide fuel for the mines at Kimberley.

On the most stony soils the vegetation is dense bushland. The principal species is *Tarchonanthus camphoratus*. Its woody associates include: *Acacia karroo*, *A. mellifera*, *A. tortilis*, *Boscia albitrunca*, *Buddleja saligna*, *Croton gratissimus*, *Diospyros lycioides*, *Ehretia rigida*, *Euclea crispa* subsp. *ovata*, *E. undulata*, *Euphorbia avasmontana*, *Grewia flava*, *Lebeckia macrantha*, *Maytenus heterophylla*, *Olea africana*, *Rhigozum obovatum*, *R. trichotomum*, *Rhus ciliata*, *R. dregeana*, *R. pyroides*, *R. lancea*, *R. undulata*, *Tarchonanthus minor* and *Ziziphus mucronata*.

The Windhoek Mountains
(mapping unit 35c)

Refs.: Giess (1971, p. 11); Volk & Leippert (1971).
Phots.: Giess (1971: 39–42).

Giess's photographs show a wooded grassland, but the original vegetation was probably denser. The flora is a mixture of Zambezian and Karoo–Namib species. The principal trees and bushes are *Acacia hereroensis*, *Combretum apiculatum*, *Acacia reficiens* subsp. *reficiens*, *A. hebeclada*, *Euclea undulata*, *Dombeya rotundifolia*, *Tarchonanthus camphoratus*, *Rhus marlothii*, *Albizia anthelmintica*, *Heeria* (*Ozoroa*) *crassinervia*, *Ficus cordata* and *F. guerichiana*. The original grass cover consisted of *Anthephora pubescens*, *Brachiaria nigropedata*, *Cymbopogon* spp., *Heteropogon contortus*, *Hyparrhenia hirta*, and others, but these grasses are now sparse in many parts because of overgrazing. The Karoo–Namib grasses, *Stipagrostis obtusa*, *S. uniplumis*, and *Panicum lanipes*, also occur. *Phaeoptilum spinosum* and species of *Aptosimum*, *Eriocephalus*, *Galenia*, *Pentzia*, *Plinthus*, *Salsola* and *Tetragonia* represent the Karoo–Namib flora among small shrubs and forbs.

The Kalahari/Karoo–Namib transition
(mapping unit 56)

Refs.: Coetzee & Werger (1975, p. 549–50); Giess (1971, p. 13); Leistner (1967); Leistner & Werger (1973); Walter (1971, p. 256–8).
Phots.: Coetzee & Werger (1975: 20); Giess (1971: 65–7); Leistner (1967: 1, 2, 8, 18, 36, 39–42); Leistner & Werger (1973: 2, 4).
Profile: Walter (1971: 148).

This is an area of wind-blown sand which occurs as fixed dunes in the form of long parallel ridges. Sand covers 90 per cent of the surface. In undisturbed areas, the lower slopes of dunes are largely consolidated by vegetation but the upper slopes and crests are subjected to erosion by strong winds and the cover is much sparser.

The vegetation is a mosaic of lightly wooded grassland on the dune crests, pure grassland in shallow depressions between the dunes, and *Rhigozum trichotomum* shrubby grassland in deeper hollows where the underlying calcrete is near the surface.

In undisturbed places the grasses are mostly perennials including *Asthenatherum glaucum*, *Stipagrostis uniplumis*, *Eragrostis lehmanniana*, *Stipagrostis ciliata*, and, on the dune crests, *Stipagrostis amabilis*. A common pioneer is *Megaloprotachne albescens*, whereas in disturbed areas in drier regions the annual *Schmidtia kalahariensis* is dominant. The commonest trees are *Acacia erioloba, Boscia albitrunca, Acacia reficiens* subsp. *reficiens, Albizia anthelmintica* and *Terminalia sericea*, which often occurs as a shrub. *Acacia haematoxylon*, which is almost confined to this region, is usually a shrub, but sometimes a tree.

Highveld grassland and associated communities
(mapping unit 58)

Refs.: Acocks (1975, p. 87–95); Coetzee & Werger (1975, p. 551–3); Mostert (1958, p. 85–161); Van Zinderen Bakker, Jr (1971, 1973); Werger (1973a, p. 113–27).

Phots.: Acocks (1957: 81–90); Coetzee & Werger (1975: 22); Van Zinderen Bakker, Jr (1973: 1).

Highveld grassland represents the climatic climax between 1220 and 2150 m on large parts of the high interior plateau in South Africa which extends west of the Drakensberg from the extreme south of the Transvaal through the Orange Free State to the Eastern Cape. The total tree flora of this region is very small and the development of woody vegetation is precluded nearly everywhere by the dry, extremely frosty winters. Although fire is almost certainly a natural ecological factor there is insufficient evidence that it is primarily responsible for the almost total absence of larger woody plants. The latter are virtually confined to riparian forest, bushland, and thicket on the few rocky hills and escarpments, and to scrub forest in sheltered ravines in the foothills of the Maluti Mts and the Drakensberg in the east.

Grassland

Acocks (1975) recognizes 10 types of Highveld grassland, which are distinguished mainly by the different proportions in which a handful of species occur. The following species are of general occurrence in one or more of his types: *Alloteropsis semialata, Andropogon amplectens, A. appendiculatus, A. schirensis, Anthephora pubescens, Aristida congesta, A. junciformis, Brachiaria serrata, Chloris virgata, Ctenium concinnum, Cymbopogon plurinodis, Cynodon dactylon, C. incompletus, Digitaria argyrograpta, D. diagonalis, D. monodactyla, D. tricholaenoides, Elionurus argenteus, Eragrostis atherstonei, E. capensis, E. chloromelas, E. gummiflua, E. lehmanniana, E. micrantha, E. obtusa, E. plana, E. racemosa (chalcantha), E. sclerantha, E. superba, Eustachys paspaloides, Harpechloa falx, Heteropogon contortus, Microchloa caffra, Monocymbium ceresiiforme, Panicum coloratum, P. natalense, Pogonarthria squarrosa, Setaria flabellata, S. nigrirostris, S. sphacelata, Sporobolus discosporus, S. fimbriatus, Themeda triandra, Trachypogon spicatus, Tragus koelerioides, T. racemosus, Trichoneura grandiglumis, Triraphis andropogonoides* and *Tristachya leucothrix (hispida)*. The absence from this list of *Hyparrhenia*, the tall species of which are so conspicuous in the Zambezian Region, is noteworthy. *Themeda triandra* is by far the most widespread and abundant species in the Highveld. It usually forms a sward 25–75 cm tall, which in summer looks dense, but the basal cover rarely exceeds 25 per cent. In winter and during droughts the grass cover is much shorter and the associated forbs and bare spaces are more conspicuous.

It has been widely assumed that *Themeda* is the natural climax species throughout most of the Highveld area, but there is some evidence (Roux, 1969) that its dominance depends on fire, and that with fire-protection *Themeda* is partly replaced by other species. In the wetter parts of the Highveld, *Themeda* is commonly associated with *Elionurus argenteus, Heteropogon contortus, Trachypogon spicatus,* and *Tristachya leucothrix*. In drier types the lower-growing species, *Aristida congesta, Eragrostis lehmanniana* and *Tragus berteronianus*, are plentiful.

Over extensive areas, because of overgrazing, *Themeda* has been largely eliminated and has been replaced by pioneer grasses such as *Aristida* spp. and *Chloris virgata* as well as invasive Karoo shrublets (*Chrysocoma tenuifolia*) and annual weeds (*Tribulus terrestis*).

The following forbs are of general occurrence in one or more of Acocks's types: *Ajuga ophrydis, Anthospermum rigidum, Asclepias multicaulis, Barleria macrostegia, Berkheya onopordifolia, B. rigida, Conyza pinnata, Crabbea acaulis, Cyperus obtusiflorus, Dicoma macrocephala, Euphorbia inaequilatera, E. striata, Felicia filifolia, F. muricata, Geigeria aspera, Gnidia kraussiana, Haplocarpha scaposa, Helichrysum dregeanum, H. latifolium, H. rugulosum, H. oreophilum, Hermannia betonicifolia, H. coccocarpa, H. depressa, Hypoxis rigidula, H. rooperi, Indigofera alternans, I. rostrata, Ipomoea crassipes, Kohautia amatymbica, Osteospermum scariosum, Oxalis depressa, Rhynchosia totta, Scabiosa columbaria, Scilla nervosa, Senecio coronatus, S. erubescens, Sonchus nanus, Stachys spathulata, Vernonia oligocephala, Walafrida densiflora* and *W. saxatilis. Ziziphus zeyherana* is a rhizomatous geoxylic suffrutex.

Several of the above belong to genera, e.g. *Berkheya, Geigeria, Gnidia, Haplocarpha, Helichrysum, Osteospermum, Oxalis,* and *Walafrida*, which have their greatest concentration of species in South Africa. The majority, however, belong to predominantly tropical or subcosmopolitan genera.

Riparian scrub

Principal species are *Acacia karoo* (7 m), *Celtis africana, Diospyros lycioides, Rhus lancea* and *Ziziphus mucronata*.

Rupicolous bushland and shrubland

In the drier lower-lying western parts the following bushes are characteristic: *Acacia karroo, Buddleja saligna, Celtis africana, Cussonia paniculata, C. spicata, Diospyros austro-africana, D. lycioides, Ehretia rigida, Euclea crispa, Grewia occidentalis, Heteromorpha arborescens, Olea africana, Osyris* sp., *Rhus ciliata, R. erosa, R.*

lancea, R. undulata, Tarchonanthus camphoratus, and *Ziziphus mucronata*. Towards the east, Afromontane species become increasingly prominent.

In the Kimberley area there are karroid communities characterized by *Chrysocoma tenuifolia, Cotyledon decussata, Eberlanzia spinosa, Eriocephalus spinescens, Euphorbia mauritanica, Pentzia sphaerocephala, Rhigozum obovatum* and *Ruschia unidens*. From about Maseru southwards *Aloe ferox* is conspicuous on northern slopes.

Even the most luxuriant communities are rarely more than 5 m tall.

Scrub forest

Patches of 10 m tall scrub forest composed predominantly of Afromontane species occur between 1500 and 1900 m in the eastern parts of the Orange Free State and adjacent parts of Lesotho. They are confined to shallow water-retaining soils overlying consolidated screes in deep ravines. Because of heavy exploitation for firewood and degradation by cattle only small fragments remain. The principal species are: *Buddleja salviifolia, Cassinopsis ilicifolia, Celtis africana, Diospyros whyteana, Euclea coriacea, E. crispa, Grewia occidentalis, Halleria lucida, Ilex mitis, Kiggelaria africana, Leucosidea sericea, Maytenus acuminata, M. heterophylla, M. undata, Myrsine africana, Olea africana, Olinia emarginata, Osyris* sp., *Pittosporum viridiflorum, Podocarpus latifolius, Rhamnus prinoides, Rhus pyroides* and *Scolopia mundii*. Of these, *Celtis* is deciduous. *Leucosidea, Kiggelaria* and *Maytenus acuminata* are semi-deciduous. The others are evergreen.

The Highveld/Karoo transition
(mapping unit 57b)

Refs.: Acocks (1975, p. 78–81); Werger (1973*a*, 1973*b*).
Phots.: Acocks (1975: 72, 73, 75, 76); Werger (1973*b*: 2–4).
Syn.: false Upper Karoo; false karroid broken veld, false Central Lower Karoo; pan turf veld invaded by Karoo; karroid *Merxmuellera* mountain veld replaced by Karoo (all of Acocks).

Apart from riparian scrub forest, and various types of shrubland and bushland on rocky slopes, the whole region is believed to have formerly been grassland similar to the surviving Highveld grassland of today.

Early travellers (summarized by Werger, 1973*a*) describe the general grassiness of the region and the absence of shrubs and bushes, besides commenting on the multitude of wild animals. It appears that at the end of the eighteenth century, the country near the present Colesberg, which was then outside the borders of the Cape Colony, had an abundance of grass, but that thirty-five years later the grass had largely been replaced by dwarf shrubs. In the area north of the Orange River, not at that time included in the colonized areas, grassland was still the dominant vegetation type.

Because of overgrazing the grassland has been converted to a secondary Karoo dwarf shrubland similar to that of the Central Upper Karoo but more grassy and floristically poorer. The principal surviving grasses are *Aristida congesta, Cynodon hirsutus, Eragrostis curvula, E. lehmanniana, E. obtusa, Themeda triandra* and *Tragus koelerioides*. The most abundant Karoo shrublets include *Chrysocoma tenuifolia, Aptosimum procumbens* (*depressum*), *Gnidia polycephala, Hermannia conocarpa, Pentzia globosa* and *Walafrida saxatilis*. This type occurs on the rather deep soils of the pediplains.

By contrast, the lithosols of mesas, kopjes, and ridges support various types of shrubland from which the shrublets of the false Karoo have been recruited, and, much more locally, bushland. The principal species of these shrubland and bushland communities are:

Large shrubs and bushes: *Buddleja saligna, Celtis africana, Cussonia paniculata, Diospyros austro-africana, D. lycioides, Ehretia rigida, Euclea crispa,* subsp. *ovata, Maytenus polyacantha, Olea africana, Osyris* sp., *Rhigozum obovatum, Rhus ciliata, R. erosa, R. undulata, Tarchonanthus camphoratus* and *Ziziphus mucronata.*

Dwarf shrubs: *Aptosimum procumbens, Eriocephalus spinescens, Hermannia candidissima, Hibiscus marlothianus, Pegolettia retrofracta* and *Pentzia sphaerocephala* as well as the ubiquitous *Chrysocoma tenuifolia* and *Pentzia globosa.*

Grasses: *Aristida diffusa, Enneapogon desvauxii, Fingerhuthia africana, Heteropogon contortus, Hyparrhenia hirta* and *Themeda triandra.*

Succulents: *Euphorbia clavarioides, Haworthia tesselata* and *Pachypodium succulentum.*

The deep sandy levees fringing the Orange River support 6–10 m tall riparian forest composed principally of *Acacia karroo, Celtis africana* and *Diospyros lycioides.*

The transition from Afromontane scrub forest to Highveld grassland
(mapping unit 20)

Refs.: Jacot Guillarmod (1971); Killick (1978*c*, p. 540–2).
Phot.: Killick (1978*c*: 10).

Published information is meagre. Although the Natal slopes of the Drakensberg are occupied by Afromontane communities between the 1280 m contour and the summit area, the corresponding slopes in Lesotho below about 2900 m are almost bereft of the most typical Afromontane elements. Hence this area must be excluded from the Afromontane Region. The Lesotho slopes are almost entirely covered by *Themeda–Festuca* grassland. There are only isolated patches of scrub dominated by *Leucosidea sericea, Buddleja corrugata, Passerina montana* and species of *Erica*. In places *Leucosidea* is less than 2 m tall and is closely pressed to the ground. On northern slopes *Themeda* prevails up to 2750 m, above which *Festuca caprina* becomes dominant. The grasses, however, are often replaced by *Chrysocoma tenuifolia* or *Felicia filifolia* as a result of overgrazing. On southern slopes *Festuca caprina* descends as a dominant to 2135 m.

The Afromontane/Tongaland–Pondoland transition
(mapping unit 24)

Ref.: Acocks (1975, p. 100–3).
Phots.: Acocks (1975: 97–9).

This transition separates the Tongaland–Pondoland Region from the Afromontane vegetation of the Natal Drakensberg north of the Tugela basin, and continues on the upper slopes of the ridge which connects the Natal and Transvaal sectors of the Drakensberg. It lies chiefly between 800 and 1700 m. The vegetation today is chiefly grassland but originally was probably bushland with scrub forest in sheltered kloofs. The woody relics are chiefly Afromontane at higher altitudes and include *Apodytes dimidiata*, *Halleria lucida*, *Leucosidea sericea*, *Pittosporum viridiflorum*, *Podocarpus latifolius*, *Rapanea melanophloeos* and *Scolopia mundii*.

'Lowland' woody species include *Acacia caffra*, *A. davyi*, *A. nilotica* subsp. *kraussiana*, *A. sieberana*, *Aloe arborescens*, *Celtis africana*, *Commiphora harveyi*, *Dalbergia obovata*, *Ekebergia pterophylla*, *Ficus capensis*, *F. sonderi* and *Syzygium cordatum*.

The most abundant grasses are *Andropogon schirensis*, *Brachiaria serrata*, *Elionurus argenteus*, *Eragrostis racemosa*, *Heteropogon contortus*, *Hyparrhenia hirta*, *Monocymbium ceresiiforme*, *Rendlia altera*, *Themeda triandra*, *Trachypogon spicatus* and *Tristachya leucothrix*.

The Zambezia/Highveld transition
(mapping unit 34)

Refs.: Acocks (1975, p. 99); White (1978a, p. 477–9).
Phots.: Acocks (1975: 96).

This area coincides with parts of Acocks's veld type 61, Bankenveld. Its original vegetation was probably bushland dominated by *Acacia caffra*, but the prevalent vegetation today is secondary grassland. There are relatively few tree species and their number diminishes rapidly towards the south. Afromontane and Zambezian species are present, surviving chiefly in bushland and scrub forest in sheltered kloofs. Afromontane species include *Calodendrum capense*, *Diospyros whyteana*, *Halleria lucida*, *Ilex mitis*, *Kiggelaria africana*, *Leucosidea sericea*, *Nuxia congesta*, *Olinia* and *Pterocelastrus*. Zambezian species include *Acacia robusta*, *A. sieberana*, *Burkea africana*, *Combretum molle*, *Dombeya rotundifolia*, *Ficus ingens*, *F. soldanella*, *Lannea discolor*, *Mimusops zeyheri*, *Ochna pulchra* and *Strychnos pungens*.

xv The Tongaland–Pondoland regional mosaic

Geographical position and area

Geology and physiography

Climate

Flora

Mapping units

Vegetation
    Tongaland–Pondoland undifferentiated forest
    Tongaland–Pondoland scrub forest
    Tongaland–Pondoland swamp forest
    Tongaland–Pondoland evergreen and semi-evergreen bushland and thicket
    Tongaland–Pondoland edaphic grassland
    Tongaland–Pondoland secondary grassland

Geographical position and area

This Region extends from the Limpopo River mouth (25° S.) to Port Elizabeth (34° S.). In the north it is up to 240 km wide, but locally in the south where mountains come close to the sea its width is no more than 8 km. Elsewhere in the south it penetrates inland along river valleys far into the interior. For most of its length it lies below the Afromontane Region or the Afromontane/Tongaland–Pondoland transition zone (page 196). (Area: 148 000 km^2.)

Geology and physiography

The coastal plain in the north is composed of Cretaceous and Tertiary marine sediments. Elsewhere, the more undulating landscape which rises locally to 1 600 m is carved out of rocks of the Basement Complex, Table Mountain Sandstone and sedimentary strata of the Karoo system.

Climate

Owing to the ameliorating effect of the warm Mozambique Current the coastal regions have a moderately high and well-distributed rainfall and, except in the extreme south, are frost-free. Further inland, however, climate changes rapidly over short distances and there is often a great contrast between xerocline and mesocline vegetation. The desiccating 'berg' winds have a profound effect on valley vegetation. The rainfall is more evenly distributed throughout the year than in most parts of the Zanzibar–Inhambane Region. Mean annual temperature diminishes from 22° C in the north to 17° C in the south. (See Fig. 20.)

Flora

There are about 3 000 species. More than 200 of the larger woody species, approximately 40 per cent of the total, are endemic. The proportion of endemic herbs and smaller woody plants is unknown but is probably less.

Endemic family. The Achariaceae is centred on this region, but is not strictly endemic.

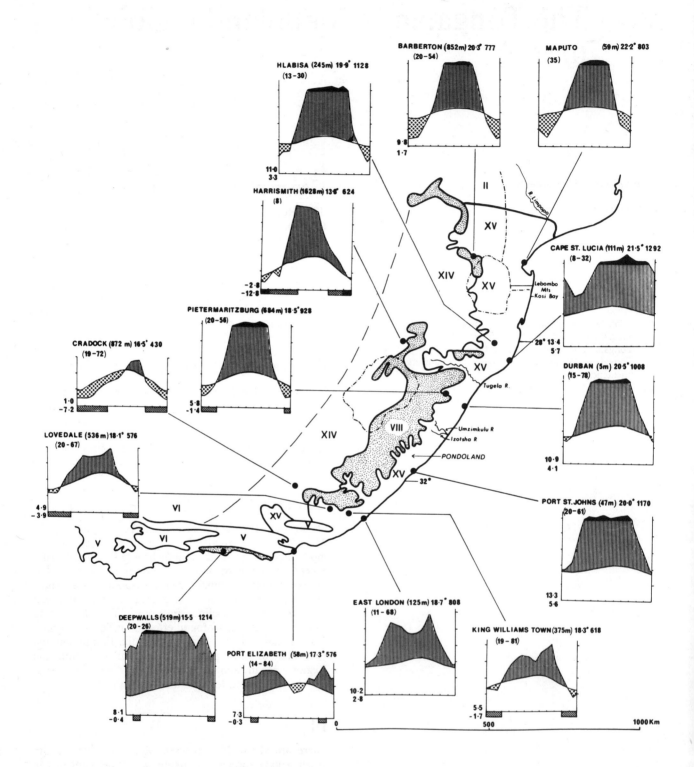

FIG. 20. Climate and topography of the Tongaland–Pondoland regional mosaic (XV)

Endemic genera. The twenty or so endemic woody genera include *Anastrabe, Bachmannia, Burchellia, Ephippiocarpa, Galpinia, Harpephyllum, Hippobromus, Jubaeopsis, Loxostylis, Pseudosalacia, Rhynchocalyx, Stangeria* and *Umtiza*. Two genera, *Atalaya* and *Protorhus*, occur nowhere else in Africa but are found in Asia and Madagascar.

Other characteristic genera. Twenty-six of the 35 South African species of *Encephalartos*, 12 of the 23 South African tree Aloes and nine of the 13 South African succulent tree Euphorbias occur in Tongaland–Pondoland.

Linking elements. Of the 500 or so larger woody plants occurring in Tongaland–Pondoland, 7.6 per cent are Zanzibar–Inhambane linking species, 20 per cent are Zambezian linking species, 8.7 per cent are Afromontane linking species, 5.1 per cent are Guineo–Congolian linking species, and 2.5 per cent are Karoo–Namib linking species. A further 1.5 per cent also occur in upland areas in tropical Africa but are not truly Afromontane. The importance of the Zambezian element decreases and that of the Karoo–Namib element increases towards the south. The representation of the Afromontane element is higher than in any other lowland phytochorion (Moll & White, 1978).

Mapping units

16c. The Tongaland–Pondoland coastal mosaic.
29e. The transition from undifferentiated Zambezian woodland to Tongaland–Pondoland bushland (see page 96 and below).
39. Tongaland–Pondoland evergreen and semi-evergreen bushland and thicket.
48. Tugela basin wooded bushland.

Vegetation

Where the vegetation has not been completely destroyed it consists of a complex mosaic of forest, scrub forest, and evergreen and semi-evergreen bushland and thicket in a matrix of secondary grassland and wooded grassland. There are small patches of woodland in the north and of edaphic grassland and swamp forest on the coastal plain. The vegetation has been described in some detail by Moll & White (1978).

Tongaland–Pondoland undifferentiated forest
(mapping unit 16c)

Refs.: Acocks (1952, 1975: mapping units 1 p.p., 2, 6 p.p.); Breen (1971); Edwards (1967, p. 82–6); Huntley (1965); Killick (1958, p. 60–72); Moll (1968b, 1968d); Moll & White (1978); Rogers & Moll (1975); Wild & Barbosa (1967: type 5, p.p.).

Phots.: Breen (1971: 1); Killick (1958: 18–23); Moll (1968b: 1–5); Rogers & Moll (1975, 2, 4–7); Moll & White (1978: 4, 5).
Profiles: Huntley (1965, fig. 4); Moll (1968b, figs. 3–4; 1968d, fig. 1).

Tongaland–Pondoland forest formerly extended as a narrow more or less continuous band along the coast. Further inland it was confined to mesocline slopes and, in regions of low relief, to soils with a high water-table throughout the year. Forest is virtually absent from the deep valleys influenced by desiccating berg winds. Variation in floristic composition is kaleidoscopic and classification is difficult.

The canopy varies in height from 10 to 30 m or slightly more. The most luxuriant stands approach rain forest in stature and structure, but since they occupy a small area and all their species occur more extensively in drier, less luxuriant types, they are not described separately.

In the tallest stands the trees are straight and well grown with long boles clear of branches for 20 m or more, but in the stunted types, although most individuals have single stems, they are often crooked and branch low down between 2 and 5 m. On steep slopes, in particular, most of the boles are often leaning.

The canopy is evergreen to semi-evergreen (E. J. Moll, pers. comm.). In the moister types as at Ngoye no more than 5 per cent of the canopy is leafless at any one time, whereas in drier forests, e.g. Gwalaweni, as much as 60 per cent may be briefly bare. Nearly all the smaller trees and shrubs other than secondary species such as *Trema orientalis* are completely evergreen.

About 120 species occur in the canopy, though normally not more than 30 would be present in any one stand. The more important species include:

— Afromontane species: *Calodendrum capense* (18 m), *Combretum kraussii* (18 m), *Zanthoxylum (Fagara) davyi, Kiggelaria africana, Nuxia congesta* (16 m), *Podocarpus falcatus* (20 m), *Podocarpus latifolius* (24 m), *Scolopia mundii* and *Xymalos monospora*.
— Upland species: *Celtis gomphophylla* (24 m), *Chrysophyllum viridifolium* (33 m), *Drypetes gerrardii* (24 m), *Heywoodia lucens* (27 m) and *Homalium dentatum* (24 m).
— Endemic species: *Atalaya natalensis* (15 m), *Anastrabe integerrima* (18 m), *Beilschmiedia natalensis* (15 m), *Brachylaena uniflora* (16 m), *Cola natalensis* (18 m), *Commiphora harveyi* (15 m), *Cordia caffra* (16 m), *Diospyros inhacaensis* (16 m), *Erythrina caffra* (18 m), *Harpephyllum caffrum* (30 m), *Manilkara concolor* (20 m), *Millettia grandis, M. sutherlandii* (30 m), *Oricia bachmanii* and *Protorhus longifolia* (21 m).
— Guineo–Congolian linking species: *Albizia adianthifolia* (18 m), *Blighia unijugata* (16 m), *Celtis mildbraedii* (21 m), *Chaetacme aristata* (15 m), *Croton sylvaticus* (21 m), *Ficus capensis, F. natalensis, Morus mesozygia* (21 m), *Phyllanthus discoideus* and *Sapium ellipticum* (24 m).

— Zanzibar–Inhambane linking species: *Bequaertiodendron natalense*, *Inhambanella henriquesii* (16 m), *Olea woodiana* (15 m), *Strychnos decussata* (16 m), *Sideroxylon inerme* (15 m) and *Vepris undulata* (24 m).

— Zambezian linking species: *Berchemia discolor* (16 m), *Clerodendrum glabrum* (21 m) and *Ziziphus mucronata* (16 m).

— Transgressors: *Celtis africana* (24 m), *Croton gratissimus* (18 m), *Ekebergia capensis* (24 m), *Euclea natalensis* (16 m), *Olea capensis* subsp. *macrocarpa* (21 m), *Ptaeroxylon obliquum* (15 m), *Strychnos madagascariensis* (15 m), *Syzygium guineense* subsp. *gerrardii*, *S. cordatum*, *Trichilia dregeana* and *T. emetica*.

All the species mentioned above reach, at least sometimes, a height of 15 m. Some of them, however, are more characteristic of various types of bushland and scrub forest, e.g. *Anastrabe*, *Commiphora harveyi*, *Cordia caffra*, *Croton gratissimus*, *Euclea natalensis*, *Mimusops obovata*, *Sideroxylon inerme* and *Strychnos madagascariensis*.

Species which enter the canopy only in shorter types of Tongaland–Pondoland forest which is 10–15 m high include *Diospyros dichrophylla*, *Dombeya cymosa*, *Euclea racemosa*, *Euphorbia grandidens*, *E. tetragona*, *E. triangularis*, *Hippobromus pauciflorus*, *Schotia latifolia* and *Umtiza listerana*. All these species are more characteristic of evergreen bushland. *Schotia latifolia* is a Karoo–Namib linking species.

Small trees and shrubs are represented by many species. A rosette tree, *Dracaena hookerana*, is locally abundant. Cycads, including *Stangeria eriopus* and *Encephalartos altensteinii*, *E. ferox* and *E. villosus*, are widespread.

Large lianes, although scarce in some of the taller denser forests, are a very conspicuous feature of shorter more open types. Some, e.g. *Dalbergia armata*, *D. obovata*, *Entada spicata*, *Oncinotis inhandensis*, *Pisonia aculeata* and *Rhoicissus tomentosa*, are of immense size with stems 15 cm or more in diameter. *Dalbergia armata*, *Entada* and *Pisonia* are armed with vicious spines. Climbers in general are probably more numerous in this type of forest, area for area, than in any other in Africa.

The abundance of vascular epiphytes varies greatly from place to place even over quite short distances. In general they are much scarcer than in the Afromontane forest of the 'mist belt' and are often almost absent. Bryophytes are also inconspicuous and epiphytic lichens are only locally significant.

Of the many variants of Tongaland–Pondoland forest one of the most extreme is the 'Sand forest' of Tongaland, characterized by *Afzelia quanzensis*, *Albizia forbesii*, *Balanites maughamii*, *Cleistanthus schlechteri*, *Dialium schlechteri*, *Erythrophleum lasianthum*, *Hymenocardia ulmoides*, *Newtonia hildebrandtii* and *Pteleopsis myrtifolia*. Afromontane species are almost absent. Nevertheless it is a remarkable sight to see, in this type, *Podocarpus falcatus* growing side by side with such characteristic bushland species as *Schotia brachypetala* and *Spirostachys africana*.

Tongaland–Pondoland scrub forest
(mapping unit 16c)

The transition from forest to bushland and thicket is often abrupt. Scrub forest consequently is of relatively restricted occurrence, but of very varied composition.

In the shelter of the first ridge of high dunes the wind-trimmed coastal thicket is replaced by scrub forest in which the tallest trees, notably *Mimusops caffra*, *Sideroxylon inerme*, *Euclea racemosa*, *Trichilia dregeana*, *Cordia caffra* and *Ekebergia capensis*, form an open canopy over a dense bushy understorey.

In places on the rocky coast of the Transkei, *Euphorbia triangularis* forms a narrow fringe of 10 m tall scrub forest at the mouths of rivers. Its associates include *Cassipourea gerrardii*, *Zanthoxylum* (*Fagara*) *capense*, *Euclea natalensis*, *Turraea obtusifolia*, *Psychotria capensis*, *Millettia grandis*, *Turraea floribunda*, *Sideroxylon inerme*, *Dracaena hookerana*, *Phoenix reclinata*, *Rapanea melanophloeos* and wind-trimmed *Diospyros natalensis*.

Scrub forest in the interior valleys is sometimes dominated by 12–15 m tall *Aloe bainesii* associated with *Casearia gladiiformis*, *Commiphora harveyi*, *Craibia zimmermannii*, *Diospyros natalensis*, *Euclea natalensis*, *Euphorbia grandidens*, *Ficus natalensis*, *Galpinia transvaalica*, *Harpephyllum caffrum*, *Manilkara discolor*, *Strychnos henningsii*, *Suregada africana*, *Teclea gerrardii*, *Trichilia emetica*, *Turraea floribunda*, *Vitellariopsis marginata* and *Ziziphus mucronata*.

Tongaland–Pondoland swamp forest
(mapping unit 16c)

This type occurs on the coastal plain as far south as 31° S., and is best developed near Kosi Bay. The canopy is up to 30 m tall. *Ficus trichopoda*, *Syzygium cordatum*, *Raphia australis*, *Voacanga thouarsii* and *Rauvolfia caffra* are characteristic species. The climbing fern *Stenochlaena tenuifolia* is conspicuous (Moll & White, 1978).

Tongaland–Pondoland evergreen and semi-evergreen bushland and thicket
(mapping units 16c, 29e, 39, and 48)

Refs.: Acocks (1975, p. 28–9 p.p., 52–8); Comins (1962, p. 12–13); Dyer (1937, p. 87–90); Edwards (1967, p. 96–100); Moll & White (1978); Story (1952, p. 53–60); White (MS, 1973).
Phots.: Acocks (1975: 15, 45–50); Comins (1962: 6); Dyer (1937: 4, 18, 27–34); Story (1952: 9, 17, 18).
Syn.: lowveld (p.p.), valley bushveld (Acocks, 1975); Fort Cox scrub, Nqhuema scrub (Story, 1952); karroid scrub (Dyer, 1937).

Where the rainfall is too low to support forest, the most widespread climax vegetation is evergreen and semi-evergreen (locally semi-deciduous) bushland and thicket. In the north this type is most extensively developed in the low-lying country between the forests of the coastal plain and the mountainous country inland. Further south it occupies deep valleys.

There are pronounced floristic and physiognomic gradients from north to south. In general, stature and deciduousness decrease towards the south while succulence, sclerophylly, and spinescence increase. In the north there are many Zambezian species but few extend much further than the Tugela basin. Some of them are trees which locally form woodland and scrub woodland (see below).

Because of its position in the heart of one of the most diversified floras in Africa, Tongaland–Pondoland bushland shows some floristic overlap with other major vegetation types, principally Afromontane forest, coastal forest, broad-leaved Zambezian woodland and Karoo shrubland.

The most widespread bushes, which grow 3–6 m tall, include *Azima tetracantha, Bauhinia natalensis, Brachylaena ilicifolia, Carissa bispinosa, Cassine aethiopica, Cussonia* spp., *Diospyros dichrophylla, D. lycioides, D. scabrida, D. simii, Ehretia rigida, Euclea* spp., *Zanthoxylum capense, Grewia occidentalis, G. robusta, Maytenus* spp., *Olea africana, Pappea capensis, Phyllanthus verrucosus, Plumbago auriculata (capensis), Rhus* spp., *Schotia* spp., *Sideroxylon inerme, Tarchonanthus camphoratus* and *Xeromphis rudis*.

Arborescent succulent species of *Aloe* and *Euphorbia* occur throughout, though not necessarily in every stand. In general they are more abundant in the south, though even in the north they locally dominate the landscape, but probably only on disturbed sites.

The larger Aloes (*A. candelabrum, ferox, marlothii, spectabilis*) have unbranched stems up to 6 m tall and 0.3 m in diameter, the lower parts of which are densely clothed with marcescent leaves. The Euphorbias (*E. evansii, grandidens, ingens, tetragona, tirucalli, triangularis*) have a single bole 1–2 m long and a spreading crown of the familiar candelabra type. Smaller succulents, e.g. species of *Crassula* and *Kalanchoe*, and *Portulacaria afra*, are also locally plentiful especially towards the south. Species of the fibrous-leaved *Sansevieria* occur throughout.

In the northern variant of Tongaland–Pondoland bushland, especially north of the Tugela basin, Zambezian tree species in addition to the Acacias mentioned above are present, but they are usually less plentiful than the bushes. They are mostly less than 9 m tall, often much less, and rarely have straight trunks more than 2 m long. Since they are usually widely spaced, the communities in which they occur can be referred to as 'wooded bushland'. The principal species are *Acacia sieberana, Afzelia quanzensis, Albizia versicolor, Berchemia discolor, Combretum apiculatum, C. collinum, C. imberbe, C. molle, C. zeyheri, Dombeya rotundifolia, Entandrophragma caudatum, Ficus sycomorus, Lannea stuhlmannii, Lonchocarpus capassa, Peltophorum africanum, Pterocarpus rotundifolius, Sclerocarya caffra, Spirostachys africana, Terminalia sericea, Trichilia emetica* and *Ziziphus mucronata*.

Very rarely, the tallest of these species, especially *Acacia nigrescens*, form small patches of woodland or wooded grassland (Werger & Coetzee, 1978, fig. 51), which may however have been derived from wooded bushland following the elimination of the smaller woody plants by fire and cultivation.

Several species of *Acacia*, including *A. borleae, burkei, caffra, davyi, gerrardii, nilotica, robusta* and *senegal*, are present in the north, but only one, *A. caffra*, other than the ubiquitous, invasive, *A. karroo*, extends to the extreme south. Acacias are deciduous, and when abundant, contribute greatly to the physiognomy. Their abundance, however, is usually due to human activity. In some places, e.g. parts of the Tugela basin, they dominate the degraded landscape. In less-disturbed places they are usually less conspicuous than the evergreen bushes. Evergreen species, including succulent Euphorbias, establish themselves in the shade of the Acacias and eventually overtop them and shade them out.

Climbers, including *Asparagus* spp., *Capparis sepiaria, Cissus quadrangularis, Dalbergia armata, Entada spicata, Scutia myrtina, Rhoicissus digitata, R. tridentata* and *Sarcostemma viminale*, are often abundant.

Corticolous, foliaceous lichens are sometimes conspicuous. A few epiphytic orchids, including *Ansellia gigantea*, occur. The field layer is mostly sparse and may include a few ferns, shade-tolerant grasses, and, especially, species of Acanthaceae.

Semi-evergreen bushland also occurs on rock outcrops in areas where the rainfall is sufficient to support forest, for instance on outcropping Table Mountain Sandstone at the lips of the gorges of the Izotsha and Umzinkulu Rivers, and at Mills Kloof and the Ngongo stream in Natal (rainfall c. 1 150 mm per year). Although there is some floristic overlap with the types just described the majority of species are different. Cactiform Euphorbias are virtually absent and arborescent Aloes are rare. Several of the species, e.g. *Apodytes dimidiata, Ekebergia capensis, Harpephyllum caffrum* and *Protorhus longifolia*, are more characteristic of forest. Some forest species, however, including *Apodytes, Halleria lucida* and *Diospyros whyteana*, can also occur at least locally in much drier types of bushland.

Only a few of the characteristic species of Tongaland–Pondoland semi-evergreen bushland, such as *Aloe speciosa, Cadaba aphylla, Carissa haematocarpa, Crassula portulacea, Euclea undulata, Euphorbia grandidens, Lycium austrinum, Montinia caryophyllacea, Maytenus linearis, Portulacaria afra, Schotia afra* and *S. latifolia*, also occur in the Karoo–Namib Region, where their most characteristic habitat is 'Spekboomveld' (page 141).

Tongaland–Pondoland edaphic grassland
(mapping unit 16c)

Badly drained grassland with scattered palms, chiefly *Hyphaene natalensis* and *Borassus aethiopum*, together with *Garcinia livingstonei* and *Syzygium cordatum*, occurs at several places along the Mozambique coast. This type also extends into Tongaland where, as in the extreme south of Mozambique, it is characterized by the abundance of a few species of geoxylic suffrutex, including *Parinari capensis* and *Diospyros galpinii*, and suffruticose variants of *Eugenia capensis*, *Diospyros lycioides*, *Syzygium cordatum* and *Salacia kraussii*.

Tongaland–Pondoland secondary grassland
(mapping unit 16c)

When Tongaland–Pondoland coastal forest is destroyed it is replaced by *Acacia karroo* wooded grassland in which the following grasses occur: *Alloteropsis semialata*, *Cymbopogon excavatus*, *C. validus*, *Digitaria* spp., *Diheteropogon amplectens*, *Eulalia villosa*, *Heteropogon contortus*, *Hyparrhenia filipendula*, *Loudetia simplex*, *Paspalum scrobiculatum* (*orbiculare*), *Themeda triandra* and *Tristachya leucothrix*. Heavy grazing encourages *Aristida junciformis* ('Ngongoni'), which is now dominant over large areas.

XVI The Sahel regional transition zone

Introduction
Geographical position and area
Geology and physiography
Climate
Flora
Mapping units
Vegetation
    Sahel wooded grassland
    Sahel semi-desert grassland and the transition to the Sahara
    Sahel deciduous bushland
    Sahelomontane scrub forest
    Sahelomontane secondary grassland
Vegetation pattern in relation to the environment
    Gross pattern in the Jebel Marra area
        Introduction
        Vegetation of the Basement Complex
            1. *Acacia mellifera, Commiphora africana* communities on indurated soils
            2. *Acacia mellifera* communities on hill soils of the Basement Complex
            3. *Anogeissus leiocarpus* woodland on Basement Complex soils
            4. *Anogeissus leiocarpus, Boswellia papyrifera* communities on hill soils of the Basement Complex
            5. *Acacia seyal, Balanites aegyptiaca* communities on clay soils
            6. *Acacia albida* and *Balanites aegyptiaca* communities on alluvial soils
            7. *Acacia senegal, Combretum glutinosum* communities on aeolian sands
            8. *Combretum glutinosum, Guiera senegalensis* communities on Nubian sandstone soils
        Vegetation of the volcanic massif
            9. *Acacia albida* communities on ash piedmont soils
            10. *Combretum glutinosum, Terminalia laxiflora* communities on ash piedmont soils
            11. *Acacia mellifera* bushland on volcanic soils
            12. *Anogeissus* communities on volcanic soils
            13. Riparian forest
            14. Sahelomontane communities
    Gross pattern in Kordofan
        1. Semi-desert grassland on aeolian sand
        2. *Acacia senegal* wooded grassland on aeolian sand
        3. *Acacia mellifera* bushland
        4. Broad-leaved woodland
    Detailed patterns in Darfur, Kordofan and the Nile valley
Vegetation change and the Great Drought of 1968–73

Introduction

Because of its geographical position on the southern fringes of the largest desert in the world, rainfall in the Sahel zone is insufficient for permanent agriculture based on rain-fed crops. Nevertheless, where the water supply permits either permanent or seasonal settlement, rain-fed crops are grown even where rainfall is as low as 200 mm per year. Success, however, is intermittent, and even in the wetter phase of the climatic cycle the crops fail on average at least once every three years. In the drier phase they fail completely. Permanent agriculture is only possible in the few places where permanent rivers, which originate in wetter regions, make water available for irrigation.

In most parts of the Sahel, livestock-raising in the form of pastoralism is the main source of livelihood and is the basis of the economy. Nearly everywhere, pastoralism involves common ownership of grazing lands, and the people are nomadic or semi-nomadic.

Rainfall in the Sahel zone shows pronounced fluctuations of a cyclical nature. During the dry phase of the cycle prolonged and severe drought can result in the death of large numbers of people and livestock, and in severe degradation of the vegetation. The most recent prolonged drought, from 1968 to 1973, received much publicity, particularly towards its end, when the cumulative effects were most pronounced. Consequently the governments concerned, and international organizations, in particular the United Nations Sudano–Sahelian Office (UNSO), are now attempting to formulate policies which should mitigate the effects of future dry phases in the climatic cycle.

This chapter includes information on the detailed pattern of vegetation in the Sahel zone in relation to environmental factors, and also on the effects of the Great Drought of 1968–73 on the natural and semi-natural vegetation. It is not possible to produce a detailed account of the influence of geology, physiography, and soils on vegetation for the entire Sahel zone, since the information either does not exist or is widely scattered and not readily accessible. It is fortunate, however, that these topics have been studied in some detail for certain representative parts of the Sahel. Sites of intensive studies in the region include:
— Richard Toll in the Fété Olé of Senegal (see Bourlière, 1978, for synthesis and references);

— the Eghazer and Azawak region of Niger (see case study presented by the Government of Niger to the UN Conference on Desertification, published in Mabbutt and Floret, 1980);
— two sites in the Sudan Republic, namely in Kordofan Province (Hunting Technical Services, 1964), and in the Jebel Marra region in Darfur Province (Hunting Technical Services, 1958, 1968, 1977; Wickens, 1977a; Wickens, pers. comm.).

The studies in the Sudan are briefly summarized in the last two sections of this chapter. The Jebel Marra Region was intensively studied shortly before the Great Drought began (1963–67) and shortly after it ended (in 1977). The study area is partly situated in the transition zone between the Sahel and the Sudan. The north-east is Sahelian and the south-west is Sudanian, whereas, with the exception of the massif itself and its immediate proximity, much of the remainder is transitional. Where there is a permanent water supply, as in the mountains and along major water courses, permanent agriculture is practised by the Fur tribesmen. Elsewhere, throughout the area and extending a short distance to the south, pastoralism is practised by nomadic and semi-nomadic Arabs. Hence, the study area is ideally situated to exemplify the effects of the Great Drought, not only on Sahelian vegetation, but also in those parts of the Sudan zone which are used by the Sahelian pastoralists during the dry season.

Geographical position and area

The Sahel zone occupies a relatively narrow band, mostly c. 400 km wide, which extends across north Africa from the Atlantic coast to the Red Sea. The massifs of Adrar des Iforas, Aïr, and Ennedi are responsible for a local increase in precipitation which permits a northerly extension of several Sahel species. Their floras, however, also include Saharan species and, with equal justification, they could have been included in the Saharan Region. (Area: 2 482 000 km².)

Geology and physiography

Most of the Sahel Region forms a flat or gently undulating landscape below 600 m. Large areas are covered with Pleistocene clays, or sand sheets which were distributed by wind action during the drier phases of the Pleistocene but were mostly derived from earlier continental deposits. In a few places older rocks form small islands of higher ground. Two southern extensions of the Saharan massif of Ahaggar, namely Adrar des Iforas (727 m) and Aïr (1 900 m), are composed of Precambrian crystalline rocks. By contrast, the Ennedi Plateau (1 450 m) is capped by horizontal Devonian sandstone. In the nearby Sudan Republic two late Tertiary volcanic mountains, Jebel Gurgeil and the main Jebel Marra massif (page 208) rise to 2 400 m and 3 057 m respectively. The Red Sea Hills on the Eritrean border and the Erkowit Plateau (1 273 m) a little further north are formed from the Precambrian basement complex. The former supports Afromontane forest.

Climate

Rainfall is unreliable and mostly between 150 and 500 mm per year but rises to more than 1 000 mm on Jebel Marra (page 208). Most rain falls in 3–4 summer months and the dry season is long and severe. Except near the coast mean annual temperature is between 26° and 30°C. Light frosts occur occasionally in some places. (See Fig. 21.)

Flora

There are about 1 200 species, of which probably fewer than 40 species (3 per cent) are strictly endemic. An additional 150 species (approximately) are more or less confined to the Sahel and other parts of Africa and Asia with a similar or drier climate (see below). The above figures exclude species which are confined to the high Sahelian mountains.

Endemic families and genera. None.

Endemic species. Endemic species include *Ammannia gracilis, Chrozophora brocchiana, Farsetia stenoptera, Indigofera senegalensis, Launaea (Sonchus) chevalieri, Nymphoides ezannoi, Panicum laetum, Rotala pterocalyx, Tephrosia gracilipes, T. obcordata* and *T. quartiniana*.

Linking elements. Six per cent of the Sahel flora is more or less confined to the Sahel Region except for extensions into Asia or the Somalia–Masai Region. This element includes *Barleria hochstetteri, Cadaba glandulosa, Crotalaria microphylla, Gossypium somalense, Heliotropium rariflorum, Indigofera cordifolia, Solanum albicaule, Tephrosia nubica* and *Vahlia geminiflora*.

Fourteen per cent of the Sahel flora belongs to the Sahara–Sahel linking element which includes *Aristida sieberana, Blepharis ciliaris, Calligonum comosum, Chascanum marrubifolium, Cleome scaposa, Cornulaca monacantha, Forsskålea tenacissima, Glossonema boveanum, Leptadenia pyrotechnica, Maerua crassifolia, Panicum turgidum, Olea laperrinei* (Sahelo- and Saharo-montane), *Stipagrostis pungens* and *Ziziphus lotus*. Many of these extend far into Asia.

Five per cent of species, the 'arid disjuncts', in Africa north of the equator are more or less confined to the Sahel or the Sahel plus the Sahara, but also occur in the drier parts of South Africa. They include *Geigeria alata, Indigofera disjuncta, Lotus arabicus, Stipagrostis ciliata, S. hirtigluma, S. uniplumis, Tragus racemosus* and *Zygophyllum simplex*.

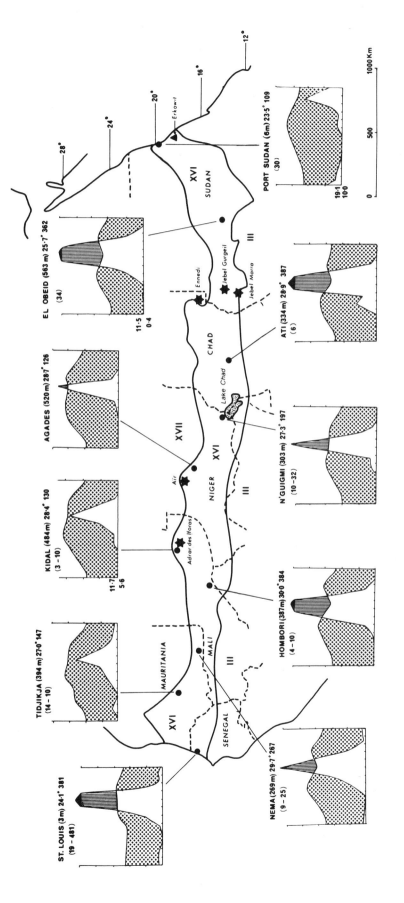

FIG. 21. Climate and topography of the Sahel transition zone (XVI)

Twenty-six per cent also occur in the Sudanian Region and are members of its characteristic vegetation types. A few, e.g. *Bauhinia rufescens* and *Piliostigma reticulatum*, are confined to the Sudanian and Sahelian Regions, but the majority are much more widespread in Africa, e.g. *Diospyros mespiliformis*, or in the tropics generally, e.g. *Abrus precatorius*.

The largest element (46 per cent) comprises pluriregional species which are not members of characteristic Sudanian vegetation although many occur in the Sudanian Region in azonal or ruderal habitats. The majority are either aquatics or semi-aquatics, e.g. *Neptunia oleracea*, or are weeds or plants of disturbed places, e.g. *Corchorus tridens*, *Urena lobata*, and *Waltheria indica*. About 30 species included in this element, e.g. *Acacia tortilis*, *Cadaba farinosa*, *Cordia sinensis*, *Schmidtia pappophoroides* and *Tephrosia uniflora*, scarcely extend south of the Sahel in West Africa, but have a much more diffuse distribution in East and Southern Africa.

Mapping units

19b. Undifferentiated Sahelomontane vegetation.
38 (p.p.). East African evergreen and semi-evergreen bushland and thicket.
43. Sahel *Acacia* wooded grassland and *Acacia* deciduous bushland.
54a. Northern Sahel semi-desert grassland and shrubland.
62 (p.p.). Mosaic of edaphic grassland and *Acacia* wooded grassland (see Chapter III and below).
64 (p.p.). Mosaic of edaphic grassland and semi-aquatic vegetation (see Chapters III and XXII).
75 (p.p.). Swamp and aquatic vegetation (see Chapter XXII).

Vegetation

The extensive sand sheets of the Sahel support wooded grassland in the south and semi-desert grassland in the north. Bushland is much more restricted and is mostly confined to rocky outcrops. The upper slopes of two high mountains, Jebel Gurgeil and Jebel Marra, were formerly covered with scrub forest, which has been largely replaced by secondary grassland, especially on the latter. Various types of scrub forest, bushland, and wooded grassland occur on the massifs on Ennedi and Aïr, but published information is sparse. Edaphic grassland and wooded grassland occurring on Pleistocene clays is described along with similar Sudanian communities in Chapter III. Evergreen and semi-evergreen bushland related to that described in Chapter IV occurs on the Erkowit Hills near Suakin, Sudan Republic (Kassas, 1956b).

Sahel wooded grassland
(mapping unit 43)

Refs.: Audry & Rossetti (1962); Harrison & Jackson (1958); Keay (1959a); Quézel (1969); Rosevear (1953); Rossetti (1962).
Phots.: Audry & Rossetti (1962: 1, 4, 23); Quézel (1969: 5, 6); Rosevear (1953: 29–35).

This is the most widespread type on sandy soils in the southern Sahel, where annual rainfall is between 250 and 500 mm.

In West Africa, of the chief woody species, namely *Acacia tortilis*, *A. laeta*, *Commiphora africana*, *Balanites aegyptiaca*, *Boscia senegalensis*, *Maerua crassifolia* and *Leptadenia pyrotechnica*, the first is by far the most abundant. In the south it occurs as a small bushy tree up to 8 m tall and with a bole up to 1.3 m long. Further north it is both shorter and more bushy, and neither this species nor its associates often exceed a height of 4 m. The density of the larger woody plants varies greatly, especially in relation to water supply and the amount of human interference. Locally the crowns are almost touching but more commonly they are several crown-diameters apart. The grass sward is more or less continuous and is no more than 60 cm tall. It mostly consists of annual species, principally *Cenchrus biflorus*, *Schoenefeldia gracilis*, *Aristida stipoides* and *Tragus racemosus*. Perennial grasses are localized, but *Andropogon gayanus* sometimes occurs in almost pure stands on deeper sands towards the south. Formerly it was more widespread, but it has been eliminated from large areas by cultivation. It is also characteristic of sandy stripes (*brousse tigrée*). In heavily grazed and trampled areas the grasses are replaced by annual weeds such as *Boerhavia coccinea* and *Tribulus terrestris*.

In the Sudan Republic sandy soils, although extensive (65 000 km^2), are relatively restricted because of the large areas occupied by black, cracking clays and rocky hills. Where rainfall is between 280 and 450 mm the most characteristic tree on sandy soils is *Acacia senegal*, which often occurs in almost pure stands. It frequently follows cultivation and in places appears to be a secondary species.

Sahel semi-desert grassland and the transition to the Sahara
(mapping unit 54a)

Refs.: Audry & Rossetti (1962); Harrison & Jackson (1958); Monod (1958); Rossetti (1962).
Phots.: Audry & Rossetti (1962: 5, 6, 21–6); Rossetti (1962: 5–7, 19).

In the northern Sahel rainfall is less than 250 mm per year and grassland is the prevalent vegetation on deep sandy soils.

It usually contains an admixture of bushes and small bushy trees, the density of which is partly determined by local conditions. The whole region has, however, been subjected to intense human activity, and the extent to which treeless areas are natural is largely conjectural.

The crown cover of the woody species is usually less than 10 per cent. Woody plants are only sufficiently numerous to constitute bushland on rocky outcrops and specially favoured, water-receiving sites.

The chief woody species are *Acacia tortilis*, *Commiphora africana*, *Balanites aegyptiaca*, *Boscia senegalensis*, *Leptadenia pyrotechnica*, *Acacia laeta* and *A. ehrenbergiana* (*flava*). They all, except the last, also occur throughout the southern Sahel where they attain larger dimensions. In the drier northern Sahel they are never more than 5 m tall and often no more than 2 m.

The most extensively occurring dominant grasses in the northern Sahel are all annual species, notably *Cenchrus biflorus*, *Schoenefeldia gracilis*, *Aristida stipoides* and *Tragus racemosus*. They are equally characteristic of the southern Sahel. In the northern Sahel, however, certain desert grasses, principally *Panicum turgidum* and *Stipagrostis pungens*, which are completely absent from the southern Sahel, are locally dominant and increase in abundance towards the north.

Within the northern Sahel the transition to desert grassland is not a gradual one closely correlated with rainfall, but is greatly modified by local edaphic factors, particularly the relief of the sandy covering. On sandy plains and low dunes this transition takes place at about 100 mm but where the dune relief is more accentuated desert grassland extends sporadically much further south. Thus, patches of desert grassland dominated by the perennial species, *Panicum turgidum* and *Aristida sieberana*, extend as far south as the 250 mm isohyet on the loose sands which form the crests of the taller dunes. They alternate with patches of typical Sahel grassland dominated by the annual species *Cenchrus biflorus* and its habitual associates, which cover the stable sand of the lower dune slopes and the intervening hollows. The shrub, *Calligonum comosum*, a characteristic Saharan species, becomes associated with *Panicum turgidum* where the rainfall is about 200 mm.

Where the rainfall is about 100 mm per year, *Panicum turgidum* becomes generally dominant except on the unstable crests of dunes where it is replaced by *Stipagrostis pungens*. Several woody species, e.g. *Acacia senegal*, *Balanites aegyptiaca* and *Commiphora africana*, drop out more or less where *Panicum turgidum* becomes dominant, but *Acacia tortilis* still occurs with *Maerua crassifolia*, *Leptadenia pyrotechnica*, *Calligonum comosum* and *Euphorbia balsamifera*. In this part of its range *A. tortilis* is normally more than 2 m high, but never more than 5 m. The cover of the woody plants rarely exceeds 3 per cent, except very locally, where *A. ehrenbergiana* forms dense stands on rocky outcrops, or *Salvadora persica* and *Tamarix* occur on brackish water-receiving soils.

Stipagrostis increases in abundance towards the north and is generally dominant at the 80 mm isohyet where the dunes become more mobile.

Sahel deciduous bushland
(mapping unit 43)

Ref.: Quézel (1969, p. 22–5, 33–4, 80–2).
Phots.: Quézel (1969: 3, 8, 12b, 13).

Rocky outcrops in the Sahel zone, such as those on the plateau of north-west Darfur and the lower slopes of the volcanic inselberg Jebel Gurgeil, at least on water-receiving sites, support bushland or thicket.

In Darfur the bushland on the plateau is usually 2–3 m tall and is often impenetrable. It is dominated by *Acacia mellifera* and *Commiphora africana*, with *Boscia senegalensis* and *Dichrostachys cinerea* usually present. On the piedmont of Jebel Gurgeil bushland and thicket, which is 3–5 m high, has persisted in the principal valleys and on basalt lava flows, but elsewhere has been replaced by wooded grassland. The most frequent thicket species are *Commiphora africana*, *Acacia mellifera* and *Euphorbia candelabrum*.

Sahelomontane scrub forest
(mapping unit 19b)

Refs.: Quézel (1969, p. 90–2); Wickens (1977a, p. 32–3).
Phots.: Quézel (1969: 15); Wickens (1977a: 7, 8).

On Jebel Gurgeil *Olea laperrinei* dominates scrub forest on rocky sheltered slopes between 1 700 and 2 200 m. Cover varies from 50 per cent to almost closed. Most individuals are 6–8 m tall and have massive stems 70–90 cm in diameter, though their form is that of large bushes rather than trees. The largest trees are 15 m tall. Regeneration is plentiful and the trees look healthy in contrast to their twisted and mutilated appearance on the Sahara mountains. The principal woody associates of the Olive are *Boscia angustifolia*, *Ficus populifolia*, *F. salicifolia*, *Maytenus senegalensis*, *Rhus vulgaris* and *Vangueria venosa*. In the lower part of the *Olea* zone *Euphorbia candelabrum* is not entirely absent and persists up to 1 900 m.

On Jebel Marra, *Olea laperrinei* has a scattered distribution between 2 300 and 3 000 m, but the communities it formerly occurred in have been destroyed.

Sahelomontane secondary grassland
(mapping unit 19b)

Refs.: Quézel (1969, p. 91–3); Wickens (1977a, p. 31–4).
Phots.: Quézel (1969: 16); Wickens (1977a: 6).
Profile: Wickens (1977a: 15).

The rolling plains and lava peaks of the upland plateau of Jebel Marra above 1 800–2 000 m are covered with montane grassland. The mountain has been occupied by man for at least 2 000 years and there can be little doubt that much of the grassland is secondary. Today the plateau is sparsely inhabited and cultivation is almost entirely restricted to the valleys, though in the past almost all the area has been cultivated, and abandoned cultivation terraces can be found up to 2 750 m. Except for a few secluded valleys fires occur at least once a year.

The original vegetation was probably *Olea laperrinei* scrub forest, but this has been almost completely destroyed for firewood and building. The secondary grasslands on Jebel Marra are briefly described on page 211.

The tabular plateau, which forms the summit of Jebel Gurgeil between 2 200 and 2 400 m, is occupied by a dense 40–60 cm high sward, which is dominated almost exclusively by perennial grasses, principally *Cymbopogon commutatus*, *Hyparrhenia papillipes*, *H. hirta*, *Andropogon distachyos*, *Heteropogon contortus*, *Themeda triandra* and *Aristida adoensis*. A few bushes are present but their cover is always less than 10 per cent. The most frequent is *Dichrostachys cinerea* but *Terminalia brownii*, *Albizia amara*, *Acacia tortilis* and *A. albida* also occur. The status of the Jebel Gurgeil grasslands is uncertain.

Vegetation pattern in relation to the environment

Gross pattern in the Jebel Marra area

Refs.: Hunting Technical Services (1958, 1968, 1977); Wickens (1977a, pers. comm.).

Introduction

The Jebel Marra area (Fig. 21) lies in that part of Africa which is more than 1 600 km from the sea. It is almost in the centre of the continent, being equidistant from the Atlantic and Indian Oceans, the Mediterranean and the Red Sea. Within this area the vegetation forms a transition from typical Sahel to typical Sudan. The most prominent physical feature is Jebel Marra itself, a dormant, late Tertiary, volcanic massif resting on a base of Archaean rocks (Basement Complex) which form the Nile–Chad watershed.

The volcanic massif is a lofty and rugged range rising to over 3 000 m, approximately 90 km long by 65 km wide. It is composed principally of basalt and trachyte lavas with pockets of pumice and ash. On the south and south-western flanks are the remnants of a far more extensive ash piedmont that rises to an elevation of 1 150 m. It is partly alluvial and colluvial in origin. The other flanks are fringed by rather broken country formed by Basement Complex hills. The slopes of the mountain rise gently at first, then steeply, to a high plateau lying between 2 300 and 2 600 m.

The Basement Complex peneplain surrounding Jebel Marra has had a complex erosional history. In many places it is surmounted by inselbergs, and it is incised by a complicated drainage pattern. The lowlands of the Basement Complex have been formed from the more easily weathered schists and gneisses, whereas the hill lands which rise to more than 1 400 m are composed of the more resistant paraschists and gneiss and represent the remnants of a higher and older land surface. Aeolian sands are of restricted extent and only occur north of 13°00′ N. There is also a very small patch of Nubian sandstone near Garsila.

The high Jebel Marra massif strongly modifies the regional climate and is responsible for an increase in precipitation. The rain-bearing winds bring moist unstable air from the South Atlantic Ocean, and the augmented rainfall to the south-west of the mountain permits a northward extension of Sudan zone vegetation. On the higher parts of the west slope of the mountain rainfall slightly exceeds 1 000 mm per year. In the south-west of the study area it is mostly between 700 and 800 mm, elsewhere between 600 and 700 mm, and in the rain-shadow area north-east of Jebel Marra it falls off to below 400 mm. Rainfall is almost confined to the period from May to September, with 60 per cent falling in July and August. Relative humidity is generally low. Jebel Marra acts as a shield to the dry northerly winds blowing from the Sahara, evidence for which is the almost total absence of aeolian sands within the survey area, except on the eastern flank of the massif and in the extreme north-west.

Jebel Marra has been occupied by man for at least 2 000 years and its vegetation has been drastically altered. The soils, however, have remained productive because of the skilled use of bench terracing going back several centuries. All the accessible slopes have been terraced, up to 2 750 m, although today cultivation rarely occurs above 2 600 m.

In the Jebel Marra area the distribution of the following fourteen vegetation types is shown to be closely related to the main environmental features.

Vegetation of the Basement Complex

1. Acacia mellifera, Commiphora africana *communities on indurated soils*. This type is found on the eastern, northern and north-western borders of the survey area on the truncated and indurated red-brown drift and brown clay soils of the peneplain.

Acacia mellifera, the dominant species, often forms dense stands. Associated tree and shrub species are *Commiphora africana*, *Acacia nubica*, *A. senegal*, *A. tortilis*, *Albizia amara*, *Dalbergia melanoxylon*, *Ziziphus* spp. and *Lannea humilis*. The rather sparse grass cover is mainly restricted to areas of rain-washed sand. It is principally composed of *Aristida adscensionis*, *A. rhiniochloa*, *Cymbopogon schoenanthus* (*proximus*), *Schoenefeldia gracilis*, *Loudetia togoensis* and *Chloris gayana*.

To the north of the western flanks of Jebel Marra this type forms a complex mosaic with *Anogeissus leiocarpus* woodland. *Acacia mellifera* dominates the interfluves and *Anogeissus* fringes the drainage lines.

2. Acacia mellifera *communities on hill soils of the Basement Complex*. This type occurs on the lower eastern

flanks of the Jebel Marra massif and in the granite hills to the north of the massif beyond the study area.

The dominant species is *Acacia mellifera*, which generally occurs in pure stands of almost impenetrable thicket. Other tree species include *Acacia nilotica* subsp. *adstringens* (*adansonii*), *A. seyal*, *Balanites aegyptiaca*, *Commiphora africana*, *Albizia amara*, *Dichrostachys cinerea*, *Boswellia papyrifera*, *Terminalia brownii* and *Anogeissus leiocarpus*. The eastern flanks of the massif show signs of extensive terracing, long ago abandoned for cultivation, although they must have been usable in the past. The high rainfall, 700 mm, and the presence of occasional *Anogeissus leiocarpus* and *Terminalia brownii* suggests that extensive soil erosion has so reduced the moisture-holding capacity of the soil that *Acacia mellifera* has invaded an area which formerly supported a more mesophytic type of vegetation.

The more frequent and characteristic grass species present, although rather sparse, are *Aristida adscensionis*, *A. rhiniochloa*, *Sporobolus festivus* and *Tetrapogon cenchriformis*.

3. *Anogeissus leiocarpus woodland on Basement Complex soils*. This is the largest and most important lowland community within the area investigated, and occurs mainly to the south and east of the Jebel Marra massif. It is best developed on the sedentary soils of foothills where *Anogeissus leiocarpus* occurs in almost pure stands. The major tree associates are *Combretum glutinosum*, *Terminalia laxiflora*, *Sclerocarya birrea*, *Dalbergia melanoxylon* and *Dichrostachys cinerea*.

The more important grasses include *Sporobolus festivus*, *Loudetia togoensis*, *L. simplex*, *Ctenium elegans*, *Hyparrhenia rufa*, *H. confinis* and *Pennisetum pedicellatum*.

In areas of severe denudation, especially near villages, *Albizia amara* becomes dominant.

4. *Anogeissus leiocarpus, Boswellia papyrifera communities on hill soils of the Basement Complex*. These types are found on the southern and western fringes of the Jebel Marra massif as well as on the Tebella massif and Kobara hills. They occur on slopes undergoing progressive secular sheet erosion accelerated by misuse of the land, and pass through the following sequence of increasingly less water-demanding species: *Anogeissus leiocarpus*, *Boswellia papyrifera*, *Terminalia brownii*, *Lannea fruticosa*, *Acacia gerrardii* and finally *Albizia amara*.

The hill crests are usually dominated by pure stands of *Boswellia papyrifera*, with *Anogeissus leiocarpus* on the flanks, and sometimes thickets of *Acacia ataxacantha* at the foot of the scree slopes.

5. *Acacia seyal, Balanites aegyptiaca communities on clay soils*. These communities, which are well represented elsewhere in the Sudan on the clay plains of the Nile and its major tributaries, are here found on the gravel clays bordering the Wadi Saleh.

Acacia seyal generally occurs in pure stands interspersed by open areas of grassland. Occasional tree associates are *Acacia gerrardii*, *Balanites aegyptiaca* and *Albizia amara*. Associated grasses are *Setaria lynesii*, *Panicum subalbidum*, *Brachiaria lata*, *Pennisetum ramosum* and *Hyparrhenia rufa*.

On the clay plain and red-brown drift soils of the upper basin of the Wadi Azum system the *Acacia seyal*, *Balanites* community forms a mosaic with *Anogeissus leiocarpus*, with *Acacia seyal* dominating the interfluves and *Anogeissus* bordering the drainage lines. *Lannea humilis* is locally dominant in otherwise pure stands of rather stunted *Acacia seyal*. The inhibited growth of the *Acacia seyal* is probably due to the shallow soils, often less than 1 m deep, overlying the Basement Complex rocks.

Sporobolus festivus is the dominant grass throughout the area at the beginning of the rains, followed on stony soils by *Microchloa kunthii*, which is in turn succeeded by *Loudetia simplex*. In shallow depressions *Setaria pallide-fusca* succeeds *Sporobolus festivus* with *Anthephora lynesii* occurring at the edges. Eventually *Hyparrhenia confinis* and *H. rufa* dominate the deeper soils, with *Eragrostis tremula* dominant on the poorer soils.

6. *Acacia albida and Balanites aegyptiaca communities on alluvial soils*. These communities occur on the terrace soils of the Wadi Azum system.

In the Upper Azum Catchment the lower terraces are dominated by pure stands of *Acacia albida*, often forming a close canopy. Minor associates are *Ficus* spp., *Kigelia africana*, *Cordia abyssinica*, *Acacia sieberana* and *A. polyacantha*. The drier soils of the upper terrace carry a more xerophytic vegetation with *Balanites aegyptiaca* as the dominant species and *Ziziphus spina-christi*, *Acacia gerrardii*, *Albizia amara* and *Combretum aculeatum* also present.

The grass cover on both the upper and the lower terraces follows the same seasonal sequence. *Sporobolus festivus* is dominant at the start of the rains, followed by *Pennisetum pedicellatum*, *Cymbopogon schoenanthus*, *Hyparrhenia* spp. and *Andropogon gayanus*.

The Middle Reaches carry a similar vegetation but include a few species such as *Celtis integrifolia* and *Combretum paniculatum* on the lower terrace soils.

In the Lower Reaches the hitherto pure stands of *Acacia albida* are gradually replaced by a mixture of tree species, although the closed or almost closed canopy is still maintained. There is less distinction between the upper and lower terraces in both elevation and vegetation.

More species typical of the Sudan zone are present. They include *A. sieberana*, *Terminalia laxiflora*, *Combretum paniculatum*, *C. collinum*, *Pterocarpus lucens*, *Pseudocedrela kotschyi*, *Tamarindus indica* and *Prosopis africana*.

Locally there are pure stands of *Borassus aethiopum*, which are protected by the farmers.

7. Acacia senegal, Combretum glutinosum *communities on aeolian sands.* These communities, which are generally widespread in the Sahel zone, are very restricted in the Jebel Marra area owing to the small extent of wind-blown sand. *Acacia senegal* and *Combretum glutinosum* are co-dominant, with *Balanites aegyptiaca, Ziziphus* spp. and *Boscia senegalensis* also present.

The major grasses are *Eragrostis tremula, Cenchrus biflorus, C. prieurii, Aristida rhiniochloa, Loudetia togoensis* and *Cymbopogon* sp.

8. Combretum glutinosum, Guiera senegalensis *community on Nubian sandstone soils.* This type, which is better represented on the more extensive sandstone outcrops further east, is of very limited occurrence in the Jebel Marra area. The sandstone hills are dominated by *Combretum glutinosum*, with *Dichrostachys cinerea* and *Grewia flavescens* locally dominant on some of the steeper and better-drained slopes. Other associates include *Strychnos spinosa, Gardenia ternifolia, Dalbergia melanoxylon, Combretum collinum, Boswellia papyrifera* and *Boscia salicifolia.*

Part of the area either is under cultivation or is lying fallow, and *Guiera senegalensis*, a typical shrub of over-cultivated fallow sandy soils, is dominant. The ground cover is very sparse, with scattered *Aristida* sp. and *Eragrostis tremula.*

Vegetation of the volcanic massif

9. Acacia albida *communities on ash piedmont soils.* These communities occupy areas which were formerly extensively cultivated and probably abandoned between fifty and a hundred years ago. At the time of abandonment probably little natural vegetation remained and *Acacia albida*, which is now dominant and often forms pure stands, has subsequently spread.

Where *Acacia albida* woodland has been severely degraded, *Balanites aegyptiaca, Ziziphus spina-christi* and *Z. abyssinica* are the dominant species, with *Albizia amara, Acacia albida, A. nilotica* subsp. *adstringens, A. seyal* and *Dichrostachys cinerea* also present, while *Acacia nubica* becomes dominant on the old fallow lands on the alluvial fans. Other tree associates of *Acacia albida* include *Azanza garckeana, Acacia sieberana, Dombeya quinqueseta* and *Cordia abyssinica.*

At the start of the rainy season the grass cover is mainly *Dactyloctenium aegyptium, Cynodon dactylon* and *Setaria pallide-fusca*, later succeeded by *Hyparrhenia filipendula, H. anthistirioides, Andropogon gayanus* and *Cymbopogon giganteus.*

10. Combretum glutinosum, Terminalia laxiflora *communities on ash piedmont soils.* In this rather open community *Combretum glutinosum* is locally dominant, as well as being generally distributed throughout the area; the seedling stages are abundantly represented. *Terminalia laxiflora*, although equally widely distributed, is more frequent on the rather poorly drained ash soils. *Azanza garckeana* is locally dominant and is regarded as an indication of former settlements; it propagates very freely by means of root suckers. *Ficus* spp., *Acacia sieberana, Dombeya quinqueseta, Albizia aylmeri, Piliostigma thonningii, Ziziphus spina-christi, Stereospermum kunthianum* and *Securidaca longepedunculata* are other minor tree associates.

In areas with a high water-table and protected from the annual grass fires, *Khaya senegalensis* is regenerating naturally. But for fire it is believed that *Anogeissus leiocarpus* would be the dominant species.

Although by no means abundant, *Elionurus hirtifolius* is very conspicuous during the dry season for it is the only grass growing and flowering at that time. *Sporobolus festivus* is the dominant grass throughout the entire area early in the rainy season, followed by the taller *Anthephora lynesii, Hyparrhenia* spp., *Ctenium newtonii, C. somalense, Andropogon gayanus, Brachiaria brizantha, Cymbopogon giganteus* and *C. excavatus.*

11. Acacia mellifera *bushland on volcanic soils.* This community is found on the drier eastern ash slopes of the massif and on flat-topped basalt hills to the east and north-east. Rainfall is between 450 and 650 mm per year, hence a more mesophytic vegetation might be expected. The area was formerly cultivated and the bench terraces are still well preserved, though the finer ash particles have been washed away and the soils today are quite unsuitable for cultivation. The dominance of *Acacia mellifera* is believed to be a secondary consequence of reduced soil moisture rather than a direct reflection of the rainfall.

Acacia mellifera often forms dense impenetrable thickets on the hill slopes. Other associates include *Acacia seyal, A. nilotica* subsp. *adstringens, Commiphora africana, Mundulea sericea, Euphorbia candelabrum, Grewia flavescens* and *Dichrostachys cinerea.* On old fallow lands *Azenza garckeana*, which reproduces vigorously from root suckers, may form quite dense stands; *Acacia seyal, Sclerocarya birrea, Terminalia brownii* and *Anogeissus leiocarpus* are other typical species of the old fallow lands, and probably represent a transitional stage before dominance by *Acacia mellifera.*

The more important grasses present include *Aristida adscensionis, A. rhiniochloa, Tetrapogon cenchriformis, Loudetia simplex, Tripogon minimus, Schizachyrium exile* and *Cymbopogon* spp.

12. Anogeissus *communities on volcanic soils.* The extensive bench terraces on Jebel Marra indicate man's widespread activity. Everywhere, except for the inaccessible gallery forest and possibly the higher peaks, the vegetation has been destroyed by man or greatly modified by fire and grazing animals. Hence it is impossible to reconstruct the original vegetation on the slopes between the upper limit of *Acacia mellifera* bushland and the lower limit of the truly montane communities, which occurs at 1800 to 2000 m. *Anogeissus leiocarpus*, however, is widespread in this zone and Wickens (1977*a*)

describes all the vegetation under the heading '*Anogeissus* hill savanna'.

On ash loams of the lower mountain slopes, *Anogeissus leiocarpus* is the dominant tree species, although growth is not as good as might be expected in view of the fertile soils and high rainfall; fire or frost could be the limiting factors. Other tree associates include *Ficus sycomorus*, *Cordia abyssinica*, *Sterculia setigera*, *Lonchocarpus laxiflorus*, *Stereospermum kunthianum*, *Khaya senegalensis* and *Albizia malacophylla*. The shrubs include the scandent *Ipomoea verbascoidea* and thickets of *Grewia flavescens*.

On the basalt outcrops *Anogeissus leiocarpus* is the dominant species, forming dense stands, often with an almost closed canopy. Other tree and shrub associates include *Khaya senegalensis*, *Ziziphus mauritiana*, *Z. spina-christi*, *Acacia polyacantha*, *A. sieberana*, *A. ataxacantha*, *Gardenia ternifolia*, *Strychnos madagascariensis*, *Commiphora africana*, *Grewia mollis*, *G. flavescens*, *G. villosa*, *Erythrina sigmoidea*, *Securidaca longepedunculata* and *Pterocarpus lucens*.

On terraced mountain soils *Anogeissus* is dominant throughout, although on the drier eastern slopes more xerophytic species, such as *Boswellia papyrifera*, are locally dominant. *Terminalia brownii* and *Combretum molle* are also important, especially on the southern slopes. The vegetation on broken lava-plateau soils and on ash loams on steep mountain slopes is similar to that just described. *Acacia seyal* locally forms pure stands on poorly drained basalt soils and *Acacia albida* occurs on the ash loams.

13. *Riparian forest.* This type occurs in deeply etched and almost inaccessible gorges. The principal trees are *Trema orientalis*, *Syzygium guineense*, *Polyscias fulva*, *Diospyros mespiliformis* and *Phoenix reclinata*, with *Teclea nobilis*, *Albizia zygia*, *Maesa lanceolata* and *Casearia barteri* occurring more rarely.

14. *Sahelomontane communities.* The rolling plains and lava peaks of the upland plateau of Jebel Marra above 1 800–2 000 m are covered with montane grassland, but most is secondary, since almost the whole area has been cultivated and, apart from a few secluded valleys, is swept annually by fires.

The original vegetation was probably *Olea laperrinei* scrub forest, but this has been almost completely destroyed for firewood and building, though relict trees occur scattered throughout, almost to the summit.

Wickens (1977a) recognizes two main types of secondary grassland. The more extensive, which occurs on the steeper slopes and eroded lands, is an open community of bunch grasses and small shrubs. *Andropogon distachyos*, the dominant grass, is frequently associated with *Themeda triandra* and *Hyparrhenia hirta*. The commonest shrubs are *Lavandula pubescens* and *Blaeria spicata*.

On the better-drained sites of the gently rolling grass plains, the second type of grassland often forms a sward no more than 5 cm high. The dominant species are *Hyparrhenia multiplex* and *Vulpia bromoides*, with *Aristida congesta*, *Festuca abyssinica*, *Panicum pusillum*, *Tripogon leptophyllus* and *Pentaschistis pictigluma* also present. The flatter ash plains are semi-waterlogged during the rainy season. It is possible that, at least in places, the grassland they support is edaphically determined.

Gross pattern in Kordofan

Ref.: Hunting Technical Services (1964).

The vegetation of that part of Kordofan Province, Sudan Republic, which lies between 12° and 14° N. and 28°30′ and 31°15′ E. was surveyed and mapped by Hunting Technical Services as part of the Kordofan Land and Water Use Survey sponsored by FAO and the United Nations Special Fund. Almost the whole of this area is in the Sahel zone, and the four vegetation types described below can be used to illustrate the way in which the distribution of Sahelian vegetation in regions of low-to-moderate relief is influenced by climate, geology, soils, and human activity. Despite the low rainfall and absence of large permanent rivers agriculture is widely practised.

In the drier northern fringes cultivation is generally scattered but becomes more intensive and concentrated around water points. The principal crops are 'dukhn' (*Pennisetum*) and water-melons, but grazing is the dominant land use. In the *Acacia senegal* wooded grassland zone the chief crops are 'dukhn', 'dura' (*Sorghum*), sesame, ground-nuts and water-melons, usually grown in rotation (gum-cultivation cycle) with a fallow of *Acacia senegal* which is the source of gum Arabic. In the sand-dune areas cultivation is normally restricted to interdune hollows. A small amount of irrigation allows the cultivation of vegetables and citrus. On the old land surface (vegetation types 3, p.p. and 4, below) cultivation is almost entirely restricted to the sandier soils, and most of the area is heavily grazed by the nomadic Baggara tribesmen. The clay plains are not used for grazing, because of seasonal waterlogging and the unpalatable nature of the grass. A small amount of cotton and 'dura' is grown.

Nomadic movements into and out of the area are determined by the rains. The camel-owning nomads move northwards out of the area during the rainy season to graze their animals in semi-desert areas to the north. By contrast, the cattle-owning nomads (Baggara) move into the area from the south during the rainy season and graze their animals on the new pastures of the old land surface.

1. Semi-desert grassland on aeolian sand

Rainfall is less than 250 mm per year. *Aristida sieberana* (*pallida*) is dominant on freely drained sands and *Cymbopogon schoenanthus* in less well-drained areas. *Aristida mutabilis* and *Eragrostis tremula* also occur, and

Panicum turgidum is dominant on unstable dunes. The shrubs *Leptadenia pyrotechnica*, *Calotropis procera* and *Ziziphus spina-christi*, when abundant, indicate severe overcultivation. Tree species are represented by *Acacia tortilis*, *A. albida*, *A. senegal*, *Balanites aegyptiaca* and *Maerua crassifolia*. They are very thinly scattered and are chiefly confined to the hollows. The shrubs *Combretum aculeatum* and *Guiera senegalensis* occur at the foot of some unstable dune ridges where moisture is increased by seepage.

2. Acacia senegal *wooded grassland on aeolian sand*

Rainfall is mostly between 250 and 400 mm per year, but the distribution of this type of vegetation is edaphically restricted in the study area and it would extend into wetter areas if the soils were suitable. *Acacia senegal* is the most important tree, but it has greatly increased in abundance because of agricultural practice. Nearly everywhere the sand-sheet soils have been intensively cultivated using the gum-cultivation cycle. In an ideal cycle gum arabic is harvested from the trees for approximately fourteen years, followed by four years of cultivation, but increasing pressure for cultivable land increases the period under crops and eventually *Acacia senegal* is dropped from the cycle, and continuous cultivation is practised until the soil is exhausted. The parasite *Striga hermonthica* is an important weed of cultivation. Where the gum-cultivation cycle is practised, *Striga* is suppressed during the fallow period, but when the fallow is shortened or abandoned it may increase sufficiently to limit the yield of grain crops, especially 'dura'.

Within the area occupied by *Acacia senegal* there is a gradual change in floristic composition from north to south.

In the drier northern parts *Acacia tortilis* is more abundant than *A. senegal*. *Lannea humilis* is locally dominant and forms thickets. *Acacia laeta* is also locally dominant, and elsewhere the trees occasionally follow square outlines probably indicating the position of former hedges (zarribas) which delimited areas of cultivation. In wetter areas, *Albizia amara* frequently outlines the field boundaries of now abandoned cultivated land (page 214). Other trees include *Acacia albida*, *A. nubica*, *Balanites aegyptiaca* and *Maerua crassifolia*. The tree cover is very open and in the north trees are often restricted to the hollows. Among grasses, *Aristida sieberana* is the most widespread dominant, with *Cenchrus biflorus* locally dominant near areas of cultivation. On old lands that have become completely exhausted and abandoned, *Panicum turgidum* may occur. In dune areas the grasses show a catenary sequence with *Aristida sieberana* dominant on the crest and sides of the dune. *Eragrostis tremula* may also occur on the sides. The change in drainage at the bottom of the slope is marked by *Aristida mutabilis*, which is successively replaced by *Stipagrostis* (*Aristida*) *acutiflora*, *Cymbopogon schoenanthus*, *Schoenefeldia gracilis* and *Aristida adscencionis*, and finally bare soil in the wettest parts of the hollow. In depressions with wells where stock are watered, trampling and dunging give rise to concentric zones of practically monospecific vegetation surrounding the drinking-trough. The bare soil at the centre is successively replaced by zones dominated by *Tribulus terrestris*, *Amaranthus graecizans*, *Solanum dubium*, *Cassia tora* and *Acacia nubica*.

In the wetter variants of *Acacia senegal* wooded grassland certain broad-leaved trees, such as *Combretum glutinosum*, *Terminalia brownii*, *Albizia amara*, *Stereospermum kunthianum*, *Sclerocarya birrea* and *Terminalia laxiflora*, which are more characteristic of type no. 4, begin to appear.

3. Acacia mellifera *bushland*

This type is most extensively developed on the alkaline clay soils of the pediplain south of El Obeid. It also occurs on dark, cracking clays. Rainfall is mostly between 400 and 500 mm per year.

The pediplain soils are usually overlain by a shallow topsoil of wind-blown sand with a moderately rapid permeability, whereas the underlying clay is only very slowly permeable. Hence the topsoil promotes water entry into the lower layers but reduces losses by evaporation and so acts as a mulch. On the old land surface south of El Obeid, *A. mellifera* often forms dense thickets, especially around shallow seasonal pools. On drier sites it is associated with *A. nubica*, *Commiphora africana* and *Boscia senegalensis*. Occasional associates on the banks of small gulleys include *Cordia sinensis*, *Dichrostachys cinerea*, *Albizia amara*, *Dalbergia melanoxylon* and *Terminalia brownii*, whereas *Adansonia digitata* prefers shallow depressions. The most abundant grasses are *Schoenefeldia gracilis*, *Eragrostis tremula*, *Sporobolus humifusus* and *Chloris virgata*, though the grass cover is thin and discontinuous and generally confined to patches of sand. In overgrazed areas the field layer is dominated by *Blepharis linariifolia* and *Zornia glochidiata* on the sandier and harder soils respectively. After cultivation *Acacia nubica* often becomes dominant.

On dark, cracking clays *Acacia mellifera* may form pure stands or is associated with *Boscia senegalensis*, *Cadaba glandulosa*, *Albizia anthelminthica*, *Balanites aegyptiaca* and *Dichrostachys cinerea*. On water-receiving sites *Acacia seyal* occurs, especially towards the south. The major grass species are *Schoenefeldia gracilis*, *Aristida funiculata*, *Tetrapogon cenchriformis*, *Hyparrhenia anthistirioides* (*pseudocymbaria*), *H. petiolata* and *Cymbopogon nervatus*.

4. Broad-leaved woodland

Broad-leaved woodland is the vegetation of the sandy pediplain. Rainfall is more than 400 mm per year. The most characteristic tree species are more typical of Sudanian than Sahelian vegetation. The area has been so much cultivated and grazed by domestic stock that much of the vegetation is now secondary scrub regrowth. The principal trees, which mostly exceed 6 m in height, are *Combretum glutinosum*, *Albizia amara*, *Terminalia brownii*, *Dalbergia melanoxylon*, *Sclerocarya birrea*, *Adansonia digitata* and *Balanites aegyptiaca*.

Detailed patterns in Darfur, Kordofan and the Nile valley

Vegetation arcs or stripes running parallel to the contours and occurring on virtually flat to gently sloping surfaces (from 1 in 500 to 1 in 50) are a distinctive feature of many arid and semi-arid regions (pages 000 and 000). Mean annual rainfall ranges from 100 to 500 mm and is often in the form of heavy showers. The vegetation of the stripes is denser, taller and physiognomically more complex than that of the intervening lanes, which are sometimes almost bare. Within the Sahel zone such patterns have been described from Niger by L. P. White (1970) and from the Sudan Republic by several authors (summarized by Wickens & Collier, 1971). In West Africa this type of pattern is often referred to as *brousse tigrée* (see Chapters 2 and 3).

According to Wickens & Collier this type of pattern occurs on a wide variety of soils developed from different parent materials and there are no significant chemical differences between the vegetated and non-vegetated areas. The soils under the vegetation, however, are more permeable to water. The plants on the arcs provide a barrier at ground level which interrupts water movement along the soil surface and traps additional water for plant growth. In the areas of the Sudan examined by Wickens & Collier, the bare indurated surfaces between the arcs have been exposed by the removal of the sandy topsoil by sheet erosion. The arcs themselves are in a delicate state of equilibrium and increased grazing pressure is likely to bring about their degeneration, accompanied by increased surface run-off, ultimately resulting in an eroded landscape which will prove difficult to recolonize except with poor-quality grasses.

Of the three patterns mentioned below, Wickens & Collier describe the first two and Worrall (1959) the third:

— 1. *Terminalia brownii* arcs in Kordofan. Rainfall 400–500 mm per year. Parent rock Nubian sandstone and conglomerate. Slope 1 in 50 to 1 in 200. Pattern: crescent-shaped groves of trees aligned along the contour are separated by more open spaces 60–120 m apart. Sheet erosion has occurred on all the soils in the inter-grove areas.

Terminalia brownii is the main tree constituent of the groves, associated with *Albizia amara*, *Dalbergia melanoxylon*, *Grewia flavescens* and *G. tenax*. Grasses included *Alloteropsis cimicina*, *Aristida sieberana*, *Erogrostis tremula* and *Schoenefeldia gracilis*.

The inter-arc area contained a few scattered shrubs of *Boscia senegalensis*, *Grewia flavescens*, *G. tenax* and *Dalbergia melanoxylon*. Apart from *Microchloa indica*, the sparse grass flora was not identifiable, because of trampling by grazing animals.

— 2. *Acacia mellifera* fingerprint pattern in Kordofan. Rainfall 400–500 mm per year. Parent rock: Basement Complex. Slope less than 1 in 200. Pattern: tightly packed 'fingerprint' whorls of bushes aligned along the contour. The pattern is clearly visible from the air, but because of the density of *Acacia mellifera* is almost impossible to discern on the ground.

The vegetation consists of pure stands of *Acacia mellifera* with occasional understorey shrubs such as *Boscia senegalensis* and *Cadaba glandulosa*. The intervening grass cover is thin and patchy and confined to wind-blown sand. It consists of *Sporobolus humifusus*, *Schoenefeldia gracilis*, *Eragrostis tremula* and *Chloris virgata*. This pattern is generally found on waterless and therefore uninhabited country.

— 3. The Butana grass pattern in the Nile valley east of Khartoum. Rainfall 100–400 mm per year. Slope 1 in 200. The grey-brown clay-loam soils have been derived by *in situ* weathering of rocks of the Basement Complex. Pattern: 8–12 m wide bands of grass alternate with bare areas about twice that width.

The dominant grasses are *Aristida* spp., chiefly *funiculata*, *Sehima ischaemoides*, *Schoenefeldia gracilis* and *Cymbopogon nervatus*. Worrall observed that the next season's vegetation occurred on the upslope side of the standing dead grass of the previous year and postulated a slow annual migration of the grass bands upslope. The sparse vegetation of the 'bare' areas consists chiefly of stunted grasses of the same species as occur in the bands.

According to Wickens & Collier, the grass arcs are being destroyed by excessive grazing and the area is being changed to a clay desert by sheet and wind erosion.

Wickens & Collier also describe the following detailed patterns which are determined by soil differences or previous land use:

— 1. *Acacia mellifera* 'frog-spawn' pattern in Kordofan. Rainfall: 200–250 mm per year. Parent material: aeolian sand forms a confused system of low, crescent-shaped, 1–3 m high dunes, separated by fluvio-lacustrine deposits in the intervening hollows. On aerial photographs the peculiar pattern produced by the sparsely vegetated dunes and the heavily vegetated hollows is reminiscent of frog spawn.

The dunes carry a sparse tree cover of *Acacia tortilis*, *A. senegal* and *Leptadenia pyrotechnica*. The dominant grass is *Aristida sieberana*, except where *Panicum turgidum* dominates in recently disturbed places. Impenetrable thickets of *Acacia mellifera* occur on the fluviolacustrine deposits.

— 2. Fossil drainage patterns in Darfur. Rainfall 800–900 mm per year. Parent material: volcanic ash of the piedmont plain bordering the western slopes of Jebel Marra. The ash has suppressed the earlier dendritic drainage pattern and there is no apparent difference in level between the fossil drainage lines and the surrounding land, nor in the soils themselves.

The fossil waterways are bare of trees but are fringed by well-grown trees of *Acacia albida* and *Khaya senegalensis*. Elsewhere, away from the former drainage lines, *Acacia albida*, *Balanites aegyptiaca* and *Ziziphus spina-christi* are dominant or co-dominant. Although the area was formerly intensively cultivated, it was depopulated during the Mahdist era, when the people fled to the safety of the hills, and has been uninhabited for forty years or more, except for a few nomadic tribesmen. Wickens & Collier suggest that seasonal flooding of the waterways might be the cause of their treelessness.

— 3. Reticulate cultivation patterns in Kordofan. Rainfall 400 mm per year. Parent rock: Nubian sandstone. Slope: up to 1 in 200. Pattern: a mosaic of tree-lined interlocking squares and rectangles indicating former field boundaries, within which there is little tree growth. The most abundant tree is *Albizia amara* with *Dichrostachys cinerea* and *Piliostigma thonningii* as minor associates. The original vegetation was probably dominated by *Terminalia brownii*, relics of which still survive around the seasonal water-holes. The area was heavily cultivated during pre-Mahdist days but became depopulated when the local inhabitants left in support of the Mahdi and were killed in battle. It seems that the *Albizia amara* trees date from this time. Wickens & Collier suggest that the brushwood fences with their accumulations of trash and wind-blown sand might have formed a suitable medium for their establishment.

Vegetation change and the Great Drought of 1968–73

The immediate consequences of the drought and possible long-term future implications were widely reported, often sensationally in the international press, and gave rise to much controversy. Much attention was focused on six West African countries, Chad, Mali, Mauritania, Niger, Senegal, and Upper Volta, whereas the Sudan Republic received less publicity, possibly because the movement of its pastoralists is almost entirely within its own frontiers.

Estimates of the death-rate among animals and of human losses have varied greatly and it seems that in the absence of reliable demographic data only the most general impressions can be formed. The people showed a remarkable capacity for survival, and most estimates of the death-rate are probably exaggerated (Konczacki, 1978).

Although there is abundant evidence for long-term oscillations in precipitation in Africa throughout the Pleistocene, existing rainfall records provide no evidence for secular climatic change within the Sudano–Sahelian region in recent times. The picture, rather, is one of short-term fluctuations. Existing records show that droughts of approximately five-year duration can be traced back to the eighteenth century. The drought of 1911–14 and those of the early 1920s and early 1940s were also severe.

A decade of above-average rainfall and good pastures before the 1968–73 drought resulted in a great increase in the livestock population, but exogenous man-made interventions played an important part. These included the elimination of warfare, the provision of veterinary services, the digging of wells and boreholes, and less extensive seasonal movement in search of new pasture. Hence the purely climatic consequences of the drought were greatly accentuated, resulting in a sudden extension of the area of desertification and a marked decrease in the productive capacity of the rangeland. Some of the consequences are briefly described below. They are chiefly based on vegetation surveys made before and after the drought in the Jebel Marra region of the Sudan Republic (Hunting Technical Services, 1977, G. E. Wickens, pers. comm.).

The principal objectives of the ecological survey undertaken by Hunting Technical Services were to compare conditions in 1977 with those encountered in 1963–67, and to identify any adverse trends resulting from the decreased rainfall of the Great Drought or from changes of land use, and to make recommendations for reversing such changes. The principal observations and conclusions were as follows:

1. Anogeissus *woodland on the Basement Complex*. Serious deterioration was encountered in the Wadi Azum basin, which formerly was regarded as the best winter grazing in the Sudan, because of its exceptionally good supply of water and winter fodder. Except for the sparsely inhabited Qoz Salsilgo, perennial grasses are now only a minor constituent of the ground flora, whereas in the past they were a major one. They have been replaced by annual grasses, but even these are confined, in the more devastated areas, to the lower-lying patches of accumulated sheet wash, with large intervening expanses of bare, indurated, 'B' horizons. These indurated areas are now unlikely to produce any worthwhile grass cover without remedial surface treatment to prevent run-off and increase water penetration.

2. Acacia mellifera *and* A. seyal *communities on clay soils*. The grazing had deteriorated and some of the trees were dying owing to drought stress.

3. *The piedmont slopes of Jebel Marra*. Except locally, the grazing of the piedmont, especially the more western parts, has deteriorated, with a noticeable decrease in perennial grasses and an increase in gully activity along the wadi banks. The southern piedmont is now much more intensively cultivated than it was twelve years ago, while the area under bush fallow has noticeably decreased.

The reasons for the deterioration are again believed to be the combined effects of the lower rainfall and increased pressure on the land. The soils are more susceptible to wind than sheet erosion, so that in areas where mechanized farming is being introduced adequate shelter belts and other conservation practices such as trash mulching during the dry season must be effectively implemented.

4. *The Jebel Marra massif.* On the massif, which is avoided by the nomads, and where grazing is mainly by the sheep and goats owned by the resident villagers, the effects from the drought years appear to be far less drastic. Soil erosion has been kept under control by the very effective terracing of the lands under cultivation. The practice of periodically rotating the cultivated lands ensures that terraces neglected during periods under fallow are eventually brought into repair.

5. Acacia albida *communities.* Because the tree has the unusual property of being leafless during the rains, crops can be grown beneath its canopy without any adverse effect due to shade. They even benefit from the enhanced fertility around the tree. The green foliage during the dry season provides a useful browse at a time when green fodder is in short supply. In addition the winter supply of pods from twelve trees has a crude protein equivalent to that from a hectare of ground-nuts. Since the stands can be as high as twenty trees per hectare, the combined return from the trees plus the crops is extremely productive and is unlikely to be exceeded by any other form of crop production in the area.

The former practice of lopping leafy branches by means of a knife attached to a long pole did not adversely affect the tree. The camel nomads, however, who do not normally visit the area, but during the drought years wintered even farther south than their traditional quarters, reduced the tree to a severely pollarded trunk. Locally they have virtually destroyed fine stands. This practice can be expected to spread into other areas as the browse reserves become depleted, unless it is prevented by legislation.

6. *Conclusions.* During the drought years there was a general degradation of the vegetation accompanied by a widespread increase in sheet erosion, with gully formation in some areas. The adverse effects of the drought on the vegetation were greatly exacerbated by the high population of domestic animals, and increased pressure on land was brought about by a general southward movement of people and their livestock because of the even more serious consequences of the drought in the lower rainfall areas further north. Although climatic conditions are beyond control, their effect can be ameliorated by adjusting livestock numbers to the carrying capacity. This is obviously a difficult and unpopular measure but it is essential for the future prosperity, and even survival, of the livestock industry.

XVII The Sahara regional transition zone

Area, geographical position, geology and climate

Flora

Mapping units

Vegetation
   Oases
   Wadis
      Tamarix communities
      Acacia communities
      Hyphaene ('Doum') communities
   Psammophilous vegetation
   Hamadas
   Regs
   Saharomontane vegetation
      Saharomontane wadi vegetation
      Saharomontane grassland
      Saharomontane dwarf shrubland
      The *Erica arborea* community
   Halogypsophilous vegetation
      Hyperhalogypsophilous vegetation
      Halogypsophilous vegetation on drier soils
      Vegetation of gypsaceous loamy sands
   Absolute desert
   The Atlantic coastal desert
   The Red Sea coastal desert

Area, geographical position, geology and climate

The Sahara is the largest desert in the world and the most extreme. It extends across North Africa from the Atlantic coast to the Red Sea. Climatically it is characterized by high temperatures, absence of frost (except in the high mountains), and, apart from the coastal fringes, a very dry atmosphere (see Fig. 22). The daily amplitude in temperature can exceed 35°C and the annual amplitude 60°C. Wind is a continual factor. (Area: 7 387 000 km^2.)

The limits of the desert are somewhat arbitrary, but the 'best fit' with biological reality is obtained if the northern limit is drawn to coincide with the 100 mm isohyet and the southern limit to coincide with the 150 mm isohyet (Quézel, 1965a). The 100 mm isohyet more or less corresponds to the northern limit of cultivation of the date palm (*Phoenix dactylifera*) and the southern limit of 'Alfa', *Stipa tenacissima*, one of the most characteristic species of the Mediterranean–Sahara transition zone. The 150 mm isohyet more or less corresponds to the southern limits of *Cornulaca monacantha*, *Stipagrostis pungens* and *Panicum turgidum*, and the northern limits of several Sahel species, notably the grass, 'Cram-Cram', *Cenchrus biflorus*, and, among woody plants, *Commiphora africana* and *Boscia senegalensis*. The southern boundary of the Sahara is much less abrupt than the northern because of the absence of pronounced features of relief. Over a wide transition zone Saharan and Sahelian elements occur in mosaic. The precise distribution of each is determined principally by local features of physiography.

Three climate zones, northern, central, and southern, can be recognized on the basis of rainfall distribution. In the northern zone the rain falls during the cold season with two maxima, in autumn and spring. Although rain falls every year there is considerable variation from one year to the next both in distribution and amount. Rainfall declines rapidly towards the south.

In the central region rain is episodic in character and low in amount. Between 18° and 30°N, the mean annual rainfall is less than 20 mm, except in the high mountains. Large areas of the Libyan desert are virtually rainless. The western Sahara is better watered but even there several years may pass without a drop of rain.

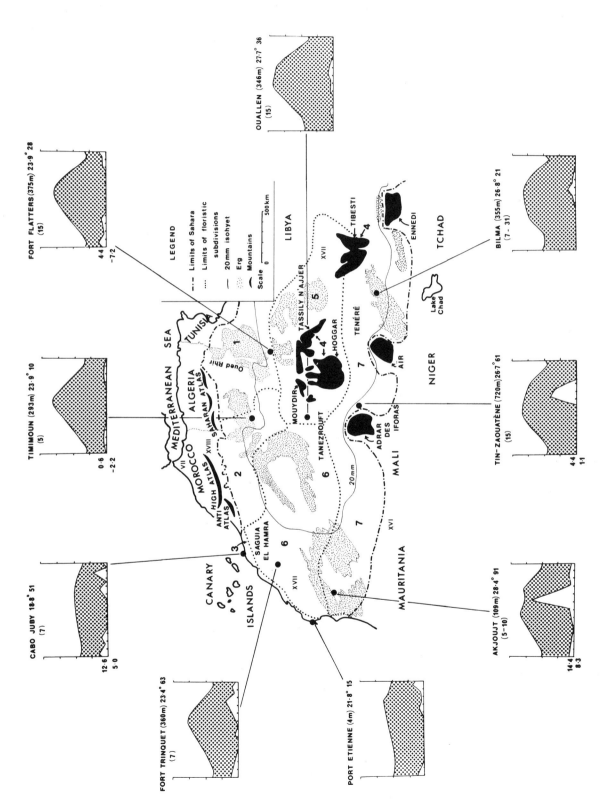

FIG. 22. Climate and topography of the western half of the Sahara regional transition zone (XVII). The floristic subdivisions of Quézel (1965a) are shown as follows: (1) Northern Sahara; (2) North-Western Sahara; (3) Atlantic Sahara; (4) Montane Sahara; (5) Central Sahara; (6) Western Sahara; (7) Southern Sahara

Further south, rainfall increases, but this is summer rain.

Three major floristic zones more or less coincide with the rainfall zones, and the dividing line between two great floristic realms, the Holarctic and the Palaeotropic, runs through the central Sahara. In the northern Sahara, elements of Mediterranean affinity predominate. The latter are almost completely absent from the southern Sahara, where the flora is overwhelmingly tropical in character. The flora of the central zone is mixed. In this zone the vegetation of the stone detritus of hamadas and of sandy beds of wadis is almost completely comprised of northern elements, while the predominant vegetation of stony and rocky wadis, characterized by *Acacia tortilis* and *Panicum turgidum*, consists essentially of tropical species. The high mountains which arise from the central Sahara are even more mixed in character.

Apart from the high mountains, which are of limited extent, the Sahara consists of several basins mostly isolated from the sea and lacking outward drainage. The substrate of most of the Sahara consists of Cretaceous and Tertiary deposits. Crystalline bedrock only breaks through to form the summits of the higher mountains, namely Tibesti (3415 m), Ahaggar (2918 m), and Jebel Uweinat (1900 m), and locally elsewhere to form granite inselbergs. The depressions are filled with Pleistocene deposits, which form either sand deserts (erg) or gravel deserts (reg). Between the depressions the stratified terraced landscape forms a stone desert (hamada) incised by dry valleys (wadis). Oases occur wherever water of low salt content issues as springs in the desert. Saline soils play only a minor role in the Sahara proper. The rainfall is so small that salts are neither removed by leaching nor accumulated in depressions.

Flora

According to Quézel (1978) there are 1620 species, of which 11.6 per cent are endemic. A further 22.7 per cent, however, are Saharan species which extend into the deserts of Arabia. Among the linking elements, 21.2 per cent of Saharan species are shared with the Mediterranean Region. The largest element, 32.3 per cent, comprises tropical species.

Endemic genera. Endemic genera are few and mostly monospecific. There are no endemic taxa above generic rank. According to Quézel (1978) there are 16 endemic genera, including *Foleyola, Monodiella, Nucularia, Tibestina* and *Warionia*. Several other widespread Saharan genera, such as *Agathophora, Anabasis, Anastatica, Neurada, Ochradenus, Rhanterium, Schouwia* and *Zilla*, extend into the deserts of south-west Asia.

Endemic species. About half are widespread; the remainder have more restricted distributions. The Atlantic coastal Sahara and the Sahara mountains are relatively rich in endemics. The Southern Sahara has very few endemic species. Many Saharomontane endemics, including *Cupressus dupreziana* and *Myrtus nivellei*, are closely related to Mediterranean species. The palm *Medemia argun* of the Nubian desert belongs to a genus with one other species in Madagascar.

Mapping units

38 (p.p.). Eastern African evergreen and semi-evergreen bushland and thicket.
67. Absolute desert.
68a. Atlantic coastal desert.
68b (p.p.). Red Sea coastal desert.
69. Desert dunes without perennial vegetation.
70. Desert dunes with perennial vegetation.
71 (p.p.). Regs, hamadas, wadis.
72. Saharomontane vegetation.
73 (p.p.). Oasis.
76 (p.p.). Halophytic vegetation.

Vegetation

Because Saharan vegetation is exiguous and is often physiognomically mixed it is most conveniently classified according to the habitats in which it occurs. The vegetation of the Atlantic and Red Sea coastal deserts, which is different from that of the interior, is described separately.

The vegetation of the western half of the Sahara is more diversified and is better known than that of the eastern Sahara. The remarkable synthesis of Quézel (1965a) extends only as far east as Chad, though his more recent (1978) floristic study covers the whole of the Sahara.

Quézel recognizes the following seven phytogeographical divisions ('domains') of the western Sahara: North-Western, Northern, Atlantic, Western, Central, Southern, Saharomontane. His classification is followed here. The names of the domains are given capital initial letters in order to avoid confusion with parts of the Sahara *sensu lato* which are similarly designated but in a different and more general sense, for which the lower case is used. To the above divisions Quézel (1978) has subsequently added two domains, the North-Eastern and the Eastern in the eastern Sahara.

Oases
(mapping unit 73)

Ref.: Walter (1971, p. 444).
Phots.: Ozenda (1958: 5, 6); Quézel (1965a: plate 1c); Stocker (1927: 1–6); Walter (1971: 262).

Places where water of low salt content issues as springs are relatively few. At the edge of the water, swamps of *Typha latifolia, Phragmites australis* or *Scirpus holoschoenus* often occur. They grade into stands of *Tamarix*[1] '*gallica*' and *T.* '*nilotica*'.

According to Walter (1971) the original vegetation of the oases consisted of Doum Palms (*Hyphaene thebaica*), species of *Acacia*, *Maerua* and *Capparis*, *Calotropis procera* and the cucurbit *Citrullus colocynthis* (*Colocynthis vulgaris*). Today this vegetation has been almost completely replaced by the Date palm, *Phoenix dactylifera*, and other cultivated plants.

Stocker describes the large depression of Wadi Natrun, WNW. of Cairo, which is fed by subterranean water from the Nile. Where ground water appears at the surface, swamp vegetation was formerly developed with *Typha*, *Phragmites* and *Juncus acutus*, and the surface between the plants was covered with a crust of white salt. The plant roots, however, occurred in deeper layers, which have only low salt concentrations. Wadi Natrun has recently been converted into an oasis with a dense population.

Wadis
(mapping unit 71)

Apart from oases these are the only desert habitats where trees and large bushes are found. There are three main vegetation types, *Tamarix* communities, *Acacia* communities and *Hyphaene* communities.

Tamarix *communities*

Refs.: Kassas (1952); Kassas & Imam (1954); Quézel (1965a, p. 81–3, 179–84, 200–3).
Phots.: Kassas (1952: 3); Quézel (1965a: 43, 44, 49).

In the Central Sahara, communities dominated by *T.* '*articulata*' occur in the larger sandy wadis wherever the water-table is within 7 or 8 m of the surface. They are particularly well developed between 300 and 1 300 m in the large wadis that radiate from the mountainous massifs. *Tamarix* communities vary greatly in luxuriance. Under favourable conditions where the water-table is near the surface for most of the year *Tamarix* may form completely closed riparian forest, in which the largest individuals are up to 10 m tall with boles 1.7 m in diameter. Where conditions are less favourable the *Tamarix* plants are more widely spaced, or, if occurring in closed communities, are no more than 2 or 3 m tall. In open *Tamarix* communities the ground between the trees is often occupied by *Stipagrostis pungens*, *Leptadenia pyrotechnica* and *Calligonum comosum*.

T. '*articulata*', which is very resistant to browsing camels and goats, provides the only joinery timber locally available. Tannin is also obtained from its galls.

1. It would appear from Baum's revision (*The genus* Tamarix, Jerusalem, Israel Academy of Sciences and Humanities, 1978) that most of the names most commonly used in the ecological literature for Saharan species of this genus have been misapplied. Since it is impossible to correct them with any degree of confidence, the names as originally published are quoted here, but with the epithets between inverted commas.

In certain wadis in the Northern and North-Western Sahara which receive their water supply from the mountains of the High Atlas or Saharan Atlas some tree species of northern affinity, e.g. *Nerium oleander*, *Populus euphratica* and *Vitex agnus-castus*, are associated with *Tamarix* '*gallica*'.

Acacia *communities*

Refs.: Léonard (1969a); Quézel (1965a, p. 124–33, 160–75).
Phots.: Léonard (1969a: 3, 8, 10); Quézel (1965a: pl. 3b, 4c; figs. 41, 42).

Acacia communities are characteristic of the rocky beds of wadis and the deposits of gravelly alluvium of outwash fans. Except for the *Tamarix* and *Hyphaene* communities and some vegetation types on the summits of the high mountains, all the larger woody plants of the Sahara belong to the *Acacia* communities. Nearly all of them are characteristic Sahel species, which in the Sahara are confined to specially favoured water-receiving sites. Of the larger woody plants, the most important are *Acacia tortilis* subsp. *raddiana*, *A. ehrenbergiana*, *Maerua crassifolia*, *Balanites aegyptiaca*, *Capparis decidua*, *Calotropis procera*, *Salvadora persica* and *Ziziphus mauritiana*. They occur in communities which, when well developed, are physiognomically similar to the bushlands and bushed grasslands of the Sahel zone.

The most widespread community is characterized by *Acacia tortilis* and the grass *Panicum turgidum*. This type is found in suitable habitats throughout the tropical Sahara below altitudes of 1 800 m wherever the annual rainfall exceeds 30 mm. It is therefore virtually absent from the hyper-arid areas such as Tanezrouft and Ténéré. It reaches its best development in the valleys radiating from the mountainous massifs, but is otherwise widely distributed from coastal Mauritania to Tibesti and beyond. It occurs characteristically on the gravelly beds of wadis and outwash fans. The most constantly present species are, among larger woody plants: *Acacia tortilis* subsp. *raddiana*, *A. ehrenbergiana*, *Maerua crassifolia*, *Balanites aegyptiaca*, and *Ziziphus lotus*; and, among herbs and subshrubs, *Cassia italica* (*aschrek*), *Caylusea hexagyna* (*canescens*), *Lavandula stricta* (*coronopifolia*) and *Seetzenia africana* (*orientalis*). The grass, *Panicum turgidum*, is constantly present.

The following variations of the *Acacia tortilis–Panicum turgidum* community are found where edaphic or climatic conditions are somewhat different from the normal:
1. Where fixed sand occurs in the beds of wadis and the water-table is near the surface, *Leptadenia pyrotechnica*, *Chrozophora brocchiana* and *Stipagrostis pungens*, among others, join the community.
2. On gentle slopes in wadis, where water stands for a long time and the soil is more silty, *Psoralea plicata* and *Hyoscyamus muticus* are especially characteristic. These sites are avoided by the caravans of the Touaregs because of the toxicity of *Hyoscyamus*.

3. Wadis on the south- and west-facing slopes of Ahaggar and Tibesti receive rain-bearing winds from the Atlantic Ocean. Rainfall is higher, more than 50 mm per year, and the *Acacia–Panicum* community is richer in species, mostly of tropical genera, such as *Abutilon*, *Hibiscus*, *Rhynchosia* and *Tephrosia*.
4. Vegetation in which *Acacia tortilis* occurs is extensive in wadis in the North-Western Sahara. *Panicum turgidum* and *Caylusea hexagyna* are usually present, but their other tropical associates are much more sporadic, and the community as a whole consists of species of Mediterranean rather than tropical affinity.
5. *Acacia tortilis* also occurs in a few wadis in the Northern Sahara. Here virtually all its associates are of Mediterranean affinity.
6. The wadis of Jebel Uweinat in the Libyan Desert also support *Acacia tortilis–Panicum turgidum* communities. Their density and composition vary greatly but they are mostly very open.

Hyphaene ('*Doum*') *communities*

The large wadis radiating from the south-western slopes of Tibesti support several tree species which locally form well-grown fringing forest. The dominant trees are *Hyphaene thebaica*, *Salvadora persica*, *Tamarix* '*articulata*', *Acacia nilotica* subsp. *adstringens* and *A. albida*. In the piedmont zone *Salvadora* and *Tamarix* are dominant and the other species are often lacking. Towards the south the water-table becomes progressively deeper and *Tamarix* becomes dominant. As one penetrates the massif, *Tamarix* and *Salvadora* gradually disappear and are replaced by *Hyphaene* and *Acacia nilotica*. In the deep gorges, where water occurs in the wadi beds, there are two strips of *Hyphaene* palms. Such communities extend along the wadis for up to 30 km. The Doum palm plays an important part in the economy of the local people. Fibres from its leaves are made into very strong ropes; its trunks are used for various constructional purposes, especially for making equipment for drying dates. The fruit is edible. The date palm can be grown on Tibesti wherever *Hyphaene* occurs, but the plantations are not extensive.

Psammophilous vegetation
(mapping units 69, 70, and 71)

Ref.: Quézel (1965a, p. 84–99).
Phots.: Quézel (1965a: 21–3).

Sand covers more than one-third of the surface of the western Sahara and rather less further east. More than half of the desert sands are without perennial vegetation, especially those in the hyper-arid zone, otherwise vegetation is always present though its degree of development varies greatly. There are two principal sandy habitats (other than the sandy floor of wadis, described above) namely desert dunes and sandy regs.

1. *Desert dunes* (*erg*). Some dunes are completely sterile. On others, after heavy rain, the vegetation cover may be as much as 50 per cent. When the wind blows from opposite directions but predominantly from one, crescent dunes ('barchans') which are mobile and devoid of vegetation are formed. Where one of the wind directions is only slightly more frequent the dunes themselves are more or less stable but their surfaces, especially on the crests, are unstable. Shifting dunes are barren of vegetation. On more stable dunes plants become established if their roots can penetrate into the damp layers. Since moisture is available throughout the year they show no seasonality in their development. Dune vegetation is poor in species. In the western Sahara two species occur throughout: the grass, *Stipagrostis pungens*, which has sclerophyllous, spine-tipped leaves; and the chenopodiaceous shrub, *Cornulaca monacantha*. In the Northern Sahara several other shrubs, including three species of *Calligonum*, *Retama retam*, *Genista saharae* and *Ephedra alata*, also occur. In the Central and Southern Sahara the shrub *Leptadenia pyrotechnica* is often present. Apart from *Malcolmia aegyptiaca*, annuals are rarely found, possibly because the constant movement of the surface sand is inimical to their establishment.

2. *Sandy regs*. These are frequent throughout the Sahara, especially in the hyper-arid zone. Their vegetation is extremely homogeneous and consists almost exclusively of therophytes. Wind erosion of the sand is greatly reduced because of a superficial covering of somewhat larger rock fragments. The most frequently occurring species are *Asthenatherum* (*Danthonia*) *forskalii*, *Plantago ciliata*, *Polycarpaea repens* (*fragilis*), *Monsonia nivea*, *Neurada procumbens*, *Ifloga spicata* and *Fagonia glutinosa*. In the driest parts of the western Sahara, such as Tanezrouft and Ténéré, episodic rainfall of 10–20 mm, which occurs in certain years, is sufficient for the development of ephemeral vegetation on sandy regs. The plants, which grow very quickly, cover 8–20 per cent of the surface.

Hamadas
(mapping unit 71)

Refs.: Kassas & Girgis (1964); Quézel (1965a, p. 101–12, 120–2, 133–43, 149–51, 187–94, 213–17, 315–21).
Phots.: Kassas & Girgis (1964: 1–7); Ozenda (1958: 9); Quézel (1965a: figs. 24, 25, 28, 33, 34, 39, 47, 48); Walter (1971: 266).
Profile: Walter (1971: 270).

All fine products of weathering are removed by the wind from the tops of the plateaux on which stone fragments about the size of a hand accumulate to form a stone pavement. The surface strata of Cretaceous and Tertiary origin form stone pavements, which occupy the largest area of the Sahara. Where the parent material is of marine origin beads of gypsum and salt-sheets may accumulate in the layer beneath the surface stones, which become coated with a blackish 'desert varnish' of iron and manganese compounds. The plateaux surfaces of hamadas are usually devoid of vegetation because of the lower water reserves in the soil and the relatively high salt content. Plants can only grow in rock crevices and water-receiving depressions. Steep rocky slopes incised by deep erosion channels support a more luxuriant vegetation.

Although the vegetation of hamadas is usually exiguous, it is relatively rich in species. Despite considerable regional variation, the vegetation of the hamadas occurring in the Northern, North-Western, Western and Central Sahara has a certain uniformity. Among widespread species the following are characteristic: *Forsskålea tenacissima, Asteriscus graveolens, Salvia aegyptiaca, Farsetia aegyptiaca, Reseda villosa, Fagonia latifolia* and other species, *Enneapogon scaber, E. desvauxii* (*brachystachyus*) and *Anastatica hierochuntica*.

In the Northern Sahara, *Moricandia arvensis, Cymbopogon schoenanthus, Fagonia microphylla* and *Haloxylon scoparium* are particularly characteristic.

In the North-Western Sahara a striking plant, *Fredolia* (*Anabasis*) *aretioides*, is often the only perennial to occur over extensive plateau surfaces. Its dense rounded cushions of tightly packed stems may be up to 50 cm in diameter and almost as high. Other characteristic species include *Limoniastrum feei* and the almost leafless umbellifer, *Pituranthos battandieri*.

In the Central Sahara the hamada flora is virtually absent except on the rocky flanks of the high mountains, where it is characterized by two perennial species, *Anabasis articulata* and *Aerva persica*, and several species of annual grass including six species of *Aristida* and *Stipagrostis* and two of *Enneapogon*.

The rocky outcrops of the Southern Sahara are virtually sterile, except for those on the south-west flanks of Tibesti, where several tropical species occur.

In the Egyptian desert the hamadas are generally without vegetation. Only a few chasmophytes such as *Erodium glaucophyllum, Reaumuria hirtella, Helianthemum kahiricum* and *Fagonia mollis* can be found growing in rock crevices. The stones may be covered with lichens.

Regs
(mapping unit 71)

Refs.: Davis (1953); Kassas (1952); Kassas & Imam (1959).
Phots.: Davis (1953: 2); Kassas (1952: 4, 5); Kassas & Imam (1959: 1–6).

Gravel deserts are formed from heterogeneous deposits, such as conglomerates or coarse alluvium, by the removal of fine material by the wind. Beneath the superficial layer of gravel, gypsum accumulates as a fine powdery mass and may form a strongly cemented hardpan at somewhat greater depth. The surface gravel is commonly so compact that roots can scarcely penetrate, and gravel desert is often without vegetation, especially in the drier parts. Such vegetation as there is is mostly confined to drainage channels.

In the Northern Sahara, *Haloxylon scoparium* is a characteristic species of regs, where it forms very diffuse communities, which have, however, often been degraded by man, since *Haloxylon* provides a useful fuel. Where there is a sandy cover the vegetation is much denser and then three species of *Stipagrostis, S. plumosa, obtusa* and *ciliata*, form a close sward, among which geophytes (*Androcymbium* and *Asphodelus*) and annuals (*Daucus* and *Ammodaucus*) occur. Communities of *Cornulaca monacantha* and *Randonia africana* are found where the sand has a high clay content.

The regs of the Central Sahara are mostly bare, though *Stipagrostis obtusa* may appear after rain.

Much further to the east, the vast pebble desert between Cairo and Suez is virtually without vegetation except in small sand-filled drainage channels, in which the density of the vegetation increases with the depth of the sand. On very shallow sand *Haloxylon salicornicum* is dominant with less than 5 per cent cover. Deeper sands support communities of *Lasiurus hirsutus, Panicum turgidum* and *Zilla spinosa*.

Saharomontane vegetation
(mapping unit 72)

Refs.: Léonard (1969a); Quézel (1965a).
Phot.: Léonard (1969a: 13).

Above aproximately 1800 m the vegetation of the high mountains of the Sahara is markedly different from that at lower altitudes. Relatively few massifs are sufficiently elevated to support distinct Saharomontane types. In the western Sahara these are found only on Ahaggar and satellites Tafedest and Mouydir, and on Tassili n'Ajjer and Tibesti, to which might be added the most northerly summits of Aïr. Saharomontane vegetation similar to that of Ahaggar and Tibesti is not represented on the less-elevated Jebel Uweinat (1900 m), which rises from the Libyan desert much further east; its summit is virtually devoid of vegetation. In the narrow sandstone gorges between 1250 and 1800 m, however, a shrubby community of Mediterranean affinity occurs. This is characterized by species of *Lavandula* and *Salvia*. That this community is not truly montane is shown by the occurrence in it of scattered trees of *Acacia tortilis*.

At least in favourable years, Saharomontane vegetation is diffuse and not confined to wadis, as is the case elsewhere in the Central Sahara. The summits of the high mountains receive much more rain than the surrounding lowlands, as much as 150 mm per year. It often falls as fine rain for several consecutive days and occurs both in summer and winter. At the highest elevation cloud is frequent. The flora is very rich and diversified. Besides several endemic species, the following elements are represented: Mediterranean, Saharan, Sahelian, and Afromontane. The Saharan element plays a subordinate role. The four main types of vegetation are described below.

Saharomontane wadi vegetation

Ref.: Quézel (1965a: p. 226–52).
Phots.: Quézel (1965a: figs. 16, 17, 51–6).

On Ahaggar relatively few larger woody species grow in this type, which begins to appear at about 1 800 m. They all occur also in the Mediterranean Region or are Saharan endemic species of Mediterranean affinity. *Olea laperrinei* is the most characteristic tree of wadis. It is often found on rock detritus associated with the perennial grasses, *Stipa parviflora*, *S. capensis* and *Oryzopsis* (*Piptatherum*) *coerulescens*. Elsewhere it occurs with the bushy *Pistacia atlantica* and *Rhus tripartita*, which locally form dense thickets 3–4 m tall. On gentle slopes of wadis *Olea* is rarer than elsewhere and the low shrub *Artemisia campestris* subsp. *glutinosa* is extensively dominant. The *Olea laperrinei* communities, which provide one of the most striking features on Ahaggar, are completely absent from Tibesti, but recur on Aïr at the southern fringe of the Sahara, and are a prominent feature of the Sahel mountains Jebel Gurgeil and Jebel Marra (page 207).

In moister wadis on Ahaggar and its satellites and on the western slopes of Tibesti, especially near 'gueltas' between 1 500 and 2 300 m, communities characterized by *Myrtus nivellei* and *Nerium oleander* are found.

In still moister situations, where the water-table is at the surface or within 1–2 m, both throughout the central massif of Ahaggar and on the northern slopes of Tibesti, *Tamarix* '*gallica* subsp. *nilotica*' occurs, often with *Nerium oleander*. Many hygrophilous species of northern affinity are also found in these moist places. They include *Scirpus holoschoenus*, *Juncus maritimus*, *J. bufonius*, *Typha australis*, *Phragmites australis* and *Equisetum ramosissimum*.

The remarkable tree, *Cupressus dupreziana*, which reaches a height of 20 m and may have a bole up to 3 m in diameter, is confined to a few rocky valleys in Tassili n'Ajjer, though dead trunks are reported on Ahaggar, and its unmistakeable large pollen grains have been widely found in fossil deposits. The wadis on the moister southern and south-western slopes of Tibesti between 1 600 and 2 300 m have a much richer woody flora than those of Ahaggar. Except for the endemic *Ficus teloukat*, all the larger woody plants are characteristic of the Sahel Zone, and some of these are also widespread even further south. *Acacia seyal* (*stenocarpa*) is the most characteristic species. Other important woody species are *Acacia albida*, *A. laeta*, *A. ehrenbergiana*, *Balanites aegyptiaca*, *Boscia salicifolia*, *Cordia sinensis*, *Ficus ingens*, *F. salicifolia*, *F. sycomorus*, *Grewia tenax*, *Maytenus senegalensis*, *Securinega virosa* and *Rhus incana*. The most abundant grass is *Chrysopogon plumulosus*. Among other smaller plants some are of Mediterranean affinity, e.g. *Globularia alypum* and *Lavandula pubescens*; others are of tropical affinity, e.g. *Abutilon fruticosum* and *Rhynchosia memnonia*.

Saharomontane grassland

Ref.: Quézel (1965a, p. 252–8, 268).
Phot.: Quézel (1965a: pl. 3a, fig. 60).

Grassland, dominated by the perennial species, *Stipagrostis obtusa* and *Aristida coerulescens*, and locally by *Eragrostis papposa*, occupies all rocky slopes between 1 800 and 2 400 m, the beds of wadis above 2 400 m, and plateaux surfaces up to 2 700 m. This grassland consists almost entirely of Saharan species and Saharo–Mediterranean linking species. Several species of low shrub also occur, including *Anabasis articulata*, *Fagonia flamandii* and *Zilla spinosa*.

A dwarf, closely grazed, grassland less than 5 cm tall, dominated by the endemic species, *Eragrostis kohorica*, is found in sheltered places on the summits of Emi Koussi in Tibesti above 3 000 m.

Saharomontane dwarf shrubland

Ref.: Quézel (1965a, p. 259–64).
Phots.: Quézel (1965a: figs. 61, 62).

The summits of Ahaggar and Tibesti above 2 600 m are covered with dwarf shrubland, which is of uniform appearance and is floristically poor. The dwarf shrubs are mostly between 20 and 50 cm (locally up to 1 m) tall and in favourable years ground cover may be as much as 60–70 per cent. On Ahaggar the dominants are *Pentzia monodiana* and *Artemisia herba-alba* (*inculta*), whereas on Tibesti they are *Pentzia monodiana*, *Artemisia tilhoana* and *Ephedra tilhoana*.

The Erica arborea *community*

Ref.: Quézel (1965a, p. 264–8).
Phot.: Quézel (1965a: fig. 64).

The giant heath, *Erica arborea*, a characteristic component of Mediterranean shrubland and Afromontane bushland and thicket, occurs on the higher peaks of Tibesti, where it is confined to narrow fissures in the basaltic lava flows. The floor of these fissures is permanently moist and is covered with a community of mosses containing as many as 24 species.

Halogypsophilous vegetation
(mapping unit 76)

Ref.: Quézel (1965a, p. 57–78).
Phots.: Quézel (1965a: 18–20).

In the Sahara, salt accumulation takes place wherever water evaporates from depressions without drainage. Salt pans are formed which are often covered with a white crust and are known as 'sebkhas' or 'chotts'. Saline soils play only a minor role in the driest parts of the Sahara. The rainfall there is so small that salts are neither removed nor deposited except locally in the largest wadis. Elsewhere, they are found in every locality where ground water wets the surface by capillary action.

The chemical composition of the deposits varies from place to place and depends on the kinds of salt dissolved in the ground water. White sheets of salt can often be seen in the depressions between dunes.

Although very large saline depressions (chotts) occur towards the northern limits of the Sahara, they are much more common outside the Sahara proper in the Mediterranean–Sahara transition zone. Their extensive development there is dependent more on the occurrence of salt-bearing strata than on climatic factors.

For the western Sahara the available information on halogypsophilous vegetation is summarized by Quézel (1965a). Halophytic and gypsophilous communities are much better developed and are more diversified and richer in species in the Northern and North-Western parts of the western Sahara and along its Atlantic fringes than in the Central and Southern parts. Their virtual absence from the Central Sahara is largely due to the low rainfall, but this explanation does not apply to the Southern Sahara. Quézel suggests that other climatic factors (higher temperatures combined with summer rainfall) and the nature of the salts (predominantly carbonates, not chlorides) are both responsible for the paucity of halophytic vegetation in the Southern Sahara.

Even in the Northern Sahara halogypsophilous vegetation occupies only small areas. The most important occurrences are the chotts of South Constantine and South Tunisia but there are numerous smaller sebkhas throughout the Northern and Western Sahara occurring in the proximity of oases and in basins without outlets to the sea.

There is usually a well-marked zonation of the vegetation surrounding the sterile salt-encrusted interior of the chotts and sebkhas. Floristic composition is related to water-regime and salt content.

In the Northern Sahara three main types of halogypsophilous vegetation can be recognized.

Hyperhalogypsophilous vegetation

This type is dominated by *Halocnemum strobilaceum*. It occurs in the sebkhas of South Constantine, Southern Tunisia, and in the Rhir wadi. *Halocnemum* is an extreme halophyte which occurs in the central part of sebkhas which dry out during the summer, or, if the central part is devoid of vegetation, forms a belt around the zone of salt encrustation. *Halocnemum* is always dominant and often occurs in pure stands. It is a low shrub which covers from 55 to 70 per cent of the surface.

Halogypsophilous vegetation on drier soils

The larger chotts at the northern edge of the desert are fringed with halogypsophilous vegetation, characterized by the low shrubs *Salsola sieberi* and *Zygophyllum cornutum*. Cover varies from 25 to 60 per cent.

Further south in similar situations in the valley of Wadi Rhir comparable vegetation is characterized by *Zygophyllum album* and *Traganum nudatum*. This type is also very widespread elsewhere in the Northern Sahara, where it occupies gypsaceous areas with only small amounts of soluble salts and often with a thin cover of wind-blown sand. Cover varies from 15 to 40 per cent. Other characteristic species include *Limoniastrum guyonianum*, *Suaeda mollis*, *Limonium pruinosum*, *Nitraria retusa* and *Salsola tetragona*.

Vegetation of gypsaceous loamy sands

These sands occur locally in the beds of wadis and on ancient alluvial terraces. They are characterized by the low shrubs *Suaeda vermiculata* and *Salsola baryosma* (*foetida*) and the larger shrub *Tamarix* '*brachystylis*'. In the North-Western Sahara there are two main types:

1. In the beds of wadis, where brackish water occurs at a depth of 50–100 cm, communities characterized by *Limoniastrum ifniense*, *Nitraria retusa*, *Suaeda ifniensis*, *Atriplex halimus* and *Tamarix* sp. are found. Where the water-table is much closer to the surface, especially in the beds of the larger wadis, *Halocnemum strobilaceum*, *Arthrocnemum indicum*, and *Salicornia arabica* appear. On less saline slopes of the valleys *Suaeda monodiana*, *Salsola baryosma*, *Nucularia perrinnii* and *Lycium intricatum* predominate.
2. The halophytic vegetation of the sebkhas of the Western Sahara is extremely poor in species and often consists of a single species of *Zygophyllum*, either *Z. gaetulum* or *Z. waterlotii*. Cover varies from 5 to 60 per cent.

Little is known of the halophilous vegetation of the Atlantic Sahara. Floristically it is relatively rich, with species of *Zygophyllum*, *Lycium*, *Salsola*, *Suaeda* and *Frankenia*. Where the surface of the soil is moist *Arthrocnemum indicum* is important.

Halophilous vegetation is very rare in the Central Sahara. Outside the mountainous massifs, because of the severity of the climate, vegetation is almost totally excluded from saline and gypsaceous depressions, which in any case are not frequent. Very small saline areas occur in some of the larger wadis. Here, various species of *Tamarix*, and *Suaeda fruticosa*, may occur. Nearly all the species which characterize halogypsophilous soils in the Northern Sahara are completely absent from the Central Sahara. The only species which remain are *Juncus maritimus*, *Cyperus laevigatus* and *Phragmites australis*.

In the Southern Sahara saline and natron soils cover a large area but are mostly devoid of vegetation. The only community in this region which can tolerate a certain degree of salinity is that characterized by *Sporobolus robustus* and *Hyphaene thebaica*.

Absolute desert
(mapping unit 67)

Absolute desert occurs where the rainfall in very low (less than 20 mm per year) and is episodic so that many consecutive years are without rain. A single shower can support annuals on shallow sand, but perennials can grow only on water-receiving sites where the soil remains damp for several years without rain. Absolute desert is much more extensive in the eastern Sahara than in the western. Walter (1971) did not see a single plant for a distance of over 200 km in the southern Libyan desert. In the Libyan desert, J. Léonard (pers. comm.) on a journey of 600 km saw no plants except the two perennial species *Stipagrostis zitellii* and *Cornulaca monacantha*, which were widely spaced. The surface sand, however, contained abundant seeds of annuals which germinated when moistened.

The Atlantic coastal desert
(mapping unit 68a)

Ref.: Guinea (1949); Quézel (1965a, p. 154–9).

The Atlantic Coastal Desert extends as a narrow band up to 40 km wide from Saguia el Hamra at the southern limit of the succulent shrubland (page 228) of the Mediterranean–Sahara transition zone in the north to the northern limit of the Sahel zone in the south. Rainfall is low but mists are frequent and permit the growth of crustaceous and even fruticose lichens, on the branches of shrubs, and a dense growth of fruticose lichens (*Ramalina* sp.) on the otherwise bare ground between the vascular plants. The vegetation cover, which is denser than that of most parts of the Sahara, is relatively rich in species. The succulent shrubs which characterize the vegetation immediately to the north almost disappear, quite suddenly, at Seguia el Hamra but some of them, e.g. *Euphorbia regis-jubae* and *Senecio anteuphorbium,* have a scattered distribution further south, and *Euphorbia echinus* extends even as far south as Vila Cisneros (Guinea, 1949).

Other Mediterranean–Saharan linking species which are common in the northern part of this coastal strip include *Lycium intricatum* and *Launaea arborescens*. In the southern part of the coastal Sahara, in Mauritania, Sahelo–Saharan linking species such as *Acacia tortilis, Salvadora persica* and *Balanites aegyptiaca* are well represented.

Other common species in this zone are mostly halophytes or Saharan endemics and near-endemics including *Anabasis articulata, Arthrocnemum glaucum, Stipagrostis pungens, Panicum turgidum, Cornulaca monacantha,* and species of *Salsola, Suaeda, Atriplex* and *Tamarix*.

The Red Sea coastal desert
(mapping unit 68b)

Ref.: Kassas & Zahran (1965).
Phots.: Kassas & Zahran (1965: 1–7).

The Red Sea coastal plain, which is 15–20 km wide, receives very little rainfall. At Hurghada the mean annual rainfall is only 3 mm and most years are rainless, though as much as 41 mm has been known to fall on a single day. Apart from halophytic communities on the littoral itself, the plain is devoid of vegetation except in the wadis. Inland from the coastal plain a chain of rugged mountains, with peaks up to 2 184 m high, runs the entire length of the Red Sea. Their summits intercept cloud moisture in the form of orographic rain or condensation; this feeds permanent springs or 'nakkat' and contributes to the water revenue of tunnels and wadis associated with the mountains.

Littoral salt marshes are characterized by *Arthrocnemum glaucum, Halocnemum strobilaceum, Zygophyllum album, Nitraria retusa* and *Suaeda monoica*. Cover varies from 5 to 100 per cent.

In the wadis of the coastal plain, saline areas have a dense growth of *Juncus arabicus* and *Tamarix 'mannifera'*. Elsewhere in the wadis, *Acacia tortilis, Zilla spinosa, Capparis decidua, Calligonum comosum, Lasiurus hirsutus, Panicum turgidum* and *Retama retam* are characteristic species.

The springs in the mountains provide a habitat for ferns (*Adiantum capillus-veneris*), bryophytes, *Ficus pseudosycomorus* and other water-loving plants such as *Phragmites australis* and *Imperata cylindrica*. The wadis in the mountains have a rich flora. Most species found in the wadis of the lowland plain also occur in the mountains. But there are several others, including the small tree, *Moringa peregrina*. Since the oil obtained from the seeds of the latter is highly prized, the plant is too valuable to be cut for fuel.

A floristically depauperate variant of East African evergreen bushland (mapping unit 38), which contains *Olea africana, Euclea racemosa* subsp. *schimperi* and *Dracaena ombet*, occurs on Gebel Elba in the south-east corner of Egypt.

XVIII The Mediterranean/Sahara regional transition zone

Area, geographical position, geology and physiography

Climate

Flora

Mapping units

Vegetation
    Sub-Mediterranean forest
    Argania spinosa scrub forest and bushland
    Acacia gummifera–Ziziphus lotus bushland
    Succulent sub-Mediterranean shrubland
        Euphorbia resinifera succulent shrubland
        Euphorbia res-jubae, E. beaumierana succulent shrubland
        Euphorbia echinus succulent shrubland
    Sub-Mediterranean grassland
    Sub-Mediterranean halophytic vegetation
    Sub-Mediterranean anthropic landscapes

Area, geographical position, geology and physiography

The eastern part of the transition zone is mostly low lying and consists principally of the Mediterranean coastal plain, extending from the so-called 'Sahel' of Tunisia eastwards to the Suez Canal. In Cyrenaica, Gebel Akdhar rises to 878 m. (Area: 107 000 km².)

The western part in the southern Maghreb is much more diversified. The largest area is occupied by the High Plateau, an undulating region, principally in Algeria, between the Tell Atlas and the Saharan Atlas. Its elevation is mostly between 750 and 1 000 m. On the plateau are extensive shallow semi-permanent salt lakes or chotts. Further west the transition zone is narrow where it occupies the lower southern slopes of the High Atlas and the Anti-Atlas. At its western end it broadens considerably to include the Atlantic coastal plain between Safi and Cabo Yubi, the lowlands of Haouz-Tadla and the Souss and the whole of the western end of the Anti-Atlas. (Area: 366 000 km².)

The lithology is diverse. The Anti-Atlas is largely composed of basement rocks and the high plateau is covered with mostly unconsolidated clay marls of Upper Tertiary age. Elsewhere the substrate consists principally of Cretaceous, Lower Tertiary, Miocene, and Quaternary sediments.

In the legend to the map and in places in the text the term 'sub-Mediterranean' is applied to this transition zone.

Climate

Except very locally, rainfall is between 100 and 250 mm per year. Precipitation is mostly concentrated in the winter months, but in the rain shadow of the Atlas Mts and on the High Plateau the main peaks are in spring and autumn, or rainfall is irregular throughout the year.

Temperature varies appreciably from place to place. At the western and eastern ends of the Maghreb mean annual temperatures are mostly between 18° and 20°C and frosts are not very severe. By contrast, on the High Plateau mean annual temperature is between 13° and 17°C, frosts are severe, and the frosty season may last up to eight months.

East of southern Tunisia the coastal belt is frost-free. (See Fig. 13.)

Flora

The flora is relatively poor. There are possibly no more than 2500 species. Only a few are endemic.

The monotypic genus *Argania* is almost confined to the western end of the transition zone in Morocco. Several other endemic or near-endemic species have a similar distribution. They include *Acacia gummifera* and the succulent Euphorbias *E. resinifera*, *E. beaumierana* and *E. echinus*. All these species belong to genera or infra-generic groupings that are absent from typical Mediterranean vegetation. Their closest relatives are found in tropical Africa or even further south. Some of their associates, including *Euphorbia regis-jubae*, *Aeonium arboreum* and *Sonchus pinnatifidus*, also occur in the Canary Islands.

Most species in the Mediterranean–Sahara transition zone are Mediterranean or Saharan wides or have even wider distributions. Apart from the western-Morocco endemics just mentioned, the only other important group of local endemics is that occurring on Gebel Akhdar in the Libyan Arab Jamahiriya. Its one hundred or so endemic species include *Arbutus pavarii*, *Crocus boulosii* and *Cyclamen rohlfsianum* (Bartolo et al., 1977).

According to Boulos (1975), 1095 species are known from the coastal strip of Egypt, but few are typical Mediterranean species. Mediterranean trees are completely absent, other than *Ceratonia* and the cultivated olive, as are many other circum-Mediterranean species such as *Rosmarinus officinalis*, *Spartium junceum*, *Calicotome villosa* and the entire genus *Cistus*. A high proportion of the species that do occur are annuals and many are widely distributed weeds.

The coastal strip of the Libyan Arab Jamahiriya is slightly richer, with about 1440 species. Most occur on Gebel Akhdar, which supports typical Mediterranean vegetation and might possibly be considered as an exclave of the Mediterranean Region proper. Although the coastal strip of Tripolitania has many more true Mediterranean species than that of Egypt, they are still relatively insignificant.

Mapping units

10 (p.p.). Mediterranean sclerophyllous forest.
49. Transition from sub-Mediterranean *Argania* scrubland to succulent semi-desert shrubland.
55. Sub-Mediterranean semi-desert grassland and shrubland.
73 (p.p.). Oasis.
76 (p.p.). Halophytic vegetation (see Chapter XXII and below).
79. Western sub-Mediterranean anthropic landscapes.
80. Eastern sub-Mediterranean anthropic landscapes.

Vegetation

There is appreciable change from west to east. In western Morocco the prevalent types are scrub forest and bushland dominated by *Argania spinosa*, and *Euphorbia*-dominated succulent shrubland.

From eastern Morocco to Tunisia the landscape is dominated by a mosaic of grassland consisting almost entirely of *Stipa tenacissima* or *Lygeum spartum*, alternating with patches of dwarf *Artemisia* shrubland. Much of this region is thought to have been formerly forest. Patches of forest consisting principally of *Pinus halepensis*, *Juniperus phoenicea* and *Quercus ilex* remain, especially in the mountains.

In two areas, one in Morocco, the other in Tunisia, the natural vegetation has been almost entirely replaced by cultivation (mapping unit 79). In the former, the Haouz–Tadla region, the original vegetation was probably *Acacia gummifera*, *Ziziphus lotus* bushland. It includes the large oasis of Marrakech, which is surrounded by extensive groves of the date palm, *Phoenix dactylifera*. Otherwise this species is almost confined to the Sahara.

Further east in the Libyan Arab Jamahiriya and Egypt the vegetation nearly everywhere has been severely degraded, and, except for an important exclave of Mediterranean forest, has been shown (mapping unit 80) as an anthropic landscape.

Saline depressions (chotts) bearing dwarf halogypsophilous shrubland occur throughout the transition zone wherever there are outcrops of salt-bearing strata.

Sub-Mediterranean forest

In the western part of the transition zone the mountains of the Saharan Atlas enjoy a more humid climate than the High Plateau and have a vegetation similar to that of the mountains to the north and west. The topography is dominated by anticlinal ridges, mainly of calcareous rock, running from south-west to north-east. The slopes are gentle, allowing easy penetration by firewood-gatherers and graziers. The surviving forests are dominated by *Quercus ilex* and *Pinus halepensis*, the latter being more resistant to forest fire and less sought after as a fuel. Associated tree species include *Juniperus oxycedrus*, *Juniperus phoenicea*, *Olea europaea* and *Pistacia atlantica*. Characteristic elements of the ground flora are *Globularia alypum*, *Rosmarinus eriocalyx*, *Leuzea conifera* and most notably *Stipa tenacissima*, which becomes the dominant plant where the forest has been destroyed (P. J. Stewart, pers. comm.).

Further east the richest forests floristically, though they are severely degraded, are those of Gebel Akhdar in Cyrenaica, which rises to 878 m and receives an annual rainfall of 300–600 mm. They contain *Arbutus pavarii*, *Ceratonia siliqua*, *Cupressus sempervirens*, *Juniperus oxycedrus*, *J. phoenicea*, *Laurus nobilis*, *Olea europaea*, *Phillyrea angustifolia*, *Pinus halepensis*, *Pistacia lentiscus* (but not *P. atlantica*), *Quercus coccifera* and *Q. ilex*. None of these species, except *Olea* and *Ceratonia*, has been recorded from the coastal plain of Egypt (Wickens, 1977b). In Tripolitania there are only a few relict stands of *Ceratonia*, *Olea*, *Pistacia atlantica* and *P. lentiscus*.

Argania spinosa scrub forest and bushland
(mapping unit 49)

Refs.: Boudy (1948, p. 142–3; 1950, p. 382–416); Emberger (1939, p. 63–6, 101); Métro (1958, p. 84–6); Sauvage (1948); White (MS, 1974).
Phots.: Boudy (1950: 47–55); Emberger (1939: 3); Métro (1958: 9).

Argania spinosa does not normally occur as a tree with a single well-defined trunk, but, at its optimum, is a 10 m tall multiple-stemmed bushy tree with massive contorted boles, or, when single-stemmed, it branches low down and has stout wide-spreading branches. Its most luxuriant stands are best regarded as scrub forest, although in drier areas it is much shorter and then forms bushland. *Argania* is usually evergreen, but sheds its leaves in very dry summers.

Argania spinosa is the only member of the tropical family Sapotaceae to occur on the mainland of Africa north of the Sahara. It takes its generic name from the village of Argana on the slopes of the High Atlas and is confined to Morocco, where it covers enormous areas in the arid and semi-arid 'étages'. It is the dominant tree in south-west Morocco, and grows on all types of soil except mobile sand. With increasing humidity it gives way to *Olea–Pistacia* bushland or *Tetraclinis* forest. This is so in the hinterland of Essaouira (Mogador), where *Argania* is replaced by *Tetraclinis* in all the hollows. On mountain slopes the altitude at which it is replaced by *Tetraclinis* is very variable and depends on humidity alone near the sea, and a combination of humidity and winter cold further inland. It does not ascend above 1400–1500 m, which corresponds with the lower limit of winter snow.

The main area occupied by *Argania* extends from Safi to Oued Noun, and penetrates far inland in the plain of the Souss and on the lower southern slopes of the High Atlas and the north-western slopes of the Anti-Atlas. South of Oued Noun it is much rarer and is practically confined to wadis, reaching its southern limit just south of Oued Drar. Well outside and to the north of the main area, there are also some small populations south-west of Rabat and in the Moulouya valley near Berkane. They testify to its former much more extensive distribution, during a drier epoch.

Virgin *Argania* forest is dense, with a shrubby understorey which is difficult to penetrate. This type has virtually disappeared, except for small remnants surrounding certain shrines. The photographs published by Boudy (1950) give an idea of its former luxuriance. Most *Argania* forest, even on stony hillsides, has been converted into an orchard of widely spaced trees between which cereals are cultivated during the rainy season. The trees are left standing for the sake of their seeds, which yield a highly prized edible oil. On uncultivated slopes most *Argania* forest has been severely degraded by flocks of goats, which not only destroy the understorey and prevent regeneration, but also climb into the crown of *Argania* and browse its branches. Most of the surviving Arganias in the semi-desert country south of Goulimine have been severely pruned by browsing camels into bizarre unnatural shapes.

Some *Argania* forests are currently undergoing severe soil erosion. In places, on the southern slopes of the High Atlas for instance, up to 0.6 m of soil has been lost within the lifetime of quite small individuals of an associated species, *Ziziphus lotus*, the roots of which have locally held the soil in place.

In the semi-arid *Argania* forests of the littoral the following associates occur: *Acacia gummifera, Clematis cirrhosa, Cytisus albidus, Ephedra altissima, Euphorbia beaumierana, Genista ferox, Helianthemum canariense, Lavandula dentata, L. multifida, Olea europaea, Periploca laevigata, Pistacia lentiscus, Rhamnus oleoides, Rhus oxyacantha, R. pentaphylla, Senecio anteuphorbium* and *Withania frutescens*. Some of these extend into the arid 'étage', but others drop out, notably *Pistacia lentiscus, Rhus pentaphylla*, and *Euphorbia beaumierana*, and are replaced by *Ziziphus lotus, Euphorbia echinus, Pistacia atlantica* and *Laburnum platycarpum*. The succulent *Euphorbia, E. echinus*, is often abundant, especially in degraded types towards the south, and sometimes persists after *Argania* has disappeared, though not all *Euphorbia* shrubland has originated in this way.

North of Agadir, on rocky coastal slopes and cliff tops, *Argania* occurs as widely spaced, severely wind-pruned individuals no more than 2 m high in a matrix of succulent *Euphorbia* shrubland (see below).

Acacia gummifera–Ziziphus lotus bushland
(mapping unit 79 p.p.)

Ref.: Emberger (1939, p. 66–7).
Phot.: Emberger (1939: 3.2.).

According to Emberger the climax vegetation of Haouz-Tadla, which has an arid Mediterranean climate, was a bushland or thicket dominated by *Ziziphus lotus, Withania frutescens, Acacia gummifera* and *Pistacia atlantica. Acacia gummifera*, which is often no more than 1–2 m high, sometimes occurs as a 5–6 m high bushy tree. It is endemic to Morocco and, besides occurring extensively in the upper basins of Oued Oum and Oued Tensift, ascends the foothills of the Central Atlas to 1100–1200 m. Here it occurs scattered in *Euphorbia resinifera* succulent shrubland, which may be largely secondary having replaced *Tetraclinis articulata* forest. It also occurs on the southern slopes of the Anti-Atlas but does not extend much further south than Goulimine. Its range does not overlap with the desert Acacias mentioned in Chapter XVII.

Most of Haouz-Tadla is cultivated and only tiny degraded remnants of the original vegetation remain, in which the trees are accompanied by *Ephedra altissima, Asparagus stipularis, Lavandula multifida, Ballota hispanica, Bryonia dioica, Peganum harmala*, and many therophytes including *Calendula algeriensis, Diplotaxis tenuisiliqua, Reseda battandieri, Ononis polysperma* and the grasses *Stipa capensis (tortilis), Lamarckia aurea,*

Bromus madritensis and *B. rubens*. In places *Stipa capensis* covers enormous areas. Saline soils are frequent, and *Acacia gummifera* and its entourage are replaced there by halophytes, of which *Atriplex halimus* and *Lycium intricatum* are dominant, and *Salsola vermiculata*, *Suaeda fruticosa* and *Sphenopus divaricatus* (*gouanii*) occur in the most saline parts.

Succulent sub-Mediterranean shrubland
(mapping units 10 p.p., 49, 55, and 79 p.p.)

Refs.: Maire & Emberger (1939); Sauvage (1948, p. 118–24; 1971, p. 60–3, 72); White (MS, 1974).
Syn.: la steppe à 'daghmous' (*Euphorbia echinus*) (Sauvage).

Three succulent shrubby species of *Euphorbia* occur in the south-western corner of Morocco, frequently as dominants. Collectively they are confined to the arid 'étage' and the drier, low-lying parts of the semi-arid 'étage', with, in the case of two species, a considerable southwards extension into the Atlantic Coastal Sahara. They reach their maximum development at the western end of the Mediterranean–Sahara transition zone, where mists are frequent and summer temperatures are ameliorated because of proximity to the sea.

The three species *E. resinifera*, *E. beaumierana* and *E. echinus* are cactoid shrubs with thick, succulent, ridged, spinous stems. They form dense clumps and normally grow 0.6–1 m tall. They replace each other geographically.

E. resinifera grows on the lower slopes of the Middle Atlas for a considerable distance to the south-west of Beni Mallal.

E. beaumierana is confined to the coastal belt northwards from the Souss to a point about 100 km S. of Essaouira.

South of the Souss, *E. beaumierana* is replaced by *E. echinus*, which is particularly abundant between Oued Noun and Oued Dra, but also extends southwards into the coastal Sahara, with diminishing frequency, as far as Cap Blanc in Mauritania.

A fourth species, *E. regis-jubae*, which is usually associated with either *E. beaumierana* or *E. echinus*, is a pachycaul treelet c. 2 m tall. Its soft semi-succulent stems show candelabra-branching and bear narrow summer-deciduous leaves c. 15 cm long. It extends from Safi to a considerable distance south of Oued Drar. It is also locally abundant in the Canary Islands.

All four species occur in degraded forest and scrub forest, especially *Argania* scrub forest, and in places it is quite clear that succulent *Euphorbia* shrubland has increased considerably at the expense of forest, following misuse of the latter. However, not all *Euphorbia* shrubland occurring in the forest regions should be regarded as secondary. On shallow rocky soils and in wind-swept places near the sea, these species represent a local edaphic or climatic climax. To the south of the *Argania* zone, there can be little doubt that *Euphorbia* shrubland represents a climatic climax wherever suitable soils occur.

Euphorbia resinifera *succulent shrubland*

On shallow, often calcareous, lithosols within the zone of *Tetraclinis articulata* forest, on the foothills of the Atlas Mts south-west of Beni Mellal, *Euphorbia resinifera* occurs as a local dominant. Where the forest has been destroyed it is more widespread. It covers about 60 per cent of the surface; elsewhere there is mostly bare rock. Bushes and small bushy trees are sometimes present but their cover is small, and *Euphorbia* is physiognomically dominant. The principal larger woody plants are *Acacia gummifera*, *Ceratonia siliqua*, *Chamaerops humilis*, *Olea europaea*, *Pistacia lentiscus*, *Rhus pentaphylla*, *Tetraclinis articulata* and *Ziziphus lotus*. Small species belong mostly to typical Mediterranean genera such as *Asphodelus*, *Ballota*, *Biscutella*, *Cistus*, *Ferula*, *Hippocrepis*, *Lavandula* and *Ruta*.

Euphorbia res-jubae, E. beaumierana *succulent shrubland*

At various places on the Atlantic coast between Safi and Agadir succulent shrubland, dominated by one or other but usually both of these species, occurs on steep rocky slopes leading down to the sea, or on flat, windswept cliff-tops covered with stone detritus. Although tree species are often present they are usually small and severely wind-trimmed, often less than 2 m tall, and contribute less to the phytomass than *Euphorbia*. On very steep rapidly eroding slopes *Euphorbia* covers less than 10 per cent of the surface. In flatter places there may be 50 per cent cover. The older stems of the Euphorbias are sometimes covered with lichens. The principal associated bushes and trees are *Acacia gummifera*, *Argania spinosa*, *Ceratonia siliqua*, *Maytenus senegalensis*, *Olea europaea*, *Phillyrea angustifolia* and *Pistacia lentiscus*. Two soft-stemmed 2 m tall pachycaul treelets, *Aeonium arboreum* and *Sonchus pinnatifidus*, which also occur in the Canary Islands, are sometimes present, as are the climbers *Periploca laevigata* and the succulent-stemmed *Senecio anteuphorbium*. Small shrubs and chamaephytes include *Helianthemum canariense*, *Lavandula dentata*, *L. maroccana*, *L. multifida* and *Withania frutescens*.

Euphorbia echinus *succulent shrubland*

This community in the region of Goulimine has been described by Sauvage (1948). The most characteristic habitat of *Euphorbia echinus* is rocky hillsides, from which it descends to lower slopes covered with pebbles and gravel. It avoids both clayey and sandy soils, and, especially, places liable to flooding. In this region the landscape is a mosaic of *Euphorbia* shrubland on permeable soils and dwarf *Haloxylon scoparium* shrubland where the soil is impermeable. According to Sauvage, a large part of the *Euphorbia* shrubland is the result of degradation of a community dominated by *Argania spinosa*. Since, however, the latter is at the southern limits of its distribution there, and the rainfall is only about 125 mm per year, it is likely that the original vegetation was no more than a very open stunted *Argania* bushland in which *Euphorbia echinus* was abundant. South of Goulimine, for a distance of about 20 km, *Argania* is found away from watercourses but only as very widely spaced bushes, which have assumed fantastic shapes as a result of browsing camels. Further south, it is very rare and is strictly confined to the banks of seasonal watercourses, reaching its southern limit just beyond Oued Drar. There can be little doubt that towards the southern limits of the range of *Argania*, the *Euphorbia* communities are not secondary and they do, of course, also extend much further south than *Argania*.

South of Goulimine, *Euphorbia echinus* is usually 0.3–0.5 m tall. In the more oceanic parts of its range its older stems are often covered with lichens and it is frequently associated with *E. regis-jubae*. The following are its principal associates:

— Shrubs and bushy trees: *Acacia gummifera* (very rare), *Argania spinosa* (only in north), *Euphorbia regis-jubae*, *Launaea arborescens*, *Lycium intricatum*.

— Climbers: *Asparagus pastorianus*, *Periploca laevigata*, *Senecio anteuphorbium* (stem succulent).

— Chamaephytes: *Anabasis aphylla*, *Artemisia herba-alba*, *Chenolea tomentosa*, *Convolvulus trabutianus*, *Frankenia corymbosa*, *Haloxylon scoparium*, *Salsola sieberi*, *S. tetragona*, *S. vermiculata*, *Suaeda ifniensis*, *S. mollis*, *Traganopsis glomerata*, *Zygophyllum gaetulum*.

— Hemicryptophyte: *Limonium fallax*.

— Parasites: *Cynomorium coccineum*, *Cistanche phelipaea*.

— Therophytes: *Aizoon canariense*, *Asphodelus tenuifolius*, *Calendula murbeckii*, *Eryngium ilicifolium*, *Fagonia cretica*, *Linaria sagittata*, *Matthiola kralikii*, *Sclerosciadium nodiflorum*.

In the northern half of its range between Bou Izakarn and the Souss, *E. echinus* is much less plentiful and mostly occurs in *Argania*-dominated communities. Here the *Argania* is growing under optimum conditions and reaches a height of 9 m and *Euphorbia* is associated with many species belonging to typical Mediterranean genera, such as *Genista* and *Lavandula*, which are absent further south.

Sub-Mediterranean grassland
(mapping unit 55)

Refs.: Boudy (1950, p. 773–818); Emberger (1939, p. 67–9); Métro (1958, p. 96–8); Ozenda (1954, p. 210–15); White (MS, 1974).

Phots.: Boudy (1950: 130, 131); Emberger (1939: 4.2); Ozenda (1954: 3); Quézel & Santa (1962: 10).

Large parts of the Mediterranean/Sahara transition zone are covered with tussock grasslands of considerable economic importance, the origin of which is still a matter of controversy. They are dominated by two species, *Stipa tenacissima*, the Esparto grass or 'Alfa', and a second grass of similar but more diffuse growth-form, *Lygeum spartum* or 'Sparte'. Two dwarf shrubby composites, *Artemisia herba-alba* ('Armoise' or 'Chih') and *A. campestris*, dominate communities which occur in mosaic with *Stipa* and *Lygeum* and form a pattern which partly reflects soil differences, partly the history of the site. Saline areas are dominated by *Atriplex halimus* and *Salsola vermiculata*, sometimes associated with *Lygeum*.

Stipa tenacissima is characteristic of the driest parts of the Maghreb. On the High Plateau of Algeria and Morocco it dominates the landscape as far as the eye can see, as it also does in the Moulouya Valley and the drier parts of the Mediterranean plain. It also occurs more sporadically in west Morocco, southern Portugal, east and southern Spain, the Balearic Islands and the Libyan Arab Jamahiriya. In North Africa its southern limit, except on specially favoured sites, more or less coincides with the 100 mm isohyet, which Quézel (1965a) uses to demarcate the northern limit of the Sahara, but further south it penetrates the northern fringes of the Sahara as far as Tilrempt, on the banks of seasonal watercourses.

Stipa tenacissima only grows where the rainfall is less than 500 mm per year. Where it is higher than this it is replaced by *Ampelodesma mauritanicum*. *Stipa* is just as intolerant of an excess of soil moisture as of high atmospheric humidity. It is particularly abundant between the 200 and 400 mm isohyets. Where the rainfall is less than 200 mm per year it becomes much rarer.

Alfa is extremely resistant to cold. It can withstand temperatures as low as $-10°$ to $-16°C$ and prolonged covering by snow. Throughout the winter the leaves function wherever the temperature reaches $3°–5°C$. Every year there are two seasons of retarded growth, each lasting about 3–4 months, one caused by summer drought, the other by winter cold. The new leaves emerge when the first rain falls.

S. tenacissima is very widely distributed in the semi-arid 'étage' in the field layer of more or less open forest dominated by *Juniperus phoenicea*, *Pinus halepensis*, *Tetraclinis articulata* and *Quercus ilex*. In this community it extends as far west as the hinterland of Essaouira (Mogador). But it is on the virtually treeless High Plateau that it achieves its maximum development. The destruction of forest in the semi-arid 'étage' favours the extension of Alfa, almost pure stands of which represent their ultimate degradation. It has usually been thought, however, that most of the Alfa grasslands of the High Plateau did not originate in this way but represent a climatic climax. Thus, Emberger believed that the treelessness of the plateau was due to a combination of winter cold and extreme aridity accentuated by strong winds. It now appears, however, that the grasslands on the High Plateau are also secondary (A. M. Monjauze, MS). Throughout the High Plateau there are outcrops of rock which support *Juniperus phoenicea*, *Olea europea*, *Rosmarinus eriocalyx* and *Globularia alypum*. These are all characteristic species of *Pinus halepensis* forest, though that species does not occur naturally on the plateau today. It has, however, been planted in several places, and has been shown to be capable of invading

Alfa grassland when the latter is fenced to exclude grazing animals.

Confirmation that the Alfa grassland of the High Plateau is secondary comes from its reproductive behaviour. Inside the forest, *Stipa tenacissima* reproduces regularly by seed and on deep forest soils c. 4 seedlings per square metre can often be found. On the plateau, by contrast, seedlings are scarcely ever found and reproduction is entirely vegetative. On balance the evidence favours the hypothesis that the climax vegetation on the High Plateau is a forest or scrub forest dominated by species of the arid and semi-arid 'étages', and that the reduction of Alfa grasslands is attributable to man and his flocks. The very poor flora of Alfa grasslands supports this view. Apart from ephemerals which are not visible every year, Alfa is often the only species present over extensive areas.

Alfa is a tussock grass with long narrow evergreen sclerophyllous leaves. Its growth-form is similar to that of the dominants of grassland in the Ericaceous and Afroalpine belts of the high mountains of tropical Africa (Chapter VIII) and Madagascar (Chapter XIX).

From the original Alfa tuft, rhizomes slowly extend radially and give rise to a ring of daughter tufts which, after the death of the mother tuft, become independent of each other. From a distance, the formation looks closed but ground cover is small and much bare soil is exposed. In the denser stands there are 3000–5000 tufts per hectare, but when degraded there may be no more than 1000–2000. The leaf-blades are 30–120 cm long and have strong nerves and a waxy covering. During moist weather the blade is flat and ribbon-like but at other times the blades are tightly inrolled. They remain photosynthetically active for 2 or 3 years, after which they persist for years as dead or dying leaves and form a grey mat from which the young leaves emerge. The leaves contain cellulose fibres and this is the basis of the important Esparto-grass industry.

The distribution of *Lygeum spartum* in the Maghreb is similar to that of Alfa but it is not quite as widespread. It, too, only penetrates the northern fringes of the Sahara. On the sandy shores of the Mediterranean it extends as far east as Egypt.

Artemisia herba-alba is a Mediterranean–Sahara linking species which is most abundant on the High Plateau of the transition zone.

The relative abundance of *Stipa tenacissima*, *Artemisia herba-alba*, and *Lygeum spartum* and the degree to which they intermingle varies from place to place. Where the relief is feeble, mixed communities may be more extensive than the pure types. Where the relief is more accentuated, however, the three dominants are often mutually exclusive. Characteristically, *Stipa* occurs on rather coarse, often gravelly, freely drained soils. By contrast, *Artemisia herba-alba* is dominant on clay soils and depressions where water accumulates. It is frequently stated that *Lygeum* occurs in similar situations to *Artemisia* but only covers a small area. Ozenda (1954), however, has shown that *Lygeum* normally occurs on a much sandier soil than *Artemisia*, and, at least on the southern border of the High Plateau eastwards from Djelfa, it covers a considerable area. In this region, taken as a whole, if mixed communities are excluded from consideration the three dominants are more or less equally abundant, though elsewhere *Lygeum* is the rarest species. Where, because of warping of the earth's crust, drainage is impeded and there is no outlet, *Artemisia* dominates the landscape, but where the drainage is slightly incised it is restricted to quite narrow channels. In the region studied by Ozenda, Alfa is characteristic of soils derived from Cretaceous rocks, both on the mountain chains and in areas of feeble relief, whereas *Lygeum* is usually found on Quaternary alluvium.

At the edges of saline depressions on Quaternary alluvium, *Lygeum* is associated with *Atriplex halimus* or *Salsola vermiculata*. Where the soil becomes more sandy its associates include *Thymelaea microphylla* and *Ferula*, in a facies which is transitional to *Stipagrostis pungens* grassland. In the Djelfa region, *Lygeum* grassland is often cultivated. When cultivation is abandoned the land is invaded, not by *Lygeum*, but by *Artemisia campestris*. Some *Lygeum* grassland is transitional towards *Stipa* grassland and then shares several associates with the latter, such as *Alyssum serpyllifolium*, *Helianthemum pergamaceum*, *Onobrychis argentea*, *Thymelaea nitida* and *Nardurus cynosuroides*, in addition to some psammophytes, including *Koeleria pubescens*, *Avena bromoides* and *K. vallesiana*, which appear to characterize this variant.

South of the Saharan Atlas, the *Stipa*, *Artemisia* and *Lygeum* communities gradually give way to those of the desert. Before it drops out, *Stipa* is joined by Saharan species such as *Anabasis oropediorum* and *Fagonia* sp. In the depressions, *Artemisia* and *Lygeum* are replaced by *Haloxylon scoparium*.

Sub-Mediterranean halophytic vegetation
(mapping unit 76)

The vegetation of the largest salt-pan or chott in Tripolitania, namely Tauorga, has been described by Berger-Landefeldt (1959). The most characteristic species include *Aeluropus lagopoides* (*repens*), *Arthrocnemum glaucum*, *Atriplex mollis*, *Bassia muricata*, *Frankenia laevis*, *Halocnemum strobilaceum*, *Limoniastrum monopetalum*, *Nitraria retusa*, *Reaumuria vermiculata* (*muricata*), *Salicornia arabica* (*fruticosa*), *Salsola longifolia*, *S. tetragona*, *Sphenopterus divaricatus*, *Limonium cymuliferum*, (*Statice cyrtostachya*), *Suaeda fruticosa*, *Salsola vermiculata* and *Zygophyllum album*.

Sub-Mediterranean anthropic landscapes
(mapping units 79 and 80)

In Haouz-Tadla in western Morocco the landscape is dominated by wheat fields and pastures for sheep and cattle. In springtime the fallows are covered with a colourful carpet of attractive annuals. Older fallows support 2 m high secondary shrubland of *Retama monosperma* and *Ziziphus lotus*. *Opuntia* is planted locally as a hedge plant and in places is extensively naturalized. There are some plantations of *Pinus halepensis*, *Acacia cyanophylla* and *Eucalyptus*. Indigenous trees are virtually absent except for a few *Acacia gummifera*, and, on the highest hills, *Quercus ilex*.

The coastal plain of Tunisia between Hammamet and Sfax is a rich agricultural area noted for its olive groves. Cereals are also grown and there are apricot and almond orchards. The indigenous forests have completely disappeared.

Further east, in the Libyan Arab Jamahiriya and Egypt, the rainfall is much lower. Nevertheless the region has been intensively cultivated since Roman times and grazed for longer. Few trees remain in the Libyan Arab Jamahiriya, and in Egypt forest has probably not existed during the present climatic era. In Egypt barley is grown in depressions 50 km from the coast and a little further north over the whole surface, but is only harvested in good years (Walter, 1971). Locally olives are grown even where the rainfall is no more than 100 mm per year, but only where the run-off from stony ridges can be diverted into the orchards. Similar systems were used in Roman times for viticulture. The vegetation in uncultivated areas has been degraded by centuries of overgrazing and only unpalatable species remain.

Madagascar and other offshore islands

Madagascar

A comprehensive review (Koechlin *et al.*, 1974) of the flora and vegetation of Madagascar, which owes much to the earlier studies of Perrier de la Bâthie (1921*a*, 1936) and Humbert (1927*a*–1959; Humbert & Cours Darne, 1965) has been published. The facts recorded in the following brief account have been selected chiefly to facilitate comparison with analogous vegetation on the African mainland.

The floras and vegetation of the East and West Malagasy Regions are so distinct that they must be treated separately. Most published information on family and generic endemism, however, refers to the island as a whole and hence is summarized collectively below.

The delimitation of the East and West Malagasy Regions follows Humbert (1955*b*). His Domains are also referred to in the text. Their boundaries are shown in Figure 23.

Flora

About 8500 species of vascular plants are known, but the actual total for the angiosperms alone may be of the order of 10 000 (Humbert, 1959). Of the 7900 species of phanerogams known to Humbert, 6400 (81 per cent) are endemic.

Endemic families. Asteropeiaceae (1 genus, 5–6 species), Didiereaceae (4 genera, 11 species), Didymelaceae (1 genus, 2 species), Diegodendraceae (monotypic), Geosidiraceae (monotypic), Humbertiaceae (monotypic), Sphaerosepalaceae (Rhopalocarpaceae, 2 genera, 14 species), Sarcolaenaceae (10 genera, 35 species). The monotypic Medusagynaceae is confined to the Seychelles.

Endemic genera. About 240 (20 per cent) of the 1200 genera are endemic. They include *Apodocephala, Ascarinopsis, Centauropsis, Chrysalidocarpus, Cuphocarpus, Dicoryphe, Dilobeia, Dypsis, Ephippiandra, Hedycaryopsis, Megistostegium, Neodypsis, Neophloga, Oncostemum, Ravenala, Ravensara, Tambourissa, Tetrapterocarpon, Tina* and *Xerosicyos*.

Linking elements. See Dejardin *et al.* (1973) and, especially for relationships with the African mainland, Leroy (1978).

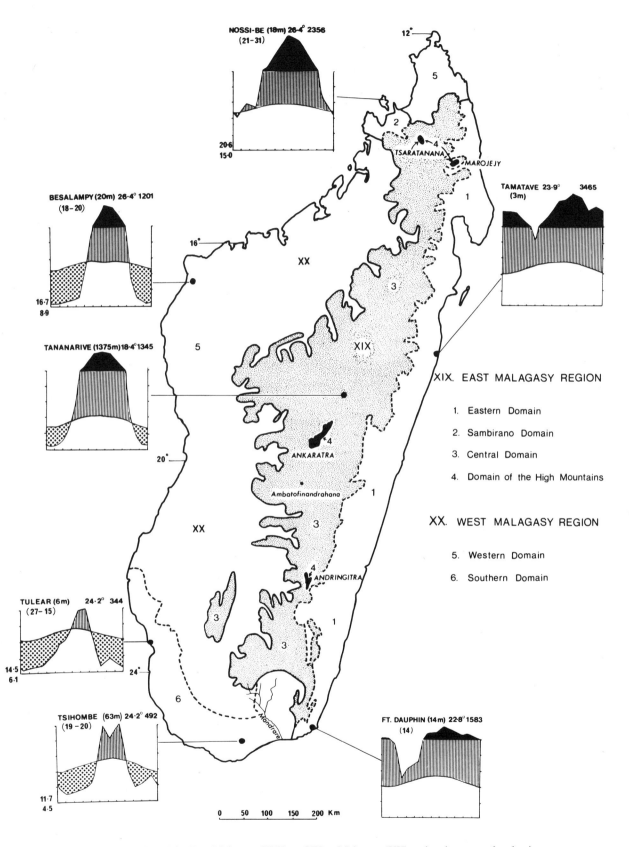

FIG. 23. Climate and topography of the East Malagasy (XIX) and West Malagasy (XX) regional centres of endemism

XIX The East Malagasy regional centre of endemism

Geographical position and area

Geology and physiography

Climate

Flora

Mapping units

Vegetation
    East Malagasy primary lowland rain forest
    East Malagasy secondary lowland rain forest
    East Malagasy moist montane forest
    East Malagasy sclerophyllous montane forest
    East Malagasy 'tapia' forest
    East Malagasy montane bushland and thicket
    East Malagasy rupicolous shrubland
    East Malagasy secondary grassland
        The coastal region
        'Tanety' grassland
        'Tampoketsa' grassland
        Grassland on the western slopes
        Grassland on mountain slopes above 2000 m

Geographical position and area

This region occupies eastern Madagascar and descends from the central highlands, which run almost the entire length of the island, to sea-level in the east and to approximately 800 m in the west. The Sambirano Domain forms a small exclave on the north-west coast. (Area: 272 000 km^2.)

Geology and physiography

Almost the whole of the East Malagasy Region is formed of the Basement Complex of igneous and metamorphic rocks. The Region is dominated by the central highlands, which are essentially formed by schists, migmatites, granites, and gneisses.

The relief of the central highlands is complex and there are several types of landscape. In most places erosion is active, but residual ancient peneplains, known as 'tampoketsa', cover large areas north and north-east of Tananarive between 900 and 1 600 m. Because of the low temperature there are few people and little cultivation. Nevertheless most of the natural vegetation has been destroyed by fire. Elsewhere there are extensive areas of rounded hills known as 'tanety'.

The highest mountains are of diverse origin. In the north the upper slopes of Tsaratanana (2 886 m) above 2 000 m are composed of basalt. In central Madagascar the volcanic massif of Ankaratra (2 643 m) south of Tananarive originated in the Upper Tertiary and vulcanism has continued into recent times. The third high mountain, Andringitra (2 650 m), is granitic.

To the east, the central highlands end in steep escarpments overlooking the coastal plain, which is rarely more than 30 km wide, except where it penetrates inland along river valleys. Numerous relatively short swift-flowing rivers reach the plain. Extensive marshes occur in the lowest part, while coastal lagoons are widespread between Tamatave and Manakara.

Climate

The Eastern Domain has a rain-forest climate. Mean annual rainfall is mostly more than 2 000 mm and in places exceeds 3 000 mm. Towards the south it falls off to 1 500 mm but remains well distributed. The mean monthly rainfall is less than 100 mm during 1–4 months, but no month receives less than 50 mm. Mean annual temperatures are from 23° to 25°C. Cyclones are not infrequent, and the rain often falls in violent showers.

The Central Domain is cooler (mean annual temperature 17°–20°C) and drier (mean annual rainfall 1 300–1 500 mm with a dry season of 5–6 months).

Flora

There are about 6 100 species of which *c.* 4 800 (78.7 per cent) are endemic. There are approximately 1 000 genera, of which *c.* 160 (16 per cent) are endemic. See also page 232.

Mapping units

1b. Malagasy lowland rain forest.
5. Malagasy moist montane forest.
11b. Mosaic of Malagasy lowland rain forest and secondary grassland.
18. Mosaic of cultivation and secondary grassland with relictual sclerophyllous montane and tapia forest.
19c. Malagasy Altimontane communities.

Vegetation

The original vegetation nearly everywhere was forest: evergreen rain forest below 800 m, and three other types of evergreen forest of lower stature above 800 m. Of the upland forests, two types, namely moist montane forest and sclerophyllous montane forest, occupy the eastern slopes, and 'tapia' forest occurs on the western slopes. Over extensive areas, especially above 800 m, the forests have been replaced by secondary grassland. Above 2 000 m the characteristic vegetation is montane thicket. Rocky outcrops both above and below that contour support distinctive rupicolous communities.

East Malagasy primary lowland rain forest
(mapping units 1b and 11b)

Refs.: Humbert & Cours Darne (1965, p. 49–57); Koechlin, Guillaumet & Morat (1974, p. 103–65); Perrier de la Bâthie (1921*a*, p. 89–105).
Phots.: Humbert & Cours Darne (1965, p. 50); Koechlin *et al.* (1974: 10–13, 16, 18, 22, 25); Perrier de la Bâthie (1921*a*, pp. 95, 97).
Syn.: forêt dense humide sempervirente de basse altitude (Koechlin *et al.*); série à Myristicacées et *Anthostema* ou forêt dense ombrophile orientale (Humbert & Cours Darne); série à Sarcolaenacées–Myristicacées et *Anthostema* ou forêt dense ombrophile du Sambirano (Humbert & Cours Darne).

Lowland tropical rain forest of uniform aspect formerly extended along the entire length of the east coast of Madagascar below an altitude of 800 m, but much has been destroyed for cultivation and replaced by secondary regrowth or secondary grassland. It cannot develop on windswept ridges, where it is replaced by shorter forest only 10–15 m tall. Mean annual rainfall is mostly more than 2 000 mm and in places exceeds 3 000 mm.

Malagasy lowland rain forest differs from Guineo–Congolian rain forest in several structural features. The main canopy is lower (25–30 m) and large emergent trees are absent. Except towards its upper altitudinal limit, no species is deciduous. The foliage is more coriaceous and xeromorphic. Palms are much better represented, especially below 200 m, and occur in all strata. Dwarf palms (*Dypsis*, *Neophloga*) play an important part in the lowest layers. Bamboos, including lianoid species, e.g. *Ochlandra capitata*, are scattered throughout. Tree ferns occur at low altitudes but are rarer than in submontane forest. Large humus-collecting epiphytic ferns (*Asplenium nidus*, *Vittaria elongata*, *Oleandra articulata* and *Platycerium*) are conspicuous on the trunks of trees.

Malagasy lowland rain forest is very rich in species, and varies greatly in floristic composition from place to place. One hundred individual plants growing side by side may belong to more than 50 species. Dominance by one or a few species does not occur. The following families are most frequently represented in the upper canopy: Euphorbiaceae, Rubiaceae, Araliaceae, Ebenaceae (*Diospyros*), Sapindaceae, Burseraceae (*Canarium*), Anacardiaceae, Elaeocarpaceae (*Echinocarpus*), Lauraceae (*Ocotea*, *Ravensara*), Guttiferae (*Mammea*, *Symphonia*), Myrtaceae, Malpighiaceae, Monimiaceae (*Tambourissa*), Flacourtiaceae, Loganiaceae, Proteaceae (*Dilobeia*) and Leguminosae (*Dalbergia*, *Cynometra*). Leguminosae are relatively much less important than in Guineo–Congolian rain forest, and Meliaceae subfamily Swietenioideae are scarcely represented. The giant monocot, *Ravenala madagascariensis*, and several species of *Dracaena* also occur in the upper canopy. Large lianes are rather rare.

The lower canopy is composed of small trees and large shrubs, the leaves of which are larger and less coriaceous than those of the upper canopy. Rubiaceae, Euphorbiaceae, Ochnaceae, Erythroxylaceae, Myrsinaceae, Celastraceae, Violaceae (*Rinorea*), Flacourtiaceae, and Tiliaceae are well represented.

Epiphytes are abundant, especially Orchidaceae, Melastomataceae, and ferns, and increase with altitude. Epiphytic ferns show a preference for the stems of tree ferns.

East Malagasy secondary lowland rain forest
(mapping units 1b and 11b)

Ref.: Koechlin *et al.* (1974, p. 399–410).

In the rain-forest region of eastern Madagascar undisturbed or little disturbed forest occupies 6 400 000 ha and degraded formations 3 600 000 ha. Very little of the latter, however, is secondary forest, which is known locally as 'savoka'. There are two reasons for this. First, Madagascar has very few indigenous species of secondary forest and they are much less vigorous than corresponding species in tropical Africa. Second, the indigenous secondary forest species are unable to compete with smaller plants such as *Panicum maximum* and *Imperata cylindrica* among grasses, *Pteridium aquilinum* and *Sticherus flagellaris* among ferns, *Aframomum angustifolium* and *Hedychium coronarium* among giant forbs, and *Solanum auriculatum* and *Lantana camara* among shrubs. Most of these species are not indigenous. The introduced trees *Psidium cattleianum* and *P. guajava* are also much more invasive than the indigenous secondary species. Pre-eminent among Malagasy secondary forest 'trees' is the Traveller's Tree, *Ravenala madagascariensis*, of the family Musaceae (Strelitziaceae). Its unbranched trunk, which can exceed a height of 20 m, ends in a tuft of giant leaves arranged like a fan in a single plane. The heliophilous *Ravenala* is widespread in eastern Madagascar between sea-level and 800 m. It is one of the few forest species capable of surviving in secondary grassland. It is also abundant in swamp forest with the screw pine, *Pandanus*, and the giant aroid, *Typhonodorum lindleyanum*, and this probably represents its original habitat. Other indigenous secondary forest species include *Harungana madagascariensis*, *Psiadia altissima*, species of *Canarium*, *Croton*, *Dombeya* and *Macaranga*, and the bamboo, *Ochlandra capitata*.

The ease with which secondary forest can become established is largely dependent on the nature of the soil. Porous soils of good structure and rich in decomposing minerals, which are found on strongly rejuvenated land surfaces, favour forest vegetation; hence bamboo thickets, clumps of *Ravenala*, or patches of secondary forest soon became established. By contrast, on compact ferrallitic soils formed from impoverished parent material, the destruction of forest is succeeded almost immediately by grassland with more or less abundant *Ravenala*, which, however, is progressively eliminated by fire.

Coastal forest on leached sand dunes is replaced by a totally different type dominated by *Philippia*, often with scattered guavas. It is highly combustible and is rapidly replaced by grassland.

East Malagasy moist montane forest
(mapping unit 5)

Refs.: Guillaumet & Koechlin (1971); Humbert & Cours Darne (1965, p. 58–9); Koechlin *et al.* (1974, p. 358–74); Perrier de la Bâthie (1921*a*, p. 133–45).

Phots.: Guillaumet & Koechlin (1971: 9a); Koechlin *et al.* (1975: 103–8); Perrier de la Bâthie (1921*a*, p. 135, 139, 141, 143–4).

Syn.: forêt à sous-bois herbacé entre 800 et 2000 m (Perrier de la Bâthie, 1921*a*); forêt à mousses et sous-bois herbacé (Perrier de la Bâthie, 1927); étage de moyenne altitude (de 800 à 1800 m)—forêt dense ombrophile (série à *Tambourissa* et *Weinmannia*) (Humbert & Cours Darne, 1965); forêt dense humide de montagne (Koechlin *et al.*, 1974).

This type is as rich in species as Malagasy lowland rain forest, but is of lower stature (main canopy up to 20–25 m), and the trees are often branched low down and their boles are rarely straight. The leaves of the canopy species are more sclerophyllous. Epiphytes, especially mosses, are more abundant. The herb layer is much better developed and includes both abundant ferns and species with large non-xeromorphic leaves. It occurs chiefly between 800 and 1300 m, but on deep, well-watered soils in sheltered places may ascend to 2000 m or more. The most abundant canopy species belong to the following genera: *Tambourissa*, *Weinmannia*, *Symphonia*, *Dombeya*, *Dilobeia*, *Dalbergia*, *Canarium*, *Vernonia*, *Diospyros*, *Eugenia*, *Protorhus*, *Grewia*, *Brachylaena*, *Schefflera* and *Cuphocarpus*. Among shrubby plants, Compositae, Rubiaceae and Myrsinaceae are well represented. Large lianes, especially species of Compositae, Rubiaceae and monocarpic bamboos, are abundant. There are several species of tree fern and *Pandanus*. The herb layer is dense and consists of many ferns and species of Labiatae, Acanthaceae, shade-tolerant Gramineae, and *Impatiens*. It is remarkable for the occurrence of species belonging to several temperate genera, namely *Ajuga*, *Plantago*, *Cardamine*, *Alchemilla*, *Rubus*, *Vaccinium*, *Ranunculus*, *Sanicula*, *Cerastium*, *Hydrocotyle* and *Viola*. It is the epiphytes that provide the most striking feature, especially the bryophytes and lichens, which cover the boles and branches with an almost continuous layer. The most abundant vascular epiphytes are ferns, orchids, (especially species of *Bulbophyllum*), *Medinilla*, *Kalanchoe*, *Rhipsalis* and *Peperomia*.

Although many species are endemic to this formation, others, such as *Podocarpus madagascariensis*, extend from sea-level to the summits of the highest mountains. Two genera of Monimiaceae, *Ephippiandra* (3 spp.) and *Hedycaryopsis* (4 spp.), are confined to moist montane forest and formations at higher altitudes.

Malagasy moist montane forest is transitional between lowland rain forest and sclerophyllous montane forest. With increasing altitude the stature of the trees diminishes, the stratification becomes simplified and the understorey more open; the field layer becomes better developed and more diversified; palms diminish in frequency, and epiphytes become increasingly well represented and cover the boles and branches of all the trees and even occur on the soil itself; the foliage of the trees becomes more sclerophyllous. There is concomitant floristic change, but this has been little studied.

East Malagasy sclerophyllous montane forest
(mapping unit 18)

Refs.: Guillaumet & Koechlin (1971); Humbert & Cours Darne (1965, p. 60–1); Koechlin *et al.* (1974, p. 374–83); Perrier de la Bâthie (1921*a*, p. 146–8); Thomasson (1977).

Phots.: Guillaumet & Koechlin (1971: 9b); Koechlin *et al.* (1974: 113–15).
Syn.: sylve des lichens (Perrier de la Bâthie, 1921*a*); étage montagnard—sylve à lichens (Humbert & Cours Darne, 1965); forêt dense sclérophylle de montagne (Koechlin *et al.*, 1974).

In general, this type occurs at higher altitudes than moist montane forest, mostly between 1300 and 2300 m, but, according to Koechlin *et al.* (1974, p. 359), its distribution is determined as much by edaphic conditions as by climate, in that it descends much lower than 1300 m on the shallow soil of exposed ridges. Compared to moist montane forest the environment is characterized by lower temperatures with greater daily and seasonal variations, stronger winds, and greater variation in humidity, which affects all strata down to ground-level.

The canopy is lower (10–12 m) than that of Malagasy moist montane forest and is less well differentiated from the lower strata since most of the trees are more richly branched and branch low down. In structure it is intermediate between forest and thicket, but closer to the former. The leaves of canopy species are smaller and are much more xeromorphic. When the canopy is relatively open, several ericoid species which are more characteristic of montane thicket (see below) occur in the understorey. Bryophytes and lichens are even more abundant than in the preceding type. Many of the trees are festooned with *Usnea*. The ground is covered with a dense layer of bryophytes and lichens often several decimetres thick, in which are present species normally occurring as epiphytes (ferns, orchids, *Peperomia*). This mat is chiefly composed of pleurocarpous mosses, but there are also mounds of *Sphagnum*. Fruticose lichens (*Cladonia*) are characteristic of more rocky situations.

Among the most characteristic species of this forest the following may be mentioned: *Dicoryphe viticoides*, *Tina isoneura*, *Alberta minor*, *Rhus taratana*, and the recently discovered *Ascarinopsis coursii*, the only member of the Chloranthaceae known from Madagascar. It is related to the New Caledonian *Ascarina*.

Families which predominate in the canopy are chiefly Compositae (*Vernonia*, *Senecio*, *Centauropsis*, *Psiadia*, *Apodocephala*), Rubiaceae (especially Psychotrieae), Lauraceae (*Ocotea*), Verbenaceae (*Clerodendrum*, *Vitex*) and Ericaceae (*Philippia*, *Agauria*). The following genera and species are also important: *Oncostemum*, *Dombeya*, *Faurea*, *Podocarpus*, *Heteromorpha*, *Aphloia*, *Nuxia*, *Symphonia*, *Myrica*, *Cussonia*, *Schefflera*, *Weinmannia*, *Vaccinium* and *Ilex mitis*. Bamboos are represented by species of *Arundinaria* and *Ochlandra*, and palms by species of *Chrysalidocarpus* and *Neodypsis*. *Arundinaria marojejyensis*, which is more characteristic of a higher zone, forms almost pure populations in rocky places. This type of forest is very susceptible to fire, which can easily spread through the thick layer of humus.

East Malagasy 'tapia' forest
(mapping unit 18)

Refs.: Guillaumet & Koechlin (1971); Humbert & Cours Darne (1965, p. 61–3); Koechlin *et al.* (1974, p. 215–42); Perrier de la Bâthie (1921*a*, p. 153–9).
Phots.: Guillaumet & Koechlin (1971: 36); Koechlin *et al.* (1974: 47–59).
Syn.: bois des pentes occidentales (Perrier de la Bâthie, 1921*a*); étages des pentes occidentales—forêt basse sclérophylle à *Uapaca bojeri* et Chlaenacées (Humbert & Cours Darne, 1965); les forêts sclérophylles de moyenne altitude (Koechlin *et al.*, 1974).

This type occurs between 800 and 1600 m on the western slopes of the upland massif which runs along almost the entire length of Madagascar. Since it occurs in the rain shadow of higher ground to the east, the climate is drier; temperatures are also higher and insolation more intense. It is particularly sensitive to fire and has largely been replaced by secondary grassland or open woodland. 'Tapia' is the vernacular name of the dominant species, *Uapaca bojeri*.

In appearance this type is similar to the Cork Oak (*Quercus suber*) forests of the Mediterranean, but more species are present in the canopy, which is at 10–12 m. The gnarled canopy trees have evergreen coriaceous leaves, often of small size. Their crowns are more or less in contact but cast only a light shade. The understorey is composed largely of ericoid shrubs. Lianes are quite frequent, but of small size. Tree ferns are lacking, and the only palm is *Chrysalidocarpus decipiens*, which is confined to the wettest places. Epiphytes are rare (a few small ferns and species of *Bulbophyllum*) and at the lowest altitudes are represented only by lichens. There is no ground layer of bryophytes.

The canopy is composed chiefly of *Uapaca bojeri*, which is very fire-resistant and often persists as the dominant of secondary open woodland long after its associates have disappeared. Its most frequent associates are three species of Sarcolaenaceae, namely *Leptolaena pauciflora*, *L. bojerana* and *Sarcolaena oblongifolia*. Other associates include *Asteropeia densiflora*, *Agauria salicifolia*, *Weinmannia* spp., *Dodonaea madagascariensis*, *Faurea forficuliflora*, *Brachylaena microphylla*, *Dicoma incana*, *Rhus taratana*, *Protorhus buxifolia*, *Schefflera bojeri*, *Alberta* spp. and *Enterospermum* spp. The shrub layer is composed chiefly of *Philippia*, Compositae (*Senecio*, *Vernonia*, *Psiadia*, *Conyza*, *Helichrysum*), Rubiaceae, *Vaccinium* and Leguminosae.

East Malagasy montane bushland and thicket
(mapping unit 19c)

Refs.: Humbert & Cours Darne (1965, p. 64–6); Koechlin *et al.* (1974, p. 383–8); Perrier de la Bâthie (1921*a*, p. 149–52).
Phots.: Humbert & Cours Darne (1965, opp. p. 66); Koechlin *et al.* (1974: 116–21).
Syn.: broussailles éricoïdes des hautes altitudes (Perrier de la Bâthie, 1921*a*); fourré dense d'altitude (Humbert & Cours Darne, 1965); fourrés de montagne (Koechlin *et al.*, 1974).

This vegetation is found on the high mountains of Madagascar wherever conditions are suitable between (1800) 2000 m and the highest summit (2886 m). It occurs above montane sclerophyllous forest, of which it can be regarded as an extremely depauperate derivative. There is only a single stratum of woody plants, which is never more than 6 m tall and is often impenetrable. It is composed almost entirely of species of ericoid habit with short twisted stems. All species are evergreen and most have ericoid, cupressoid, or myrtilloid leaves. The chief components are Ericaceae (several species of *Philippia*, *Vaccinium*), Rubiaceae, and Compositae (*Psiadia*, *Senecio*, *Vernonia*, *Stoebe*, *Stenocline*, *Helichrysum*). Somewhat rarer species of other families—e.g. Labiatae, Gentianaceae, Melastomataceae, *Thesium*—are also of ericoid habit. A few widely spaced bushy trees emerge slightly from the general canopy, e.g. *Agauria salicifolia*, *Ilex mitis*, *Schefflera bojeri*, *Alberta minor*, *Dodonaea madagascariensis*, *Tambourissa gracilis*, *Podocarpus rostratus*, *Vitex humbertii*, *Pittosporum* sp., several species of *Weinmannia*, *Faurea forficuliflora*, etc. Palms are absent from this formation, except for *Chrysalidocarpus acuminum*, which is endemic to the massif of Manongarivo. Other arborescent monocots are represented by *Dracaena reflexa* and *Pandanus alpestris*.

Lianes are almost completely absent, as are vascular epiphytes, except for a few small orchids, but epiphytic bryophytes and lichens are plentiful. There is a discontinuous ground layer of bryophytes and lichens which is lacking on better-drained sites. The field layer is also poorly developed but includes a few endemic species of Gramineae, Cyperaceae and *Impatiens*. Plants of temperate affinity are no better represented in this vegetation type than in other vegetation types of the Central Domain.

East Malagasy rupicolous shrubland
(locally occurring in several mapping units)

Ref.: Koechlin et al. (1974, p. 488–553).
Phots.: Koechlin et al. (1974: 157–69).
Syn.: pelouse à xérophytes (Perrier de la Bâthie, 1921a).

Whereas the most characteristic vegetation of large rocky outcrops on the African mainland is bushland and thicket, on similar outcrops in Madagascar the tallest plants rarely exceed 2 m in height. Although Malagasy rupicolous vegetation has some floristic similarities with African rupicolous bushland, because of its much lower stature, it is classified as shrubland in the present work.

Rupicolous communities occur on karstic outcrops in West Malagasy, but it is in the Central Domain and the Domain of High Mountains in East Malagasy that they are most frequent. They also occur in the Oriental Domain but their flora is poor and not very well known.

Most of the inselbergs are made of granite but some are of sandstone or quartzite. The plants are rooted in crevices or, more often, in mats of coarse shallow soil on the less steep slopes. These mats of vegetation are unstable and are liable to be washed away during heavy rain. In the region of Fort Dauphin the development of vegetation is severely retarded by the violence of the precipitations.

The shallow soils dry out rapidly between the rains and the plants growing on them are exposed to strong sunshine by day and, because of rapid heat-loss by radiation, low temperatures at night. Strong wind, also, is often an adverse factor. Most species show adaptations to drought.

Although the Malagasy rupicolous flora is not very rich in species it shows a wide diversity of growth forms.

Reviviscent pteridophytes are represented by the moss-like *Selaginella echinata* and species of *Pellaea*, *Actiniopteris* and *Notholaena*. There are several endemic species of the leaf-succulent genus *Aloe*. Another monocotyledonous leaf-succulent is *Cyanotis nodiflora*. There are three species of *Xerophyta* of which *X. dasylirioides* is the most widespread—normally 20–60 cm tall, exceptionally it attains 2 m. Its pseudodichotomous stems are thickly covered with an adventitious root system which permits rapid absorption of water after rain when the leaves, which fold up and are dingy in dry weather, soon become verdant. Among grasses, species of *Loudetia*, *Aristida*, *Heteropogon* and *Hyparrhenia* are often present, and it is possible that it is from habitats such as this that they have spread to dominate the secondary grasslands which are so widespread today. Four other grasses, *Redfieldia hitchcockii* and three species of *Isalus*, are confined to inselbergs. The tufted sedge, *Coleochloa setifera*, an important pioneer, is less vigorous than on the African mainland. Terrestrial orchids are represented by species of *Cynorkis* and *Angraecum*, while *Bulbophyllum leptostachyum* is epiphytic on *Xerophyta*.

Among dicots, there are many endemic species belonging to the leaf-succulent genus *Kalanchoe*.

Both cactoid and coralliform species of *Euphorbia* are conspicuous; other species of *Euphorbia* are spartioid or have large subterranean structures and a fugaceous rosette of leaves borne at ground-level. The leguminous *Mundulea phylloxylon* has cladodes. The reviviscent small shrub *Myrothamnus moschatus* is closely related to the African *M. flabellifolius*. There are several stem succulents of diverse form in the genus *Pachypodium* which are described in detail by Koechlin et al. There are also a few stem-succulent asclepiads. *Ceropegia dimorpha* has a short perennial fleshy stem which in the rainy season gives rise to a deciduous leafy and flowering twining stem. There are several species of *Senecio* with succulent leaves and a number of rupicolous *Helichrysum* species with ericoid or densely tomentose leaves.

Above 2000 m, rock outcrops are characterized by a sharp impoverishment of the flora. Certain genera, very abundant at lower altitudes (*Pachypodium*, *Euphorbia*, *Myrothamnus*, *Selaginella*, etc.), disappear completely above the 2000 m contour, and the importance of some species like *Coleochloa setifera* is greatly diminished. Genera with a wider ecological amplitude which ascend to high altitudes include *Kalanchoe*, *Aloe*, *Senecio*,

Helichrysum, *Xerophyta* and *Cheilanthes*. The first three are often represented by endemic species on each mountain. A few genera are confined to high-altitude rupicolous shrubland. *Sedum*, for instance, is represented there by the small shrub, *S. madagascariense*. High-altitude rupicolous shrubland is also characterized by the abundance of bryophytes and lichens, including the fruticulose *Cladonia pycnoclada*.

East Malagasy secondary grassland
(mapping units 1b, 11b, 18, and 19c)

Ref.: Koechlin et al. (1974, p. 434–57).
Phots.: Koechlin et al. (1974: 138–44).

Secondary grassland covers enormous areas, particularly in the Central Domain. The following major variants have been recognized.

The coastal region

After the destruction of forest regrowth by fire, *Imperata cylindrica* and *Hyparrhenia rufa* invade, but are soon eliminated because of soil erosion caused by the heavy and violent rainfall. *Aristida similis* then becomes dominant. Associates include *Panicum luridum*, *P. dregeanum*, *Digitaria humbertii*, *Setaria sphacelata*, *Eragrostis lateritica*, *Andropogon eucomus*, *Sporobolus subulatus*, *Hyparrhenia nyassae* and *Cymbopogon plicatus*. On certain soils, if fire is excluded this type can revert to forest.

'Tanety' grassland

The tanety is a region of hills between 1 200 and 1 500 m. The short grassland consists of widely spaced tufts of low ground cover. The infertile soil is very hard and compact and almost impermeable. It is often covered between the tufts with a layer of lichens and blue-green algae. On the slopes, *Aristida rufescens* is dominant and in the most degraded places it is virtually the only species present. Normally, however, it is associated with other species such as *Ctenium concinnum*, *Elionurus tristis*, *Alloteropsis semialata*, *Cymbopogon plicatus*, *Craspedorachis africana*, *Sporobolus centrifugus*, *Panicum luridum* and *Urelytrum squarrosum*. On deeper soils and where fire is less intense, *Aristida* is replaced by grassland of *Hyparrhenia rufa*, *H. newtonii* and *Heteropogon contortus*, but its total area is small.

'Tampoketsa' grassland

The tampoketsa comprise the plateaux situated north and north-east of Tananarive. Their slightly undulating surface at an altitude of 1 600–1 900 m represents the end-Tertiary peneplain. The soils are ferrallitic. There are few relict forest patches, but most of the surface is covered with grassland. *Loudetia simplex* subsp. *stipoides*, which is endemic to Madagascar, is dominant throughout. On gentle slopes it is associated with *Elionurus tristis* and *Trachypogon spicatus*, and on steeper slopes and more degraded soils with *Aristida similis*. Tampoketsa grassland is very uniform floristically and has a total of only 34 species. Species other than the dominants play only a minor role.

Grassland on the western slopes

Between 800 and 1 600 m the western slopes of the central spine of Madagascar were formerly covered with tapia forest. This has almost entirely been replaced by grassland, which is intermediate between the short grasslands of the plateau described above and the taller grasslands of the West Malagasy Region (page 242). At higher altitudes, species of the tampoketsa such as *Loudetia simplex* subsp. *stipoides* and *Aristida rufescens* are still important constituents. Lower down, *Hyparrhenia rufa* and *Heteropogon contortus* increasingly become dominant towards the west, and western elements such as *Hyperthelia dissoluta* enter the community. Nevertheless, the zone as a whole is floristically poor and very homogeneous.

Grassland on mountain slopes above 2 000 m

In most places the original montane bushland and thicket have been almost entirely replaced by grassland which is regularly grazed and burnt. It seems that this transformation occurred relatively recently. In 1777, Ankaratra was covered with 'forest'. Up till the end of the last century Andringitra served as a refuge for the local people and their flocks in times of trouble and its 'forests' had thus escaped conversion into pasture. Since it became a reserve there has been a pronounced succession back towards forest.

Above 2 000 m the soils are rich in humus, and peat readily forms at the surface. The grass species are completely different from those occurring at lower altitudes and mostly belong to genera which also figure prominently in secondary grassland in the Ericaceous and Afroalpine belts on the African mountains. On Ankaratra the principal constituents of montane grassland are *Pentaschistis perrieri*, *P. humbertii*, *Andropogon trichozygus*, *Anthoxanthum madagascariense*, *Digitaria ankaratrensis*, *Agrostis elliotii*, *Merxmuellera* (*Danthonia*) *macowanii*, *Brachypodium perrieri*, *Poa madecassa*, *P. ankaratrensis* and *Festuca camusiana*. Many species of secondary montane grassland are also found in bogs.

xx The West Malagasy regional centre of endemism

Geographical position, geology, physiography and area

Climate

Flora

Mapping units

Vegetation
    West Malagasy dry deciduous forest
    West Malagasy deciduous thicket
    West Malagasy grassland

Geographical position, geology, physiography and area

This region occupies the western side of the island up to about the 800 m contour. Towards the east, where the land rises to meet the central highlands, there are outcrops of crystalline Precambrian rocks, but the greater part of the region is underlain by sediments of Triassic, Jurassic, Cretaceous, and Tertiary age. The flat plains of the west coast are wider than those along the east coast. (Area: 322 000 km^2.)

Climate

This region lies in the rain shadow of the SE. monsoon, which arrives desiccated and heated after losing its moisture further east. The dry season, which lasts seven months or more, is very severe, but in most years there is a small amount of precipitation.

In the Western Domain mean annual rainfall increases from 500 mm in the south to 2 000 mm in the north. Mean annual temperature is mostly between 25° and 27°C.

The Southern Domain is the driest part of the island, with a rainfall of only 300–500 mm per year. Most rain falls in summer as local heavy showers. The dry season normally lasts at least eight months, but some rain may fall in any month of the year. On the other hand, droughts can last from 12 to 18 months.

Flora

There are about 2 400 species, of which 1 900 (79.2 per cent) are endemic, and about 700 genera, of which *c.* 140 (20 per cent) are endemic. (See also page 232.)

Mapping units

7. Malagasy dry deciduous forest (see below).
22b. Mosaic of Malagasy dry deciduous forest and secondary grassland (see below).
41. Malagasy deciduous thicket (see below).
46. Mosaic of Malagasy deciduous thicket and secondary grassland (see below).

Vegetation

There are two main types of primary vegetation, dry deciduous forest and deciduous thicket, but the most extensive vegetation is secondary grassland.

West Malagasy dry deciduous forest
(mapping units 7, 22b, 41, and 46)

Refs.: Guillaumet & Koechlin (1971); Humbert & Cours Darne (1965, p. 68–72); Koechlin *et al.* (1974, p. 167–213); Perrier de la Bâthie (1921*a*, p. 204–23).
Phots.: Guillaumet & Koechlin (1971: 1, 2, 3a); Humbert & Cours Darne (1965: opp. p. 70); Koechlin (1972: 3); Koechlin *et al.* (1964: 30–40, 42–6); Perrier de la Bâthie (1921*a*, p. 212–13, 215, 217–19, 221–2).

This is the characteristic vegetation of the Western Domain, which lies below 800 m in the rain shadow of the SE. monsoon. During the wet season, rain occurs in the form of storms brought by winds from the north or west, augmented by those from the north-east. Some 30 to 80 mm of rain may fall in the dry season.

This type is less dense and less rich floristically than most of the moister forests of the east. Nevertheless, its flora is large and varied. No single species or small group of species is ever dominant. The upper canopy, which is rather open, occurs at 12–15 m with scattered taller trees up to 25 m high. Lianes are abundant and the shrub layer is well developed. The soil is mostly bare except for small patches of Acanthaceous subshrubs which die back in the dry season. There are very few vascular epiphytes (only a few small orchids in the wetter types), no bryophytes, and very few lichens. Ferns and palms are absent. The trees of the main canopy are always deciduous but the length of deciduousness varies greatly. Some species retain their foliage for scarcely longer than four months, whereas others lose the last of their old leaves when the new ones unfold. Every intermediate occurs. Several trees, e.g. species of *Adansonia*, *Dalbergia*, and *Cassia*, flower precociously a few weeks before the appearance of the new leaves. The latter appear suddenly in all species, precocious and otherwise, immediately after the first rains. Certain herbs (*Kalanchoe*, *Plectranthus*) have large membranaceous leaves during the rainy season. These are replaced by smaller leaves in the dry season. In the moister types some of the lianes and some shrubs of the understorey retain their leaves. In the drier types evergreen species are almost absent. Excluding riparian forest, there are three main types, which occur in different substrates.

1. *On lateritic clays.* These soils, which are developed from basalt and gneiss, support the most luxuriant of the dry deciduous forests. The humus layer is deeper than that of the moister forests of the east. The larger tree species include *Dalbergia*, *Stereospermum euphorioides*, *Givotia madagascariensis*, *Xylia hildebrandtii*, *Ravensara* and *Cordyla madagascariensis*. The majority of lianes belong to the Asclepiadaceae and the following genera: *Dichapetalum*, *Salacia*, *Combretum*, *Landolphia* and *Tetracera*. Rubiaceae, Euphorbiaceae, and Leguminosae dominate the understorey. There is a single species of *Dracaena* and one bamboo, which has deciduous leaves; these types are lacking from the other dry deciduous forests.

2. *On sandy soils.* These are derived from Liassic, Jurassic, and Cretaceous sandstones. The forest is similar to that on lateritic soils but is shorter. It varies according to the depth and moisture content of the soil. On dry soils the larger trees drop out, the distinction between the upper and lower canopy is obscured, and forest passes gradually into thicket. On the moister soils *Tamarindus indica* is frequent. This type covers large areas and varies little from north to south. On drier soils in the southern part of the domain, deciduous forest is characterized by the cactoid *Euphorbia*, *E. enterophora*, which reaches a height of 15–20 m, in association with *Broussonetia* (*Chlorophora*) *greveana*, *Securinega seyrigii*, *Hernandia voyroni*, *Protorhus deflexa*, *Flacourtia indica* and *Adansonia grandidieri*.

3. *On calcareous plateaux.* This type is similar to 1 and 2 but generally of lower stature. There are fewer lianes and fewer evergreen species. The latter contribute less than 2 per cent to the phytomass. Trees and shrubs with swollen stems, e.g. *Adansonia*, *Bathiaea* and *Harpagophytum*, are relatively more abundant. A taller type formerly occurred on deeper soils, but only a few vestiges remain. In rocky situations, which are less affected by fires, the shorter variant survives more plentifully. In deep fissures, trees can reach a large size. This is the preferred habitat of *Diospyros perrieri*, which was formerly exploited to yield a valuable ebony. But among the rocks themselves, the height falls off very quickly, the upper and lower canopy are no longer distinct, spinous lianes and shrubs appear, succulents and plants with swollen stems become more numerous, and one passes almost without transition to thicket.

The following are important in the canopy: *Albizia* spp., *Protorhus humbertii*, *P. perrieri*, *Erythrophysa* and *Sideroxylon collinum*. One of the most widely cultivated tropical ornamental trees, *Delonix regia*, belongs to this community, but it is exceedingly rare. A few larger trees occur as emergents: *Adansonia za*, *A. rubrostipa*, and species of *Diospyros* and *Acacia*. The understorey is made up principally of Euphorbiaceae, Leguminosae, Acanthaceae, and Rubiaceae. Lianes belong chiefly to Asclepiadaceae, Passifloraceae, and Leguminosae.

Dry deciduous forest also occurs, though rarely, in the Southern Domain, where it is confined to moister sites. The prevalent vegetation in the Southern Domain is thicket, to which the endemic family Didiereaceae is virtually confined. It is almost always present. Didiereaceae are normally absent from dry deciduous forest, though they do occur in such forest locally in the Southern Domain, for instance in the forest of *Didierea madagascariensis* and *Adansonia fony* near Tuléar and in the forest of *Alluaudia procera* and *A. ascendens* (Guillaumet & Koechlin, 1971, phot. 2) in the Mandrare basin. These forests are stratified and the Didiereaceae occur in the upper canopy mixed with other species. Plants which have microphyllous leaves on short shoots are rare in these forests in contrast to their abundance in thicket. It would appear that this type of forest is transitional to thicket and is only of local significance.

West Malagasy deciduous thicket
(mapping units 41 and 46)

Refs.: Guillaumet & Koechlin (1971); Humbert & Cours Darne (1965, p. 72–5); Koechlin (1972, p. 171–80); Koechlin *et al.* (1974, p. 243–341); Perrier de la Bâthie (1921a, p. 245–54).
Phots.: Guillaumet & Koechlin (1971: 5–8); Humbert & Cours Darne (1965, opp. p. 73, 74, 75); Koechlin (1972: 6–10); Koechlin *et al.* (1974: 60–9, 71–6, 80–3, 85, 86, 88–91, 93–8).
Profile.: Koechlin (1972: 5, 6); Koechlin *et al.* (1974: 21–2).

This is the characteristic but still very imperfectly known vegetation of the Southern Domain of Madagascar, which is the driest part of the island.

The irregular distribution of rainfall, combined with high relative humidity throughout the year, may account for the fact that a higher (though still small) proportion of the flora is evergreen than in the dry deciduous forests described above.

The soils are usually shallow, and are often stony.

Height and density of the thicket vary greatly in relation to the amount of rainfall and soil moisture. Shorter and more open types are mostly confined to rocky situations. Most undamaged stands are impenetrable or almost so. At one extreme there is a gradual transition to dry deciduous forest, and at the other, on shallow rocky soils, the canopy is less than 2 m tall.

These thickets are commonly between 3 and 6 m tall, and may have a very discontinuous stratum of emergent trees, exceptionally up to 8 or 10 m tall. Otherwise there is no stratification and the thicket consists of a complex mixture of plants of different sizes. Physiognomically the most distinctive element is provided by the Didiereaceae and arborescent species of *Euphorbia*, which are usually present in, and almost restricted to, this vegetation type.

The Didiereaceae is a small endemic family of bushy or arborescent pachycauls of distinctive ascending branching habit, and with very small, somewhat persistent leaves borne in fascicles scattered along the main stems. There are four genera and 12 species, namely the two species of *Didierea*, *D. madagascariensis* and *D. trollii*, the genus *Alluaudia* with six species, the two species of *Alluaudiopsis*, and the monotypic *Decaryia* (*D. madagascariensis*). *Alluaudia procera* and *A. ascendens* reach a height of 8 m or more, but most species are smaller.

There are several species of *Euphorbia* which have caducous leaves and green fleshy stems; the latter sometimes bear paired spines, as in *E. stenoclada*, one of the most abundant species. Some species reach a height of 10 m or more.

Other emergent species are *Adansonia za*, *A. fony*, *Tetrapterocarpon geayi*, *Dicoma incana*, *D. carbonaria*, *Gyrocarpus americanus*, *Maerua filiformis* and *Ficus marmorata*.

The thicket itself is rich in species and varies greatly in floristic composition. Important woody plants include: *Acacia*, *Commiphora monstruosa*, *Grewia*, *Dichrostachys*, *Iphiona*, *Uncarina*, *Jatropha*, *Gardenia*, *Rhigozum madagascariensis*, *Cadaba*, *Megistostegium*, *Sclerocarya*, *Diospyros latispathulata* and *Terminalia subserrata*.

Lianes are numerous but rather small, and include several Asclepiadaceae, leafless species of *Cissus* and *Adenia*, and species of *Xerosicyos* (Cucurbitaceae) with thick fleshy leaves.

The ground flora is sparse and consists of isolated tufts of an endemic grass, *Humbertochloa bambusiuscula*, together with reviviscent species of ferns and *Selaginella*, species of *Xerophyta*, scattered Acanthaceae and other herbs, and succulent species of *Aloe*, *Kalanchoe*, *Euphorbia*, *Senecio* and *Notonia*.

The photosynthetic organs are very variable in structure and behaviour. Some species have large leaves, which appear suddenly after heavy rainfall and are shed equally suddenly. In other species the more fugacious leaves are produced very irregularly. Many shrubs have narrow, greyish leaves which persist much longer and may still occur on some shoots while others on the same plant are covered with a new crop of leaves. A few rare shrubs are evergreen. Many species have green photosynthetic stems, and may or may not also produce fugacious leaves. In some of these, e.g. certain cactiform Euphorbias, species of *Cissus* and Asclepiadaceae, the photosynthetic stems themselves are caducous. A large number of woody species have small narrow leaves in fascicles on short shoots of very limited growth.

Several of the taller species have distended water-storing stems of characteristic, often bottle-like appearance. They include *Adansonia*, *Moringa*, *Delonix adansonioides*, *Gyrocarpus americanus*, *Pachypodium lamerei* and *P. geayi*. Many species are spinous.

West Malagasy grassland
(mapping units 22b and 46)

Refs.: Koechlin *et al.* (1974, p. 457–86); Morat (1973).
Phots.: Koechlin *et al.* (1974, p. 145–56); Morat (1973: 5, 13–24).
Profiles: Morat (1973: 16–27).

More than 80 per cent of the surface of the West Malagasy Region is covered with secondary grassland or wooded grassland which is burnt each year. In general, the dominant species are taller than those of east Madagascar and have wider, flat, ribbon-like leaves which contain less sclerenchyma. Only *Aristida rufescens* in the grasslands of the north-west and *A. congesta* in the grasslands of the south-west possess narrow, tightly rolled, sclerenchymatous leaves. The dominants, which in addition to *Aristida* include *Heteropogon contortus, Loudetia simplex* subsp. *stipoides, L. filifolia* subsp. *humbertiana, Themeda quadrivalvis, Hyparrhenia rufa, H. schimperi, H. cymbaria, Panicum maximum* and *Hyperthelia dissoluta*, are hemicryptophytes, well adapted to withstand the annual destruction of their subaerial parts by fire.

Therophytes are numerous but physiognomically unimportant; among them, grasses and sedges include *Bulbostylis xerophila, B. firingalavensis, Brachiaria ramosa, B. nana, Eragrostis lateritica,* and *Tragus berteronianus* in the lower herb layer, and *Perotis* aff. *patens, Pogonarthria squarrosa, Digitaria biformis, Chloris virgata* and *Aristida adscensionis* in the upper herb layer. *Imperata cylindrica* is a geophyte. Other geophytes, which include several orchids, are virtually confined to protected places where fires are infrequent. Chamaephytes are few in species and never abundant. Many belong to the Papilionoideae (*Rothia, Eriosema, Crotalaria, Indigofera, Otoptera*).

Above all, it is the presence of trees and bushes which distinguishes the secondary grasslands of the West from those of the East. Tall trees are rare and are either relics from the forest or, like the palm *Medemia nobilis*, otherwise occur on hydromorphic soils. Most trees are no more than 8–12 m tall (*Sclerocarya caffra, Maytenus linearis, Acridocarpus excelsus, Hyphaene shatan, Dicoma incana, D. oleifolia, Erythroxylum platycladum*). Forest species which also occur in grassland (*Stereospermum variabile, S. euphorioides, Tamarindus indica*) are always of lower stature in the latter. Apart from an ability to coppice after fire, the trees show few adaptations to fire, an argument in favour of a recent origin of the communities in which they occur.

In Western Madagascar the woody vegetation along streams and in damp hollows is frequently dominated by species of *Pandanus* spp., which are very conspicuous (e.g. phot. in *Webbia*, 28, 1973, p. 42). In the higher areas they are associated with the introduced *Cosmos*. The grasslands of the southern plateaux about Ambatofinandrahana are characterized by the presence of caulescent *Aloe capitata* var. *cipolinicola*. The other varieties of this species are almost stemless but this plant seems to be fire-resistant, the area being a 'pachycaul' grassland (D. J. Mabberley, pers. comm.).

Despite their great extent and very varied climatic and edaphic conditions, the secondary grasslands of West Madagascar are floristically poor. At most there are 300 species, and of these more than half are transient ruderals. If those species which only grow in the shade of trees, or are confined to swampy places, are also excluded, then only 84 heliophilous species which occur on well-drained soil and can withstand annual burning are left.

Some species, e.g. *Themeda quadrivalvis, Erythroxylum platycladum, Dicoma oleifolia* and *Medemia nobilis*, are confined to or are more abundant in the North-West. Others, e.g. *Tragus berteronianus, Aristida congesta, Loudetia filifolia* subsp. *humbertiana* and *Terminalia seyrigii* characterize the South-West. Nevertheless, the majority of species occur throughout the Western Region, and several extend into the Eastern Region, some as far as the coast.

The majority of species occurring in the grasslands of the West are neither characteristic nor faithful. This is a reflection of their origin. The majority have been introduced from other countries, or are forest species which survive the destruction of forest without showing notable modifications. Of the 84 typical grassland species mentioned above, 31 are adventive. Of the others, 42 are certainly indigenous and 11 probably so. Twenty-four indigenous species have had a forest origin, four have been recruited from the deciduous thicket of the South, and two species of palm, *Medemia nobilis* and *Borassus madagascariensis*, have originated from riparian forest. Eighteen other species come from dry deciduous forest. They include the following, which are among the most frequent species in the western grasslands: *Tamarindus indica, Cassine aethiopica, Erythroxylum platycladum, Stereospermum variabile, Fernandoa (Kigelianthe) madagascariensis* and *Terminalia seyrigii*.

Only 18 endemic species are confined to grassland. Four are phanerophytes (*Hyphaene shatan, Acridocarpus excelsus, Dicoma incana* and *D. oleifolia*) and eight are hemicryptophytes, including *Aristida rufescens* and *Loudetia simplex* subsp. *stipoides* (the typical subsp. is widespread on the African mainland). Certain other species, which are confined to grassland in Madagascar, also occur in Africa (*Sclerocarya caffra, Maytenus linearis, Sporobolus festivus*) or Asia (*Leptadenia reticulata*). Morat convincingly argues that the presence of the heliophilous endemics is evidence for the occurrence in Madagascar before the appearance of man of small open communities which perhaps occupied stations least favourable to forest, such as compact soils and rock outcrops. In such situations the forest was probably stunted and had an open canopy, thus permitting the persistence of species intolerant of shade.

XXI Other offshore islands

Introduction

Macaronesia
    The Azores
    Madeira
    The Canary Islands
    The Cape Verde Islands

Islands in the Gulf of Guinea
    São Tomé
    Príncipe
    Annobon

South Atlantic Islands
    Ascension
    St Helena

Socotra

The Comoro Islands

The Seychelles

The Mascarenes
    Mauritius
    Réunion
    Rodrigues

Aldabra and other coral islands of the western Indian Ocean

Introduction

This chapter deals with all the main islands lying between the African mainland and the mid-oceanic ridges in the Atlantic and Indian Oceans, except for Madagascar, which is the subject of the previous chapter, and islands situated on the continental shelf close to the mainland such as Bioko (Fernando Po) and Zanzibar.

The position of the islands is shown on Figures 24–27. The vegetation of the larger islands is often complex and often is markedly different from corresponding vegetation on the mainland. On some islands it has been almost entirely destroyed. For these reasons it is difficult to show at a scale of 1:5 000 000. For some islands, namely the Canaries (Fig. 25), the Cape Verde Islands (Fig. 26) and Socotra (Fig. 8—see page 112), the vegetation has been shown separately at a larger scale.

Macaronesia

Refs.: Allorge *et al.* (1946); Dansereau (1966); Engler (1910, p. 816–70); Eriksson *et al.* (1974); Humphries (1979); Sunding (1979).

The islands of Macaronesia, which comprise the five archipelagos of the Azores, Madeira, the Salvage Islands, the Canary Islands, and the Cape Verde Islands, are situated in the Atlantic Ocean between 39° and 15° N. and at distances from the European and African continents of 115 to 1 600 km. Their total area is about 14 400 km². In formal, hierarchical, chorological classifications, such as those of Engler (1964), Good (1974), and Takhtajan (1969), Macaronesia is recognized as a floristic Region.

The archipelagos vary considerably in size and diversity from the Salvage Islands with a land area of less than 15 km² and a maximum elevation of 183 m to the Canary Islands with an area of more than 7 000 km² and an altitudinal range of more than 3 700 m. The islands are principally composed of Tertiary and more recent volcanic rocks, but there are also Jurassic and Cretaceous sedimentary rocks in the Cape Verde and Canary Islands respectively. Whether the various island groups were ever connected to each other or to the mainland, and if so when, still remains controversial, though Dietz & Sproll consider the two eastern of the

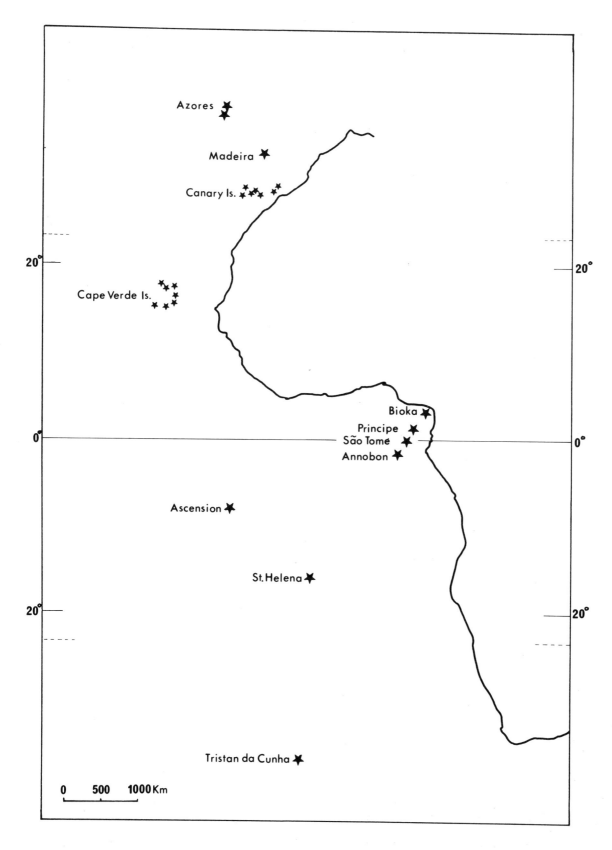

FIG. 24. Islands of the eastern Atlantic Ocean

Canary islands as continental fragments with volcanic overlays.

There is a wide range of climates. The Azores are colder and wetter than the more southerly archipelagos. By comparison, Madeira is much warmer and, although there is a north–south gradient in rainfall, the vegetation zones are relatively uniform all around the island. The climate of the Canaries, by contrast, shows much wider extremes and this is reflected in its richer flora and more diverse vegetation. Cloud belts are restricted to northerly or north-easterly slopes, while the southern slopes are more arid. Because of the high altitude of some islands several altitudinal zones of vegetation can be recognized. The tropical Cape Verde Islands are uniformly arid except in the mountains and show only slight zonation. Throughout Macaronesia the vegetation has been severely degraded by man, and several endemic species are believed to be threatened with extinction (Lucas & Synge, 1978). Although there are many publications dealing with the floristics and plant taxonomy of Macaronesia, detailed ecological studies are few and there are no syntheses.

The total flora of Macaronesia comprises approximately 3 200 species of flowering plants, of which c. 680 (20 per cent) are said to be endemic (Humphries, 1979). Included in the former total, however, is a large number of introduced species, so that in fact the endemics are proportionally more numerous. Generic endemism is relatively weak, amounting to only 31 (12.4 per cent) out of a total of 251 genera. Eighteen are confined to a single archipelago (17 in the Canaries, one in Madeira). The majority of endemic genera are mono- or oligo-specific, only four (*Aichryson, Argyranthemum, Monanthes, Sinapidendron*) having more than five species. Only three endemic genera (*Picconia, Pleiomeris* and *Visnea*) include large trees.

Most of the non-endemic genera are also poor in species, but a few genera such as *Aeonium* (36 species), *Sonchus* (29 species), *Echium* (28 species), *Lotus* (27 species) and *Argyranthemum* (endemic, 22 species) have undergone a remarkable adaptive radiation, especially in the Canary Islands.

A distinctive feature of the Macaronesian flora is the large number of arborescent, often pachycaul, species in otherwise predominantly herbaceous genera, such as *Echium, Sonchus, Limonium, Plantago* and *Sanguisorba*.

Most of the tree species are endemic, notably *Apollonias barbujana, Arbutus canariensis, Clethra arborea, Cytisus (Teline) stenopetalus, Dracaena draco, Erica scoparia* subsp. *azorica, Euphorbia tuckeyana, Heberdenia excelsa (bahamensis), Ilex canariensis, I. perado, I. perado* subsp. *platyphylla, Juniperus brevifolia, J. cedrus, Laurus azorica, Ocotea foetens, Persea indica, Phoenix atlantica, P. canariensis, Picconia (Notelaea) azorica, P. excelsa, Pinus canariensis, Pittosporum coriaceum, Pleiomeris canariensis, Sideroxylon marmulano* and *Visnea mocanera*.

The floristic relationships of Macaronesia are extremely diverse, including affinities with America. The largest elements, however, consist of Mediterranean linking species, and of endemic species of Mediterranean affinity. The floras of the arid lowlands in the Canaries and Cape Verde Islands are closely related to those of the nearby African mainland. Other linking elements involving more distant parts of Africa include:

1. *Erica arborea*: Canaries, Madeira, Mediterranean Region, Sahara Mts (Tibesti), Mts of E. Africa from Ethiopia to southern Tanzania.
2. *Myrsine africana*: Azores, African mainland from the Red Sea Hills southwards to the Cape, westwards from Tanzania to Angola, and eastwards to China.
3. *Canarina canariensis*: Canaries; *C. abyssinica* and *C. eminii* on E. African mountains from Ethiopia to Tanzania.
4. *Ocotea foetens*: Madeira, Canaries; *O. gabonensis*, Gabon, Congo Republic; *O. bullata* (including *O. kenyensis*), Ethiopia to the Cape.
5. *Dracaena draco*: Canaries, Cape Verde Islands. *D. ombet*: Egypt south to Ethiopia. *D. cinnabari*: Socotra.
6. *Visnea mocanera*: Canaries. *Balthasaria mannii*: São Tomé. *B. schliebenii*: East African Mts.

There can be little doubt that the history of the Macaronesian flora is complex, and many unresolved problems remain. It is widely believed that the 'Laurel' forest represents a relic of a humid subtropical flora which was widespread in southern Europe and parts of North Africa during the Upper Tertiary.

The Azores

Refs.: Dansereau (1966); Guppy (1917); Marler & Boatman (1952); Sjögren (1973); Tutin (1953).

The Azores are a group of nine islands and a number of rocks lying between about 37° and 39° N. and 25° and 32° W. Their total land area is about 1 800 km². The island of Fayal in the middle of the group is 450 km from Lisbon and 1 900 km from Newfoundland. The islands are all volcanic and of recent origin. Pico, the highest island, rises to 2 300 m.

The climate is extremely oceanic, characterized by a moderate rainfall spread evenly throughout the year, high relative humidity, and a small temperature range. Frost occurs at high elevations.

Of the 700 or so species of flowering plants at least 200 have been introduced. About 40 are endemic. There are no endemic genera. The affinity of the flora is overwhelmingly European, though the trees of the 'Laurel forest' are all, with the exception of *Myrica faya* and *Persea indica*, endemic species or varieties, nearly all of which appear to be closely allied to Madeiran species.

1. Coastal vegetation

The most abundant species include *Solidago sempervirens, Juncus acutus, Euphorbia azorica* and the grasses *Cynodon dactylon, Agrostis azorica* and *Polypogon monspeliensis*. A lava flow on the eastern side of Pico produced by an eruption in 1718 which was still uncultivated in 1929 gave an indication of the natural vegetation. *Myrica faya* (2–3 m high) was dominant, though the bushes were deformed by the wind. There were scattered individuals of *Erica scoparia* subsp. *azorica* and *Calluna vulgaris*.

2. *'Laurel forest'*

This type, which consists almost entirely of broad-leaved trees, appears to represent the climax up to *c.* 600 m. It is dominated by *Laurus* (*Persea*) *azorica* and *Myrica faya*, which reach a height of 6–7 m in favourable localities. Other species in the tree layer include *Rhamnus latifolia*, *Ilex perado* subsp. *azorica*, *Erica scoparia* subsp. *azorica*, *Viburnum tinus*, *Vaccinium cylindraceum*, *Persea indica* and *Picconia azorica*. The shrub layer is dominated by *Myrsine africana*. According to Guppy the forests were originally much taller than at present and *Myrica faya*, *Laurus azorica* and *Erica scoparia* subsp. *azorica* reached heights of 15 m, 15 m and 11 m respectively.

3. *'Ericetum azoricae'*

From 600 to 1500 m; frost occurs above 760 m. *Erica scoparia* subsp. *azorica*, which grows to a height of 4.5–6 m, is dominant, with *Juniperus brevifolia* subdominant. *Juniperus*, which has suffered greatly from felling for its valuable timber, was probably formerly dominant, and was considerably larger. Guppy states that *Taxus baccata* formerly grew in the lower part of the zone in considerable quantity but became extinct through excessive felling.

4. *'Callunetum'*

From 1500 m to the summit of Pico (2300 m); between 1500 and 1800 m *Calluna* is mixed with *Erica scoparia* subsp. *azorica*. True *Callunetum*, which is an open community on steep and often rather unstable slopes of volcanic debris and little-weathered lava, is floristically very poor. The only common species are *Calluna vulgaris*, *Daboecia azorica* and *Thymus caespititius*.

Madeira

Ref.: Cockerell (1928); Hansen (1969); Sjögren (1973, 1974, 1978); Vahl (1905).

The Madeira islands lie approximately 560 km from the African coast and 450 km N. of the Canaries. Madeira, the main island, reaches an elevation of *c.* 2000 m.

About 1140 species of flowering plants and ferns have been recorded from the islands. Of these at least 250, and probably many more, have been introduced. About 120 species are believed to be endemic. The monotypic *Chamaemeles* is the only endemic genus.

Accounts of the vegetation are conflicting and difficult to reconcile. Much of the vegetation has been destroyed by man, but, according to Dansereau (1966), on steep escarpments on the northern side of the island extensive Laurel forests nearly 30 m high still survive. The 'relict Tertiary' flora is represented by *Pittosporum coriaceum* (endemic), *Visnea mocanera* (Canaries), *Clethra arborea* (endemic), *Sideroxylon marmulano* (Canaries, Cape Verde), *Heberdenia excelsa* (Canaries), *Picconia excelsa* (Canaries), *Persea indica* (Azores, Canaries), *Appollonias barbujana* (Canaries), *Ocotea foetens* (Canaries) and *Dracaena draco* (Canaries, Cape Verde).

The Canary Islands

Refs.: Börgensen (1924); Bramwell (1976); Burchard (1929); Ceballos & Ortuño (1951); Ciferri (1962); Dansereau (1968); Follmann (1976); Kämmer (1974, 1976); Künkel (1971, 1976); Lems (1960); Lindinger (1926); Schenck (1907); Schmidt (1954, 1976); Sunding (1970, 1972, 1973a).

The Canaries are a group of seven islands situated approximately 28°N. of the Equator (Fig. 25). Their total land area is 7273 km^2 and the highest peak, on Tenerife, reaches 3718 m. Ecologically the islands can be divided into two groups. In the first, the eastern islands of Lanzarote and Fuerteventura, which lie little more than 100 km from the African coast and do not exceed an elevation of 650 m, have an arid climate. By contrast, the western islands (Gran Canaria, Tenerife, Gomera, Hierro, and La Palma) are situated between 200 and 360 km from the mainland and have a more oceanic climate.

In general the Canaries have hot, dry summers and warm, wet winters. Moisture is brought by the northeast trade winds, which are responsible for a cloud zone between 800 and 1500 m on the north side of all the western islands. It is widely believed that fog precipitation has a great influence on vegetation, but according to Kämmer the distribution of forest types on Tenerife is not greatly influenced by this factor. The southern sectors of the islands are in a rain shadow and are mostly without a forest zone at medium altitudes. The extreme dryness of the eastern islands and the south of Gran Canaria is partly due to the hot dry Sahara wind, the Harmattan, which sometimes blows for up to a week at a time. The eastern islands are too low to intercept the moisture-laden winds except at their highest points. The trade wind is responsible for differences of over 10°C between the north and south coasts of the larger islands. On Tenerife above 1900 m snow lies for about five months of the year.

The flora of the Canaries amounts to *c.* 1800 species including many introduced species. Some 460 species (25.5 per cent) are endemic, as are 17 genera. All 13 of the Macaronesian genera occurring in more than one archipelago are found in the Canaries.

Of recent studies on the vegetation of the Canaries, Ceballos & Ortuño (1951) and Dansereau (1968) have described the Laurel forests of the western islands, and

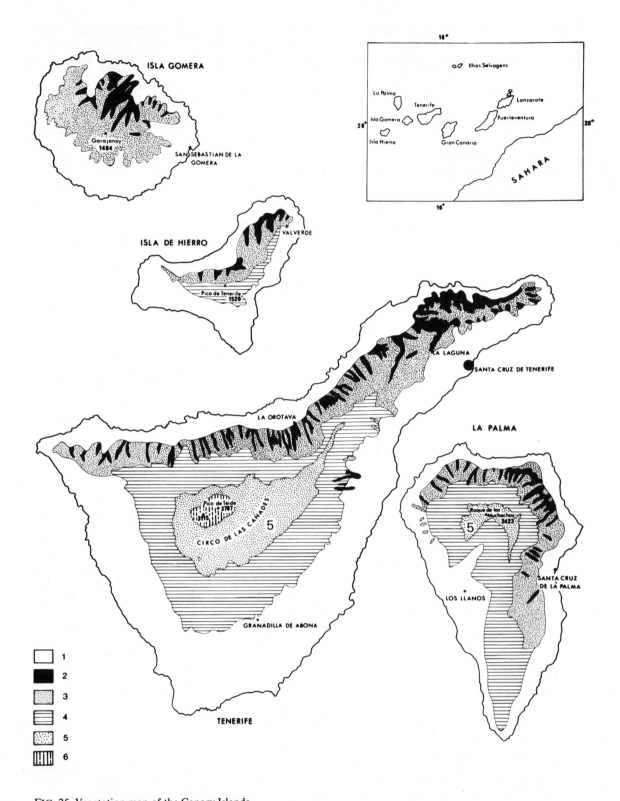

FIG. 25. Vegetation map of the Canary Islands
1. Xerophytic scrub with small islands of *Juniperus phoenicea*. 2. Laurel forest. 3. 'Fayal–Brezal' communities characterized by *Myrica faya* and *Erica arborea*. 4. *Pinus canariensis* forest with small areas of *Cytisus proliferus* shrubland. 5. 'Retama–Codeso' shrubland characterized by *Spartocytisus nubigenus* and *Adenocarpus viscosus*.
6. Altimontane communities

Dansereau (1966) has written a brief account of zonation on Tenerife, as has Kämmer (1974). The following generalized zones are recognized by Bramwell & Bramwell (1974):
1. *Xerophytic scrub zone.* 0–700 m. Occupies the lower slopes of all islands. Stem and leaf succulents, especially species of *Euphorbia*, *Aeonium* and Compositae, predominate. Scrub forest dominated by *Erica arborea* and *Juniperus phoenicea* occurs on some southern slopes towards the upper limit.
2. *Evergreen forest zone.* 400–1300 m. Laurel forest is mostly confined to the northern slopes of the western islands, but is occasionally found on southern slopes. The main dominants are *Laurus azorica*, *Apollonias barbujana*, *Ocotea foetens* and *Persea indica*. On Gran Canaria only 1 per cent of the original forest survives, and on Tenerife less than 10 per cent.
3. *Pine 'forest' zone.* 1200–1900 m. *Pinus canariensis*, which can grow up to 30 m tall, usually forms open stands with sparse field and shrub layers. The most common shrubs are *Adenocarpus foliolosus*, *Cistus symphytifolius*, *Daphne gnidium* and species of *Micromeria*.
4. *Montane zone.* Above 1900 m. Open shrubland dominated by Leguminosae with many endemic species belonging to several families.

The Cape Verde Islands

Refs.: Barbosa (1968a, 1968c); Chevalier (1935); Humphries (1979); Saraiva (1961); Sunding (1973b, 1974, 1977, 1979); Teixeira & Barbosa (1958).

The Cape Verde Archipelago, which comprises ten islands and eight islets (Fig. 26), is situated in the Atlantic Ocean 445 km from the African mainland, and 1400 km SSW. of the Canaries. It lies between latitudes 14°48′ and 17°12′N. and 22°44′ and 25°22′W. The highest altitudes of the islands are: Fogo 2829 m, Santo Antão 1979 m, Santiago 1392 m, São Nicolau 1304 m, Brava 976 m, São Vicente 725 m, Maio 436 m, Sal 406 m, Santa Luzia 395 m, Boa Vista 387 m.

The islands, which were uninhabited when they were discovered by the Portuguese in 1460–62, now have a population of *c.* 160000. They are almost entirely volcanic in origin and are principally composed of basalt and phonolite. Volcanic activity has probably not entirely ceased and there are extensive areas of ash, pumice, and relatively recent lava. The caldera of Fogo erupted as recently as 1857 and produced lava flows which are still covered mainly with lichens. It erupted again in 1951.

Where sufficient water is available the soils are very fertile, but elsewhere the surface is stony and unproductive and the characteristic soils are alkaline pedocals. Saline soils are also extensive and locally, in the arid lowlands, there are large areas of sand dunes.

In general, because of the geographical position of the islands, the climate is dry. Precipitation adequate for permanent cultivation is confined to certain favoured slopes. In much of the lowlands mean annual rainfall is less than 250 mm. In the capital, Praia, for instance, it is 213 mm (183 mm July to October, 30 mm during the rest of the year), and the mean annual temperature is 25°C.

The archipelago is situated at the same latitude as the Sahel zone of the African mainland and is influenced by the same wind systems, and experiences the same problems due to cycles of drought years. Basically it has a similar semi-desert climate. In the archipelago, however, humidity is profoundly modified locally by aspect and altitude.

The prevailing winds are the trade winds, which blow regularly from the NNE. for most of the year. They are responsible for an appreciable amount of condensation, often in the form of mist on NNE.-facing slopes, especially between 400 and 1300 m. Outside the regions favourable for condensation, however, they exert a damaging, desiccating influence on the vegetation, and most of the trees are deformed by wind-action.

The SSW.-facing slopes of the islands are subjected to a totally different rainfall regime. They are characterized by rare and sporadic but violent rainfall confined to a short period of the year. Such precipitation occurs when the warm and humid monsoon from the southern Atlantic arrives at the islands from the W. to SSW. In good years rainfall is adequate for a rich harvest of maize and beans, and supports lush pastures, but often these rains fail completely, sometimes for years on end.

At low altitudes the islands are always arid and the lower they are the most desert-like are the conditions. At higher altitudes there are more indigenous species, and the islands with the largest high-altitude area exposed to the NNE. are those which show least oscillation in crop production. This is the case with Santiago, which has the greatest length along the direction ESE.–WNW. In position it is thus the most favoured of the islands.

Although rainfall is irregular, it seems to follow a cycle of approximately ten dry years alternating with a similar number of wetter years. Teixeira & Barbosa (1958) give figures which show that during the first half of this century the population increased during the moist phases and declined by emigration and starvation during the dry phases. Thus, between 1920 and 1930 it declined by 13376. During the next decade it increased by 34987, only to fall again by 23243 during the next three years.

Flora

At present *c.* 650 species of vascular plants have been recorded from the archipelago (Sunding, 1973b, 1974) of which many were deliberately or accidentally introduced by man. In general, species with tropical affinities occur in the lowlands, whereas species otherwise confined to Macaronesia, or with Mediterranean affinities, occupy the mountains, but there are many exceptions to this. It seems that, at most, only 17 large woody species form part of the original flora, and the status of even some of

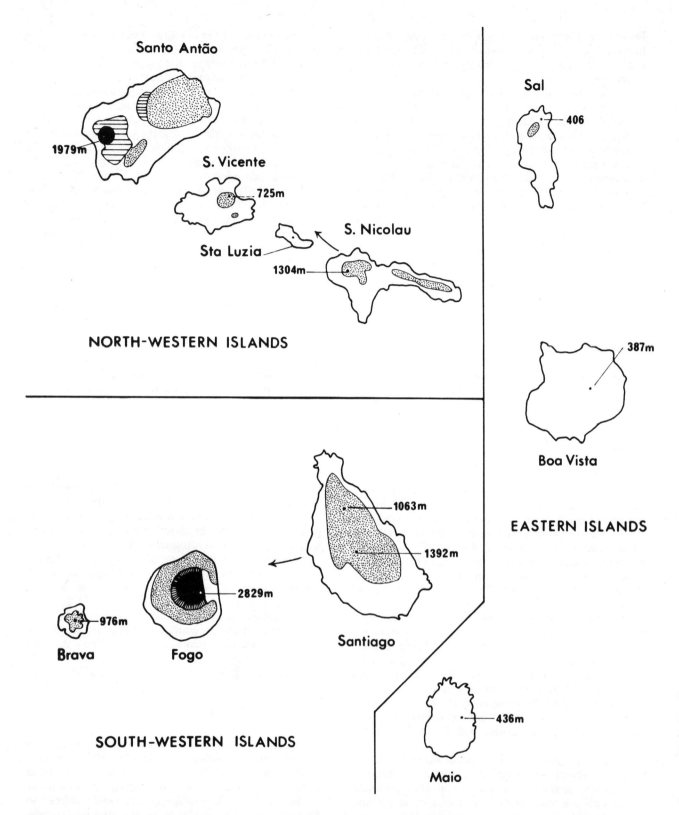

FIG. 26. Cape Verde Islands, showing the four main agroclimatic zones: (1) (stippled), agricultural crops, plantations and subhumid pastures; (2) (striped), upland arid pastures; (3) (white), lowland arid pastures; (4) (black), rocky high mountain peaks.
(The relative position of the islands within the three groups is shown accurately, but, for reasons of scale, the distances between the groups have been telescoped)

these is doubtful. The information summarized below was obtained from Chevalier (1935) and L. A. Grandvaux Barbosa (pers. comm., and from his collections described by various authors in *Garcia de Orta*, vols. 2–4, 1975–79):

1. *Euphorbia tuckeyana*. Candelabra-shaped pachycaul 'tree' up to 4 m high. Endemic but with close relatives in the Canaries. The most characteristic plant of Cape Verde vegetation. Sometimes widely spaced, sometimes it forms almost impenetrable thickets mixed with giant *Echium*, and *Sideroxylon*, and entwined with *Periploca laevigata* and *Sarcostemma daltonii*. It is most abundant above 1000 m but descends locally almost to sea-level. The stems are used for firewood and nearly everywhere it has been destroyed.
2. *Echium hypertropicum*. Candelabra-shaped pachycaul up to 2 m high. Endemic and confined to Fogo. From 500 to 800 m. Used for firewood, and in danger of extinction.
3. *Phoenix atlantica*. Palm up to 10 high. Endemic. From 200 to 300 m. Leaves and stems fed to goats. Fruits eaten by people and swine.
4. *Sideroxylon marmulano*. Shrub or small tree up to 6 m high. Also in Madeira and the Canaries. Up to 800 m. Bark used for tanning. Fruits eaten. In regression but survives on inaccessible rocky escarpments.
5. *Cytisus stenopetalus*. Shrub or small tree up to 5 m high. Also in Canaries, and with a related species in Madeira. Formerly common on the mountains of São Antão and Fogo. Formerly much sought after for carpentry, and now almost extinct.
6. *Rhus albida*. Shrub. Also in Morocco and the Canaries. Rocky places near the sea.
7. *Tamarix canariensis*. Shrub or small tree up to 8 m high. Also in the Canaries and western Mediterranean. Near the sea.
8. *Dracaena draco*. Candelabra-shaped tree up to 10 m high. Also in Canaries and Madeira. Almost extinct in Cape Verde. Sometimes cultivated, but elsewhere indigenous and surviving on almost vertical rocky faces (Chevalier).
9. *Olea europaea*. Small tree up to 9 m, often stunted. Widespread in Mediterranean and beyond. From near sea-level to 1000 m. Indigenous (Chevalier). Introduced and normally not fruiting (Barbosa).
10. *Acacia albida*. Tree 7–12 m high, often deformed by wind, sometimes occurring as very spiny bushes 2–3 m high. Widespread in the drier parts of tropical Africa and beyond. In the Cape Verde Islands, together with *Ficus sycomorus*, it is now the most characteristic species of pastures up to c. 1000 m. It was probably very abundant when the Archipelago was discovered, but it has often been destroyed by man and damaged by goats which climb into its branches (Chevalier). Possibly introduced, but if so now thoroughly naturalized (Barbosa).
11. *Dichrostachys cinerea*. Shrub or small tree up to 5 m high. Widespread in the drier parts of Africa and beyond. Locally abundant in the lowlands. Certainly indigenous (Chevalier).
12. *Tamarindus indica*. Tree up to 10 m high. Widespread in the drier parts of Africa and beyond. Possibly introduced, but if so now thoroughly naturalized.
13. *Ficus sycomorus*. Small tree. Widespread in the drier parts of Africa and beyond. In arid, uncultivated places. One of the most characteristic spontaneous species of the Archipelago from sea-level to 1000 m, and occurring on all the islands (Chevalier).
14. *Grewia villosa*. Shrub. Widespread in the drier parts of Africa and beyond. In dry, uncultivated places. One of the most widespread and characteristic species in the Archipelago. Probably indigenous.
15. *Calotropis procera*. Pachycaul 'tree' up to 4 m high. Widespread in the drier parts of Africa and beyond. Throughout the Archipelago below 1000 m, especially on the drier islands. Probably indigenous.
16. *Dodonaea viscosa*. Shrub. Pan-tropical. In ravines near the sea on Fogo. Almost certainly indigenous.
17. *Ficus capensis*. Tree up to 10 m. Widespread in the wetter parts of tropical Africa. In moist valleys in the most humid mountainous parts of the Archipelago. Almost certainly indigenous.

Vegetation

Nearly everywhere the original vegetation has been almost totally destroyed or altered beyond recognition, and a very high proportion of the contemporary flora has been deliberately or accidentally introduced by man. Today these exotics are much more conspicuous than what remains of the original flora, and it is sometimes difficult to decide whether a species is indigenous or not. For these reasons it is impossible to prepare maps of the potential vegetation. For each island, however, Teixeira & Barbosa (1958) have published very detailed agroclimatic maps in colour at scales of 1:50000 to 1:100000. Barbosa, has subsequently (1968a) produced a simpler classification including some details of the indigenous flora.

In the following account, which is based on the aforementioned publications, the ten mapping units of Teixeira & Barbosa have been reduced to four. Their approximate extent is shown diagrammatically in Figure 26.

1. *Areas of agricultural crops, plantations and subhumid pastures* ('Sequeiro húmido, sub-húmido, semi-árido, pastagens sub-húmidas de altitude' of Teixera & Barbosa).

On the well-watered NNE.-facing slopes, the original vegetation has virtually disappeared and has been replaced by crops on the intensively cultivated and carefully terraced steep terrain. The more important products are sugar-cane (especially under irrigation), bananas, *Coffea arabica*, potatoes, sweet potatoes (widely planted: the true potato, 'bateta inglesa', is much less frequent), tobacco, *Carica papaya*, *Mangifera indica*, *Psidium guajava*, *Cicca disticha*, *Cajanus cajan*, *Colocasia esculenta* (*antiquorum*) etc. The soils are so fertile that more or less continuous cultivation is possible for thirty years or more. Fallows are characterized by abundant Compositae including *Tagetes patula*, *Bidens pilosa*, *Acanthospermum hispidum*, *Ageratum conyzoides* etc. There are also extensive thickets of *Lantana camara*. A few indigenous species survive by streamsides, notably *Pteris vittata*, *Dryopteris parasitica*, *Equisetum ramosissimum*, *Melinis minutiflora* and *Elvira biflora*. In deep ravines *Hyparrhenia hirta* plays an important role in binding the soil.

In drier areas, which are transitional between the croplands just described and the arid pastures, maize (*Zea mays*) and various pulses, principally *Dolichos lablab* (*Lablab niger*, *L. purpureus*), *Vigna unguiculata*, *Phaseolus lunatus* and *P. vulgaris*, are the most important food plants, though manioc and the sweet potato are also grown, as is cotton. Elsewhere there are patches of grazed wooded grassland with *Hyparrhenia hirta*, *Heteropogon contortus* and *Acacia albida*. *Ziziphus mauritiana*, probably introduced for its edible fruits, also occurs, as do *Desmanthus virgatus*, *Desmodium tortuosum*, *Crotalaria retusa*, *Panicum maximum* and *Rhynchelytrum repens*, which were also introduced.

2. *Upland arid pastures* ('Pastagens áridas de altitude', Teixeira & Barbosa, 1958).

This type occurs mostly above 1 400 m. On well-developed soil there is a mixture of grasses, including *Hyparrhenia hirta*, *Pennisetum polystachyon* and *Melinis minutiflora*, and dwarf shrubs such as *Lavandula dentata*, *L. rotundifolia* and *Micromeria forbesii*.

In more rocky places the following species occur: *Campylanthus salsoloides*, *Verbascum capitis-viridis* (*Celsia insularis*), *Cytisus* (*Teline*) *stenopetalus*, *Echium stenosiphon*, *E. vulcanorum*, *Globularia* (*Lytanthus*) *amygdalifolia*, *Erysimum caboverdeanum* and *Sonchus daltonii*.

Euphorbia tuckeyana, *Globularia amygdalifolia* and *Artemisia gorgonum* are characteristic of ancient lava flows, whereas *Helianthemum gorgoneum* is dominant on pumice.

In sheltered places where there is sufficient soil and an adequate water-supply, fruit-trees characteristic of the Mediterranean region are grown but usually only in small patches. They include *Cydonia oblonga*, *Ceratonia siliqua*, *Punica granatum* and *Ficus carica*, and more rarely *Prunus persica*, *Malus domestica* (*sylvestris*) and *Eriobotrya japonica*. In drier places *Ricinus communis* is often planted.

3. *Lowland arid pastures* ('Pastagens áridas de baixa altitude, pastagens muito áridas', Teixeira & Barbosa, 1958).

These pastures support large numbers of goats. The most important grasses are *Aristida adscensionis*, *A. cardosoi*, *A. funiculata*, *Schmidtia pappophoroides* and *Elionurus royleanus*. On the better soils trees, principally *Acacia albida*, *Ficus sycomorus*, *Tamarindus indica* and *Ziziphus mauritiana*, are conspicuous.

In places forbs are dominant. They include *Aerva persica*, *Boerhavia repens*, *Commicarpus verticillatus*, *Cleome viscosa*, *Lotus glinoides*, *Corchorus* spp. and several Malvaceae and Sterculiaceae. In the driest places *Sclerocephalus arabicus* and *Zygophyllum simplex* occur. The principal bushes, in addition to the possibly indigenous *Calotropis procera*, are: *Acacia farnesiana*, *A. nilotica*, *Gossypium hirsutum*, *Jatropha gossypiifolia*, *Nicotiana glauca* and *Parkinsonia aculeata*. They are all introduced. *Jatropha curcas* was formerly extensively planted for the purgative oil, extracted from its seeds, which was exported.

Chenopodiaceae are dominant in saline depressions, and *Sporobolus spicatus* occurs on mobile dunes.

4. *Rocky high mountain peaks*. On the rocky summits of Fogo and Santo Antão higher plants are relatively inconspicuous.

Islands in the Gulf of Guinea

Refs.: Chevalier (1938); Exell (1944, 1952, 1973); Mildbraed (1922); Monod (1960).

The four islands of Bioko, Príncipe, São Tomé and Annobon lie along a line of volcanic activity in a direction approximately NE.–SW. which is continued to the north-east by the massif of Mt Cameroun. Bioko (69 × 32 km) reaches an altitude of 2 850 m and is only 32 km from the mainland. Príncipe (17 × 8 km) is 948 m high and is 210 km SSW. of Bioko and about the same distance from the continent. São Tomé (47 × 27 km) rises to 2 024 m and is about 135 km from Príncipe and 275 km from the African coast. Annobon (7 × 2.5 km) reaches a height of 655 m and lies 180 km SSW. of São Tomé and 340 km from the nearest mainland (Gabon).

The islands are of comparatively recent origin (probably Tertiary) and are mainly composed of basalts and phonolites, which form a fertile soil, usually red in colour. The equator passes just to the south of São Tomé, which has a typical equatorial climate, hot, but not excessively so, with considerable rainfall. Near sea-level mean annual precipitation ranges from less than 1 000 mm in the north-east to more than 4 000 mm in the south-west. In the drier northern half of the island precipitation increases with altitude to 2 600 mm at 700 m and even higher in the mountains, where mist condensation is also an important factor, although, according to Monod, the summit is frequently above the clouds and its climate is drier than is commonly supposed. Annobon is rather drier than São Tomé, but no statistics are available. On São Tomé, at least in the lowlands, little or no rain falls in July and August, and June and September are also fairly dry.

At the time of the discovery of the islands by the Portuguese in 1470–71, Bioko was peopled by Africans of the Bubi tribe, but the other three islands had apparently never been inhabited. Since Bioko is situated on the Continental shelf it is not considered further.

São Tomé

Dense forest formerly covered almost the whole of the island but nearly everywhere this has been destroyed and replaced by plantations, mainly cocoa (below 800 m). Exell (1944) recognizes the following zones:
1. *Littoral zone.* The principal species on sand dunes are *Ipomoea pes-caprae, Canavalia rosea, Cynodon dactylon* and *Sporobolus virginicus.* The small patches of mangrove are dominated by *Rhizophora harrisonii* and *Avicennia marina*, with *Conocarpus erectus* and *Dalbergia ecastaphyllum.*
2. *Lower rain forest zone.* From 0 to 800 m. Now almost completely under cultivation. The principal trees were: *Anisophyllea cabole* (endemic), *Ceiba pentandra, Celtis gomphophylla, C. mildbraedii, C. prantlii, Chlorophora excelsa, Chrysophyllum albidum, Cynometra mannii, Dacryodes edulis, Dialium guineense, Drypetes glabra* (endemic), *Funtumia africana, Heisteria parvifolia, Mammea africana, Mesogyne henriquesii* (endemic), *Monodora myristica, Musanga cecropioides, Pentaclethra macrophylla, Polyscias quintasii* (endemic), *Pseudospondias microcarpa, Tetrapleura tetraptera, Treculia africana* and *Zanthoxylum gilletii.*
3. *Mountain rain forest zone.* From 800 to 1400 m. The principal trees are: *Craterispermum montanum, Discoclaoxylon occidentale* (endemic), *Maesa lanceolata, Olea capensis, Pseudagrostistachys africana, Sapium ellipticum, Symphonia globulifera, Tabernaemontana stenosiphon* (endemic) and *Trichilia grandifolia* (endemic). Rubiaceae and Euphorbiaceae are abundant. Exell did not see a single leguminous tree in undisturbed forest. The canopy is dense and the trees are festooned with rope-like lianes. The trunks of the trees are almost hidden by a dense covering of epiphytic bryophytes, ferns, orchids and species of *Begonia* and *Peperomia.*
4. *Mist forest region.* From 1400 to 2024 m. The principal trees are: *Balthasaria mannii, Cassipourea gummiflua, Peddiea thomensis* (endemic), *Prunus africana, Nuxia congesta, Podocarpus mannii* (endemic), *Schefflera mannii* and *Syzygium guineense* subsp. *bamendae.* The mist forest ascends to the summit, but at that altitude the trees are small and the canopy is not dense, and two non-forest species, namely *Philippia thomensis* and *Lobelia barnsii*, which have close relatives in more open communities in the Cameroun highlands, also occur.

Príncipe

Dense forest formerly clothed the island but in the more accessible regions it has been extensively replaced by cocoa and coffee. Much of what remained of the forest was destroyed during the campaign against sleeping sickness, about 1906, though there has been considerable regeneration of secondary forest since then. Rubiaceae, Euphorbiaceae, Connaraceae, and Orchidaceae are abundant in the natural vegetation whereas Leguminosae and Compositae are poorly represented. The principal forest trees in Príncipe include: *Anthostema aubryanum, Ceiba pentandra, Celtis prantlii, Chlorophora excelsa, Cola digitata, Croton stelluliferus, Dialium guineense, Drypetes principum, Funtumia africana, Heisteria parvifolia* (very abundant), *Irvingia gabonensis, Mammea africana, Monodora myristica, Neoboutonia mannii, Pentaclethra macrophylla, Sterculia tragacantha, Xylopia aethiopica* and *Zanthoxylum gilletii.* Nothing is known of the vegetation of the Pico do Príncipe (948 m), but the narrow, exposed rocky summit of the Pico Papagaio (680 m) is covered by bushes, and has few or no trees.

Annobon

Mildbraed recognizes five formations of which the following are the most important:
1. *Lowland vegetation.* 'Savanna'-like with scattered bushes and patches of cultivated land. Principal species: *Ficus annobonensis, Mucuna sloanei, Rauvolfia vomitoria, Turraea glomeruliflora, Vernonia amygdalina* and *Ximenia americana.*
2. *Dry forest.* Composed principally of *Olea capensis* and *Lannea welwitschii* with *Cavacoa quintasii, Ceiba pentandra, Celtis prantlii, Chaetacme aristata, Discoglypremna caloneura, Pseudospondias microcarpa, Trilepisium madagascariense* and many ferns.
3. *Mist forest.* From 500 m upwards. The principal woody species include *Agelaea* spp. *Cassipourea annobonensis, Craterispermum montanum, Heisteria parvifolia, Rubus pinnatus, Schefflera mannii* and *Strombosia* sp. There is a rich development of epiphytes.

South Atlantic Islands

Ascension

Refs.: Duffy (1964); Hemsley (1885); P. James (pers. comm.).

Ascension is an isolated peak on the mid-Atlantic ridge, and is entirely volcanic except for some patches of beach material. It is probably not more than 10000 years old. The African coast lies 1536 km to the north-east, and due west is the island of Fernando de Noronha, 2048 km away. The island rises to 860 m on an east/west ridge, the Green Mountains. Rainfall increases from 132 mm near sea-level to 645 mm in the Green Mountains, which are frequently capped with clouds. Below 600 m semi-desert conditions prevail.

The indigenous vascular flora is very poor, comprising no more than seven species of flowering plants (three endemic) and 12 species of pteridophytes (three endemic). Non-vascular cryptogams are represented by c. 34 species of moss, 10 hepatics, and 270 species of lichens, of which a few are endemic and the majority reproduce asexually.

Of the indigenous flora the endemic *Euphorbia origanoides* is confined to the coastal desert area, where *Ipomoea pes-caprae*, *Aristida adscensionis*, *Digitaria* cf. *adscendens* and *Portulaca oleracea* also occur. Of the remaining endemic phanerogams, *Sporobolus durus* is now extremely rare and *Hedyotis adscensionis* is believed to be extinct. They, together with a possibly introduced *Wahlenbergia*, were confined to the Green Mountains, where, however, almost all the natural vegetation is composed of cryptogams. Most of the vegetation to be seen today, even far from the settlements, consists of species either deliberately or accidentally introduced by man, including *Setaria verticillata*, *Enneapogon cenchroides*, *Melinis minutiflora*, *Argemone mexicana*, *Opuntia*, *Psidium guajava*, an *Acacia*, and a tall bamboo, which occurs near the summit.

St Helena

Refs.: Hemsley (1885); Henry (1974); Kerr (1971); Mabberley (1975b); Melliss (1875); Turrill (1949).

St Helena is situated 1120 km south-east of Ascension and lies 1760 km from the African coast and 2880 km from the nearest part of America. It is about 16 km long and 13 km broad and is wholly volcanic. The surface is rugged and mountainous and reaches a height of 825 m above sea-level.

The prevailing winds bring rain from the south-east but the coastal belt is arid with a rainfall, in places, scarcely exceeding 200 mm per year. Precipitation increases rapidly inland reaching more than 1000 mm per year in the interior highlands, where mist is also an important factor.

The indigenous flora is small, though there are more than 1000 introduced species, many of which are extensively naturalized. There are c. 39 indigenous species of flowering plants (38 endemic) belonging to 28 genera, of which eight, namely *Commidendrum*, *Lachanodes*, *Melanodendron*, *Petrobium* and *Pladaroxylon* (all Compositae), *Nesiota* (Rhamnaceae), *Trimeris* (Campanulaceae) and *Mellissia* (Solanaceae), are confined to St Helena. There are 27 indigenous species of vascular cryptogams in 13 genera; 12 species are endemic, including the tree-fern *Dicksonia arborescens*. The majority of flowering plants are small trees or shrubs. Several are pachycauls. Some flowering plants have become extinct within historic times. Most of the remainder are now very rare and their future is problematical. Nearly all the non-endemic genera also occur on the African mainland.

Before its discovery in 1502 most of the island was covered with scrub forest. Nearly everywhere the original vegetation has been completely destroyed, either by man for fuel and timber, and during clearing for agricultural operations, or by goats, which were introduced in 1513. Today, the drier parts of the island are virtually desert. Elsewhere, the non-cultivated parts are occupied by invasive exotic plants, notably *Phormium tenax*, *Ulex europaeus* and species of *Solanum* and *Rubus*.

Melliss and others, notably Kerr (reported by Henry), have suggested that the indigenous species occurred in three distinct zones as follows:

1. An outer and lower zone, which is a rocky belt extending all round the coast: *Trochetia* (*Melhania*) *melanoxylon*, *Commidendrum rugosum*, *Mellissia begoniifolia*, *Frankenia potulacifolia*, *Plantago robusta*, *Pelargonium cotyledonis*, *Mesembryanthemum cryptanthum* (*Hydrodea cryptantha*), *Pharnaceum acidum*.
2. An intermediate zone, less rocky than the first: *Phylica ramosissima*, *Commidendrum robustum*, *C. spurium*, *Trochetia erythroxylon*.
3. The Central Highlands zone which has deep soil and originally supported dense vegetation: *Melanodendron integrifolium*, *Pladaroxylon* (*Senecio*) *leucadendron*, *Lachanodes arborea*, (*S. prenanthiflorus*, *S. redivivus*), *Petrobium arboreum*, *Hedyotis arborea*, *Dicksonia arborescens*, *Trochetia erythroxylon*, *Nesiota elliptica*, *Trimeris scaevolifolia*, *Sium helenianum*, *Wahlenbergia angustifolia*, *W. linifolia*.

Socotra

Refs.: Gwynne (1968); Pichi-Sermolli (1957); Popov (1957).

The island of Socotra lies on the African continental shelf 225 km east of Cape Guardafui. Although imperfectly known, it appears that the geological structure is simple and resembles that of neighbouring parts of Africa and Arabia. Socotra is 115 km long and 35 km wide. It is bordered by wide alluvial plains of recent origin. The interior consists largely of an undulating plateau of Eocene limestones, which varies in altitude from 300 to 400 m, locally rising to 900 m. Near the north-east coast a granitic intrusion forms the Hagghier massif, which rises to a height of over 1500 m.

The climate is influenced by the NE. and SW. monsoons. The former bring the main rains, whereas the latter, which are very strong and desiccating, rarely bring rain. There are no long-term rainfall records, but the fragmentary figures available and the vegetation indicate that much of the plateau probably receives 125–200 mm per year, rising to 600 mm or more in the highlands, where heavy mists occur.

Floristically, Socotra belongs to the Somalia–Masai Regional Centre of Endemism. The majority of its species also occur on the African mainland, but a sufficient number are confined to the island to make Socotra stand out as an important local centre of endemism.

Apart from some small patches of *Avicennia* mangrove, a narrow littoral zone of herbaceous halophytic communities, and a large area of almost bare sand dunes on the south coast, the vegetation of the coastal plains consists mainly of semi-desert dwarf shrubland and grassland, sometimes with scattered bushes or dwarf trees.

The vegetation of the limestone plateau is very variable. Surfaces exposed to the desiccating SW. monsoon support sparse grassland with scattered bushes of *Jatropha unicostata*, *Croton socotranus* and *Aloe perryi* with an occasional *Dracaena cinnabari*. In sheltered valleys at about 900 m altitude there are dense thickets of *Acacia pennivenia*, *Ruellia insignis*, *Psiadia schweinfurthii*, *Rhus thyrsiflora*, *Ficus socotrana* etc. On the slopes of the Hamadera Hills there is a remarkable community dominated by *Dracaena cinnabari* (page 115). This type also occurs on the southern slopes of the Hagghier massif. Another spectacular community is the succulent shrubland of the limestone cliffs and valley slopes, especially the rugged slopes of the north. Characteristic species include: *Dendrosicyos socotranus*, *Adenium socotranum*, *Euphorbia arbuscula*, *E. spiralis*, *Dorstenia gigas*, *Kleinia scottii*, *Kalanchoe robusta* and *Aloe perryi*.

On the granite massif of Hagghier, evergreen bushland and thicket (page 115) occurs on the lower slopes, while grasslands dominated by *Themeda quadrivalvis*, *Hyparrhenia hirta* and *Arthraxon lancifolius* cover large areas of the watershed between the granite peaks. They form the main pastures for the Socotran cattle. Rocky outcrops are thickly covered with lichens.

The Comoro Islands

Ref.: Legris (1969).

The Comoro archipelago, which comprises the four volcanic islands of Anjouan, Mayotte, Mohéli and Grand Comore, is situated in the middle of the Mozambique Channel, 300 km from the African mainland and a similar distance from the north-west tip of Madagascar (see Fig. 27). Very little is known about the vegetation but a brief account has been published for Grand Comore, the largest and most recent island. Grand Comore is 62 km long and 24 km wide and has an area of 1 148 km^2. It is the only island which is still volcanically active and the conical peak of Karthala reaches an altitude of 2 355 m. The island consists entirely of basalt. Most of the soils are very immature and only 35 per cent of the surface is cultivable. Despite the high rainfall, much of the water is not available to the vegetation and crops because it rapidly descends to a deep water-table which emerges below sea-level at some distance from the shore.

In the extreme north-east and south-east of the island rainfall is less than 1 500 mm per year and the dry season lasts for 3–6 months. Elsewhere rainfall is between 1 500 and 3 000 mm or more per year and the dry season is of 0–3 months.

According to Voeltzkow (reported by Renvoize, 1979), 935 vascular plants, of which 416 are indigenous, occur in the Comoros; 136 are said to be endemic.

Natural vegetation only survives on any scale in the mountains. The most extensive and luxuriant forests, which are 20–30 m high, occur on the southern and eastern slopes of Karthala between the upper limit of cultivation at 500–800 m and 1 300–1 800 m. Principal species are *Ocotea comoriensis*, *Khaya comorensis*, which sometimes forms up to 80 per cent of the canopy, *Olea* sp., *Chrysophyllum boivinianum*, *Prunus africana* and *Filicium decipiens*.

Recent lava flows are colonized by *Nuxia pseudodentata*, *Breonia* sp., *Weinmannia* sp., *Apodytes dimidiata* and *Olea* sp. *Nuxia pseudodentata* is the pioneer species *par excellence*. It colonizes all lava flows from sea-level to the upper limit of the forest and establishes itself at the same time as the saxicolous lichens and pteridophytes. *Philippia comorensis* is also sometimes found on lava flows down to 600 m.

With increasing altitude the forest becomes shorter and at 1 900 m is replaced by 6–8 m high thickets of *Philippia comorensis*. The anthropic landscapes of the north and east of the island are characterized by scattered trees of *Adansonia madagascariensis*, *Tamarindus indica*, *Jatropha curcas* etc. On uncultivable volcanic soils there are thickets dominated by *Erythroxylum lanceum*, *Phyllanthus comorensis* and *Diospyros comorensis*.

The Seychelles

Refs.: Gibson (1938); Jeffrey (1963, 1968); Procter (1974); Sauer (1967); Sörlin (1957); Swabey (1961, 1970); Vesey-Fitzgerald (1940).

The Seychelles archipelago comprises some 77 islands with a total area of 260 km^2, scattered over 388 500 km^2 of the Indian Ocean and extending in a south-westerly direction for nearly 1 000 km (see Fig. 27). Within this administrative unit most of the islands belong to three distinct groups: the Seychelles proper (a group of mountainous islands of mainly Precambrian granitic and syenitic igneous rocks lying on the Seychelles Bank between 4° and 5° S. and 55° and 56° E.); the coralline Aldabra islands; and the sand-cay islands of the Amirantes. These last two groups are included in the final section of this chapter.

The igneous group of islands is a fragment of Gondwanaland. Mahé, the largest and most mountainous island, is 905 m high and is situated nearly 1 200 km NW. of Madagascar. The principal smaller islands are Silhouette, Praslin, La Digue and Curieuse. The amount and distribution of rainfall varies greatly from year to year and from island to island. At Victoria, at sea-level on Mahé, it is 2 250 mm per year, whereas in the uplands it rises to more than 4 000 mm per year. Relative humidity is high, averaging 70–80 per cent throughout the year.

The indigenous flora comprises 233 species, of which at least 72 are said to be endemic. Subsequent to the settlement of the islands in 1770 many weeds have been

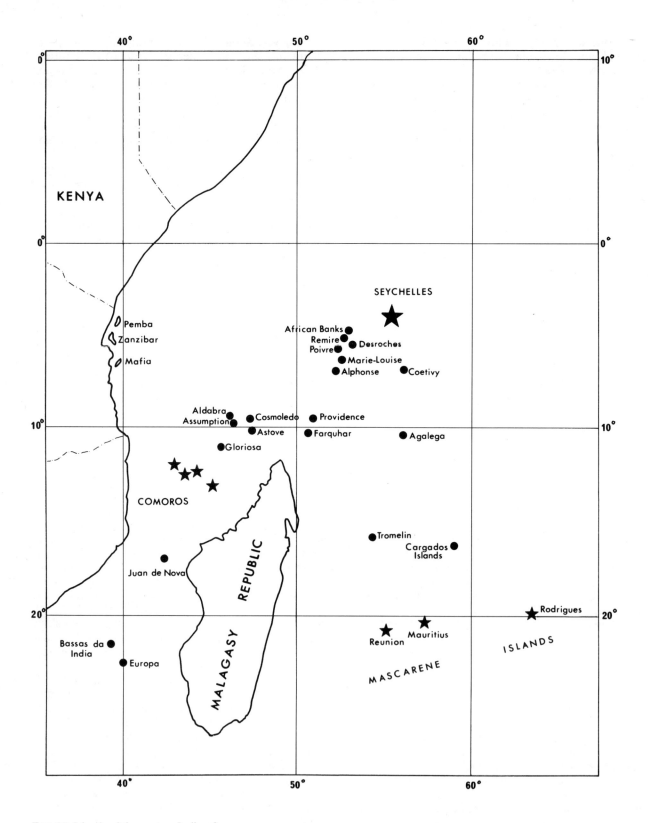

FIG. 27. Islands of the western Indian Ocean
A single large star designates the, mainly Precambrian, Seychelles archipelago. Smaller stars represent the four volcanic islands of the Comoro group and the three volcanic islands of the Mascarenes. Solid circles indicate low islands, which are either raised coral-reef limestone islands or sand cays on sea-level reefs

introduced and other species have escaped from cultivation, so that the adventive flora (247 species) now exceeds the indigenous. One family (Medusagynaceae) is endemic, as are the following ten genera: *Vateria, Geopanax, Indokingia, Protarum, Deckenia, Lodoicea, Nephrosperma, Phoenicophorium, Roscheria* and *Verschaffeltia*, the last six of which belong to Palmaceae.

The original vegetation has been profoundly modified and over extensive areas totally destroyed. Vesey-Fitzgerald and Jeffrey have reconstructed the original vegetation as follows:

1. *Coastal formations*. In addition to mangrove and *Ipomoea pes-caprae* formations, the coast was fringed with coconuts (*Cocos nucifera*), with which were associated such typical sublittoral species as *Scaevola* sp., *Cordia subcordata, Hibiscus tiliaceus, Hernandia ovigera* and *Tournefortia argentea*.
2. *Lowland rain forest*. Up to 300 m on Mahé and Silhouette. Canopy at about 30 m. Dominated by *Imbricaria seychellarum* and *Calophyllum inophyllum*. Associates: *Dillenia ferruginea, Intsia bijuga, Vateria seychellarum* (Mahé only), *Medusagyne oppositifolia* (in deep clefts of granite masses, almost extinct).
3. *Intermediate forest*. From 300 to 550 m on Mahé and Silhouette. Canopy at about 18 m. Dominated by *Dillenia ferruginea* and *Northea seychellana*. Associates: *Soulamea terminalioides, Colea seychellarum, Campnosperma seychellarum, Riseleya griffithii, Aphloia theiformis, Pandanus hornei*. Most of the endemic species apparently occurred in this community.
4. *Mossy montane forest*. Above 550 m on Mahé. Canopy at 12 m or lower. Dominated by *Northea seychellana*. Associates: *Roscheria melanochaetes, Timonius seychellensis, Nepenthes pervillei*.
5. *Drier forest*. In drier parts of Mahé, Silhouette, and Praslin. Dominant: *Dillenia ferruginea*. Prominent associates include *Diospyros seychellarum, Dodonaea viscosa* and *Memecylon eleagni*, and several endemic palms including *Lodoicea maldivica* (Praslin and Curieuse only, the famous Coco-de-Mer), *Verschaffeltia splendida* and *Deckenia nobilis*.

The lowland rain forest no longer exists and the other forest types survive only as small relict patches. Most of the land is occupied by plantations or secondary communities, of which the following are the most important: coppice of *Cinnamomum zeylanicum*, secondary *Albizia falcata* forest, and thickets of the fern *Dicranopteris linearis* on worn-out, eroded, and denuded land.

The Mascarenes

This group of three volcanic islands is situated on the southernmost part of the Seychelles–Mauritius ridge (see Fig. 27). The nearest large land mass is Madagascar, 680 km north-west of Réunion. The prevailing winds are the south-east trades which blow throughout the year, but more or less fitfully from December to April.

The earlier Floras of the Mascarene islands are out of date and not always reliable, so that only very approximate figures on endemism and floristics can be given. A new Flora (Bosser et al., 1976), however, is being prepared, and the vegetation of Réunion has been recently described in considerable detail (Cadet, 1980).

Mauritius

Refs.: Brouard (1963); Sauer (1961, 1962); Vaughan (1968); Vaughan & Wiehe (1937–47).

Mauritius is 62 km long and 46 km wide and has a total area of 1 865 km². The land rises from the coastal plains to a rugged and precipitous central plateau (305–730 m) on which are several extinct craters, and peaks which reach an altitude of 826 m. Vaughan & Wiehe recognize two ecological zones, lowland and upland, which correspond to the coastal plains and the central plateau respectively.

In the lowlands mean annual rainfall varies from 890 mm on the leeward side of the island to 1 905 mm on the south-east coast. In the uplands it ranges from 2 540 to 4 445 mm. Cyclones are frequent and devastate crops but do little damage to the indigenous forests, possibly because of the efficient anchorage of the trees. Torrential rains which accompany cyclones, however, frequently cause landslides in the montane forests which are rapidly colonized by exotic species.

The flora of Mauritius is relatively small. Baker records 869 species for Mauritius and the Seychelles combined.

The indigenous vegetation has disappeared from most of Mauritius. Even where it has not been destroyed it is threatened by more vigorous, invasive, exotic species such as *Furcraea foetida, Ligustrum robustum, Ravenala madagascariensis, Leucaena leucocephala* (*glauca*), *Albizia lebbeck* and *Psidium cattleianum*, which prevent natural regeneration of the native species.

The lowland forests have been virtually destroyed. From the accounts of early explorers it appears that palm stands consisting of *Latania lontaroides, Dictyosperma album* and *Hyophorbe* sp. occurred where rainfall is less than 1 000 mm per year. The moister forests were probably dominated by *Diospyros tesselaria* (endemic) and *Elaeodendron orientale* (also in Réunion and Rodrigues) associated with *Foetidia mauritiana, Stadmannia oppositifolia, Hornea mauritiana* (endemic) and *Terminalia bentzoe* (Réunion, Rodrigues).

The upland communities are somewhat better preserved. The following are the more important ones described by Vaughan & Weihe:

1. *Swamp forest*. Dominated by five endemic species of *Pandanus*.
2. *Sideroxylon thicket*. The dominant, *S. cinereum* (endemic), forms an 8–10 m high open canopy over a dense closed lower stratum of phanerophytes belonging to ninety species.
3. *Upland forest*. The main canopy at 18–21 (25) m is

composed of *Calophyllum eputamen, Canarium paniculatum* (*mauritianum*), *Mimusops maxima* (endemic), *M. petiolaris* (endemic), *Nuxia verticillata* (also in Réunion), *Sideroxylon cinereum* and *S. majus* (*Calvaria major*, endemic). The boles are short and stout, about 1 m in diameter and branch at 10–15 m. There is an intricate system of large roots on the surface of the soil sometimes covering an area three or four times greater than the spreading canopy of the tree. The Sapotaceae have well-developed buttresses.

4. *Mossy forest.* This occurs on Mt Cocotte (744 m). Rainfall is more than 4000 mm per year. Clouds and mists are frequent. The 8–15 m canopy is very irregular, and is composed of *Nuxia verticillata, Eugenia* sp., *Molinaea* sp., *Tambourissa* sp., *Aphloia theiformis, Turraea* (*Quivisia*) *oppositifolia* etc. The trunks and branches are covered with a great variety of filmy ferns, mosses and hepatics.

5. *Philippia thicket.* The thicket, 4 m tall, occupies a small part of the plateau at 610–670 m, where it occurs on little-altered lava. It is dominated by *Philippia abietina* (endemic) associated with *Phylica nitida* (*mauritiana*, also on Réunion) and *Helichrysum yuccifolium* (Réunion).

Réunion

Refs.: Cadet (1980); Rivals (1952, 1968).

Réunion, which is 75 km long and 70 km wide, is situated 780 km E. of Madagascar and 200 km SW. of Mauritius. The central massif culminates in Piton des Neiges at 3069 m. Rainfall at sea-level ranges from 425 mm per year on the dry side of the island to 4290 mm per year on the south-east coast.

Of the 630 species constituting the flora c. 480 are indigenous; 160 species belong to the Orchidaceae, but three-quarters of them are believed to be extinct. Of the remaining species. c. fifty are endemic to Réunion, but a much larger number is shared with the other Mascarene islands.

Vegetation zones are not clear-cut and some tree species have very wide ecological amplitudes. Thus, *Aphloia theiformis* and *Nuxia verticillata* range from sea-level to 2000 m and experience a mean annual rainfall from 800 to 7500 mm, and mean annual temperatures from 10° to 25°C. *Agauria salicifolia* grows from sea-level to 1000 m, above which it is replaced by *A. buxifolia* (also in Madagascar) up to 2500 m.

Rivals recognizes the following plant communities:

1. *Coastal.* Mangrove, halophytic, and *Ipomoea pes-caprae* and *Scaevola* communities.
2. *Dry, megathermic forest* (now destroyed). Below 400 m on the dry side of the island. Important species were *Elaeodendron orientale, Terminalia bentzoe* (*benzoin*), *Mimusops petiolaris, Diospyros melanida* and *Ocotea obtusata*.
3. A complex of *moister forests* formerly ascending from sea-level to 1800 (2000) m on the wet side of the island and occurring between 400 and 1200 m on the dry side. These forests are remarkable for their low stature, with few trees exceeding 15 m in height. Characteristic species include *Calophyllum tacamahaca, Grangeria borbonica* and *Pittosporum senacia*, in addition to those mentioned under 2 above, which, in the moister forests, only occur at low altitudes. On the windward slopes the forest diminishes in stature and at 1700 m is replaced by 4–5 m high elfin thicket dominated by *Forgesia borbonica*.
4. '*Tamarin*' (*Acacia heterophylla*) *scrub forest.* This type occurs in rain-shadow areas between 1200 and 2000 m. The gnarled 8–10 m high Acacias are strongly dominant. Apart from a bamboo, *Nastus borbonicus*, other woody plants are said to be infrequent. Tamarin forest is less dense than the moister forests and is very susceptible to fire. It only regenerates on bare soil, notably following fires. Although young plants are resistant to cyclones, the adults are easily blown over and up to 80 per cent of a stand may be destroyed in this way.
5. *Ericoid communities.* Between 2000 and 2500 m the vegetation of slopes is dominated by *Philippia montana*, and above 2500 m by *Stoebe passerinoides*. Associates include *Hypericum revolutum* (*lanceolatum*), *Phylica nitida* (*leucocephala*) and species of *Psiadia* and *Philippia*.

Rodrigues

Refs.: Balfour *et al.* (1879); Vaughan (1968); Weihe (1949).

Rodrigues, the smallest (109 km^2) of the Mascarene group, is situated 584 km due east of Mauritius and rises to a height of 395 m above sea-level. The climate, which is relatively uniform, is not unlike that of the north-western lowlands of Mauritius. Mean annual rainfall on the north-east coast is 1325 mm. Of the 145 indigenous species, 35 are endemic.

The natural plant communities have disappeared and only individual plants remain. These are confined to the upper reaches of some mountain streams or to sites where peculiar edaphic conditions have restricted cultivation, and alien vegetation has not yet gained a stranglehold.

The original vegetation, apparently, was a low forest 10–15 m high with *Sideroxylon galeatum* (*Calvaria galeata*), *Elaeodendron orientale, Mathurina penduliflora* (endemic genus, Turneraceae), *Diospyros diversifolia, Terminalia bentzoe, Foetidia rodriguesiana* and the palm *Dictyosperma album*.

Palm stands, dominated by *Latania verschaffeltii* and *Hyophorbe verschaffeltii*, and the drought-resisting screw-pine, *Pandanus heterocarpus*, seem to have been the characteristic vegetation of the coral plain on the drier eastern coast. This plain is exposed to the south-east trade winds, which exert a profound influence on the growth form of many species. Shrubs—such as *Carissa xylopicron*—assume a cushion habit and rarely exceed a height of 40 cm.

Aldabra and other coral islands of the western Indian Ocean

Refs.: Gwynne & Wood (1969); Renvoize (1975, 1979); Stoddart (1970).

The scattered low islands of the West Indian Ocean are of two kinds: (a) raised reef-limestone islands best exemplified by the Aldabra group, which are mostly 3.5–8 m above mean low-tide level; (b) sand cays on sea-level reefs including the Farquhar group and African Banks. In recent years the flora and vegetation of Aldabra, the largest island (97 km²), have been studied in great detail in connection with an investigation of the 150 000 or so giant tortoises (*Geochelone gigantea*) which occur on the island. Comparative studies of some of the other islands have also been made. Although no comprehensive synthesis has yet appeared, this work has resulted in an extensive botanical literature, which is summarized below, together with floristic details, where known.

Islands of the Mozambique Channel
— Juan de Nova: Bosser (1952); Capuron (1966); Perrier de la Bâthie (1921*b*).
— Europa: Bosser (1952); Perrier de la Bâthie (1921*b*).

Amirantes group: 72 indigenous species; 25 introduced species. Gwynne & Wood (1969); Vesey-Fitzgerald (1942).
— African Banks: Fosberg & Renvoize (in Stoddart, 1970); Stoddart & Poore (in Stoddart, 1970).
— Remire: Fosberg & Renvoize, op. cit.; Stoddart & Poore, op. cit.
— Desroches: Fosberg & Renvoize, op. cit.; Stoddard & Poore, op. cit.

Aldabra group: *c.* 175 indigenous species of terrestrial vascular plants (*c.* 30 endemic); *c.* 85 introduced species. Fosberg & Renvoize (1980); Renvoize (1971); Vesey-Fitzgerald (1942); Wickens (1979).
— Aldabra: Fosberg (1971); Hnatiuk & Merton (1979); Hnatiuk, Woodell & Bourn (1976); MacNae (1971); Merton, Bourn & Hnatiuk (1976); Stoddart & Wright (1967).
— Assumption: Fosberg & Renvoize, op. cit.; Stoddart (1967).
— Cosmoledo: Bayne *et al.* (in Stoddart, 1970); Fosberg & Renvoize, op. cit.; Stoddart, op. cit.
— Aslove: Bayne *et al.*, op. cit.; Fosberg & Renvoize, op. cit.; Stoddart, op. cit.

Farquhar group: 59 indigenous species. 13 introduced species.
— Farquhar: Fosberg & Renvoize, op. cit.; Stoddart & Poore, op. cit.
— St Pierre: Stoddart (1967); Vesey-Fitzgerald (1942).
— Providence: Stoddart (1967).

Cargados Carajos: 48 km²; 17 indigenous species; 24 introduced species. Staub & Gueho (1968).

Isolated islands
— Coetivy: 49 indigenous species; 16 introduced species. Gwynne & Wood (1969).
— Gloriosa: 20 species. Battistini & Cremers (1972); Stoddart (1967).
— Agalega: *c.* 60 species (J. Procter, pers. comm. in Renvoize, 1979).
— Tromelin: 6 species. Staub (in Stoddart, 1970).

On many of the islands the primary vegetation has been destroyed, either to make way for coconut plantations, as on the Amirantes, or by guano-diggers, as on St Pierre and the islands of the Aldabra group excluding Aldabra itself. The vegetation on Aldabra, which has never been permanently settled, is little disturbed by man, although in places it has been profoundly modified by the resident tortoise population.

The more important vegetation types on Aldabra are as follows:
1. *Sclerodactylon macrostachyum* tussock grassland and *Sporobolus virginicus* turf. Near the coast.
2. Mangrove. About 1–10 m or more high. Dominated by *Avicennia marina*, *Bruguiera gymnorrhiza*, *Ceriops tagal* and *Rhizophora mucronata*, with less abundant *Sonneratia albida*, *Xylocarpus granatum* and *X. molucensis*.
3. *Pemphis acidula* thicket. About 0.5–6 m high, usually on very rocky limestone.
4. Coconut (*Cocos nucifera*) groves. Introduced by man, locally naturalized.
5. *Casuarina equisetifolia* groves. Status uncertain. Possibly indigenous.
6. Mixed scrub. Widespread throughout the atoll, mostly 3–5 m tall, occasionally up to 12 m. No general dominants but the following are common: *Apodytes dimidiata*, *Canthium bibracteatum*, *Cassine aethiopica*, *Erythroxylum acranthum*, *Euphorbia pyrifolia*, *Ficus* spp., *Maytenus senegalensis*, *Ochna ciliata*, *Polysphaeria multiflora*, *Sideroxylon inerme*, *Terminalia boivinii*.
7. Tortoise turf. When intensely grazed 1–2 cm high, otherwise up to 15 cm. Composed principally of *Dactyloctenium pilosum*, *Eragrostis decumbens*, *Panicum aldabrense*, *Sporobolus testudinum*, and *Bulbostylis basalis* and other dwarf sedges.

Azonal vegetation

XXII Mangrove, halophytic and fresh-water swamp vegetation

Mangrove

Herbaceous fresh-water swamp and aquatic vegetation

Halophytic vegetation

Mangrove
(mapping unit 77)

Refs.: Barbosa (1970, p. 121–5); Boughey (1957a, p. 679–80); Chapman (1977); Cole (1968, p. 72–5); Dale (1939, p. 7–8); Giglioli & Thornton (1965); Gledhill (1963); Gossweiler & Mendonça (1939, p. 70–2); Graham (1929); Hédin (1928); Hemming (1961, p. 64); G. Jackson (1964); Kassas (1957, p. 194–6); Kassas & Zahran (1965, p. 167–8); Keay (1953; 1959a, p. 11–13); Koechlin, Guillaumet & Morat (1974, p. 583–91); Letouzey (1968a, p. 239–40); Lugo & Snedaker (1974); Macnae (1963; 1968); Macnae & Kalk (1962a); Moll & Werger (1978); Naurois & Roux (1965); Pichi-Sermolli (1957); Pynaert (1933); Rabinowitz (1978); Rosevear (1947); Savory (1953); Trochain (1940); Walter (1971, p. 150–66); Walter & Steiner (1936); White (MS, 1975–76); White & Werger (1978).

Phots.: Barbosa (1970: 14A.1); Boughey (1957a: 11–12); Gossweiler & Mendonça (1939: 4); Hemming (1961: 1); Karsten & Schenck (1915: 43); Kassas (1957: 1); Kassas & Zahran (1965: 4); Koechlin *et al.* (1974: 186–8); Macnae & Kalk (1962a: 1–4); Naurois & Roux (1965: 1–5); Pynaert (1933: 57–9); Walter (1971: 88–89, 98); Walter & Steiner (1936: 4–5, 8–11, 13–14, 21–2).

Profiles: Giglioli & Thornton (1965: 5); Gledhill (1963: 4); Naurois & Roux (1965: 5); Walter (1971: 87, 91); Walter & Steiner (1936: 7).

Mangrove occurs only on shores where the vigour of the surf is broken by sand-bars or coral reefs or islands. It is most extensively developed on the deltas of large rivers but also occurs in quite small bays and lagoons. Along rivers it can penetrate far inland. In West Africa, for example, it extends for 190 km along the banks of the River Gambia. The seedlings of mangroves require quiet water on accrescent shores for their establishment.

At the mouths of rivers the sea-water is diluted, sometimes for a considerable distance in a seaward direction, as in the case of large rivers like the Niger. By contrast the coastal mangroves growing away from the river inlets, in the shelter of coral reefs, are subjected to undiluted sea-water with an osmotic concentration of 24 atm.

Mangrove attains its best development under a rainforest climate. In Africa the tallest stands, at the mouth of the Niger delta, are 45 m tall (Rosevear, 1947). On some shores, however, it extends far beyond the equatorial zone and even beyond the Tropics of Cancer and Capricorn, but it cannot tolerate frost. On the west side of Africa mangrove is confined to the tropics. The northernmost stand occurs near Tidra in Mauritania at 19°50′N., (Naurois & Roux, 1965), and the southernmost is near Benguela at 12°30′S. (Barbosa, 1970). At both places the mean annual rainfall is very low, *c.* 100 mm and 150 mm, respectively. On the east side of Africa mangrove extends as far north as the Gulf of Aqaba (30°N.) and the Gulf of Suez (28°N.). The southernmost occurrences are near East London (33°S.).

The species occurring in West Africa are completely different from those in East Africa. There are five species in West Africa, namely *Rhizophora mangle, R. harrisonii, R. racemosa, Avicennia germinans (A. africana, A. nitida)* and *Laguncularia racemosa*. All five are also widely distributed on the east coast of tropical America and neighbouring islands.

The mangrove flora of East Africa is more diversified, with nine species, namely *Rhizophora mucronata, Avicennia marina, Sonneratia alba, Ceriops tagal, Bruguiera gymnorrhiza, Xylocarpus granatum, X. moluccensis, Lumnitzera racemosa* and *Heritiera littoralis*. All these extend far towards the east and most reach the western Pacific Ocean.

True mangrove species have either pneumatophores, which are exposed at low tide, or are viviparous or almost so. Most African species show both these features. Species growing with mangroves which lack these characteristics, e.g. *Conocarpus erectus* and *Barringtonia racemosa*, are referred to as 'mangrove associates'.

In *Rhizophora* the stilt roots function as pneumatophores. *Bruguiera, Ceriops* and *Lumnitzera* have knee roots. In *Xylocarpus* and *Heritiera* the main roots are laterally compressed and ribbon-like and their upper surface projects above the surface of the soil. In *Avicennia* and *Sonneratia* erect pneumatophores arise from the main subterranean roots. In *Avicennia* they resemble *Asparagus* shoots. The very large pneumatophores of *Sonneratia* may be 75 cm high and 25 cm in diameter; they are used as floats for fishing nets.

Rhizophora, Ceriops and *Bruguiera* are viviparous. The embryo develops precociously and ruptures the testa and pericarp, after which the hypocotyl undergoes enormous development. It contains the reserve food of the seedling. In *Rhizophora mucronata* it reaches a length of almost one metre. When the seedlings fall they may find anchorage beneath the parent plant, or may be dispersed by ocean currents. Although not truly viviparous, the embryos of *Avicennia, Laguncularia* and *Xylocarpus* are well developed when they fall and emerge as seedlings soon afterwards, from the fruit (in *Avicennia* and *Laguncularia*, Gledhill, 1963; Rosevear, 1947) or seed (*Xylocarpus*, F. White, pers. obs.). The indehiscent fruits of *Sonneratia* are dispersed by bats (van der Pijl, 1957) and those of *Lumnitzera* and *Heritiera* by ocean currents. More information on the dispersal and establishment of water-borne propagules of mangrove species has been published by Rabinowitz (1978).

According to Walter, mangroves are obligate halophytes in that they accumulate sodium chloride in their cell sap. They can, however, in the absence of competition, be grown on non-saline soil. Their leaves are succulent and have water-storage tissue. The transpiration rate is very low. Their roots are able to desalinate seawater to a high degree, but insufficiently so to prevent some salt accumulation in their tissues. Among African genera only *Avicennia* is salt-excreting. In the dry climate of East Africa the undersurface of the leaves of *A. marina* is densely covered with sodium chloride crystals. At night the crystals absorb water hygroscopically from the atmosphere and so dissolve. In more humid climates sodium chloride crystals are not formed. The salt is washed off by the rain.

The three main environmental factors that influence the occurrence and abundance of individual mangrove species in relation to others are:
1. Frequency and duration of flooding with sea-water.
2. Consistency of the soil; sandy or clayey.
3. The degree of admixture with fresh-water at river mouths and the concentration of brackish water. Rainfall influences salt content, particularly that of the rarely flooded higher regions.

Mangroves frequently show a well-defined zonation of species, but owing to the great variation from place to place in the factors just mentioned and in floristic composition, and the wide climatic tolerance of most mangrove species, there is no generally applicable zonation.

Mangrove in West Africa

In the mangrove of the Niger delta, *Rhizophora 'mangle'* (= *R. racemosa*?, see below) covers 99 per cent of the area and is said to attain a maximum height of 45 m and a girth of 2.5 m above the 5 m long stilt roots (Rosevear, 1947). The latter do not penetrate the soil in the manner of normal roots, but, immediately below the surface of the mud, divide into innumerable rootlets of the thickness of a piece of fine twine; and it is the vast development of these that completely alters the nature of the soil by displacement of the soft mud. The tree therefore stands upon a system of arches supported by a thick felt raft of its own making and to which it is securely anchored as it were by multitudinous tie-ropes. According to Rosevear this raft of rootlets, which is often a metre or more thick, spells the doom of the mangrove because *Rhizophora* can only reach its optimum size where it grows on newly deposited soft mud. On mature mangrove sites the root systems of seedlings cannot develop properly and nothing more than a low shrubby tangle results. Many mangrove islands in the creeks consist of an outer fringe of tall trees on freshly deposited mud surrounding an inner core of this low tangle, which represents secondary or tertiary growth on the altered soil. Rosevear states that the dominants of the low and high mangrove are precisely the same tree. Savory (1953) and Keay (1953, 1959a), however, recognize three species of *Rhizophora* in the Niger delta and elsewhere on the coast of West Africa and eastern tropical America where they form separate communities within the

Rhizophora zone. The tall species, *R. racemosa*, is the commonest species and is the pioneer at the edge of the alluvial salt swamp to the exclusion of other species. *R. harrisonii*, which does not exceed a height of 6 m in Nigeria, is dominant in the middle areas, and the even smaller shrubby *R. mangle* is found only on the drier inner limit of the *Rhizophora* zone. Other workers have doubted the existence of as many as three distinct biological species. Gledhill (1963) suspected a high degree of interfertility and hybridization between *R. mangle* and *R. harrisonii*. According to Breteler (1969), *R. harrisonii* is probably a hybrid between *R. racemosa* and *R. mangle* and sets little ripe fruit. J. B. Hall (in litt. 24 February 1977) shares this opinion and confirms that *R. harrisonii* scarcely forms any fruit in Ghana. In view of the uncertainty of field identifications the generic name alone is used below.

Behind the *Rhizophora* zone is the zone of *Avicennia germinans* with the grass *Paspalum vaginatum*, which lies between the limits of the normal high tides and the spring tides, and is flooded twice monthly. *Avicennia* is normally a small tree which in the Niger delta rarely occurs within 24 km of the sea. It can, however, in places grow in pure stands up to 30 m tall on the sea-board, as on Soden Island in the estuary of the Rio del Rey. This seems to happen when sand from the sea rather than silt from the river is deposited. *Laguncularia* bushes, which are almost totally submerged at high spring tides, appear to be the first colonizers in this situation and are subsequently suppressed by *Avicennia*. *Avicennia* can, like *Rhizophora*, occur in a stunted form, and the two are often found together in scrub mangrove, though it is exceedingly rare to find even a single specimen of this tree as a component of the main *Rhizophora* community of the open creeks. There is some evidence, however, that when *Avicennia* establishes itself as a pioneer on sandbanks as described above, it causes deposition of mud which favours invasion by *Rhizophora* and hence the development of transitional mixed communities.

The mangrove vegetation of Ghana, which occurs principally in lagoons, has been described by Boughey (1957a). West of Takorodi, where the rainfall is more than 1 250 mm per year, all the lagoons are *open*, since they have a permanent outlet to the sea, and are thus inundated with sea-water by each tide. Most of the lagoons east of Takorodi, where the rainfall is lower, are *closed*, at least for the greater part of the year. They are open only for one or two months between June and September, the main rainy season, and not necessarily every year. Mangrove in open lagoons is dominated by *Rhizophora*, which is virtually confined to this habitat. *Pandanus candelabrum* also occurs and, in open places, the fern *Acrostichum aureum*. In closed lagoons at the high-water level reached during the seasonal flooding there is a fringe of scrubby *Avicennia germinans* associated with *Laguncularia racemosa* and *Conocarpus erectus*. Channels and pools of water which persist throughout the year in closed lagoons are fringed with *Avicennia*. The dried-out bed of the lagoon between the channels is usually covered by a dense sward of *Sesuvium portulacastrum* mixed with *Philoxerus vermicularis*.

The mangrove occurring on the Gambia River near Keneba 80 km inland has been described by Giglioli and Thornton (1965). Mean annual rainfall is 1 125 mm and falls mostly between June and October. The inland swamps of the Gambia, unlike most littoral mangrove swamps, are not composed of regular belts of *Rhizophora* and *Avicennia*, backed by grasslands. The flatness of the land and the slow silting of the ancient Gambia flood plain have combined to form large areas of swamp, which ramify deeply into the surrounding country but nevertheless are reached by tidal water through a maze of meandering creeks. Each large or small watercourse is characteristically bordered, up to the limit of daily tidal flooding, by a gallery of *Rhizophora*, which in turn is usually surrounded to the mean limit of inundation by spring tides by open bushland of *Avicennia germinans*. *Laguncularia racemosa* is rarely seen. *Rhizophora* is usually the pioneer species. In some localities along the Gambia River, but not along its tributaries and associated swamps, *Avicennia* is the pioneer species. This happens only where sedimentation has been extremely rapid and a new high-shelving bank has been produced, which is barely covered by the daily tides. Such areas are usually immediately downstream of the mouth of a tributary. Behind or between the mangroves, there are isolated areas or continuous stretches of barren mud-flats which account for a quarter of the total area. They are 15 to 30 cm higher than the mangrove soil-level and are usually entirely without vegetation because they are desiccated during the dry season and their soils contain a high concentration of soluble salts, mainly chlorides and sulphates. Where the mud-flats are low enough to receive periodic flooding during the dry season from the high spring tides, they often, but not always, support perennial lawns of *Sesuvium portulacastrum*, and more rarely, at the end of the rains, seasonal lawns of *Eleocharis* spp.

There are well-defined gradients in the depth of the water in the flood plains of the creeks, so that pure and highly zoned plant communities are produced. *Rhizophora* and *Avicennia* usually occur in remarkably pure stands. Mixing occurs only in the narrow ribbon-like ecotone between them, in which *Rhizophora* is being invaded and replaced by *Avicennia*.

In West Africa *Rhizophora* reaches its northern limits just north of Saint-Louis in Senegal, and *Avicennia* at Tidra in Mauritania, where it forms an open bushland 2–2.5 m tall.

South of Cameroun, mangrove is poorly developed, except at the mouth of the Zaire R., where it shows the full floristic complement of West African species.

Mangrove in East Africa

All nine of the East African mangroves occur in Kenya, Tanzania, and Mozambique, but only *Heritiera littoralis* is confined to this part of the East African coast. The other species, however, thin out rapidly to the north and south. *Avicennia marina*, *Bruguiera gymnorrhiza* and *Rhizophora mucronata* have the widest ranges and extend from the Red Sea to the Eastern Cape in South Africa. *Ceriops tagal*, *Sonneratia alba* and *Xylocarpus granatum* occur as far north as Somalia but reach their southern limits in Mozambique or the extreme north of Natal (*Ceriops*). *Lumnitzera racemosa* extends from Kenya to Natal.

The zonation of mangroves in East Africa, which is much more complicated than in West Africa, has been described for the Tanga region of Tanzania by Walter & Steiner (1936). They emphasize the fact that zonation shows many irregularities and that the occurrence of the species is not directly related to distance from the outer and inner limits of the mangrove swamp but to the depth and salinity of the water and the texture of the alluvium. Sandbanks often occur at the outer limit of mangrove and here the water is very shallow. Elsewhere deep channels, which are only briefly uncovered by the receding tides, may extend a considerable distance inland. The ground level is not always flat and a difference in micro-relief of only 20 cm, which is imperceptible to the eye, may be responsible for a change in dominance. For these reasons only a generalized zonation can be described.

In the deepest water at the outer edge of the mangrove *Sonneratia alba* is dominant and is so dense that it cannot be penetrated by a small boat. Inside the *Sonneratia* zone a *Rhizophora mucronata* zone is usually well developed. At river mouths *Sonneratia*, which is more closely associated with undiluted sea-water than other species, disappears almost completely from the outer zone and is replaced there by *Rhizophora*. Inside the *Rhizophora* zone there is usually a narrow zone of *Ceriops tagal* along small channels far into the innermost zone, the *Avicennia marina* zone. On the sides of such channels, which are caused by erosion, zonation is telescoped into a few metres and *Rhizophora*, *Ceriops*, and *Avicennia* replace each other in rapid succession.

In the wide innermost zone, which is not flooded daily with sea-water, but only at spring tides, *Avicennia* occurs almost pure as a small bush only a few metres tall. Towards the inner margin of the mangrove swamp *Avicennia* gets smaller and smaller until it occurs only as juvenile individuals in thick stands along scarcely perceptible channels with only scattered individuals on the intervening flats, which are only a few centimetres more elevated. Small hummocks in the *Avicennia* zone are colonized by *Arthrocnemum indicum* and *Sporobolus virginicus*.

On the landward side of the mangrove there is a vegetation-free sand-flat. This is related to the dry climate. The barren sand-flat is flooded only twice a year for a few days during the equinoctial spring tides. Subsequently the salt concentration increases considerably through evaporation. During the rainy season the salt is leached out. It appears that neither halophytes nor glycophytes can grow under these conditions. At the lower limit of the barren zone, *Suaeda monoica* is often frequent, or there may be scattered bushes of *Lumnitzera racemosa*. In the more open parts of the *Avicennia* fringe, *Suaeda monoica* is often associated with *Arthrocnemum indicum*, *Sesuvium portulacastrum* and the grasses *Sporobolus virginicus*, *Paspalum vaginatum* and *Dactyloctenium geminatum*.

In the Tanga region *Bruguiera gymnorrhiza* does not form a distinct zone but occurs scattered with *Rhizophora* and *Ceriops*. It also penetrates a considerable distance inland along rivers.

The fern *Acrostichum aureum* is often abundant near river mouths but only in the inner zones where the water is brackish. It is completely absent from mangrove bathed with pure sea-water.

In the innermost mangrove zone *Avicennia* is always bushy. It can also occur as a large tree at the outer edge of mangrove, though it never forms a continuous zone there. It seems that in this situation it always occurs on sandy alluvium.

Mangrove species also grow on coral rock but because of the extreme irregularity of the surface they show no zonation there.

In Kenya the behaviour of the mangrove species is similar to that just described for Tanga in Tanzania, though their relative abundance varies greatly according to local conditions. *Rhizophora mucronata* is the most abundant species, covering 70 per cent of the area. It often occurs pure or with a few scattered *Bruguiera*. The most extensive and luxuriant mangrove swamp occurs in the Lamu archipelago at the mouth of the Tana River.

Further south in Kenya in Gazi Creek at the mouth of the very much smaller Kidogoweni River, the alluvium is mostly sandy, and *Rhizophora* is largely confined to the muddy banks of the main channels, which receive a copious supply of fresh-water; *Xylocarpus granatum* is locally abundant and there are scattered individuals of *Bruguiera*. Smaller channels which have sandy banks and receive less fresh-water are fringed by a narrow zone of *Ceriops* and *Lumnitzera* with scattered *Avicennia* up to 11 m tall, and *Heritiera*. The extensive sandy flats between the smaller channels support open *Avicennia* bushland 2–5 m in height.

In Eritrea, mangrove dominated by *Avicennia* occurs on sandy mud overlying coral rock in shallow bays which are partly land-locked and sheltered from the full force of the tides by coral reefs (Hemming, 1961).

In the Sudan *Avicennia marina* is the principal species. Where the water is shallow and the substrate compact, camels browse the leaves and shoots and the vegetation is noticeably thinned. In deeper water *Avicennia* forms dense thickets. In the extreme south of the Sudan, *Avicennia* is associated with *Rhizophora mucronata* and *Bruguiera gymnorrhiza*.

Nearly all the East African mangrove species occur at the mouth of the Zambezi delta in Mozambique, but mangrove swamp is relatively restricted there and extends only 15 km along the main channel, although salt-water goes up much further (Macnae, 1968).

On Inhaca Island at the extreme southern end of Mozambique, mangrove occurs on the sheltered shores. Zonation occurs along creeks but not in long-shore mangrove. On open slopes and at the mouths of estuaries *Avicennia marina* is the principal species probably because of the sandy soil. Upstream it is replaced by *Rhizophora mucronata*. The greater part of the mangrove swamp is made up of *Ceriops tagal* in the drier parts, and *Bruguiera gymnorrhiza* where the water-table is at or near the surface. On the landward fringe of the *Ceriops* zone, *Lumnitzera racemosa* is frequent and *Xylocarpus granatum* occurs rarely. *Avicennia* reappears locally in the form of low stunted bushes at the landward edge of the swamp; elsewhere there may be bare flats with saline efflorescence (Macnae & Kalk, 1962a).

On all Inhaca shores, *Avicennia marina* is the principal colonist. It grows on sandy beaches and on muddy ones in which drainage is improved by the inclusion of shell debris. Once established the pneumatophores cause the accumulation of silt around them, followed by the invasion of *Ceriops* and *Bruguiera*, the seedlings of which require shade for their establishment. Further accumulation of mud and the formation of mangrove peat causes waterlogging of the soil. *Avicennia* dies back under these conditions. *Rhizophora* on the other hand will only grow into a tree in waterlogged soil, but only if the salinity is below that of normal sea-water (Macnae & Kalk, 1962a).

In South Africa, mangrove is not extensively developed but has a scattered distribution along the east coast. It is absent from the mouths of rivers which run to the sea through deep gorges and have only a narrow flood plain. Mangrove is also absent from the mouths of rivers which become cut off from the sea by sand-bars during the dry season. Such estuaries are dominated by *Barringtonia racemosa* and *Hibiscus tiliaceus* (Macnae, 1963).

Five species, *Avicennia marina*, *Bruguiera gymnorrhiza*, *Rhizophora mucronata*, *Ceriops tagal* and *Lumnitzera racemosa*, occur at Kosi Bay in Zululand. *Ceriops* and *Lumnitzera* reach their southern limits there. The behaviour of *Lumnitzera* in unusual. Elsewhere it occurs at the landward margin of mangrove swamps and its roots are submerged only at the highest spring tides. At Kosi Bay it occurs with *Ceriops* and *Bruguiera* in all thickets and is to be found on the seaward fringe as well as on the landward. Its occurrence at low intertidal levels may be explained by the absence of mud and the good drainage provided by a sandy substrate. The same factor may account for the abundance of *Avicennia* in Natal and the rarity of *Rhizophora*. *Avicennia* is normally the pioneer species in southern Africa and is capable of colonizing stable sandy shores. *Bruguiera* is the species most tolerant of fresh-water and reaches its most luxuriant development in areas of low salinity. *Rhizophora* is more dependent on mud and only occurs abundantly in those estuaries where mud-banks are well consolidated and border creeks or channels.

Avicennia, *Bruguiera* and *Rhizophora* extend as far south as the mouth of the Kei River.

In Madagascar mangrove covers 217600 hectares (Koechlin *et al.*, 1974). Nearly all of this occurs on the western coast. Mangrove is extremely localized on the exposed eastern coast, where it is confined to a few sheltered estuaries. All of the nine species of mangrove which are found in East Africa also occur in Madagascar, where their ecology appears to be similar.

Mangrove is of considerable economic importance. It provides poles and planks for house-building and small planks for boat-building. In East Africa poles of *Heritiera littoralis* are much used for the masts of Arab dhows. Mangrove is also an important source of fuel. In Nigeria the coastal towns were formerly dependent on it. In East Africa large quantities of *Rhizophora mucronata* bark were formerly exported for tannin, while the local people still use mangrove tannin for preserving their fishing nets, ropes and sails. Mangrove swamps in West Africa also provide breeding-places for the salt-water vector of malaria, *Anopheles gambiae melas*, but after appropriate reclamation measures they can be used for growing rice and other crops.

Herbaceous fresh-water swamp and aquatic vegetation
(mapping units 64 and 75)

Refs.: Boughey (1963a); Eggeling (1935); Germain (1965, p. 217–44); Greenway (1973); Howard-Williams & Walker (1974); Léonard (1952; 1969b); Lind (1956b); Lind & Morrison (1974, p. 102–27); Lind & Visser (1962); Mitchell (1978); Seagrief (1962); Thompson (1976); Van der Ben (1959); Van Meel (1952, 1953, 1966); Wild (1961); Wild & Barbosa (1968, p. 64).
Phots.: Eggeling (1935: 1–4); Germain (1965: 1–8); Lind (1956b: 1, 2); Lind & Morrison (1974: 30–5); Seagrief (1962: 1, 2); Thompson (1976: 28); Van der Ben (1959: 1–10); Van Meel (1952: 14, 15, 18; 1953: 1–13).
Profiles: Lind & Morrison (1974: 3, 4); Thompson (1976: 26).

Throughout the wetter parts of tropical and subtropical Africa water accumulates in depressions, where it gives rise to swamps and lakes. In the Guineo–Congolian Region most swampy areas are covered with swamp forest (page 82). By contrast, reed-swamp and aquatic communities are relatively restricted.

Outside the Guineo–Congolian Region most of the shallower lakes, except those which are strongly saline, have a wide belt of reed-swamp, of which the main constituent is *Cyperus papyrus* or 'papyrus', the tallest member of the Cyperaceae.

Reed-swamp also occurs less extensively in sheltered bays on the shores of deeper lakes and in the backwaters and lagoons of the larger rivers. Conditions favourable for the development of reed-swamp are widespread in East and South tropical Africa, where they have developed following warping of the earth's crust and other tectonic movements, in places associated with rift formation.

In Uganda, swamp occupies 6 per cent of the total land surface, of which the largest areas are produced by the Nile, especially where the Victoria Nile flows through Lake Kioga. Further north, in the Sudan, the White Nile with its tributary the Bar el Ghazal forms the largest swamp in the world, the Sudd, which covers an area of 150000 km^2 and extends for more than 600 km from north to south and a similar distance from east to west.

The largest swamp areas in West Africa are on the shores of Lake Chad and in the valley of the Upper Niger south of Tombouctou.

There are also many swamps in the Zambezian Region, principally the Okavango, Busanga, and Lukanga swamps, and those associated with Lakes Upemba, Mweru, Mweru Wantipa, Bangweulu, Shirwa and Chiuta. Smaller swamps fringe the flood plains of the Zambezi and Kafue Rivers.

According to Debenham (1952) the vegetation growing in a swamp is not merely a concomitant of the water conditions but sometimes is the cause of them. In some swamps there was no real ponding by physical features until the growth of the vegetation itself caused sufficient obstruction to hold back the water.

The dominants of reed-swamp are normally rooted in the soil beneath the water, but some, especially papyrus and the grass *Vossia cuspidata*, also extend outwards as a floating mat at the edge of the swamp, which frequently becomes detached and forms floating islands. In deeper water beyond the reed-swamp, communities of submerged and free-floating aquatics occur.

Cyperus papyrus, the commonest dominant of reed-swamp, is widely distributed in tropical and south Africa and in Madagascar. It does not occur above 2300 m. Formerly there were extensive growths in the Egyptian Nile, where for centuries in dynastic time it provided the basis of a paper industry and was widely used for boat-making, cordage, matting, food, and medicine. It has long been thought to be extinct there, but has been rediscovered (El Hadidi, 1971).

Cyperus papyrus is a vigorous species which usually forms almost pure stands. It is very fast-growing and can attain its maximum size of 5 m in 10 weeks. Other reed-swamp dominants are much more localized. In East Africa, *Phragmites australis* and *P. mauritianus* are commonest in silted areas and in lakes of volcanic origin. *Typha australis*, *T. latifolia* and *Cladium mariscus* locally replace papyrus at higher altitudes. In some places in East and South tropical Africa belts of the grasses *Loudetia phragmitoides* and *Miscanthus violaceus* occur in shallow water on the landward side of papyrus swamp.

The principal associates of papyrus are *Cyperus haspan*, *Dissotis rotundifolia*, *Hibiscus diversifolius*, *Impatiens irvingii*, *Ipomoea* spp., *Ludwigia erecta*, *L. leptocarpa*, *L. octovalvis*, *L. stolonifera*, *Limnophyton obtusifolium*, *Melanthera scandens*, *Melastomastrum segregatum*, *Mikania cordata*, *Polygonum pulchrum*, *P. salicifolium*, *P. strigosum*, *Thelypteris striata* and *Vigna luteola*.

The grass *Vossia cuspidata* is the most characteristic pioneer of reed-swamp and is usually abundant at the outer edge of the papyrus zone. Its stems, which are up to 6 m long, lie on the water, and with other prostrate species such as *Ludwigia* and the fern *Thelypteris striata* form strong floating mats which will bear the weight of a man. The stout rhizomes of *Cyperus papyrus* push out into this mat and so extend the area of reed-swamp. Other species rooted in the *Vossia* mat include the grasses *Echinochloa pyramidalis*, *E. scabra* (*stagnina*), *Oryza longistaminata* (*perennis*) and *Paspalidium geminatum*, and the sedges, *Eleocharis acutangula* and *Scirpus inclinatus*.

In deeper water beyond the reed-swamp there are normally two communities, the submerged community and the floating-leaf community.

The most characteristic members of the submerged community are *Ceratophyllum demersum*, *Hydrilla verticillata*, *Lagarosiphon* spp., *Najas* spp., *Ottelia ulvifolia*, *Potamogeton schweinfurthii*, *Utricularia* spp. including *U. foliosa*, *Vallisneria aethiopica*, *V. spiralis* and the charophyte *Nitella*. They show a clearly marked zonation in relation to water depth. *Ceratophyllum*, for instance, grows in water up to 8 m deep, while *Potamogeton schweinfurthii* rarely tolerates more than 3 m.

There are two types of plants with floating leaves, namely those which are rooted in the mud and those which are free-floating. Among the former are species of *Nymphaea*, including *N. caerulea* and *N. lotus*, *Potamogeton richardii* and *Nymphoides indica*. The principal free-floating species are *Azolla africana*, *Lemna* spp. including *perpusilla*, *Trapa natans*, *Wolffia arrhiza*, *Eichhornia crassipes*, *Pistia stratiotes* and *Salvinia molesta*. These species are readily dispersed by water-movements and by wind. When they are transported to favourable conditions they increase explosively, and all seven can become troublesome, especially *Eichhornia*, *Pistia* and *Salvinia*.

Eichhornia crassipes, the Water Hyacinth, is indigenous to tropical America but was introduced to Africa about 100 years ago. It is now widespread and is a serious pest in many places since it interferes with navigation in large rivers, blocks irrigation canals, and fills dams. Because of its rapid vegetative reproduction it is difficult to eradicate, and expensive control measures in Zaire and the Sudan have been unsuccessful. An indigenous species, *E. natans*, which is distributed from Senegal to Angola and Zimbabwe, is not normally troublesome but is regarded as a potential pest in the Gambia.

Pistia stratiotes, the Water Lettuce or Nile Cabbage, has a pan-tropical distribution and has been known in Egypt since the time of Pliny (A.D. 77). It is widespread in tropical Africa but is not so dangerous as *Eichhornia*. It is said to be troublesome in the Gambia, the Niger Delta, the Upper Nile, parts of Kenya, and the littoral regions of Angola and Mozambique.

Salvinia molesta, a free-floating water-fern, has been confused with *S. auriculata*. The former is now thought to be a sterile triploid hybrid which probably originated in cultivation and may have been introduced to Africa as an aquarium plant. Little was known about it before the construction of the Kariba dam in the Zambezi Valley in 1959, following which it rapidly became an extremely serious weed on Lake Kariba where it covered large areas with a blanket 25 cm thick. The latter provided a substrate for sudd-forming species such as *Vossia cuspidata* and *Scirpus cubensis*. *Salvinia* has subsequently declined somewhat but is still a troublesome pest. Its explosive development is described by Boughey (1963a) and Mitchell (1978).

In Uganda *Miscanthus violaceus* occurs in the inner part of the papyrus zone but it also forms a distinct zone in shallower water from which papyrus is absent. In the inner part of the *Miscanthus* zone another tall tussock grass, *Loudetia phragmitoides*, is co-dominant. Wet hollows alternate with the grass tussocks.

The associates of *Miscanthus* are more diversified than those of papyrus and include *Cyperus haspan*, *Dissotis incana* (*canescens*), *Fuirena umbellata*, *Hypericum lalandii*, *Leersia hexandra*, *Paspalum scrobiculatum*, *Polygonum* spp., *Scleria nyasensis*, *S. nutans*, *Smithia elliotii*, *Thelypteris confluens*, *Tristemma incompletum* and *Utricularia gibba*.

Towards the landward margin of reed-swamp a narrow zone of shrubs and small trees adapted to swamp conditions often occurs. The principal species are *Aeschynomene elaphroxylon*, *A. pfundii*, *Kotschya africana*, *Mimosa pigra* and *Sesbania sesban*, often with scattered juveniles of swamp-forest trees such as *Syzygium cordatum*, *Ficus verruculosa* and *Ficus congensis*. Reed-swamp fringing larger lakes, however, is not normally replaced by swamp forest, probably because the succession is arrested by the great changes in water-level which periodically occur.

Sphagnum is normally absent from papyrus swamps, but is often found in *Miscanthus* swamps, especially those at higher altitudes.

In the Zaire basin permanent swamp along all the major rivers is almost exclusively dominated by *Vossia cuspidata*. Its principal associates, except for *Polygonum acuminatum*, are grasses, namely *Brachiaria mutica*, *Panicum subalbidum*, *Echinochloa pyramidalis*, *Leersia hexandra*, *Echinochloa scabra* and *Panicum parvifolium*. Most openings in the swamp forest are dominated by *Cyrtosperma senegalense*, a giant aroid, which largely replaces *Cyperus papyrus* in the Zaire basin.

There have been several plans for the draining and cultivation of African reed-swamp or the harvesting of their natural growth for manufacturing paper or hardboard. The dangers involved in exploiting such a fragile ecosystem are stressed by Thompson (1976).

Halophytic vegetation
(mapping unit 76)

Refs.: Emberger (1939, p. 147–8); Giess (1971, p. 9); Greenway (1973, p. 57); Lind & Morrison (1974, p. 174–5); Seagrief & Drummond (1957, p. 110–11); Symoens (1953); Vesey-FitzGerald (1955a; 1963, p. 261–3; 1970); Walter (1971); Wild & Barbosa (1967, p. 61–2, 67–8).
Phots.: Giess (1971: 46–9); Vesey-FitzGerald (1963: 4, 7).

Saline soils are frequently found in arid and semi-arid regions where rainfall is insufficient to transport salts formed, during weathering, to the sea. Their distribution, however, is also partly determined by geology, in that locally they occur in wet regions around springs which bring soluble salts to the surface from salt-containing strata. An intermediate condition occurs extensively under a mean annual rainfall of 250–1000 mm in parts of East Africa where the salts which are deposited in lake basins and river valleys are derived from volcanic deposits rich in sodium.

Only a relatively small number of plant species, known as halophytes, grow on saline soils. The most typical halophytes absorb soluble salts, especially sodium chloride, which reaches high concentrations in the cell sap of the leaves. Non-halophytes cannot tolerate high internal concentrations of sodium and die.

A comprehensive description of the vegetation of coastal saline soils that are influenced by sea-water is beyond the scope of this book. Some types, however, are briefly mentioned in the section on mangrove.

The halophytic vegetation of the Sahara desert is dealt with in Chapter XVII and that of the Karoo–Namib Region in Chapter VI. In parts of the Karoo, especially the 'vloere', the vegetation consists essentially of halophytes, but halophytes are also widely distributed elsewhere in the Karoo. Indeed, species of the genus *Mesembryanthemum sensu lato*, which occur throughout the Karoo, always have high osmotic values and accumulate chloride even on non-saline soils. The vegetation on saline soils in the Somalia–Masai Region is briefly mentioned in Chapter IV. The vegetation on inland saline soils is physiognomically varied and includes grassland, wooded grassland, shrubland, and bushland.

In the Maghreb halophytic vegetation occurs chiefly in the arid and semi-arid 'étages'. The principal species are *Atriplex halimus*, *Lycium intricatum*, *Suaeda fruticosa*, *Salsola longifolia* (*oppositifolia*), *S. vermiculata*, *Asparagus stipularis*, *Anabasis aphylla*, *Peganum harmala*, *Artemisia herba-alba*, *Asphodelus fistulosus* and species of *Frankenia*, *Mesembryanthemum*, *Sphenopus*, *Lepturus* and *Aïzoon*. Non-halophilous therophytes are also plentiful. In more permanently moist depressions the following are frequent: *Tamarix* spp., *Juncus maritimus*, *J. acutus*, *Statice* spp., *Scirpus holoschoenus*, *Spergularia maritima* and *Plantago coronopus* (see also page 230).

On the eastern side of tropical Africa halophytic vegetation occurs in most of the lake basins in the Eastern Rift, principally Lakes Turkana, Hannington,

Nakuru, Elementeita, Magadi, Natron, Manyara, Eyasi and Rukwa. Two lakes, however, L. Baringo and L. Naivasha, are much less saline and probably owe their freshness to subterranean outlets. Lake Mweru Wantipa in north-east Zambia also lies in a down-faulted depression with internal drainage, which supports halophytic vegetation. Some of the lakes mentioned above are surrounded by extensive salt deposits that are the basis of an important industry.

The chief plants around the saline lakes in Kenya and Uganda are *Cyperus laevigatus*, *Sporobolus spicatus* and *Dactyloctenium* sp.

The halophytic vegetation in the Lake Rukwa basin in Tanzania is chiefly grassland (Vesey-FitzGerald, 1963). It covers extensive level plains which are liable to be inundated by the lake during periods of high water, but which themselves extend over the dry bed of the lake when the latter dries up. The soil is highly alkaline (pH 8.0–9.6) and the surface is impregnated with soda. The grassland can be divided into the following three zones:

1. *Beach zone.* The fringe of the lake at its maximum extent is marked by a pure stand of 1–2 m tall tussocks of *Sporobolus robustus*. It never grows in water. Under dry conditions it persists for an indefinite period, making a clearly defined but obsolete beach zone which is often remote from the contemporary perimeter of the water in the lake.

2. *Alkaline swamp.* Because the beds of alkaline lakes are very flat, extensive areas are shallowly flooded when the basin is full. This alkaline swamp is colonized by a single species, *Diplachne fusca*, which forms a dense and even mat up to 2 m in height and extends for many kilometres. If the swamp dries up, even for a period of years, as often happens, the *D. fusca* mat remains in occupation of the ground for an indefinite period, but only grows to a height of 50 cm. When growing in water *D. fusca* remains green throughout the year but when the lake has receded it dries up and becomes burnt during the dry season. If man-made fires do not consume the dry mat, lightning frequently ignites it at the onset of the rains, but even if burnt every year the *Diplachne* mat still persists.

3. *Alkaline flats.* The lake bed itself (as opposed to the lake swamps just described) shows a cyclical succession. The two grass species that are principally involved replace each other as the bed is successively flooded or dries out. When the lake dries out a crust of saline soil is left on the surface but moist mud remains below the surface mulch. This newly exposed surface is colonized by *Sporobolus spicatus*, which spreads rapidly by runners and produces a perennial sward. The latter persists during the most prolonged dry periods and is not replaced by any other species. Another grass of similar growth-form, *Psilolemma* (*Odyssea*) *jaegeri*, may occur with *S. spicatus* but its tufts do not spread when the ground is dry. When, however, the lake is rising this species colonizes vast areas of flats that are shallowly flooded with warm water in which the soda crust is dissolved. Under these conditions the *S. spicatus* mat rots away and *Psilolemma* replaces it.

In the Wembere depression south of Lake Eyasi in the East African Rift Valley, alkaline swamp is dominated by *Diplachne fusca* grassland. Scattered *Tamarix* and 'salt bushes' (Chenopodiaceae) occur along the edges of small drainage lines.

Many of the flat valleys in the drier parts of Tanzania have alkaline soils. This is particularly true of the Pangani River, the headwaters of which rise from the volcanic deposits of Mts Meru and Kilimanjaro, which weather rapidly and release large amounts of salt into the drainage water. Saline soils on the flood plain are dominated by grasses and *Sesbania sesban* and the tree *Acacia xanthophloea* (but see page 30). Prominent halophytes include *Salvadora persica*, *Sporobolus robustus*, *Suaeda monoica* and *Triplocephalum holstii*.

In South tropical Africa saline soils are more localized than in East Africa. The principal occurrences are in the Etosha depression in Namibia, the Makarikari basin in Botswana, and in the Changane valley in Mozambique.

The Etosha pan itself is quite barren but it has a fringe of halophytic vegetation consisting principally of *Suaeda articulata*, *Atriplex vestita*, *Sporobolus spicatus*, *S. tenellus*, *S. virginicus* and *Odyssea paucinervis*, surrounded by a dwarf-shrub zone characterized by *Acacia nebrownii*, *Monechma tonsum*, *M. genistifolium*, *Leucosphaera bainesii*, *Petalidium engleranum* and *Salsola tuberculata*.

The Makarikari pan is surrounded by a narrow fringe of grassland dominated by *Sporobolus spicatus* and *Odyssea paucinervis*.

Halophytic communities are widespread in the valley of the Changane, a tributary of the Limpopo. Rainfall amounts to 400–600 mm per year. In moderately saline zones there are grasslands of *Acacia nilotica* subsp. *kraussiana*. When the salinity increases, the grasses *Eriochloa meyerana*, *Sporobolus nitens* and *Aristida adscensionis* form discontinuous patches with extensive bare areas in between. Near the river itself salinity is high and species of *Arthrocnemum*, *Salicornia*, *Atriplex* and *Suaeda* predominate.

Symoens (1953) has described the Mwashya salt-flats in the Lufira Valley 30 km SW. of Lukafu in Upper Shaba. They are dominated by *Juncus maritimus* and *Sporobolus* cf. *virginicus*, two coastal species which have a very sporadic distribution in the interior of Africa.

Glossary and index of vernacular names of vegetation types and habitat

Note. Definitions, where given, are generally brief, since the meaning of most terms should be apparent from the text (to which reference is made).

ALFA, 216, 229ff. Esparto grass, *Stipa tenacissima*.

BAN, 116. A treeless plain in Somalia.
BARCHAN, 220. A crescent-shaped mobile sand dune devoid of vegetation.
BATEKE. *See* TEKE
BATHA, 159. A type of spiny dwarf shrubland in the Mediterranean Region.
BOWAL (pl. BOWE), 108. Ironstone outcrop covered with open, seasonal marsh vegetation.
BROKEN VELD. Applied by South African botanists to a landscape dominated by dwarf shrubs, grasses, and succulents with scattered larger bushes and occasional small trees, usually occurring on flattish, gravelly plains or rugged, rocky mountains. Walter (1971, p. 391), however, defines it as grassland interrupted by areas with a different vegetation.
BROUSSE TIGRÉE, 23, 27, 213. A pattern of arcs or stripes of vegetation alternating with bare areas in arid and semi-arid regions.

CHIPYA, 60, 96. Fiercely burning wooded grassland on Central African Plateau (from Bemba 'cipya').
CHOTT. *See* SHATT
CITEMENE, 92. A type of slash-and-burn shifting cultivation practised mainly in miombo woodland on the Central African Plateau.
CRAM-CRAM, 216. *Cenchrus biflorus*.

DAMBO, 61, 99ff. A seasonally waterlogged depression on the Central African Plateau covered with grassland.

ELFIN THICKET, 83, 168. The term 'elfin woodland' was introduced by the translators of Schimper's *Pflanzengeographie* (Schimper, 1898, 1903) as the English equivalent of 'Krummholz'. Tropical examples of the latter were described by Schimper as montane woody communities dominated by dwarf trees with short, thick, gnarled trunks which often are almost horizontal or, at least, heavily leaning. In English, the term has subsequently been applied to various types of stunted, montane, cloud 'forest' with abundant epiphytes, especially liverworts and mosses. African examples, however, fit the definition of thicket, rather than woodland, as used in the present work. In Africa, elfin thicket occurs on the summits of rather low mountains at lower altitudes than the general upper limit of the forest belt. Ericaceous bushland and thicket are classified separately. Elfin thicket does not comfortably fit any of the categories of conventional classifications of vegetation. Hence its continued use seems appropriate, notwithstanding its somewhat whimsical connotation.
ERG, 218, 220. A sand desert usually in the form of dunes.

FADAMA, 107. A grassy flood plain in West Africa.
FIRKI, 108. Seasonally flooded wooded grassland in the Chad basin.
FYNBOS, 32, 41, 48, 49, 132ff, 159. The characteristic sclerophyllous vegetation of the Cape Region. Most fynbos is shrubland, less often bushland or thicket.

GARRIC. *See* GARRIGUE
GARRIGUE, 49, 159. A type of short Mediterranean shrubland.
GUELTA, 222. A pool, more or less temporary, in the bed of a wadi, usually sheltered in a canyon.

HALFA. *See* ALFA
HAMADA, 218, 220. A stone desert incised by dry valleys (wadis).
HIGHVELD, 51, 64, 194. An Afrikaans word applied to the plateaux surfaces of the interior of southern Africa. In this work restricted to the eastern part of the transition zone between the Zambezian and Karoo–Namib Regional Centres of Endemism.

ITIGI THICKET, 48, 97.

KARAL. *See* FIRKI
KAROO, 137ff. A Hottentot word meaning bald, applied to the essentially treeless semi-desert parts of South Africa which are mainly dominated by succulent communities and dwarf malacophyllous shrublands.

MABWATI, 172. A type of dry evergreen forest in Zaire and Angola.
MACCHIA. *See* MAQUIS
MAQUIS, 41, 49, 159. A type of Mediterranean sclerophyllous shrubland, usually tall, often impenetrable. The term is frequently misapplied to sclerophyllous shrublands (e.g. fynbos, q.v.) in other parts of the world.
MATESHI, 96. A type of Zambezian dry evergreen thicket.
MATTORAL, 149. All woody non-forest vegetation in the Mediterranean Region.
MAVUNDA, 90. Zambezian dry evergreen forest dominated by *Cryptosepalum pseudotaxus*.
MBUGA, 116. Water-receiving depressions in East Africa covered with grassland and *Acacia*-wooded grassland on seasonally saturated, black, cracking clays. Mostly occurring at lower altitudes and under a drier and hotter climate than dambos.

MIKWATI, 173. A type of (chiefly secondary) wooded grassland in Zaire.

MIOMBO, 54, 57, 60, 61, 92, 181. A kind of woodland in the Zambezian Region dominated by species of *Brachystegia* and related genera.

MOPANE, 54, 61, 62, 93, 143, 191. A vernacular name applied to *Colophospermum mopane* and vegetation dominated by it.

MUHULU, 91. A kind of Zambezian, dry evergreen forest.

MUTEMWA, 90. The shrub layer of dry deciduous forest dominated by *Baikiaea plurijuga*. It usually forms a 5–8 m high thicket.

NOORSVELD, 141. Vegetation in the Eastern Cape intermediate between bushland and shrubland, and dominated by *Euphorbia coerulescens* (Noors).

PAPYRUS, 264. The giant sedge *Cyperus papyrus* and vegetation dominated by it.

PÁRAMO, 169.

PENGBELE. *See* BOWAL

PSEUDOSTEPPE, 173.

QOZ, 102. A consolidated sand dune.

REG, 218, 220, 221. A gravel desert.

RENOSTERVELD. *See* RHENOSTERBOSVELD

RHENOSTERBOSVELD, 132, 135. Cape shrubland dominated by Rhenosterbos, *Elytropappus rhinocerotis*.

SABKHA (also transliterated SEBKHA or SEBKRA), 222. The classical Arabic term for saline or swampy land. It is used for the salt lakes of inland drainage basins, which are periodically filled but usually dry. 'Sebkha' is the form most often used in botanical literature. *See also* SHATT.

SAHEL. The original meaning in Arabic is seashore, but it was subsequently extended to gravel plains and to the northern margins of the Sahara (P. J. Stewart, pers. comm.). It was first applied in its phytogeographical sense by Chevalier (1900) to the southern fringes of the Sahara, but he gives no explanation of this usage. According to Monod (in litt. 7.X.1974), in parts of Mauritania, in Hodd, Azaouad, Timbuktoo, etc. 'Sahel' just means north. He suggests that Chevalier, who reached the future Sahel zone from the south, adopted, perhaps inadvertently, the name of a geographical direction for a botanical zone. It was a mere coincidence that the zone is situated along the southern fringes or 'coast' of the desert. As applied in the present work it extends as a narrow east–west band, from Senegal to the Red Sea, where rainfall is 150–500 mm p.a. See Figures 1 *and* 21.

SAVANE STEPPIQUE, 173.

SAVANNA, 45, 50ff, 113. This term, in the present work, is used only in a general sense for certain tropical landscapes in which both trees and grasses are conspicuous. No attempt is made to use it in a precise classificatory sense.

SAVOKA, 235–6. Malagasy secondary rain forest.

SEBKHA. *See* SABKHA

SHATT (also transliterated CHOTT), 222. A classical Arabic term for river bank or water's edge. In some modern dialects it is used as a synonym for 'sabkha'. 'Chott' is the form most frequently used in botanical literature.

SPARTE, 229. *Lygeum spartum*.

SPEKBOOMVELD, 141, 201.

STEPPE, 45, 50, 149, 173. In this work the term steppe is not used for African vegetation.

SUDD, 265, 266. The swamp region of the Upper Nile. Sometimes applied to similar swamps elsewhere.

TAMPOKETSA GRASSLAND, 239.

TANDOS, 189. Seasonally flooded grassland on the coastal plain of Mozambique.

TANETY GRASSLAND, 239.

TAPIA, 237. A vernacular name applied to *Uapaca bojeri* and a forest type dominated by it.

TEKE, 173. The Kalahari Sand-covered plateaux in Congo north of Brazzaville.

VELD. An Afrikaans word used by South African botanists to signify vegetation.

VLOERE, 136, 139, 266. Brackish seasonal swamps in South Africa.

WADI. A desert valley, usually dry at the surface except after heavy rainfall.

YAÉRÉ, 108. Grassland in the Chad basin subjected to prolonged flooding.

Geographical bibliography

The references cited below are not meant to be exhaustive, but include the more important works. In addition, many of the publications mentioned in Chapters 3, 4, and 5 include information on vegetation. References to the offshore islands, other than Bioko, the Malagasy Republic, and Zanzibar, are given under the individual islands and archipelagos in Chapter XXI.

ALGERIA (see also MAGHREB and SAHARA). Barry et al. (1970). Barry & Faurel (1973). Cannon (1913). Guinet (1958). Guinochet & Quézel (1954). Hochreutiner (1904). Humbert (1928a). Killian (1961). Lemée (1953). Leredde (1957). Maire (1916). Monjauze (1958, 1968). Monjauze, Faurel & Schotter (1955). Ozenda (1954). Peyerimhoff (1941). Pons & Quézel (1955). Quézel (1954, 1956, 1957b). Quézel & Santa (1962–63). Rikli & Schröter (1912). Simonneau (1954a, 1954b).

ANGOLA. Airy Shaw (1947). Azancot de Menezes (1969). Barbosa (1970). Diniz (1973). Diniz & Aguiar (1969a, 1969b, 1972, 1973). Gossweiler & Mendonça (1939, 1941). Humbert (1940a, 1940b). Matos (1970). Matos & Sousa (1970). Mendes (1962). Mendonça (1961). Monteiro (1962, 1965, 1970). Redinha (1961). Teixeira (1968a, 1968b). Teixeira & Corrêa de Pinho (1961). Teixeira & Matos (1967). Teixeira, Matos & Sousa (1967). Warburg (1903). Whellan (1965).

BENIN. Adjanohoun (1965, 1968). Aubréville (1937a). FAO (1980b, 1980d). Paradis (1975a, 1975b, 1976). Paradis, De Souza & Houghnon (1978).

BIOKO. Adams (1957). Guinea (1968). Mildbraed (1933b).

BOTSWANA. Bawden (1965). Bawden & Stobbs (1963). Blair Rains & McKay (1968). Blair Rains & Yalala (1972). Bremekamp (1935). Campbell (1977). Cole & Brown (1976). Lang & Bremekamp (1935). Miller (1939, 1946). Pole Evans (1948a). Seagrief & Drummond (1958). Seiner (1911). Simpson (1975). Tinley (1966). Weare & Yalala (1971). Wild (1968b).

BURUNDI. Deuse (1963, 1966). Lebrun (1956). Lewalle (1968, 1972, 1975). Pahaut & Van der Ben (1962). Reekmans (1980a, 1980b). Van der Ben (1961).

CAMEROUN. Aubréville (1948a). FAO (1980c, 1980d). Guillaume (1968). John B. Hall (1973). Hawkins & Brunt (1965). Hédin (1930). Jacques-Félix (1945, 1971). Keay (1955, 1959d). Ledermann (1912). Letouzey (1957, 1958, 1960, 1966, 1968a, 1968b, 1969, 1975, 1977). Letouzey et coll. (1978, 1979). Maitland (1932). Mildbraed (1932, 1933a). Paulian & Gèze (1940). Portères (1946). Preuss (1892). Richards (1963a, 1963b). Vaillant (1945).

CENTRAL AFRICAN REPUBLIC. Aubréville (1964). Chevalier (1951). Guigonis (1968). Lanly (1966). Sillans (1951, 1952a, 1952b, 1952c, 1954, 1958).

CHAD (see also SAHARA). Bruneau de Miré & Quézel (1959). Cavalho & Gillet (1960). Depierre & Gillet (1971). Gillet (1957, 1958, 1959a, 1959b, 1960, 1961a, 1961b, 1961c, 1962b, 1963, 1964, 1968a, 1968b). Grondard (1964). Maire & Monod (1950). Murat (1937). Pias (1970). Quézel (1958, 1959). Quézel, Bruneau de Miré & Gillet (1964).

CONGO. Begué (1965, 1967). Descoings (1972a). Farron (1968). Koechlin (1957, 1961). Makany (1972, 1976). Rollet (1963, 1964).

DJIBOUTI. Chedeville (1972). Chevalier (1939). Verdcourt (1968).

EGYPT. Abdel Rahman & Batanouny (1959a, 1959b, 1959c, 1966). Ayyad (1973, 1976). Ayyad & Elghonemey (1976). Batanouny (1964, 1966, 1973). Batanouny & Zaki (1973). Boulos (1966). Davis (1953). El Hadidi (1971). El Hadidi & Ayyad (1975). El Hadidi & Kosinova (1971). El-Sharwaki & Fayad (1975). Hassib (1952). Kassas (1952, 1953a, 1953b). Kassas & El-Abyad (1962). Kassas & Girgis (1964, 1965, 1970). Kassas & Imam (1953, 1959). Kassas & Zahran (1962, 1965, 1967). Long (1955). Migahid (1947). Migahid et al. (1955–75). Stocker (1926, 1927). Tadros (1953, 1956). Tadros & Atta (1958a, 1958b).

EQUATORIAL GUINEA. Guinea (1946).

ETHIOPIA (including Eritrea). Beals (1968). Chaffey (1979). Cufodontis (1940). Engler (1906a). Gillett (1941). Hedberg (1971, 1975). Hemming (1961). Klötzli (1975). Logan (1946). Mooney (1963). Negri (1913). Pichi-Sermolli (1938, 1939, 1940). Posnett & Reilly (1977). Scott (1952a, 1952b, 1955, 1958). Seebald (1972).

GABON. Aubréville (1962, 1967a, 1967b). Caballé (1978). Catinot (1978). Descoings (1974, 1976b). Gloriod (1974). Hallé & Le Thomas (1968). Hallé, Le Thomas & Gazel (1967). Hladik (1974). Le Testu (1938). Saint-Aubin (1961, 1963). Villiers (1973).

GAMBIA. Giglioli & Thornton (1965). Rosevear (1937).

GHANA. Ahn (1958, 1959, 1961). Asare (1962). Aubréville (1959). Boughey (1957a). Brookman-Amissah et al. (1980). Chipp (1927). Douglas (1948). Foggie (1947a, 1947b). Hall & Jeník (1968). Hall & Lawson (1972). Hall & Popple (1968). Hall & Swaine (1974, 1976, 1981). Jeník & Hall (1966, 1976). Jeník & Lawson (1968). Lawson (1956, 1968). Lawson, Armstrong-Mensah & Hall (1970). Lawson & Jeník (1967). Lawson, Jeník & Armstrong-Mensah (1968). Morton (1957). Ramsay & Rose Innes (1963). Swaine & Hall (1974). Taylor (1952, 1960). Vigne (1936).

GUINEA. Adam (1947, 1948, 1950, 1958b, 1968b). Adam & Jaeger (1976). Jaeger & Adam (1947, 1950). Killian (1951). Killian & Schnell (1947). Lamotte, Aguesse & Roy (1962). Schnell (1950a, 1950c, 1950e, 1952a, 1952c, 1957, 1960, 1961, 1968).

GUINEA-BISSAU. Malato Beliz (1963). Malato Beliz & Alves Pereira (1965). Sousa (1958).

IVORY COAST. Adjanohoun (1962, 1964, 1965). Adjanohoun & Aké Assi (1967). Adjanohoun, Aké Assi & Guillaumet (1968). Alexandre (1977, 1978). Aubréville (1957–58). Begué (1937). Bellier (1969). Bernard-Reversat, Huttel & Lemée (1978). César & Menault (1974). Chevalier (1908). Descoings (1972b). Dugerdil (1970). Emberger, Mangenot & Miège (1950b). Guillaumet (1967). Guillaumet & Adjanohoun (1971). Huttel (1975b). Lamotte (1967, 1979). Lanly (1969). Latham & Dugerdil (1970). Mangenot (1950, 1955b, 1971). Mangenot, Miège & Aubert (1948). Menaut & César (1979). Miège (1954, 1955, 1966). Monnier (1968). Paullian (1947). Poissonet & César (1972). Portères (1950). Roland (1967). Roland & Heydacker (1963). Rougerie (1957). Sarlin (1969). Schmidt (1973). Schnell (1957). Spichiger & Pamard (1973). Vuattoux (1968, 1970).

KENYA. Agnew (1968). Ament (1975). Barkham & Rainy (1976). Birch (1963). Bogdan (1958). Coe (1967). Darling (1960b). Edwards (1935, 1940). Faden (1970). Fries (1925). Fries & Fries (1948). Glover (1966). Glover & Trump (1970). Glover, Trump & Wateridge (1964). Glover & Wateridge (1968). Graham (1929). Greenway (1969). Hedberg (1969b). Hemming (1972). Herlocker (1979a). Holland & Hove (1975). Isaac & Isaac (1968). Kerfoot (1964a). Lucas (1968). Mabberley (1975a). Moomaw (1960). Sauer (1965). Staples & Hudson (1938). Synnott (1979a). Trapnell et al. (1966, 1969). Tweedie (1976). Verdcourt (n.d.). Wimbush (1937).

LESOTHO (see also SOUTH AFRICA: AFROMONTANE REGION). Bawden & Carroll (1968).

LIBERIA. Adam (1970, 1971a, 1971b). Berger-Landefeldt (1959). Cooper & Record (1931). Jaeger & Adam (1975). Künkel (1962, 1964, 1966a, 1966b). Voerhoeve (1964, 1968).

LIBYA (see also SAHARA). Bartolo et al. (1977). Boulos (1972, 1975). Léonard (1969a, 1971). Nègre (1974). Scholz (1971). Thomas (1921).

MAGHREB (see also ALGERIA, LIBYA, MOROCCO, and TUNISIA). Boudy (1948, 1950). Blaun-Blanquet (1928). Emberger (1955b). Métro (1970). Quézel (1957a, 1976–77). Rikli (1943–48).

MALAGASY REPUBLIC. Battistini & Richard-Vindard (1973). Dejardin, Guillaumet & Mangenot (1973). Granier (1979). Guillaumet & Koechlin (1971). Humbert (1927a, 1927b, 1928b, 1955a, 1955b, 1959). Humbert & Cours Darne (1964–65). Keraudren (1968). Koechlin (1968, 1972). Koechlin, Guillaumet & Morat (1974). Leroy (1978). Morat (1973). Paulian et al. (1971, 1973). Perrier de la Bâthie (1921a, 1936). Rauh (1973). Segalen & Moureaux (1949). Straka (1960). Thomasson (1974, 1976, 1977).

MALAWI. Brass (1953). Brown & Young (1974). Chapman (1962, 1968). Chapman & White (1970). Hall-Martin (1975). Hall-Martin & Fuller (1975). Howard-Williams (1975a, 1977). Howard-Williams & Walker (1974). Jackson (1954, 1968, 1969). Kalk, McLachlan & Howard-Williams (1979).

MALI REPUBLIC. Adam (1959). Audry & Rossetti (1962). Begué (1958). Duong-Huu-Thoi (1950a, 1950b). Hagerup (1930). Jaeger (1950, 1956, 1959, 1965b, 1968). Jaeger & Jarovoy (1952). Jaeger & Winkoun (1962). Raynal & Raynal (1961). Rossetti (1962).

MAURITANIA (see also SAHARA). Adam (1962c, 1965a, 1968c). Audry & Rossetti (1962). Monod (1952a, 1954a, 1954b). Naegélé (1958a, 1958b, 1959a, 1959b, 1960). Roberty (1958). Sauvage (1946).

MOROCCO (see also MAGHREB and SAHARA). Braun-Blanquet (1928). Braun-Blanquet & Maire (1924). Cavassilas (1963). Dahlgren & Lassen (1972). Destremau (1974). Emberger (1925, 1932, 1936, 1939, 1948). Frödin (1923). Guinet & Sauvage (1954). Humbert (1924). Ionesco & Sauvage (1962, 1965–69). Ionesco & Stefanesco (1967). Killian (1941). Lecompte (1973). Maire (1924). Mathez (1973). Métro (1958). Nègre (1952a, 1952b, 1953, 1956a, 1956b, 1959). Nègre & Peltier (1976). Peltier (1971). Peyre (1973). Quézel (1952). Sauvage (1948, 1961, 1963, 1971). Theron & Vindt (1960). Vindt (1959).

MOZAMBIQUE. Amico (1967). Barbosa (1952, 1968b). Gomes e Sousa (1967). Macedo (1970). Macnae & Kalk (1962a, 1962b, 1969). Mendonça (1952). Myre (1960, 1962, 1964, 1971). Noel (1959). Pedro & Barbosa (1955). Wild (1953).

NAMIBIA. Curson (1947). Dinter (1912, 1921). Giess (1962, 1968a, 1968b, 1969, 1970, 1971). Giess & Tinley (1968). Keet (1950). Marloth (1909). Nordenstam (1970, 1974). Pearson (1907). Rennie (1936). Rutherford (1972). Tinley (1969, 1971). Volk (1966a, 1966b). Volk & Leippert (1971). Walter (1936). Walter & Volk (1954).

NIGER (see also SAHARA). Aubréville (1937b, 1973). Collier & Dundas (1937). Dundas (1938). Fairbairn (1943). Peyre de Fabregues & Lebrun (1976). Pitot (1950a). L. P. White (1970).

NIGERIA. Adejuwon (1970, 1971a, 1971b, 1971c). Ainslie (1926). Aubréville (1973). Bawden & Tuley (1966). Buxton (1935). Charter (1968). Charter & Keay (1960). Clayton (1957, 1958a, 1958b, 1958c, 1961, 1963, 1966). Collier & Dundas (1937). Cook (1968). Fairbairn (1939). Forest Department, Nigeria (1948). Golding & Gwynn (1939). Hall, John B. (1971, 1977). Hall, John B. & Medler (1975a, 1975b). Hall, John B. & Okali (1979). Hambler (1964). Hepper (1965, 1966). Hopkins (1962, 1965a, 1965b, 1965d, 1966, 1968). Jackson (1964). Jones, A.D.P. (1950). Jones, E. W. (1950, 1955–56, 1963a, 1963b). Keay (1947, 1948, 1949, 1951, 1952, 1959a, 1960, 1962, 1979). Keay & Onochie (1947). Kemp (1963). Kershaw (1968). Killick (1959). Kinako (1977). Lawton (1978a). MacGregor (1934). Monod (1952b). Onochie (1961). Ramsay (1964). Ramsay & De Leeuw (1964, 1965a, 1965b). Redhead (1966). Richards (1939, 1957). Rosevear (1947, 1953, 1954). Ross (1954). Sanford (1968, 1969, 1974). Tuley (1966). Tuley & Jackson (1971).

RWANDA. Bouxin (1974, 1975a, 1975b, 1975c, 1976). Deuse (1963, 1966, 1968). Frankart & Liben (1956). Hendrickx (1944). Lebrun (1955, 1956, 1961). Liben (1965). Renier (1954). Spinage (1972). Spinage & Guiness (1971, 1972). Troupin (1966).

SAHARA (see also individual countries). Bruneau de Miré & Quézel (1961). Capot-Rey (1953). Chipp (1930b). Cloudsley-Thompson (1974). Gram (1935). Guinea (1945, 1949). Kruger (1967). Lavauden (1927). Lebrun (1977, 1979). Léonard (1980). Maire (1933, 1938, 1940). Maire et al. (1925). Massart (1898). Monod (1938, 1958). Murat (1944). Ozenda (1958, 1977). Quézel (1965a, 1971). Quézel & Simonneau (1960, 1962). Schiffers (1971). Schulz (1979). Stocker (1926). Zolotarevsky & Murat (1938).

SENEGAL. Adam (1953, 1956, 1958a, 1961a, 1961b, 1961c, 1962a, 1962b, 1964, 1965b, 1968a, 1968d). Aubréville (1948b). Bille & Poupon (1972). Bourlière (1978). Devois (1948). Doumbia (1966). Jaeger (1949). Miège, Bodard & Carrère (1966). Miège, Hainard & Tchérémissinoff (1976). Naegélé (1959c). Pitot (1950b). Pitot & Adam (1954, 1955). Raynal, A. (1963). Raynal, J. (1964, 1968). Trochain (1940).

SIERRA LEONE. Cole (1967, 1968a, 1968b, 1973). Cole & Jarrett (1969). Fox (1968a, 1968b, 1968c, 1970). Gledhill (1963, 1970). Ifan-Dakar (1971). Jaeger (1965a, 1966, 1969, 1976). Jaeger & Adam (1967, 1971, 1972). Jaeger, Lamotte & Roy (1966, 1971). Jordan (1964). Morton (1968).

SOMALI REPUBLIC. Bally (1968, 1976). Boaler & Hodge (1962, 1964). Ciferri (1939). Collenette (1931). Engler (1904). Gillett (1941). Gilliland (1952). Hemming (1965, 1966, 1968). Macfadyen (1950). Senni (1935).

SOUTH AFRICA (REPUBLIC OF)
GENERAL. Acocks (1953, 1964, 1971, 1975, 1977, 1979). Adamson (1938a, 1938b). Aitken & Gale (1921). Bayer, Bigalke & Crass (1968). Bews (1912, 1913, 1916a, 1916b, 1917a, 1917b, 1918, 1925). Chipindall (1955). Coetzee & Werger (1975). Comins (1962). Dyer (1937). Edwards (1967). Goldblatt (1978). Hutchinson (1946). Killick (1968). Kruger (1979). Laughton (1937). Macnae (1963). Marloth (1887, 1908). Martin (1960a, 1960b). Meredith (1955). Moll (1968c). Muir (1929). Phillips (1971). Pole Evans (1936). Roberts (1968). Rycroft (1968). Scheepers (1978). Scott (1951). Story (1952). Von Breitenbach (1972). Weintroub (1933). Wellington (1955). Werger (1978a, 1978c). West (1945, 1951).

AFROMONTANE REGION (including the transition to the Highveld). Granger & Schulze (1977). Herbst & Roberts (1974). Jacot Guillarmod (1962, 1963, 1968, 1969, 1971). Killick (1963, 1978a, 1978b, 1978c, 1979). Moll (1966, 1968a, 1972a). Moll & Haigh (1966). Phillips (1928a, 1928b, 1931a). Roberts (1961, 1966, 1969). Rycroft (1944). Taylor, H. C. (1962). Van Zinderen Bakker, E. M., Jr (1971, 1973). Van Zinderen Bakker, E. M., Sr (1955, 1965). Van Zinderen Bakker, E. M., Sr & Werger (1974).

CAPE REGION. Adamson (1927, 1934, 1935, 1959). Boucher (1977, 1978). Boucher & Jarman (1977). Campbell & Moll (1977). Campbell, Gubb & Moll (1980). Day et al. (1979). Duthie (1929). Kruger (1977a, 1977b, 1977c). Kruger & Taylor (1979). Marloth (1902, 1923, 1929). McLachlan, Moll & Hall (1980). Milewski (1977). Milewski & Esterhuysen (1977). Taylor (1953, 1961a, 1963, 1972a, 1972b, 1977, 1978, 1979, 1980). Werger, Kruger & Taylor (1972a, 1972b).

KALAHARI-HIGHVELD TRANSITION ZONE. Bredenkamp (1975). Bredenkamp & Lambrechts (1979). Bredenkamp & Theron (1976, 1978, 1980). Leistner (1959, 1961a, 1967). Leistner & Werger (1973). Louw (1951). Mostert (1958). Potts & Tidmarsh (1937). Werger (1973a, 1973b, 1978d). Werger & Coetzee (1977). Werger & Leistner (1975).

KAROO-NAMIB REGION. Acocks (1964). Compton (1929a, 1929b). Levyns (1950). Marloth (1909). Werger (1978b). Werger & Coetzee (1977).

TONGALAND–PONDOLAND REGIONAL MOSAIC. Archibald (1955). Bayer (1938). Bews (1920). Breen (1971). Downing (1980). Furness & Breen (1980). Henkel, Ballenden & Bayer (1936). Huntley (1965). Killick (1959). Martin (1965, 1966). Moll (1968b, 1968d, 1972b, 1972c). Moll & Morris (1968). Moll & White (1978). Musil, Grunow & Bornman (1973). Penzhorn, Robertse & Olivier (1974). Rogers & Moll (1975). Varmeijer (1966). Van der Walt (1968). Venter (1976). Weisser (1978). Weisser & Marques (1979).

ZAMBEZIAN REGION. Brynard (1964). Coetzee (1974, 1975). Coetzee et al. (1976). Galpin (1927). Gilliland (1962). Glover & Van Rensburg (1938). Grunow (1967). Schweickerdt (1933). Van der Meulen (1978, 1979). Van der Meulen & Westfall (1979). Verdoorn (1929). Wells (1964).

SUDAN. Adams (1967). Andrews (1945, 1948). Bari (1968). Begué (1958). Bruneau de Miré (1960). Bunting & Lea (1962). Chipp (1929, 1930a). Eyre, Ramsay & Jewitt (1953). Gay (1960). Gay & Berry (1959). Good (1924). Halwagy (1961, 1962a, 1962b, 1963). Hancock (1944). Harrison & Jackson (1958). Hunting Technical Services (1958, 1964, 1968). Jackson (1950, 1951, 1956). Jenkin et al. (1977). Jonglei Investigation Team (1954). Kassas (1956a, 1956b, 1957). Lamprey (1975). Mahmoud & Obeid (1971). Migahid (1947). Morison, Hoyle & Hope-Simpson (1948). Obeid & Mahmoud (1971). Obeid & Seif el Din (1971a, 1971b). Quézel (1969, 1970). Radwanski & Wickens (1967). Ramsay (1958). Ruxton & Berry (1960). Schweinfurth (1968). Smith (1949). Wickens (1977a). Wickens & Collier (1971). Willimot (1957). Worrall (1959, 1960, 1960b).

TANZANIA (mainland). Albrecht (1964). Anderson & Herlocker (1973). Anderson & Talbot (1965). Backlund (1956). Bjørnstad (1976). Boaler (1966). Boaler & Sciwale (1966). Brunnthaler (1914). Buchwald (1896). Burtt (1942). Clutton-Brock & Gillett (1979). Dean (1967). Engler (1894, 1900, 1903). Gillman (1949). Goetze & Engler (1902). Greenway (1933, 1955, 1965). Greenway & Vesey-FitzGerald (1969). Herlocker (1975). Herlocker & Dirschl (1972). Jeffers & Boaler (1966). Kerfoot (1964b). Klötzli (1958). Lamprey (1963, 1964, 1979). Leippert (1968). Milne (1947). Moreau (1935a). Pearsall (1957). Phillips (1930, 1931b). Pielou (1952). Pitt-Schenkel (1938). Pócs (1974, 1976a, 1976b, 1976c). Polhill (1968). Rodgers & Homewood (1979). Rodgers & Ludanga (1973). Salt (1951, 1954). Schmidt (1975a, 1975b). Scott (1934). Tobler-Wolff & Tobler (1915). Vageler (1910). Vesey-FitzGerald (1955a, 1973a, 1974b). Volkens (1897). Welch (1960). Welsh & Denny (1978). Werth (1915). Wood (1965).

TOGO. Aubréville (1937a). Busse (1907). Ern (1979). FAO (1980a, 1980d).

TUNISIA. Burollet (1927). Gaussen & Vernet (1958). Knapp (1968b). Lavauden (1928). Le Houérou (1959, 1962, 1967, 1969). Long (1954). Peyerimhoff (1941). Quézel & Bounaga (1975). Vanden Bergen (1977, 1979a, 1979b, 1980).

UGANDA. Bishop (1959). Buechner & Dawkins (1961). Dale (1954). Dawkins (1954). Denny (1971, 1973). Eggeling (1935, 1938, 1947). Harrington & Ross (1974). Jackson & Gartlan (1965). Kerfoot (1965). Lang Brown & Harrop (1962). Langdale-Brown (1959a, 1959b, 1960a, 1960b, 1960c). Langdale-Brown, Osmaston & Wilson (1964). Laws (1970b). Leggat (1965). Lind (1956a, 1956b). Lind & Visser (1962). Lock (1973, 1977a, 1977b). Loveridge (1968). Osmaston (1968). Ross (1955a, 1955b). Snowden (1933, 1953). Thomas, A. S. (1941, 1943, 1945, 1946). Wilson (1962). Wood (1960).

ZAIRE. Aubréville (1957). Balle (1953). Bamps (1975). Bernard (1945). Bouillenne, Moureau & Deuse (1955). Bourbeau et al. (1955). Bourguignon, Streel & Calembert (1960). Chambon & Leruth (1954). Colonval-Elenkov & Malaisse (1975). Compère (1970). Cornet D'Elzius (1964). Delevoy (1933). Delevoy & Robert (1935). Delvaux (1958). Demaret (1958). Denisoff & Devred (1954). De Saeger (1954). Desenfans (1950). Deuse (1960). Devred (1956, 1957, 1958). Devred, Sys & Berce (1958). De Wildeman (1932, 1934). Dieterlen (1978). Diels (1915). Dubois (1955). Duvigneaud (1949a, 1949b, 1950, 1952, 1953, 1958, 1959). Duvigneaud & Denaeyer-de Smet (1960, 1963). Duvigneaud & Symoens (1951). Évrard (1957, 1965, 1968). Focan & Mullenders (1949, 1955). Frankart & Liben (1956). Freson, Goffinet & Malaisse (1974). Gérard (1960).

Germain (1945, 1949, 1952, 1965, 1968). Germain, Croegaert & Sys (1955). Germain & Évrard (1956). Gilson *et al.* (1957). Gilson, Van Wambeke & Gutzwiller (1956). Hauman (1933). Hendrickx (1944, 1946). Holowaychuk *et al.* (1954). Jongen *et al.* (1960). Lebrun (1935, 1936a, 1936b, 1942, 1947, 1954, 1955, 1957, 1959, 1960a, 1960b, 1960d, 1968, 1969). Lebrun & Gilbert (1954). Léonard, A. (1959, 1962). Léonard, J. (1947, 1950, 1951, 1952a, 1952b, 1953, 1954). Liben (1958, 1962). Louis (1947a, 1947b, 1947c). Malaisse (1975, 1976a, 1976b). Malaisse & Anastassiou-Socquet (1977). Malaisse & Gregoire (1978). Meessen (1951). Mullenders (1953, 1954, 1955). Nanson & Gennart (1960). Pahaut & Van der Ben (1962). Pecrot & A. Léonard (1960). Peeters (1964). Pierlot (1966). Pynaert (1933). Robyns (1932, 1936, 1937, 1941, 1948a, 1948b, 1950). Schmitz (1950, 1952a, 1952b, 1962, 1963a, 1963b, 1971, 1977). Streel (1962, 1963). Symoens (1953, 1963). Symoens & Ohoto (1973). Sys & Schmitz (1959). Taton (1949a, 1949b). Taton & Risopoulos (1955). Thomas (1941). Van der Ben (1959). Vanderyst (1932, 1933). Van Meel (1952, 1953, 1966). Van Wambeke & Évrard (1954). Van Wambeke & Liben (1957).

ZAMBIA. Astle (1965a, 1965b, 1969). Astle, Webster & Lawrance (1969). Balon & Coche (1974). Boughey (1964). Cole (1963a). Cottrell & Loveridge (1966). Debenham (1952). Drew & Reilly (1972). Edmonds (1976). Fanshawe (1961, 1968, 1969). Fanshawe & Savory (1964). Fries (1913, 1915, 1921). Horscroft (1961). Kornaś (1977, 1978, 1979). Lawton (1963, 1964, 1967a, 1967b, 1972, 1978b). Martin (1940, 1941). Mitchell (1969). Seagrief (1962). Trapnell (1953, 1959). Trapnell & Clothier (1937). Trapnell *et al.* (1976). Trapnell, Martin & Allan (1950). Verboom (1965, 1966). Verboom & Brunt (1970). Vesey-FitzGerald (1955a). White (1968).

ZANZIBAR. Robins (1976). Werth (1901).

ZIMBABWE. Anderson & Walker (1974). Atwell (1970). Banks (1976). Barclay-Smith (1964). Boughey (1961, 1963a, 1963b). Crook (1956). Dye & Walker (1980). Farrell (1968a, 1968b). Gilliland (1938). Goldsmith (1976). Goodier & Phipps (1961, 1962). Guy (1977). Henkel (1931). Ingram (1960). Jacobsen (1967, 1968, 1970, 1973). Kelly & Walker (1976). Kennan (1972). Kennan, Staples & West (1955). Lang (1952). Magadza (1970). Mitchell, B. L. (1961b). Mitchell, D. S. (1969). Phipps & Goodier (1962). Proctor & Craig (1978). Rattray (1957, 1961). Rattray & Wild (1955). Strang (1974). Thomas, Walker & Wild (1977). Werger, Wild, & Drummond (1978a, 1978b). West (1958). Wild (1952a, 1953, 1955, 1964b, 1965, 1968c, 1968d, 1968e, 1970, 1974a, 1974b, 1974c, 1974d, 1974e, 1975).

Alphabetical bibliography

Abbreviations employed:

AETFAT	Association pour l'Étude Taxonomique de la Flore d'Afrique Tropicale/Association for the Taxonomic Study of the African Flora
ASGA	Association des Services Géologiques Africains/Association of African Geological Surveys
CCTA	Commission de Coopération Technique en Afrique au Sud du Sahara/Commission for Technical Co-operation in Africa South of the Sahara
CNRS	Centre National de la Recherche Scientifique (15 Quai Anatole-France, 75700 Paris, France)
CSA	Conseil Scientifique pour l'Afrique au Sud du Sahara/Scientific Council for Africa South of the Sahara
FAO	Food and Agriculture Organization of the United Nations/Organisation des Nations Unies pour l'Alimentation et l'Agriculture (Rome, Italy)
FULREAC	Fondation de l'Université de Liège pour les Recherches en Afrique Centrale
IEMVT	Institut d'Élevage et de Médecine Vétérinaire des Pays Tropicaux (10 Rue Pierre-Curie, 94700 Maisons-Alfort, France)
IFAN	Institut Français d'Afrique Noire (up to 1966), Institut Fondamental d'Afrique Noire (after 1966) (Dakar, Senegal)
INEAC	Institut National pour l'Étude Agronomique du Congo (Publications obtainable from SERDAT, q.v.)
IPAL	Integrated Project on Arid Lands/Projet Intégré sur les Terres Arides
IUCN	International Union for Conservation of Nature and Natural Resources (Morges, Switzerland)
MAB	Man and the Biosphere Programme (of Unesco)
ORSTOM	Office de la Recherche Scientifique et Technique Outre-Mer (70–74 Route d'Aulnay, 93 Bondy, France)
SERDAT	Service de Documentation en Agronomie Tropicale (Rue Defacqz 1, 1050 Brussels, Belgium)
UNDP	United Nations Development Programme
Unesco	United Nations Educational, Scientific and Cultural Organization/Organisation des Nations Unies pour l'Éducation, la Science et la Culture (7 Place de Fontenoy, 75700 Paris, France)

ABDEL RAHMAN, A. A.; BATANOUNY, K. H. 1959a. Seasonal variations in the desert vegetation along Cairo–Suez road. *Bull. Inst. Désert Égypte*, 9, p. 1–10.

——; ——. 1959b. The phenology of the desert vegetation in relation to environment. *Bull. Inst. Désert Égypte*, 9, p. 11–19.

——; ——. 1959c. Root development and establishment of plants under desert conditions. *Bull. Inst. Désert Égypte*, 9, p. 41–50.

——; ——. 1965. Vegetation and root distribution in the different microhabitats in wadi Hof. *Bull. Inst. Désert Égypte*, 15, p. 55–66.

ACOCKS, J. P. H. 1953. Veld types of South Africa. *Mem. bot. Surv. S. Afr.*, 28, p. 1–192, with coloured vegetation map 1:1500000.

——. 1964. Karoo vegetation in relation to the development of deserts. In: Davis, D. H. S. (ed.), p. 100–12.

——. 1971. The distribution of certain ecologically important grasses in South Africa. *Mitt. Bot. Staatssamml. Münch.*, 10, p. 149–60.

——. 1975. Veld types of South Africa. 2nd ed. *Mem. bot. Surv. S. Afr.*, 40, p. 1–128, with coloured vegetation map 1:1500000.

——. 1977. Riverine vegetation of the semi-arid and arid regions of South Africa. *J. S. Afr. biol. Soc.*, 17, p. 21–35.

——. 1979. The flora that matched the fauna. *Bothalia*, 12, p. 673–709.

ADAM, J. G. 1947. La végétation de la région de la source du Niger. *Annls Géogr.*, 56, p. 192–200.

——. 1948. Les reliques boisées et les essences des savanes dans la zone préforestière en Guinée française. *Bull. Soc. bot. Fr.*, 95, p. 22–6.

——. 1950. Les formations végétales ligneuses secondaires de Guinée française. *Conf. int. Afr. Occid. em Bissau, 1947*, 2 (1a), p. 225–41.

——. 1953. Notes sur la végétation des Niayes de la presqu'île du Cap Vert (Dakar, AOF). *Bull. Soc. bot. Fr.*, 100, p. 153–8.

——. 1956. La végétation de l'extrémité occidentale de l'Afrique. La pointe des Almadies aux environs de Dakar (Sénégal). *Bull. IFAN*, sér. A, 18, p. 685–702.

——. 1958a. Flore et végétation de la réserve botanique de Noflaye (environs de Dakar, Sénégal). *Bull. IFAN*, sér. A, 20, p. 809–68.

——. 1958b. *Éléments pour l'étude de la végétation des hauts plateaux du Fouta Djalon (Secteur des Timbis), Guinée française. I. La flore et ses groupements.* Dakar, Gouvernement Général de l'AOF, Bureau des Sols. 80 p., with coloured vegetation map 1:50000.

——. 1959. Contribution à l'étude floristique des pâturages du Soudan français. In: Charreau, C., et al. (eds.), *Études des pâturages tropicaux de la zone soudanienne*, p. 49–75, with map.

——. 1961*a*. La végétation du bois sacré d'Oussouye (Casamance) et quelques intrusions du domaine de la forêt dense en basse Casamance. *Bull. IFAN*, sér. A, 23, p. 1–10.

——. 1961*b*. Florule et végétation de la grande Mamelle de Dakar (Phare). *Bull. IFAN*, sér. A, 23, p. 406–22.

——. 1961*c*. Flore et végétation de l'île de la Madeleine (Dakar). *Bull. IFAN*, sér. A, 23, p. 708–15.

——. 1962*a*. Contribution à l'étude de la flore et de la végétation de l'Afrique occidentale. La Basse-Casamance (Sénégal). *Bull. IFAN*, sér. A, 24, p. 116–53.

——. 1962*b*. Éléments pour l'étude des groupements végétaux de la presqu'île du Cap-Vert (Dakar). La série du massif de N'Diass. *Bull. IFAN*, sér. A, 24, p. 154–67.

——. 1962*c*. Itinéraires botaniques en Afrique occidentale; flore et végétation d'hiver de la Mauritanie occidentale; les pâturages; inventaire des plantes signalées en Mauritanie. *J. Agric. trop. Bot. appl.*, 9, p. 85–200.

——. 1964. Contribution à l'étude de la végétation du lac de Guiers (Sénégal). *Bull. IFAN*, sér. A, 26, p. 1–72.

——. 1965*a*. La végétation du delta du Sénégal en Mauritanie. *Bull. IFAN*, sér. A, 27, p. 121–38, with small vegetation map.

——. 1965*b*. Généralités sur la flore et la végétation du Sénégal. *Étud. sénégal.*, 9 (3), p. 155–214.

——. 1968*a*. La flore et la végétation du Parc National du Niokolo–Koba (Sénégal). *Adansonia*, sér. 2, 8, p. 439–59.

——. 1968*b*. Flore et végétation de la lisière de la forêt dense en Guinée. *Bull. IFAN*, sér. A, 30, p. 920–52, with vegetation map 1:25 000.

——. 1968*c*. La Mauritanie. In: Hedberg, I.; Hedberg, O. (eds.), p. 49–51, with small vegetation map.

——. 1968*d*. Sénégal. In: Hedberg, I.; Hedberg, O. (eds.), p. 65–9.

——. 1970. État actuel de la végétation des monts Nimba au Libéria et en Guinée. *Adansonia*, sér. 2, 10, p. 193–211.

——. 1971*a*. La végétation littorale aux environs de Buchanan (Libéria). *Bull. IFAN*, sér. A, 32, p. 995–1018.

——. 1971*b*. Aperçu sur la flore et la végétation des Monts Nimba au Libéria. In: Flore descriptive des Monts Nimba, 1. *Mém. Mus. natn. Hist. nat., Paris*, n.s., sér. B (Bot.), 20, p. 23–144.

ADAM, J. G.; JAEGER, P. 1976. Suppression de la floraison consécutive à la suppression des feux dans les savanes et prairies de la Guinée (Afrique occidentale). *C.r. hebd. Séanc. Acad. Sci., Paris*, 282, p. 637–9.

ADAMS, C. D. 1957. Observations on the fern flora of Fernando Po. I. A description of the vegetation with particular reference to the Pteridophyta. *J. Ecol.*, 45, p. 479–94.

ADAMS, M. E. 1967. A study of the ecology of *Acacia mellifera*, *A. seyal*, and *Balanites aegyptiaca* in relation to land clearing. *J. appl. Ecol.*, 4, p. 221–37.

ADAMSON, R. S. 1927. The plant communities of Table Mountain: preliminary account. *J. Ecol.*, 15, p. 278–309.

——. 1934. The vegetation and flora of Robben Island. *Trans. R. Soc. S. Afr.*, 22, p. 279–96.

——. 1935. The plant communities of Table Mountain. III. A six years' study of regeneration after burning. *J. Ecol.*, 23, p. 44–55.

——. 1938*a*. *The vegetation of South Africa*. London, British Emp. Veg. Comm. 235 p., with 3 small vegetation maps.

——. 1938*b*. Notes on the vegetation of the Kamiesberg. *Mem. bot. Surv. S. Afr.*, 18, p. 1–25.

——. 1959. Notes on the phytogeography of the flora of the Cape Peninsula. *Trans. R. Soc. S. Afr.*, 35, p. 443–62.

ADEJUWON, J. O. 1970. The ecological status of coastal savannas in Nigeria. *J. trop. Geogr.*, 30, p. 1–10.

——. 1971*a*. Savanna patches within forest areas in Western Nigeria: a study of the dynamics of forest savanna boundary. *Bull. IFAN*, sér. A, 33, p. 327–44.

——. 1971*b*. The ecological status of savannas associated with inselbergs in the forest areas of Nigeria. *Trop. Ecol.*, 12, p. 51–65.

——. 1971*c* [1973]. The ecological status of fresh water swamp savannas in the forest zone of Nigeria. *J. W. Afr. Sci. Ass.*, 16, p. 133–54.

ADJANOHOUN, E. 1962. Étude phytosociologique des savanes de basse Côte d'Ivoire (savanes lagunaires). *Vegetatio*, 11, p. 1–38.

——. 1964. Végétation des savanes et des rochers découverts en Côte d'Ivoire centrale. *Mém. ORSTOM*, 7, p. 1–250.

——. 1965. Comparaison entre les savanes côtières de Côte d'Ivoire et du Dahomey. *Ann. Univ. Abidjan, Fac. Sci.*, 1, p. 1–20.

——. 1968. Le Dahomey. In: Hedberg, I.; Hedberg, O. (eds.), p. 86–91, with small vegetation map.

ADJANOHOUN, E.; AKÉ ASSI, L. 1967. Inventaire floristique des forêts claires subsoudanaises et soudanaises en Côte d'Ivoire septentrionale. *Ann. Univ. Abidjan, Fac. Sci.*, 3, p. 89–147.

ADJANOHOUN, E.; AKÉ ASSI, L.; GUILLAUMET, J. L. 1968. La Côte d'Ivoire. In: Hedberg, I.; Hedberg, O. (eds.), p. 76–81, with small vegetation map.

AGNEW, A. D. Q. 1968. Observations on the changing vegetation of Tsavo National Park (East). *E. Afr. Wildl. J.*, 6, p. 75–80.

AHLGREN, I. F.; AHLGREN, C. E. 1960. Ecologic effects of forest fires. *Bot. Rev.*, 26, p. 483–533.

AHN, P. M. 1958. Regrowth and swamp vegetation in the western forest areas of Ghana. *J. W. Afr. Sci. Ass.*, 4, p. 163–73.

——. 1959. The savanna patches of Nzima, South-Western Ghana. *J. W. Afr. Sci. Ass.*, 5, p. 10–25.

——. 1961. Soil–vegetation relationships in the western forest areas of Ghana. In: *Humid tropics research. Tropical soils and vegetation*, p. 75–84. See Unesco, 1961.

——. 1970. *West African soils*. London, Oxford Univ. Press. 332 p.

——. 1974. Some observations on basic and applied research in shifting cultivation. *FAO Soils Bull.*, 24, p. 123–54.

AINSLIE, J. R. 1926. The physiography of Southern Nigeria and its effect on the forest flora of the country. *Oxf. For. Mem.*, 5, p. 1–36. Oxford, Clarendon Press.

AIRY SHAW, H. K. 1947. The vegetation of Angola. *J. Ecol.*, 35, p. 23–48.

AITKEN, R. D.; GALE, G. W. 1921. Botanical survey of Natal and Zululand. *Mem. bot. Surv. S. Afr.*, 2, p. 1–19.

ALBRECHT, F. O. 1964. Natural changes in grass zonations in a Red Locust outbreak centre in the Rukwa Valley, Tanganyika. *S. Afr. J. agr. Sci.*, 7, p. 123–30.

ALEXANDRE, D. Y. 1977. Régénération naturelle d'un arbre caractéristique de la forêt équatoriale de Côte d'Ivoire: *Turraeanthus africanus* Pellegr. *Oecol. Plant.*, 12, p. 241–62.

——. 1978. Le rôle disséminateur des éléphants en forêt de Tai, Côte d'Ivoire. *La Terre et la Vie*, 32, p. 47–72.

ALLAN, W. 1965. *The African husbandman*. Edinburgh, London, Oliver & Boyd. 505 p.

——. 1968. Soil resources and land use in tropical Africa. In: Hedberg, I.; Hedberg, O. (eds.), p. 9–13.
ALLORGE, P. & V., et al. 1946. Contribution à l'étude du peuplement des Îles Atlantides. *Mém. Soc. Biogéogr.*, 8, p. 1–500.
AMENT, J. G. 1975. The vascular plants of Meru National Park, Kenya. Part 1. A preliminary survey of the vegetation. *J. E. Afr. nat. Hist. Soc.*, 154, p. 1–10, with large-scale vegetation map.
AMICO, A. 1967. Contributo alla conoscenza della flora della Zambesia inferiore (Mozambico). 1. Itinerario botanico e considerazioni fitogeografiche. *Webbia*, 22, p. 469–526, with small vegetation map.
AMPOFO, S. T.; LAWSON, G. W. 1972. Growth of seedlings of *Afromosia elata* Harms in relation to light intensity. *J. appl. Ecol.*, 9, p. 301–6.
ANDERSON, G. D.; HERLOCKER, D. J. 1973. Soil factors affecting the distribution of the vegetation types and their utilization by wild animals in Ngorongoro crater, Tanzania. *J. Ecol.*, 61, p. 627–51.
ANDERSON, G. D.; TALBOT, L. M. 1965. Soil factors affecting the distribution of the grassland types and their utilization by wild animals on the Serengeti Plains, Tanganyika. *J. Ecol.*, 53, p. 33–56.
ANDERSON, G. D.; WALKER, B. H. 1974. Vegetation composition and elephant damage in the Sengwa Wildlife Research Area, Rhodesia. *J. S. Afr. Wildl. Mgmt Ass.*, 4, p. 1–14.
ANDREWS, F. W. 1945. Water plants in the Gezira canals. *Ann. appl. Biol.*, 32, p. 1–14.
——. 1948. The vegetation of the Sudan. In: Tothill, J. D. (ed.), *Agriculture in the Sudan*, p. 32–61. London, Oxford University Press.
ANON. 1976. *World atlas of agriculture*, vol. 4. Novara (Italy), Inst. Geogr. Agostini. 761 p.
——. 1977. *Desertification: its causes and consequences.* Compiled and edited by the Secretariat of the United Nations Conference on Desertification, Nairobi, Kenya, 29 August to 9 September 1977. Oxford, Pergamon Press. 448 p.
ARCHIBALD, E. E. A. 1955. An ecological survey of the Addo Elephant National Park. *J. S. Afr. Bot.*, 20, p. 137–54.
ASARE, E. O. 1962. A note on the vegetation of the transition zone of the Tain basin in Ghana. *Ghana J. Sci.*, 2, p. 60–73.
ASGA–UNESCO. 1963. *Carte géologique de l'Afrique/Geological map of Africa.* Paris, ASGA–Unesco. 9 sheets in colour (1:5 000 000). *See also* Furon & Lombard, 1964.
ASTLE, W. L. 1965a. The grass cover of the Chambeshi Flats, Northern Province, Zambia. *Kirkia*, 5, p. 37–48, with vegetation map 1:250 000.
——. 1965b. The edaphic grasslands of Zambia. *Proc. 9th Int. Grassl. Congr.*, vol. 1, p. 363–73. São Paulo, Brazil.
——. 1969. The vegetation and soils of Chishinga Ranch, Luapula Province, Zambia. *Kirkia*, 7, p. 73–102, with large-scale vegetation map.
ASTLE, W. L.; WEBSTER, R.; LAWRANCE, C. J. 1969. Land classification for management planning in the Luangwa Valley of Zambia. *J. appl. Ecol.*, 6, p. 143–69.
ATWELL, R. I. G. 1970. Some effects of Lake Kariba on the ecology of a flood-plain of the mid–Zambezi Valley of Rhodesia. *Biol. Conserv.*, 2, p. 189–96.
AUBRÉVILLE, 1937a. Les forêts du Dahomey et du Togo. *Bull. Com. Étud. hist. scient. Afr. occid. fr.*, 20, p. 1–112.
——. 1937b. *The Niger Colony forestry expedition September to December 1935.* Ibadan, Nigeria, For. Dep. 83 p.

——. 1938. La forêt coloniale: les forêts de l'Afrique occidentale française. *Annls Acad. Sci. colon.*, 9, p. 1–244.
——. 1947a. Les brousses secondaires en Afrique équatoriale. *Bois Forêts Trop.*, 2, p. 24–49.
——. 1947b. Érosion et 'bovalisation' en Afrique noire française. *Agron. trop.*, 2, p. 339–57.
——. 1948a. Étude sur les forêts de l'Afrique équatoriale française et du Cameroun. *Agron. trop., Bull. sci.*, 2, p. 1–132. *Also republished in* Aubréville, 1948c.
——. 1948b. La Casamance. *Agron. trop., Bull. sci.*, 3, p. 25–52. *Also republished in* Aubréville, 1948c.
——. 1948c. *Richesses et misères des forêts de l'Afrique noire française.* Paris. 251 p.
——. 1949a. *Climats, forêts et désertification de l'Afrique tropicale.* Paris, Soc. Éd. Géogr. Marit. Colon. 351 p.
——. 1949b. Ancienneté de la destruction de la couverture forestière primitive de l'Afrique tropicale. *Bull. Agric. Congo belge*, 40, p. 1347–52.
——. 1951. Le concept d'association dans la forêt dense équatoriale de la basse Côte d'Ivoire. *Mém. Soc. bot. Fr.*, 1950–51, p. 145–58.
——. 1955. La disjonction africaine dans la flore forestière tropicale. *C.r. somm. Séanc. Soc. Biogéogr.*, 278, p. 42–9.
——. 1957. Echos du Congo belge. Climax yangambiens, Muhulus, termitières fossiles géantes et forêt claire katangiens. *Bois Forêts Trop.*, 51, p. 28–39.
——. 1957–58. A la recherche de la forêt en Côte d'Ivoire. *Bois Forêts Trop.*, 56, p. 17–32 (1957), 57, p. 12–27 (1958).
——. 1959. Les fourrés alignés et les savanes à termitières buissonnantes des plaines de Winneba et d'Accra (Ghana). *Bois Forêts Trop.*, 67, p. 21–4.
——. 1962. Position chorologique du Gabon. In: *Flore du Gabon*, 3, p. 3–11.
——. 1963. Classification des formes biologiques des plantes vasculaires en milieu tropical. *Adansonia*, sér. 2, 3, p. 221–6.
——. 1964. La forêt dense de la Lobaye. *Cah. Maboké*, 2 (1), p. 5–9.
——. 1965. Principes d'une systématique des formations végétales tropicales. *Adansonia*, sér. 2, p. 153–96.
——. 1967a. Les étranges mosaïques forêts–savane du sommet de la boucle de l'Ogooué au Gabon. *Adansonia*, sér. 2, 7, p. 13–22.
——. 1967b. La forêt primaire des montagnes de Bélinga. *Biol. gabon.*, 3, p. 95–112.
——. 1970. Vocabulaire de biogéographie appliquée aux régions tropicales. *Adansonia*, sér. 2, 10, p. 439–97.
——. 1971. The destruction of forests and soils in the tropics. *Adansonia*, sér. 2, 11, p. 5–39.
——. 1973. Rapport de la mission forestière Anglo-Française Nigeria–Niger (déc. 1936–févr. 1937). *Bois Forêts Trop.*, 148, p. 3–26.
AUBRÉVILLE, A.; DUVIGNEAUD, P.; HOYLE, A. C.; KEAY, R. W. J.; MENDONÇA, F. A.; PICHI-SERMOLLI, R. E. G. 1959. *Vegetation map of Africa south of the tropic of Cancer/Carte de la végétation de l'Afrique au sud du tropique du Cancer.* 1:10 000 000. *See also* Keay, 1959.
AUDRY, P.; ROSSETTI, C. 1962. *Observations sur les sols et la végétation en Mauritanie du sud-est et sur la bordure adjacente du Mali (1959 et 1961).* Rome, FAO. 267 p. (Projet Pélerin, Rapp. no. UNSF/DL/ES/3.)
AUSTEN, B. 1972. The history of veld burning in the Wankie National Park, Rhodesia. *Proc. Ann. Tall Timbers Fire Ecol. Conf.*, 11, p. 277–96.
AYYAD, M. A. 1973. Vegetation and environment of the

western Mediterranean coastal land of Egypt. I. The habitat of sand dunes. *J. Ecol.*, 61, p. 509–23.
——. 1976. Vegetation and environment of the western Mediterranean coastal land of Egypt. IV. The habitat of non-saline depressions. *J. Ecol.*, 64, p. 713–22.
AYYAD, M. A.; AMMAR, M. Y. 1973. Relationship between local physiographic variations and the distribution of common Mediterranean desert species. *Vegetatio*, 27, p. 163–76.
AYYAD, M. A.; ELGHONEMY, A. A. 1976. Phytosociological and environmental gradients in a sector of western desert of Egypt. *Vegetatio*, 31, p. 93–102.
AZANCOT DE MENEZES, O. J. 1969. *Estudo fito-ecológico da Região do Mucope e Carta da Vegetação (1:200000).* Luanda, Inst. Invest. Cient. Angola. 50 p.
BACKLUND, H. O. 1956. Aspects and successions of some grassland vegetation in the Rukwa Valley, a permanent breeding area of the Red Locust. *Oikos*, Suppl. 2, p. 1–132.
BAGNOULS, F.; GAUSSEN, H. 1957. Les climats biologiques et leur classification. *Annls Géogr.*, 355, p. 193–220.
BALFOUR, I. B. 1888. Botany of Socotra. *Trans. R. Soc. Edin.*, 31, p. 1–446.
BALFOUR, I. B. *et al.* 1879. Botany of Rodriguez. Transit of Venus Expedition. *Proc. R. Soc. Lond.*, 168 (extra volume), p. 302–419.
BALLANTYNE, A. O. 1968. *Soils of Zambia.* Lusaka, Mount Makulu Res. Sta., Dep. Agric.
BALLE, S. 1953. La végétation du Ruwenzori. *Naturalistes belg.*, 34, p. 75–83.
BALLY, P. R. O. 1968. Somali Republic South. In: Hedberg, I.; Hedberg, O. (eds.), p. 145–8, with small vegetation map.
——. 1976. The vegetation of Somalia. *Boissiera*, 24, p. 447–50.
BALON, E. K.; COCHE, A. G. (eds.). 1974. *Lake Kariba, a man-made tropical ecosystem in Central Africa.* The Hague, Junk. 767 p. (Monogr. biol. 24.)
BAMPS, P. 1975. Glossaire des dénominations indigènes désignant les paysages végétaux au Zaïre. *Bull. Jard. bot. nat. Belg.*, 45, p. 137–47.
BANDS, D. P. 1977. Prescribed burning in Cape fynbos. In: Mooney, H. A.; Conrad, C. E. (eds.), p. 245–56.
BANKS, P. F. 1976. Chirinda forest. *Rhod. Sci. News*, 10, p. 39–40.
BARBOSA, L. A. Grandvaux. 1952. Esboço da vegetação da Zambézia. *Doc. Moç.*, 69, p. 5–65, with coloured vegetation map 1:1000000.
——. 1968*a*. L'archipel du Cap-Vert. In: Hedberg, I.; Hedberg, O. (eds.), p. 94–7.
——. 1968*b*. Moçambique. In: Hedberg, I.; Hedberg, O. (eds.), p. 224–32.
——. 1968*c*. Vegetation. In: Bannerman, D. A. & W. M., *History of the birds of the Cape Verde Islands*, p. 58–61. Edinburgh, Oliver & Boyd. (Birds of the Atlantic Islands, vol. 4.)
——. 1970. *Carta fitogeográfica de Angola.* Luanda, Inst. Invest. Cient. Angola. 323 p., with coloured vegetation map 1:2500000.
BARCLAY-SMITH, R. W. 1964. A report on the ecology and vegetation of the Great Dyke within the Horseshoe Intensive Conservation Area. *Kirkia*, 4, p. 25–34, with small vegetation map.
BARI, E. A. 1968. Sudan. In: Hedberg, I.; Hedberg, O. (eds.), p. 59–64.
BARKHAM, J. P.; RAINY, M. E. 1976. The vegetation of the Samburu–Isiolo Game Reserve. *E. Afr. Wildl. J.*, 14, p. 297–329.
BARRY, J. P.; BÉLIN, B.; CELLES, J. Cl.; DUBOST, D.; FAUREL, L.; HETHENER, P. 1970. Essai de monographie du *Cupressus dupreziana* A. Camus, Cyprès endémique du Tassili des Ajjer. *Bull. Soc. Hist. nat. Afr. N.*, 61, p. 95–178.
BARRY, J. P.; FAUREL, L. 1973. Notice de la feuille de Ghardaia. Carte de la végétation de l'Algérie au 1:500000. *Mém. Soc. Hist. nat. Afr. N.*, n.s., 11, p. 1–125, 1 carte h.t.
BARTHA, R. 1970. *Fodder plants in the Sahel zone of Africa.* München, Weltforum Verlag. 306 p. (IFO-Inst. Wirtschaftsforsch. Münch. Afr.-Studienstelle, no. 48.)
BARTLETT, H. H. 1955–57. *Fire in relation to primitive agriculture and grazing in the tropics: annotated bibliography.* Vol. 1, 1955, p. 1–568, vol. 2, 1957, p. 1–873. Ann Arbor, Mich., Univ. of Michigan Botanical Gardens.
——. 1956. Fire, primitive agriculture and grazing in the tropics. In: Thomas. W. L. (ed.), p. 692–720.
BARTOLO, G.; BRULLO, S.; GUGLIELMO, A.; SCALIA, C. 1977. Considerazioni fitogeografiche sugli endemismi della Cirenaica settentrionale. *Arch. bot. biogeogr. ital*, 53, p. 131–54.
BATANOUNY, K. H. 1964. Sand dune vegetation of El Arish area. *Bull. Fac. Sci., Cairo Univ.*, 39, p. 11–23.
——. 1966. Vegetation and root development in the different microhabitats in wadi Hoff. *Bull. Inst. Désert Égypte*, 15, p. 55–66.
——. 1973. Habitat features and vegetation of deserts and semi-deserts in Egypt. *Vegetatio*, 27, p. 181–99.
BATANOUNY, K. H.; ABU EL SOUOD, S. 1972. Ecological and phytosociological study of a sector in the Libyan desert. *Vegetatio*, 25, p. 335–56.
BATANOUNY, K. H.; ZAKI, M. A. F. 1973. Range potentialities of a sector in the Mediterranean coastal region in Egypt. *Vegetatio*, 27, p. 115–30.
BATTISTINI, R.; CREMERS, G. 1972. Geomorphology and vegetation of Îles Glorieuses. *Atoll Res. Bull.*, 159, p. 1–25.
BATTISTINI, R.; RICHARD-VINDARD, G. (eds.). 1972. *Biogeography and ecology in Madagascar.* The Hague, Junk. 765 p. (Monogr. biol. 21.)
BAWDEN, M. G. 1965. A reconnaissance of the land resources of Eastern Bechuanaland. *J. appl. Ecol.*, 2, p. 357–65.
BAWDEN, M. G.; CARROLL, D. M. 1968. *The land resources of Lesotho.* Tolworth Tower, Surbiton, Surrey, Directorate of Overseas Surveys. 89 p., with vegetation map 1:1000000. (Land resource study 3.)
BAWDEN, M. G.; STOBBS, A. R. 1963. *The land resources of eastern Bechuanaland.* Tolworth Tower, Surbiton, Surrey, Directorate of Overseas Surveys. 75 p., with small coloured vegetation map.
BAWDEN, M. G.; TULEY, P. 1966. *The land resources of Southern Sardauna and Southern Adamawa Provinces, Northern Nigeria.* Tolworth Tower, Surbiton, Surrey, Directorate of Overseas Surveys. 120 p., with coloured map 1:500000 showing land systems and vegetation. (Land resource study 2.)
BAX, P. N. *See* Napier Bax, P.
BAYER, A. W. 1938. An account of the plant ecology of the Coastbelt and Midlands of Zululand. *Ann. Natal Mus.*, 8 (3), p. 371–454.
BAYER, A. W.; BIGALKE, R. C.; CRASS, R. S. 1968. Natal. In: Hedberg, I.; Hedberg, O. (eds.), p. 243–7.
BEADLE, L. C. 1974. *The inland waters of tropical Africa. An introduction to tropical limnology.* London, Longman. 365 p.

BEADLE, N. C. W. 1966. Soil phosphate and its role in molding segments of the Australian flora and vegetation, with special reference to xeromorphy and sclerophylly. *Ecology*, 47, p. 992–1007.
——. 1968. Some aspects of the ecology and physiology of Australian xeromorphic plants. *Aust. J. Sci.*, 30, p. 348–55.
BEALS, E. W. 1968. Ethiopia. In: Hedberg, I.; Hedberg, O. (eds.), p. 137–40.
BEARD, J. S. 1967. Some vegetation types of tropical Australia in relation to those of Africa and America *J. Ecol.*, 55, p. 271–90.
——. 1978. The physiognomic approach. In: Whittaker, R. H. (ed.), *Classification of plant communities*, p. 33–64. The Hague, Junk.
BEGUÉ, L. 1937. Contribution à l'étude de la végétation forestière de la Haute Côte d'Ivoire. *Publs Com. Étude Hist. scient. Afr. occ. fr.*, sér. B, 4, p. 1–127.
——. 1958. Les forêts de la République du Soudan. *Bois Forêts Trop.*, 62, p. 3–19.
——. 1965. Les savanes du Sud de la République du Congo (Brazzaville). *Bois Forêts Trop.*, 99, p. 52–8; 100, p. 58–63.
——. 1967. Les forêts du Nord de la République du Congo (Brazzaville). *Bois Forêts Trop.*, 111, p. 63–76.
BELL, R. H. V. 1970. The use of the herb layer by grazing ungulates in the Serengeti. In: Watson, A. (ed.), *Animal populations in relation to their food resources*, p. 111–24. Oxford, Blackwell.
——. 1971. A grazing ecosystem in the Serengeti. *Sci. Am.*, 225, p. 86–93.
BELL, R. H. V.; GRIMSDELL, J. J. R. 1973. The persecuted black lechwe of Zambia. *Oryx*, 12, p. 77–92.
BELLIER, L.; GILLON, D.; GILLON, Y.; GUILLAUMET, J. L.; PERRAUD, A. 1969. Recherches sur l'origine d'une savane incluse dans le bloc forestier du Bas-Cavally (Côte d'Ivoire) par l'étude des sols et de la biocoenose. *Cahiers ORSTOM*, sér. Biol., 10, p. 65–94.
BELLOUARD, P. 1950. Le rônier en Afrique occidentale française. *Bois Forêts Trop.*, 14, p. 117–26.
BEN SAI, S. 1950. Note sur la végétation des hautes vallées du Sénégal et du Niger. *C.r. Première Conf. int. Afr. Ouest*, 1, p. 407–31.
BERGER-LANDEFELDT, U. 1959. Beiträge zur Ökologie der Pflanzen nordafrikanischer Salzpfannen. IV. Vegetation. *Vegetatio*, 9, p. 1–47.
BERNARD, E. 1945. Le climat écologique de la Cuvette centrale congolaise. *Publs INEAC*, Coll. in 4°, p. 1–240.
BERNHARD-REVERSAT, F.; HUTTEL, C.; LEMÉE, G. 1978. Structure and functioning of evergreen rain forest ecosystems of the Ivory Coast. In: Unesco/UNEP/FAO, *Tropical forest ecosystems*, p. 557–74.
BEWS, J. W. 1912. The vegetation of Natal. *Ann. Natal Mus.*, 2 (3), p. 253–331.
——. 1913. An oecological survey of the midlands of Natal with special reference to the Pietermaritzburg District. *Ann. Natal Mus.*, 2 (4), p. 485–545, with large-scale vegetation map.
——. 1916a. An account of the chief types of vegetation in South Africa with notes on the plant succession. *J. Ecol.*, 4, p. 129–59.
——. 1916b. The growth-forms of Natal plants. *Trans. R. Soc. S. Afr.*, 5, p. 605–36.
——. 1917a. The plant ecology of the Drakensberg Range. *Ann. Natal Mus.*, 3 (3), p. 511–65.
——. 1917b. The plant succession in the thornveld. *S. Afr. J. Sci.*, 15, p. 153–72.
——. 1918. *The grasses and grasslands of South Africa*. Pietermaritzburg, Davis. 161 p.
——. 1920. The plant ecology of the coast belt of Natal. *Ann. Natal Mus.*, 4 (2), p. 367–469.
——. 1925. *Plant forms and their evolution in South Africa*. London, Longman Green. 199 p.
BIGALKE, R. C. 1961. Some observations on the ecology of the Etosha Game Park, South West Africa. *Ann. Cape Prov. Mus.*, 1, p. 49–67.
——. 1978. Mammals. In: Werger, M. J. A. (ed.), p. 981–1048.
BILLE, J. C.; POUPON, H. 1972. Recherches écologiques sur une savane sahélienne du Ferlo septentrional, Sénégal. Description de la végétation. *La Terre et la Vie*, 26, p. 351–65.
BIRCH, W. R. 1963. Observations on the littoral and coral vegetation of the Kenya coast. *J. Ecol.*, 51, p. 603–15.
BISHOP, W. W. 1959. Raised swamps of Lake Victoria. *Rec. Geol. Surv. Uganda 1955–1956*, p. 33–42.
BJØRNSTAD, A. 1976. The vegetation of Ruaha National Park, Tanzania. 1, Annotated check-list of the plant species. *Serengeti Res. Inst. Publ.*, 215, p. 1–61, with small vegetation map.
BLAIR RAINS, A.; MCKAY, A. D. 1968. *The Northern State Lands, Botswana*. Tolworth Tower, Surbiton, Surrey, Directorate of Overseas Surveys. 125 p., with coloured vegetation map 1:500 000. (Land resource study 5.)
BLAIR RAINS, A.; YALALA, A. M. 1972. *The Central and Southern State Lands, Botswana*. Tolworth Tower, Surbiton, Surrey, Directorate of Overseas Surveys. 115 p., with coloured vegetation map 1:500 000. (Land resource study 11.)
BOALER, S. B. 1966. Ecology of a miombo site, Lupa North Forest Reserve, Tanzania. II. Plant communities and seasonal variation in the vegetation. *J. Ecol.*, 54, p. 465–79.
BOALER, S. B.; HODGE, C. A. H. 1962. Vegetation stripes in Somaliland. *J. Ecol.*, 50, p. 465–74.
——; ——. 1964. Observations on vegetation arcs in the Northern Region, Somali Republic. *J. Ecol.*, 52, p. 511–44.
BOALER, S. B.; SCIWALE, K. C. 1966. Ecology of a miombo site, Lupa North Forest Reserve, Tanzania. III. Effects on the vegetation of local cultivation practices. *J. Ecol.*, 54, p. 577–87.
BÖCHER, T. W. 1977. Convergence as an evolutionary process. *J. Linn. Soc. Bot.*, 75, p. 1–19.
BOGDAN, A. V. 1958. Some edaphic vegetational types at Kiboko, Kenya. *J. Ecol.*, 46, p. 115–26.
BOLUS, H. 1875. Extract from a letter of Harry Bolus, Esq., F.L.S., to J. D. Hooker. *J. Linn. Soc. Bot.*, 14, p. 482–4.
BOND, W. J. 1980. Periodicity in fynbos of the non-seasonal rainfall belt. *J. S. Afr. Bot.*, 46, p. 343–54.
BÖRGENSEN, F. 1924. Contributions to the knowledge of the vegetation of the Canary Islands (Teneriffe and Gran Canaria). *K. danske Vidensk. Selsk. Skr. Naturv. Math.*, Afd. 8, 6 (3), p. 283–399.
BORNMAN, C. H.; BOTHA, C. E. J.; NASH, L. J. 1973. *Welwitschia mirabilis*: observations on movement of water and assimilates under föhn and fog conditions. *Madoqua*, ser. 2, 2, p. 25–31.
BORNMAN, C. H.; ELSWORTHY, J. A.; BUTLER, V.; BOTHA, C. E. J. 1972. *Welwitschia mirabilis*: observations on general habit, seed, seedlings, and leaf characteristics. *Madoqua*, ser. 2, 1, p. 53–66.

BOSCH, O. J. K. 1978. Vergelyking tussen die plantegroei en habitatte van die gronde van die Estcourt-en Sterkspruitvorms in die suidoostelike Oranje–Vrystaat. *Bothalia*, 12, p. 499–511.

BOSSER, J. 1952. Note sur la végétation des Îles Europa et Juan de Nova. *Le Naturaliste Malgache*, 4, p. 41–2.

BOSSER, J.; CADET, T.; JULIEN, H. R.; MARAIS, W. (eds.). 1976–. *Flore des Mascareignes: La Réunion, Maurice, Rodrigues*. Mauritius, Sug. Ind. Res. Inst.; Kew, R. Bot. Gdns; Paris, ORSTOM.

BOUCHER, C. 1977. Cape Hangklip area. I. The application of association analysis, homogeneity functions and Braun-Blanquet techniques in the description of south-western Cape vegetation. *Bothalia*, 12, p. 293–300.

———. 1978. Id. II. The vegetation. *Bothalia*, 12, p. 455–97, with large-scale vegetation map.

———. 1980. Notes on the use of the term 'Renosterveld'. *Bothalia*, 13, p. 237.

BOUCHER, C.; JARMAN, M. L. 1977. The vegetation of the Langebaan area, South Africa. *Trans. R. Soc. S. Afr.*, 42, p. 241–72, with large-scale vegetation map.

BOUDET, G. 1972. Désertification de l'Afrique tropicale sèche. *Adansonia*, sér. 2, 12, p. 505–24.

BOUDY, P. 1948. *Économie forestière nord-africaine*. 1. *Milieu physique et milieu humain*. Paris, Larose. 686 p.

———. 1950. *Économie forestière nord-africaine*. 2. *Monographie et traitements des essences forestières*. Paris, Larose. 878 p.

BOUGHEY, A. S. 1955a. The vegetation of the mountains of Biafra. *Proc. Linn. Soc. Lond.*, 165, p. 144–50.

———. 1955b. The nomenclature of the vegetation zones on the mountains of tropical Africa. *Webbia*, 11, p. 413–23.

———. 1956a. The lowland rain forest of tropical Africa. *Proc. Trans. Rhod. sci. Ass.*, 44, p. 36–52.

———. 1956b. The vegetation types of the Federation. *Proc. Trans. Rhod. sci. Ass.*, 45, p. 73–91.

———. 1957a. Ecological studies of tropical coast-lines. I. The Gold Coast, West Africa. *J. Ecol.*, 45, p. 665–87.

———. 1957b. The physiognomic delimitation of West African vegetation types. *J. W. Afr. Sci. Ass.*, 3, p. 148–65.

———. 1958. The plant colonisation of the islands in the Gulf of Guinea. *6a. Conf. Int. Afr. Occid.*, vol. 3, p. 69–76.

———. 1961. The vegetation types of Southern Rhodesia: a reassessment. *Proc. Trans. Rhod. sci. Ass.*, 49, p. 54–98.

———. 1963a. The explosive development of a floating weed vegetation on Lake Kariba. *Adansonia*, sér. 2, 3, p. 49–61.

———. 1963b. Interaction between animals, vegetation and fire in Southern Rhodesia. *Ohio J. Sci.*, 63, p. 193–209.

———. 1964. Deciduous thicket communities in Northern Rhodesia. *Adansonia*, sér. 2, 4, p. 239–61.

BOUILLENNE, R.; MOUREAU, J.; DEUSE, P. 1955. Esquisse écologique des faciès forestiers et marécageux des bords du lac Tumba. *Mém. Acad. r. Sci. colon., Cl. Sci. nat. méd., 8°*, n.s., 3 (1), p. 1–44, with small vegetation map.

BOULOS, L. 1966. A natural history study of Kurkur Oasis, Libyan Desert, Egypt. IV. The vegetation. *Postilla*, 100, p. 1–22.

———. 1972. Our present knowledge on the flora and vegetation of Libya; bibliography. *Webbia*, 26, p. 365–400.

———. 1975. The Mediterranean element in the flora of Egypt and Libya. In: *La flore du bassin méditerranéen. Coll. Int. CNRS*, 235, p. 119–24.

BOULVERT, Y. 1980. Végétation forestière des savanes centrafricaines. *Bois Forêts Trop.*, 191, p. 21–45.[1]

BOURBEAU, G.; SYS, C.; FRANKART, R.; MICHEL, G.; REED, J. 1955. *Carte Sols Vég. Congo belge*, 5, *Mosso* (Urundi). A, Sols (1:50000; 1:100000). B, Végétation (1:200000). *Notice explicative*, p. 1–40. Brussels, INEAC.

BOURGUIGNON, P.; STREEL, M.; CALEMBERT, J. 1960. *Prospection pédo-botanique des plaines supérieures de la Lufira (Haut-Katanga)*. Liège, Ed. FULREAC, Univ. de Liège. 111 p., with coloured vegetation map 1:45000.

BOURLIÈRE, F. 1965. Densities and biomasses of some ungulate populations in Eastern Congo and Rwanda, with notes on population structure and lion/ungulate ratios. *Zool. Afr.*, 1, p. 199–207.

———. 1978. La savane sahélienne de Fété Olé, Sénégal. In: Lamotte, M.; Bourlière, F. (eds.), *Structure et fonctionnement des écosystèmes terrestres*, p. 187–229. Paris, Masson.

BOURLIÈRE, F.; HADLEY, M. 1970. The ecology of tropical savannas. *Ann. Rev. Ecol. Syst.*, 1, p. 125–52.

BOURLIÈRE, F.; VERSCHUREN, J. 1960. *Introduction à l'écologie des ongulés du Parc National Albert. Exploration du Parc National Albert: Mission F. Bourlière et J. Verschuren*, fasc. 1. Brussels, Inst. Parc Nat. Congo belge. 158 p., 59 pl.

BOURNE, R. 1931. *Regional survey*. Oxford, Clarendon Press. 169 p. (Oxf. For. Mem., 13.)

BOURREIL, P.; GILLET, H. 1971. Synthèse des connaissances et des recherches nouvelles sur *Aristida rhinochloa*, graminée africaine amphitropicale. *Mitt. Bot. Staatssamml. Münch.*, 10, p. 309–40.

BOUXIN, G. 1974. Distribution des espèces dans la strate herbacée au sud du parc national de l'Akagera (Rwanda, Afrique centrale). *Oecol. Plant.*, 9, p. 315–31.

———. 1975a. Ordination and classification in the savanna vegetation of the Akagera Park (Rwanda, Central Africa). *Vegetatio*, 29, p. 155–67.

———. 1975b. Action des feux saisonniers sur la strate ligneuse dans le Parc National de l'Akagera (Rwanda, Afrique centrale). *Vegetatio*, 30, p. 189–96.

———. 1975c. Ordination of quantitative and qualitative data in a savanna vegetation (Rwanda, Central Africa). *Vegetatio*, 30, p. 197–200.

———. 1976. Ordination and classification in the upland Rugege forest (Rwanda, Central Africa). *Vegetatio*, 32, p. 97–115.

BRAMWELL, D. 1976. The endemic flora of the Canary Islands; distribution, relationships and phytogeography. In: Künkel, G. (ed.), p. 207–40.

———. (ed.). 1979a. *Plants and islands*. London, Academic Press. 459 p.

———. 1979b. A local botanic garden (Canary Islands): its role in plant conservation. In: Synge, H.; Townsend, H. (eds.), *Survival or extinction*, p. 47–52. Kew, England, R. Bot. Gdns.

BRAMWELL, D.; BRAMWELL, Z. I. 1974. *Wild flowers of the Canary Islands*. London and Burford, Stanley Thornes. 261 p.

BRASS, L. J. 1953. Vegetation of Nyasaland. Report of the Vernay Nyasaland expedition of 1946. *Mem. N.Y. bot. Gdn*, 8, p. 161–90.

BRAUN-BLANQUET, J. 1928. Zur Kenntnis der Vegetationsverhältnisse des Grossen Atlas. *Vjschr. Naturf. Ges. Zürich*, Jg. 73, Beibl. 15, p. 334–57.

1. This important publication, which includes detailed information on the distribution of woody vegetation types in the Central African Republic, unfortunately appeared too late to be taken into account in the compilation of the map.

BRAUN-BLANQUET, J.; MAIRE, R. 1924. Études sur la végétation et la flore marocaines. *Mém. Soc. Sci. nat. Maroc*, 8, p. 1–244.
BREDENKAMP, G. J. 1975. Plant communities of the Suikerbosrand Nature Reserve, Transvaal. *S. Afr. J. Sci.*, 71, p. 30–1.
BREDENKAMP, G. J.; LAMBRECHTS, A. v. W. 1979. A check list of ferns and flowering plants of the Suikerbosrand Nature Reserve. *J. S. Afr. Bot.*, 45, p. 25–47.
BREDENKAMP, G. J.; THERON, G. K. 1976. Vegetation units for management of the grasslands of the Suikerbosrand Nature Reserve. *S. Afr. J. Wildl. Res.*, 6, p. 113–22.
——; ——. 1978. A synecological account of the Suikerbosrand Nature Reserve. I. The phytosociology of the Witwatersrand geological system. *Bothalia*, 12, p. 513–29.
——; ——. 1980. Id. II. The phytosociology of the Ventersdorp Geological System. *Bothalia*, 13, p. 199–216.
BREEN, C. M. 1971. An account of the plant ecology of the dune forest at Lake Sibayi. *Trans. R. Soc. S. Afr.*, 39, p. 223–34.
BREMAN, H.; CISSÉ, A. M. 1977. Dynamics of Sahelian pastures in relation to drought and grazing. *Oecologia* (Berl.), 28, p. 301–15.
BREMEKAMP, C. E. B. 1935. The origin of the flora of the Central Kalahari. *Ann. Transv. Mus.*, 16, p. 443–55.
BRENAN, J. P. M. 1978. Some aspects of the phytogeography of tropical Africa. *Ann. Mo. Bot. Gdn*, 65, p. 437–78.
BRETELER, F. J. 1969. The Atlantic species of *Rhizophora*. *Acta bot. neerl.*, 18, p. 434–41.
BROOKMAN-AMISSAH, J.; HALL, J. B.; SWAINE, M. D.; ATTAKORAH, J. Y. 1980. A re-assessment of a fire-protection experiment in North-Eastern Ghana savanna. *J. appl. Ecol.*, 17, p. 85–99.
BROUARD, N. R. 1963. *A history of woods and forests in Mauritius*. Port Louis, Mauritius, Govt Printer. 86 p.
BROWN, L. H. 1971. The biology of pastoral man as a factor in conservation. *Biol. Conserv.*, 3, p. 93–100.
BROWN, L. H.; COCHEMÉ, J. 1969. *A study of the agroclimatology of the highlands of Eastern Africa*. Rome, FAO. 330 p.
BROWN, P.; YOUNG, A. 1974. *The physical environment of central Malawi with special reference to soils and agriculture*. Zomba, Govt Printer. 93 p.
BROWN, W. L. 1960. Ants, acacias and browsing animals. *Ecology*, 41, p. 587–92.
BRUNEAU DE MIRÉ, P. 1960. Note préliminaire sur l'étage culminal du Djebel Marra (Republic of the Sudan) et ses affinités avec les hauts sommets du Tibesti. *C.r. somm. Séanc. Soc. Biogéogr.*, 37 (321–2), p. 11–18.
BRUNEAU DE MIRÉ, P.; QUÉZEL, P. 1959. Sur quelques aspects de la flore résiduelle du Tibesti: les fumerolles du Toussidé et les lappiaz volcaniques culminaux de l'Emi Koussi. *Bull. Soc. Hist. nat. Afr. N.*, 50, p. 126–45.
——; ——. 1961. Remarques taxonomiques et biogéographiques sur la flore des montagnes de la lisière méridionale du Sahara et plus spécialement du Tibesti et du Djebel Marra. *J. Agric. trop. Bot. appl.*, 8, p. 110–33.
BRUNNTHALER, J. 1914. Vegetationsbilder aus Deutsch-Ostafrika. Regenwald von Usambara. *Vegetationsbilder*, 11, t. 43–8.
BRYNARD, A. M. 1964. The influence of veld burning on the vegetation and game of the Kruger National Park. In: Davis, D. H. S. (ed.), p. 371–93, with small vegetation map.
BUCHWALD, J. 1896. Beitrag zur Gliederung der Vegetation von West-Usambara. *Mitt. dt. Schutzgeb.*, 9, p. 213–33.

BUECHNER, H. K.; DAWKINS, H. C. 1961. Vegetation change induced by elephants and fire in Murchison Falls National Park, Uganda. *Ecology*, 42, p. 752–66.
BULTOT, F. 1950. *Régimes normaux et cartes de précipitations dans l'est du Congo belge (Long.: 26° à 31° Est, Lat.: 4° Nord à 5° Sud) pour la période 1930 à 1946*. Brussels, INEAC. (Communication n° 1 du Bureau climatologique.) (Publ. INEAC, Coll. in-4°.)
——. 1971–74. *Atlas climatique du Bassin congolais*. 1. *Les composantes du bilan de rayonnement* (1971). 2. *Les composantes du bilan d'eau* (1971). 3. *Température et humidité de l'air, rosée, température du sol* (1972). Brussels, SERDAT. (Publ. INEAC, h.s., without pagination.)
——. 1977. *Atlas climatique du Bassin zaïrois*. 4. *Pression atmosphérique, vent en surface et en altitude, température et humidité de l'air en altitude, nébulosité et visibilité, classifications climatiques, propriétés chimiques de l'air et des précipitations*. Brussels, SERDAT. (Publ. INEAC, h.s., without pagination.)
BUNTING, A. H.; LEA, J. D. 1962. The soils and vegetation of the Fung, east central Sudan. *J. Ecol.*, 50, p. 529–58.
BURCHARD, O. 1929. Beiträge zur Ökologie und Biologie der Kanarenpflanzen. *Biblioth. bot.*, 24 (98), p. 1–262.
BURKE, K. C.; DEWEY, J. F. 1972. Orogeny in Africa. In: Dessauvage, T. F. J.; Whiteman, A. J. (eds.), *African geology*, p. 583–608. Univ. of Ibadan, Nigeria, Dep. of Geology.
BUROLLET, P. A. 1927. *Le Sahel de Sousse. Monographie phytogéographique*. Tunis. 276 p. (Thesis.)
BURTT, B. D. 1929. A record of fruits and seeds dispersed by mammals and birds from the Singida District of Tanganyika Territory. *J. Ecol.*, 17, p. 351–5.
——. 1942. Some East African vegetation communities. *J. Ecol.*, 30, p. 65–146.
BURTT DAVY, J. 1931. The forest vegetation of South Central Africa. *Emp. For. J.*, 10, p. 73–85.
——. 1935. A sketch of the forest vegetation and flora of tropical Africa. *Emp. For. J.*, 14, p. 191–201.
——. 1938. The classification of tropical woody vegetation-types. *Inst. Pap. Imp. For. Inst.*, 13, p. 1–85.
BUSSE, W. 1907. Das südliche Togo. *Vegetationsbilder*, 4, t. 7–12.
——. 1908. Die periodische Grasbrände im tropischen Afrika, ihr Einfluss auf die Vegetation und ihre Bedeutung für die Landeskultur. *Mitt. dt. Schutzgeb.*, 21, p. 113–39.
BUXTON P. A. 1935. Seasonal changes in vegetation in the north of Nigeria. *J. Ecol.*, 23, p. 134–9.
BYSTRÖM, K. 1960. *Dracaena draco* in the Cape Verde Islands. *Acta Hort. Gotoburgensis*, 22, p. 179–214.
CABALLÉ, G. 1978. Essai sur la géographie forestière du Gabon. *Adansonia*, sér. 2, 17, p. 425–40.
——. 1980*a*. Caractéristiques de croissance et multiplication végétative en forêt dense du Gabon de la 'liane à eau' *Tetracera alnifolia* Willd. (Dilleniaceae). *Adansonia*, sér. 2, 19, p. 465–75.
——. 1980*b*. Caractères de croissance et déterminisme chorologique de la liane *Entada gigas* (L.) Fawcett & Rendle (Leguminosae: Mimosoideae) en forêt dense du Gabon. *Adansonia*, sér. 2, 20, p. 309–20.
CADET, L. J. T. (1980). *La végétation de l'Île de la Réunion: étude phytoécologique et phytosociologique*. Saint-Denis de la Réunion, Impr. Cazal. 312 p., with small vegetation map. (Thesis.)
CAHEN, L. 1954. *Géologie du Congo belge*. Liège, Vaillant-Carmanne. 577 p.

CAIN, S. A. 1947. Characteristics of natural areas and factors in their development. *Ecol. Monogr.*, 17, p. 185–200.
——. 1950. Life-forms and phytoclimate. *Bot. Rev.*, 16, p. 1–32.
CAMPBELL, A. (ed.). 1977. *The Okavango delta and its future utilization. Proceedings of a symposium held at the National Museum, Gaborone, Botswana, 30 August–2 September, 1976.* Gaborone, Botswana Society.
CAMPBELL, B.; GUBB, A.; MOLL, E. 1980. The vegetation of the Edith Stephenson Cape Flats Flora Reserve. *J. S. Afr. Bot.*, 46, p. 435–44.
CAMPBELL, B. M.; MOLL, E. J. 1977. The forest communities of Table Mountain, South Africa. *Vegetatio*, 34, p. 105–15.
CAMPBELL, B. M.; VAN DER MEULEN, F. 1980. Patterns of plant species diversity in fynbos vegetation, South Africa. *Vegetatio*, 43, p. 43–7.
CANNON, W. A. 1913. Botanical features of the Algerian Sahara. *Publs Carnegie Instn*, 178, p. 1–81.
——. 1924. General and physiological features of the vegetation of the more arid portions of Southern Africa, with notes on the climatic environment. *Publs Carnegie Instn*, 354, p. 1–159.
CAPOT-REY, R. 1953. *Le Sahara français.* Paris, Presses Universitaires de France. 564 p.
CAPURON, R. 1966. Rapport succinct sur la végétation et la flore de l'île Europa. *Mém. Mus. natn. Hist. nat., Paris*, sér. A (Zool.), 41, p. 19–21.
CARVALHO, G.; GILLET, H. 1960. Catalogue raisonné et commenté des plantes de l'Ennedi (Tchad septentrional). *J. Agric. trop. Bot. appl.*, 7, p. 49–96, 193–240, 317–78.
CASEBEER, R. L.; KOSS, G. G. 1970. Food habits of wildebeest, zebra, hartebeest and cattle in Kenya Masailand. *E. Afr. Wildl. J.*, 8, p. 25–36.
CATINOT, R. 1978. The forest ecosystems of Gabon: an overview. In: Unesco/UNEP/FAO, *Tropical forest ecosystems*, p. 575–9.
CAUGHLEY, G. 1976. The elephant problem. An alternative hypothesis. *E. Afr. Wildl. J.*, 14, p. 265–84.
CAVASSILAS, Y. 1963. Étude morphologique, écologique et floristique du bassin d'El Haroura (Maroc). *Mém. Soc. Sci. nat. phys. Maroc. Bot.*, n.s., 3, p. 1–154.
CCTA/CSA. 1956. *Phytogéographie/Phytogeography.* CCTA/CSA. 33 p. (Publ. no. 53.)
——. 1959. *Open forests/Forêts claires.* CCTA/CSA. 126 p. (Publ. no. 52.)
CEBALLOS, L.; ORTUÑO, F. 1951. *Vegetación y la flora forestal de las Canarias occidentales.* Madrid, Minist. Agric. 465 p.
CÉSAR, J.; MENAUT, J. C. 1974. Analyse d'un écosystème tropical humide: la savane de Lamto (Côte d'Ivoire). 2. Le peuplement végétal. *Bull. Liais. Cherch. Lamto*, num. spéc. 1974, fasc. 2, p. 1–161, with vegetation map 1:35000. N'Douci, Côte d'Ivoire, Stn Écol. Trop. Lamto.
CHAFFEY, D. R. 1979. *South-west Ethiopia forest inventory project. A reconnaissance inventory of forest in south-west Ethiopia.* Tolworth Tower, Surbiton, Surrey, Directorate of Overseas Surveys. 316 p., with 7 vegetation maps 1:250000. (Land Resources Development Centre, Project Rep. 31.)
CHAMBON, R.; LERUTH, A. 1954. Monographie des Bena Muhona. Territoire de Kongolo, District du Tanganyika. *Bull. Agric. Congo belge*, 45, p. 519–98.
CHAPIN, J. P. 1932. The birds of the Belgian Congo. Pt 1. *Bull. Am. Mus. nat. Hist.*, 65, p. 1–756.
——. 1939. Id. Pt 2, op. cit., 75, p. 1–632.
CHAPMAN, J. D. 1962. *The vegetation of the Mlanje Mountains, Nyasaland.* Zomba, Govt Printer. 78 p.
——. 1968. Malawi. In: Hedberg, I.; Hedberg, O. (eds.), p. 215–24.
CHAPMAN, J. D.; WHITE, F. 1970. *The evergreen forests of Malawi.* Oxford, Comm. For. Inst. 190 p.
CHAPMAN, V. J. 1960. *Salt marshes and salt deserts of the world.* London, Leonard Hill; New York, Interscience Publishers. 292 p.
——. 1974. Id. 2nd ed. Lehre, Cramer. Reprint of 1st ed. with suppl. of 104 p.
——. 1976. *Mangrove vegetation.* Vaduz, Cramer. 447 p.
——. (ed.). 1977. *Wet coastal ecosystems.* Amsterdam, Oxford, New York, Elsevier Scientific. 428 p. (Ecosystems of the world, vol. 1.)
CHARTER, J. R. 1968. Nigeria. In: Hedberg, I.; Hedberg, O. (eds.), p. 91–4.
CHARTER, J. R.; KEAY, R. W. J. 1960. Assessment of the Olokemeji fire-control experiment (investigation 254) 28 years after institution. *Nig. For. Inf. Bull.*, n.s., 3, p. 1–32.
CHEDEVILLE, E. 1972. La végétation du Territoire français des Afars et des Issas. *Webbia*, 26, p. 243–66.
CHEEMA, M. S. Z. A.; QADIR, S. A. 1973. Autecology of *Acacia senegal* (L.) Willd. *Vegetatio*, 27, p. 131–62.
CHEVALIER, A. 1900. Les zones et les provinces botaniques de l'Afrique occidentale française. *C.r. hebd. Séanc. Acad. Sci., Paris*, 130, p. 1205–8.
——. 1908. La forêt vierge de la Côte d'Ivoire. *La Géographie*, 17, p. 201–10.
——. 1933. Le territoire géobotanique de l'Afrique tropicale nord-occidentale et ses subdivisions. *Bull. Soc. bot. Fr.*, 80, p. 4–26.
——. 1935. Les îles du Cap Vert. Géographie, biogéographie, agriculture. Flore de l'Archipel. *Rev. Bot. appl. Agric. trop.*, 15, p. 733–1090.
——. 1938. La végétation de l'île de San Thomé. *Bol. Soc. Brot.*, sér. 2, 13, p. 101–16.
——. 1939. La Somalie française. Sa flore et ses productions végétales. *Rev. Bot. appl. Agric. trop.*, 19, p. 663–87.
——. 1951. Sur l'existence d'une forêt vierge sèche sur de grandes étendues aux confins des bassins de l'Oubangui, du Haut-Chari et du Nil (Bahr-el Ghazal). *Rev. int. Bot. appl. Agric. trop.*, 31, p. 135–6.
——. 1953*a*. La négation de la notion des associations végétales telles qu'elles sont admises par le système de J. Braun-Blanquet pour les pays tempérés et par les auteurs récents pour la grande forêt tropicale d'Afrique. *C.r. hebd. Séanc. Acad. Sci., Paris*, 236, p. 1520–3.
——. 1953*b*. Le remplacement des associations végétales par la notion des biotopes pour désigner les groupements végétaux. *C.r. hebd. Séanc. Acad. Sci., Paris*, 236, p. 1621–4.
CHIPP, T. F. 1927. The Gold Coast forest: a study in synecology. *Oxf. For. Mem.*, 7, p. 1–94. Oxford, Clarendon Press.
——. 1929. The Imatong Mountains, Sudan. *Kew Bull.*, p. 177–97.
——. 1930*a*. Forest and plants of the Anglo–Egyptian Sudan. *Geogrl J.*, 75, p. 123–43, with small map.
——. 1930*b*. The vegetation of the central Sahara. *Geogrl J.*, 76, p. 126–37, with 2 small maps.
CHIPPINDALL, L. K. A. 1955. A guide to the identification of grasses in South Africa. In: Meredith, D. (ed.), p. 1–527.
CIFERRI, R. 1939. Le associazioni del litorale marino della Somalia meridionale. *Riv. Biol. colon.*, 2, p. 5–42.
CLAYTON, W. D. 1957. The swamps and sand dunes of Hadejia.

Nig. geogr. J., 1, p. 31–7, with large-scale vegetation map.
—. 1958a. Secondary vegetation and the transition to savanna near Ibadan, Nigeria. *J. Ecol.*, 46, p. 217–38.
—. 1958b. A tropical moor forest in Nigeria. *J. W. Afr. Sci. Ass.*, 4, p. 1–3.
—. 1958c. Erosion surfaces in Kabba Province, Nigeria. *J. W. Afr. Sci. Ass.*, 4, p. 141–9.
—. 1961. Derived savanna in Kabba Province, Nigeria. *J. Ecol.*, 49, p. 595–604.
—. 1963. The vegetation of Katsina Province, Nigeria. *J. Ecol.*, 51, p. 345–51.
—. 1966. Vegetation ripples near Gummi, Nigeria. *J. Ecol.*, 54, p. 415–17.
CLEMENTS, J. B. 1933. The cultivation of finger millet (*Eleusine coracana*) and its relation to shifting cultivation in Nyasaland. *Emp. For. J.*, 12, p. 16–20.
CLOS-ARCEDUC, M. 1956. Études sur photographies aériennes d'une formation végétale sahélienne: la brousse tigrée. *Bull. IFAN*, sér. A, 18, p. 677–84.
CLOUDSLEY-THOMPSON, J. L. 1969. *The zoology of tropical Africa.* London, Weidenfeld & Nicolson. 355 p.
—. 1974. The expanding Sahara. *Environ. Conserv.*, 1, p. 5–13.
CLUTTON-BROCK, T. H. (ed.). 1977. *Primate ecology: Studies of feeding and ranging behaviour in lemurs, monkeys and apes.* London, New York, San Francisco, Academic Press. 631 p.
CLUTTON-BROCK, T. H.; GILLETT, J. B. 1979. A survey of forest composition in the Gombe National Park, Tanzania. *Afr. J. Ecol.*, 17, p. 131–58.
COCHEMÉ, J.; FRANQUIN, P. 1967. *Une étude d'agroclimatologie de l'Afrique sèche au sud du Sahara en Afrique occidentale.* Rome, FAO–Unesco–WMO.
COCKERELL, T. D. A. 1928. Aspects of the Madeira flora. *Bot. Gaz.*, 85, p. 66–73.
CODY, M. L.; MOONEY, H. A. 1978. Convergence versus nonconvergence in Mediterranean-type ecosystems. *Ann. Rev. Ecol. Syst.*, 9, p. 265–321.
COE, M. J. 1967. *The ecology of the alpine zone of Mount Kenya.* The Hague, Junk. 136 p. (Monogr. biol. 17.)
COE, M. J.; CUMMING, D. H.; PHILLIPSON, J. 1976. Biomass and production of large African herbivores in relation to rainfall and primary production. *Oecologia* (Berl.), 22, p. 341–54.
COETZEE, B. J. 1974. A phytosociological classification of the vegetation of the Jack Scott Nature Reserve. *Bothalia*, 11, p. 329–47.
—. 1975. A phytosociological classification of the Rustenburg Nature Reserve. *Bothalia*, 11, p. 561–80.
COETZEE, B. J.; VAN DER MEULEN, F.; ZWANZIGER, S.; GONSALVES, P.; WEISSER, P. J. 1976. A phytosociological classification of the Nylsvley Nature Reserve. *Bothalia*, 12, p. 137–60.
COETZEE, B. J.; WERGER, M. J. A. 1975. A west–east vegetation transect through Africa south of the Tropic of Capricorn. *Bothalia*, 11, p. 539–60.
COLE, M. M. 1963a. Vegetation and geomorphology in Northern Rhodesia. *Geogrl J.*, 129, p. 290–310.
—. 1963b. Vegetation nomenclature and classification with particular reference to the savannas. *S. Afr. Geogrl J.*, 45, p. 3–14.
COLE, M. M.; BROWN, R. C. 1976. The vegetation of the Ghanzi area of western Botswana. *J. Biogeogr.*, 3, p. 169–96.
COLE, N. H. Ayodele. 1967. Ecology of the montane community at the Tingi Hills in Sierra Leone. *Bull. IFAN*, sér. A, 29, p. 904–24.
—. 1968a. *The vegetation of Sierra Leone (incorporating a field guide to common plants).* Njala Univ. College Press. 198 p.
—. 1968b. Ecology of a moist semi-deciduous forest on Kogia Hill in Sierra Leone. *Bull. IFAN*, sér. A, 30, p. 100–13.
—. 1973. Soil conditions, zonation and species diversity in a seasonally flooded, tropical, grass-herb swamp in Sierra Leone. *J. Ecol.*, 61, p. 831–47.
COLE, N. H. Ayodele; JARRETT, H. O. 1969. Tropical plant communities of limited occurrence. *J. W. Afr. Sci. Ass.*, 14, p. 95–102.
COLLENETTE, C. L. 1931. North-eastern British Somaliland. *Kew Bull.*, p. 401–14.
COLLIER, F. S.; DUNDAS, J. 1937. The arid regions of northern Nigeria and the French Niger Colony. *Emp. For. J.*, 16, p. 184–94.
COLONVAL-ELENKOV, E.; MALAISSE, F. 1975. Remarques sur l'écomorphologie de la flore termitophile du Haut-Shaba (Zaïre). *Bull. Soc. r. Bot. Belg.*, 108, p. 167–81.
COMINS, D. M. 1962. The vegetation of the Districts of East London and King William's Town, Cape Province. *Mem. bot. Surv. S. Afr.*, 33, p. 1–32, with coloured vegetation map 1:125000.
COMPÈRE, P. 1970. *Carte Sols Vég. Congo, Rwanda, Burundi*, 25, *Bas-Congo. B, Végétation* (1:250000). *Notice explicative*, p. 1–35. Brussels, INEAC.
COMPTON, R. H. 1929a. The vegetation of the Karoo. *J. bot. Soc. S. Afr.*, 15, p. 13–21.
—. 1929b. The flora of the Karoo. *S. Afr. J. Sci.*, 26, p. 160–5.
—. 1966. An annotated check list of the flora of Swaziland. *J. S. Afr. Bot.*, Suppl. 6, p. 1–191.
COOK, C. D. K. 1968. The vegetation of the Kainji Reservoir site in Northern Nigeria. *Vegetatio*, 15, p. 225–43.
COOMBE, D. E. 1960. An analysis of the growth of *Trema guineensis*. *J. Ecol.*, 48, p. 219–31.
COOMBE, D. E.; HADFIELD, W. 1962. An analysis of the growth of *Musanga cecropioides*. *J. Ecol.*, 50, p. 221–34.
COOPER, G. P.; RECORD, S. J. 1931. The evergreen forests of Liberia. *Yale Univ. Sch. For. Bull.*, 31, p. 1–153.
CORFIELD, T. F. 1973. Elephant mortality in Tsavo National Park, Kenya. *E. Afr. Wildl. J.*, 11, p. 339–69.
CORNET D'ELZIUS, C. 1964. *Évolution de la végétation dans la plaine au sud du Lac Édouard.* Brussels, Inst. Parcs Nat. Congo, Rwanda. 23 p., with coloured vegetation map.
COTTRELL, C. B.; LOVERIDGE, J. P. 1966. Observations on the *Cryptosepalum* forest of the Mwinilunga District of Zambia. *Proc. Trans. Rhod. sci. Ass.*, 51, p. 79–120.
COWLING, R. M.; CAMPBELL, B. M. 1980. Convergence in vegetation structure in the Mediterranean communities of California, Chile and South Africa. *Vegetatio*, 43, p. 191–7.
CREMERS, G. 1973. Architecture de quelques lianes d'Afrique tropicale. *Candollea*, 28, p. 249–80.
—. 1974. Id. 2. *Candollea*, 29, p. 57–110.
CROOK, A. O. 1956. A preliminary vegetation map of the Melsetter Intensive Conservation Area, Southern Rhodesia. *Rhod. agric. J.*, 53, p. 3–25, with small vegetation map.
CUFODONTIS, G. 1940. La vegetazione. In: Zavattari, R.; Cufodontis, G. (eds.), *Missione biologica nel Paese dei Borana. I. Condizioni biogeografiche e antropiche*, p. 141–255. Rome, R. Acad. Ital. Centro Studi AOI.

CURRY-LINDAHL, K. 1968. Zoological aspects of the conservation of vegetation in tropical Africa. In: Hedberg, I.; Hedberg, O. (eds.), p. 25–32.
——. 1974. Conservation problems and progress in equatorial African countries. *Environ. Conserv.*, 1, p. 119–20.
CURSON, H. H. 1947. Notes on eastern Caprivi Strip. *S. Afr. J. Sci.*, 43, p. 124–57.
DAHLGREN, R.; LASSEN, P. 1972. Studies in the flora of Northern Morocco. I. Some poor fen communities and notes on a number of northern and atlantic plant species. *Bot. Notiser*, 125, p. 439–64.
DALBY, D.; HARRISON-CHURCH, R. J. (eds.). 1973. *Drought in Africa.* London, School of Oriental and African Studies. 124 p. (Rep. 1973 Symp.).
DALBY, D.; HARRISON-CHURCH, R. J.; BEZZAZ, F. (eds.). 1977. *Drought in Africa/Sécheresse en Afrique*, vol. 2. London, International African Institute/Institut Africain International. 200 p. (Afr. Environ. spec. rep. no. 6.)
DALE, I. R. 1939. The woody vegetation of the Coast Province of Kenya. *Inst. Pap. Imp. For. Inst.*, 18, p. 1–28.
——. 1954. Forest spread and climatic change in Uganda during the Christian era. *Emp. For. Rev.*, 33, p. 23–9.
DANCETTE, C.; POULAIN, J. F. 1968 (1969). Influence de l'*Acacia albida* sur les facteurs pédoclimatiques et les rendements des cultures. *Sols Afr.*, 13, p. 197–239.
DANDELOT, P. 1965. Distribution de quelques espèces de Cercopithecidae en relation avec les zones de végétation de l'Afrique. *Zool. Afr.*, 1, p. 167–76.
DANSEREAU, P. 1951. Description and recording of vegetation upon a structural basis. *Ecology*, 32, p. 172–229.
——. 1966. Études macaronésiennes. III. La zonation altitudinale. *Naturaliste can.*, 93, p. 779–95.
——. 1968. Macaronesian studies. II. Structure and functions of the laurel forest in the Canaries. *Collecteana bot.*, 7, p. 227–80.
DARLING, F. Fraser. 1960a. *Wild life in an African territory.* London, Oxford Univ. Press. 160 p.
——. 1960b. An ecological reconnaissance of the Mara Plains in Kenya Colony. *Wildl. Monogr.*, 5, p. 1–41.
DASMANN, R. F. 1964. *African game ranching.* Oxford, London, Paris, Frankfurt, Pergamon Press; New York, Macmillan. 75 p.
——. 1972. Towards a system for classifying natural regions of the world and their representation by national parks and reserves. *Biol. Conserv.*, 4, p. 247–55.
——. 1973a. A system for defining and classifying natural regions for purposes of conservation. *IUCN occ. Pap.*, 7, p. 1–47.
——. 1973b. Biotic provinces of the world. *IUCN occ. Pap.*, 9. Morges.
DAUBENMIRE, R. 1968. Ecology of fire in grassland. *Adv. ecol. Res.*, 5, p. 209–66.
DAVIDGE, C. 1977. Baboons as dispersal agents for *Acacia cyclops*. *Zool. Afr.*, 12, p. 249–50.
DAVIDSON, R. L. 1962. The influence of edaphic factors on the species composition of early stages of the subsere. *J. Ecol.*, 50, p. 401–10.
——. 1964. An experimental study of succession in the Transvaal highveld. In: Davis, D. H. S. (ed.), p. 113–25.
DAVIS D. H. S. (ed.). 1964. *Ecological studies in Southern Africa.* The Hague, Junk. 415 p. (Monogr. biol. 14.)
DAVIS, P. H. 1953. The vegetation of the deserts near Cairo. *J. Ecol.*, 41, p. 157–73.
DAVIS, T. A. W.; RICHARDS, P. W. 1933. The vegetation of Moraballi Creek, British Guiana; an ecological study of a limited area of tropical rain forest. Pt 1. *J. Ecol.*, 21, p. 350–84.
——; ——. 1934. Id. Pt 2. *J. Ecol.*, 22, p. 106–34.
DAVY, J. BURTT. *See* Burtt Davy, J.
DAWKINS, H. C. 1954. Timu and the vanishing forests of North-East Karamoja. *E. Afr. agric. J.*, 19, p. 164–7.
DAY, J.; SIEGFRIED, W. R.; LOUW, G. N.; JARMAN, M. L. (eds.). 1979. *Fynbos ecology: a preliminary synthesis.* Pretoria, CSIR. 166 p. (S. Afr. Nat. Sci. Progr. Rep., no. 40.)
DEAN, G. J. W. 1967. Grasslands of the Rukwa valley. *J. appl. Ecol.*, 4, p. 45–57.
DEBENHAM, F. 1952. *Study of an African swamp.* London, HMSO. 88 p.
DEJARDIN, J.; GUILLAUMET, L.; MANGENOT, G. 1973. Contribution à la connaissance de l'élément non endémique de la flore malgache (végétaux vasculaires). *Candollea*, 28, p. 325–91.
DELANY, M. J.; HAPPOLD, D. C. D. 1979. *Ecology of African mammals.* London, New York, Longman. 434 p.
DE LEEUW, P. N. 1965. The role of savanna in nomadic pastoralism: some observations from western Bornu, Nigeria. *Neth. J. Agric. Sci.*, 13, p. 178–89.
DELEVOY, G. 1933. Contribution à l'étude de la végétation forestière de la vallée de la Lukuga. *Mém. Inst. r. colon. belg. Sect. Sci. nat. méd.* 8°, vol. 1 (8), p. 1–124.
DELEVOY, G.; ROBERT, M. 1935. Le milieu physique du Centre Africain Méridional et la phytogéographie. *Mém. Inst. r. colon. belg. Sect. Sci. nat. méd.* 8°, vol. 3 (4), p. 1–104.
DELVAUX, J. 1958. Effets mesurés des feux de brousse sur la forêt claire et les coupes à blanc dans la région d'Elisabethville (1950–1951 à octobre 1955). *Bull. Agric. Congo belg.*, 49, p. 683–714.
DELWAULLE, J. C. 1973. Désertification de l'Afrique au Sud du Sahara. *Bois Forêts Trop.*, 149, p. 3–20.
DEMARET, F. 1958. Aperçu sur la flore et la végétation des forêts à *Hagenia abyssinica* (Bruce) Gmel. du Ruwenzori occidental. *Bull. Jard. bot. État Brux.*, 28, p. 331–36.
DE NAUROIS, R.; ROUX, F. 1965. Les mangroves d'*Avicennia* les plus septentrionales de la côte occidentale d'Afrique. *Bull. IFAN*, sér. A, 27, p. 843–54.
DENISOFF, I.; DEVRED, R. 1954. *Carte Sols Vég. Congo belge*, 2, Mvuazi (Bas-Congo). A, Sols (1:50000), B, Végétation (1:50000). *Notice explicative*, p. 1–40. Brussels, INEAC.
DENNY, P. 1971. Zonation of aquatic macrophytes around Habukara Island, Lake Bunyonyi, S.W. Uganda. *Hidrobiologia*, 12, p. 249–57.
——. 1973. Lakes of south-western Uganda. II. Vegetation studies on Lake Bunyonyi. *Freshwat. Biol.*, 3, p. 123–35.
DEPIERRE, D.; GILLET, H. 1971. Désertification de la zone sahélienne au Tchad. *Bois Forêts Trop.*, 139, p. 3–25.
DE RHAM, P. 1970. L'azote dans quelques forêts, savanes et terrains de culture d'Afrique tropicale humide (Côte d'Ivoire). *Veröff. geobot. Inst., Zürich*, 45, p. 1–124.
DE SAEGER, H. 1954. Végétation. In: *Exploration du Parc National de la Garamba*, 1, p. 17–22. Brussels, Inst. Parcs Nat. Congo belge.
DESCOINGS, B. 1971. Méthode de description des formations herbeuses intertropicales par la structure de la végétation. *Candollea*, 26, p. 223–57.
——. 1972a. Notes de phytoécologie équatoriale. Les steppes Loussékés du Plateau Batéké (Congo). *Adansonia*, sér. 2, 12, p. 569–84.
——. 1972b. Note sur la structure de quelques formations herbeuses de Lamto (Côte d'Ivoire). *Ann. Univ. Abidjan*, sér. E (Écologie) 5 (1), p. 7–30.

——. 1973. Les formations herbeuses africaines et les définitions de Yangambi considérées sous l'angle de la structure de la végétation. *Adansonia*, sér. 2, 13, p. 391–421.

——. 1974. Notes de phyto-écologie équatoriale. 2. Les formations herbeuses du Moyen Ogooué (Gabon). *Candollea*, 29, p. 13–37, with small vegetation map.

——. 1976*a*. Pour une conception structurale et ouverte des classifications phytogéographiques. *Adansonia*, sér. 2, 16, p. 93–105.

——. 1976*b*. Notes de phyto-écologie équatoriale. 3. Les formations herbeuses de la vallée de la Nyanga (Gabon). *Adansonia*, sér. 2, 15, p. 307–29, with small vegetation map.

——. 1978. Les formations herbeuses dans la classification phytogéographique de Yangambi. *Adansonia*, sér. 2, 18, p. 243–56.

DESENFANS, R. 1950. La cartographie des groupements végétaux du degré carré de Sokele. *C.r. Congr. sci. Elisabethville*, 1950, 4, p. 42–51.

DESTREMAU, D. X. 1974. Précisions sur les aires naturelles des principaux conifères marocains en vue de l'individualisation de provenances. *Ann. Rech. For. Maroc*, 1974, p. 3–90.

DEUSE, P. 1960. Étude écologique et phytosociologique de la végétation des Esobe de la région Est du lac Tumba (Congo belge). *Mém. Acad. r. Sci. Outre-Mer, Cl. Sci. nat. méd.* 8°, n.s., 11, p. 1–115.

——. 1963. Marais et tourbières au Rwanda et au Burundi. *Publs Univ. Elisabethville*, 6, p. 69–80.

——. 1966. Contribution à l'étude des tourbières du Rwanda et du Burundi. *Inst. Nat. Rech. Sci. Butare, Rép. Rwandaise*, 4, p. 53–115.

——. 1968. Rwanda. In: Hedberg, I.; Hedberg, O. (eds.), p. 125–7.

DEVOIS, J. C. 1948. Peuplements forestiers de la basse Casamance. *Bull. IFAN*, 10, p. 182–211.

DE VOS, A. 1975. *Africa, the devastated continent?* The Hague, Junk. 240 p. (Monogr. biol. 26.)

DEVRED, R. 1956. Les savanes herbeuses de la région de Mvuazi (Bas-Congo). *Publs INEAC*, sér. sci., 65, p. 1–115.

——. 1957. Limite phytogéographique occidento-méridionale de la Région Guinéenne au Kwango. *Bull. Jard. bot. État Brux.*, 27, p. 417–31, with small vegetation map.

——. 1958. La végétation forestière du Congo belge et du Ruanda-Urundi. *Bull. Soc. r. for. Belg.*, 65, p. 409–68, with vegetation map.

DEVRED, R.; SYS, C.; BERCE, J. M. 1958. *Carte Sols Vég. Congo belge*, 10, Kwango. A, Sols (1:1000000). B, Végétation (1:1000000). *Notice explicative*, p. 1–64. Brussels, INEAC.

DE WILDEMAN, E. 1932. La forêt équatoriale congolaise et ses problèmes biologiques. *Bull. Acad. r. Belg. Cl. Sci.*, sér. 5, 17, p. 1475–514.

——. 1934. Remarques à propos de la forêt équatoriale congolaise. *Mém. Inst. r. colon. belg. Sect. Sci. nat. méd.* 8°, 2 (2), p. 1–120.

DE WINTER, 1966. Remarks on the distribution of some desert plants in Africa. *Palaeoecol. Afr.*, 1, p. 188–9.

——. 1971. Floristic relationship between the northern and southern arid areas in Africa. *Mitt. Bot. Staatssamml. Münch.*, 10, p. 424–37.

D'HOORE, J. 1959. Pedological comparisons between tropical South America and tropical Africa/Comparaisons pédologiques entre les continents sud-américain et africain (zones intertropicales). *African soils/Sols africains*, 4 (3), p. 5–19.

D'HOORE, J. L. 1964*a*. *La carte des sols d'Afrique au 1:5000000. Mémoire explicatif/Explanatory monograph of the soil map of Africa, scale 1:5000000*. Lagos, CCTA. 209 p. (Projet conjoint/Joint project no. 11, Publ. no. 93.)

—— (co-ordinator). 1964*b*. *Carte des sols d'Afrique/Soils map of Africa*. Lagos, CCTA. 7 sheets in colour, 1:5000000. (Serv. pédol. interafr./Interafr. Soils Serv., Projet conjoint/Joint project no. 11.)

——. 1968. The classification of tropical soils. In: Moss, R. P. (ed.), p. 7–28.

DI CASTRI, F.; MOONEY, H. A. (eds.). 1973. *Mediterranean-type ecosystems*. London, Chapman & Hall; Berlin, Heidelberg, New York, Springer-Verlag. 405 p.

DIELS, L. 1915. Vegetationstypen vom untersten Kongo. *Vegetationsbilder*, 12, t. 43–8.

——. 1922. Beiträge zur Kenntnis der Vegetation und Flora der Seychellen. In: Chun, C. (ed.), *Wiss. Ergeb. dt. Tiefsee Exped. 'Valdivia' 1898–1899*, vol. 2 (1), p. 409–66. Jena, Gustav Fischer.

DIELS, L.; MILDBRAED, J.; SCHULZE-MENZ, G. K. 1963. *Vegetationskarte von Afrika* (1:15000000, in colour). *Willdenowia*, Beih. 1. See also Domke, W.

DIETERLEN, F. 1978. *Zur Phänologie des äquatorialen Regenwaldes im Ost-Zaïre (Kivu) nebst Pflanzenliste und Klimadaten*. Vaduz, Cramer. 111 p. (Diss. bot. no. 47.)

DINIZ, A. Castanheira. 1973. *Características mesologicas de Angola*. Nova Lisboa, Miss. Inq. Agric. Angola. 482 p., with many small vegetation maps.

DINIZ, A. Castanheira; AGUIAR, F. Q. de Barros. 1969*a*. Zonagem agro-ecológica do Cuanza-Sul. *Inst. Invest. Agron. Angola*, sér. Cient., 6, p. 1–5.

——; ——. 1969*b*. Regiões naturais de Angola. 3rd ed. *Inst. Invest. Agron. Angola*, sér. Cient., 7, p. 1–6.

——; ——. 1972. Os solos e a vegetação do Planalto occidental da Cela. *Inst. Invest. Agron. Angola*, sér. Cient., 26, p. 1–25, with coloured vegetation map 1:50000.

——; ——. 1973. Recursos em terras com aptidão para o regardio na Bacia do Cubango. *Inst. Invest. Agron. Angola*, sér. Técn., 33, p. 1–27.

DINTER, K. 1912. *Die vegetabilische Veldkost Deutsch-Südwest-Afrikas*. Okahandja, privately pub. 47 p.

——. 1921. Botanische Reisen in Deutsch-Südwest-Afrika. *Beih. Repert. Spec. nov. Regni veg.*, 3, p. 1–169.

DOMKE, W. 1963. Bemerkungen zu der von L. Diels, J. Mildbraed und G. K. Schulze-Menz 1939–1942 bearbeiteten Vegetationskarte von Afrika. *Willdenowia*, Beih. 1, p. 1–4. See also Diels et al., 1963.

——. 1966. Grundzüge der Vegetation des tropischen Kontinental-Afrika von Johannes Mildbraed, herausgegeben und revidiert von Walter Domke. *Willdenowia*, Beih. 2, p. 1–253.

DOUGLAS, H. A. 1948. The vegetation of the Afram plains. *Farm and Forest*, 9, p. 32–40.

DOUGLAS-HAMILTON, I. 1973. On the ecology and behaviour of the Lake Manyara elephants. *E. Afr. Wildl. J.*, 11, p. 401–3.

DOUMBIA, F. I. 1966. Étude des forêts de Basse Casamance au sud de Ziguinchor. *Annls Fac. Sci. Univ. Dakar*, 19, sér. Sci. Végétales, n° 3, p. 61–100.

DOWNING, B. H. 1980. Changes in the vegetation of Hluhluwe Game Reserve, Zululand, as regulated by edaphic and biotic factors over 36 years. *J. S. Afr. Bot.*, 46, p. 225–31.

DRAR, M. 1955. Egypt, Eritrea, Libya and the Sudan. In: *Plant ecology/Écologie végétale*, p. 151–94. Paris, Unesco. (Arid zone research/Recherches sur la zone aride, 6.)

DREW, A.; REILLY, C. 1972. Observations on copper tolerance in the vegetation of a Zambian copper clearing. *J. Ecol.*, 60, p. 439–44.

DRUDE, O. 1913. *Die Ökologie der Pflanzen.* Braunschweig. (Die Wiss., Bd 50.)

DUBOIS, L. 1955. La jacinthe d'eau au Congo belge. *Bull. Agric. Congo belge*, 46, p. 893–900.

DUGERDIL, M. 1970. Recherches sur le contact forêt–savane en Côte d'Ivoire. I. Quelques aspects de la végétation et de son évolution en savane préforestière. *Candollea*, 25, p. 11–19. II. Note floristique sur des îlots de forêt semidécidue, id., p. 235–43.

DUNDAS, J. 1938. Vegetation types of the Colonie du Niger. *Inst. Pap. Imp. For. Inst.*, 15, p. 1–10, with small-scale vegetation map.

DUONG-HUU-THOI. 1950a. Introduction à l'étude de la végétation du Soudan français. *Conf. int. Afr. occid. 2a. Conf. Bissau, 1947*, 2 (1). *Trabalhos apresentados à 2a Secção (Meio biologico)*, p. 7–51, with vegetation map.

——. 1950b. Étude préliminaire de la végétation du delta central nigérien. Id., p. 53–156.

DU RIETZ, G. E. 1931. Life-forms of terrestrial flowering plants. *Acta phytogeogr. suec.*, 3, p. 1–95.

DUTHIE, A. V. 1929. Vegetation and flora of the Stellenbosch Flats. *Annale Univ. Stellenbosch*, 7 (sect. A, no. 4), p. 1–59 (with vegetation map).

DU TOIT, A. L. 1954. *Geology of South Africa.* 3rd ed. Edinburgh, Oliver & Boyd. 611 p.

DUVIGNEAUD, P. 1949a. Voyage botanique au Congo belge à travers le Bas-Congo, le Kwango, le Kasai et le Katanga. De Banana à Kasenga. *Bull. Soc. r. Bot. Belg.*, 81, p. 15–34.

——. 1949b [1953]. Les savanes du Bas-Congo. Essai de phytosociologie topographique. *Lejeunia*, Mém., 10, p. 1–192.

——. 1950. Sur la véritable identité du *Parinari* sp. 'Mafuca' de Gossweiler et sur l'existence d'une laurisilve de transition Guinéo–Zambézienne. *Bull. Soc. r. Bot. Belg.*, 83, p. 105–10.

——. 1952 [1953]. La flore et la végétation du Congo méridional. *Lejeunia*, 16, p. 95–124, with small-scale coloured vegetation map.

——. 1953. Les formations herbeuses (savanes et steppes) du Congo méridional. *Naturalistes belg.*, 34, p. 66–75.

——. 1955. Études écologiques de la végétation en Afrique tropicale. *Les divisions écologiques du monde*, p. 131–48. (Colloques int. CNRS, 59.) Also published in: *Année biol.*, sér. 3, 31, p. 375–92.

——. 1958. La végétation du Katanga et de ses sols métallifères. *Bull. Soc. r. Bot. Belg.*, 90, p. 127–286.

——. 1959. Plantes 'cobaltophytes' dans le Haut-Katanga. *Bull. Soc. r. Bot. Belg.*, 91, p. 111–34.

——. (ed.). 1971. *Productivity of forest ecosystems. Proceedings of the Brussels Symposium, 1969.* Paris, Unesco. 707 p.

DUVIGNEAUD, P.; DENAYER-DE SMET, S. 1960. Action de certains métaux lourds du sol (Cu-Co-Mn-Ur) sur la végétation dans le Haut-Katanga. In: *Colloque sur les rapports sol–végétation sous la dir. de Viennot-Bougin*, p. 121–39. Paris, Masson.

——; ——. 1963. Cuivre et végétation au Katanga. *Bull. Soc. r. Bot. Belg.*, 96, p. 93–231.

DUVIGNEAUD, P.; SYMOENS, J. J. 1951. Contribution à l'étude des associations tourbeuses du Bas-Congo. Le *Rhynchosporetum candidae* à l'étang de Kibambi. *Verh. Int. Verein. Limnol.*, 11, p. 100–4.

DYE, P. J.; WALKER, B. H. 1980. Vegetation–environment relations on sodic soils of Zimbabwe–Rhodesia. *J. Ecol.*, 68, p. 589–606.

DYER, R. A. 1937. The vegetation of the Divisions of Albany and Bathurst. *Mem. bot. Surv. S. Afr.*, 17, p. 1–138.

——. 1955. Angola, South-West Africa, Bechuanaland and the Union of South Africa. In: *Plant ecology/Écologie végétale*, p. 195–218. Paris, Unesco. (Arid zone research/Recherches sur la zone aride, 6.)

EDMONDS, A. C. R. 1976. *The Republic of Zambia vegetation map 1:500000* (in colour). Frankfurt a./M (Federal Republic of Germany), Institut für Angewandte Geodäsie/Govt Repub. Zambia.

EDROMA, E. L. 1974. Copper pollution in Rwenzori National Park, Uganda. *J. appl. Ecol.*, 11, p. 1043–56.

——. 1977. Outbreak of the African army worm (*Spodopteca exempta* Walk.) in Rwenzori National Park, Uganda. *E. Afr. Wildl. J.*, 15, p. 157–8.

EDWARDS, D. 1967. A plant ecological survey of the Tugela River basin. *Mem. bot. Surv. S. Afr.*, 36, p. 1–285, with coloured vegetation map 1:250000 in 6 sheets.

EDWARDS, D. C. 1935. The grasslands of Kenya. *Emp. J. exp. Agric.*, 3, p. 153–9.

——. 1940. A vegetation map of Kenya with particular reference to grassland types. *J. Ecol.*, 28, p. 377–85, with vegetation map 1:4000000.

——. 1942. Grass burning. *Emp. J. exp. Agr.*, 10, p. 219–31.

——. 1945. *Horn of Africa (including Kenya): vegetation* (map 1:3000000). Nairobi, Govt Printer.

——. 1951. The vegetation in relation to soil and water conservation in East Africa. *Bull. Commonw. Bur. Past. Fld Crops*, 41, p. 28–43.

EDWARDS, D. C.; BOGDAN, A. V. 1951. *Important grassland plants of Kenya.* Nairobi, London, Pitman. 124 p., with small-scale vegetation map.

EDWARDS, K. A.; FIELD, C. R.; HOGG, I. G. G. 1979. *A preliminary analysis of climatological data from the Marsabit District of Northern Kenya.* Nairobi, UNEP–MAB Integrated Project in Arid Lands. 44 p. (IPAL tech. Rep., no. B-1.)

EGGELING, W. J. 1935. The vegetation of Namanve Swamp, Uganda. *J. Ecol.*, 23, p. 422–35.

——. 1938. The savannah and mountain forests of South Karamoja, Uganda. *Inst. Pap. Imp. For. Inst.*, 11, p. 1–14.

——. 1947. Observations on the ecology of the Budongo rain forest, Uganda. *J. Ecol.*, 34, p. 20–87.

EIG, A. 1931. Les éléments et les groupes phytogéographiques auxiliaires dans la flore palestinienne. *Beih. Repert. Spec. nov. Regni veg.*, 63, p. 1–201.

EL HADIDI, M. N. 1971. Distribution of *Cyperus papyrus* L. and *Nymphaea lotus* L. in inland waters of Egypt. *Mitt. Bot. Staatssamml. Münch.*, 10, p. 470–5.

EL HADIDI, M. N.; AYYAD, M. A. 1975. Floristic and ecological features of Wadi Habis (Egypt). In: *La flore du bassin méditerranéen*, p. 247–58. (Coll. Int. CNRS, 235.)

EL HADIDI, M. N.; KOSINOVA, J. 1971. Studies on the weed flora of cultivated land in Egypt. *Mitt. Bot. Staatssamml. Münch.*, 10, p. 354–67.

ELLENBERG, H.; MUELLER-DOMBOIS, D. 1966. Tentative physiognomic–ecological classification of plant formations of the earth. *Ber. geobot. Forsch. Inst. Rübel*, 37, p. 21–55.

ELLIS, B. S. 1950. A guide to some Rhodesian soils. II: A note on mopani soils. *Rhod. agric. J.*, 47, p. 49–61.

——. 1958. Soil genesis and classification. *Soils Fertil.*, 21, p. 145–7.

EL-SHARKAWI, H. M.; FAYED, A. A. 1975. Vegetation of inland desert wadis in Egypt. I: Wadi Bir-El-Ain. *Feddes Rep.*, 86, p. 589–94.

ELTRINGHAM, S. K. 1976. The frequency and extent of uncontrolled grass fires in the Rwenzori National Park, Uganda. *E. Afr. Wildl. J.*, 14, p. 215–22.

EMBERGER, L. 1925. Les limites naturelles climatiques de l'Arganier. *Bull. Soc. Sci. nat. Maroc*, 5, p. 94–7.
——. 1930. La végétation de la région méditerranéenne. Essai d'une classification des groupements végétaux. *Rev. gén. Bot.*, 43, p. 641–62, 705–21.
——. 1932. Recherches botaniques et phytogéographiques dans le Grand Atlas oriental (Massifs du Ghat et du Mgoun). *Mém. Soc. Sci. nat. Maroc*, 33, p. 1–49.
——. 1936. Remarques critiques sur les étages de végétation dans les montagnes marocaines. *Bull. Soc. bot. suisse*, 46, p. 614–31.
——. 1939. Aperçu général sur la végétation du Maroc. In: Rübel, E.; Ludi, W. (eds.), Ergebnisse der internationalen pflanzengeographischen Excursion durch Marokko und Westalgerian 1936. *Veröff. geobot. Inst., Zürich*, 14, p. 40–157, with coloured vegetation map 1:1500000.
——. 1948. La flore de l'horizon culminal des montagnes marocaines. *Vol. jubilaire Soc. Sci. nat. Maroc, 1920–1945*, p. 95–105.
——. 1955*a*. Une classification biogéographique des climats. *Recl Trav. Labs. Bot. Géol. Zool. Univ. Montpellier*, sér. bot., 7, p. 3–43.
——. 1955*b*. Afrique du Nord-Ouest. In: *Plant ecology/Écologie végétale*, p. 219–49. Paris, Unesco. (Arid zone research/Recherches sur la zone aride, 6.)
EMBERGER, L.; MANGENOT, G.; MIÈGE, J. 1950*a*. Existence d'associations végétales typiques dans la forêt dense équatoriale. *C.r. hebd. Séanc. Acad. Sci., Paris*, 231, p. 640–2.
——; ——; ——. 1950*b*. Caractères analytiques et synthétiques des associations de la forêt équatoriale de Côte d'Ivoire. *C.r. hebd. Séanc. Acad. Sci., Paris*, 231, p. 812–14.
ENGLER, A. 1891 [1892]. Über die Hochgebirgsflora des tropischen Afrika. *Phys. Abh. K. Akad. Wiss. Berl.*, 2, p. 1–461.
——. 1894. Über die Gliederung der Vegetation von Usambara und der angrenzenden Gebiete. *Phys. Abh. K. Akad. Wiss. Berl.*, p. 1–86.
——. 1895. Die Pflanzenwelt Ost-Afrikas und der Nachbargebiete. Teil A. Grundzüge der Pflanzenverbreitung in Deutsch-Ost-Afrika und den Nachbargebieten. *Deutsch-Ost-Afrika*, Bd 5. Berlin, Dietrich Reimer. 154 p.
——. 1900. Über die Vegetationsverhältnisse des Ulugurugebirges in Deutsch-Ostafrika. *Sitzber. preuss. Akad. Wiss.*, 16, p. 191–211.
——. 1903. Über die Vegetationsformationen Ost-Afrikas auf Grund einer Reise durch Usambara zum Kilimandscharo. *Z. Ges. Erdk. Berl.*, p. 254–79, 398–421.
——. 1904. Über die Vegetationsverhältnisse des Somalilandes. *Sitzber. preuss. Akad. Wiss.*, 10, p. 355–416.
——. 1906*a*. Über die Vegetationsverhältnisse von Harar und des Gallahochlandes auf Grund der Expedition von Freiherrn von Erlanger und Hrn. Oscar Neumann. *Sitzber. preuss. Akad. Wiss.*, 40, p. 726–47.
——. 1906*b*. Beiträge zur Kenntnis der Pflanzenformationen von Transvaal und Rhodesia. *Sitzber. preuss. Akad. Wiss.*, 52, p. 866–906.
——. 1908. Pflanzengeographische Gliederung von Afrika. *Sitzber. preuss. Akad. Wiss.*, 38, p. 781–837.
——. 1910. *Die Pflanzenwelt Afrikas, inbesondere seiner tropischen Gebiete.* Bd 1, *Allgemeiner Überblick über die Pflanzenwelt Afrikas und ihre Existenzbedingungen*, p. 1–1029, with coloured vegetation maps of East Africa, South-West Africa, Cameroun (all 1:6000000), and Togo (1:250000). In: Engler, A.; Drude, O. (eds.), *Die Vegetation der Erde*, 9. Leipzig, Engelmann.
——. 1925. *Die Pflanzenwelt Afrikas, inbesondere seiner tropischen Gebiete*, Bd 5, Tl 1, *Ausführliche Schilderung der Vegetationsverhältnisse des tropischen Afrika*, p. 1–341, with small-scale vegetation map. In: Engler, A; Drude, O. (eds.), *Die Vegetation der Erde*, 9. Leipzig, Engelmann.
——. 1964. Übersicht über die Florenreiche und Florengebiete der Erde. In: *Syllabus der Pflanzenfamilien*, 12th ed. (rev. H. Melchior), vol. 2, p. 626–9.
ENGLER, A.; DRUDE, O. See Engler, A., 1910 and 1925.
ENTI, A. A. 1968. Distribution and ecology of *Hildegardia barteri* (Mast.) Kosterm. *Bull. IFAN*, sér. A, 30, p. 881–95.
ERIKSSON, J. 1964. Botanical notes from the Somali Plateau in Southern Ethiopia. I. Some general observations on flora and vegetation. *Bot. Notiser*, 117, p. 1–9.
ERIKSSON, O.; HANSEN, A.; SUNDING, P. 1974. *Flora of Macaronesia. Check-list of vascular plants.* Univ. of Umeå. 66 p.
ERN, H. 1979. Die Vegetation Togos. Gliederung, Gefährdung, Erhaltung. *Willdenowia*, 9, p. 295–311.
ERNST, W. 1971. Zur Ökologie der Miombo-Wälder. *Flora*, 160, p. 317–31.
——. 1975. Variation in the mineral contents of leaves of trees in miombo woodland in South Central Africa. *J. Ecol.*, 63, p. 801–7.
ERNST, W.; WALKER, B. H. 1973. Studies on the hydrature of trees in miombo woodland in South Central Africa. *J. Ecol.*, 61, p. 667–73.
ERNST, W. H. O. 1974. *Schwermetallvegetation der Erde.* Stuttgart, Gustav Fischer. 194 p. (Geobot. sel. 5.)
EVANS, G. C. 1939. Ecological studies on the rain forest of Southern Nigeria. II. The atmospheric environmental conditions. *J. Ecol.*, 27, p. 436–82.
ÉVRARD, C. 1957. L'association à *Aneulophus africanus* Benth. Forêt périodiquement inondée sur podzol humique au Congo belge. *Bull. Jard. bot. État Brux.*, 27, p. 335–49.
——. 1965. Données préliminaires à une statistique phytogéographique de la flore du secteur forestier central congolais. *Webbia*, 19, p. 619–26.
——. 1968. Recherches écologiques sur le peuplement forestier des sols hydromorphes de la Cuvette centrale congolaise. *Publs INEAC*, sér. sci., 110, p. 1–295.
EWUSIE, J. Yanney. See Yanney Ewusie, J.
EXELL, A.W. 1944. *Catalogue of the vascular plants of S. Tomé* (includes an account of the vegetation of Príncipe, S. Tomé and Annobon). London, British Museum (Nat. Hist.). 428 p.
——. 1952 [1953]. The vegetation of the islands of the Gulf of Guinea. *Lejeunia*, 16, p. 57–66.
——. 1957. La végétation de l'Afrique tropicale australe. *Bull. Soc. r. Bot. Belg.*, 89, p. 101–6.
——. 1968. Príncipe, S. Tomé and Annobon. In: Hedberg, I.; Hedberg, O. (eds.), p. 132–4.
EYRE V. E. F.; RAMSAY, D. M.; JEWITT, T. N. 1953. Agriculture, forests and soils of the Jur ironstone country of the Bahr el Ghazal Province, Sudan. *Bull. Minist. Agric. Sudan*, 9, p. 1–40.
FADEN, R. B. 1970. *A preliminary report on the Kakamega forest in Kenya.* Nairobi, East African Herbarium. 15 p. (Mimeo.)
FAIRBAIRN, W. A. 1939. Ecological succession due to biotic factors in northern Kano and Katsina Provinces of northern Nigeria. *Inst. Pap. Imp. For. Inst.*, 22, p. 1–32.
——. 1943. Classification and description of the vegetation types of the Niger Colony, French West Africa. *Inst. Pap. Imp. For. Inst.*, 23, p. 1–38.

FANSHAWE, D. B. 1961. Evergreen forest relics in Northern Rhodesia. *Kirkia*, 1, p. 20–4.
——. 1968. The vegetation of Zambian termitaria. *Kirkia*, 6, p. 169–79.
——. 1969. The vegetation of Zambia. *For. Res. Bull.*, Kitwe, Zambia, 7, p. 1–67.
——. 1972*a*. The biology of the reed *Phragmites mauritianus* Kunth. *Kirkia*, 8, p. 147–50.
——. 1972*b*. The bamboo, *Oxytenanthera abyssinica*. Its ecology, silviculture and utilization. *Kirkia*, 8, p. 157–66.
FANSHAWE, D. B.; SAVORY, B. M. 1964. *Baikiaea plurijuga* dwarf shell forests. *Kirkia*, 4, p. 185–90.
FAO. See also UNDP/FAO.
——. 1971. *Range development in Marsabit District*. Nairobi, UNDP/FAO. (AGP:SF/KEN.11.) (Wkg pap. 9.)
——. 1977. Les systèmes pastoraux sahéliens. *Étude FAO: Production végétale et protection des plantes*, 5. Rome, FAO. 389 p.
FAO. 1980*a*. *Système mondial de surveillance continue de l'environnement. Projet pilote sur la surveillance continue de la couverture forestière tropicale. Togo. Cartographie du couvert végétal et étude de ses modifications*. Rome, FAO. 117 p., with coloured vegetation map 1:500000. (UN32/6.1102-75-005.) (Rapp. tech. 1.)
FAO. 1980*b*. Id. *Bénin*. Rome, FAO. 73 p., with coloured vegetation map 1:500000. (UN32/6.1102-75-005). (Rapp. tech. 2.)
FAO. 1980*c*. Id. *Cameroun*. Rome, FAO. 84 p., with coloured vegetation map 1:1000000. (UN32/6.1102-75-005.) (Rapp. tech. 3.)
FAO. 1980*d. Global environment monitoring system. Pilot project on tropical forest cover monitoring. Benin, Cameroon, Togo. Project implementation: methodology, results and conclusions*. Rome, FAO. 99 p., with coloured vegetation maps 1:500000 (Benin, Togo) and 1:1000000 (Cameroon). (UN32/6.1102-75-005.) (Proj. Rep. 4.)
FAO–UNESCO. 1974. *Soil map of the world, 1:5000000*, vol. I, *General legend*. Paris, Unesco. 59 p., with legend sheet.
——. 1977. *Soil map of the world, 1:5000000*, vol. VI, *Africa*. Paris, Unesco. 299 p., with coloured map in 3 sheets.
FARRELL, J. A. K. 1968*a*. Preliminary notes on the vegetation of the lower Sabi-Lundi basin, Rhodesia. *Kirkia*, 6, p. 223–48, with 2 vegetation maps 1:250000.
——. 1968*b*. Preliminary notes on the vegetation of southern Gokwe District, Rhodesia. *Kirkia*, 6, p. 249–57, with 2 vegetation maps 1:250000.
FARRON, C. 1968. Congo-Brazzaville. In: Hedberg, I.; Hedberg, O. (eds.), p. 112–15.
FIELD, C. R. 1968*a*. The food habits of some wild ungulates in relation to land use and management. *E. Afr. agric. For. J.*, 33 (spec. issue), p. 159–62.
——. 1968*b*. A comparative study of the food habits of some wild ungulates in the Queen Elizabeth National Park, Uganda. Preliminary report. *Symp. zool. Soc. Lond.*, 21, p. 135–51.
——. 1970. A study of the feeding habits of the hippopotamus (*Hippopotamus amphibius* Linn.) in the Queen Elizabeth National Park, Uganda, with some management implications. *Zool. Afr.*, 5, p. 71–86.
——. 1972. The food habits of wild ungulates in Uganda by analyses of stomach contents. *E. Afr. Wildl. J.*, 10, p. 17–42.
——. 1975. Climate and food habits of ungulates on Galana Ranch. *E. Afr. Wildl. J.*, 13, p. 203–20.

FIELD, C. R.; ROSS, I. C. 1976. The savanna ecology of Kidepo Valley National Park. II. Feeding ecology of elephant and giraffe. *E. Afr. Wildl. J.*, 14, p. 1–15.
FISHWICK, R. W. 1970. Sahel and Sudan zone of Northern Nigeria, North Cameroons and the Sudan. In: Kaul, R. N. (ed.), p. 59–85.
FOCAN, A.; MULLENDERS, W. 1949. Communication préliminaire sur un essai de cartographie pédologique et phytosociologique dans le Haut-Lomami (Congo belge). *Bull. Agric. Congo belg.*, 40, p. 511–32.
——; ——. 1955. *Carte Sols Vég. Congo belge*, 1, *Kaniama (Haut-Lomami)*. Feuille 1, Carte de Reconnaissance, A, Sols (1:100000), B, Végétation (1:50000). Feuille 2, Carte semi-détaillée de la Région de Tshibonde (1:25000). A, Sols, B, Végétation, C, Utilisation des Sols. *Notice explicative*, p. 1–53. Brussels, INEAC.
FOGGIE, A. 1947*a*. On the definition of forest types in the closed forest zone of the Gold Coast. *Farm and Forest*, 8, p. 50–5.
——. 1947*b*. Some ecological observations on a tropical forest type in the Gold Coast. *J. Ecol.*, 34, p. 88–106.
FOLLMANN, G. 1976. Lichen flora and lichen vegetation of the Canary Islands. In: Künkel, G. (ed.), p. 267–86.
FORD, J. 1971. *The role of trypanosomiasis in African ecology. A study of the tsetse fly problem*. Oxford, Clarendon Press. 576 p.
FOREST DEPARTMENT, NIGERIA. 1948. *The vegetation of Nigeria. Descriptive terms*. Lagos, Govt Printer. 34 p.
FOSBERG, F. R. 1961. A classification of vegetation for general purposes. *Trop. Ecol.*, 2, p. 1–28.
——. 1971. Preliminary survey of Aldabra vegetation. *Phil. Trans. R. Soc. Lond.*, ser. B, 260, p. 215–25.
FOSBERG, F. R.; RENVOIZE, S. A. 1980. The flora of Aldabra and neighbouring islands. *Kew Bull. add. Ser.*, 7, p. 1–358.
FOSTER, J. B. 1966. The giraffe of Nairobi National Park: home range, sex ratios, the herd and food. *E. Afr. Wildl. J.*, 4, p. 139–48.
FOSTER, J. B.; DAGG, A. I. 1972. Notes on the biology of the giraffe. *E. Afr. Wildl. J.*, 10, p. 1–16.
FOURCADE, H. G. 1889. *Report on the Natal forests*. Pietermaritzburg, Natal Govt. 192 p.
FOX, J. E. D. 1968*a*. Derived coastal savanna at Bradford, Sierra Leone. *J. W. Afr. Sci. Ass.*, 13, p. 173–83.
——. 1968*b*. Exploitation of the Gola Forest. *J. W. Afr. Sci. Ass.*, 13, p. 185–210.
——. 1968*c*. *Didelotia idae* in the Gola Forest, Sierra Leone. *Econ. Bot.*, 22, p. 338–46.
——. 1970. Natural regeneration of the Kambui Hills forest in eastern Sierra Leone. Pt I. Ecological status of the *Lophira/Heritiera* rain forest. *Trop. Ecol.*, 11, p. 169–85.
FRANC DE FERRIÈRE, J.; JACQUES-FÉLIX, H. 1936. Le marais à *Raphia gracilis* de Guinée française. Valeur et utilisation agricoles. *Rev. Bot. appl. Agric. trop.*, 16, p. 105–23.
FRANCKE, A. 1942. Zur Gliederung der forstlich wichtigen Vegetationsformationen des tropischen Afrikas. *Kolonialforstl. Mitt.*, 5, p. 1–44.
FRANKART, R. 1960. *Carte Sols Vég. Congo belge*, 14, *Uele*. A, Sols (1:40000). *Notice explicative*, p. 1–128. Brussels, INEAC.
——. 1967. *Carte Sols Vég. Congo, Rwanda, Burundi*, 21, *Paysannat Babua*. A, Sols (1:40000). *Notice explicative*, p. 1–88. Brussels, INEAC.
FRANKART, R.; LIBEN, L. 1956. *Carte Sols Vég. Congo belge*, 7, *Bugesera-Mayaga (Ruanda)*. A, Sols (1:100000), B,

Végétation (1:100000). *Notice explicative*, p. 1–57. Brussels, INEAC.

FRESON, R.; GOFFINET, G.; MALAISSE, F. 1974. Ecological effects of the regressive succession muhulu–miombo–savannah in Upper Shaba (Zaïre). In: *Proc. 1st Int. Congr. Ecol.* (The Hague, September 1974), p. 365–71. Wageningen, Centre for Agricultural Publishing and Documentation. 414 p.

FRIES, R. E. 1913. Die Vegetation des Bangweolo-Gebietes. *Svensk bot. Tidskr.*, 7, p. 233–57, with coloured vegetation map 1:500000.

——. 1915. Vegetationsbilder aus dem Bangweologebiet. *Vegetationsbilder*, 12, t. 1–6.

——. 1921. Zur Kenntnis der Vegetation Termitenhügels in Nord-Rhodesia. In: *Botanische Untersuchungen, Wiss. Ergebnisse Schwed. Rhodesia–Kongo Expedition 1911–1912*, 1, p. 30–9. Stockholm, Aftonbladet.

——. 1925. Vegetationsbilder von den Kenia- und Aberdare-Bergen. *Vegetationsbilder*, 16, t. 37–42.

FRIES, R. E.; FRIES, Th. C. E. 1948. Phytogeographical researches on Mt. Kenya and Mt. Aberdare, British East Africa. *K. svenska Vetensk. Handl.*, ser. 3, 25 (5), p. 1–83.

FRÖDIN, J. 1923. Recherches sur la végétation du Haut Atlas. *Lunds Univ. Årsskr.*, N. F., Avd. 2, Bd 19, Nr 4, p. 1–24.

FURNESS, H. D.; BREEN, C. M. 1980. The vegetation of seasonally flooded areas of the Pongola River floodplain. *Bothalia*, 13, p. 217–30.

FURON, R. 1963. *The geology of Africa*. Edinburgh, London, Oliver & Boyd. 377 p. (English translation by A. Hallam & L. A. Stevens of *Géologie de l'Afrique*, 2nd ed.)

——. 1968. *Géologie de l'Afrique*. 3rd ed. Paris, Payot. 376 p.

FURON, R.; LOMBARD, J. 1964. *Notice explicative, Carte géologique de l'Afrique (1:5000000)/Explanatory note, Geological map of Africa (1:5000000)*, p. 1–39. Paris, Unesco. See also ASGA–Unesco, 1964. (Recherches sur les ressources naturelles/Natural resources research, 3.)

GAFF, D. F. 1977. Desiccation-tolerant vascular plants of southern Africa. *Oecologia* (Berlin), 31, p. 95–109.

GALLAIS, J. 1967. Le delta intérieur du Niger. Étude de géographie régionale. *Mém. IFAN*, 79, p. 1–621.

——. 1975. *Pasteurs et paysans du Gourma. La condition sahélienne*. Paris, CNRS, Mémoires du Centre d'Études de Géographie Tropicale (CEGET). 239 p.

GALPIN, E. E. 1927. Botanical survey of the Springbok Flats. *Mem. bot. Surv. S. Afr.*, 12, p. 1–100, with large-scale vegetation map.

GAUSSEN, H. 1952. Les résineux d'Afrique du Nord. Écologie, reboisements. *Rev. int. Bot. appl. Agric. trop.*, 32, p. 505–32.

——. 1955. Expression des milieux par des formules écologiques; leur représentation cartographique. *Les divisions écologiques du monde*, p. 13–25. (Colloques Int. CNRS, 59.) Also published in: *Année biol.*, sér. 3, 31, p. 257–69.

——. 1958. L'emploi des couleurs en cartographie. *Bull. Serv. Carte phytogéogr.*, sér. A, 3 (1), p. 5–19.

GAUSSEN, H.; DE PHILIPPIS, A. 1955. *La limite euméditerranéenne et les contrées de transition*. Carte au 1:5000000 en couleur. Rome, FAO, Sous-commission de Coordination des Questions Forestières Méditerranéennes.

GAUSSEN, H.; VERNET, A. 1958. *Carte internationale du tapis végétal. Feuille de Tunis–Sfax* (1:1000000, in colour). Tunisian Govt.

GAY, P. A. 1960. Ecological studies of *Eichhornia crassipes* Solms. in the Sudan. I. Analysis of spread in the Nile. *J. Ecol.*, 48, p. 183–91.

GAY, P. A.; BERRY, L. 1959. The water hyacinth. A new problem on the Nile. *Geogrl J.*, 125, p. 89–91.

GELDENHUYS, J. N. 1976. Physiognomic characteristics of wetland vegetation in South African shelduck habitat. *S. Afr. J. Wildl. Res.*, 6, p. 75–8.

GERAKIS, P. A.; TSANGARAKIS, C. Z. 1970. The influence of *Acacia senegal* on the fertility of a sand sheet ('goz') soil in the Central Sudan. *Plant and Soil*, 33, p. 81–6.

GÉRARD, P. 1960. Étude écologique de la forêt dense à *Gilbertiodendron dewevrei* dans la région de l'Uele. *Publs INEAC*, sér. sci., 87, p. 1–159.

GERMAIN, R. 1945. Note sur les premiers stades de la reforestation naturelle des savanes du Bas-Congo. *Bull. Agric. Congo belge*, 36, p. 16–25.

——. 1949. Reconnaissance géobotanique dans le nord du Kwango. *Publs INEAC*, sér. sci., 43, p. 1–22.

——. 1952. Les associations végétales de la plaine de la Ruzizi (Congo belge) en relation avec le milieu. *Publs INEAC*, sér. sci., 52, p. 1–321.

——. 1957. Un essai d'inventaire de la flore et des formes biologiques en forêt équatoriale congolaise. *Bull. Jard. bot. État Brux.*, 27, p. 563–76.

——. 1965. Les biotopes alluvionnaires herbeuses et les savanes intercalaires du Congo équatorial. *Mém. Acad. r. Sci. Outre-Mer, Cl. Sci. nat. méd.*, n.s., 8°, 15 (4), p. 1–399.

——. 1968. Congo-Kinshasa. In: Hedberg, I.; Hedberg, O. (eds.), p. 121–5.

GERMAIN, R.; CROEGAERT, J.; SYS, C. 1955. *Carte Sols Vég. Congo belge*, 3, *Vallée de la Ruzizi*. A, Sols (1:1000000), B, Végétation (1:50000). *Notice explicative*, p. 1–48. Brussels, INEAC.

GERMAIN, R.; ÉVRARD, C. 1956. Étude écologique et phytosociologique de la forêt à *Brachystegia laurentii*. *Publs INEAC*, sér. sci., 67, p. 1–105.

GETAHUN, A. 1974. The role of wild plants in the native diet in Ethiopia. *Agro-Ecosyst.*, 1, p. 45–56.

GIBSON, H. S. 1938. *A report on the forests of the granitic islands of the Seychelles*. Mahé, Seychelles, Govt Printer. 50 p.

GIESS, W. 1962. Some notes on the vegetation of the Namib desert, with a list of plants collected in the area. *Cimbebasia*, 2, p. 3–35.

——. 1968a. A short report on the vegetation of the Namib coastal area from Swakopmund to Cape Frio. *Dinteria*, 1, p. 13–29.

——. 1968b. Die Gattung *Rhigozum* Burch. und ihre Arten in Südwestafrika. *Dinteria*, 1, p. 31–51.

——. 1969. *Welwitschia mirabilis* Hook. f. *Dinteria*, 3, p. 3–55.

——. 1970. Ein Beitrag zur Flora des Etoscha Nationalparks. *Dinteria*, 5, p. 19–55.

——. 1971. A preliminary vegetation map (1:3000000, coloured) of South West Africa. *Dinteria*, 4, p. 5–114.

GIESS, W.; TINLEY, K. L. 1968. South West Africa. In: Hedberg, I.; Hedberg, O. (eds.), p. 250–3, with small vegetation map.

GIGLIOLI, M. E. C.; THORNTON, I. 1965. The mangrove swamps of Keneba, Lower Gambia River Basin. I. Descriptive notes on the climate, the mangrove swamps and the physical composition of their soils. *J. appl. Ecol.*, 2, p. 81–103.

GILLET, H. 1957. Une enclave floristique soudanienne dans le massif de l'Ennedi (Nord-Tchad). *C.r. somm. Séanc. Soc. Biogéogr.*, 34 (301), p. 172–7.

——. 1958. Sur quelques plantes relictes du massif de l'Ennedi (Nord-Tchad). *C.r. somm. Séanc. Soc. Biogéogr.*, 35 (307), p. 63–8.

——. 1959a. Découverte de nouvelles plantes relictes dans le massif de l'Ennedi (Nord-Tchad). *C.r. somm. Séanc. Soc. Biogéogr.*, 36 (312), p. 27–34.

——. 1959b. Une mission scientifique dans l'Ennedi (Nord-Tchad) et en Oubangui. *J. Agric. trop. Bot. appl.*, 6, p. 505–73.

——. 1960. Étude des pâturages du Ranch de l'Ouadi Rimé (Tchad). *J. Agric. trop. Bot. appl.*, 7, p. 465–528.

——. 1961a. Ibid., p. 615–708.

——. 1961b. Flore sahélienne tchadienne: un essai d'analyse biogéographique. *C.r. somm. Séanc. Soc. Biogéogr.*, 37 (330), p. 7–21.

——. 1961c. Pâturages sahéliens. Le ranch de l'ouadi Rimé. *J. Agric. trop. Bot. appl.*, 8, p. 557–692.

——. 1962a. Variations de la flore sahélienne en fonction de l'importance de la saison des pluies. *C.r. somm. Séanc. Soc. Biogéogr.*, 39 (341), p. 13–23.

——. 1962b. Végétation, agriculture et sol du Centre Tchad. Feuilles de Mongo, Melfi, Bohoro, Guera. *J. Agric. trop. Bot. appl.*, 9, p. 451–501.

——. 1963. Végétation, agriculture et sol du Centre et du Sud Tchad. Feuilles de Miltou, Dagela, Koumra, Moussafoyo. *J. Agric. trop. Bot. appl.*, 10, p. 53–160.

——. 1964. Agriculture, végétation et sol du Centre et du Sud Tchad. Feuilles de Miltou, Dagéla, Koumra, Moussafoyo. Fort-Lamy, ORSTOM–CRT. 108 p. Also published in *J. Agric. trop. Bot. appl.*, 10.

——. 1967. Essai d'évaluation de la biomasse végétale en zone sahélienne. *J. Agric. trop. Bot. appl.*, 14, p. 123–58.

——. 1968a. Le peuplement végétal du massif de l'Ennedi (Tchad). *Mém. Mus. natn. Hist. nat.*, Paris, n.s., sér. B (Bot.), 17, p. 1–206.

——. 1968b. Tchad et Sahel tchadien. In: Hedberg, I.; Hedberg, O. (eds.), p. 54–8.

GILLETT, J. B. 1941. The plant formations of Western British Somaliland and the Harar Province of Abyssinia. *Kew Bull.*, p. 37–199, with small-scale vegetation map.

——. 1955. The relation between the highland floras of Ethiopia and British East Africa. *Webbia*, 11, p. 459–66.

GILLHAM, M. E. 1963. Some interactions of plants, rabbits and seabirds on South African islands. *J. Ecol.*, 51, p. 275–94.

GILLI, A. 1975. Pflanzensoziologische Beobachtungen aus Ostafrika. *Feddes Rep.*, 86, p. 233–52.

GILLILAND, H. B. 1938. The vegetation of Rhodesian Manicaland. *J. S. Afr. Bot.*, 4, p. 73–99.

——. 1952. The vegetation of Eastern British Somaliland. *J. Ecol.*, 40, p. 91–124.

GILLILAND, M. 1962. The flora of the Merensky Nature Reserve. *Fauna Flora*, Pretoria, 13, p. 41–9.

GILLMAN, C. 1949. A vegetation-types map (1:2000000 in colour) of Tanganyika Territory. *Geogrl Rev.* (New York), 39, p. 7–37.

GILSON, P.; FRANÇOIS, P. 1969. *Carte Sols Vég. Congo, Rwanda, Burundi*, 23, *Zone de la Haute Lulua*. A, Sols (1:50000; 1:100000; 1:200000). *Notice explicative*, p. 1–22. Brussels, INEAC.

GILSON, P.; JONGEN, P.; VAN WAMBEKE, A.; LIBEN, L. 1957. *Carte Sols Vég. Congo belge*, 6, *Yangambi (3), Lilanda*. A, Sols (1:50000), B, Végétation, (1:50000). *Notice explicative*, p. 1–32. Brussels, INEAC.

GILSON, P.; LIBEN, L. 1960. *Carte Sols Vég. Congo belge*, 15, *Kasai*. A, Sols (1:50000; 1:200000). *Notice explicative*, p. 1–76. Brussels, INEAC.

GILSON, P.; VAN WAMBEKE, A.; GUTZWILLER, R. 1956. *Carte Sols Vég. Congo belge*, 6, *Yangambi (2), Yangambi*. A, Sols (1:50000), B, Végétation (1:50000). *Notice explicative*, p. 1–35. Brussels, INEAC.

GLEDHILL, D. 1963. The ecology of the Aberdeen Creek mangrove swamp. *J. Ecol.*, 51, p. 693–703.

——. 1970. The vegetation of superficial ironstone hardpans in Sierra Leone. *J. Ecol.*, 58, p. 265–74.

GLORIOD, G. 1974. La forêt de l'Est du Gabon. *Bois Forêts Trop.*, 155, p. 35–57.

GLOVER, J. 1963. The elephant problem at Tsavo. *E. Afr. Wildl. J.*, 1, p. 30–9.

GLOVER, J.; GWYNNE, M. D. 1962. Light rainfall and plant survival in East Africa. I. Maize. *J. Ecol.*, 50, p. 111–18.

GLOVER, P. E. 1937. A contribution to the ecology of the Highveld flora. *S. Afr. J. Sci.*, 34, p. 224–59.

——. 1950–51. The root systems of some British Somaliland plants. *E. Afr. agric. J.*, 16, p. 98–113 (1950), p. 154–73 (1951), p. 205–17 (1951), 17, p. 38–50 (1951).

——. 1966. *An ecological survey of the Narok District of Kenya Masailand 1961–1965* (pt 1). Kenya National Parks. 72 p. (Mimeo.)

——. 1968. The role of fire and other influences on the savannah habitat with suggestions for further research. *E. Afr. Wildl. J.*, 6, p. 131–7.

——. 1972. Nature, wildlife and the habitat with a discussion on fire and other influences. *Proc. Ann. Tall Timbers Fire Ecol. Conf.*, 11, p. 319–36.

GLOVER, P. E.; GLOVER, J.; GWYNNE, M. D. 1962. Light rainfall and plant survival in East Africa. II: Dry grassland vegetation. *J. Ecol.*, 50, p. 199–206.

GLOVER, P. E.; GWYNNE, M. D. 1961. The destruction of Masailand. *New Scientist*, 249, p. 450–3.

GLOVER, P. E.; TRUMP, E. C. 1970. *An ecological survey of the Narok District of Kenya Masailand*. Pt 2. *Vegetation*. Nairobi, Kenya National Parks. 157 p., with vegetation map. (Mimeo.)

GLOVER, P. E.; TRUMP, E. C.; WATERIDGE, L. E. D. 1964. Termitaria and vegetation patterns on the Loita Plains of Kenya. *J. Ecol.*, 52, p. 367–77.

GLOVER, P. E.; VAN RENSBURG, H. J. 1938. A contribution to the ecology of the highveld grassland at Frankenwald, in relation to grazing and burning. *S. Afr. J. Sci.*, 35, p. 247–79.

GLOVER, P. E.; WATERIDGE, L. E. D. 1968. Erosion terraces in the Loita-Aitong areas of Kenya Masailand. *E. Afr. Wildl. J.*, 6, p. 125–9.

GODDARD, J. 1968. Food preferences of two black rhinoceros populations. *E. Afr. Wildl. J.*, 6, p. 1–18.

——. 1970. Food preferences of black rhinoceros in the Tsavo National Park. *E. Afr. Wildl. J.*, 8, p. 145–61.

GOETZE, W.; ENGLER, A. 1902. *Vegetationsansichten aus Deutschostafrika insbesondere aus der Khutusteppe, dem Ulugurugebirge, Uhehe, dem Kingagebirge, vom Rungwe, dem Kondeland und der Rukwasteppe*. Leipzig, Engelmann. 50 p., 64 phots.

GOLDBLATT, P. 1978. An analysis of the flora of Southern Africa, its characteristics, relationships and origins. *Ann. Mo. Bot. Gdn*, 65, p. 369–436.

GOLDING, F. D.; GWYNN, A. M. 1939. Notes on the vegetation of the Nigerian shore of Lake Chad. *Kew Bull.*, p. 631–43.

GOLDSMITH, B. 1976. The trees of Chirinda forest. *Rhod. Sci. News*, 10, p. 41–50.

GOMES E SOUSA, A. 1967. Formações florestais. In: *Dendrologia de Moçambique*, Estudo Geral, 1, p. 35–175, with coloured vegetation map 1:5000000. Privately pub.

GOOD, R. 1924. The geographical affinities of the flora of Jebel Marra. *New Phytol.*, 23, p. 266–81.

———. 1974. *The geography of the flowering plants.* 4th ed. London, Longman. 557 p.

GOODALL, D. W.; PERRY, R. A. 1979. *Arid-land ecosystems: structure, functioning and management.* Vol. 1. Cambridge, Cambridge Univ. Press. 881 p.

GOODIER, R. 1968. Nature conservation and forest clearance with special reference to some ecological implications of tsetse control. In: Hedberg, I.; Hedberg, O. (eds.), p. 20–5.

GOODIER, R.; PHIPPS, J. B. 1961. A revised check-list of the vascular plants of the Chimanimani Mountains. *Kirkia*, 1, p. 44–66.

———; ———. 1962. A vegetation map (large-scale, in colour) of the Chimanimani National Park. *Kirkia*, 3, p. 2–7.

GOSSWEILER, J.; MENDONÇA, F. A. 1939. *Carta fitogeográfica de Angola.* Luanda, Gov. Geral de Angola. 242 p., with coloured vegetation map 1:2 000 000.

———; ———. 1941. The grasslands of Angola. *Herb. Abstr.*, 11, suppl., p. 61–5.

GOUDIE, A. S. 1973. *Duricrusts in tropical and subtropical landscapes.* Oxford, Clarendon Press. 174 p. (Oxf. Res. Stud. Geogr.)

GRAHAM, R. M. 1929. Notes on the mangrove swamps of Kenya. *J. E. Afr. Uganda nat. Hist. Soc.*, 36, p. 157–64.

GRAM, K. 1935. *Karplantenvegetationen i Mouydir (Emmidir) i Central Sahara.* Copenhagen. 168 p.

GRANGER, J. E.; SCHULZE, R. E. 1977. Incoming solar radiation patterns and vegetation response—examples from the Natal Drakensberg. *Vegetatio*, 35, p. 47–54.

GRANIER, P. 1979. The grazing land ecosystems of the Mid-West of Madagascar. In: Unesco/UNEP/FAO, p. 602–11.

GRANT, P. M. n.d. Trace element deficiencies in Kalahari sands. *Fedl Minist. Agric. Bull.*, 2172, p. 1–3.

GRAY, C. (ed.). 1971. *Symposium on the geology of Libya.* Faculty of Science, Univ. of Libya. 522 p.

GREENWAY, P. J. 1933. The vegetation of Mpwapwa, Tanganyika Territory. *J. Ecol.*, 21, p. 28–43.

———. 1943. *Second draft report on vegetation classification for the approval of the Vegetation Committee, Pasture Research Conference, Nairobi.* (Mimeo.)

———. 1955. Ecological observations on an extinct East African volcanic mountain. *J. Ecol.*, 43, p. 544–63.

———. 1965. The vegetation and flora of Mt. Kilimanjaro. *Tanganyika Notes Rec.*, 64, p. 97–107.

———. 1969. A check-list of plants recorded in Tsavo National Park, East. *J. E. Afr. nat. Hist. Soc.*, 27, p. 169–209.

———. 1973. A classification of the vegetation of East Africa. *Kirkia*, 9, p. 1–68, with small-scale vegetation map.

GREENWAY, P. J.; VESEY-FITZGERALD, D. F. 1969. The vegetation of Lake Manyara National Park. *J. Ecol.*, 57, p. 127–49.

GREIG-SMITH, P.; CHADWICK, M. J. 1965. Data on pattern within plant communities. III. Acacia–Capparis semi-desert scrub in the Sudan. *J. Ecol.*, 53, p. 465–74.

GREWE, E. 1941. Afrikanische Mangrove-Landschaften, Verbreitung und wirtschaftsgeographische Bedeutung. *Wissensch. Veröffentl. Dt. Müs. f. Landerk.*, N.F., 9, p. 105–77.

GRIFFITHS, J. F. 1962. The climate of East Africa. In: Russell, E. W. (ed.), p. 77–87. Revised account in: Morgan, W. T. W. (ed.), p. 106–25.

———. (ed.). 1972. *Climates of Africa.* In: Landsberg, H. E. (ed.), *World survey of climatology*, 10. Amsterdam, London; New York, Elsevier. 604 p.

GRIFFITHS, J. F.; HEMMING, C. F.; 1963. *A rainfall map of Eastern Africa and Southern Arabia.* Nairobi, Meteorological Dep., E.A. Common Services Organization. 42 p. (Mimeo.)

GRIMSDELL, J. J. R.; FIELD, C. R. 1976. Grazing patterns of buffaloes in the Rwenzori National Park, Uganda. *E. Afr. Wildl. J.*, 14, p. 339–44.

GRONDARD, A. 1964. La végétation forestière au Tchad. *Bois Forêts Trop.*, 93, p. 15–34.

GROVE, A. T. 1969. Landforms and climatic change in the Kalahari and Ngamiland. *Geogrl J.*, 135, p. 190–212.

———. 1977. The geography of semi-arid lands. In: Hutchinson, J. B. (ed.), p. 457–75.

———. 1978. *Africa.* 3rd ed. London, Oxford University Press. 337 p.

GROVE, A. T.; WARREN, A. 1968. Quaternary landforms and climate on the south side of the Sahara. *Geogrl J.*, 134, p. 194–208.

GRUBB, P. J. 1977. Control of forest growth and distribution on wet tropical mountains with special reference to mineral nutrition. *Ann. Rev. Ecol. Syst.*, 8, p. 83–107.

GRUBB, P. J.; TANNER, E. V. J. 1976. The montane forests and soils of Jamaica. A reassessment. *J. Arn. Arb.*, 57, p. 313–68.

GRUNOW, J. O. 1967. Objective classification of plant communities. A synecological study in the Sourish Mixed Bushveld of Transvaal. *J. Ecol.*, 55, p. 691–710.

GUICHARD, K. M. 1955. Habitats of the Desert Locust in western Libya and Tibesti. *Anti-Locust Bull.*, 21, p. 1–33.

GUICHON, A. 1960. La superficie des formations forestières de Madagascar. *Revue for. fr.*, 6, p. 408–11.

GUIGONIS, G. 1968. République Centrafricaine. In: Hedberg, I.; Hedberg, O. (eds.), p. 107–11, with small vegetation map.

GUILLAUME, G. M. D. 1968. Quelques considérations sur les biotopes forestiers de la province de Victoria (Cameroun occidental) en relation avec les facteurs du milieu. *Bull. IFAN*, sér. A, 30, p. 896–919.

GUILLAUMET, J. L. 1967. Recherches sur la végétation et la flore de la région du Bas-Cavally (Côte d'Ivoire). *Mém. ORSTOM*, 20, p. 1–247, with vegetation map 1:1 000 000.

GUILLAUMET, J. L.; ADJANOHOUN, E. 1971. La végétation. In: Le milieu naturel de la Côte d'Ivoire. *Mém. ORSTOM*, 50, p. 157–263, with coloured vegetation map 1:500 000.

GUILLAUMET, J. L.; KOECHLIN, J. 1971. Contribution à la définition des types de végétation dans les régions tropicales (exemple de Madagascar). *Candollea*, 26, p. 263–77.

GUILLOTEAU, J. 1957. The problem of bush fires and burns in land development and soil conservation in Africa south of the Sahara. *Afr. Soils*, 4, p. 64–102.

GUINEA, E. 1945. *Aspecto forestal del desierto. La vegetación leñosa y los pastos del Sahara Español.* Madrid, Instituto Forestal de Investigaciones y Experiencias. 152 p., with coloured vegetation map.

———. 1946. *Ensayo geobotánico de la Guinea continental española.* Madrid, Dirección de Agricultura de los Territorios Españoles del Golfo de Guinea. 389 p.

———. 1949. Geobotánica. In: Hernández-Pacheco et al. (eds.), *El Sahara Español: Estudio Geológico, Geográfico y Botánico*, p. 631–806. Madrid, Instituto Estudios Africanos.

———. 1968. Fernando Po. In: Hedberg, I.; Hedberg, O. (eds.), p. 130–2, with small vegetation map.

GUINET, P. 1958. Notice détaillée de la feuille de Beni-Abbès (coupure spéciale de la carte de la végétation de l'Algérie au 1:200 000). *Bull. Serv. Carte phytogéogr.*, sér. A. *Carte de la végétation*, 3, p. 21–96. Paris, CNRS.

GUINET, P.; SAUVAGE, C. 1954. Les hamadas sud-marocaines. Botanique. *Trav. Inst. sci. chérif.*, sér. gén., 2, p. 73–167.

GUINOCHET, M.; QUÉZEL, P. 1954. Reconnaissance phytosociologique autour du Grand Erg Occidental. *Trav. Inst. Rech. sahar.* (Alger), 12, p. 11–27.

GUPPY, H. B. 1914. Notes on the native plants of the Azores as illustrated on the slopes of the mountain of Pico. *Kew Bull.*, p. 305–21.

GUY, P. R. 1976. The feeding behaviour of elephant (*Loxodonta africana*) in the Sengwa Area, Rhodesia. *S. Afr. J. Wildl. Res.*, 6, p. 55–63.

——. 1977. Notes on the vegetation types of the Zambezi Valley between the Kariba and Mpata gorges. *Kirkia*, 10, p. 543–57, with vegetation map 1:250000.

GWYNNE, M. D. 1968. Socotra. In: Hedberg, I.; Hedberg, O. (eds.), p. 179–85, with small vegetation map.

——. 1969. The nutritive values of *Acacia* pods in relation to *Acacia* seed distribution by ungulates. *E. Afr. Wildl. J.*, 7, p. 176–8.

GWYNNE, M. D.; BELL, R. H. V. 1968. Selection of vegetation components by grazing ungulates in the Serengeti National Park. *Nature*, 220, p. 390–3.

GWYNNE, M. D.; WOOD, D. 1969. Plants collected on islands in the western Indian Ocean during a cruise of the M.F.R.V. 'Manihine', Sept.–Oct. 1967. *Atoll Res. Bull.*, 134, p. 1–15.

HADAC, E. 1976. Species diversity of mangrove and Continental Drift. *Folia geobot. phytotax., Praha*, 11, p. 213–16.

HAGERUP, O, 1930. Étude des types biologiques de Raunkiaer dans la flore autour de Tombouctou. *Biol. Meddel. k. Danske Vidensk. Selsk.*, 9 (4), p. 1–116.

HALL, A. V.; RYCROFT, H. B. 1979. South Africa: the conservation policy of the National Botanic Gardens and its Regional Gardens. In: Synge, H.; Townsend, H. (eds.), *Survival or extinction*, p. 125–34. Royal Botanic Gardens, Kew, England.

HALL, J. B.; JENÍK, J. 1968. Contribution towards the classification of savanna in Ghana. *Bull. IFAN*, sér. A, 30, p. 84–99.

HALL, J. B.; LAWSON, G. W. 1972. *Talbotielletum gentii*: a single dominant forest association in Ghana. *Ghana J. Sci.*, 12, p. 28.

HALL, J. B.; OKALI, D. U. U. 1974. Phenology and productivity of *Pistia stratioides* L. on the Volta Lake, Ghana. *J. appl. Ecol.*, 11, p. 709–25.

HALL, J. B.; POPLE, W. 1968. Recent vegetational changes in the Lower Volta River. *Ghana J. Sci.*, 8, p. 24–9.

HALL, J. B.; SWAINE, M. D. 1974. *Classification and ecology of forests in Ghana*. Legon, Dep. of Botany, Univ. of Ghana. 27 p. (Mimeo.)

——; ——. 1976. Classification and ecology of closed-canopy forest in Ghana. *J. Ecol.*, 64, p. 913–51.

——; ——. 1981. *Distribution and ecology of vascular plants in a tropical rain forest: forest vegetation in Ghana*. The Hague, Boston, London, Junk. 383 p. (Geobot. 1.)

HALL, John B. 1971. Environment and vegetation on Nigeria's highlands. *Vegetatio*, 23, p. 339–59.

——. 1973. Vegetation zones on the southern slopes of Mount Cameroun. *Vegetatio*, 27, p. 49–69.

——. 1977. Forest types in Nigeria. An analysis of pre-exploitation forest-enumeration data. *J. Ecol.*, 65, p. 187–99.

HALL, John B.; BADA, S. O. 1979. The distribution and ecology of Obeche (*Triplochiton scleroxylon*). *J. Ecol.*, 67, p. 543–64.

HALL, John B.; MEDLER, J. A. 1975*a*. Botanical exploration of the Obudu Plateau area. *Nig. Fld*, 15, p. 101–17.

——; ——. 1975*b*. Highland vegetation in South-Eastern Nigeria and its affinities. *Vegetatio*, 29, p. 191–8.

HALL, John B.; OKALI, D. U. U. 1979. A structural and floristic analysis of woody fallow vegetation near Ibadan, Nigeria. *J. Ecol.*, 67, p. 321–46.

HALLÉ, F.; HALLÉ, N. 1965. Présentation de quelques formes ligneuses simples de la forêt de Bélinga (Gabon). *Biol. gabon.* 1, p. 247–55.

HALLÉ, F.; OLDEMAN, R. A. A. 1970. *Essai sur l'architecture et la dynamique de croissance des arbres tropicaux*. Paris, Masson. 178 p.

——; ——. 1975. *An essay on the architecture and dynamics of growth of tropical trees* (transl. B. C. Stone). Kuala Lumpur, Malaysia, Penerbit Universiti Malaya. 156 p.

HALLÉ, F.; OLDEMAN, R. A. A.; TOMLINSON, P. B. 1978. *Tropical trees and forests. An architectural analysis*. Berlin, Heidelberg, New York, Springer-Verlag. 441 p.

HALLÉ, N.; LE THOMAS, A. 1968. Gabon (note préliminaire). In: Hedberg, I.; Hedberg, O. (eds.), p. 111–12.

HALLÉ, N.; LE THOMAS, A.; GAZEL, M. 1967. Trois relevés botaniques dans les forêts de Bélinga (N.E. du Gabon). *Biol. gabon.*, 3, p. 3–16.

HALL-MARTIN, A. J. 1975. Classification and ordination of forest and thicket vegetation of the Lengwe National Park, Malawi. *Kirkia*, 10, p. 131–84.

HALL-MARTIN, A. J.; FULLER, N. G. 1975. Observations on the phenology of some trees and shrubs of the Lengwe National Park, Malawi. *Jl S. Afr. Wildl. Mgmt Ass.*, 5, p. 83–6.

HALWAGY, R. 1961. The vegetation of the semi-desert northeast of Khartoum, Sudan. *Oikos*, 12, p. 87–110.

——. 1962*a*. The impact of man on semi-desert vegetation in the Sudan. *J. Ecol.*, 50, p. 263–73.

——. 1962*b*. The incidence of the biotic factor in northern Sudan. *Oikos*, 13, p. 97–117.

——. 1963. Studies on the succession of vegetation on some islands and sand banks in the Nile near Khartoum, Sudan. *Vegetatio*, 11, p. 217–34.

HAMBLER, D. J. 1964. The vegetation of granitic outcrops in Western Nigeria. *J. Ecol.*, 52, p. 573–94.

HAMILTON, A. C.; PERROTT, R. A. 1980. *The vegetation of Mt. Elgon, East Africa*. New Univ. of Ulster, Coleraine, Northern Ireland. 73 p. (Mimeo.)

HANCOCK, G. M. 1944. The grass-*Acacia* cycle in the Blue Nile Province, south of Sennar, and proposals for its management. *Sudan Govt Rep. Soil Conserv. Comm.*, p. 78–86.

HANSEN, A. 1969. Check-list of the vascular plants of the Archipelago of Madeira. *Bol. Mus. Munic. Funchal*, 24, p. 1–62.

HANSEN, A.; SUNDING, P. 1979. *Flora of Macaronesia*. Checklist of vascular plants. 2nd ed. Pt 1, 93 p. Pt 2, *Synonym index*, 55 p. Oslo, Botanical Garden and Museum, Univ. of Oslo. See also Erikson et al., 1974.

HARRINGTON, G. N. 1974. Fire effects on a Ugandan savanna grassland. *Trop. Grassl.*, 8, p. 87–101.

HARRINGTON, G.N.; ROSS, I.C. The savanna ecology of Kidepo Valley National Park. 1. The effects of burning and browsing on the vegetation. *E. Afr. Wildl. J.*, 12, p. 93–105.

HARRIS, D. R. (ed.). 1980. *Human ecology in savanna environments*. London, etc., Academic Press.

HARRISON, M. N.; JACKSON, J. K. 1958. Ecological classification of the vegetation of the Sudan. *Forests Bull. Sudan*, n.s., 2, p. 1–45, with coloured vegetation map 1:4000000.

HARROY, J. P. 1949. *Afrique, terre qui meurt*. Brussels, Marcel Hayez. 557 p., with small-scale vegetation map.

HASSELO, H. N.; SWARBRICK, J. T. 1960. The eruption of the Cameroon Mountain in 1959. *J. W. Afr. Sci. Ass.*, 6, p. 96–101.

HASSIB, H. 1952. Distribution of plant communities in Egypt. *Bull. Fac. Sci. Cairo Univ.*, 29, p. 59–261.

HAUGHTON, S. H. 1963. *Stratigraphic history of Africa south of the Sahara*. Edinburgh, Oliver & Boyd.

——. 1969. *Geological history of southern Africa*. Geological Society of South Africa. 535 p.

HAUMAN, L. 1933. Esquisse de la végétation des hautes altitudes sur le Ruwenzori. *Bull. Acad. r. Belg. Cl. Sci.*, sér. 5, 19, p. 602–16, 702–17, 900–17.

——. 1955. La 'Région Afroalpine' en phytogéographie Centro-africaine. *Webbia*, 11, p. 467–9.

HAWKINS, P.; BRUNT, M. 1965. *Report to the Government of Cameroun on the soils and ecology of West Cameroun: a broad reconnaissance survey with special reference to the Bamenda area.* 1, p. 1–285; 2, p. 286–516, with large-scale vegetation map. Rome, FAO. (FAO Expanded Program of Technical Assistance Report no. 2083. Project CAM/TE/LA.)

HEADY, H. F. 1960. *Range management in East Africa*. Nairobi, Govt Printer. 125 p.

——. 1966. Influence of grazing on the composition of *Themeda triandra* grassland, East Africa. *J. Ecol.*, 54, p. 705–27.

HEDBERG, I.; HEDBERG, O. (eds.). 1968. Conservation of vegetation in Africa south of the Sahara. *Acta phytogeogr. suec.*, 54, p. 1–320.

HEDBERG, O. 1951. Vegetation belts of the East African mountains. *Svensk bot. Tidskr.*, 45, p. 140–202.

——. 1955. Altitudinal zonation of the vegetation on the East African Mountains. *Proc. Linn. Soc. Lond.*, 165, p. 134–6.

——. 1957. Afroalpine vascular plants. *Symb. bot. upsal.*, 15 (1), p. 1–411.

——. 1961. The phytogeographical position of the Afroalpine flora. *Rec. Adv. Bot.*, 1, p. 914–19.

——. 1964. Features of Afroalpine plant ecology. *Acta phytogeogr. suec.*, 49, p. 1–144.

——. 1965. Afroalpine flora elements. *Webbia*, 19, p. 519–29.

——. 1969a. Evolution and speciation in a tropical high mountain flora. *Biol. J. Linn. Soc.*, 1, p. 135–48.

——. 1969b. Taxonomic and ecological studies on the Afroalpine flora of Mt. Kenya. *Hochgebirgsforschung*, 1, p. 171–94. Innsbruck, München, Universitätsverlag Wagner.

——. 1970. Evolution of the Afroalpine flora. *Biotropica*, 2, p. 16–23.

——. 1971. The high mountain flora of the Galama Mountain in Arussi Province, Ethiopia. *Webbia*, 26, p. 101–28.

——. 1975. Studies of adaptation and speciation in the Afroalpine flora of Ethiopia. *Boissiera*, 24, p. 71–4.

HEDIN, L. 1930. *Étude sur la forêt et les bois du Cameroun*. Paris, Librairie Larose. 230 p.

HEINZ, H. J.; MAGUIRE, B. 1974. The ethnobiology of the Kö Bushmen. Their ethnobotanical knowledge and plant lore. *Occ. Pap., Botswana Soc.*, 1, p. 1–53.

HEMMING, C. F. 1961. The ecology of the coastal area of northern Eritrea. *J. Ecol.*, 49, p. 55–78.

——. 1965. Vegetation arcs in Somaliland. *J. Ecol.*, 53, p. 57–67.

——. 1966. The vegetation of the northern region of the Somali Republic. *Proc. Linn. Soc. Lond.*, 177, p. 173–250.

——. 1968. Somali Republic North. In: Hedberg, I.; Hedberg, O. (eds.), p. 141–5, with small vegetation map.

——. 1972. The south Turkana expedition. 8. The ecology of south Turkana. *Geogrl J.*, 138, p. 15–40.

HEMMING, C. F.; SYMMONS, P. M. 1969. The germination and growth of *Schouwia purpurea* (Forsk.) Schweinf. and its role as a habitat of the Desert Locust. *Anti-Locust Bull.*, 46, p. 1–38.

HEMMING, C. F.; TRAPNELL, C. G. 1957. A reconnaissance classification of the soils of the South Turkana desert. *J. Soil Sci.*, 8, p. 167–83.

HEMSLEY, W. B. 1885. Report on the present state of knowledge of various insular floras. In: Thomson, C. W.; Murray, J. (eds.), Report of the scientific results of the voyage of H.M.S. *Challenger*, Botany, vol. 1, p. 1–75.

HENDRICKX, F. L. 1944. Esquisse de la végétation des Mfumbiru. *Bull. Ass. col. Ingén. Inst. agron. Gembloux*, 1944, p. 1–12.

——. 1946. Esquisse de la végétation du Kahusi. *Rev. Agron. col., Bukavu*, 3ᵉ trim. 1946, p. 25–32.

HENKEL, J. S. 1931. Types of vegetation in Southern Rhodesia. *Proc. Rhod. Sci. Ass.*, 30, p. 1–23, with coloured vegetation map 1:2 000 000.

HENKEL, J. S.; BALLENDEN, S. St C.; BAYER, A. W. 1936. An account of the plant ecology of the Dukuduku Forest Reserve and adjoining areas of the Zululand Coast Belt. *Ann. Natal Mus.*, 8, p. 95–125.

HENRY, P. W. T. 1974. *Forestry on St. Helena*. Tolworth Tower, Surbiton, Surrey, Directorate of Overseas Survey. 160 p. (Land resource rep. 2.)

HEPPER, F. N. 1965. The vegetation and flora of the Vogel Peak massif, Northern Nigeria. *Bull. IFAN*, sér. A, 27, p. 413–513.

——. 1966. Outline of the vegetation and flora of Mambila Plateau, Northern Nigeria. *Bull. IFAN*, sér. A, 28, p. 91–127.

HERBST, S. N.; ROBERTS, B. R. 1974. The alpine vegetation of the Lesotho Drakensberg. A study in quantitative floristics at Oxbow. *J. S. Afr. Bot.*, 40, p. 257–67.

HERLOCKER, D. 1975. *Woody vegetation of the Serengeti National Park*. Caesar Kleberg Research Program in Wildlife Ecology, The Texas Agricultural Experiment Station, The Texas A. & M. University System, College Station, Texas 77843. 31 p., with coloured vegetation map 1:250 000.

——. 1979a. *Vegetation of southwestern Marsabit District, Kenya*. Nairobi, UNEP–MAB Integrated Project in Arid Lands. 68 p., with coloured vegetation map 1:500 000. (IPAL Tech. Rep. no. D-1.)

——. 1979b. *Implementing forestry programmes for local community development, Southwestern Marsabit District, Kenya*. Nairobi, UNEP–MAB Integrated Project in Arid Lands. (IPAL Tech. Rep. no. D-2c.)

HERLOCKER, D. J.; DIRSCHL, H. J. 1972. *Vegetation of the Ngorongoro Conservation Area, Tanzania*. Ottawa, Canadian Wildlife Service. 37 p., with coloured vegetation map 1:125 000. (Rep. ser. no. 19.)

HERRE, H. 1961. The age of *Welwitschia bainesii* (Hook. f.) Carr.: C^{14} research. *J. S. Afr. Bot.*, 27, p. 139.

HESSE, P. R. 1955. A chemical and physical study of the soils of termite mounds in East Africa. *J. Ecol.*, 43, p. 449–61.

HLADIK, A. 1974. Importance des lianes dans la production foliaire de la forêt équatoriale du Nord-Est du Gabon. *C.r. hebd. Séanc. Acad. Sci., Paris*, sér. D, 278, p. 2527–30.

HLADIK, C. M.; HLADIK, A. 1967. Observations sur le rôle des primates dans la dissémination des végétaux de la forêt gabonaise. *Biol. gabon.*, 3, p. 43–57.

HNATIUK, R. J.; MERTON, L. F. H. 1979. A perspective of the vegetation of Aldabra. *Phil. Trans. R. Soc. Lond.*, ser. B, 286, p. 79–84.

HNATIUK, R. J., WOODELL, S. R. J.; BOURN, D. M. 1976. Giant tortoise and vegetation interactions on Aldabra atoll. II. Coastal. *Biol. Conserv.*, 9, p. 305–16.

HOCHREUTINER, B. P. G. 1904. Le Sud-oranais. Études floristiques et phytogéographiques. *Annu. Conserv. Jard. bot. Genève*, 7–8, p. 22–276.

HOCKING, B. 1970. Insect association with the swollen thorn acacias. *Trans. R. Entom. Soc. London*, 122, p. 211–55.

——. 1975. Ant–plant mutualism: evolution and energy. In: L. E. Gilbert; P. H. Raven (eds.), *Coevolution of animals and plants,* p. 78–90. Austin, London, Univ. of Texas Press.

HOLLAND, P. G.; HOVE, A. R. T. 1975. The distribution of *Euphorbia candelabrum* in the Southern Rift Valley, Kenya. *Vegetatio*, 30, p. 49–54.

HOLOWAYCHUK, N.; DENISOFF, I.; GILSON, P.; CROEGAERT, J.; LIBEN, L.; SPERRY, T. 1954. *Carte Sols Vég. Congo belge,* 4, *Nioka* (Ituri). A, Sols (1:50000), B, Végétation (1:10000; 1:50000). *Notice explicative,* p. 1–31. Brussels, INEAC.

HOPKINS, B. 1962. Vegetation of the Olokemeji Forest Reserve, Nigeria. I. General features of the Reserve and the Research Sites. *J. Ecol.*, 50, p. 559–98.

——. 1963. The role of fire in promoting the sprouting of some savanna species. *J. W. Afr. Sci. Ass.*, 7, p. 154–62.

——. 1965a. Vegetation of the Olokemeji Forest Reserve, Nigeria. II. The climate with special reference to its seasonal changes. *J. Ecol.*, 53, p. 109–24.

——. 1965b. Id. III. The microclimates with special reference to their seasonal changes. *J. Ecol.*, 53, p. 125–38.

——. 1965c. *Forest and savanna. An introduction to tropical plant ecology with special reference to West Africa.* 2nd ed. Ibadan, London, Heinemann. 154 p.

——. 1965d. Observations on savanna burning in the Olokemeji Forest Reserve, Nigeria. *J. appl. Ecol.*, 2, p. 367–81.

——. 1966. Vegetation of the Olokemeji Forest Reserve, Nigeria. IV. The litter and soil with special reference to their seasonal changes. *J. Ecol.*, 54, p. 687–703.

——. 1968. Id. V. The vegetation on the savanna site with special reference to its seasonal changes. *J. Ecol.*, 56, p. 97–115.

HORSCROFT, F. D. M. 1961. Vegetation. In: Mendelsohn, F. (ed.), *The geology of the Northern Rhodesian Copper Belt,* p. 73–80. London, Macdonald.

HOWARD-WILLIAMS, C. 1970. The ecology of *Becium homblei* in Central Africa with special reference to metalliferous soils. *J. Ecol.*, 58, p. 745–63.

——. 1975a. Seasonal and spatial changes in the composition of the aquatic and semi-aquatic vegetation of Lake Chilwa, Malawi. *Vegetatio*, 30, p. 33–9.

——. 1975b. Vegetation changes in a shallow African lake: response of the vegetation to a recent dry period. *Hydrobiol.*, 47, p. 381–98.

——. 1977. A check-list of the vascular plants of Lake Chilwa, Malawi, with special reference to the influence of environmental factors on the distribution of taxa. *Kirkia*, 10, p. 563–79.

HOWARD-WILLIAMS, C.; WALKER, B. H. 1974. The vegetation of a tropical African lake. Classification and ordination of the vegetation of Lake Chilwa (Malawi). *J. Ecol.*, 62, p. 831–54, with small-scale vegetation map.

HUMBERT, H. 1924. Végétation du Grand Atlas marocain oriental. Exploration botanique de l'Ari Ayachi. *Bull. Soc. Hist. nat. Afr. N.*, 15, p. 147–234.

——. 1927a. Sur la flore des hautes montagnes de Madagascar. *C.r. somm. Séanc. Soc. Biogéogr.*, 4 (31), p. 84–9.

——. La destruction d'une flore insulaire par le feu. Principaux aspects de la végétation à Madagascar. *Mém. Acad. malgache*, 5, p. 1–78.

——. 1928a. Végétation de l'Atlas saharien occidental et additions à l'Étude botanique de l'Ari Ayachi. *Bull. Soc. Hist. nat. Afr. N.*, 19, p. 204–40.

——. 1928b. Végétation des hautes montagnes de Madagascar. *Mém. Soc. Biogéogr.*, 2, p. 195–220.

——. 1938. Les aspects biologiques du problème des feux de brousse et la protection de la nature dans les zones intertropicales. *Inst. Roy. Colon. Belg. Bull. Séanc.*, 9, p. 811–35.

——. 1940a. Zones et étages de végétation dans le sud-ouest de l'Angola. I. Généralités. Zones littorale, désertique et de transition. *C.r. somm. Séanc. Soc. Biogéogr.*, 17 (147), p. 47–51.

——. 1940b. Id. II. Zones forestières. Tom. cit., p. 54–7.

——. 1955a. Une merveille de la nature à Madagascar. Première exploration botanique du massif du Marojejy et de ses satellites. *Mém. Inst. scient. Madagascar,* sér. B (Biol. vég.), 6, p. 1–210.

——. 1955b. Les territoires phytogéographiques de Madagascar. *Les divisions écologiques du monde,* p. 195–204. (Colloques Int. CNRS, 59.) Also published in: *Année biol.*, sér. 3, 31, p. 439–48, with coloured vegetation map, 1:3500000.

——. 1959. Origines présumées et affinités de la flore de Madagascar. *Mém. Inst. scient. Madagascar,* sér. B (Biol. vég.), 9, p. 149–87.

HUMBERT, H.; COURS DARNE, G. 1964–65. *Carte internationale du tapis végétal à 1:1 000000 en 3 feuilles de Madagascar.* Trav. Sect. Sci. Techn. Inst. Franç. Pondichéry, h.s., 6. (In colour.)

——; ——. 1965. *Notice de la carte Madagascar. Carte internationale du tapis végétal à 1:1000000.* Trav. Sect. Sci. Techn. Inst. Franç. Pondichéry, h.s., 6, p. 1–164.

HUMPHRIES, C. J. 1979. Endemism and evolution in Macaronesia. In: Bramwell, D. (ed.), p. 171–99.

HUNTING TECHNICAL SERVICES. 1958. *Jebel Marra investigations. Report on phase I studies.* London, Hunting Technical Services Ltd. 112 p.

——. 1964. *Land and water use survey in Kordofan Province of the Republic of the Sudan. Report on the survey of geology, geomorphology and soils, vegetation and present land use.* Athens, Doxiadis Associates. 349 p., with coloured vegetation map at 1:250000.

——. 1968. *Land and water resources survey of the Jebel Marra area. Reconnaissance vegetation survey.* Rome, FAO. With coloured vegetation map at 1:250000. (LA:SF/SUD/17.)

HUNTLEY, B. J. 1965. A preliminary account of the Ngoye Forest Reserve, Zululand. *J. S. Afr. Bot.*, 31, p. 177–205.

——. 1978. Ecosystem conservation in southern Africa. In: Werger, M. J. A. (ed.), p. 1333–84.

HUTCHINSON, J. 1946. *A botanist in Southern Africa.* London, Gawthorn. 686 p.

HUTCHINSON, J. B. (ed.). 1977. Resource development in semi-arid lands. *Phil. Trans. R. Soc. Lond.*, sér. B, 278, p. 439–55.

HUTTEL, C. 1969. Répartition verticale des racines dans une forêt dense humide sempervirente de basse Côte d'Ivoire. *J. W. Afr. Sci. Ass.*, 14, p. 65–72.

——. 1975a. Root distribution and biomass in three Ivory

Coast rain forest plots. In: Golley, F. B.; Medina, E. (eds.), *Tropical ecological systems: trends in terrestrial and aquatic research*, p. 123–30. Berlin, Heidelberg, New York, Springer-Verlag. 398 p. (Ecol. stud., no. 11.)

——. 1975*b*. Recherches sur l'écosystème de la forêt subéquatoriale de Basse Côte d'Ivoire. III. Inventaire et structure de la végétation ligneuse. *La Terre et la Vie*, 2, p. 178–91.

HUTTEL, C.; BERNHARD-REVERSAT, F. 1975. Recherches sur l'écosystème de la forêt subéquatoriale de Basse Côte d'Ivoire. V. Biomasse végétale et productivité primaire. Cycle de la matière organique. *La Terre et la Vie*, 2, p. 203–28.

HUXLEY, J. 1958. *The conservation of wild life and natural habitats in Central and East Africa*. Paris, Unesco. 113 p.

IFAN–DAKAR. 1971. Le massif des Monts Loma. 1. Recherches géographiques, botaniques et zoologiques. *Mém. IFAN*, 86, p. 1–417.

INGRAM, P. 1960. A preliminary account of the ecology of the vegetation of the South Matopos. I.C.A. *Rhod. agric. J.*, 57, p. 319–30, with large-scale vegetation map.

INNIS, A. C. 1958. The behaviour of the giraffe, *Giraffa camelopardalis*, in the eastern Transvaal. *Proc. Zool. Soc. Lond.*, 131, p. 254–78.

IONESCO, T.; SAUVAGE, C. 1962. Les types de végétation du Maroc. Essai de nomenclature et de définition. *Rev. Géogr. maroc.*, 1–2, p. 75–83.

——; ——. (1965–69). Fichier des espèces climax. *Al Awamia*, 16, p. 1–21; 20, p. 103–23; 32, p. 105–24.

IONESCO, T.; STEFANESCO, E. 1967. La cartographie de la végétation de la région de Tanger. *Al Awamia*, 22, p. 17–147, with 4 vegetation maps 1 : 100000.

IRBY, L. R. 1977. Food habits of Chanler's mountain reedbuck in a Rift valley ranch. *E. Afr. Wildl. J.*, 15, p. 289–94.

ISAAC, W. E.; ISAAC, F. M. 1968. Marine botany of the Kenya coast. 3. General account of the environment, flora and vegetation. *J. E. Afr. nat. Hist. Soc.*, 27, p. 7–28.

IUCN. 1964. *The ecology of man in the tropical environment/ L'écologie de l'homme dans le milieu tropical*. Morges, Switzerland. 355 p. (IUCN Publs, n.s., vol. 4.)

——. 1973. *A working system for classification of world vegetation prepared by the IUCN Secretariat with the guidance of the IUCN Commission on Ecology*. Morges, Switzerland. 20 p. (IUCN Occ. Pap. 5.)

JACKSON, G. 1954. Preliminary ecological survey of Nyasaland. *Proc. 2nd Inter-Afr. Soils Conf.*, p. 679–90.

——. 1964. Notes on West African vegetation. I. Mangrove vegetation at Ikorodu, Western Nigeria. *J. W. Afr. Sci. Ass.*, 9, p. 98–110.

——. 1968. The vegetation of Malawi. II. The *Brachystegia* woodlands. X. *Brachystegia* with evergreen understorey. *Soc. Malawi J.*, 21, p. 11–19.

——. 1969. The grasslands of Malawi. *Soc. Malawi J.*, 22, p. 7–77, 18–25, 73–81.

JACKSON, G.; GARTLAN, J. S. 1965. The flora and fauna of Lolui Island, Lake Victoria. A study of vegetation, men and monkeys. *J. Ecol.*, 53, p. 573–97.

JACKSON, I. J. 1977. *Climate, water and agriculture in the tropics*. London, New York, Longman. 248 p.

JACKSON, J. K. 1950. The Dongotona Hills, Sudan. *Emp. For. Rev.*, 29, p. 139–42. Also issued as *Sudan Govt Mem. For. Div.*, 1 (1950).

——. 1951. Mount Lotuke, Didinga Hills. *Sudan Notes Rec.*, 32, p. 339–41. Also issued as *Sudan Govt Mem. For. Div.*, 3 (1951).

——. 1956. The vegetation of the Imatong Mountains, Sudan. *J. Ecol.*, 44, p. 341–74.

JACKSON, J. K.; SHAWKI, M. K. 1950. Shifting cultivation in the Sudan. *Sudan Notes Rec.*, 31, p. 210–22. Reissued as: *Sudan Govt Mem. For. Div.*, 2.

JACKSON, S. P. 1962. *Atlas climatologique de l'Afrique*. Lagos, Nairobi, CCTA/CSA.

JACOBSEN, N. H. G.; DU PLESSIS, E. 1976. Observations on the ecology and biology of the Cape fruit bat *Rousettus aegyptiacus leachi* in the eastern Transvaal. *S. Afr. J. Sci.*, 72, p. 270–3.

JACOBSEN, W. B. G. 1967. The influence of the copper content of the soil on trees and shrubs of Molly South Hill, Mangula. *Kirkia*, 6, p. 63–84, with large-scale vegetation map.

——. 1968. The influence of the copper content of the soil on the vegetation at Silverside North, Mangula Area. *Kirkia*, 6, p. 259–77, with large-scale vegetation map.

——. 1970. Further notes on the vegetation of copper-bearing soils at Silverside. *Kirkia*, 7, p. 285–90.

——. 1973. A check-list and discussion of the flora of a portion of the Lomagundi District, Rhodesia. *Kirkia*, 9, p. 139–207.

JACOT GUILLARMOD, A. 1962. The bogs and sponges of the Basutoland mountains. *S. Afr. J. Sci.*, 58, p. 179–82.

——. 1963. Further observations on the bogs of the Basutoland mountains. *S. Afr. J. Sci.*, 59, p. 115–18.

——. 1968. Lesotho. In: Hedberg, I.; Hedberg, O. (eds.), p. 253–6.

——. 1969. The effects of land usage on aquatic and semi-aquatic vegetation at high altitudes in southern Africa. *Hydrobiologia*, 34, p. 3–13.

——. 1971. *Flora of Lesotho* (Basutoland). Lehre, Cramer. 474 p.

JACQUES-FÉLIX, H. 1945. Une réserve botanique à prévoir au Cameroun. Le sommet des Monts Bambutos. *Bull. Mus. natn. Hist. nat., Paris*, sér. 2, 17, p. 506–13.

——. 1971. Compte rendu d'un voyage au Cameroun. Reconnaissance d'un étage montagnard méconnu de l'Adamaoua. *Mitt. Bot. Staatssamml. Münch.*, 10, p. 341–53.

JAEGER, P. 1949. La végétation. In: La presqu'île du Cap-Vert. *Étud. sénégal.*, 1, p. 93–140.

——. 1950. Aperçu sommaire de la végétation du massif de Kita (Soudan français). *Rev. int. Bot. appl. Agric. trop.*, 30, p. 501–6.

——. 1956. Contribution à l'étude des forêts reliques du Soudan occidental. *Bull. IFAN*, sér. A, 18, p. 993–1053.

——. 1959. Les plateaux gréseux du Soudan occidental. Leur importance phytogéographique. *Bull. IFAN*, sér. A, 21, p. 1147–59.

——. 1965*a*. Espèces végétales de l'étage altitudinal des monts Loma (Sierra Leone). *Bull. IFAN*, sér. A, 27, p. 34–120.

——. 1965*b*. Sur l'endémisme dans les plateaux soudanais ouest-africains. *C.r. somm. Séanc. Soc. Biogéogr.*, 42 (368), p. 38–48.

——. 1966. La prairie d'altitude des monts Loma (Sierra Leone). Distribution et structure. *C.r. hebd. Séanc. Acad. Sci., Paris*, sér. D, 263, p. 1089–91.

——. 1968. Mali. In: Hedberg, I.; Hedberg, O. (eds.), p. 51–3.

——. 1969. Première esquisse d'une étude bioclimatique des monts Loma (Sierra Leone). *Bull. IFAN*, sér. A, 31, p. 1–21.

——. 1976. Le massif des monts Loma (Sierra Leone). Son importance phytogéographique. *Boissiera*, 24, p. 473–5.

JAEGER, P.; ADAM, J. G. 1947. Aperçu sommaire sur la végétation de la région occidentale de la dorsale Loma–

Man. La galerie forestière de la source du Niger. *Bull. Soc. bot. Fr.*, 94, p. 323–37.

——; ——. 1950. Sur les îlots forestiers du Haut-Niger. *Conf. Int. Afr. Occid. em Bissau, 1947*, 2 (1*a*), p. 217–21.

——; ——. 1967. Sur la présence en piedmont ouest des monts Loma (Sierra Leone) d'un groupement forestier rélictuel à *Tarrietia utilis* Sprague (Sterculiacées). *C.r. hebd. Séanc. Acad. Sci., Paris*, sér. D, 265, p. 1627–9.

——; ——. 1971. Aperçu succinct sur la flore et la végétation de l'étage culminal des monts Loma (Sierra Leone). *Mitt. Bot. Staatssamml. Münch.*, 10, p. 478–83.

——; ——. 1972. Contribution à l'étude de la végétation des Monts Loma (Sierra Leone). *C.r. somm. Séanc. Soc. Biogéogr.*, 424, p. 77–103.

——; ——. 1975. Les forêts de l'étage culminal du Nimba libérien. *Adansonia*, sér. 2, 15, p. 177–88.

JAEGER, P.; JAROVOY, M. 1952. Les grès de Kita (Soudan occidental); leur influence sur la répartition du peuplement végétal. *Bull. IFAN*, 14, p. 1–18.

JAEGER, P.; LAMOTTE, M.; ROY, R. 1966. Les richesses floristiques et faunistiques des monts Loma (Sierra Leone). Urgence de leur protection intégrale. *Bull. IFAN*, sér. A, 28, p. 1149–90.

——; ——; ——. (eds.). 1971. Le massif des Monts Loma (Sierra Leone), fasc. 1. *Mém. IFAN*, 86, p. 1–417.

JAEGER, P.; WINKOUN, D. 1962. Premier contact avec la flore et la végétation du plateau de Bandiagara. *Bull. IFAN*, sér. A, 24, p. 69–111.

JAMAGNE, M. 1965. *Carte Sols Vég. Congo, Rwanda, Burundi*, 19, *Maniema*. A, Sols (1:50000; 1:250000). *Notice explicative*, p. 1–132. Brussels. INEAC.

JAMESON, J. D. (ed.) 1970. *Agriculture in Uganda*. London, Oxford Univ. Press. 395 p.

JANZEN, D. H. 1972. Protection of *Barteria* (Passifloraceae) by *Pachysima* ants (Pseudomyrmecinae) in a Nigerian rainforest. *Ecology*, 53, p. 885–92.

JARMAN, P. J. 1971. Diets of large mammals in the woodlands around Lake Kariba, Rhodesia. *Oecologia* (Berl.), 8, p. 157–78.

——. 1976. Damage to *Acacia tortilis* seeds eaten by Impala. *E. Afr. Wildl. J.*, 14, p. 223–6.

JEANNEL, R. 1950. *Hautes montagnes d'Afrique*. Publs. Mus. natn. Hist. nat., Paris, Suppl. 1. 253 p.

JEFFERS, J. N. R.; BOALER, S. B. 1966. Ecology of a miombo site, Lupa North Forest Reserve, Tanzania. I. Weather and plant growth, 1962–64. *J. Ecol.*, 54, p. 447–63.

JEFFREY, C. 1963. *The botany of the Seychelles*. London, Dept Techn. Co-op. 22 p.

——. 1968. Seychelles. In: Hedberg, I.; Hedberg, O. (eds.), p. 275–9.

JENÍK, J. 1968. Wind action and vegetation in tropical West Africa. In: Misra, R.; Gopal, B. (eds.), *Proc. Symp. recent Adv. trop. Ecol.*, vol. 1, p. 108–13. Pub. International Society for Tropical Ecology, Dep. Botany, Banares Hindu University.

——. 1970. The pneumatophores of *Voacanga thouarsii* Roem. & Schult. (Apocynaceae). *Bull. IFAN*, sér. A, 32, p. 986–94.

——. 1971. Root structure and underground biomass in equatorial forests. In: Duvigneaud, P. (ed.), p. 323–9.

JENÍK, J.; HALL, J. B. 1966. The ecological effects of the harmattan wind in the Djebobo Massif (Togo Mountains, Ghana). *J. Ecol.*, 54, p. 767–79.

——; ——. 1969. The dispersal of *Detarium microcarpum* by elephants. *Nig. Fld*, 34, p. 39–42.

——; ——. 1976. Plant communities of the Accra Plains, Ghana. *Folia geobot. phytotax., Praha*, 11, p. 163–212.

JENÍK, J.; LAWSON, G. W. 1968. Zonation of microclimate and vegetation on a tropical shore in Ghana. *Oikos*, 19, p. 198–205.

JENKIN, R. N.; HOWARD, W. J.; THOMAS, P.; ABELL, T. M. B.; DEANE, G. C. 1977. Vegetation. In: *Forestry development prospects in the Imatong Central Forest Reserve, Southern Sudan*, p. 61–78. With vegetation map 1:50000. Tolworth Tower, Surbiton, Surrey, Directorate of Overseas Surveys. (Land resource study 28, vol. 2.)

JOHANSSON, D. 1974. Ecology of vascular epiphytes in West African rain forest. *Acta phytogeogr. suec.*, 59, p. 1–129.

JONES, A. P. D. 1950. *The Natural Forest Inviolate Plot, Akure Forest Reserve, Nigeria*. Ibadan, Nigerian Govt Forest Dept. 33 p.

JONES, E. W. 1950. Some aspects of natural regeneration in the Benin rain forest. *Emp. For. Rev.*, 29, p. 108–24.

——. 1955–56. Ecological studies on the rain forest of Southern Nigeria. IV. The plateau forest of the Okomu Forest Reserve. *J. Ecol.*, 43, p. 564–94 (1955); 44, p. 83–117 (1956).

——. 1963*a*. The forest outliers in the Guinea zone of Northern Nigeria. *J. Ecol.*, 51, p. 415–34.

——. 1963*b*. The Cece Forest Reserve, Northern Nigeria. *J. Ecol.*, 51, p. 461–6.

JONES, M. J.; WILD, A. 1975. *Soils of the West African savanna*. Harpenden, Commonwealth Bureau of Soils. 246 p. (Tech. Comm., 55.)

JONGEN, P. 1968. *Carte Sols Vég. Congo, Rwanda, Burundi*, 22, *Ubangi*. A, Sols (1:50000; 1:100000). *Notice explicative*, p. 1–39. Brussels, INEAC.

JONGEN, P.; JAMAGNE, M. 1966. *Carte Sols Vég. Congo, Rwanda, Burundi*, 20, *Région Tshuapa–Équateur*. A, Sols (1:50000; 1:100000). *Notice explicative*, p. 1–82. Brussels, INEAC.

JONGEN, P.; LECLERCQ, J.; SOBERON, M. 1970. *Carte Sols Vég. Congo, Rwanda, Burundi*, 26, *Nord-Kivu et Région du Lac Édouard*. A, Sols (1:50000; 1:200000). *Notice explicative*, p. 1–78. Brussels, INEAC.

JONGEN, P.; VAN COSTEN, M.; ÉVRARD, C.; BERCE, J.-M. 1960. *Carte Sols Vég. Congo belge*, 11, *Ubangi*. A, Sols (1:50000). B, Végétation (1:100000). *Notice explicative*, p. 1–82. Brussels, INEAC.

JONGLEI INVESTIGATION TEAM. 1954. *The Equatorial Nile Project and its effects in the Anglo–Egyptian Sudan*. Khartoum, Govt Printer.

JORDAN, H. D. 1964. The relation of vegetation and soil to development of mangrove swamps for rice growing in Sierra Leone. *J. appl. Ecol.*, 1, p. 209–12.

KAHN, F. 1977. Analyse structurale des systèmes racinaires des plantes ligneuses de la forêt tropicale dense humide. *Candollea*, 32, p. 321–58.

——. 1980. Comportements racinaire et aérien chez les plantes ligneuses de la forêt tropicale humide (sud-ouest de la Côte d'Ivoire). *Adansonia*, sér. 2, 19, p. 413–27.

KALK, M.; MCLACHLAN, A. G.; HOWARD-WILLIAMS, C. (eds.). 1979. *Lake Chilwa: studies of change in a tropical ecosystem*. The Hague, Junk. 462 p. (Monogr. biol. 35.)

KÄMMER, F. 1974. Klima und Vegetation auf Tenerife, besonders im Hinblick auf den Nebelniederschlag. *Scripta geobot.* (Göttingen), 7, p. 1–78.

——. 1976. The influence of man on the vegetation of the island of Hierro (Canary Islands). In: Künkel, G. (ed.), p. 327–46.

KASSAS, M. 1952. Habitat and plant communities in the Egyptian desert. I. Introduction. *J. Ecol.*, 40, p. 342–51.
——. 1953*a*. Landforms and plant cover in the Egyptian desert. *Bull. Soc. Géogr. Égypte*, 26, p. 193–205.
——. 1953*b*. Habitat and plant communities in the Egyptian desert. II. The features of a desert community. *J. Ecol.*, 41, p. 248–56.
——. Land forms and plant cover in the Omdurman Desert, Sudan. *Bull. Soc. Géogr. Égypte*, 29, p. 43–58.
——. 1956*b*. The Mist Oasis of Erkowit, Sudan. *J. Ecol.*, 44, p. 180–94.
——. 1957. On the ecology of the Red Sea coastal land. *J. Ecol.*, 45, p. 187–203.
KASSAS, M.; EL-ABYAD, M.S. 1962. On the phytosociology of the desert vegetation of Egypt. *Ann. Arid Zone*, 1, p. 54–83.
KASSAS, M.; GIRGIS, W. A. 1964. Habitat and plant communities in the Egyptian desert. V. The limestone plateau. *J. Ecol.*, 52, p. 107–19.
——; ——. 1965. Id. VI. The units of a desert ecosystem. *J. Ecol.*, 53, p. 715–28.
——; ——. 1970. Id. VII. Geographical facies of plant communities. *J. Ecol.*, 58, p. 335–50.
KASSAS, M.; IMAM, M. 1954. Id. III. The wadi bed ecosystem. *J. Ecol.*, 42, p. 424–41.
——; ——. 1959. Id. IV. The gravel desert. *J. Ecol.*, 47, p. 289–310.
KASSAS, M.; ZAHRAN, M. A. 1962. Studies on the ecology of the Red Sea coastal land. I. The district of Gebel Ataqa and El-Galada El-Bahariya. *Bull. Soc. Géogr. Égypte*, 35, p.129–75.
——; ——. 1965. Id. II. The district from El-Galaba El-Qibliya to Hurghada. *Bull. Soc. Géogr. Égypte*, 38, p. 155–93.
——; ——. 1967. On the ecology of the Red Sea littoral salt marsh, Egypt. *Ecol. Monogr.*, 37, p. 297–315.
KAUL, R. N. (ed.). 1970. *Afforestation in arid zones*. The Hague, Junk. 435 p. (Monogr. biol., 20.)
KEAY, R. W. J. 1947. Notes on the vegetation of Old Oyo Forest Reserve. *Farm and Forest*, 8, p. 36–46.
——. 1948–59*a. An outline of Nigerian vegetation*. Lagos, Govt Printer. 52 p., with coloured vegetation map 1:3 000 000 (1948). 2nd ed., 55 p. (1953). 3rd ed. (minor corrections only), 1959*a*.
——. 1949. An example of Sudan zone vegetation in Nigeria. *J. Ecol.*, 37, p. 335–64.
——. 1951. Some notes on the ecological status of savanna vegetation in Nigeria. *Bull. Commonw. Bur. Past. Fld Crops*, 41, p. 57–68.
——. 1952 [1953]. *Isoberlinia* woodlands in Nigeria and their flora. *Lejeunia*, 16, p. 17–26.
——. 1955. Montane vegetation and flora in the British Cameroons. *Proc. Linn. Soc. Lond.*, 165, p. 140–3, with small vegetation map.
——. 1957. Wind-dispersed species in a Nigerian forest. *J. Ecol.*, 45, p. 471–8.
——. 1959*b. Vegetation map of Africa south of the tropic of Cancer. Explanatory notes*. With French transl. by A. Aubréville. Oxford Univ. Press. 24 p., with coloured map 1:10000000. See also Aubréville et al., 1959.
——. 1959*c*. Derived savanna—derived from what? *Bull. IFAN*, sér. A, 21, p. 427–38.
——. 1959*d*. Lowland vegetation on the 1922 lava flow, Cameroons Mountain. *J. Ecol.*, 47, p. 25–9.
——. 1960. An example of Northern Guinea Zone vegetation in Nigeria. *Nig. For. Inf. Bull.*, n.s., 1, p. 1–46, with large-scale coloured vegetation map.
——. 1962. Yankari Game Reserve, Bauchi Province, Northern Nigeria. *Tech. Note*, 17, p. 1–19. Ibadan, Nigeria, Dep. For. Res.
——. 1979. A botanical study of the Obudu Plateau and Sonkwala Mountains. *Nig. Fld*, 44, p. 106–19.
KEAY, R. W. J.; ONOCHIE, C. F. A. 1947. Some observations on the Sobo Plains. *Farm and Forest*, 8, p. 71–80.
KEET, J. D. M. 1950. Forests of the Okavango Native Territory. *J. S. Afr. For. Ass.*, 19, p. 77–88.
KELLY, R. D.; WALKER, B. H. 1976. The effects of different forms of land use on the ecology of a semi-arid region in South-Eastern Rhodesia. *J. Ecol.*, 64, p. 553–76.
KEMP, R. H. 1963. Growth and regeneration of open savanna woodland in Northern Nigeria. *Commonw. For. Rev.*, 42, p. 200–6.
KENNAN, T. C. D. 1972. The effects of fire on two vegetation types at Matopos, Rhodesia. *Proc. Ann. Tall Timbers Fire Ecol. Conf.*, 11, p. 53–98.
KENNAN, T.C.D.; STAPLES, R.R.; WEST, O. 1955. Veld management in Southern Rhodesia. *Rhod. Agr. J.*, 52, p. 4–21.
KERAUDREN, M. 1968. Madagascar. In: Hedberg, I.; Hedberg, O. (eds.), p. 261–5.
KERAUDREN-AYMONIN, M. 1971. Quelques remarques à propos des Cucurbitacées des flores sèches. *Mitt. Bot. Staatssamml. Münch.*, 10, p. 449–57.
KERFOOT, O. 1963. The root systems of tropical forest trees. *Commonw. For. Rev.*, 42, p. 19–26.
——. 1964*a*. The vegetation of the South-West Mau Forest. *E. Afr. agric. For. J.*, 29, p. 295–318.
——. 1964*b*. A preliminary account of the vegetation of the Mbeya Range, Tanganyika. *Kirkia*, 4, p. 191–206.
——. 1965. The vegetation of an experimental catchment in the semi-arid ranchland of Uganda. *E. Afr. agric. For. J.*, 30, p. 227–45.
——. 1968. Mist precipitation on vegetation. *For. Abstr.* 29, p. 8–20.
KERR, N. 1971. *Report on a preliminary Nature Conservation Project, Island of St. Helena, July/August 1970*. (IBP/4 (71).) (Mimeo.)
KERS, L. E. 1967. The distribution of *Welwitschia mirabilis* Hook. f. *Svensk bot. Tidskr.*, 61, p. 97–125.
KERSHAW, K. A. 1968*a*. A survey of the vegetation in Zaria Province, N. Nigeria. *Vegetatio*, 15, p. 244–68.
——. 1968*b*. Classification and ordination of Nigerian savanna vegetation. *J. Ecol.*, 56, p. 467–82.
KILEY, M. 1966. A preliminary investigation into the feeding habits of the waterbuck by faecal analysis. *E. Afr. Wildl. J.*, 4, p. 153–7.
KILLIAN, C. 1941. Sols et plantes indicatrices dans les parties non irriguées des oasis de Figuig et de Beni-Ounif. *Bull. Soc. Hist. nat. Afr. N.*, 32, p. 301–14.
——. 1942. Sols de forêt et sols de savane en Côte d'Ivoire. Leurs caractères pédologiques, chimiques, physiques et microbiologiques. *Annls agron.*, 12, p. 600–52.
——. 1951. Mesures écologiques sur des végétaux types du Fouta Djallon (Guinée Française) et sur leur milieu, en saison sèche. *Bull. IFAN*, 13, p. 601–81.
KILLIAN, C.; SCHNELL, R. 1947. Contribution à l'étude des formations végétales et des sols humifères correspondants des massifs du Benna et du Fouta-Djallon (Guinée française). *Revue can. Biol.*, 6, p. 379–435.
KILLIAN, J. 1961. Contribution à l'étude phytosociologique du Grand Erg oriental. *Terres et Eaux*, 37, p. 46–64.

KILLICK, D. J. B. 1959. An account of the plant ecology of the Table Mountain area of Pietermaritzburg, Natal. *Mem. bot. Surv. S. Afr.*, 32, p. 1–133, with vegetation map 1:12500.
——. 1963. An account of the plant ecology of the Cathedral Peak area of the Natal Drakensberg. *Mem. bot. Surv. S. Afr.*, 34, p. 1–178.
——. 1968. Transvaal. In: Hedberg, I.; Hedberg, O. (eds.), p. 239–43.
——. 1978a. Notes on the vegetation of the Sani Pass area of the southern Drakensberg. *Bothalia*, 12, p. 537–42.
——. 1978b. Further data on the climate of the Alpine vegetation belt of eastern Lesotho. *Bothalia*, 12, p. 567–72.
——. 1978c. The Afroalpine Region. In: Werger, M. J. A. (ed.), p. 515–60.
——. 1979. African mountain heathlands. In: Specht, R. L. (ed.), p. 97–116.
KILLICK, H. J. 1959. The ecological relationships of certain plants in the forest and savanna of Central Nigeria. *J. Ecol.*, 47, p. 115–27.
KINAKO, P. D. S. 1977. Conserving the mangrove forest of the Niger Delta. *Biol. Conserv.*, 11, p. 35–9.
KING, J. M.; HEATH, B. R. 1976. Game domestication for animal production in Kenya. Experiences at the Galana Ranch. *Wld Anim. Rev.*, 16, p. 23–30.
KING, L. C. 1967a. *South African scenery*. 3rd ed. Edinburgh, Oliver & Boyd.
——. 1967b. *The morphology of the earth*. 2nd ed. Edinburgh, Oliver & Boyd. 726 p.
——. 1978. The geomorphology of central and southern Africa. In: Werger, M. J. A. (ed.), p. 1–17.
KINGDON, J. 1971–77. *East African mammals: an atlas of evolution in Africa*, vol. 1, p. 1–446 (1971); vol. 2a, p. 1–341 + i–l (1974); vol. 2b, p. 342–704 + i–lvi (1974); vol. 3a, p. 1–476 (1977). London, New York, San Francisco, Academic Press.
KINLOCH, J. B. 1939. The use of the term 'deciduous' as applied to tropical forests. *Emp. For. J.*, 18, p. 247–51.
KLAUS, D.; FRANKENBERG, P. 1979. Statistical relationships between floristic composition and mean climatic conditions in the Sahara. *J. Biogeogr.*, 6, p. 391–405.
KLÖTZLI, F. 1958. Zur Pflanzensoziologie des Südhanges der alpinen Stufe des Kilimandscharo. *Ber. geobot. Forsch. Inst. Rübel.*, 1957, p. 33–58.
——. 1975. Zur Waldfähigkeit der Gebirgssteppen Hoch-Semiens (Nordäthiopien). *Beitr. naturk. Forsch. Südw. Dtl.*, 34, p. 131–47.
KNAPP, R. 1968a. Probleme der Vegetationskartierung in den Tropen mit Beispielen aus Kamerun, Äthiopien und Madagaskar. *Ber. oberhess. Ges. Nat. u. Heilk.*, N.F. (Naturw. Abt.), 36, p. 81–94.
——. 1968b. Vegetation und Landnutzung in Süd-Tunesien. *Ber. oberhess. Ges. Nat. u. Heilk.*, N.F. (Naturw. Abt.), 36, p. 103–24.
——. 1973. *Die Vegetation von Afrika*. 626 p., with several small-scale vegetation maps. Stuttgart, Gustav Fischer. (Vegetationsmonogr. einzelnen Grossraüme (ed. H. Walter), vol. 3.)
KOECHLIN, J. 1957. Morphoscopie des sables et végétation dans la région de Brazzaville. *Bull. Inst. Étud. centrafr.*, n.s., 13–14, p. 39–48.
——. 1961. La végétation des savanes dans le sud de la République du Congo (Capitale Brazzaville). *Mém. ORSTOM*, 1, p. 1–310, with vegetation map 1:1000000.
——. 1968. Sur la signification des formations graminéennes à Madagascar et dans le monde tropical. *Ann. Univ. Madagascar*, sér. Sci. Techn., 6, p. 211–34.
——. 1972. Flora and vegetation of Madagascar. In: Battistini, R.; Richard-Vindard, G. (eds.), p. 145–89, with small vegetation map.
KOECHLIN, J.; GUILLAUMET, J. L.; MORAT, P. 1974. *Flore et végétation de Madagascar*. Vaduz, Cramer. 687 p.
KOMAREK, E. V., Sr. 1964. The natural history of lightning. *Proc. Ann. Tall Timbers Fire Ecol. Conf.*, 3, p. 139–83.
——. 1972. Lightning and fire ecology in Africa. *Proc. Ann. Tall Timbers Fire Ecol. Conf.*, 11, p. 473–511.
KONCZACKI, Z. A. 1978. *The economics of pastoralism: a case study of sub-Saharan Africa*. London, Frank Cass. 185 p.
KORIBA, K. 1958. On the periodicity of tree growth in the tropics, with reference to the mode of branching, the leaf-fall, and the formation of the resting bud. *Gdns Bull.* (Singapore), 17, p. 11–81.
KORNAŚ, J. 1977. Life-forms and seasonal patterns in the pteridophytes in Zambia. *Acta Soc. Bot. Poloniae*, 46, p. 669–90.
——. 1978. Fire-resistance in the pteridophytes of Zambia. *Fern Gaz.*, 11, p. 373–84.
——. 1979. *Distribution and ecology of the pteridophytes in Zambia*. Warsaw, Krakow, Państwowe Wydawnictwo Naukowe. 207 p.
KORTLANDT, A. 1976. Tree destruction by elephants in Tsavo National Park and the role of man in African ecosystems. *Neth. J. Zool.*, 26, p. 449–51.
KOWAL, J. M.; KASSAM, A. H. 1978. *Agricultural ecology of savanna. A study of West Africa*. Oxford, Clarendon Press. 403 p.
KOZLOWSKI, T. T.; AHLGREN, C. E. 1974. *Fire and ecosystems*. New York, San Francisco, London, Academic Press. 542 p.
KRUGER, C. (ed.). 1967. *Sahara*. Vienna, Munich, Anton Schroll. 184 p.
KRUGER, F. J. 1977a. Ecology of Cape fynbos in relation to fire. In: Mooney, H. A.; Conrad, C. E. (eds.), p. 230–44.
——. 1977b. Ecological reserves in the Cape fynbos—towards a strategy for conservation. *S. Afr. J. Sci.*, 73, p. 81–5.
——. 1977c. A preliminary account of the aerial plant biomass in fynbos communities of the Mediterranean-type climate zone of the Cape Province. *Bothalia*, 12, p. 301–7.
——. 1979. South African heathlands. In: Specht, R. L. (ed.), p. 19–80.
KRUGER, F. J.; TAYLOR, H. C. 1979. Plant species diversity in Cape fynbos: gamma and delta diversity. *Vegetatio*, 41, p. 85–93.
KÜCHLER, A. W. 1947. A geographic system of vegetation. *Geogrl Rev.* (New York), 37, p. 233–40.
——. 1949. A physiognomic classification of vegetation. *Annls Ass. Am. Geogr.*, 39, p. 201–10.
——. 1950. Die physiognomische Kartierung der Vegetation. *Petermanns geogr. Mitt.*, 94, p. 1–6.
——. 1951. The relation between classification and mapping vegetation. *Ecology*, 32, p. 275–83.
——. 1960. Vegetation mapping in Africa. *Annls Ass. Am. Geogr.*, 50, p. 74–84.
——. 1967. *Vegetation mapping*. New York, N.Y., Ronald Press. 472 p.
——. 1970. International bibliography of vegetation maps. 4, *Africa, South America and World maps*. Lawrence, Kansas, University of Kansas Libraries. 561 p.
——. 1973. Problems in classifying and mapping vegetation for ecological regionalization. *Ecology*, 54, p. 512–23.
KÜNKEL, G. 1962. Aufzeichnungen über die Vegetation des Mt.

Wutivi-Gebiets (Nordwest-Liberia), unter besonderer Berücksichtigung der Pteridophyten. *Willdenowia*, 3, p. 151–68.

——. 1964. Die Vegetationsverhältnisse am Cape Mount (Liberia, Westafrika). *Willdenowia*, 3, p. 641–52.

——. 1966*a*. Über die Struktur und Sukzession der Mangrove Liberias und deren Rand-formationen. *Ber. schweiz. bot. Ges.*, 75, p. 20–40.

——. 1966*b*. Anmerkungen über Sekundärbusch und Sekundärwald in Liberia (Westafrika). *Vegetatio*, 13, p. 233–48.

——. 1971. On some floristic relationships between the Canary Islands and neighbouring Africa. *Mitt. Bot. Staatssamml. Münch.*, 10, p. 368–74.

——. (ed.). 1976. *Biogeography and ecology in the Canary Islands*. The Hague, Junk. 511 p. (Monogr. biol. 30.)

LAINS E SILVA, H. 1958. *São Tomé e Príncipe e a cultura do Café*. Lisbon. 499 p., with coloured vegetation maps, 1:100000 (S. Tomé), 1:200000 (Príncipe). (Mems Junta Invest. Ultram., vol. 1.)

LAMOTTE, M. 1967. Recherches écologiques dans la savane de Lamto (Côte d'Ivoire). Présentation du milieu et du programme de travail. *La Terre et la Vie*, 114, p. 197–215.

——. 1975*a*. Observations préliminaires sur les flux d'énergie dans un écosystème herbacé tropical, la savane de Lamto (Côte d'Ivoire). *Geo-Eco-Trop.*, 1, p. 45–63.

——. 1975*b*. The structure and function of a tropical savanna ecosystem. In: Golley, F. B.; Medina, E. (eds.), *Tropical ecological systems: trends in terrestrial and aquatic research*, p. 179–222. Berlin, Heidelberg, New York, Springer Verlag. 398 p. (Ecol. stud., no. 11.)

——. 1979. Structure and functioning of the savanna ecosystems of Lamto (Ivory Coast). In: Unesco/UNEP/FAO, p. 511–61.

LAMOTTE, M.; AGUESSE, P.; ROY, R. 1962. Données quantitatives sur une biocoenose ouest-africaine. La prairie montagnarde du Nimba (Guinée). *La Terre et la Vie*, 4, p. 351–70.

LAMPREY, H. 1963. Ecological separation of the large mammal species in the Tarangire Game Reserve, Tanganyika. *E. Afr. Wildl. J.*, 1, p. 63–92.

——. 1964. Estimation of the large mammal densities, biomass and energy exchange in the Tarangire Game Reserve and the Masai Steppe in Tanganyika. *E. Afr. Wildl. J.*, 2, p. 1–46.

——. 1967. Notes on the dispersal and germination of some tree seeds through the agency of mammals and birds. *E. Afr. Wildl. J.*, 5, p. 179–80.

——. 1975. *Report on the Desert Encroachment Reconnaissance in Northern Sudan, November 1975*. Unesco/UNEP. 16 p.

——. 1978. The integrated project on arid lands (IPAL)/Le projet intégré sur les terres arides (IPAL). *Nature and Resources/Nature et ressources*, 14, no. 4, p. 2–12.

——. 1979. Structure and functioning of the semi-arid grazing land ecosystem of the Serengeti region (Tanzania). In: Unesco/UNEP/FAO, p. 562–601.

LAMPREY, H.; GLOVER, P. E.; TURNER, M.I.M.; BELL, R.H.V. 1967. Invasion of the Serengeti National Park by elephants. *E. Afr. Wildl. J.*, 5, p. 151–66.

LAMPREY, H.; HALEVY, G.; MAKACHA, S. 1974. Interactions between *Acacia*, bruchid seed beetles and large herbivores. *E. Afr. Wildl. J.*, 12, p. 81–5.

LANG, H.; BREMEKAMP, C. E. B. 1935. Views of the vegetation of the Central Kalahari. *Ann. Transv. Mus.*, 16, p. 457–8 (with plates 12–21).

LANG, P. D. 1972. The vegetation of the Insiza and Shangani Intensive Conservation Areas. *Rhod. agric. J.*, 49, p. 346–51, with small vegetation map.

LANG BROWN, J. R.; HARROP, R. F. 1962. The ecology and soils of the Kibale grasslands, Uganda. *E. Afr. agric. For. J.*, 27, p. 264–72.

LANGDALE-BROWN, I. 1959*a*. The vegetation of the Eastern Province of Uganda. *Mem. Res. Div.*, ser. 2, 1, p. 1–154. Dep. Agric., Uganda. (Mimeo.)

——. 1959*b*. The vegetation of Buganda. *Mem. Res. Div.*, ser. 2, 2, p. 1–84. Dep. Agric., Uganda. (Mimeo.)

——. 1960*a*. The vegetation of the West Nile, Acholi, and Lango Districts of the Northern province of Uganda. *Mem. Res. Div.*, ser. 2, 3, p. 1–106. Dep. Agric., Uganda. (Mimeo.)

——. 1960*b*. The vegetation of the Western Province of Uganda. *Mem. Res. Div.*, ser. 2, 4, p. 1–111. Dep. Agric., Uganda. (Mimeo.)

——. 1960*c*. The vegetation of Uganda (excluding Karamoja). *Mem. Res. Div.*, ser. 2, 6, p. 1–45. Dep. Agric., Uganda. (Mimeo.)

LANGDALE-BROWN, I.; OSMASTON, H. A.; WILSON, J. G. 1964. *The vegetation of Uganda and its bearing on land-use*. Entebbe, Govt Printer. 159 p., with coloured vegetation map 1:500000.

LANGLANDS, B.W. 1967. Burning in Eastern Africa with particular reference to Uganda. *E. Afr. Geogr Rev.*, 5, p. 21–37.

LANLY, J. P. 1966. La forêt dense centrafricaine. *Bois Forêts Trop.*, 108, p. 43–55.

——. 1969. Régression de la forêt dense en Côte d'Ivoire. *Bois Forêts Trop.*, 127, p. 45–59.

LATHAM, M.; DUGERDIL, M. 1970. Contribution à l'étude de l'influence du sol sur la végétation au contact forêt–savane dans l'ouest et le centre de la Côte d'Ivoire. *Adansonia*, sér. 2, 10, p. 553–76.

LAUGHTON, F. S. 1937. The silviculture of the indigenous forests of the Union of South Africa with special reference to the forests of the Knysna region. *Sci. Bull. Dep. Agric. For. Un. S. Afr.*, 157, p. 1–168.

LAVAUDEN, L. 1927. Les forêts du Sahara. *Rev. Eaux Forêts*, 65, p. 265–77, 329–41.

——. 1928. La forêt de Gommiers du Bled Talha (Sud-tunisien). *Rev. Eaux Forêts*, 66, p. 699–713.

LAVRANOS, J. J. 1975. Note on the northern temperate element in the flora of the Ethio–Arabian Region. *Boissiera*, 24, p. 67–9.

——. 1978. On the Mediterranean and western-Asiatic floral element in the area of the Gulf of Aden. *Bot. Jb. Syst.*, 99, p. 152–67.

LAWS, R. M. 1970*a*. Elephants as agents of habitat and landscape change in East Africa. *Oikos*, 21, p. 1–15.

——. 1970*b*. Elephants and habitats in North Bunyoro, Uganda. *E. Afr. Wildl. J.*, 8, p. 163–80.

LAWS, R. M.; PARKER, I. S. C.; JOHNSTONE, R. C. B. 1975. *Elephants and their habitats*. Oxford, Clarendon Press. 376 p.

LAWSON, G. W. 1956. Rocky shore zonation on the Gold Coast. *J. Ecol.*, 44, p. 153–70.

——. 1966. *Plant life in West Africa*. London, Accra, Ibadan, Oxford Univ. Press. 150 p.

——. 1968. Ghana. In: Hedberg, I.; Hedberg, O. (eds.), p. 81–6, with small vegetation map.

LAWSON, G. W.; ARMSTRONG-MENSAH, K. O.; HALL, J. B. 1970. A catena in tropical moist semi-deciduous forest near Kade, Ghana. *J. Ecol.*, 58, p. 371–98.

LAWSON, G. W.; JENÍK, J. 1967. Observations on microclimate and vegetation inter-relationships on the Accra Plains (Ghana). *J. Ecol.*, 55, p. 773–85.

LAWSON, G. W.; JENÍK, J.; ARMSTRONG-MENSAH, K. O. 1968. A study of a vegetation catena in Guinea savanna at Mole Game Reserve (Ghana). *J. Ecol.*, 56, p. 505–22.

LAWTON, R. M. 1963. Palaeoecological and ecological studies in the Northern Province of Northern Rhodesia. *Kirkia*, 3, p. 46–77.

——. 1964. The ecology of the *Marquesia acuminata* (Gilg) R.E. Fr. evergreen forests and the related chipya vegetation types of North-Eastern Rhodesia. *J. Ecol.*, 52, p. 467–79.

——. 1967a. The conservation and management of the riparian evergreen forests of Zambia. *Commonw. For. Rev.*, 46, p. 223–32.

——. 1967b. Bush encroachment in Zambia. *Pest Artic. News Summ.*, Sect. C, 13, p. 335–53.

——. 1972. An ecological study of miombo and chipya woodland with particular reference to Zambia. Bodleian Library, Oxford Univ. 145 p. (Unpub. D.Phil. thesis.)

——. 1978a. The management and regeneration of some Nigerian high forest ecosystems. In: Unesco/UNEP/FAO, p. 580–88.

——. 1978b A study of the dynamic ecology of Zambian vegetation. *J. Ecol.*, 66, p. 175–98.

LEBRUN, J. 1935. Les essences forestières des régions montagneuses du Congo oriental. *Publs INEAC*, sér. sci., 1, p. 1–264, with large-scale vegetation map.

——. 1936a. Répartition de la forêt équatoriale et des formations végétales limitrophes. Brussels, Ministère des Colonies. 195 p., with one coloured vegetation map 1:2 000 000 and one at 1:4 000 000 and 7 small black-and-white maps.

——. 1936b. La forêt équatoriale congolaise. *Bull. Agric. Congo belge*, 27, p. 163–92, with 2 coloured vegetation maps at 1:4 000 000 and 1:2 000 000.

——. 1942. La végétation de Nyiragongo. Aspects de végétation des Parcs nationaux du Congo belge. Sér. 1. Parc National Albert, 1 (3–5), p. 1–121.

——. 1947. La végétation de la plaine alluviale au sud du lac Édouard. Exploration du Parc National Albert. Mission J. Lebrun (1937–38), pt 1 (2 vols.). Brussels, Inst. Parcs Nat. Congo belge. 800 p.

——. 1954. Sur la végétation du Secteur littoral du Congo belge. *Vegetatio*, 5–6, p. 157–60.

——. 1955. *Esquisse de la végétation du Parc National de la Kagera.* Exploration du Parc National de la Kagera. Mission J. Lebrun (1937–38), vol. 2, p. 1–89. Brussels, Inst. Parcs Nat. Congo belge.

——. 1956. La végétation et les territoires botaniques du Ruanda–Urundi. *Naturalistes belg.*, 37, p. 230–56.

——. 1957. Sur les éléments et groupes phytogéographiques de la flore du Ruwenzori (versant occidental). *Bull. Jard. bot. État Brux.*, 27, p. 453–78.

——. 1958. Les orophytes africains. *Comm. 6a Sess. Conf. Int. Afr. Occid.*, 3, Bot., p. 121–31.

——. 1959. Sur les processus de colonisation végétale des champs de lave des Virunga (Kivu, Congo belge). *Bull. Acad. r. Belg. Cl. Sci.*, sér. 5, 45, p. 759–76.

——. 1960a. Sur une méthode de délimitation des horizons et étages de végétation des montagnes du Congo oriental. *Bull. Jard. bot. État Brux.*, 30, p. 75–94.

——. 1960b. Sur les horizons et étages de végétation de divers volcans du massif des Virunga (Kivu–Congo). *Bull. Jard. bot. État Brux.*, 30, p. 255–77.

——. 1960c. Sur la richesse de la flore de divers territoires africains. *Bull. Séanc. Acad. r. Sci. Outre-Mer*, n.s., 6, p. 669–90.

——. 1960d. *Études sur la flore et la végétation des champs de lave au nord du lac Kivu (Congo belge).* Exploration du Parc National Albert. Mission J. Lebrun (1937–38), vol. 2, p. 1–352. Brussels, Inst. Parcs Nat. Congo belge.

——. 1961. Les deux flores d'Afrique tropicale. *Mém. Acad. r. Belg. Cl. Sci.*, 8°, 32 (6), p. 1–81.

——. 1962. Le 'couloir littoral' atlantique, voie de pénétration de la flore sèche en Afrique guinéenne. *Bull. Séanc. Acad. r. Sci. Outre-Mer*, n.s., 8, p. 719–35.

——. 1968. A propos des formations 'sclérophylles' du littoral congolais. *Bull. Soc. r. Bot. Belg.*, 102, p. 89–100.

——. 1969. La végétation psammophile du littoral congolais. *Mém. Acad. r. Sci. Outre-Mer, Cl. Sci. nat. méd.*, n.s., 8°, 18 (1), p. 1–166.

LEBRUN, J.; GILBERT, G. 1954. Une classification écologique des forêts du Congo. *Publs INEAC*, sér. sci., 63, p. 1–89, with small-scale vegetation map.

LEBRUN, J. P. 1971a. Les activités botaniques de l'Institut d'Élevage et de Médecine Vétérinaire des Pays Tropicaux de 1961 à 1970. *Mitt. Bot. Staatssamml. Münch.*, 10, p. 86–90.

——. 1971b. Quelques phanérogames africaines à aire disjointe. *Mitt. Bot. Staatssamml. Münch.*, 10, p. 438–48.

——. 1975. Quelques aires remarquables de phanérogames africaines des zones sèches. *Boissiera*, 24, p. 91–105.

——. 1976. La contribution de l'IEMVT à la connaissance de la flore africaine (fin 1970–début 1974). *Boissiera*, 24, p. 529–34.

——. 1977. *Éléments pour un atlas des plantes vasculaires de l'Afrique sèche*, 1. Maisons-Alfort, IEMVT. 265 p. (Étude botanique, n° 4.)

——. 1979. Id. 2. 255 p.

——. 1980. *Les bases floristiques des grandes divisions chorologique de l'Afrique sèche.* Univ. Pierre et Marie Curie, Paris 6. 483 p. (Thesis.)

LEBRUN, J. P.; STORK, A. L. 1977. *Index 1935–1976 des cartes de répartition des plantes vasculaires d'Afrique.* Geneva, Conservatoire et Jardin Botaniques. 138 p.

LECOMPTE, M. 1973. Aperçu sur la végétation dans le Rif Occidental calcaire (Massif de Talassemtane). *Trav. RCP 249*, 1, p. 89–104. Paris, CNRS.

LEDERMANN, C. 1912. Eine botanische Wanderung nach Deutsch-Adamaua. *Mitt. dt. Schutzgeb.*, 25, p. 20–55, with coloured vegetation map 1:1 000 000.

LEE, K. E.; WOOD, J. G. 1971. *Termites and soils.* London, New York, Academic Press. 251 p.

LEE, R. B. 1965. *Subsistence ecology of !Kung Bushmen.* Ann Arbor, Mich., University Microfilms Inc. 209 p. (Ph.D. thesis, University of California, Berkeley.)

LEE, R. B.; DE VORE, I. (eds.). 1976. *Kalahari hunter-gatherers. Studies of the !Kung San and their neighbours.* Cambridge, Mass., Harvard Univ. Press. 408 p.

LEGGAT, G. J. 1965. The reconciliation of forestry and game preservation in Western Uganda. *E. Afr. agric. For. J.*, 30, p. 355–66.

LEGRIS, P. 1969. *La Grande Comore. Climats et végétation.* Trav. Sect. Sci. Techn. Inst. Franç. Pondichéry, 3 (5), p. 1–28, with coloured vegetation map 1:100 000.

LE HOUÉROU, H. N. 1959. *Recherches écologiques et floristiques sur la végétation de la Tunisie méridionale.* Algiers, Institut de Recherches Sahariennes, Université d'Alger. Mém. n° 6, vol. 1: *Les milieux naturels, la végétation*, 281 p.; vol. 2: *La flore*, 229 p.

——. 1962. Recherches écologiques et floristiques sur la végétation de la Tunisie méridionale. Les milieux naturels, la végétation. *Mém. Inst. Rech. sahar.* (Alger), 6, p. 1–323.
——. 1967. *Carte phytoécologique de la Tunisie méridionale.* 2 feuilles au 1:500000 en couleurs. Tunis, Institut National de Recherche Agronomique de Tunisie.
——. 1969. La végétation de la Tunisie steppique (avec référence aux végétations analogues du Maroc, d'Algérie et de Libye). *Annls Inst. natn. Rech. agron., Tunisie*, 42 (5), p. 1–622.
——. 1975. Étude préliminaire sur la compatibilité des flores nord-africaines et palestinienne. In: *La flore du bassin méditerranéen.* Coll. Int. CNRS, 235, p. 345–50.
——. 1977. Fire and vegetation in North Africa. In: Mooney, H. A.; Conrad, C. E. (eds.), p. 334–41.
LEIPPERT, H. 1968. *Pflanzenökologische Untersuchungen im Masai-Land Tanzanias.* Munich, IFO-Inst. Wirtschaftsforschung. 184 p.
LEISTNER, O. A. 1959. Notes on the vegetation of the Kalahari Gemsbok National Park with special reference to its influence on the distribution of antelopes. *Koedoe*, 2, p. 128–51.
——. 1961a. Zur Verbreitung und Ökologie der Bäume der Kalaharidünen. *J. S. W. Afr. scient. Soc.*, 15, p. 35–40.
——. 1961b. On the dispersal of *Acacia giraffae* by game. *Koedoe*, 4, p. 101–4.
——. 1967. The plant ecology of the Southern Kalahari. *Mem. bot. Surv. S. Afr.*, 38, p. 1–172.
——. 1979. Southern Africa. In: Perry, R. A.; Goodall, D. W. (eds.), *Arid-land ecosystems: structure, functioning and management,* 1, p. 109–43. Cambridge Univ. Press.
LEISTNER, O. A.; WERGER, M. J. A. 1973. Southern Kalahari phytosociology. *Vegetatio*, 28, 353–99.
LEMÉE, G. 1953. Les associations à thérophytes des dépressions sableuses et limoneuses non salées et des rocailles aux environs de Béni-Ounif. *Vegetatio*, 4, p. 137–54.
——. 1954. L'économie de l'eau chez quelques graminées vivaces du Sahara septentrional. *Vegetatio*, 5–6, p. 534–41.
LEMON, P. C. 1968a. Effects of fire on an African plateau grassland. *Ecology*, 49, p. 316–22.
——. 1968b. Fire and wildlife grazing on an African plateau. *Proc. Ann. Tall Timbers Fire Ecol. Conf.*, 8, p. 71–88.
LEMS, K. 1960. Floristic botany of the Canary Islands. *Sarracenia*, 5, p. 1–94.
LÉONARD, A. 1959. Contribution à l'étude de la colonisation des laves du volcan Nyamuragira par les végétaux. *Vegetatio*, 8, p. 250–8.
——. 1962. Les savanes herbeuses du Kivu. *Publs INEAC*, sér. sci., 95, p. 1–87.
LÉONARD, J. 1947. Contribution à l'étude des formations ripicoles arbustives et arborescentes de la région d'Eala. *C.r. Semaine agricole de Yangambi*, 2, p. 863–77. Brussels. (Publs Inst. natn. Étude agron. Congo belge, h.s.)
——. 1950. Botanique du Congo belge. Les groupements végétaux. *Encycl. Congo belge*, 1, p. 345–89.
——. 1951. Contribution à l'étude de la végétation des bains d'éléphants au Congo belge. Le *Rhynchosporeto–Cyperetum longibracteati. Bull. Soc. r. Bot. Belg.*, 84, p. 13–27.
——. 1952a. Aperçu préliminaire des groupements végétaux pionniers dans la région de Yangambi. *Vegetatio*, 3, p. 279–97.
——. 1952b. [1953]. Les divers types de forêts du Congo belge. *Lejeunia*, 16, p. 81–93.
——. 1953. Les forêts du Congo belge. *Naturalistes belg.*, 34, p. 53–65.
——. 1954. La végétation pionnière des pentes sableuses sèches dans la région de Yangambi–Stanleyville (Congo belge). *Vegetatio*, 5–6, p. 97–104.
——. 1969a. Expédition scientifique belge dans le désert de Libye (Jebel Uweinat 1968–1969). *Africa–Tervuren*, 15, p. 101–36.
——. 1969b. *Aperçu sur la végétation. Complément au chapitre 1 de la 'Monographie hydrologique du Lac Tchad'.* Paris, ORSTOM. 11 p., with vegetation map 1:1000000. (Mimeo.)
——. 1971. Aperçu de la flore et de la végétation du Jebel Uweinat (Désert de Libye). Résumé. *Mitt. Bot. Staatssamml. Münch.*, 10, p. 476–7.
——. 1980. *Noms de plantes et de groupements végétaux cités dans Pierre Quézel: 'La végétation du Sahara, du Tchad à la Mauritanie'.* Meise, Jardin Botanique National de Belgique. 45 p. (Mimeo.)
LEREDDE, C. 1957. Étude écologique et phytogéographique du Tassili n'Ajjer. *Trav. Inst. Rech. sahar.* (Alger), sér. du Tassili, t. 2, p. 1–455. (Also issued as *Trav. Lab. for. Toulouse*, t. 5, sect. 3, vol. 3.)
LE ROUX, A.; MORRIS, J. W. 1977. Effects of certain burning and cutting treatments and fluctuating annual rainfall on seasonal variation in grassland basal cover. *Proc. Grassl. Soc. S. Afr.*, 12, p. 55–8.
LEROY, J. F. 1978. Composition, origin and affinities of the Madagascan vascular flora. *Ann. Mo. Bot. Gdn*, 65, p. 535–89.
LE TESTU, G. 1938. Note sur la végétation dans le bassin de la Nyanga et de la Ngounyé au Gabon. *Mém. Soc. linn. Normandie,* n.s., 1, p. 83–108.
LETOUZEY, R. 1957. La forêt à *Lophira alata* de la zone littorale camerounaise. *Bois Forêts Trop.*, 53, p. 9–20.
——. 1958. Phytogéographie camerounaise. In: *Atlas du Cameroun,* with coloured map 1:2000000. Paris, ORSTOM.
——. 1960. La forêt à *Lophira alata* Banks du littoral camerounais. Hypothèses sur ses origines possibles. *Bull. Inst. Étud. centrafr.,* n.s., 19–20, p. 219–40.
——. 1966. Étude phytogéographique du Cameroun. *Adansonia*, 2, 6, p. 205–15, with small vegetation map.
——. 1968a. *Étude phytogéographique du Cameroun.* Paris, Lechevalier. 511 p., with small-scale vegetation map.
——. 1968b. Cameroun. In: Hedberg, I.; Hedberg, O. (eds.), p. 115–21, with small vegetation map.
——. 1969. Observations phytogéographiques concernant le plateau africain de l'Amadoua. *Adansonia*, sér. 2, 9, p. 321–37.
——. 1975. Premières observations concernant au Cameroun la forêt sur cordons littoraux sablonneux. *Adansonia*, sér. 2, 14, p. 529–42.
——. 1977. Présence de *Ternstroemia polypetala* Melchior (Théacées) dans les montagnes camerounaises. *Adansonia*, sér. 2, 17, p. 5–10.
LETOUZEY, R. et coll. 1978. *Flore du Cameroun. Documents phytogéographiques,* n° 1, Introduction (14 pages + 8 calques) et 130 fiches avec cartes au 1:5000000 [of tree species of which the generic name begins with the letter 'A']. Paris, Association de Botanique Tropicale, Muséum National d'Histoire Naturelle.
——. 1979. Id. no. 2 [116 tree species of which the generic names begin with the letter 'B'].
LEUTHOLD, B. M.; LEUTHOLD, W. 1972. Food habits of giraffe

in Tsavo National Park, Kenya. *E. Afr. Wildl. J.*, 10, p. 129–41.
LEUTHOLD, W. 1970. Preliminary observations on food habits of gerenuk in Tsavo National Park, Kenya. *E. Afr. Wildl. J.*, 8, p. 73–84.
——. 1971. Studies on food habits of lesser kudu in Tsavo National Park, Kenya. *E. Afr. Wildl. J.*, 9, p. 35–45.
——. 1972. Home range, movements and food of a buffalo herd in Tsavo National Park. *E. Afr. Wildl. J.*, 10, p. 237–43.
——. 1977a. *African ungulates. A comparative review of their ethology and behavioural ecology.* Berlin, Heidelberg, New York, Springer-Verlag. 307 p.
——. 1977b. Changes in tree populations of Tsavo East National Park, Kenya. *E. Afr. Wildl. J.*, 15, p. 61–9.
LEUTHOLD, W.; LEUTHOLD, B. M. 1976. Density and biomass of ungulates in Tsavo East National Park, Kenya. *E. Afr. Wildl. J.*, 14, p. 49–58.
LEUTHOLD, W.; SALE, J. B. 1973. Movements and patterns of habitat utilization of elephants in Tsavo National Park, Kenya. *E. Afr. Wildl. J.*, 11, p. 369–84.
LEVYNS, M. R. 1929a. Veld-burning experiments at Ida's Valley, Stellenbosch. *Trans. R. Soc. S. Afr.*, 17, p. 61–92.
——. 1929b. The problem of the Rhenoster Bush. *S. Afr. J. Sci.*, 26, p. 166–9.
——. 1950. The relations of the Cape and Karroo floras near Ladismith, Cape. *Trans. R. Soc. S. Afr.*, 32, p. 235–46.
——. 1964. Migrations and origin of the Cape flora. *Trans. R. Soc. S. Afr.*, 37, p. 85–107.
LEWALLE, J. 1968. Burundi. In: Hedberg, I.; Hedberg, O. (eds.), p. 127–30.
——. 1972. Les étages de végétation du Burundi occidental. *Bull. Jard. bot. nat. Belg.*, 42, p. 1–247.
——. 1975. Endémisme dans une haute vallée du Burundi. *Boissiera*, 24, p. 85–9.
LEWIS, I. M. (ed.). 1975. *Abaar, the Somali drought.* London, International African Institute.
LEWIS, J. G. 1977. *Report of a short-term consultancy on the grazing ecosystem in the Mt Kulal region, Northern Kenya.* Nairobi, UNEP–MAB Integrated Project on Arid Lands. 62 p. (IPAL tech. rep., no. E-3.)
LIBEN, L. 1958. Esquisse d'une limite phytogéographique Guinéo–Zambézienne au Katanga occidental. *Bull. Jard. bot. État Brux.*, 28, p. 299–305.
——. 1961. Les bosquets xérophiles du Bugesera (Ruanda). *Bull. Soc. r. bot. Belg.*, 93, p. 93–111.
——. 1962. Nature et origine du peuplement végétal (Spermatophytes) des contrées montagneuses du Congo oriental. *Mém. Acad. r. Belg. Cl. Sci.*, 4°, sér. 2, 15 (3), p. 1–195.
——. 1965. Note sur quelques mammifères du Bugesera (Rwanda) et leurs relations avec les paysages végétaux. *Naturalistes belg.*, 46, p. 141–56.
LIEBERMAN, D.; HALL, J. B.; SWAINE, M. D.; LIEBERMAN, M. 1979. Seed dispersal by baboons in the Shai Hills, Ghana. *Ecology*, 60, p. 65–75.
LIND, E. M. 1956a. The natural vegetation of Buganda. *Uganda J.*, 20, p. 13–16.
——. 1956b. Studies in Uganda swamps. *Uganda J.*, 20, p. 166–76.
LIND, E. M.; MORRISON, M. E. S. 1974. *East African vegetation.* London, Longman. 257 p., with small-scale vegetation map.
LIND, E. M.; VISSER, S. A. 1962. A study of a swamp at the north end of Lake Victoria. *J. Ecol.*, 50, p. 599–613.

LINDINGER, L. 1926. Beiträge zur Kenntnis von Vegetation und Flora der kanarischen Inseln. *Abh. Gebiet Auslandsk.*, Bd 21, Reihe C, Naturw., Bd 8, p. 1–350. Hamburg Univ.
LIVINGSTONE, D. A. 1967. Postglacial vegetation of the Ruwenzori Mountains in equatorial Africa. *Ecol. Monogr.*, 37, p. 25–52.
——. 1975. Late Quaternary climatic change in Africa. *Ann. Rev. Ecol. Syst.*, 6, p. 249–80.
LOCK, J. M. 1972a. The effects of hippopotamus grazing on grasslands. *J. Ecol.*, 60, p. 445–67.
——. 1972b. Baboons feeding on *Euphorbia candelabrum*. *E. Afr. Wildl. J.*, 10, p. 73–6.
——. 1973. The aquatic vegetation of Lake George, Uganda. *Phytocoenologia*, 1, p. 250–62.
——. 1977a. Preliminary results from fire and elephant exclusion plots in Kabalaga National Park, Uganda. *E. Afr. Wildl. J.*, 15, p. 229–32.
——. 1977b. The vegetation of Rwenzori National Park, Uganda. *Bot. Jb. Syst.*, 98, p. 372–448.
LOCK, J. M.; MILBURN, T. R. 1971. The seed biology of *Themeda triandra* Forsk. in relation to fire. *Symp. Brit. ecol. Soc.*, 11, p. 337–49.
LOGAN, W. E. M. 1946. An introduction to the forests of central and southern Ethiopia. *Inst. Pap. Imp. For. Inst.*, 24, p. 1–65, with small-scale vegetation map.
LONG, G. 1954. Contribution à l'étude de la végétation de la Tunisie centrale. *Annu. Serv. Bot. Agron. Tunisie*, 27, p. 1–388, with coloured vegetation maps at 1:12500.
LONG, G. A. 1955. The study of the natural vegetation as a basis for pasture improvement in the western desert of Egypt. *Bull. Inst. Désert Égypte*, 5, p. 18–45.
LONGMAN, K. A.; JENÍK, J. 1974. *Tropical forest and its environment.* London, Longman. 196 p.
LOUIS, J. 1947a. Contribution à l'étude des forêts équatoriales congolaises. *C.r. Semaine agric. de Yangambi*, 2, p. 902–15. Brussels. (Publs INEAC, h.s.)
——. 1947b. La phytosociologie et le problème des jachères au Congo. Tom. cit., p. 916–23.
——. 1947c. L'origine et la végétation des îles du fleuve dans la région de Yangambi. Tom. cit., p. 924–33.
LOUW, W. J. 1951. An ecological account of the vegetation of the Potchefstroom area. *Mem. bot. Surv. S. Afr.*, 24, p. 1–105.
LOVELESS, A. R. 1961. A nutritional interpretation of sclerophylly based on differences in the chemical composition of sclerophyllous and mesophytic leaves. *Ann. Bot.*, n.s., 25, p. 168–84.
——. 1962. Further evidence to support a nutritional interpretation of sclerophylly. *Ann. Bot.*, n.s., 26, p. 551–61.
LOVERIDGE, J. P. 1968. Plant ecological investigations in the Nyamagasani Valley, Ruwenzori Mountains, Uganda. *Kirkia*, 6, p. 153–68, with large-scale vegetation map.
LOWE, R. G. 1968. Periodicity of a tropical rain forest tree, *Triplochiton scleroxylon* K. Schum. *Comm. For. Rev.*, 47, p. 150–63.
LUBINI, A. 1980. Étude analytique du groupement messicole à *Spermacoce latifolia* dans la région de Kisangani (Zaïre). *Bull. Jard. bot. nat. Belg.*, 50, p. 123–33.
LUCAS, G. L. 1968. Kenya. In: Hedberg, I.; Hedberg, O. (eds.), p. 152–63, with small vegetation map.
LUCAS, G. L.; SYNGE, H. 1978. *The IUCN plant red data book, comprising red data sheets on 250 selected plants threatened on a world scale.* Morges, Switzerland, IUCN. 540 p.
LUGO, A. E.; SNEDAKER, S. C. 1974. The ecology of mangroves. *Ann. Rev. Ecol. Syst.*, 5, p. 39–64.

LUNDGREN, B. 1978. *Soil conditions and nutritional cycling under natural and plantation forests in Tanzanian highlands.* Uppsala, Swedish Univ. of Agricultural Sciences. 429 p. (Rep. For. Ecol. For. Soils, 31.)

MABBERLEY, D. J. 1973. Evolution in the giant Groundsels. *Kew Bull.*, 28, p. 61–96.

——. 1974. The pachycaul Lobelias of Africa and St. Helena. *Kew Bull.*, 29, p. 535–84.

——. 1975a. Notes on the vegetation of the Cherangani Hills, N.W. Kenya. *J. E. Afr. nat. Hist. Soc.*, 150, p. 1–11.

——. 1975b. The pachycaul senecio species of St Helena. 'Cacalia paterna' and 'Cacalia materna'. *Kew Bull.*, 30, p. 413–20.

——. 1976. [1977]. The origin of the Afroalpine pachycaul flora and its implications. *Gdns Bull.* (Singapore), 29, p. 41–55.

MABBUT, J. A.; FLORET, C. (eds.) 1980. *Case studies on desertification.* Prepared by Unesco/UNEP/UNDP. Paris, Unesco. 279 p.

MACEDO, J. de Aguiar. 1970. Carta de vegetação da Serra de Gorongosa. *Comm. Inst. Investig. Agron. Moçambique*, 50, p. 1–75, with coloured vegetation map 1:75 000. Lourenço Marques.

MCCUSKER, A. 1977. Seedling establishment in mangrove species. *Geo-Eco-Trop.*, 1, p. 23–33.

MACFADYEN, W A. 1950. Vegetation patterns in the semi-desert plains of British Somaliland. *Geogrl J.*, 116, p. 199–211.

MCFARLANE, M. J. 1976. *Laterite and landscape.* London, New York, San Francisco, Academic Press. 151 p.

MACGREGOR, W. D. 1934. Silviculture of the mixed deciduous forests of Nigeria. *Oxf. For. Mem.*, 18, p. 1–108.

——. 1937. Forest types and succession in Nigeria. *Emp. For. J.*, 16, p. 234–42.

MÄCKEL, R. 1974. Dambos: a study in morphodynamic activity on the plateau regions of Zambia. *Catena*, 1, p. 327–66.

MCLACHLAN, D.; MOLL, E. J.; HALL, A. V. 1980. Re-survey of the alien vegetation in the Cape Peninsula. *J. S. Afr. Bot.*, 46, p. 127–46.

MCCLURE, F. A. 1966. *The bamboos, a fresh perspective.* Cambridge, Mass., Harvard Univ. Press. 347 p.

MACNAE, W. 1957. The ecology of the plants and animals in the intertidal regions of the Zwartkops estuary near Port Elizabeth, South Africa. *J. Ecol.*, 45, p. 113–31, 361–87.

——. 1963. Mangrove swamps in South Africa. *J. Ecol.*, 51, p. 1–25.

——. 1968. A general account of the fauna and flora of mangrove swamps and forests in the Indo-West-Pacific Region. *Adv. Mar. Biol.*, 6, p. 73–270.

——. 1971. Mangroves on Aldabra. *Phil. Trans. R. Soc. Lond.*, ser. B, 260, p. 237–47.

MACNAE, W.; KALK, M. 1962a. The ecology of the mangrove swamps at Inhaca Island, Moçambique. *J. Ecol.*, 50, p. 19–34.

——; ——. 1962b. The flora and fauna of sand flats at Inhaca Island, Moçambique. *J. Anim. Ecol.*, 31, p. 93–128.

——, ——. (eds.). 1969. *A natural history of Inhaca Island, Moçambique.* 2nd ed. Johannesburg, Witwatersrand Univ. Press. 163 p., with large-scale vegetation map.

MADGE, D. S. 1965. Leaf fall and litter disappearance in a tropical forest. *Pedobiologia*, 5, p. 273–88.

MAGADZA, C. H. D. 1970. A preliminary survey of the vegetation of the shore of Lake Kariba. *Kirkia*, 7, p. 253–67.

MAGUIRE, B. 1978. *The food plants of the !Khû Bushmen of North-Eastern South West Africa.* Witwatersrand University. 539 p. (M.Sc. thesis.) (Mimeo.)

MAHMOUD, A.; OBEID, M. 1971. Ecological studies in the vegetation of the Sudan. I. General features of the vegetation of Khartoum Province. *Vegetatio*, 23, p. 153–76.

MAIRE, R. 1916. La végétation des montagnes du Sud Oranais. *Bull. Soc. Hist. nat. Afr. N.*, 7, p. 210–92.

——. 1924. Études sur la végétation et la flore du Grand Atlas et du Moyen Atlas marocains. *Mém. Soc. Sci. nat. Maroc*, 7, p. 1–220.

——. 1938. La flore et la végétation du Sahara occidental. *Mém. Soc. Biogeogr.*, 6, p. 325–33.

——. 1933, 1940. Études sur la flore et la végétation du Sahara central. *Mém. Soc. Hist. nat. Afr. N.*, 3, p. 1–272 (1933); p. 273–433 (1940).

MAIRE, R.; BATTANDIER, A.; LAPIE, G.; PEYERIMHOFF, P. de; TRABUT, L. 1925. *Carte phytogéographique d'Algérie et de Tunisie.* 1:1 500 000, in colour. Alger.

MAIRE, R.; MONOD, T. 1950. Études sur la flore et la végétation du Tibesti. *Mém. IFAN*, 8, p. 1–140.

MAITLAND, T. D. 1932. The grassland vegetation of the Cameroons Mountain. *Kew Bull.*, p. 417–25.

MAKANY, L. 1972. Une tourbière tropicale à sphaignes sur les contreforts des plateaux batéké (Congo). *Annls Univ. Brazzaville*, 8, sér. C, p. 93–104.

——. 1976. Végétation des plateaux Teke (Congo). *Trav. Univ. Brazzaville*, 1, p. 1–301.

MALAISSE, F. 1974. Phenology of the Zambezian woodland area, with emphasis on the miombo ecosystem. In: Lieth, H. (ed.), *Phenology and seasonality modelling*, p. 269–86. New York, Heidelberg, Berlin, Springer-Verlag.

——. 1975. Carte de la végétation du bassin de la Luanza. In: Symoens, J. J. (ed.), *Expl. hydrobiol. bassin du lac Bangweolo et du Luapula*, vol. 18 (2), p. 1–41, with coloured vegetation map 1:40 000. Brussels, Cercle Hydrobiologique de Bruxelles.

——. 1976a. De l'origine de la flore termitophile du Haut-Shaba (Zaïre). *Boissiera*, 24, p. 505–13.

——. 1976b. Écologie de la rivière Luanza. In: Symoens, J. J. (ed.), *Expl. hydrobiol. bassin du lac Bangweolo et du Luapula*, vol. 17 (2), p. 1–151. Brussels, Cercle Hydrobiologique de Bruxelles.

——. 1978a. The miombo ecosystem. In: Unesco/UNEP/FAO, p. 589–606.

——. 1978b. High termitaria. In: Werger, M. J. A. (ed.), p. 1279–300.

MALAISSE, F.; ANASTASSIOU-SOCQUET, F. 1977. Contribution à l'étude de l'écosystème forêt claire (miombo). 24. Phytogéographie des hautes termitières du Shaba méridional (Zaïre). *Bull. Soc. r. Bot. Belg.*, 110, p. 85–95.

MALAISSE, F.; FRESON, R.; GOFFINET, G.; MALAISSE-MOUSSET, M. 1975. Leaf fall and litter breakdown in miombo. In: Medina, E.; Golley, F. B. (eds.), *Tropical ecological systems. Trends in terrestrial and aquatic research*, p. 137–52. Berlin, New York, Springer-Verlag.

MALAISSE, F.; GRÉGOIRE, J. 1978. Contribution à la phytogéochimie de la Mine de l'Étoile (Shaba, Zaïre). *Bull. Soc. r. Bot. Belg.*, 111, p. 252–60 (1978).

MALAISSE. F.; MALAISSE-MOUSSET, M. 1970. Contribution à l'étude de l'écosystème forêt claire (miombo). Phénologie de la défoliation. *Bull. Soc. r. Bot. Belg.*, 103, p. 115–24.

MALATO BELIZ, J. 1963. Aspectos da investigação geobotánica na Guiné Portuguesa. *Estud. agron.*, 4 (1), p. 1–20.

MALATO BELIZ, J.; ALVES PEREIRA, J. 1965. Constituição e ecologia das pastagens naturais da Guiné Portuguesa. *Rev. Junta Invest. Ultram.*, 13, p. 1–7.

MANGENOT, G. 1950. Essai sur les forêts denses de la Côte d'Ivoire. *Bull. Soc. bot. Fr.*, 97, p. 159–62.
——. 1951. Une formule simple permettant de caractériser les climats de l'Afrique intertropicale dans leurs rapports avec la végétation. *Rev. gén. Bot.*, 58, p. 353–69.
——. 1955a. Écologie et représentation cartographique des forêts équatoriales et tropicales humides. *Les divisions écologiques du monde*, p. 149–56. (Colloques int. CNRS, 59.) Also published in: *Année biol.*, sér. 3, 31, p. 393–400.
——. 1955b. Étude sur les forêts des plaines et plateaux de la Côte d'Ivoire. *Étud. éburn.*, 4, p. 5–61.
——. 1971. Une nouvelle carte de la végétation de la Côte d'Ivoire. *Mitt. Bot. Staatssamml. Münch.*, 10, p. 116–21, with vegetation map 1:4000000.
MANGENOT, G.; MIÈGE, J.; AUBERT, G. 1948. Les éléments floristiques de la basse Côte d'Ivoire et leur répartition. *C.r. somm. Séanc. Soc. Biogéogr.*, 25 (214), p. 30–4.
MARLER, P.; BOATMAN, D. J. 1952. An analysis of the vegetation of the northern slopes of Pico—the Azores. *J. Ecol.*, 40, p. 143–55.
MARLOTH, R. 1887. Das südöstliche Kalahari-Gebiet. Ein Beitrag zur Pflanzen-Geographie Süd-Afrikas. *Bot. Jb.*, 8, p. 247–60.
——. 1902. Notes on the occurrence of Alpine types in the vegetation of the higher peaks of the south-western region of the Cape. *Trans. S. Afr. Phil. Soc.*, 11, p. 161–8.
——. 1907. On some aspects in the vegetation of South Africa which are due to the prevailing winds. *Rep. S. Afr. Ass. Adv. Sci. 1905 & 1906*, p. 215–18.
——. 1908. Das Kapland, insonderheit das Reich der Kapflora, das Waldgebiet und die Karroo pflanzengeographisch dargestellt. In: Chun, C. (ed.), *Wiss. Ergeb. dt. Tiefsee-Exped. 'Valdivia' 1898–1899*, 2 (3), p. 1–436. Jena, Gustav Fischer Verlag.
——. 1909. The vegetation of the Southern Namib. *S. Afr. J. Sci.*, 6, p. 80–7.
——. 1923. Observations on the Cape flora. Its distribution on the line of contact between the south-western districts and the Karoo. *S. Afr. J. Nat. Hist.*, 4, p. 335–44.
——. 1929. Remarks on the Realm of the Cape flora. *S. Afr. J. Sci.*, 26, p. 154–9.
MARTIN, A. R. H. 1960a. The ecology of Groenvlei, a South African fen. I. The primary communities. *J. Ecol.*, 48, p. 55–71.
——. 1960b. Id. II. The secondary communities. *J. Ecol.*, 48, p. 307–29.
——. 1965. Plant ecology of the Grahamstown Nature Reserve. I. Primary communities and plant succession. *J. S. Afr. Bot.*, 31, p. 1–54.
——. 1966. Id. II. Some effects of burning. *J. S. Afr. Bot.*, 32, p. 1–39.
MARTIN, J. D. 1940. The *Baikiaea* forests of Northern Rhodesia. *Emp. For. J.*, 19, p. 8–18.
——. 1941. *Report of forestry in Barotseland*. Lusaka, Govt Printer. 66 p.
MASEFIELD, G. B. 1948. Grass burning. Some Uganda experience. *E. Afr. agric. J.*, 13, p. 135–8.
MASSART, J. 1898. Voyage botanique au Sahara. *Bull. Soc. r. Bot. Belg.*, 37, p. 203–339.
MATHEZ, J. 1973. Nouveaux matériaux pour la Flore du Maroc. Fasc. 2. Contribution à l'étude de la flore de la région d'Ifni. *Trav. RCP,* 249, 1, p. 105–20. Paris, CNRS.
MATOS, G. Cardoso de. 1970. A vegetação do Parque nacional do Iona. *Bol. Soc. Brot.*, sér. 2, p. 245–7.
MATOS, G. Cardoso de; SOUSA, J. N. Baptista de. 1970. *Reserva parcial de Moçâmedes. Carta de vegetação e memoria descritiva*, p. 1–32, with vegetation map 1:500000. Nova Lisboa, Inst. Invest. Agron. Angola.
MEESSEN, J. M. T. 1951. 'Botanique'. In: *Ituri, histoire, géographie, économie*, p. 87–106. Brussels, Min. Col. (Publ. Dir. Agr. For. Élev. Colon.)
MEGGERS, B. J.; AYENSU, E. S.; DUCKWORTH, W. D. 1973. (eds.). *Tropical forest ecosystems in Africa and South America. A comparative review*. Washington, D.C., Smithsonian Instn. 350 p.
MEIKLEJOHN, J. 1965. Microbiological studies on large termite mounds. *Rhod. Zambia J. agric. Res.*, 3, p. 67–79.
MELLISS, J. C. 1875. *St. Helena: a physical, historical, and topographical description of the island, including its geology, fauna, flora and meteorology*. London, L. Reeve. 426 p.
MENAUT, J. C. 1974. Chute de feuilles et apport au sol de litière par les ligneux dans une savane préforestière de Côte d'Ivoire. *Bull. Ecol.*, 5, p. 27–39.
MENAUT, J. C.; CÉSAR, J. 1979. Structure and primary productivity of Lamto savannas, Ivory Coast. *Ecology*, 60, p. 1197–210.
MENDES, E. J. 1962. Preliminary report on a botanical journey to the Bié–Cuando–Cubango District, Angola, 1959–60. *C.r. IVe Réunion plén. AETFAT (1960)*, p. 333–6. Lisbon, Junta Invest. Ultram.
MENDONÇA, F. A. 1952 [1953]. The vegetation of Mozambique. *Lejeunia*, 16, p. 127–35.
——. 1961. Indices fitocorológicos da vegetação de Angola. *Garcia de Orta*, 9, p. 479–83.
MENTIS, M. T.; DUKE, R. R. 1976. Carrying capacities of natural veld in Natal for large wild herbivores. *S. Afr. J. Wildl. Res.*, 6, p. 65–74.
MEREDITH, D. (ed.). 1955. *The grasses and pastures of South Africa*. Central News Agency, South Africa. 771 p.
MERTON, L. F. H.; BOURN, D. M.; HNATIUK, R. J. 1976. Giant tortoise and vegetation interactions on Aldabra atoll. I. Inland. *Biol. Conserv.*, 9, p. 293–304.
MÉTRO, A. 1958. *Atlas du Maroc. Notice explicative*. Sect. 6. Biogéographie. Forêts et ressources végétales. Planche 19a, Forêts (1:1000000 in colour), p. 1–157. Rabat, Institut Scientifique Chérifien.
——. 1970. The Maghreb of Africa north of the Sahara. In: Kaul, R. N. (ed.), p. 37–58.
MEULENBERGH, J. 1974. La mangrove zaïroise. *Mém. Acad. r. Sci. Outre-Mer, Cl. Sci. Techn.*, n.s., 8°, vol. 17 (8), p. 1–86, with large-scale vegetation map.
MEUSEL, H. 1952. Über Wuchsformen, Verbreitung und Phylogenie einiger mediterran–mitteleuropäischer Angiospermen-Gattungen. *Flora*, 139, p. 333–93.
MICHEL, P.; NAEGELÉ, A.; TOUPET, C. 1969. Contribution à l'étude biologique du Sénégal septentrional. 1. Le milieu naturel. *Bull. IFAN*, sér. A, 31, p. 757–839.
MICHELL, M. R. 1922. Some observations on the effects of a bush fire on the vegetation of Signal Hill. *Trans. R. Soc. S. Afr.*, 10, p. 213–32.
MICHELMORE, A. P. G. 1939. Observations on tropical African grasslands. *J. Ecol.*, 27, p. 282–312.
MICHON, 1973. Le Sahara avance-t-il vers le Sud? *Bois Forêts Trop.*, 150, p. 3–14.
MIÈGE, J. 1954. La végétation entre Bia et Comoé (Côte d'Ivoire orientale). *Bull. IFAN*, sér. A, 16, p. 973–89.
——. 1955. Les savanes et forêts claires de Côte d'Ivoire. *Étud. éburn.*, 4, p. 62–83.
——. 1966. Observations sur les fluctuations des limites

savanes–forêts en basse Côte d'Ivoire. *Annls Fac. Sci. Univ. Dakar*, 19, sér. Sci. végétales, n° 3, p. 149–66, with small vegetation map.
MIÈGE, J.; BODARD, M.; CARRÈRE, P. 1966. Évolution floristique des végétations de jachère en fonction des méthodes culturales à Darou (Sénégal). *Inst. Rech. Huiles Oléagineux*, sér. Sci., 14, p. 1–58.
MIÈGE, J.; HAINARD, P.; TCHÉRÉMISSINOFF, G. 1976. Aperçu phytogéographique sur la Basse-Casamance. *Boissiera*, 24, p. 461–71, with large-scale vegetation map.
MIGAHID, A. M. 1947. An ecological study of the 'sudd' swamps of the Upper Nile. *Proc. Égypt. Acad. Sci.*, 3, p. 57–86.
MIGAHID, A. M.; ABDEL RAHMAN, A. A.; EL SHAFEI ALI, M.; HAMMOUDA, M. A. 1955. Types of habitat and vegetation at Ras El Hikma. *Bull. Inst. Désert Égypte*, 5 (2), p. 107–90.
MIGAHID, A. M.; BATANOUNY, K. H.; EL-SHARKAWI, H. M.; SHALABY, A. F. 1975. Phytosociological and ecological studies of Maktila Sector of Sidi-Barrani. III. A vegetation map. *Feddes Rep.*, 86, p. 93–8.
MIGAHID, A. M.; BATANOUNY, K. H.; ZAKI, M. A. F. 1971. Phytosociological and ecological study of a sector in the Mediterranean coastal region in Egypt. *Vegetatio*, 23, p. 113–34.
MIGAHID, A. M.; EL-SHARKAWI, H. M.; BATANOUNY, K. H.; SHALABY, A. F. 1974. Phytosociological and ecological studies of Maktila Sector of Sidi-Barrani. I. Sociology of the communities. *Feddes Rep.*, 84, p. 747–60.
——; ——; ——; ——. 1975. Id. IV. Range potentialities of the communities. *Feddes Rep.*, 86, p. 579–87.
MIGAHID, A. M.; SHALABY, A. F.; BATANOUNY, K. H.; EL-SHARKAWI, H. M. 1975. Id. II. Ecology of the communities. *Feddes Rep.*, 86, p. 83–91.
MILDBRAED, J. 1909. Die Vegetationsverhältnisse der zentralafrikanischen Seenzone vom Viktoria-See bis zu den Kiwu-Vulkanen. *Sitzber. preuss. Akad. Wiss.*, 39, p. 989–1017.
——. 1913. Botanische Beobachtungen in Kamerun und im Kongogebiet während der II. Afrika-Expedition des Herzogs Adolf Friedrich zu Mecklenburg. *Verh. bot. Ver. Prov. Brandenberg*, 54, p. 38–57.
——. 1914. *Wissenschaftliche Ergebnisse der Deutschen Zentral-Afrika-Expedition 1907–1908*, 2, Botanik. Leipzig, Klinkhardt & Biermann. 718 p.
——. 1922. *Wissenschaftliche Ergebnisse der zweiten deutschen Zentral-Afrika-Expedition 1910–1911*, 2, Botanik. Leipzig, Klinkhardt & Biermann. 202 p.
——. 1923. Das Regenwaldgebiet im äquatorialen Afrika. *Notizbl. bot. Gart. Mus., Berl.-Dahlem*, 8, p. 574–99.
——. 1932. Zur Kenntnis der Vegetationsverhältnisse Nord-Kameruns. *Bot. Jb.*, 65, p. 1–52.
——. 1933*a*. Ein botanischer Ausflug in das 'Grassland' des Kamerungebirges. *Kolon. Rdsch.*, 25, p. 139–47.
——. 1933*b*. Ein Hektar Regenwald auf Fernando Poo. *Notizbl. Bot. Gart. Mus., Berl.-Dahlem*, 11, p. 946–50.
——. 1966. See Domke, W.
MILES, J. 1979. *Vegetation dynamics*. London, Chapman & Hall. 80 p.
MILEWSKI, A. V. 1977. Habitat of Restionaceae endemic to the South-Western Cape coastal flats. *J. S. Afr. Bot.*, 43, p. 243–61.
MILEWSKI, A. V.; ESTERHUYSEN, E. 1977. Habitat of Restionaceae endemic to Western Cape coastal flats. *J. S. Afr. Bot.*, 43, p. 233–41.
MILLER, O. B. 1939. The Mukusi forests of the Bechuanaland Protectorate. *Emp. For. J.*, 18, p. 193–201.
——. 1946. Southern Kalahari. *Emp. For. Rev.*, 25, p. 225–9.
MILNE, G. 1935. Some suggested units of classification and mapping, particularly for East African soils. *Soil Res.*, 4, p. 183–98.
——. 1936. *A provisional soil map of East Africa*. London, Crown Agents. 34 p., with coloured soil map 1:2 000 000.
——. 1947. A soil reconnaissance journey through parts of Tanganyika Territory, December 1935 to February 1936. *J. Ecol.*, 35, p. 192–265.
MILNE, G.; CALTON, W. E. 1944. Soil salinity related to the clearing of natural vegetation. *E. Afr. agric. J.*, 10, p. 7–11.
MITCHELL, B. L. 1961*a*. Ecological aspects of game control measures in African wilderness and forested areas. *Kirkia*, 1, p. 120–8.
——. 1961*b*. Some notes on the vegetation of a portion of the Wankie National Park. *Kirkia*, 2, p. 200–9.
MITCHELL, D. S. 1969. The ecology of vascular hydrophytes on Lake Kariba. *Hydrobiologia*, 34, p. 448–64.
MOGG, A. O. D. 1963. A preliminary investigation of the significance of salinity on the zonation of species in salt-marsh and mangrove swamp associations. *S. Afr. J. Sci.*, 59, p. 81–6.
MOLL, E. J. 1966. A report on the Xumeni Forest. Natal. *For. S. Afr.*, 7, p. 99–108.
——. 1968*a*. A quantitative ecological investigation of the Krantzkloof forest, Natal. *J. S. Afr. Bot.*, 34, p. 15–25.
——. 1968*b*. An account of the plant ecology of the Hawaan Forest, Natal. *J. S. Afr. Bot.*, 34, p. 61–76.
——. 1968*c*. A plant ecological reconnaissance of the Upper Mgeni catchment. *J. S. Afr. Bot.*, 34, p. 401–20.
——. 1968*d*. Some notes on the vegetation of Mkuzi Game Reserve. *Lammergeyer*, 8, p. 25–30.
——. 1972*a*. The current status of Mistbelt mixed *Podocarpus* forest in Natal. *Bothalia*, 10, p. 595–8.
——. 1972*b*. A preliminary account of the dune communities at Pennington Park, Mtunzini, Natal. *Bothalia*, 10, p. 615–26.
——. 1972*c*. The distribution, abundance and utilization of the Lala Palm, *Hyphaene natalensis*, in Tongaland, Natal. *Bothalia*, 10, p. 627–36.
MOLL, E. J.; CAMPBELL, B. M.; PROBYN, T. A. 1976. A rapid statistical method of habitat classification using structural and physiognomic characteristics. *S. Afr. J. Wildl. Res.*, 6, p. 45–50.
MOLL, E. J.; HAIGH, H. 1966. A report on the Xumeni forest, Natal. *For. S. Afr.*, 7, p. 99–108.
MOLL, E. J.; MCKENZIE, B.; MCLACHLAN, D. 1977. Present management problems and strategies on Table Mountain, South Africa. In: Mooney, H. A.; Conrad, C. E. (eds.), p. 470–5.
MOLL, E. J.; MORRIS, J. W. 1968. Notes on a cycad community and associated vegetation in Natal. *J. S. Afr. Bot.*, 34, p. 331–43.
MOLL, E. J.; WHITE, F. 1978. The Indian Ocean Coastal Belt. In: Werger, M. J. A. (ed.), p. 561–98.
MOLL, E. J.; WOODS, D. B. 1971. The rate of forest tree growth and a forest ordination at Xumeni, Natal. *Bothalia*, 10, p. 451–60.
MONJAUZE, A. 1958. Le groupement à micocoulier (*Celtis australis* L.) en Algérie. *Mém. Soc. Hist. nat. Afr. N.*, n.s., 2, p. 1–76.
——. 1968. Répartition et écologie de *Pistacia atlantica* Desf. en Algérie. *Bull. Soc. Hist. nat. Afr. N.*, 56, p. 5–128.
MONJAUZE, A.; FAUREL, L.; SCHOTTER, G. 1955. Note préliminaire sur un itinéraire botanique dans la steppe et le

Sahara septentrional algérois. *Bull. Soc. Hist. nat. Afr. N.*, 46, p. 206–30.

MONNIER, Y. 1968. Les effets des feux de brousse sur une savane préforestière de Côte d'Ivoire. *Étude éburn.*, 9, p. 1–260, with large-scale vegetation map.

MONOD, T. 1938. Notes botaniques sur le Sahara occidental et ses confins sahéliens. *Mém. Soc. Biogéogr.*, 6, p. 351–74.

——. 1947. Notes biogéographiques sur l'Afrique de l'Ouest. *Port. Acta biol.*, sér. B, 2 (3), p. 208–85.

——. 1952*a*, 1954*a*. Contribution à l'étude du peuplement de la Mauritanie. Notes botaniques sur l'Adrar (Sahara occidental). *Bull. IFAN*, 14, p. 405–49 (1952); 16, p. 1–48 (1954).

——. 1952*b*. Notes sur la flore du Plateau Bautchi (Nigeria). *Mém. IFAN*, 18, p. 11–37.

——. 1954*a*. See Monod, T., 1952*a*.

——. 1954*b*. Mauritanie. In: *Notices botaniques et itinéraires commentés publiés à l'occasion du VIIIe Congrès international de botanique Paris–Nice 1954*, V (3), p. 1–32, with small vegetation map.

——. 1957. *Les grandes divisions chorologiques de l'Afrique.* CCTA/CSA. 147 p., with chorological map 1:34 000 000. (Publ. no. 24.)

——. 1958. Majâbat al-Koubrâ. Contribution à l'étude de l'Empty Quarter ouest-saharien. *Mém. IFAN*, 52, p. 1–407, with map 1:2 000 000.

——. 1960. Notes botaniques sur les îles de São Tomé et de Príncipe. *Bull. IFAN*, sér. A, 22, p. 19–83.

——. 1963. Après Yangambi (1956). Notes de phytogéographie africaine. *Bull. IFAN*, sér. A, 25, p. 594–619.

——. 1968. La conservation des habitats: problèmes de définitions et de choix. In: Hedberg, I.; Hedberg, O. (eds.), p. 32–5.

——. 1971. Remarques sur les symétries floristiques des zones sèches nord et sud en Afrique. *Mitt. Bot. Staatssamml. Münch.*, 10, p. 375–423.

——. 1973. *Les déserts.* Paris, Horizons de France. 247 p.

——. 1974. Spectres de modes de dissémination dans l'Adrar mauritanien (Sahara occidental). *Candollea*, 29, p. 401–25.

——. (ed.). 1975. *Pastoralism in tropical Africa. Thirteenth Int. Afr. Semin.* (*Niamey, 13–21 December 1972*). London, International African Institute/Oxford Univ. Press. 502 p.

MONOD, T.; SCHMITT, C. 1968. Contribution à l'étude des pseudogalles formicaires chez quelques Acacias africains. *Bull. IFAN*, sér. A. 30, p. 953–1027.

MONTEIRO, R. F. Roméro. 1962. Le massif forestier du Mayumbe angolais. *Bois Forêts Trop.*, 82, p. 3–17.

——. 1965. Correlação entre as florestas do Maiombe e dos Dembos. *Bol. Inst. Invest. cient. Angola*, 1, p. 257–65.

——. 1970. *Estudo da flora e da vegetação das florestas abertas do planalto do Bié.* Luanda, Inst. Invest. Cient. de Angola. 352 p., with coloured vegetation map 1:500 000.

MONTEITH, J. L. (ed.). 1976. *Vegetation and the atmosphere.* 1, *Principles*, 278 p., 2, *Case studies*, 439 p. London, Academic Press.

MOOLMAN, E.; BREYTENBACH, G. J. 1976. Stomach contents of the Chacma Baboon, *Papio ursinus*, from the Loskop Dam area, Transvaal, South Africa. *S. Afr. J. Wildl. Res.*, 61, p. 41–3.

MOOMAW, J. C. 1960. *A study of the plant ecology of the Coast Region of Kenya Colony.* Nairobi, Govt Printer. 54 p., with large-scale vegetation map.

MOONEY, H. 1963. An account of two journeys to the Araenna Mountains in Balé Province (south-east Ethiopia), 1958 and 1959–60. *Proc. Linn. Soc. Lond.*, 174, p. 127–52.

MOONEY, H. A.; CONRAD, C. E. (eds.). 1977. *Proc. Symp. Environ. Consequences Fire Fuel Mgmt Mediterr. Ecosyst.* Washington, D.C., U.S. Dept Agr. For. Serv. 498 p. (Gen. Tech. Rep. WO-3.)

MOORE, A. W. 1960. The influence of annual burning on a soil in the derived savanna zone of Nigeria. *Trans. 7th Int. Congr. Soil Sci.*, 4, p. 257–65.

MORAT, P. 1973. Les savanes du Sud-Ouest de Madagascar. *Mém. ORSTOM*, 68, p. 1–235, with vegetation map 1:500 000.

MOREAU, R. E. 1935*a*. A synecological study of Usambara, Tanganyika Territory, with particular reference to birds. *J. Ecol.*, 23, p. 1–43.

——. 1935*b*. Some eco-climatic data for closed evergreen forest in tropical Africa. *J. Linn. Soc. Zool.*, 39, p. 285–93.

——. 1938. Climatic classification from the standpoint of East African biology. *J. Ecol.*, 26, p. 467–96.

——. 1952. Africa since the Mesozoic with particular reference to certain biological problems. *Proc. Zool. Soc. Lond.*, 121, p. 869–913.

——. 1966. *The bird faunas of Africa and its islands.* New York, London, Academic Press. 424 p.

MORGAN, W. T. W. (ed.). 1972. *East Africa: its people and resources.* Nairobi, London, New York, Oxford Univ. Press. 312 p.

MORISON, C. G. T.; HOYLE, A. C.; HOPE-SIMPSON, J. F. 1948. Tropical soil-vegetation catenas and mosaics. *J. Ecol.*, 36, p. 1–84.

MORRIS, J. W. 1969. An ordination of the vegetation of Ntshongweni, Natal. *Bothalia*, 10, p. 89–120.

——. 1976. Automatic classification of the highveld grassland of Lichtenburg, south-western Transvaal. *Bothalia*, 12, p. 267–92.

MORTON, J. K. 1957. Sand-dune formation on a tropical shore. *J. Ecol.*, 45, p. 495–7.

——. 1962. The upland floras of West Africa. *C.r. IVe Réun. plén. AETFAT (1960)*, p. 391–409. Lisbon, Junta Invest. Ultram.

——. 1968. Sierra Leone. In: Hedberg, I.; Hedberg, O. (eds.), p. 72–4.

MOSS, R. P. (ed.). 1968. *The soil resources of tropical Africa.* Cambridge, Cambridge Univ. Press. 226 p.

MOSS, R. P.; MORGAN, W. B. 1970. Soils, plants and farmers in West Africa. In: Garlick, E. P.; Keay, R. W. J. (eds.), *Human ecology in the tropics*, p. 1–31. Oxford, Pergamon Press.

MOSTERT, J. W. C. 1958. Studies of the vegetation of parts of the Bloemfontein and Brandfort Districts. *Mem. bot. Surv. S. Afr.*, 31, p. 1–226, with large-scale vegetation map.

MOURGUES, G. 1950. Le nomadisme et le déboisement dans les régions sahéliennes de l'Afrique occidentale française. *C.r. 1re Conf. int. Afr. de l'Ouest, Dakar, 1945*, 1, p. 139–67.

MUIR, J. 1929. The vegetation of the Riversdale area. *Mem. bot. Surv. S. Afr.*, 13, p. 1–82.

MULLENDERS, W. 1953. Contribution à l'étude des groupements végétaux de la contrée de Goma–Kisenyi (Kivu-Ruanda). *Vegetatio*, 4, p. 73–83.

——. 1954. La végétation de Kaniama (entre Lubishi–Lubilash, Congo belge). *Publs INEAC*, sér. sci., 61, p. 1–499.

——. 1955. The phytogeographical elements and groups of the Kaniama District (High Lomami, Belgian Congo) and the analysis of the vegetation. *Webbia*, 11, p. 497–517, with small-scale vegetation map.

MURAT, M. 1937. Végétation de la zone prédésertique en

Afrique centrale (région du Tchad). *Bull. Soc. Hist. nat. Afr. N.*, 28, p. 19–83.
——. 1944 [1945]. Esquisse phytogéographique du Sahara occidental. *Mém. Off. natn. anti-acrid., Alger*, 1, p. 1–31.
MURDOCH, G.; WEBSTER, R.; LAWRANCE, C. J. 1971. *A land system atlas of Swaziland.* Christchurch, Hants, Military Engineering Experimental Establishment. 49 p.
MURRAY, J. M. 1938. An investigation of the interrelationships of the vegetation, soils and termites. *S. Afr. J. Sci.*, 35, p. 288–97.
MUSIL, C. F.; GRUNOW, J. O.; BORNMAN, C. H. 1973. Classification and ordination of aquatic macrophytes in the Pongolo River pans, Natal. *Bothalia*, 11, p. 181–90.
MYERS, N. 1973. Tsavo National Park, Kenya, and its elephants: an interim appraisal. *Biol. Conserv.*, 5, p. 123–32.
——. 1980. *Conversion of tropical moist forests.* Washington, D.C., National Academy of Sciences. 205 p.
MYRE, M. 1960. Os principais componentes das pastagens expontâneas do sul da Província de Moçambique. I. Estudo especial e geral florístico-ecologico das espéciez pascícolas, evidenciandose as gramíneas que existen na Província. *Mems Junta Invest. Ultram.* (Lisbon), sér. 2, 20, p. 1–307.
——. 1962. A grassland type of the south of the Mozambique Province. *C.r. IVe Réun. plén. AETFAT (1960)*, p. 337–61. Lisbon, Junta Invest. Ultram.
——. 1964. A vegetação do extremo sul de Província de Moçambique. *Estud., Ens. Docum.*, 110, p. 1–145. Lisbon, Junta Invest. Ultram.
——. 1971. As pastagens da região do Maputo. *Mems Inst. Invest. Agron. Moç.*, 3, p. 1–181.
NAEGÉLÉ, A. 1958a. Contributions à l'étude de la flore et des groupements végétaux de la Mauritanie. I. Notes sur quelques plantes récoltées à Chinguetti (Adrar Tmar). *Bull. IFAN*, sér. A, 20, p. 293–305.
——. 1958b. Id. II. Plantes recueillies par Mlle Odette du Puigaudeau en 1950. Tom. cit., p. 876–903.
——. 1959a. Id. III. Les parcelles protégées IFAN–Unesco de la Région d'Atar. Op. cit., 21, p. 1195–1204.
——. 1959b. La végétation de la zone aride. Les parcelles protégées d'Atar. *La Nature*, p. 72–6.
——. 1959c. Note préliminaire sur la flore et la végétation du cordon littoral ou avant-dune au Sénégal. *Bull. IFAN*, sér. A, 21, p. 1177–94.
——. 1960. Contributions à l'étude de la flore et des groupements végétaux de la Mauritanie. IV. Voyage botanique dans la presqu'île du Cap Blanc (première note). *Bull. IFAN*, sér. A, 22, p. 1231–47.
NÄNNI, U. W. 1969. Veld management in the Natal Drakensberg. *S. Afr. For. J.*, 68, p. 5–15.
NANSON, A.; GENNART, M. 1960. Contribution à l'étude du climax et en particulier du pédoclimax en forêt équatoriale congolaise. *Bull. Inst. Agron. de Gembloux*, 28, p. 287–342.
NAPIER BAX, P.; SHELDRICK, D. L. W. 1963. Some preliminary observations on the food of elephant in the Tsavo Royal National Park (East) of Kenya. *E. Afr. Wildl. J.*, 1, p. 40–53.
NASH, T. A. M. 1969. *African bane. The tsetse fly.* London, Glasgow, Collins. 224 p.
NAVEH, Z. 1966a. The development of Tanzania Masailand: a sociological and ecological challenge. *Afr. Soils*, 11, p. 499–517.

——. 1966b. Le développement du pays Masai de Tanzanie: discussion sociologique et écologique. *Afr. Soils*, 11, p. 519–39.
——. 1974. Effects of fire in the Mediterranean Region. In: Kozlowski, T. T.; Ahlgren, C. E. (eds.), p. 401–34.
NÈGRE, R. 1952a. Observations phytosociologiques et écologiques sommaires sur la cédraie de Kissarit (Brigade forestière d'Ain Leuh, Moyen-Atlas central). *Phyton*, 4, p. 59–71.
——. 1952b. Les associations végétales du Jebel Saa (Moyen-Atlas d'Itzer). *Bull. Soc. Sci. nat. Maroc*, 32, p. 139–65, with small vegetation map.
——. 1953. Id., op. cit., 33, p. 27–38.
——. 1956a. Carte [1:50000, in colour] des groupements végétaux du Sedd el Mejnoun. *Trav. Inst. sci. chérif.*, sér. Bot., 7, p. 1–35.
——. 1956b. Recherches phytosociologiques sur le Sedd el Messjoun. *Trav. Inst. sci. chérif.*, sér. Bot., 10, p. 1–190.
——. 1959. Recherches phytogéographiques sur l'étage de végétation méditerranéen aride (sous-étage chaud) au Maroc occidental. *Trav. Inst. sci. chérif.*, sér. Bot., 13, p. 1–385, with coloured vegetation map 1:500000.
——. 1974. Les pâturages de la région de Syrte (Libye). Projet de régénération. *Feddes Rep.*, 85, p. 185–243.
NÈGRE, R.; PELTIER, J. P. 1976. Premières observations sur la végétation du bassin d'Argana (Maroc). *Feddes Rep.*, 87, p. 49–81.
NEGRI, G. 1913. *Appunti di una escursione botanica nell'Etiopia meridionale (marzo–agosto 1909).* Rome, Ministero delle Colonie. 177 p. (Rapp. Monogr. colon., 4.)
NIEUWOLT, S. 1972. Rainfall variability in Zambia. *J. trop. Geogr.*, 34, p. 44–57.
——. 1977. *Tropical climatology: an introduction to the climates of the low latitudes.* London, John Wiley & Sons. 207 p.
NJOKU, E. 1959. An analysis of plant growth in some West African species. I. Growth in full daylight. *J. W. Afr. Sci. Ass.*, 5, p. 37–56.
——. 1963. Seasonal periodicity in the growth and development of some forest trees in Nigeria. I. Observations on mature trees. *J. Ecol.*, 51, p. 617–24.
——. 1964. Id. II. Observations on seedlings. *J. Ecol.*, 52, p. 19–26.
NOEL, A. R. A. 1959. The vegetation of the freshwater swamps of Inhaca Island. *J. S. Afr. Bot.*, 25, p. 189–205.
——. 1961. A preliminary account of the effect of grazing upon species of *Helichrysum* in the Amatole Mountains. *J. S. Afr. Bot.*, 27, p. 81–5.
NORDENSTAM, B. 1970. Notes on the flora and vegetation of Etosha Pan, South West Africa. *Dinteria*, 5, p. 3–18.
——. 1974. The flora of the Brandenberg. *Dinteria*, 11, p. 3–67.
NYE, P. H. 1954–55. Some soil forming processes in the humid tropics. I, *J. Soil Sci.*, 5, p. 7–21 (1954); II–IV, op. cit., 6, p. 51–83.
NYE, P. H.; GREENLAND, D. J. 1960. The soil under shifting cultivation. *Tech. Commun. Commonw. Bur. Soils*, 51, p. 1–156.
OBEID, M.; MAHMOUD, A. 1971. Ecological studies in the vegetation of the Sudan. II. The ecological relationships of the vegetation of Khartoum Province. *Vegetatio*, 23, p. 177–98.
OBEID, M.; SEIF EL DIN, A. 1971a. Ecological studies of the vegetation of the Sudan. I. *Acacia senegal* (L.) Willd. and its natural regeneration. *J. appl. Ecol.*, 7, p. 507–18.
——; ——. 1971b. Ecological studies of the vegetation of the Sudan. III. The effect of simulated rainfall distribution at

different isohyets on the regeneration of *Acacia senegal* (L.) Willd. on clay and sandy soils. *J. appl. Ecol.*, 8, p. 203–9.

OKAFOR, J. C. 1977. Development of forest tree crops for food supplies in Nigeria. *For. Ecol. Mgmt*, 1, p. 235–47.

OKALI, D. U. U. 1971. Tissue water relations of some woody species of the Accra Plains, Ghana. *J. Ecol.*, 59, p. 89–101.

OKALI, D. U. U.; HALL, J. B. 1974. Die-back of *Pistia stratiotes* on Volta-lake, Ghana. *Nature*, 248, p. 452–3.

OKALI, D. U. U.; HALL, J. B.; LAWSON, G. W. 1973. Root distribution under a thicket clump on the Accra Plains, Ghana. *J. Ecol.*, 61, p. 439–54.

OLDEMAN, R. A. A. 1974. L'architecture de la forêt Guyanaise. *Mém. ORSTOM*, 73, p. 1–204.

OLINDO, P. M. 1972. Fire and conservation of the habitat in Kenya. *Proc. Ann. Tall Timbers Fire Ecol. Conf.*, 11, p. 243–57.

OLIVIER, M. C. 1979. An annotated systematic check list of the Angiospermae of the Worcester Veld Reserve. *J. S. Afr. Bot.*, 45, p. 49–62.

OLIVIER, R. C. D.; LAURIE, W. A. 1974. Habitat utilization by hippopotamus in the Mara River. *E. Afr. Wildl. J.*, 12, p. 249–71.

OLLIER, C. D.; LAWRANCE, C. J.; BECKETT, P. H. T.; WEBSTER, R. 1969. *Land systems of Uganda: terrain classification and data storage*. Christchurch, Hants, Military Engineering Experimental Establishment. 234 p. (MEXE Rep. 859.)

ONOCHIE, C. F. A. 1961. A report on the fire-control experiment in Anara Forest Reserve. *Tech. Note*, 14, p. 1–12. Dep. For. Res. Nigeria.

ORMEROD, W. E. 1976. Ecological effects of control of African trypanosomiasis. *Science*, 191, p. 815–21.

OSMASTON, H. A. 1968. Uganda. In: Hedberg, I.; Hedberg, O. (eds.), p. 148–51.

OWAGA, M. L. 1975. The feeding ecology of wildebeest and zebra in Athi-Kaputei plains. *E. Afr. Wildl. J.*, 13, p. 375–83.

OWEN, D. F. 1976. *Animal ecology in tropical Africa*. 2nd ed. London, New York, Longman. 132 p.

OWEN, R. E. A. 1970. Some observations on the sitatunga in Kenya. *E. Afr. Wildl. J.*, 8, p. 181–95.

OWOSEYE, J. A.; SANFORD, W. W. 1972. An ecological study of *Vellozia schnitzleinia*, a drought-enduring plant of Northern Nigeria. *J. Ecol.*, 60, p. 807–17.

OZENDA, P. 1954. Observations sur la végétation d'une région semi-aride. Les Hauts-Plateaux du Sud-Algérois. *Bull. Soc. Hist. nat. Afr. N.*, 45, p. 189–223.

——. 1958. *Flore du Sahara septentrional et central*. Paris, CNRS. 486 p.

——. 1971. Sur une extension de la notion de zone et d'étage subméditerranéens. *C.r. somm. Séanc. Soc. Biogéogr.*, 47 (415), p. 92–103.

——. 1977. *Flore du Sahara*. Paris, CNRS. 622 p.

PAGE, C. N. 1979. Macaronesian heathlands. In: Specht, R. L. (ed.), p. 117–23.

PAHAUT, P.; VAN DER BEN, D. 1962. *Carte Sols Vég. Congo, Rwanda, Burundi*, 18, *Bassin de la Karuzi*. A, Sols (1:50000). B, Végétation (1:50000). *Notice explicative*, p. 1–48. Brussels, INEAC.

PAPADAKIS, J. 1970. *Agricultural potentialities of world climates*. Buenos Aires, Papadakis. 70 p.

PARADIS, G. 1975a. Observations sur les forêts marécageuses du Bas-Dahomey: localisation, principaux types, évolution au cours du Quaternaire récent. *Ann. Univ. Abidjan*, sér. E (Écologie), 8 (1), p. 281–315.

——. 1975b. Physionomie, composition floristique et dynamisme des formations végétales d'une partie de la Basse Vallée de l'Ouémé (Dahomey). *Ann. Univ. Abidjan*, sér. C (Sciences), 9, p. 65–101.

——. 1976. Contribution à l'étude de la flore et de la végétation littorales du Dahomey. *Bull. Mus. natn. Hist. nat., Paris*, 383 (Bot., 26), p. 33–67.

——. 1980. Un cas particulier de zones dénudées dans les mangroves d'Afrique de l'Ouest: celles dues à l'extraction de sel. *Bull. Mus. natn. Hist. nat., Paris*, sér. 4, 2, sect. B, p. 227–61.

PARADIS, G.; DE SOUZA, S.; HOUNGNON, P. 1978. Les stations à *Lophira lanceolata* dans la mosaïque forêt–savane du Sud-Bénin (ex Sud-Dahomey). *Bull. Mus. natn. Hist. nat., Paris*, 521 (Bot., 35), p. 39–58.

PARKER, I. S. C.; GRAHAM, A. D. 1971. The ecological and economic basis for game ranching in Africa. *Symp. Brit. ecol. Soc.*, 11, p. 393–404.

PARRIS, R.; CHILD, G. 1973. The importance of pans to wildlife in the Kalahari and the effect of human settlement in these areas. *J. S. Afr. Wildl. Mgmt Assoc.*, 3, p. 1–8.

PATON, T. R. 1961. Soil genesis and classification in Central Africa. *Soils Fertil.*, 24, p. 249–51.

PAULIAN, R. 1947. *Observations écologiques en forêt de basse Côte d'Ivoire*. Paris, Lechevalier. 148 p.

PAULIAN, R.; BETSCH, J. M.; GUILLAUMET, J. L.; BLANC, C.; GRIVEAUD, P. 1971. Études des écosystèmes montagnards dans la région malgache. I. Le massif de l'Andringitra. Géomorphologie, climatologie et groupements végétaux. *Bull. Soc. Écol.*, 2, p. 189–266.

PAULIAN, R.; BLANC, C.; GUILLAUMET, J. L.; BETSCH, J. M.; GRIVEAUD, P.; PEYRIERAS, A. 1973. Études des écosystèmes montagnards dans la région malgache. II. Les chaînes anosyennes. Géomorphologie, climatologie et groupements végétaux. *Bull. Mus. natn. Hist. nat., Paris*, sér. 3, 118, p. 1–40.

PAULIAN, R.; GÈZE, B. 1940. Les étages de végétation sur les massifs volcaniques du Cameroun occidental. *C.r. somm. Séanc. Soc. Biogéogr.*, 17 (148), p. 57–61.

PEARSALL, W. H. 1957. Report on an ecological survey of the Serengeti National Park, Tanganyika, November and December 1956. *Oryx*, 4, p. 71–136.

PEARSON, H. H. W. 1907. Some notes on a journey from Walfish Bay to Windhuk. *Kew Bull.*, p. 339–60.

PECROT, A.; LÉONARD, A. 1980. *Carte Sols Vég. Congo belge*, 16. *Dorsale du Kiwu*. A, Sols (1:50000; 1:500000). B, Végétation (1:500000). *Notice explicative*, p. 1–124. Brussels, INEAC.

PEDRO, J. Gomes; BARBOSA, L. A. Grandvaux. 1955. A Vegetação. In: *Esboço do Reconhecimento Ecológico-Agricola de Moçambique. Mems Trab. Cent. Invest. cient. algod.*, no. 23, vol. 2, p. 67–224, with coloured vegetation map 1:2000000.

PEETERS, L. 1964. Les limites forêt–savane dans le nord du Congo en relation avec le milieu géographique. *Rev. Belg. Géogr.*, 88, p. 239–70.

PELTIER, J. P. 1971. Contribution à la flore du Haouz oriental. *Bull. Soc. Sci. nat. Maroc*, 50, p. 27–35.

PENNYCUICK, L. 1975. Movements of the migratory wildebeest population in the Serengeti area between 1960 and 1973. *E. Afr. Wildl. J.*, 13, p. 65–87.

PENNYCUICK, L.; NORTON-GRIFFITHS, M. 1976. Fluctuations in the rainfall of the Serengeti ecosystem, Tanzania. *J. Biogeogr.*, 3, p. 125–40.

PENZHORN, B. L.; ROBBERTSE, P. J.; OLIVIER, M. C. 1974. The

influence of the African elephant on the vegetation of the Addo Elephant National Park. *Koedoe*, 17, p. 137–58.

PEREIRA, H. C. (ed.). 1962. Hydrological effects of changes in land use in some East African catchment areas. *E. Afr. agric. For. J.*, 27 (spec. issue), p. 1–131.

——. 1973. *Land use and water resources.* Cambridge, Cambridge Univ. Press.

——. 1977. Land-use in semi-arid southern Africa. In: Hutchinson, J. B. (ed.), p. 555–63.

PERRAUD, A. 1971. Les sols. In: Le milieu naturel de la Côte d'Ivoire. *Mém. ORSTOM*, 50, p. 265–391. See also Guillaumet & Adjanohoun, 1971.

PERRIER DE LA BÂTHIE, H. 1921a. La végétation malgache. *Annls Mus. colon. Marseille*, sér. 3, 9, p. 1–268.

——. 1921b. Note sur la constitution géologique et la flore des îles Chesterfield, Juan-de-Nova, Europa et Nosy-Trozona. *Bull. Éc. Mad.*, p. 170–6.

——. 1936. *Biogéographie des plantes de Madagascar.* Paris, Soc. Éd. Géogr. Mar. et Col. 156 p.

PETERSEN, J. C. B.; CASEBEER, R. L. 1971. A bibliography relating to the ecology and energetics of East African large mammals. *E. Afr. Wildl. J.*, 9, p. 1–23.

PETRIDES, G. A. 1963. Ecological research as a basis for wildlife management in Africa. In: Watterson, G. (ed.), *Conservation of nature and natural resources*, p. 284–93.

——. 1974. The overgrazing cycle as a characteristic of tropical savannas and grasslands in Africa. In: *Proc. 1st Int. Congr. Ecol. (The Hague, September 1974)*, p. 86–91. Wageningen, Centre for Agricultural Publishing and Documentation. 414 p.

PEYERIMHOFF, P. de. 1941. *Carte forestière de l'Algérie et de la Tunisie* (1:1 500 000, in colour). *Notice explicative*, 70 p. Alger, Service des Forêts.

PEYRE, C. 1973. Quelques aspects de la végétation du massif du Bou-Ibane. *Trav. RCP*, 249, 1, p. 129–55. Paris, CNRS.

PEYRE DE FABRÈGUES, B.; LEBRUN, J. P. 1976. *Catalogue des plantes vasculaires du Niger.* Maisons-Alfort, IEMVT. 431 p. (Étud. bot. 3.)

PHILLIPS, J. F. V. 1926a. General biology of the flowers, fruits and young regeneration of the more important species of the Knysna forests. *S. Afr. J. Sci.*, 23, p. 366–417.

——. 1926b. Wild pig (*Potamochoerus choeropotamus*) at the Knysna. *S. Afr. J. Sci.*, 23, p. 655–60.

——. 1927. The role of the bushdove, *Columba arquatrix* T. & K. in fruit dispersal in the Knysna forests. *S. Afr. J. Sci.*, 24, p. 435–40.

——. 1928a. The principal forest types in the Knysna region. An outline. *S. Afr. J. Sci.*, 25, p. 181–201.

——. 1928b. Plant indicators in the Knysna region. *S. Afr. J. Sci.*, 25, p. 202–24.

——. 1930. Some important vegetation communities in the Central Province of Tanganyika Territory. *J. Ecol.*, 18, p. 193–234.

——. 1931a. Forest succession and ecology in the Knysna region. *Mem. bot. Surv. S. Afr.*, 14, p. 1–327.

——. 1931b. A sketch of the floral regions of Tanganyika Territory. *Trans. R. Soc. S. Afr.*, 19, p. 363–72.

——. 1956. Aspects of the ecology and productivity of some of the more arid regions of southern and eastern Africa. *Vegetatio*, 7, p. 38–68.

——. 1959. *Agriculture and ecology in Africa.* London, Faber & Faber. 424 p.

——. 1965. Fire as master and servant. Its influence in the bioclimatic regions of Trans-Saharan Africa. *Proc. Ann. Tall Timbers Fire Ecol. Conf.*, 4, p. 7–109.

——. 1968. The influence of fire in trans-Saharan Africa. In: Hedberg, I.; Hedberg, O. (eds.), p. 13–20.

——. 1971. *Physiognomic classification of the more common vegetation types in South Africa, including Moçambique.* Johannesburg, Loxton, Hunting & Ass. 30 p. (Mimeo.)

——. 1972. Fire in Africa. A brief re-survey. *Proc. Ann. Tall Timbers Fire Ecol. Conf.*, 11, p. 1–7.

——. 1974. Effects of fire in forest and savanna ecosystems of sub-Saharan Africa. In: Kozlowski, T. T.; Ahlgren, C. E. (eds.), p. 435–81.

PHILLIPSON, J. 1975. Rainfall, primary production and 'carrying capacity' of Tsavo National Park (East), Kenya. *E. Afr. Wildl. J.*, 13, p. 171–201.

PHIPPS, J. B.; GOODIER, R. 1962. A preliminary account of the plant ecology of the Chimanimani Mountains. *J. Ecol.*, 50, p. 291–319.

PIAS, J. 1970. *La végétation du Tchad. Ses rapports avec les sols. Variations paléobotaniques au Quaternaire.* ORSTOM. 47 p., with coloured vegetation map 1:1 500 000. (Trav. et docum. 6.)

PICHI-SERMOLLI, R. E. G. 1938. Ricerche botaniche nella regione del Lago Tano e nel Semièn. In: Danielli, G., et al., *Missione di Studio al Lago Tana, 1. Relazioni preliminari*, p. 77–103. Rome, R. Accad. Ital. Centro Studi AOI.

——. 1939. Aspetti del paesaggio vegetale nell'alto Semièn (Africa Orientale Italiana). *Nuovo G. bot. ital.*, n.s., 45, p. CXV–CXXIV.

——. 1940. Osservazioni sulla vegetazione del versante occidentale dell'Altipiano Etiopico. *Nuovo G. bot. ital.*, n.s., 47, p. 609–23.

——. 1955. Tropical East Africa (Ethiopia, Somaliland, Kenya, Tanganyika). In: *Plant ecology/Écologie végétale*, p. 302–60. Paris, Unesco. (Arid zone research/Recherches sur la zone aride, 6.)

——. 1957. Una carta (1:5 000 000) geobotanica dell'Africa orientale (Eritrea, Ethiopia, Somalia). *Webbia*, 13, p. 15–132.

PIELOU, E. C. 1952. Notes on the vegetation of the Rukwa Rift Valley, Tanganyika. *J. Ecol.*, 40, p. 383–92.

PIENAAR, U. de V. 1963. The large mammals of the Kruger National Park—their distribution and present-day status. *Koedoe*, 6, p. 1–37.

PIERLOT, R. 1966. Structure et composition de forêts denses d'Afrique centrale, spécialement celles du Kivu. *Mém. Acad. r. Sci. Outre-Mer, Cl. Sci. nat. méd.*, n.s., 8°, vol. 16 (4), p. 1–367.

PITOT, A. 1950a. Botanique. In: Contribution à l'étude de l'Aïr. Contribution à l'étude de la flore. *Mém. IFAN*, 10, p. 31–81.

——. 1950b. L'action de l'alizé sur quelques espèces ligneuses dans la presqu'île du Cap Vert. *Conf. int. Afr. occid. 2a. Conf. Bissau, 1947*, 2 (1). *Trabalhos apresentados à 2a. Secção (Meio biologico)*, p. 285–300.

——. 1953. Feux sauvages, végétation et sols en Afrique occidentale française. *Bull. IFAN*, 15, p. 1369–83.

——. 1954. Végétation et sols et leurs problèmes en A.O.F. *Annls Éc. sup. Sci. Dakar*, 1, p. 129–39.

PITOT, A.; ADAM, J. G. 1954–55. Sénégal, Mauritanie. In: *Notices botaniques et itinéraires commentés publiés à l'occasion du VIIIe Congrès international de botanique Paris–Nice 1954*, V (3). Republished in 1955 in: *Annls Éc. sup. Sci. Dakar*, 2, p. 21–139, with map 1:500 000.

PITT-SCHENKEL, C. J. W. 1938. Some important communities of warm temperate rain forest at Magamba, West Usambara, Tanganyika Territory. *J. Ecol.*, 26, p. 50–81.

PLAISANCE, G. 1959. *Les formations végétales et paysages ruraux. Lexique et guide bibliographique.* Paris, Gauthier-Villars. 423 p.

PÓCS, T. 1974. Bioclimatic studies in the Uluguru Mountains (Tanzania, East Africa). I. *Acta bot. hung.*, 20, p. 115–35.

——. 1975. Affinities between the bryoflora of East Africa and Madagascar. *Boissiera*, 24, p. 125–8.

——. 1976a. Bioclimatic studies in the Uluguru Mountains (Tanzania, East Africa). II. Correlations between orography, climate and vegetation. *Acta bot. hung.*, 22, p. 163–83.

——. 1976b. Vegetation mapping in the Uluguru Mountains (Tanzania, East Africa). *Boissiera*, 24, p. 477–98, with coloured vegetation map 1:50000.

——. 1976c. The rôle of the epiphytic vegetation in the water balance and humus production of the rain forests of the Uluguru Mountains, East Africa. *Boissiera*, 24, p. 499–503.

——. 1977. Epiphyllous communities and their distribution in East Africa. *Bryophyt. Biblioth.*, 13, p. 681–713.

POISSONET, J.; CÉSAR, J. 1972. Structure spécifique de la strate herbacée dans la savane à palmier ronier de Lamto (Côte d'Ivoire). *Ann. Univ. Abidjan*, sér. E (Écologie), 5, p. 577–601.

POLE EVANS, I. B. 1936. A vegetation map (1:3000000, in colour) of South Africa. *Mem. bot. Surv. S. Afr.*, 15, p. 1–23.

——. 1948a. A reconnaissance trip through the Eastern portion of the Bechuanaland Protectorate, April 1931, and an expedition to Ngamiland, June–July, 1937. *Mem. bot. Surv. S. Afr.*, 21, p. 1–203.

——. 1948b. Roadside observations on the vegetation of East and Central Africa. *Mem. bot. Surv. S. Afr.*, 22, p. 1–305.

POLHILL, R. M. 1968. Tanzania. In: Hedberg, I.; Hedberg, O. (eds.), p. 166–78, with small-scale vegetation map.

PONS, A.; QUÉZEL, P. 1955. Contribution à l'étude de la végétation des rochers maritimes du littoral de l'Algérie centrale et occidentale. *Bull. Soc. Hist. nat. Afr. N.*, 46, p. 48–80.

POORE, M. E. D. 1962. The method of successive approximation in descriptive ecology. *Adv. ecol. Res.*, 1, p. 35–68.

——. 1963. Problems in the classification of tropical rain forests. *J. trop. Geogr.*, 17, p. 12–19.

POPOV, G. B. 1957. The vegetation of Socotra. *J. Linn. Soc. Bot.*, 55, p. 706–35.

PORTÈRES, R. 1946. Climat et végétation sur la chaîne des Bambuttos (Cameroun). *Bull. Soc. bot. Fr.*, 93, p. 352–60.

——. 1950. Problèmes sur la végétation de la Basse Côte d'Ivoire. *Bull. Soc. bot. Fr.*, 97, p. 153–6.

——. 1957. Paysages floristiques des parcours culturaux en Afrique tropicale. *C.r. somm. Séanc. Soc. Biogéogr.*, 34 (294), p. 16–20.

POSNETT, N. W.; REILLY, P. M. 1977. *Ethiopia.* 194 p. Tolworth Tower, Surbiton, Surrey, Land Resources Division, Ministry of Overseas Development. (Land resource bibliography 9.)

POTTS, G.; TIDMARSH, C. E. 1937. An ecological study of a piece of Karroo-like vegetation near Bloemfontein. *J. S. Afr. Bot.*, 3, p. 51–92, with vegetation map 1:20000.

PRATT, D. J. 1966. Control of *Disperma* in semi-desert dwarf shrub grassland. *J. appl. Ecol.*, 3, p. 277–91.

PRATT, D. J.; GREENWAY, P. J.; GWYNNE, M. D. 1966. A classification of East African rangeland, with an appendix on terminology. *J. appl. Ecol.*, 3, p. 369–82, with 3 maps 1:3000000.

PRATT, D. J.; GWYNNE, M. D. (eds.). 1977. *Rangeland management and ecology in East Africa.* London, Hodder & Stoughton. 310 p.

PREUSS, P. 1892. Bericht über eine botanische Exkursion in die Urwald- und-Grasregion des Kamerungebirges und auf den Kamerun-Pic. *Mitt. dt. Schutzgeb.*, 5, p. 28–44.

PROCTER, J. 1974. The endemic flowering plants of the Seychelles: an annotated list. *Candollea*, 29, p. 345–87.

PROCTOR, J.; CRAIG, G. C. 1978. The occurrence of woodland and riverine forest on the serpentine of the Great Dyke. *Kirkia*, 11, p. 129–32.

PUFF, C. 1978. Zur Biologie von *Myrothamnus flabellifolius* Welw. (Myrothamnaceae). *Dinteria*, 14, p. 1–20.

PYNAERT, L. 1933. La mangrove congolaise. *Bull. Agric. Congo belge*, 24, p. 185–207.

QUÉZEL, P. 1952. Contribution à l'étude phytogéographique et phytosociologique du Grand Atlas calcaire. *Mém. Soc. Sci. nat. Maroc*, 50, p. 1–56.

——. 1954. Contribution à l'étude de la flore et de la végétation du Hoggar. *Inst. Rech. sahar. Univ. Alger, Monogr. Rég.*, 2, p. 1–164.

——. 1956. Contribution à l'étude des forêts de chênes à feuilles caduques d'Algérie. *Mém. Soc. Hist. nat. Afr. N.*, n.s., 1, p. 1–57.

——. 1957a. *Peuplement végétal des hautes montagnes de l'Afrique du Nord.* Paris, Lechevalier. 463 p. (Encycl. Biogéogr., vol. 10.)

——. 1957b. Les groupements végétaux du massif du Tefedest. *Trav. Inst. Rech. sahar., Alger*, 15, p. 43–57.

——. 1958. Mission botanique au Tibesti. *Mém. Inst. Rech. sahar., Alger*, 4, p. 1–357.

——. 1959. La végétation de la zone Nord-Occidentale du Tibesti. *Trav. Inst. Rech. sahar., Alger*, 18, p. 75–107.

——. 1960. Flore et palynologie sahariennes. Quelques aspects de leur signification biogéographique et paléoclimatique. *Bull. IFAN*, sér. A, 22, p. 353–60.

——. 1965a. *La végétation du Sahara, du Tchad à la Mauritanie.* Stuttgart, Gustav Fischer; Paris, Masson. 333 p. (Geobot. sel., 2.) *See also* Léonard, J., 1980.

——. 1965b. Contribution à l'étude de l'endémisme chez les phanérogames sahariens. *C.r. somm. Séanc. Soc. Biogéogr.*, 41 (359), p. 89–103.

——. 1965c. L'endémisme dans la flore de l'Algérie. *C.r. somm. Séanc. Soc. Biogéogr.*, 41 (361), p. 137–49.

——. 1967. Signification biogéographique et paléoclimatique de quelques représentants de la flore saharienne. *Palaeoecol. Afr.*, 2, p. 62–7.

——. 1969 *Flore et végétation des plateaux du Darfur Nord-occidental et du jebel Gourgeil (Rép. du Soudan).* CNRS. 146 p. (Dossier n° 5 de la Recherche Coopérative sur Programme n° 45, Populations anciennes et actuelles des confins Tchado–Soudanais.)

——. 1970. A preliminary description of the vegetation in the Sahel region of North Darfur. *Sudan Notes Rec.*, 51, p. 119–25.

——. 1971. Flora und Vegetation der Sahara. In: Schiffers, H. (ed.), *Die Sahara und ihre Randgebiete*, 1, p. 429–75. Munich, Weltforum. 674 p.

——. 1976–77. Les forêts du pourtour méditerranéen. *Notes techniques du MAB 2*, 1976, p. 9–33/Forests of the Mediterranean basin. *MAB technical notes 2*, 1977, p. 9–32. Paris, Unesco.

——. 1978. Analysis of the flora of Mediterranean and Saharan Africa. *Ann. Mo. Bot. Gdn*, 65, p. 479–534.

QUÉZEL, P.; BOUNAGA, D. 1975. Aperçu sur la connaissance actuelle de la flore d'Algérie et de Tunisie. In: *La flore du bassin méditerranéen*, p. 125–39. Paris, CNRS. (Coll. Int., 235.)

QUÉZEL, P.; BRUNEAU DE MIRÉ, P.; GILLET, H. 1964. *Carte internationale du tapis végétal. Feuille de Largeau* (1:1 000 000, in colour). Chad Govt.

QUÉZEL, P.; SANTA, S. 1962–63. *Nouvelle flore de l'Algérie et des régions désertiques méridionales*, 1, p. 1–565; 2, p. 566–1170. Paris, CNRS.

QUÉZEL, P.; SIMONNEAU, P. 1960. Note sur la végétation halophile au Sahara occidental. *Bull. Res. Counc. Israel*, 8, D (Botany), p. 253–62.

——; ——. 1962. Contribution à l'étude phytosociologique du Sahara Occidental. L'action des irrigations sur la végétation spontanée. *Annls agron.*, 13, p. 221–53.

RABINOWITZ, D. 1978. Dispersal properties of mangrove propagules. *Biotropica*, 10, p. 47–57.

RADWANSKI, S. A.; OLIVIER, C. D. 1959. A study of an East African catena. *J. Soil. Sci.*, 10, p. 149–68.

RADWANSKI, S. A.; WICKENS, G. E. 1967. The ecology of *Acacia albida* on mantle soils in Zalingei, Jebel Marra, Sudan. *J. appl. Ecol.*, 4, p. 569–79.

RAINEY, R. C. 1977. Rainfall: scarce resource in 'opportunity country'. In: Hutchinson, J. B. (ed.), p. 439–55.

RAINEY, R. C.; WALOFF, Z.; BURNETT, G. F. 1957. The behaviour of the Red Locust (*Nomadacris septemfasciata* Serville) in relation to the topography, meteorology and vegetation of the Rukwa Rift Valley, Tanganyika. *Anti-Locust Bull.*, 26, p. 1–96.

RAMSAY, D. McC. 1958. The forest ecology of central Darfur. *Sudan Govt For. Bull.*, n.s., 1, p. 1–80.

——. 1964. An analysis of Nigerian savanna. II. An alternative method of analysis and its application to the Gombe sandstone vegetation. *J. Ecol.*, 52, p. 457–66.

RAMSAY, D. McC.; DE LEEUW, P. N. 1964. Id. I. The survey area and the vegetation developed over Bima sandstone. *J. Ecol.*, 52, p. 233–54.

——; ——. 1965*a*. Id. III. The vegetation of the Middle Gongola region by soil parent materials. *J. Ecol.*, 53, p. 643–60.

——; ——. 1965*b*. Id. IV. Ordination of vegetation developed on different parent materials. *J. Ecol.*, 53, p. 661–77.

RAMSAY, J. M.; ROSE INNES, R. 1963. Some quantitative observations on the effects of fire on the Guinea savanna vegetation of Northern Ghana over a period of eleven years. *Afr. Soils*, 8, p. 41–85.

RATTRAY, J. M. 1957. The grasses and grass associations of Southern Rhodesia. *Rhod. agric. J.*, 54, p. 197–234, with small-scale coloured vegetation map.

——. 1960. *The grass cover of Africa*. Rome, FAO. 168 p., with coloured vegetation map 1:10 000 000. (FAO Agric. Stud., 49.)

——. 1961. Vegetation types of Southern Rhodesia. *Kirkia*, 2, p. 68–93, with coloured vegetation map 1:2 500 000.

RATTRAY, J. M.; WILD, H. 1955. Report on the vegetation of the alluvial basin of the Sabi valley and adjacent areas. *Rhod. agric. J.*, 52, p. 484–501.

——; ——. 1961. Vegetation map (1:2 500 000, in colour) of the Federation of Rhodesia and Nyasaland. *Kirkia*, 2, p. 94–104.

RAUH, W. 1973. Über die Zonierung und Differenzierung der Vegetation Madagaskars. *Tropische und subtropische Pflanzenwelt*, 1. Mainz, Akad. Wiss. Lit. 146 p.

——. 1975. Morphologische Beobachtungen an Dornsträuchern des Mediterrangebietes. In: *La flore du bassin méditerranéen*, p. 261–71. Paris, CNRS. (Coll. Int. 235.)

——. 1979. Problems of biological conservation in Madagascar. In: Bramwell, D. (ed.), p. 405–21.

RAYNAL, A. 1963. Flore et végétation des environs de Kayar (Sénégal): de la côte au Lac Tanma. *Annls Fac. Sci. Univ. Dakar*, 9, p. 121–231, with coloured vegetation map 1:10 000.

——. 1971. Répartition géographique des *Nymphoides* (Menyanthaceae) d'Afrique. *Mitt. Bot. Staatssamml. Münch.*, 10, p. 122–34.

RAYNAL, A.; RAYNAL. J. 1961. Observations botaniques dans la région de Bamako. *Bull. IFAN*, 23, sér. A, p. 994–1021.

RAYNAL, J. 1964. *Étude botanique des pâturages du Centre de Recherches Zoo-techniques de Dahra-Djoloff (Sénégal)*, I, p. 1–99; II, *Carte* (1:20 000, in colour). Paris, ORSTOM. (Sect. Bot.)

——. 1968. Groupements herbacés et phytosociologie au Sénégal. *Rév. gén. Sci.*, 74, p. 349–56.

——. 1971. Répartition géographique des *Rhynchospora* africains et malgaches. *Mitt. Bot. Staatssamml. Münch.*, 10, p. 135–48.

REDHEAD, J. F. 1966. The Nigerian savanna: review of the history and study of the vegetation. *Obeche*, 2, p. 21–34.

REDINHA, J. 1961. Nomenclaturas nativas para as formações botânicas do Nordeste de Angola. *Agron. angol.*, 13, p. 55–78.

REEKMANS, M. 1980*a*. La végétation de la basse Rusizi (Burundi). *Bull. Jard. bot. nat. Belg.*, 50, p. 401–44.

——. (1980*b*.) La flore vasculaire de l'Imbo (Burundi) et sa phénologie. *Lejeunia*, n.s., 100, p. 1–53.

REILLY, P. M. 1978. *Ethiopia*. Tolworth Tower, Surbiton, Surrey, Land Resources Division, Ministry of Overseas Development. 280 p. (Land resource bibliography, 10.)

RENIER, H. J. 1954. L'aménagement des forêts naturelles au Ruanda. *Bull. Agric. Congo belge*, 45, p. 1489–96.

RENNIE, J. V. L. 1936. On the flora of a high mountain in South-West Africa. *Trans. R. Soc. S. Afr.*, 23, p. 259–63.

RENVOIZE, S. A. 1971. The origin and distribution of the flora of Aldabra. *Phil. Trans. R. Soc. Lond.*, ser. B, 260, p. 227–36.

——. 1975. A floristic analysis of the western Indian Ocean coral islands. *Kew Bull.*, 30, p. 133–52.

——. 1979. The origin of Indian Ocean island floras. In: Bramwell, D. (ed.), p. 107–29.

RICHARDS, P. W. 1939. Ecological studies on the rain forest of Southern Nigeria. I. The structure and floristic composition of primary forest. *J. Ecol.*, 27, p. 1–61.

——. 1952. *The tropical rain forest*. Cambridge, Cambridge Univ. Press. 450 p.

——. 1957. Ecological notes on West African vegetation. I. The plant communities of the Idanre Hills, Nigeria. *J. Ecol.*, 45, p. 563–77.

——. 1963*a*. Id. II. Lowland forest of the Southern Bakundu Forest Reserve. *J. Ecol.*, 51, p. 123–49.

——. 1963*b*. Id. III. The upland forests of Cameroons Mountain. *J. Ecol.*, 51, p. 529–54.

RICHARDS, P. W.; TANSLEY, A. G.; WATT, A. S. 1940. The recording of structure, life-form and flora of tropical forest communities as a basis for their classification. *J. Ecol.*, 28, p. 224–39. Also published as: *Inst. Pap. Imp. For. Inst.*, no. 19 (1939).

RIKLI, M. 1943–48. *Das Pflanzenkleid der Mittelmeerländer*, 1, p. 1–436 (1943); 2, p. 437–1093 (1946); 3, p. 1094–418 (1948). Bern, Hans Huber.

RIKLI, M.; SCHRÖTER, C. 1912. Vom Mittelmeer zum Nordrand der algerischen Sahara. Pflanzengeographische Excursionen. *Vjschr. Naturf. Ges. Zürich*, 57, p. 3–210.

RIVALS, P. 1952. *Études sur la végétation naturelle de l'Île de la Réunion*. Trav. Lab. For. Toulouse, T. 5, Géographie forestière du Monde, sect. 3, L'Afrique, 1 (2), p. 1–214.

——. 1968. La Réunion. In: Hedberg, I.; Hedberg, O. (eds.), p. 272–5.

ROBERTS, B. R. 1961. Preliminary notes on the vegetation of Thaba 'Nchu. *J. S. Afr. Bot.*, 27, p. 241–51.

——. 1963. Ondersoek in die plantegroei van die Willem Pretorius-Wildtuin. *Koedoe*, 6, p. 137–64.

——. 1966. Observations on the temperate affinities of the vegetation of Hangklip Mountain near Queenstown, C.P. *J. S. Afr. Bot.*, 32, p. 243–60.

——. 1968. The Orange Free State. In: Hedberg, I.; Hedberg, O. (eds.), p. 247–50.

——. 1969. The vegetation of the Golden Gate Highlands National Park. *Koedoe*, 12, p. 15–28.

ROBERTY, G. 1940. Contribution à l'étude phytogéographique de l'Afrique Occidentale Française. *Candollea*, 8, p. 83–150.

——. 1946. Les associations végétales de la vallée moyenne du Niger. *Veröff. geobot. Inst. Zürich*, 22, p. 1–168, with coloured vegetation map 1:61500.

——. 1952a. Les cartes de la végétation ouest-africaine à l'échelle du 1:1000000. *Bull. IFAN*, 14, p. 686–94.

——. 1952b. La végétation du Ferlo. *Bull. IFAN*, 14, p. 777–98.

——. 1955. *Carte de la végétation de l'Afrique occidentale française: Diafarabé*. Paris, ORSTOM. (1:250000, in colour.)

——. 1958. Végétation de la guelta de Soungout (Mauritanie méridionale) en mars 1955. *Bull. IFAN*, sér. A, 20, p. 869–75.

——. 1960. Les régions naturelles de l'Afrique tropicale occidentale. *Bull. IFAN*, sér. A, 22, p. 95–136.

——. 1962–63. *Carte de la végétation de l'Afrique tropicale occidentale à l'échelle de 1:1000000* (in colour). Feuille ND 28 Dakar (1962); Feuille NC 28 Conakry (1962); Feuille NB 28 Bonthe (1963). Paris, ORSTOM.

——. 1964. *Carte de la végétation de l'Afrique tropicale occidentale à l'échelle de 1:1000000. I. Introduction et glossaires. II. Notes de route.* 1. Feuilles NB 28 (Bonthe) et NC 28 (Freetown, Conakry, Bissau). 5. Feuille ND 28 (Dakar) et sa marge nord, NE 28. 02 à 06 (Saint-Louis). Paris, ORSTOM.

ROBINS, R. J. 1976. The composition of the Josani forest, Zanzibar. *J. Linn. Soc. Bot.*, 72, p. 223–34.

ROBYNS, W. 1932. La colonisation végétale des laves récentes du volcan Rumoka (Laves de Kateruzi). *Mém. Inst. r. colon. belge, Sect. Sci. nat. méd.*, 8°, 1, p. 3–33.

——. 1936. Contribution à l'étude des formations herbeuses du District Forestier Central du Congo belge. *Mém. Inst. r. colon. belge, Sect. Sci. nat. méd.*, 4°, vol. 5, p. 1–151.

——. 1937. *Aperçu général de la végétation*. Aspects de végétation des Parcs nationaux du Congo belge. Sér. 1, Parc National Albert, 1 (1–2), p. 1–42.

——. 1938. Considérations sur les aspects biologiques du problème des feux de brousse au Congo belge et au Ruanda–Urundi. *Inst. Roy. Colon. Belg. Bull. Séanc.*, 9, p. 383–420.

——. 1941. Note écologique sur quelques bains d'éléphants au Congo belge. *Inst. Roy. Colon. Belg. Bull. Séanc.*, 12, p. 318–27.

——. 1942. Le concept des phytocénoses biotiques, principalement dans les régions intertropicales. *Bull. Jard. bot. État, Brux.*, 16, p. 413–33.

——. 1948a. *Flore des spermatophytes du Parc National Albert*. 1, *Gymnospermes et Choripétales*: Introduction, p. XV–LV, with small vegetation map in colour. Brussels, Inst. Parcs Nat. Congo belge.

——. 1948b. *Les territoires biogéographiques du Parc National Albert*. Brussels, Inst. Parcs Nat. Congo belge. 51 p.

——. 1950. Botanique du Congo belge. II. La flore. III. La végétation. IV. Les territoires phytogéographiques. *Encycl. Congo belge*, 1, p. 390–424.

RODGERS, W. A. 1976. Seasonal diet preferences of impala from southeast Tanzania. *E. Afr. Wildl. J.*, 14, p. 331–33.

RODGERS, W. A.; HOMEWOOD, K. M. 1979. *The conservation of the East Usambara Mountains, Tanzania. A review of biological values and land use pressures*. Dar es Salaam, Tanzania, Zoology Dept, Univ. of Dar es Salaam. 105 p. (Mimeo.)

RODGERS, W. A.; LUDANGA, R. I. 1973. *The vegetation of the Eastern Selous Game Reserve*. Dar es Salaam, Tanzania, Miombo Research Centre. 67 p., with coloured vegetation map 1:125000. (Mimeo.)

ROGERS, D. J. 1979. *A bibliography of African ecology*. Westport, Conn.; London, England, Greenwood Press. 499 p.

ROGERS, D. J.; MOLL, E. J. 1975. A quantitative description of some coast forests of Natal. *Bothalia*, 11, p. 523–37.

ROLAND, J. C. 1967. Recherches écologiques dans la savane de Lamto (Côte d'Ivoire). Données préliminaires sur le cycle annuel de la végétation herbacée. *La Terre et la Vie*, 114, p. 228–48.

ROLAND, J. C.; HEYDACKER, F. 1963. Aspects de la végétation dans la savane de Lamto (Côte d'Ivoire). *Rev. gén. Bot.*, 70, p. 605–20.

ROLLET, B. 1963. *Carte au 1:1000000 des types de la végétation de la cuvette congolaise au nord de l'équateur*. Brazzaville, Rome, FAO.

——. 1964. *Nord Congo: Introduction à l'inventaire forestier*. Rome, FAO. 116 p. (N° 1782.)

ROOT, A. 1972. Fringe-eared oryx digging for tubers in the Tsavo National Park (East). *E. Afr. Wildl. J.*, 10, p. 155–7.

ROSE INNES, R. 1972. Fire in West African vegetation. *Proc. Ann. Tall Timbers Fire Ecol. Conf.*, 11, p. 147–73.

——. 1977. *A manual of Ghana grasses*. Tolworth Tower, Surbiton, Surrey, Land Resources Division, Ministry of Overseas Development. 265 p.

ROSEVEAR, D. R. 1937. Forest conditions of the Gambia. *Emp. For. J.*, 16, p. 217–26.

——. 1947. Mangrove swamps. *Farm and Forest*, 8, p. 23–30.

——. 1953. *Checklist and atlas of Nigerian mammals*, with a foreword on vegetation. Lagos, Govt Printer. 131 p., with coloured vegetation map of Nigeria 1:3000000 and of West and Central Africa 1:12000000.

——. 1954. Vegetation. In: *The Nigeria handbook*, 2nd ed., p. 139–73, with coloured vegetation map. Lagos, Govt Printer.

ROSS, I. C.; HARRINGTON, G. 1969. The practical aspects of implementing a controlled burning scheme in the Kidepo Valley National Park (second year of operation). *E. Afr. Wildl. J.*, 7, p. 39–42.

ROSS, R. 1954. Ecological studies of the rain forest of Southern Nigeria. III. Secondary succession in the Shasha Forest Reserve. *J. Ecol.*, 42, p. 259–82.

——. 1955a. Some aspects of the vegetation of the sub-alpine zone of Ruwenzori. *Proc. Linn. Soc. Lond.*, 165, p. 136–40.

——. 1955b. Some aspects of the vegetation of Ruwenzori. *Webbia*, 11, p. 451–7.

ROSSETTI, C. 1962. *Observations sur la végétation au Mali*

oriental (*1959*). Rome, FAO. 68 p. (Projet Pélerin. Rapp. n° UNSF/DL/ES/4.)

ROUGERIE, G. 1957. Les pays Agni du sud-est de la Côte d'Ivoire forestière. *Étud. éburn.*, 6, p. 1–242. (With large-scale vegetation map.)

ROUX, E. 1969. *Grass. A story of Frankenwald.* Cape Town, London, New York, Oxford University Press. 212 p.

ROUX, P. W. 1966. Die uitwerking van seisoensreënval en beweiding op gemengde Karooveld. *Proc. Grassl. Soc. S. Afr.*, 1, p. 103–10.

RUBEL, E. 1930. *Pflanzengesellschaften der Erde.* Bern–Berlin, Hans Huber. 464 p., with coloured vegetation map 1:90 000 000.

RUSSELL, E. W. (ed.). 1962. *The natural resources of East Africa.* Nairobi, D. A. Hawkins. 144 p. *See also* Morgan, W. T. W., 1972.

RUTHERFORD, M. C. 1972. Notes on the flora and vegetation of the Omuverume Plateau-Mountain, Waterberg, South West Africa. *Dinteria*, 8, p. 1–55.

——. 1978*a*. Primary production ecology in southern Africa. In: Werger, M. J. A. (ed.), p. 621–59.

——. 1978*b*. Karoo–fynbos biomass along an elevation gradient in the western Cape. *Bothalia*, 12, p. 555–60.

RUXTON, B. P.; BERRY, L. 1960. The Butana grass patterns. *J. Soil Sci.*, 11, p. 61–2.

RYCROFT, H. B. 1944. The Karkloof forest, Natal. *J. S. Afr. For. Ass.*, 11, p. 14–25.

——. 1968. Cape Province. In: Hedberg, I.; Hedberg, O. (eds.), p. 235–9.

RZÓSKA, J. 1974. The Upper Nile swamps: a tropical wetland study. *Freshwat. Biol.*, 4, p. 1–30.

——. (ed.). 1976. *The Nile. Biology of an ancient river.* The Hague, Junk. 417 p. (Monogr. biol. 29.)

SABOUREAU, P. 1959. Propos sur les cyclones et inondations à Madagascar en février et mars 1959. *Bois Forêts Trop.*, 67, p. 3–12.

SAGGERSON, E. P. 1962*a*. Physiography in East Africa. In: Russell, E. W. (ed.), p. 48–51.

——. 1962*b*. The geology of East Africa. In: Russell, E. W. (ed.), p. 52–66. Revised account in: Morgan, W. T. W. (ed.), p. 67–94.

SAINT-AUBIN, G. de. 1961. Aperçu sur la forêt du Gabon. *Bois Forêts Trop.*, 78, p. 3–17.

——. 1963. Les formations végétales et composition de la forêt. In: *La forêt du Gabon*, p. 13–30 (with small-scale vegetation map). Nogent-sur-Marne, Centre Technique Forestier Tropical. 208 p.

SALE, J. B. 1965. The feeding behaviour of rock hyraces (genera *Procavia* and *Heterohyrax*) in Kenya. *E. Afr. Wildl. J.*, 3, p. 1–18.

SALT, G. 1951. The Shira Plateau of Kilimanjaro. *Geogrl J.*, 117, p. 150–64.

——. 1954. A contribution to the ecology of upper Kilimanjaro. *J. Ecol.*, 42, p. 375–423.

SANFORD, W. W. 1968. Distribution of epiphytic orchids in semi-deciduous tropical forest in southern Nigeria. *J. Ecol.*, 56, p. 697–705.

——. 1969. The distribution of epiphytic orchids in Nigeria in relation to each other and to geographical location and climate, type of vegetation and tree species. *Biol. J. Linn. Soc.*, 1, p. 247–85.

——. 1974. The use of epiphytic orchids to characterize vegetation in Nigeria. *J. Linn. Soc. Bot.*, 68, p. 291–301.

SARAIVA, A. C. Coutinho. 1961. *Conspectus da Entomofauna Cabo-Verdiana*, pt 1. Estudos, Ensaios e Documentos, no. 83, p. 1–189 (with coloured 'biogeographical' map). Lisbon, Junta Invest. Ultram.

SARLIN, P. 1969. Répartition des espèces forestières de la Côte d'Ivoire. *Bois Forêts Trop.*, 126, p. 3–14.

SAUER, E. G. F. 1971. Zur Biologie der wilden Strausse Südwestafrikas. *Z. Kölner Zoo*, 14, p. 43–64.

SAUER, J. D. 1961. Coastal plant geography of Mauritius. *Cstl Stud. Ser.*, 5, p. 1–153. Baton Rouge, La, Louisiana State Univ. Press.

——. 1962. Effects of recent tropical cyclones on the coastal vegetation of Mauritius. *J. Ecol.*, 50, p. 275–90.

——. 1965. Notes on seashore vegetation of Kenya. *Ann. Mo Bot. Gdn*, 52, p. 438–43.

——. 1967. *Plants and man on the Seychelles coast.* Madison, Milwaukee, London, Univ. of Wisconsin Press. 132 p.

SAUVAGE, C. 1946. Notes botaniques sur le Zemmour oriental (Mauritanie septentrionale). *Mém. off. natn. anti-acrid., Alger*, 2, p. 1–46.

——. 1948. Les environs de Goulimine, carrefour botanique. *Vol. jubilaire Soc. Sci. nat. Maroc 1920–1945*, p. 106–46.

——. 1949. Les reliques de la flore tropicale au Maroc. *Bull. Soc. Sci. nat. Maroc*, 29, p. 117–30.

——. 1961. Recherches géobotaniques sur les subéraies marocaines. *Trav. Inst. sci. chérif.*, sér. bot., 21, p. 1–462 (with coloured vegetation map 1:500).

——. 1963. *Atlas du Maroc. Notice explicative.* Sect. 2. Physique du globe et météorologie. Planche 6*b*, Étages bioclimatiques (1:2 000 000), p. 1–44. Rabat, Inst. Sci. Chérif.

——. 1971. Excursion botanique au Maroc (8–21 mai 1965). *Al Awamia*, 40, p. 1–100; 41, p. 105–214.

SAVORY, B. M. 1962. Boron deficiency in Eucalypts in Northern Rhodesia. *Commonw. For. Rev.*, 41, p. 118–26.

——. 1963. Site quality and tree root morphology in Northern Rhodesia. *Rhod. J. agric. Res.*, 1, p. 55–64.

SAVORY, J. H. 1953. A note on the ecology of *Rhizophora* in Nigeria. *Kew Bull.*, p. 127–8.

SCAËTTA, H. 1933. Les précipitations dans le bassin du Kivu et dans les zones limitrophes du fossé tectonique (Afrique centrale équatoriale). *Mém. Inst. r. colon. belge*, sect. sci. nat. méd., 4°, vol. 2 (2), p. 1–106.

——. 1934. Le climat écologique de la dorsale Congo–Nil. Op. cit., vol. 3, p. 1–335.

——. 1936. Les pâturages de haute montagne en Afrique centrale. *Bull. Agric. Congo belge*, 27, p. 323–78.

——. 1941. Les prairies pyrophiles de l'Afrique occidentale française. II. Les clairières à Graminées de la forêt humide subéquatoriale et de la forêt sèche tropicale. *Rev. Bot. appl. Agric. trop.*, 21, p. 221–40.

SCHEEPERS, J. C. 1978. The vegetation of Westfalia Estate on the north-eastern Transvaal Escarpment. *Mem. bot. Surv. S. Afr.*, 42, p. 1–230 (with large-scale vegetation map).

SCHEFFLER, G. 1900. Über die Beschaffenheit des Usambara-Urwaldes und über den Laubwechsel an Bäumen desselben. *Notizbl. bot. Gart. Mus., Berl.-Dahlem*, 3, p. 139–66.

SCHENCK, H. 1907. Beiträge sur Kenntnis der Vegetation der Canarischen Inseln. In: Chun, C. (ed.), *Wiss. Ergebn. dt. Tiefsee-Exped. 'Valdivia' 1898–1899*, 2 (1), p. 225–406. Jena, Gustav Fischer.

SCHIFFERS, H. 1971. *Die Sahara und ihr Randgebiete. Darstellung eines Naturgrossraumes*, 1. Physiographie. Munich, Weltforum. 674 p.

SCHIMPER, A. F. W. 1898. *Pflanzengeographie auf physiologischer Grundlage.* Jena, Fischer. 877 p.

——. 1903. *Plant geography upon a physiological basis* (transl.

W. R. Fischer). Ed. P. Groom & I. B. Balfour. Oxford, Clarendon Press. 839 p.

——. 1935. *Pflanzengeographie auf physiologischer Grundlage.* 3rd ed. Rev. F. C. von Faber. Jena, Fischer. 2 vols., 1613 p.

SCHMID, E. 1949. Prinzipien der natürlichen Gliederung der Vegetation des Mediterrangebietes. *Ber. schweiz. bot. Ges.*, 59, p. 169–200 (with coloured vegetation map 1:15 000 000).

——. 1950. Zur Vegetationsanalyse numidischer Eichenwälder. *Ber. geobot. Forsch. Inst. Rübel 1949*, p. 23–39.

——. 1952. Natürliche Vegetationsgliederung am Beispiel des Spanischen Rif. *Ber. geobot. Forsch. Inst. Rübel 1951*, p. 55–79.

——. 1954. Beiträge zur Flora und Vegetation der Kanarischen Inseln. *Ber. geobot. Forsch. Inst. Rübel 1953*, p. 28–49.

——. 1976. The laurisilva of Hierro. In: Künkel, G. (ed.), p. 241–8.

SCHMIDT, W. 1973. Vegetationskundliche Untersuchungen im Savannenreservat Lamto (Elfenbeinküste). *Vegetatio*, 28, p. 145–200.

——. 1975a. The vegetation of the northeastern Serengeti National Park, Tanzania. *Phytocoenologia*, 3, p. 30–82.

——. 1975b. Plant communities on permanent plots of the Serengeti Plains. *Vegetatio*, 30, p. 133–45.

SCHMITZ, A. 1950. Principaux types de végétation forestière dans le Haut-Katanga. *C.r. Congr. sci. Elisabethville 1950*, 4, p. 276–304.

——. 1952a. Essai sur la délimitation des régions naturelles dans le Haut-Katanga. *Bull. Agric. Congo belge*, 43, p. 697–734.

——. 1952b. Note sur l'expérimentation forestière portant sur l'effet du feu dans le Haut-Katanga (Congo belge). *CCTA. Première Conf. For. Interafr., Abidjan, 1951*, p. 401–12.

——. 1962. Les muhulu du Haut-Katanga méridional. *Bull. Jard. bot. État Brux.*, 32, p. 221–99.

——. 1963a. Aperçu sur les groupements végétaux du Katanga. *Bull. Soc. r. Bot. Belg.*, 96, p. 233–447.

——. 1963b. Climax et forêts claires du Parc national de l'Upemba. *Publs Univ. Elisabethville*, 6, p. 57–68.

——. 1971. La végétation de la plaine de Lubumbashi (Haut-Katanga). *Publs INEAC*, sér. sci., 113, p. 1–388.

——. 1977. *Atlas des formations végétales du Shaba (Zaïre).* Fondation Arlon, Belgium, Univ. Luxemb. 96 p. (Sér. doc., 4.)

SCHNELL, R. 1950a. Contribution préliminaire à l'étude botanique de la Basse-Guinée française. *Étud. guiné.*, 6, p. 31–76.

——. 1950b. *La forêt dense. Introduction à l'étude botanique de la région forestière d'Afrique occidentale.* Paris, Lechevalier. 330 p.

——. 1950c. Note sur le peuplement végétal des montagnes de l'Afrique occidentale et particulièrement du massif du Nimba. *C.r. 1re Conf. intern. Afr. Ouest*, 1, p. 496–504.

——. 1950d. Études préliminaires sur la végétation et la flore des hauts plateaux du Mali (Fouta-Djalon). *Bull. IFAN*, 12, p. 905–26.

——. 1950e. Esquisse de la végétation côtière de la basse Guinée française. *Conf. int. Afr. occid. 2a Conf. Bissau, 1947*, 2 (1). *Trab. apresentados à 2a Secção (Meio biol.)*, p. 201–14.

——. 1950f. État actuel des recherches sur la végétation de l'Afrique intertropicale française. *Vegetatio*, 2, p. 331–40.

——. 1952a. Végétation et flore de la région montagneuse du Nimba. *Mém. IFAN*, 22, p. 1–604.

——. 1952b. Contribution à une étude phytosociologique et phytogéographique de l'Afrique occidentale. Les groupements et les unités géobotaniques de la région guinéenne. *Mém. IFAN*, 18, p. 45–234.

——. 1952c. Végétation et flore des monts Nimba. *Vegetatio*, 3, p. 350–406.

——. 1957. Remarques sur les forêts des montagnes ouest-africaines (Guinée et Côte d'Ivoire) et leur individualisation floristique. *Bull. Jard. bot. État Brux.*, 27, p. 279–87.

——. 1960. Notes sur la végétation et la flore des plateaux gréseux de la moyenne Guinée et de leurs abords. *Rev. gén. Bot.*, 67, p. 325–98.

——. 1961. Contribution à l'étude botanique de la chaîne de Fon (Guinée). *Bull. Jard. bot. État Brux.*, 31, p. 15–54.

——. 1968. Guinée. In: Hedberg, I.; Hedberg, O. (eds.), p. 69–72.

——. 1970–71. *Introduction à la phytogéographie des pays tropicaux: les problèmes généraux.* Vol. 1. *Les flores—les structures*, p. 1–499 (1970). Vol. 2. *Les milieux—les groupements végétaux*, p. 500–951 (1971). Paris, Gauthier-Villars.

——. 1977. *La flore et la végétation de l'Afrique tropicale.* Paris, Gauthier-Villars. Vol. 1, 459 p.; vol. 2, 378 p.

SCHOLZ, H. 1971. Einige botanische Ergebnisse einer Forschungsreise in die libysche Sahara (April 1970). *Willdenowia*, 6, p. 341–69.

SCHULZ, E. 1979. Zur Flora und Vegetation der Randgebiete des Murzuk-Beckens (Fezzan–Libyen und Nord-Niger). *Willdenowia*, 9, p. 239–59.

SCHULZE, E. D. *et al.* 1976. Environmental control of Crassulacean acid metabolism in *Welwitschia mirabilis* Hook. fil. in its range of natural distribution in the Namib Desert. *Oecologia* (Berl.), 24, p. 323–34.

SCHULZE, R. E.; MCGEE, O. S. 1978. Climatic indices and classifications in relation to the biogeography of southern Africa. In: Werger, M. J. A. (ed.), p. 19–52.

SCHÜTTE, K. H. 1960. Trace element deficiencies in Cape vegetation. *J. S. Afr. Bot.*, 26, p. 45–9.

SCHWARTZ, O. 1939. Flora des tropischen Arabien. *Mitt. Inst. allg. Bot., Hamburg*, 10, p. 1–393.

SCHWEICKERDT, H. C. 1933. A preliminary account of the vegetation in the neighbourhood of the Zoutpan in the Northern Transvaal. *S. Afr. J. Sci.*, 30, p. 270–9.

SCHWEINFURTH, G. 1868. Pflanzengeographische Skizze des gesamten Nilgebiets und der Uferländer des Roten Meeres. *Petermanns geogr. Mitt.*, 14, p. 113–29, 155–69, 244–8.

——. 1891. Über die Florengemeinschaft von Südarabien und Nord-abessinien. *Verh. Ges. Erdk. Berl.*, 9–10, p. 1–20.

SCOTT, G. D. 1967. Studies of the lichen symbiosis. 3. The water relations of lichens on granite kopjes in Central Africa. *Lichenologist*, 3, p. 368–85.

SCOTT, H. 1952a. Journey to the Gughé Highlands (Southern Ethiopia) 1948–1949. *Proc. Linn. Soc. Lond.*, 163, p. 85–189.

——. 1952b. [1953]. La végétation de la haute Éthiopie centrale et méridionale. *Lejeunia*, 16, p. 67–80.

——. 1955. Journey to the High Simien District, Northern Ethiopia 1952–1953. *Webbia*, 11, p. 425–50.

——. 1958. Biogeographical research in High Simien (Northern Ethiopia), 1952–1953. *Proc. Linn. Soc. Lond.*, 170, p. 1–91.

SCOTT, J. D. 1934. Ecology of certain plant communities of the Central Province, Tanganyika Territory. *J. Ecol.*, 22, p. 177–229.

——. 1951. Conservation of vegetation in South Africa. *Bull. Commonw. Bur. Past. Fld Crops*, 41, p. 9–27.

―――. 1972. Veld burning in Natal. *Proc. Ann. Tall Timbers Fire Ecol. Conf.*, 11, p. 33–51.
SCOTT, R. M. 1962. The soils of East Africa. In: Russell, E. W. (ed.), p. 67–76. Revised account in Morgan, W. T. W. (ed.), p. 95–105.
SCOTT, R. M.; WEBSTER, R.; LAWRANCE, C. J. 1971. *A land system atlas of western Kenya.* Christchurch, Hants, Military Engineering Experimental Establishments. 363 p., with map 1:500 000.
SCULTZ, J. 1976. *Land use in Zambia.* Pt I: *The basically traditional land use systems and their regions.* Pt II: *Land use map.* Munich, Weltforum Verlag. 215 p., with coloured map in 4 sheets, 1:750 000.
SEAGRIEF, S. C. 1962. The Lukanga swamps, Northern Rhodesia, *J. S. Afr. Bot.*, 28, p. 3–7.
SEAGRIEF, S. C.; DRUMMOND, R. B. 1958. Some investigations on the vegetation of the north-eastern part of Makarikari salt pan, Bechuanaland. *Proc. Trans. Rhod. scient. Ass.*, 46, p. 103–33.
SEBALD, O. 1972. Bericht über botanische Studien und Sammlungen bei Lalibela, am Tana-See und im Awash-Tal (Äthiopien). *Stuttgarter Beitr. Naturk.*, 236, p. 1–32.
SEDDON, G. 1974. Xerophytes, xeromorphs and sclerophylls. The history of some concepts in ecology. *Biol. J. Linn. Soc.*, 6, p. 65–87.
SÉGALEN, P.; MOUREAUX, C. 1949. La végétation de la région de Befandriana (Bas Mangoky). *Mém. Inst. sci. Madagascar*, sér. B, *Biol. vég.*, 2, p. 141–57.
SEIF EL DIN, A.; OBEID, M. 1971*a*. Ecological studies of the vegetation of the Sudan. II. The germination of seeds and establishment of seedlings of *Acacia senegal* (L.) Willd. under controlled conditions in the Sudan. *J. appl. Ecol.*, 8, p. 191–201.
―――; ―――. 1971*b*. Id. IV. The effect of simulated grazing on the growth of *Acacia senegal* (L.) Willd. seedlings. *J. appl. Ecol.*, 8, p. 211–16.
SEINER, F. 1911. Pflanzengeographische Beobachtungen in der Mittel-Kalahari. *Bot. Jb.*, 46, p. 1–50.
SENNI, L. 1935. *Gli alberi e le formazioni legnose della Somalia.* Florence, Inst. Agric. Colon. Ital. 305 p.
SHACHORI, A. Y.; MICHAELI, A. 1965. Water yields of forests, maquis and grass covers in semi-arid regions. A literature review. In: *Methodology of plant eco-physiology. Proceedings of the Montpellier Symposium/Méthodologie de l'éco-physiologie végétale. Actes du colloque de Montpellier*, p. 467–77. Paris, Unesco. (Arid zone research/Recherches sur la zone aride, XXV.)
SHANTZ, H. L.; MARBUT, C. F. 1923. *The vegetation and soils of Africa.* American Geogr. Soc., Res. Ser., 13, p. 1–263, with coloured vegetation map 1:10 000 000.
SHANTZ, H. L.; TURNER, B. L. 1958. *Photographic documentation of vegetational changes in Africa over a third of a century.* Univ. of Arizona, Coll. of Agriculture. 158 p. (Rep. 169.)
SHAW, J. 1875. On the changes going on in the vegetation of South Africa through the introduction of the Merino sheep. *J. Linn. Soc. Bot.*, 14, p. 202–8.
SHEPPE, W.; OSBORNE, T. 1971. Patterns of use of a flood plain by Zambian mammals. *Ecol. Monogr.*, 41, p. 179–205.
SHEWRY, P. R.; WOOLHOUSE, H. W.; THOMPSON, K. 1979. Relationships of vegetation to copper and cobalt in the copper clearings of Haut Shaba, Zaire. *Bot. J. Linn. Soc.*, 79, p. 1–35.
SILLANS, R. 1951–52*a*. Contribution à l'étude phytogéographique des savanes du Haut-Oubangui. Note préliminaire sur la composition floristique de quelques 'kagas' (rochers). *Bull. Mus. natn. Hist. nat., Paris*, sér. 2, vol. 23, p. 542–7 (1951); II. Le Kaga Mbrès, vol. 23, p. 685–91 (1951); III. Les Kagas des Mbrès à la rivière Kuku, vol. 24, p. 108–13 (1952*a*).
―――. 1952*b*. Contribution à l'étude phytogéographique des savanes du Haut-Oubangui. Note préliminaire sur la végétation des termitières géantes. *Bull. Soc. bot. Fr.*, 99, p. 2–4.
―――. 1952*c*. Contribution à l'étude phytogéographique des savanes du Haut-Oubangui. Note préliminaire sur la végétation de quelques formations rocheuses du N-W oubanguien. *Bull. Mus. natn. Hist. nat., Paris*, sér. 2, 24, p. 382–91.
―――. 1954. Étude préliminaire de la végétation du Haut-Oubangui et du Haut-Chari. *Bull. IFAN*, sér. A, 16, p. 637–773.
―――. 1958. *Les savanes de l'Afrique centrale.* Paris, Lechevalier. 423 p.
SIMONNEAU, P. 1954*a*. La végétation des sols salés d'Oranie. Sur quelques modifications de l'association à *Suaeda fruticosa* et *Sphenopus divaricatus*. *Annls agron.*, sér. A, 5, p. 91–117.
―――. 1954*b*. Id. Les groupements à *Atriplex* dans les plaines sublittorales. *Annls agron.*, sér. A, 5, p. 225–57.
SIMPSON, C. D. 1975. A detailed vegetation study on the Chobe River in North-East Botswana. *Kirkia*, 10, p. 185–227.
SINCLAIR, A. R. E. 1974. The natural regulation of buffalo populations in East Africa. *E. Afr. Wildl. J.*, 12, p. 135–54, 169–83, 185–200, 291–311.
―――. 1977. *The African buffalo.* Chicago, London, University of Chicago Press. 355 p.
SINCLAIR. A. R. E.; GWYNNE, M. D. 1972. Food selection and competition in the East African buffalo (*Syncerus caffer* Sparrman). *E. Afr. Wildl. J.*, 10, p. 77–89.
SINCLAIR, A. R. E.; NORTON-GRIFFITHS, M. (eds.). 1979. *Serengeti: dynamics of an ecosystem.* Chicago, London, Univ. of Chicago Press. 389 p.
SJÖGREN, E. (1973). Plant communities of the natural vegetation of Madeira and the Azores. *Monogr. Biol. Canar.*, 4, p. 107–11.
―――. (1974). Local climatic conditions and zonation of vegetation on Madeira. *Agron. Lusit.*, 36, p. 95–139.
―――. (1978). Bryophyte vegetation in the Azores Islands. *Mem. Soc. Brot.*, 26, p. 1–283.
SMITH, J. 1949. Distribution of tree species in the Sudan in relation to rainfall and soil texture. *Bull. Minist. Agric. Sudan*, 4, p. 1–63.
SNOWDEN, J. D. 1933. A study in altitudinal zonation in South Kigezi and on Mounts Muhavura and Mgahinga, Uganda. *J. Ecol.*, 21, p. 7–27.
―――. 1953. *The grass communities and mountain vegetation of Uganda.* London, Crown Agents. 94 p., with small-scale vegetation map.
SOBANIA, N. W. 1979. *Background history of the Mount Kulal Region of Kenya.* Nairobi, Kenya, UNEP. (IPAL Tech. Rep. A-1, UNEP–MAB Integrated Project on Arid Lands.)
SOCIÉTÉ DE BIOGÉOGRAPHIE. 1946. See Allorge, P. & V. et al., 1946.
SÖRLIN, A. 1957. Om vegetationen på Seychellerna. *Sv. bot. Tidskr.*, 51, p. 135–58.
SOURIE, R. 1954. Contribution à l'étude écologique des côtes rocheuses du Sénégal. *Mém. IFAN*, 38, p. 1–342.

SOUSA, E. P. 1958. Observações acerca da distribuição e área das espécies consideradas mais significativas da flora da Guiné Portuguesa. *6a Conf. Int. Afr. Occid.*, 3, p. 139–53.

SPECHT, R. L. (ed.). 1979. Heathlands and related shrublands: descriptive studies. *Ecosystems of the world*, 9a. Amsterdam, Oxford, New York, Elsevier Scientific. 497 p.

SPENCE, D. H. N.; ANGUS, A. 1971. African grassland management—burning and grazing in Murchison Falls National Park, Uganda. *Symp. Brit. ecol. Soc.*, 11, p. 319–31.

SPICHIGER, R.; PAMARD, C. 1973. Recherches sur le contact forêt–savane en Côte d'Ivoire. Étude du recrû forestier sur des parcelles cultivées en lisière d'un îlot forestier dans le sud du pays baoulé. *Candollea*, 28, p. 21–37.

SPINAGE, C. A. 1972. The ecology and problems of the Volcano National Park, Rwanda. *Biol. Cons.*, 4, p. 194–204.

SPINAGE, C. A.; GUINESS, F. E. 1971. Tree survival in the absence of elephants in the Akagera National Park, Rwanda. *J. appl. Ecol.*, 8, p. 723–8.

——; ——. 1972. Effects of fire in the Akagera National Park and Mutara Hunting Reserve, Rwanda. *Rev. Zool. Bot. Afr.*, 86, p. 302–36.

STAPF, O. 1904. Die Gliederung der Gräserflora von Südafrika. Eine pflanzengeographische Skizze. In: *Festschrift zu P. Ascherson's siebstigsten Geburtstage*, p. 391–412. Berlin.

STAPLES, R. R.; HUDSON, W. K. 1938. *An ecological survey of the mountain areas of Basutoland*. London, Crown Agents. 68 p., with coloured vegetation map 1:500000.

STAUB, F.; GUEHO, J. 1968. The Cargados Carajos Shoals. *Proc. Roy. Soc. Arts Sci. Maurit.*, 3, p. 7–46.

STEPHEN, I.; BELLIS, E.; MUIR, A. 1956. Gilgai phenomena in tropical black clays of Kenya. *J. Soil Sci.*, 7, p. 1–9.

STEWART, D. R. M. 1971a. Diet of *Lepus capensis* and *L. crawshayi*. *E. Afr. Wildl. J.*, 9, p. 161–2.

——. 1971b. Food preferences of *Pronolagus*. *E. Afr. Wildl. J.*, 9, p. 163.

——. 1971c. Seasonal food preferences of *Lepus capensis* in Kenya. *E. Afr. Wildl. J.*, 9, p. 163–6.

——. 1971d. Food preferences of an impala herd. *J. Wildl. Mgmt*, 35, p. 86–93.

STEWART, D. R. M.; STEWART, J. 1971. Comparative food preferences of five East African ungulates at different seasons. *Symp. Brit. ecol. Soc.*, 11, p. 351–66.

STEWART, M. M. 1972. Relation of fire to habitat preference of small mammals on the Nyika Plateau, Malawi. *Soc. Malawi J.*, 25, p. 33–42.

STOCKER, O. 1926. Die ägyptisch–arabische Wüste. *Vegetationsbilder*, 17, t. 25–36.

——. 1927. Das Wadi Natrun. *Op. cit.*, 18, t. 1–6.

STODDART, D. R. 1967. Summary of the ecology of coral islands north of Madagascar (excluding Aldabra). *Atoll Res. Bull.*, 118, p. 53–61.

——. (ed.). 1970. Coral islands of the Western Indian Ocean. *Atoll Res. Bull.*, 136, p. 1–224.

STODDART, D. R.; WRIGHT, C. A. 1967. Geography and ecology of Aldabra atoll. *Atoll Res. Bull.*, 118, p. 11–52.

STORY, R. 1952. A botanical survey of the Keiskammahoek District. *Mem bot. Surv. S. Afr.*, 27, p. 1–84, with vegetation map 1:54000.

——. 1958. Some plants used by the Bushmen in obtaining food and water. *Mem. bot. Surv. S. Afr.*, 30, p. 1–115.

STRAKA, H. 1960. Über Moore und Torf auf Madagaskar und den Maskarenen. *Erdkunde*, 14, p. 81–98.

STRANG, R. M. 1973. Bush encroachment and veld management in South-central Africa: the need for a re-appraisal. *Biol. Conserv.*, 5, p. 96–104.

——. 1974. Some man-made changes in successional trends on the Rhodesian highveld. *J. appl. Ecol.*, 11, p. 249–63.

STREEL, M. 1962. Les savanes boisées à *Acacia* et *Combretum* de la Lufira moyenne. *Bull. Séanc. Acad. r. Sci. Outre-Mer*, n.s., 8 (2), p. 229–55.

——. 1963. La végétation tropophylle des plaines alluviales de la Lufira moyenne (Katanga méridional). Relations du complexe végétation–sol avec la géomorphologie. Liège, Ed. FULREAC, Univ. de Liège. 242 p., with 2 vegetation maps 1:100000.

SUMMERHAYES, V. S. 1931. An enumeration of the angiosperms of the Seychelles. *Trans. Linn. Soc. London*, 2, Zool., 19, p. 261–99.

SUNDING, P. 1970. Elementer i Kanariøenes flora, og teorier til forklaring av floraens opprinnelse. *Blyttia*, 4, p. 229–59.

——. 1972. The vegetation of Gran Canaria. *Norske Vid-Akad. Oslo*, I. Mat.-Naturv. Klasse, N.S., no. 29, p. 1–186, with 2 vegetation maps.

——. 1973a. *A botanical bibliography of the Canary Islands*. 2nd ed. Oslo, Botanical Garden, Univ. of Oslo. 46 p.

——. 1973b. *Check-list of the vascular plants of the Cape Verde Islands*. Oslo, Botanical Garden, Univ. of Oslo. 36 p.

——. 1974. Additions to the vascular flora of the Cape Verde Islands. *Garcia de Orta*, sér. bot., 2, p. 5–30.

——. 1977. A botanical bibliography of the Cape Verde Islands. *Bol. Mus. Munic. Funchal*, 31, p. 100–9.

——. 1979. Origins of the Macaronesian flora. In: Bramwell, D. (ed.), p. 13–40.

SWABEY, C. 1961. *Forestry in the Seychelles*. Mahé, Seychelles, Govt Printer. 34 p.

——. 1970. The endemic flora of the Seychelles Islands and its conservation. *Biol. Conserv.*, 2, p. 171–77.

SWAINE, M. D.; HALL, J. B. 1974. Ecology and conservation of upland forests in Ghana. *Proc. Ghana Scope's Conf. Environ. Dev. West Afr.*, p. 151–8.

SWAMI, K. 1973. *Moisture conditions in the savanna region of West Africa*. Montreal, McGill University. 106 p. (Savanna Res. Ser. no. 18.)

SWIFT, J. 1973. Disaster and a Sahelian nomad economy. In: Dalby, D.; Harrison-Church, J. (eds.), *Drought in Africa*, p. 71–8. London, School of Oriental and African Studies. 124 p.

SWYNNERTON, C. F. M. 1917. Some factors in the replacement of the ancient east African forest by wooded pasture land. *S. Afr. J. Sci.*, 14, p. 493–518.

SYMOENS, J. J. 1953. Note sur la végétation des salines de Mwashya (Katanga). *Bull. Soc. r. Bot. Belg.*, 86, p. 113–21.

——. 1963. Le Parc national de l'Upemba. Son histoire—son intérêt. *Publs Univ. Elisabethville*, 6, p. 43–56.

SYMOENS, J. J.; OHOTO, E. 1973. Les éléments phytogéographiques de la flore macrophytique aquatique et semi-aquatique du Haut-Katanga. *Verh. Int. Verein. Limnol.*, 18, p. 1385–94.

SYNNOTT, T. J. 1977. *Monitoring tropical forests. A review with special reference to Africa*. London, Monitoring and Assessment Research Centre, Chelsea College, Univ. of London. 45 p. (MARC rep. no. 5.)

——. 1979a. *A report on the status, importance and protection of the montane forests*. Nairobi, UNEP–MAB Integrated Project on Arid Lands. 57 p. (IPAL Tech. Rep., no. D-2a.)

——. 1979b. *A report on prospects, problems and proposals for tree planting*. Nairobi, UNEP–MAB Integrated Project on Arid Lands. 41 p. (IPAL Tech. Rep., no. D-2b.)

SYS, C. 1955. The importance of termites in the formation of

latosols in the region of Elisabethville. *Sols Afr.*, 3, p. 393–5.
——. 1960. *Carte des sols du Congo belge et du Rwanda–Urundi (1:5 000 000)*. Notice explicative, p. 1–84. Brussels, INEAC.
Sys, C.; Hubert, P. 1969. *Carte Sols Vég. Congo, Rwanda, Burundi, 24, Mahagi*. A, Sols (1:200 000). *Notice explicative*, p. 1–50. Brussels, INEAC.
Sys, C.; Schmitz, A. 1959. *Carte Sols Vég. Congo belge, 9, Région d'Elisabethville (Haut-Katanga)*. A, Sols (1:60 000). B, Végétation (1:25 000; 1:60 000). *Notice explicative*, p. 1–70. Brussels, INEAC.
Tadros, T. M. 1953. A phytosociological study of the halophilous communities from Mareotis (Egypt). *Vegetatio*, 4, p. 102–24.
——. 1956. An ecological survey of the semi-arid coastal strip of the western desert of Egypt. *Bull. Inst. Désert Égypte*, 6(2), p. 28–56.
Tadros, T. M.; Atta, B. A. M. 1958a. Further contribution to the study of the sociology and ecology of the halophilous communities of Mareotis (Egypt). *Vegetatio*, 8, p. 137–60.
——; ——. 1958b. The plant communities of barley fields and uncultivated desert areas of Mareotis (Egypt). *Vegetatio*, 8, p. 161–75.
Takhtajan, A. 1969. *Flowering plants: origin and dispersal*. Edinburgh, Oliver & Boyd. 310 p.
Talbot, L. M.; Talbot, M. H. 1962. Food preferences of some East African wild ungulates. *E. Afr. agric. For. J.*, 27, p. 131–8.
——; ——. 1963. The wildebeest in western Masailand, East Africa. *Wildl. Monogr.*, 12, p. 1–88.
Tansley, A. G.; Chipp, T. F. (eds.). 1926. *Aims and methods in the study of vegetation*. London, Crown Agents. 383 p.
Taton, A. 1949a. La colonisation des roches granitiques de la région de Nioka (Haut-Ituri, Congo belge). *Vegetatio*, 1, p. 317–32.
——. 1949b. Les principales associations herbeuses de la Région de Nioka et leur valeur agrostologique. *Bull. Agric. Congo belg.*, 40, p. 1884–900.
Taton, A.; Risopoulos, S. 1955. Contribution à l'étude des principales formations marécageuses de la région de Nioka (Distr. du Kibali–Ituri). *Bull. Soc. r. Bot. Belg.*, 87, p. 5–19.
Taylor, C. J. 1952. The vegetation zones of the Gold Coast. *For. Dep. Bull.*, 4, p. 1–12, with coloured vegetation map 1:1 500 000. Accra, Govt Printer.
——. 1960. Vegetation. In: *Synecology and silviculture in Ghana*, p. 31–73. London, Thomas Nelson.
——. 1962. *Tropical forestry with particular reference to West Africa*. London, Oxford Univ. Press. 162 p.
Taylor, H. C. 1953. Forest types and floral composition of Grootvadersbosch. *J. S. Afr. For. Ass.*, 23, p. 33–46.
——. 1961a. Ecological account of a remnant coastal forest near Stanford, Cape Province. *J. S. Afr. Bot.*, 27, p. 153–65.
——. 1961b. The Karkloof forest: a plea for its protection. *For. S. Afr.*, 1, p. 123–30.
——. 1962. A report on the Nxamalala forest. *For. S. Afr.*, 2, p. 29–51.
——. 1963. A bird's eye view of the Cape mountain vegetation. *J. bot. Soc. S. Afr.*, 49, p. 17–19.
——. 1972a. Notes on the vegetation of the Cape Flats. *Bothalia*, 10, p. 637–46.
——. 1972b. Fynbos. *Veld and Flora*, 2, p. 68–75.
——. 1977. Aspects of the ecology of the Cape of Good Hope Nature Reserve in relation to fire and conservation. In: Mooney, H. A.; Conrad, C. E. (eds.), p. 483–6.

——. 1978. Capensis. In: Werger, M. J. A. (ed.), p. 171–229.
——. 1979. Observations on the flora and phytogeography of Rooiberg, a dry fynbos mountain in the Southern Cape Province (Little Karoo), South Africa. *Phytocoenologia*, 6, p. 524–31.
——. 1980. Phytogeography of fynbos. *Bothalia*, 13, p. 231–5.
Teixeira, A. J. da Silva; Barbosa, L. A. Grandvaux. 1958. A agricultura do Arquipélago de Cabo Verde. *Mems Junta Invest. Ultram.*, sér. 2, 2, p. 1–178. With 10 agroclimatic maps in colour (1:50 000, 1:75 000, and 1:100 000). Lisbon, Minist. Ultram.
Teixeira, J. B. 1968a. *Parque nacional do Bicuar. Carta da vegetação e memória descritiva*. Nova Lisboa, Inst. Invest. Agron. Angola. 29 p., with large-scale vegetation map.
——. 1968b. Angola. In: Hedberg, I.; Hedberg, O. (eds.), p. 193–7.
Teixeira, J. B.; Corrêa de Pinho, M. I. 1961. Subsídos para o estudo e caracterização das forragens da Reserva Pastoril de Vila Arriaga. *Agron. Angol.*, 13, p. 3–51.
Teixeira, J. B.; Matos, G. Cardoso de. 1967. *Memória descritiva e carta da vegetação do Centro de Estudos da Chianga*. Nova Lisboa, Inst. Invest. Agron. Angola. 7 p., with large-scale vegetation map.
Teixeira, J. B.; Matos, G Cardoso de; Sousa, J. N. Baptista de. 1967. *Parque Nacional de Quiçama. Carta de Vegetação e Memória descritiva*. Nova Lisboa, Inst. Invest. Agron. Angola. 14 p., with large-scale vegetation map.
Theron, A.; Vindt, J. 1960. *Carte de la végétation du Maroc, Feuille Rabat, Casablanca*. 1:200 000 (in colour). Toulouse, Service Carte Internationale du Tapis Végétal, Faculté des Sciences.
Thomas, A. S. 1941. The vegetation of the Sese Islands, Uganda. An illustration of edaphic factors in tropical ecology. *J. Ecol.*, 29, p. 330–53.
——. 1942. The wild *Arabica* coffee on the Boma Plateau, Anglo–Egyptian Sudan. *J. Exp. Agr.*, 10, p. 207–12.
——. 1943. The vegetation of the Karamoja District, Uganda. An illustration of biological factors in tropical ecology. *J. Ecol.*, 31, p. 149–77.
——. 1945. The vegetation of some hillsides in Uganda. Pt I. *J. Ecol.*, 33, p. 10–43 (1945).
——. 1946. Id. II. tom. cit., p. 153–72.
Thomas, D. B.; Pratt, D. J. 1967. Bush-control studies in the drier areas of Kenya. IV. Effects of controlled burning on secondary thicket in upland *Acacia* woodland. *J. appl. Ecol.*, 4, p. 325–35.
Thomas, H. H. 1921. Some observations on plants in the Libyan desert. *J. Ecol.*, 9, p. 75–89.
Thomas, P. I.; Walker, B. H.; Wild, H. 1977. Relationships between vegetation and environment on an amphibolite outcrop near Nkai, Rhodesia. *Kirkia*, 10, p. 503–41, with vegetation map 1:26 000.
Thomas, R. 1941. *Carte forestière du domaine 1:1 000 000. Commentaire*, p. 1–39. Comité National du Kivu.
Thomas, W. L. (ed.). 1956. *Man's role in changing the face of the earth*. Chicago, Ill., Univ. of Chicago Press.
Thomasson, M. 1974. Essai sur la physionomie de la végétation des environs de Tuléar (sud-ouest malgache). *Bull. Mus. natn. Hist. nat., Paris*, sér. 3, 250 (*Écologie générale*, 22), p. 1–27.
——. 1976. Le fourré d'Orangéa (nord-ouest malgache). *Adansonia*, sér. 2, 15, p. 481–9.
——. 1977. La forêt dense sclérophylle de montagne du Tsiafajavona (Madagascar). *Adansonia*, sér. 2, 16, p. 487–92.

THOMPSON, B. W. 1955. *The climate of Africa*. Nairobi, London, New York, Oxford Univ. Press. 15 p. + 132 maps.

THOMPSON, J. G. 1960. A description of the growth habits of mopani in relation to soil and climatic conditions. *Proc. First Fed. Sci. Congr. Salisbury*, p. 181–6.

——. 1965. The soils of Rhodesia and their classification. *Techn. Bull. Rhod. agric. J.*, 6, p. 1–66.

THOMPSON, K. 1976. Swamp development in the head waters of the White Nile. In: Rzóska, J. (ed.), p. 177–96.

THOMPSON, K.; SHEWRY, P. R.; WOOLHOUSE, H. W. 1979. Papyrus swamp development in the Upemba basin, Zaire: studies of population structure in *Cyperus papyrus* stands. *J. Linn. Soc. Bot.*, 78, p. 299–316.

THOMSON, P. J. 1975. The role of elephants, fire and other agents in the decline of a *Brachystegia boehmii* woodland. *J. S. Afr. Wildl. Mgmt Ass.*, 5, p. 11–18.

TINLEY, K. L. 1966. *An ecological reconnaissance of the Moremi Wildlife Reserve, Northern Okovango Swamps, Botswana*. Johannesburg, Okovango Wildlife Society. 146 p.

——. 1969. Dikdik (*Madoqua kirki*) in South West Africa: notes on distribution, ecology and behaviour. *Madoqua*, 1, p. 7–33, with small vegetation map.

——. 1971. Etosha and the Kaokoveld. Suppl. to *Afr. Wildl.*, 25 (1), p. 1–16.

TOBLER-WOLFF, G.; TOBLER, F. 1915. Vegetationsbilder vom Kilimandscharo. *Vegetationsbilder*, 12, t. 7–18.

TOMASELLI, R. 1976–77. La dégradation du maquis méditerranéen. *Notes techniques du MAB 2*, 1976, p. 35–76/*MAB technical notes 2*, 1977, p. 33–69. Paris, Unesco.

TOMLINSON, P. B.; ZIMMERMAN, M. H. (eds.). 1978. *Tropical trees as living systems*. Cambridge Univ. Press. 675 p.

TOMLINSON, T. E. 1957. Relationship between mangrove vegetation, soil texture and reaction of surface soil after empoldering saline swamps in Sierra Leone. *Trop. Agric., Trin.*, 34, p. 41–50.

TOTHILL, J. D. (ed.). 1940. *Agriculture in Uganda*. London, Oxford Univ. Press. 551 p.

——. (ed.). 1948. *Agriculture in the Sudan*. London, Oxford Univ. Press. 974 p.

TRAPNELL, C. G. 1953. *The soils, vegetation and agriculture of North-Eastern Rhodesia*. 2nd ed. Lusaka, Govt Printer. 146 p.

——. 1959. Ecological results of woodland burning experiments in Northern Rhodesia. *J. Ecol.*, 47, p. 129–68.

TRAPNELL, C. G.; BIRCH, W. R.; BRUNT, M. A.; PRATT, D. J. 1966. *Kenya vegetation. Sheet 1* (map 1:250 000 in colour). Tolworth Tower, Surbiton, Surrey, Directorate of Overseas Surveys.

TRAPNELL, C. G.; BRUNT, M. A.; BIRCH, W. R.; TRUMP, E. C. 1969. *Kenya vegetation. Sheet 3* (map 1:250 000 in colour). Tolworth Tower, Surbiton, Surrey, Directorate of Overseas Surveys.

TRAPNELL, C. G.; CLOTHIER, J. N. 1937. *The soils, vegetation and agricultural systems of North-Western Rhodesia*. Lusaka, Govt Printer. 87 p., with coloured vegetation–soil map 1:2 000 000.

TRAPNELL, C. G.; FRIEND, M. T.; CHAMBERLAIN, G. T.; BIRCH, H. F. 1976. The effects of fire and termites on a Zambian woodland soil. *J. Ecol.*, 64, p. 577–88.

TRAPNELL, C. G.; GRIFFITHS, J. F. 1960. The rainfall–altitude relation and its ecological significance in Kenya. *E. Afr. agric. J.*, 25, p. 207–13.

TRAPNELL, C. G.; LANGDALE-BROWN, I. 1962. The natural vegetation of East Africa. In: Russell, E. W. (ed.), p. 92–102; with coloured vegetation map 1:4 000 000. Revised account in: Morgan, W. T. W. (ed.), p. 127–39.

TRAPNELL, C. G.; MARTIN, J. D.; ALLAN, W. 1950. *Vegetation–soil map of Northern Rhodesia*. Lusaka, Govt Printer. 20 p., map 1:1 000 000, in colour.

TROCHAIN, J. 1940. Contribution à l'étude de la végétation du Sénégal. *Mém. IFAN*, 2, p. 1–433.

——. 1951. Nomenclature et classification des types de végétation en Afrique noire française (deuxième note). *Bull. Inst. Étud. centrafr.*, n.s., 2, p. 9–18.

——. 1952. Les territoires phytogéographiques de l'Afrique noire française d'après leur pluviométrie. *Recl Trav. Labs Bot. Géol. Zool. Univ. Montpellier*, sér. bot., 5, p. 113–24.

——. 1955. Nomenclature et classification des milieux végétaux en Afrique noire française. *Les divisions écologiques du monde*, p. 73–90. (Colloques int. CNRS, 59.) Also published in: *Année biol.*, sér. 3, 31, p. 317–34.

——. 1957. Accord interafricain sur la définition des types de végétation de l'Afrique tropicale. *Bull. Inst. Étud. centrafr.*, n.s., 13–14, p. 55–93.

——. 1969. Les territoires phytogéographiques de l'Afrique noire francophone d'après la trilogie: climat, flore et végétation. *C.r. somm. Séanc. Soc. Biogéogr.*, 45/46 (402), p. 139–57.

——. 1980. *Écologie végétale de la zone intertropicale non désertique*. Toulouse, Université Paul Sabatier. 468 p.

TROCHAIN, J.; KOECHLIN, J. 1958. Les pâturages naturels du sud de l'A.E.F. *Bull. Inst. Étud. centrafr.*, n.s., 15–16, p. 59–83.

TROLL, C. 1935a. Wüstensteppen und Nebeloasen im südnubischen Küstengebirge. *Z. Ges. Erdk. Berl.*, p. 241–81.

——. 1935b. Bericht über eine Forschungsreise durch das östliche Afrika. *Kolon. Rdsch.*, 27, p. 1–34.

——. 1936. Termitensavannen. In: *Länderkundliche Forschung. Festschrift für N. Krebs*, p. 275–312. Stuttgart, Engelhorns Nachf.

——. 1948. Der asymmetrische Aufbau der Vegetationszonen und Vegetationsstufen auf der Nord- und Südhalbkugel. *Ber. geobot. Forsch. Inst. Rübel* (1947), p. 46–83.

——. 1958. Zur Physiognomik der Tropengewächse. *Jber. Ges. Freunden u. Förderern d. Univ. Bonn*, 1958, p. 1–75.

——. 1959. Die tropischen Gebirge. Ihre dreidimensionale klimatische und pflanzengeographische Zonierung. *Bonner geogr. Abh.*, 25, p. 1–93.

——. 1970. Die naturraumliche Gliederung N-Äthiopiens. *Erdkunde*, 24, p. 249–68.

TROLLOPE, W. S. W. 1972. Fire as a method of eradicating macchia vegetation in the Amatole mountains of South Africa.... *Proc. Ann. Tall Timbers Fire Ecol. Conf.*, 11, p. 99–120.

——. 1974. Role of fire in preventing bush encroachment in the eastern Cape. *Proc. Grassl. Soc. S. Afr.*, 9, p. 67–72.

TROUPIN, G. 1966. *Étude phytocénologique du Parc National de l'Akagera et du Rwanda Oriental*. Inst. nat. Rech. sci. Butare, Rép. Rwandaise. 293 p. (Publs 2.)

TULEY, P. 1966. The Obudu Plateau. Utilization of high altitude tropical grassland. *Bull. IFAN*, sér. A, 28, p. 899–911.

TULEY, P.; JACKSON, J. K. 1971. The vegetation of Chappal Waddi (Gangirwal) of the Cameroon Republic/Nigeria border. *Nig. Fld*, 36, p. 3–19.

TURNER, M. I. M.; WATSON, R. M. 1965. An introductory study on the ecology of hyrax (*Dendrohyrax brucei* and *Procavia johnstonii*) in the Serengeti National Park. *E. Afr. Wildl. J.*, 3, p. 49–60.

TURRILL, W. B. 1949. On the flora of St. Helena. *Kew Bull.*, 1948, p. 358–62.

TUTIN, T. G. 1953. The vegetation of the Azores. *J. Ecol.*, 41, p. 53–61.
TWEEDIE, E. M. 1976. Habitats and check-list of plants on the Kenya side of Mount Elgon. *Kew Bull.*, 31, p. 227–57.
TYSON, P. D. 1964. Berg winds of South Africa. *Weather*, 19, p. 7–11.
——. 1978. Rainfall changes over South Africa during the period of meteorological record. In: Werger, M. J. A. (ed.), p. 53–69.
UDVARDY, M. D. F. 1975. A classification of the biogeographical provinces of the world. *IUCN Occ. Pap.*, 18, p. 1–48.
——. 1976. *World biogeographical provinces.* Sausalito, Calif., The Co-Evolution Quarterly. (Coloured map 1:40 000 000.)
UNESCO. 1955. *Plant ecology. Review of research/Écologie végétale. Compte rendu de recherches.* Paris, Unesco. 379 p. (Arid zone research/Recherches sur la zone aride, 6.)
——. 1961. *Tropical soils and vegetation. Proceedings of the Abidjan symposium/Sols et végétation des régions tropicales.* Travaux du colloque d'Abidjan. Paris, Unesco. 115 p. (Humid tropics research/Recherches sur la zone tropicale humide.)
——. 1973. *International classification and mapping of vegetation/Classification internationale et cartographie de la végétation/Clasificación internacional y cartografía de la vegetación.* Paris, Unesco. 93 p. (Ecology and conservation, 6.)
——. 1975. *The Sahel: ecological approaches to land use.* Paris, Unesco. 99 p. (MAB Tech. notes 1.)
——. 1977. *Man and the environment in Marsabit District. Proceedings of a symposium held at Mt. Kulal, December 1966.* Nairobi, Unesco Regional Office for Science and Technology for Africa. (Mimeo.)
——. 1979. *Map of the world distribution of arid regions. Explanatory note.* Paris, Unesco. 54 p., map 1:25 000 000 in colour. (MAB Tech. notes 7.)
UNESCO–FAO. 1963. *Bioclimatic map of the Mediterranean zone. Explanatory notes.* Paris, Unesco; Rome, FAO. 58 p. (Arid zone research, 21.) *Carte bioclimatique (1:5 000 000) de la zone méditerranéenne. Notice explicative.* Paris, Unesco; Rome, FAO. 60 p. (Recherches sur la zone aride, 21.)
——. 1968. *Carte de la végétation de la région méditerranéenne. Notice explicative/Vegetation map of the Mediterranean zone. Explanatory notes.* Paris, Unesco; Rome, FAO. Map 1:5 000 000, in colour + 90 p. notes. (Recherches sur la zone aride/Arid zone research, 30.)
UNESCO/UNEP/FAO. 1978. *Tropical forest ecosystems: a state-of-knowledge report.* Paris, Unesco. 683 p. (Natural resources research, XIV.)
——. 1979. *Tropical grazing land ecosystems: a state-of-knowledge report.* Paris, Unesco. 655 p. (Natural resources research, XVI.)
UNITED KINGDOM. IMPERIAL AGRICULTURAL BUREAUX. 1947. *The use and misuse of shrubs and trees as fodder.* Aberystwyth, Imperial Agricultural Bureaux. 232 + xxxv p. (Joint Publ. no. 10.)
UNITED NATIONS. 1977. See Anon, 1977.
UNITED STATES OF AMERICA. DEPARTMENT OF AGRICULTURE. 1965. *Radiant energy in relation to forest.* (Tech. bull. 1344.)
——. NATIONAL ACADEMY OF SCIENCES. NATIONAL RESEARCH COUNCIL, BOARD ON SCIENCE AND TECHNOLOGY FOR INTERNATIONAL DEVELOPMENT. 1979. *Tropical legumes: resources for the future*, Washington, D.C. 331 p.
VAGELER, P. 1910. Die Mkattaebene. Beiträge zur Kenntnis der ostafrikanischen Alluvialböden und ihrer Vegetation. *Beih. Tropenpfl.*, 11, p. 251–395. With schematic vegetation map 1:300 000.

VAHL, M. 1905. Über die Vegetation Madeiras. *Bot. Jb.*, 36, p. 253–349.
VAHRMEIJER, J. 1966. Notes on the vegetation of northern Zululand. *Afr. Wildl.*, 20, p. 151–61.
VAILLANT, A. 1945. La flore méridionale du lac Tchad. *Bull. Soc. Étud. Cameroun*, 9, p. 13–98.
VANDEN BERGHEN, C. 1977. Observations sur la végétation de l'Île de Djerba (Tunisie méridionale). 1. Introduction et végétation des dunes mobiles. *Bull. Soc. r. Bot. Belg.*, 110, p. 217–27.
——. 1979a. Id. 2. Les dunes fixées. L'association à *Imperata cylindrica* et *Ononis angustissima.* Op. cit., 111, p. 227–36.
——. 1979b. Id. 3. Les dépressions dans les dunes littorales. Op. cit., 112, p. 45–63.
——. 1980. Id. 4. La végétation adventice des Moissons. Op. cit., 113, p. 33–4.
VAN DER BEN, D. 1959. *Exploration hydrobiologique des Lacs Kivu, Édouard et Albert (1952–1954). Résultats scientifiques*, vol. 4, fasc. 1, *La végétation des rives des Lacs Kivu, Édouard et Albert*, p. 1–191. Brussels, Inst. R. Sci. Nat. Belg.
——. 1961. Phytosociologie. In: Pahaut, P.; Van der Ben, D.; Bodeux, A.; Damiean, G., Archives de la mission de la Karuzi (*Urundi*), 1, p. 109–33, 2(B), Carte de Reconnaissance, Végétation (1:50 000). Brussels, Min. Affaires Étrangères.
VAN DER MEULEN, F. 1978. Progress with vegetation studies in the Sourish Mixed Bushveld of the western Transvaal. *Bothalia*, 12, p. 531–6.
——. 1979. *Plant sociology of the western Transvaal Bushveld, South Africa. A syntaxonomic and synecological study.* Vaduz, Cramer. 192 p. (Diss. bot., 49.)
VAN DER MEULEN, F.; WESTFALL, R. H. 1979. A vegetation map of the western Transvaal Bushveld. *Bothalia*, 12, p. 731–5. With vegetation map 1:250 000.
——; ——. 1980. Structural analysis of Bushveld vegetation in Transvaal, South Africa. *J. Biogeogr.*, 7, p. 337–48.
VAN DER PIJL, L. 1957. The dispersal of plants by bats (Chiropterochory). *Acta bot. neerl.*, 6, p. 291–315.
VAN DER SCHIJFF, H. P. 1958. Inleidende verslag oor veldbrandnavorsing in die Nasionale Krugerwildtuin. *Koedoe*, 1, p. 60–93.
——. 1963. A preliminary account of the vegetation of the Mariepskop complex. *Fauna Flora* (Pretoria), 14, p. 42–53.
——. 1964. Die ekologie en verwantskappe van die Sandveldflora van die Nasionale Krugerwildtuin. *Koedoe*, 7, p. 56–76.
——. 1968a. The affinities of the flora of the Kruger National Park. *Kirkia*, 7, p. 109–20.
——. 1968b. Die topografie, geologie en grondsoorte van die Nasionale Krugerwildtuin met verwysing na plantgemeenskappe wat op die verskillende grandsoorte voorkom. *Tydskr. wet. Kuns*, 1, p. 32–50.
VAN DER SCHIJFF, H. P.; SCHOONRAAD, E. 1971. The flora of the Mariepskop complex. *Bothalia*, 10, p. 461–500.
VAN DER WALT, P. T. 1968. A plant ecological survey of the Noorsveld. *J. S. Afr. Bot.*, 34, p. 215–34.
VANDERYST, H. 1932. Introduction à la phytogéographie agrostologique de la province Congo-Kasai. Les formations et associations. *Mém. Inst. r. colon. belge*, sect. sci. nat. méd., 4°, vol. 1 (3), p. 1–154.
——. 1933. Introduction générale à l'étude agronomique du Haut-Kasai. Les domaines, districts, régions et sous-régions géoagronomiques du vicariat apostolique du Haut-Kasai. Op. cit., vol. 1 (7), p. 3–82.

VAN MEEL, L. 1952. Le milieu végétal. In: Leloup, E., *et al.* (eds.), *Exploration hydrobiologique du Lac Tanganyika (1946–1947). Résultats scientifiques*, vol. 1, p. 51–61. Brussels, Inst. R. Sci. Nat. Belg. 165 p.

——. 1953. *Exploration du Parc National de l'Upemba. Contribution à l'étude du lac Upemba.* A: *Le milieu physico-chimique*, fasc. 9, p. 1–190. Brussels, Inst. Parcs Nat. Congo belge.

——. 1966. Le milieu végétal. In: Witte, G. F. de (ed.), *Exploration du Parc National de l'Upemba*, p. 39–120. Mission G. F. de Witte, fasc. 1, *Introduction*. Brussels, Inst. Parc. Nat. Congo. 122 p.

VAN RENSBURG, H. J. 1952. Grass-burning experiments in the Msima River Stock Farm, Southern Highlands, Tanganyika. *E. Afr. agric. J.*, 17, p. 119–29.

——. 1972. Fire: its effects on grasslands, including swamps—Southern, Central and Eastern Africa. *Proc. Ann. Tall Timbers Fire Ecol. Conf.*, 11, p. 175–99.

VAN ROOYEN, M. W.; THERON, G. K.; GROBBELAAR, N. 1979*a*. Phenology of the vegetation in the Hester Malan Nature Reserve in the Namaqualand Broken Veld. 1. General observations. *J. S. Afr. Bot.*, 45, p. 279–93.

VAN ROOYEN, M. W.; THERON, G. H.; GROBBELAAR, N. 1979*b*. Id. 2. The therophyte population. Tom. cit., 45, p. 433–52.

VAN WAMBEKE, A. 1958. *Carte Sols Vég. Congo belge*, 12, *Bengamisa*. A, Sols (1:100000). *Notice explicative*, p. 1–47. Brussels, INEAC.

——. 1959. *Carte Sols Vég. Congo belge*, 13, *Région du Lac Albert.* A, Sols (1:100000). *Notice explicative*, p. 1–50. Brussels, INEAC.

——. 1960. *Carte Sols Vég. Congo, Ruanda–Urundi*, 17, *Région de Yanonge–Yatolema.* A, Sols (1:100000). *Notice explicative*, p. 1–28. Brussels, INEAC.

——. 1963. *Carte des associations des sols du Rwanda et du Burundi* (1:1000000). *Notice explicative*, p. 1–67. Brussels, INEAC.

VAN WAMBEKE, A.; ÉVRARD, C. 1954. *Carte Sols Vég. Congo belge*, 6, *Yangambi (1), Weko.* A, Sols (1:50000). B, Végétation (1:50000). *Notice explicative*, p. 1–23. Brussels, INEAC.

VAN WAMBEKE, A.; LIBEN, L. 1957. *Carte Sols Vég. Congo belge*, 6, *Yangambi (4), Yambaw.* A, Sols (1:25000). B, Végétation (1:25000). *Notice explicative*, p. 1–47. Brussels, INEAC.

VAN WAMBEKE, A.; VAN OOSTEN, M. F. 1956. *Carte Sols Vég. Congo belge*, 8, *Vallée de la Lufira (Haut-Katanga).* A, Sols (1:100000). *Notice explicative*, p. 1–71. Brussels, INEAC.

VAN WYK, P. 1972. Veld burning in the Kruger National Park. An interim report on some aspects of research. *Proc. Ann. Tall Timbers Fire Ecol. Conf.*, 11, p. 9–31.

VAN WYK, P.; FAIRALL, N. 1969. The influence of the African elephant on the vegetation of the Kruger National Park. *Koedoe*, 12, p. 57–89.

VAN ZINDEREN BAKKER, E. M., Jr. 1971. *Ecological investigations on ravine forests of the eastern Orange Free State (South Africa).* Bloemfontein, Univ. of the Orange Free State. 123 p. (Thesis.)

——. 1973. Ecological investigations of forest communities in the eastern Orange Free State and the adjacent Natal Drakensberg. *Vegetatio*, 28, p. 299–334.

VAN ZINDEREN BAKKER, E. M., Sr. 1955. A preliminary survey of the peat bogs on the alpine belt of northern Basutoland. *Acta geogr.* (Helsingfors), 14, p. 413–22.

——. 1965. Über Moorvegetation und den Aufbau der Moore in Süd- und Ostafrika. *Bot. Jb.*, 84, p. 215–31.

VAN ZINDEREN BAKKER, E. M., Sr.; WERGER, M. J. A. 1974. Environment, vegetation and phytogeography of the high-altitude bogs of Lesotho. *Vegetatio*, 29, p. 37–49.

VAN ZYL, J. H. M. 1965. The vegetation of the S.A. Lombard Nature Reserve and its utilisation by certain antelope. *Zool. Afr.*, 1, p. 55–71.

VAUGHAN, R. E. 1968. Mauritius and Rodriguez. In: Hedberg, I.; Hedberg, O. (eds.), p. 265–72.

VAUGHAN, R. E.; WIEHE, P. O. 1937. Studies on the vegetation of Mauritius. I. A preliminary survey of the plant communities. *J. Ecol.*, 25, p. 289–343.

——; ——. 1939. Id. II. The effect of environment on certain features of leaf structure. *J. Ecol.*, 27, p. 263–81.

——; ——. 1941. Id. III. The structure and development of the upland climax. *J. Ecol.*, 29, p. 127–60.

——; ——. 1947. Id. IV. Some notes on the internal climate of the upland climax forest. *J. Ecol.*, 34, p. 126–36.

VENTER, H. J. T. 1976. An ecological study of the dune forest at Mapelana, Cape St. Lucia, Zululand. *J. S. Afr. Bot.*, 42, p. 211–30.

VERBOOM, W. C. 1965. The use of aerial photographs for vegetation surveys in relation with tsetse control and grassland surveys. *Int. train. Cent. aerial Surv.*, Publs ser. B, 28, p. 1–21.

——. 1966. The grassland communities of Barotseland. *Trop. Agric.* (Trin.), 43, p. 107–15.

VERBOOM, W. C.; BRUNT, M. A. 1970. *An ecological survey of Western Province, Zambia, with special reference to fodder resources.* Vol. 1, 95 p., *The environment.* Vol. 2, 133 p., *The grasslands and their development.* Tolworth Tower, Surbiton, Surrey, Directorate of Overseas Surveys.

VERDCOURT, B. 1952. Observations on the ecology of the land and freshwater Mollusca of North-East Tanganyika. *Tanganyika Notes Rec.*, 33, p. 67–82.

——. 1968. French Somaliland. In: Hedberg, I.; Hedberg, O. (eds.), p. 140–1.

——. 1969. The arid corridor between the north-east and south-west areas of Africa. *Palaeoecol. Afr.*, 4, p. 140–4.

——. n.d. The vegetation of the Nairobi Royal National Park. In: Heriz-Smith, S., *The wild flowers of the Nairobi Royal National Park*, p. 38–56, with large-scale vegetation map. Nairobi, Hawkins.

VERDOORN, I. C. 1929. Notes on the vegetation of the Fountains Valley, Pretoria. *S. Afr. J. Sci.*, 26, p. 190–4.

VESEY-FITZGERALD, D. F. 1940. On the vegetation of Seychelles. *J. Ecol.*, 28, p. 465–83.

——. 1942. Further studies of the vegetation on islands in the Indian Ocean. *J. Ecol.*, 30, p. 1–16.

——. 1955*a*. The vegetation of the outbreak areas of the Red Locust in Tanganyika and Northern Rhodesia. *Anti-Locust Bull.*, 20, p. 1–31, with small vegetation map.

——. 1955*b*. Vegetation of the Red Sea coast south of Jedda, Saudi Arabia. *J. Ecol.*, 43, p. 477–89.

——. 1957*a*. The vegetation of the Red Sea coast north of Jedda, Saudi Arabia. *J. Ecol.*, 45, p. 547–62.

——. 1957*b*. The vegetation of central and eastern Arabia. *J. Ecol.*, 45, p. 779–98.

——. 1960. Grazing succession among East African game animals. *J. Mammal.*, 41, p. 161–72.

——. 1963. Central African grasslands. *J. Ecol.*, 51, p. 243–73.

——. 1964. Ecology of the Red Locust. In: Davis, D. H. S. (ed.), p. 255–68.

——. 1965*a*. The utilisation of natural pastures by wild animals in the Rukwa Valley, Tanganyika. *E. Afr. Wildl. J.*, 3, p. 38–48.

——. 1965*b*. Lechwe pastures. *Puku*, 3, p. 143–7.
——. 1966. The habits and habitats of small rodents in the Congo River catchment region of Zambia and Tanzania. *Zool. Afr.*, 2, p. 111–22.
——. 1969. Utilization of the habitat by buffalo in Lake Manyara National Park. *E. Afr. Wildl. J.*, 7, p. 131–45.
——. 1970. The origin and distribution of valley grasslands in East Africa. *J. Ecol.*, 58, p. 51–75.
——. 1972. Fire and animal impact on vegetation in Tanzania National Parks. *Proc. Ann. Tall Timbers Fire Ecol. Conf.*, 11, p. 297–317.
——. 1973*a*. The dynamic aspects of the secondary vegetation in Arusha National Park, Tanzania. *E. Afr. agric. For. J.*, 38, p. 314–27.
——. 1973*b*. Animal impact on vegetation and plant succession in Lake Manyara National Park, Tanzania. *Oikos*, 24, p. 314–24.
——. 1973*c*. Browse production and utilization in Tarangire National Park. *E. Afr. Wildl. J.*, 11, p. 291–305.
——. 1974*a*. Utilization of the grazing resources by buffaloes in the Arusha National Park, Tanzania. *E. Afr. Wildl. J.*, 12, p. 107–34.
——. 1974*b*. The changing state of *Acacia xanthophloea* groves in Arusha National Park, Tanzania. *Biol. Conserv.*, 6, p. 40–7.
VIEWEG, G. H.; ZIEGLER, H. 1969. Zur Physiologie von *Myrothamnus flabellifolius*. *Ber. dt. bot. Ges.*, 82, p. 29–36.
VIGNE, C. 1936. Forests of the Northern Territories of the Gold Coast. *Emp. For. J.*, 15, p. 210–13.
VIGUIER, P. 1946. Les techniques de l'agriculture soudanaise et les feux de brousse. *Rev. int. Bot. appl. Agric. trop.*, 26, p. 42–51.
VILLIERS, J. F. 1973. Étude floristique et phytosociologique d'une mangrove atlantique sur substrat rocheux du littoral gabonais. *Ann. Fac. Sci., Cameroun*, 14, p. 3–46.
VINCENT, V.; THOMAS, R. G. 1961. *An agricultural survey of Southern Rhodesia. I. Agro-ecological survey.* Salisbury, Govt Printer.
VINDT, J. 1959. Notice détaillée de la feuille Rabat–Casablanca de la Carte de la Végétation du Maroc au 1:200000. *Bull. Serv. Carte phytogéogr. Sér. A, Carte de la végétation*, 4, p. 51–147. Paris, CNRS. See also Theron & Vindt, 1960.
VINE, H. 1968. Developments in the study of soils and shifting agriculture in tropical Africa. In: Moss, R. P. (ed.), p. 89–119.
VOLK, O. H. 1964. Die afro-meridional-occidentale Floren-Region in S.W. Afrika. In: Kreeb, K. (ed.), *Beiträge zur Phytologie. Prof. Dr. Heinrich Walter zum 65. Geburtstag gewidmet*, p. 1–16. Stuttgart, Ulmer. (Arb. Landw. Hochsch. Hohenheim, 30.)
——. 1966*a*. Einfluss von Mensch und Tier auf die natürliche Vegetation im tropischen Südwest-Afrika. In: Buchwald, K.; Lendholt, W.; Meyer, K. (eds.), *Beitr. Landespflege*, 2, p. 108–31.
——. 1966*b*. Die Florengebiete von Südwestafrika. *J. S.W. Afr. sci. Soc.*, 20, p. 25–58.
VOLK, O. H.; LEIPPERT, H. 1971. Vegetationsverhältnisse im Windhoeker Bergland, Südwestafrika. Op. cit., 25, p. 5–44.
VOLKENS, G. 1887. *Die Flora der Ägyptisch–Arabischen Wüste auf Grundlage anatomischer–physiologischer Forschungen.* Berlin, Borntraeger. 156 p.
——. 1897. *Der Kilimandscharo.* Berlin.
VON BREITENBACH, F. 1972. Indigenous forests of the Southern Cape. *J. bot. Soc. S. Afr.*, 58, p. 18–47.

VON RICHTER, W.; OSTERBERG, R. 1977. The nutritive values of some major food plants of lechwe, puku and waterbuck along the Chobe River, Botswana. *E. Afr. Wildl. J.*, 15, p. 91–7.
VOORHOEVE, A. G. 1964. Some notes on the tropical rain forest of the Yoma-Gola National Forest near Bomi Hills, Liberia. *Comm. For. Rev.*, 43, p. 17–24.
——. 1965. *Liberian high forest trees.* Wageningen, Centre Agric. Publs Doc. 416 p.
——. 1968. Liberia. In: Hedberg, I.; Hedberg, O. (eds.), p. 74–6.
VUATTOUX, R. 1968. Le peuplement du palmier Rônier (*Borassus aethiopum*) d'une savane de Côte d'Ivoire. *Ann. Univ. Abidjan*, sér. E (écologie), 1 (1), p. 1–138.
——. 1970. Observations sur l'évolution des strates arborée et arbustive dans la savane de Lamto (Côte d'Ivoire). Op. cit., 3, p. 285–315.
WADE, N. 1974. Sahelian drought: no victory for Western aid. *Science*, 185, p. 234–7.
WALTER, H. 1936. Die ökologischen Verhältnisse in der Namib-Nebelwüste (Südwestafrika) unter Auswertung der Aufzeichnungen des Dr. G. Boss (Swakopmund). *Jb. wiss. Bot.*, 84, p. 58–222.
——. 1939. Grasland, Savanne und Busch der ariden Teile Afrikas in ihrer ökologischen Bedingtheit. *Jb. wiss. Bot.*, 85, p. 750–860.
——. 1943. *Die Vegetation Osteuropas.* Berlin, Paul Parey. 180 p.
——. 1954. Die Verbuschung, eine Erscheinung der subtropischen Savannengebiete, und ihre ökologischen Ursachen. *Vegetatio*, 5–6, p. 6–10.
——. 1955*a*. Le facteur eau dans les régions arides et sa signification pour l'organisation de la végétation dans les contrées subtropicales. *Les divisions écologiques du monde*, p. 27–39. (Colloques int. CNRS, 59.) Also published in: *Année biol.*, sér. 3, 31, p. 271–83.
——. 1955*b*. Die Klimadiagramme als Mittel zur Beurteilung der Klimaverhältnisse für ökologische, vegetationskundliche und landwirtschaftliche Zwecke. *Ber. dt. bot. Ges.*, 68, p. 331–44.
——. 1958. Klimatypen dargestellt durch Klimadiagramme. *Geogr. Taschenbuch*, 1958, p. 540–4.
——. 1961. Über die Bedeutung des Grosswilds für die Ausbildung der Pflanzendecke. *Stuttgarter Beitr. Naturk.*, 69, p. 1–6.
——. 1962, 1964, 1973. *Die Vegetation der Erde in öko-physiologischer Betrachtung. I: Die tropischen und subtropischen Zonen*, 538 p. (1962). 2nd ed., 592 p. (1964). 3rd ed., 743 p. (1973). Jena, Fischer.
——. 1963. Climatic diagrams as a means to comprehend the various climatic types for ecological and agricultural purposes. In: Rutter, A. J.; Whitehead, F. H. (eds.), *The water relations of plants. Ecol. Soc. Symp.*, 3, p. 3–9.
——. 1968. *Die Vegetation der Erde in öko-physiologischer Betrachtung. II. Die gemässigten und arktischen Zonen.* Jena, Fischer. 1001 p.
——. 1971. *Ecology of tropical and subtropical vegetation* (transl. D. Mueller-Dombois), ed. J. H. Burnett. New York, Van Nostrand Reinhold. 539 p.
——. 1976*a*. *Die ökologischen Systeme der Kontinente (Biogeosphäre). Prinzipien ihrer Gliederung mit Beispielen.* Stuttgart, New York, Fischer. 131 p.
——. 1976*b*. Vegetationszonen und Klima. Kurze Darstellung in kausaler und globaler Sicht. Stuttgart, Eugen Ulmer. 232 p. Transl. from the 3rd ed. by J. Wieser (1979) as:

Vegetation of the earth and ecological systems of the geobiosphere. New York, Heidelberg, Berlin, Springer-Verlag. 274 p.

WALTER, H.; BOX, E. 1976. Global classification of natural terrestrial ecosystems. *Vegetatio*, 32, p. 75–81.

WALTER, H.; HARNICKELL, E.; MUELLER-DOMBOIS, D. 1975. *Klimadiagramm-Karten der einzelnen Kontinente und die ökologische Klimagliederung der Erde/Climate-diagram maps of the individual continents and the ecological climatic regions of the earth*. Berlin, Heidelberg, New York, Springer-Verlag. 36 p., 9 maps (incl. Africa 1: 7 000 000).

WALTER, H.; LIETH, H. 1960–67. *Klimadiagramm-Weltatlas*. Jena, Fischer.

WALTER, H.; STEINER, M. 1936. Die Ökologie der ostafrikanischen Mangroven. *Z. Bot.*, 30, p. 65–193.

WALTER, H.; VOLK, O. H. 1954. *Grundlagen der Weidewirtschaft in Südwestafrika*. Stuttgart, Ulmer. 281 p.

WALTER, H.; WALTER, E. 1953. Einige allgemeine Ergebnisse unserer Forschungsreise nach Südwestafrika 1952/53. Das Gesetz der relativen Standortskonstanz; das Wesen der Pflanzengemeinschaften. *Ber. dt. bot. Ges.*, 66, p. 228–36.

WARBURG, O. (ed.). 1903. *Kunene-Sambesi-Expedition, H. Baum*. Berlin, Kol. Wirtsch. Kom. 593 p.

WATSON, J. P. 1962*a*. The soil below a termite mound. *J. Soil Sci.*, 13, p. 46–51.

——. 1962*b*. Leached, pallid soils of the African plateau. *Soils Fertil.*, 25, p. 1–4.

——. 1964. A soil catena on granite in Southern Rhodesia. *J. Soil Sci.*, 15, p. 238–57.

——. 1965. Soil catenas. *Soils Fertil.*, 28, p. 307–10.

——. 1967. A termite mound in an Iron Age burial ground in Rhodesia. *J. Ecol.*, 55, p. 663–9.

——. 1969. Water movement in two termite mounds in Rhodesia. *J. Ecol.*, 57, p. 441–51.

——. 1974*a*. Termites in relation to soil formation, groundwater, and geochemical prospecting. *Soils Fertil.*, 37, p. 111–14.

——. 1974*b*. Calcium carbonate in termite mounds. *Nature*, 247, p. 74.

——. 1975. The composition of termite (*Macrotermes* spp.) mounds on soil derived from basic rock in three rainfall zones of Rhodesia. *Geoderma*, 14, p. 147–58.

WATSON, R. M.; BELL, R. H. V. 1969. The distribution, abundance and status of elephant in the Serengeti region of northern Tanzania. *J. appl. Ecol.*, 6, p. 115–32.

WATSON, R. M.; GRAHAM, A. D.; PARKER, I. S. C. 1969. A census of the large mammals of the Loliondo Controlled Area, Northern Tanzania. *E. Afr. Wildl. J.*, 7, p. 43–59.

WATSON, R. M.; KERFOOT, O. 1964. A short note on the intensity of grazing of the Serengeti Plains by plains-game. *Z. Saugetierk.*, 29, p. 317–20.

WEARE, P. R.; YALALA, A. 1971. Provisional vegetation map of Botswana. *Botswana Notes Rec.*, 3, p. 131–47, with vegetation map, in colour.

WEBB, D. A. 1954. Is a classification of plant communities either possible or desirable? *Bot. Tidsskr.*, 51, p. 363–70.

WEBB, J. S.; MILLMAN, A. P. 1951. Heavy metals in vegetation as a guide to ore. A biogeochemical reconnaissance in West Africa. *Trans. Inst. Min. metall. Engrs*, 60, p. 473–504.

WEBSTER, R. 1960. Soil genesis and classification in Central Africa. *Soils Fertil.* 23, p. 77–9.

——. 1965. A catena of soils on the Northern Rhodesia plateau. *J. Soil Sci.*, 16, p. 31–43.

WEIMARCK, H. 1941. Phytogeographical groups, centres and intervals within the Cape flora. *Lunds Univ. Årsskr.*, N.F. Avd. 2, Bd. 37, Nr 5, p. 1–143.

WEINTROUB, D. 1933. A preliminary account of the aquatic and subaquatic vegetation and flora of the Witwatersrand. *J. Ecol.*, 21, p. 44–57.

WEIR, J. S. 1969. Chemical properties and occurrence on Kalahari sand of salt licks created by elephants. *J. Zool.* (Lond.), 158, p. 293–310.

WEISSER, P. J. 1978. Conservation priorities in the dune area between Richards Bay and Mfolozi Mouth based on a vegetation survey. *Natal Tn Reg. Plann. Rep.*, 38, p. 1–64, with 6 large-scale vegetation maps.

WEISSER, P. J.; MARQUES, F. 1979. Gross vegetation changes in the dune area between Richards Bay and the Mfolozi River, 1937–1974. *Bothalia*, 12, p. 711–21.

WELCH, J. R. 1960. Observations on deciduous woodland in the Eastern Province of Tanganyika. *J. Ecol.*, 48, p. 557–73.

WELLINGTON, J. H. 1946. A physiographic regional classification of South Africa. *S. Afr. geogrl J.*, 28, p. 64–86.

——. 1955. *Southern Africa. A geographical study*. 1, *Physical geography*. Cambridge Univ. Press. 528 p., with vegetation map in colour.

WELLS, M. J. 1964. The vegetation of the Jack Scott Nature Reserve. *Fauna Flora* (Pretoria), 15, p. 17–25.

WELSH, R. P. H.; DENNY, P. 1978. The vegetation of Nyumba ya Mungu reservoir, Tanzania. *Biol. J. Linn. Soc.*, 10, p. 67–92.

WERGER, M. J. A. 1973*a*. *Phytosociology of the Upper Orange River Valley, South Africa. A syntaxonomical and synecological study*. Nijmegen, Holland. 222 p. (Thesis.)

——. 1973*b*. An account of the plant communities of Tussen die Riviere Game Farm, Orange Free State. *Bothalia*, 11, p. 165–76.

——. 1973*c*. Notes on the phytogeographical affinities of the Southern Kalahari. *Bothalia*, 11, p. 177–80.

——. 1977. Environmental destruction in southern Africa: the role of overgrazing and trampling. In: Miyawaki, A.; Tüxen, R. (eds.), *Vegetation Sci. Environ. Prot.: Proc. Int. Symp. Tokyo 1974*, p. 301–5. Tokyo, Maruzen.

——. 1978*a*. Biogeographical division of southern Africa. In: Werger, M. J. A. (ed.), p. 145–70.

——. 1978*b*. The Karoo-Namib Region. In: Werger, M. J. A. (ed.), p. 231–99.

——. (ed.). 1978*c*. *Biogeography and ecology of Southern Africa*. The Hague, Junk. 1439 p. (Monogr. biol. 31.)

——. 1978*d*. Vegetation structure in the Southern Kalahari. *J. Ecol.*, 66, p. 933–41.

WERGER, M. J. A.; COETZEE, B. J. 1977. A phytosociological and phytogeographical study of Augrabies Falls National Park, Republic of South Africa. *Koedoe*, 20, p. 11–51.

——; ——. 1978. The Sudano–Zambezian Region. In: Werger, M. J. A. (ed.), p. 301–462.

WERGER, M. J. A.; KRUGER, F. J.; TAYLOR, H. C. 1972*a*. A phytosociological study of the Cape fynbos and other vegetation at Jonkershoek, Stellenbosch. *Bothalia*, 10, p. 599–614.

——; ——; ——. 1972*b*. Pflanzensoziologische Studien der Fynbos-Vegetation am Kap der guten Hoffnung. *Vegetatio*, 24, p. 71–89.

WERGER, M. J. A.; LEISTNER, O. A. 1975. Vegetationsdynamik in der südlichen Kalahari. In: Schmidt, W. (ed.). *Ber. Int. Symp. Int. Verein. Vegetationskde, Rinteln*, p. 135–58. Lehre, Cramer.

WERGER, M. J. A.; WILD, H.; DRUMMOND, R. B. 1978a. Vegetation structure and substrate of the northern part of the Great Dyke, Rhodesia. I. Environment and plant communities. *Vegetatio*, 37, p. 79–90.
——; ——; ——. 1978b. Id. II. Gradient analysis and dominance–diversity relationships. Tom. cit., p. 151–61.
WERTH, E. 1901. Die Vegetation der Insel Sansibar. *Mitt. Semin. orient. Sprach.* (Univ. Berl.), 4, p. 1–97. (Thesis, Berne.)
——. 1915. *Das Deutsch-Ostafrikanische Kustenland und die vorgelagerten Inseln*, 1, 334 p., 2, 265 p. Berlin, Dietrich Reimer.
WEST, O. 1945. Distribution of mangroves in the Eastern Cape Province. *S. Afr. J. Sci.*, 41, p. 238–42.
——. 1951. The vegetation of Weenen County, Natal. *Mem. bot. Surv. S. Afr.*, 23, p. 1–183.
——. 1958. Bush encroachment, veld burning and grazing management. *Rhod. agric. J.*, 55, p. 407–25.
——. 1965. *Fire in vegetation and its use in pasture management with special reference to tropical and subtropical Africa.* Hurley, Berks, Commonw. Bur. Pastures Fld Crops. 53 p. (Mimeo publ. no. 1/1965.)
——. 1972. Fire, man and wildlife as interacting factors limiting the development of climax vegetation in Rhodesia. *Proc. Ann. Tall Timbers Fire Ecol. Conf.*, 11, p. 121–45.
WESTERN, D.; SINDIYO, D. M. 1972. The status of the Amboseli rhino population. *E. Afr. Wildl. J.*, 10, p. 43–57.
WESTERN, D.; VAN PRAET, C. 1973. Cyclical changes in the habitat and climate of an East African ecosystem. *Nature*, 241, p. 104–6.
WETTSTEIN, R. von. 1906. Sokotra. *Vegetationsbilder*, 3, t. 25–30.
WHEATER, R. J. 1972. Problems of controlling fires in Uganda National Parks. *Proc. Ann. Tall Timbers Fire Ecol. Conf.*, 11, p. 259–75.
WHELLAN, J. A. 1965. The habitat of *Welwitschia bainesii* (Hook. f.) Carr. *Kirkia*, 5, p. 33–5.
WHITE, F. 1965. The savanna woodlands of the Zambezian and Sudanian Domains. An ecological and phytogeographical comparison. *Webbia*, 19, p. 651–81.
——. 1968. Zambia. In: Hedberg, I.; Hedberg, O. (eds.), p. 208–15.
——. 1971. The taxonomic and ecological basis of chorology. *Mitt. Bot. Staatssamml. Münch.*, 10, p. 91–112.
——. 1976a. The vegetation map of Africa: the history of a completed project. *Boissiera*, 24, p. 659–66.
——. 1976b. The taxonomy, ecology and chorology of African Chrysobalanaceae (excluding *Acioa*). *Bull. Jard. bot. nat. Belg.*, 46, p. 265–350.
——. 1976c. Chrysobalanaceae. *Distr. Pl. Afr.*, 10, maps 281–334.
——. 1976d [1977]. The underground forests of Africa: a preliminary review. *Gdns Bull.* (Singapore), 29, p. 55–71.
——. 1978a. The Afromontane Region. In: Werger, M. J. A. (ed.), p. 463–513.
——. 1978b. The taxonomy, ecology and chorology of African Ebenaceae. I. The Guineo–Congolian species. *Bull. Jard. bot. nat. Belg.*, 48, p. 245–358.
——. 1978c. The Guineo–Congolian species of *Diospyros*. *Distr. Pl. Afr.*, 14, maps 440–94.
——. 1979. The Guineo–Congolian Region and its relationships to other phytochoria. *Bull. Jard. bot. nat. Belg.*, 49, p. 11–55.
——. 1981. The history of the Afromontane archipelago and the scientific need for its conservation. *Afr. J. Ecol.*, 19, p. 33–54.

WHITE, F.; WERGER, M. J. A. 1978. The Guineo–Congolian transition to southern Africa. In: Werger, M. J. A. (ed.), p. 599–620.
WHITE, L. P. 1970. 'Brousse tigrée' patterns in southern Niger. *J. Ecol.*, 58, p. 549–53.
——. 1971. Vegetation stripes on sheet wash surfaces. *J. Ecol.*, 59, p. 615–62.
WHITEMAN, A. J. 1971. *The geology of the Sudan Republic.* Oxford, Clarendon Press. 290 p.
WHITMORE, J. C. 1975. *Tropical rain forests of the Far East.* Oxford, Clarendon Press. 288 p.
WHYTE, R. O. 1974. *Tropical grazing lands, communities and constituent species.* The Hague, Junk. 220 p.
WICHT, C. L. 1971. The influence of vegetation in South African mountain catchments on water supplies. *S. Afr. J. Sci.*, 67, p. 201–9.
WICKENS, G. E. 1976. Speculations on long distance dispersal and the flora of Jebel Marra, Sudan Republic. *Kew Bull.*, 31, p. 105–50.
——. 1977a. The flora of Jebel Marra (Sudan Republic) and its geographical affinities. *Kew Bull. Add. Ser.*, 5, p. 1–368.
——. 1977b. Some of the phytogeographical problems associated with Egypt. *Publ. Cairo Univ. Herb.*, 7/8, p. 223–30.
——. 1979. Speculations on seed dispersal and the flora of the Aldabra archipelago. *Phil. Trans. R. Soc. Lond.*, ser. B, 286, p. 85–97.
WICKENS, G. E.; COLLIER, F. W. 1971. Some vegetation patterns in the Republic of the Sudan. *Geoderma*, 6, p. 43–59.
WIEHE, P. O. 1949. The vegetation of Rodrigues Island. *Mauritius Inst. Bull.*, 2, p. 280–304.
WIESER, J. *See* Walter, 1976b.
WILD, H. 1952a. The vegetation of Southern Rhodesian termitaria. *Rhod. agric. J.*, 49, p. 280–92.
——. 1952b. A guide to the flora of the Victoria Falls. In: Clark, J. D. (ed.), *The Victoria Falls*, p. 121–60. Comm. Preservation Nat. Hist. Monuments and Relics, Zambia.
——. 1953. Vegetation survey of the Changara (Portuguese East Africa) Mkota Reserve (S. Rhodesia) area. *Rhod. agric. J.*, 50, p. 407–19, with large-scale vegetation map.
——. 1955. Observations on the vegetation of the Sabi–Lundi Junction area. *Rhod. agric. J.*, 52, p. 533–46.
——. 1956. The principal phytogeographical elements of the Southern Rhodesian flora. *Proc. Trans. Rhod. scient. Ass.*, 44, p. 53–62.
——. 1961. Harmful aquatic plants in Africa and Madagascar. *Kirkia*, 2, p. 1–66. Also published as *Joint CCTA/CSA Project*, 14, p. 1–66.
——. 1964a. *Les plantes aquatiques nuisibles en Afrique et à Madagascar* (transl. R. Germain). Projet conjoint CCTA/CSA, 14, p. 1–63.
——. 1964b. The endemic species of the Chimanimani Mountains and their significance. *Kirkia*, 4, p. 125–57.
——. 1965. The flora of the Great Dyke of Southern Rhodesia with special reference to the serpentine soils. *Kirkia*, 5, p. 49–86.
——. 1968a. Phytogeography in South Central Africa. *Kirkia*, 6, p. 197–222.
——. 1968b. Bechuanaland Protectorate. In: Hedberg, I.; Hedberg, O. (eds.), p. 198–202.
——. 1968c. Rhodesia. In: Hedberg, I.; Hedberg, O. (eds.), p. 202–7.
——. 1968d. Geobotanical anomalies in Rhodesia. 1. The vegetation of copper-bearing soils. *Kirkia*, 7, p. 1–71.

———. 1968e. Id. 2. A geobotanical anomaly occurring in graphitic soils. *Port. Acta biol.*, sér. B, 9, p. 291–9.

———. 1970. Id. 3. The vegetation of nickel-bearing soils. *Kirkia*, 7, Suppl., p. 1–62.

———. 1974a. Id. 4. The vegetation of arsenical soils. *Kirkia*, 9, p. 243–64.

———. 1974b. Variations in the serpentine floras of Rhodesia. *Kirkia*, 9, p. 209–32.

———. 1974c. Indigenous plants and chromium in Rhodesia. *Kirkia*, 9, p. 233–41.

———. 1974d. Arsenic tolerant plant species established on arsenical mine dumps in Rhodesia. *Kirkia*, 9, p. 265–78.

———. 1974e. The natural vegetation of gypsum bearing soils in South Central Africa. *Kirkia*, 9, p. 279–92.

———. 1975. Termites and the serpentines of the Great Dyke of Rhodesia. *Trans. Rhod. Sci. Ass.*, 57, p. 1–11.

———. 1978. The vegetation of heavy metal and other toxic soils. In: Werger, M. J. A. (ed.), p. 1301–32.

WILD, H.; BARBOSA, L. A. Grandvaux. 1967 [1968]. *Vegetation map (1:2500000 in colour) of the Flora Zambesiaca area. Descriptive memoir.* Salisbury, Rhodesia, Collins. 71 p. (Supplement to *Flora Zambesiaca*.)

WILD, H.; BRADSHAWE, A. D. 1977. The evolutionary effects of metalliferous and other anomalous soils in South Central Africa. *Evolution*, 31, p. 282–93.

WILLAN, R. G. M. 1957. Some notes on the cold spell in Nyasaland in August 1955. *Nyasald J.*, 10, p. 7–10.

WILLIAMS, L. 1969. Forest and agricultural resources of Dahomey, West Africa. *Econ. Bot.*, 23, p. 352–72.

WILLIAMS, M. A. J. 1968. A dune catena on the clay plains of the west central Gezira, Republic of the Sudan. *J. Soil Sci.*, 19, p. 367–78.

WILLIMOTT, S. G. 1957. Soils and vegetation of the Boma plateau and Eastern District, Equatoria. *Sudan Notes Rec.*, 38, p. 10–20.

WILLS, J. B. (ed.). 1962. *Agriculture and land use in Ghana.* London, Accra, New York, Oxford Univ. Press. 503 p.

WILSON, A. T. 1956. *Report of a soil and land-use survey, Copperbelt, Northern Rhodesia.* Lusaka, Dep. Agriculture. 190 p. and 5 maps.

WILSON, J. G. 1962. The vegetation of Karamoja District, Northern Province of Uganda. *Mem. Res. Div.*, ser. 2, 5, p. 1–182. Dep. Agric., Uganda. (Mimeo.)

WILSON, V. J. 1965. Observations on the greater kudu, *Tragelephus strepsiceros* Pallas from a tsetse control hunting scheme in Northern Rhodesia. *E. Afr. Wildl. J.*, 3, p. 27–37.

———. 1966. Notes on the food and feeding habits of the common duiker, *Sylvicarpa grimmia*, in Eastern Zambia. *Arnoldia*, 14, p. 1–19.

WILSON, V. J.; CLARKE, J. E. 1962. Observations on the common duiker, *Sylvicarpa grimmia*, in Eastern Zambia. *Proc. Zool. Soc.* (Lond.), 138, p. 487–96.

WIMBUSH, S. H. 1937. Natural succession in the Pencil Cedar forest of Kenya Colony. *Emp. For. J.*, 16, p. 49–53.

WING, L. D.; BUSS, I. O. 1970. Elephants and forests. *Wildl. Monogr.*, 19, p. 1–92.

WOOD, G. H. S. 1960. A study of the plant ecology of Busoga District, Uganda Protectorate. *Inst. Pap. Imp. For. Inst.*, 35, p. 1–69, with large-scale vegetation map.

WOOD, P. J. 1965. The forest glades of west Kilimanjaro. *Tanganyika Notes Rec.*, 64, p. 108–11.

WOODHEAD, T. 1970. A classification of East African rangeland. II. The water balance as a guide to site potential. *J. appl. Ecol.*, 7, p. 647–52.

WORRALL, G. A. 1959. The Butana grass patterns. *J. Soil Sci.*, 10, p. 34–53.

———. 1960a. Patchiness in vegetation in the Northern Sudan. *J. Ecol.*, 48, p. 107–15.

———. 1960b. Tree patterns in the Sudan. *J. Soil Sci.*, 11, p. 63–7.

WORTHINGTON, E. B. 1961. *The wild resources of East and Central Africa.* London, HMSO. 26 p.

YANGAMBI. 1956. Classification. See CCTA/CSA, 1956.

YANNEY EWUSIE, J. 1968. Preliminary studies on the phenology of some woody species of Ghana. *Ghana J. Sci.*, 8, p. 126–33.

YOUNG, A. 1968. Slope form and the soil catena in savanna and rain forest environments. *Br. Geomorph. Res. Grp, Occ. Pap.*, 5, p. 3–12.

———. 1976. *Tropical soils and soil survey.* Cambridge Univ. Press. 468 p.

YOUNG, A.; BROWN, P. 1962. *The physical environment of northern Nyasaland with special reference to soils and agriculture.* Zomba, Govt Printer. 107 p.

YOUNG, A.; STEPHEN, I. 1965. Rock weathering and soil formation on high-altitude plateaux of Malawi. *J. Soil Sci.*, 16, p. 322–33.

ZEMKE, E. 1939. Anatomische Untersuchungen an Pflanzen der Namibwüste (Deutsch–Südwestafrika). *Flora*, N.F., 33, p. 365–416.

ZOHARY, M. 1973. *Geobotanical foundations of the Middle East.* Stuttgart, Gustav Fischer. Vol. 1, p. 1–340; vol. 2, p. 341–739. (Geobot. sel., 3.)

———. 1975. The phytogeographical delimitation of the Mediterranean Region towards the East. In: *La flore du bassin méditerranéen. Coll. Int. CNRS*, 235, p. 329–43.

ZOLOTAREVSKY, B.; MURAT, M. 1938. Divisions naturelles du Sahara et sa limite méridionale. *Mém. Soc. Biogéogr.*, 6, p. 335–50.

Index of plant names

Important synonyms are included in this index, but no attempt is made to indicate which are 'correct'. This is often a matter of opinion and sometimes of controversy. Page references are given after the name which is 'preferred' by the author. It should not be inferred, however, that the preferred name will prove to be taxonomically 'correct' or otherwise generally acceptable. Some names are preferred because they are so well known in the ecological literature that a wise society would choose to conserve them. Some names, which are not preferred, might eventually prove to be taxonomically acceptable, but they have been proposed in incomplete or otherwise controversial revisions, and at present there is no guarantee that they are here to stay. Synonyms in the index are cross-referenced to the preferred name. In the text, synonymy is usually given only where the name is first mentioned, or in connection with important source materials where the names are used.

Abies alba Miller (Pinaceae), 156
    A. numidica Delannoy ex Carrière, 60, 146, 149, 150, 152, 155, 156
    A. pinsapo Boiss., 60, 146, 155, 156
        ss. *marocana* (Trabut) Ceballos, 149, 150, 152, 156
Abrus precatorius L. (Leguminosae: Papilionoideae), 206
Abutilon (Malvaceae), 220
    A. fruticosum Guill. & Perr., 222
Acacia (Leguminosae: Mimosoideae), 31, 62, 63, 64, 65, 89, 98, 103, 107, 108, 110, 111, 113, 115, 116, 125, 128, 144, 181, 182, 191, 201, 206, 216, 219, 220, 241, 242, 254
    A. albida Del., 28, 55, 91, 95, 105, 141, 144, 203, 208, 209, 210, 211, 212, 214, 215, 220, 222, 251, 252
    A. ataxacantha DC., 90, 209, 211
    A. borleae Burtt Davy, 201
    A. brevispica Harms, 129
    A. burkei Benth., 201
    A. bussei Harms ex Sjöstedt, 114, 116, 188
    A. caffra (Thunb.) Willd., 96, 196, 201
    A. clavigera E. Meyer: see *A. robusta* ss. *clavigera*
    A. cyanophylla Lindley, 135, 231
    A. cyclops A. Cunn. ex G. Don, 135
    A. davyi N. E. Br., 96, 196, 201
    A. drepanolobium Harms ex Sjöstedt, 52, 115, 116, 121, 128
    A. dudgeonii Craib ex Holland, 105
    A. ehrenbergiana Hayne, 207, 219, 222
    A. elatior Brenan, 117
    A. erioloba E. Meyer, 31, 90, 95, 97, 137, 140, 141, 144, 191, 193, 194
    A. erubescens Welw. ex Oliver, 95
    A. etbaica Schweinf., 116, 120, 121
    A. farnesiana (L.) Willd., 252
    A. flava (Forssk.) Schweinf. not of Spreng. ex DC.: see *A. ehrenbergiana*
    A. fleckii Schinz, 90, 193
    A. galpinii Burtt Davy, 91
    A. gerrardii Benth, 96, 128, 129, 182, 201, 209
    A. gillettiae Burtt Davy: see *A. luederitzii*
    A. giraffae Willd.: see *A. erioloba*
    A. gourmaensis A. Chev., 105
    A. gummifera Willd., 149, 225, 226, 227, 228, 229, 231
    A. haematoxylon Willd., 191, 194
    A. hebeclada DC., 191, 193
    A. hereroensis Engl., 193
    A. heterophylla Willd., 258
    A. hockii De Wild., 87, 105, 115, 128, 129, 174, 182
    A. horrida (L.) Willd., 120
    A. kamerunensis Gandoger, 81
    A. karroo Hayne, 137, 141, 159, 193, 194, 195, 201
    A. kirkii Oliver, 95, 115
        ss. *mildbraedii* (Harms) Brenan, 181, 182
    A. laeta R. Br. ex Benth., 206, 207, 212, 222
    A. lahai Steud. & Hochst. ex Benth., 130
    A. luederitzii Engl., 96, 193
    A. macrostachya Reichenb. ex DC., 105
    A. macrothyrsa Harms, 105
    A. malacocephala Harms, 116
    A. melanoxylon R. Br., 135
    A. mellifera (Vahl) Benth., 108, 114, 116, 120, 121, 126, 128, 188, 191, 193, 203, 207, 208, 209, 210, 212, 213, 214
        ss. *detinens* (Burchell) Brenan, 137, 140
    A. montis-usti Merxm. & A. Schreiber, 140
    A. nebrownii Burtt Davy, 267
    A. nigrescens Oliver, 94, 95, 201
    A. nilotica (L.) Willd. ex Del., 96, 128, 188, 252
        ss. *adansonii* (Guill. & Perr.) Brenan: see ss. *adstringens*
        ss. *adstringens* (Schumach. & Thonn.) Roberty, 105, 209, 210, 220
        ss. *kraussiana* (Benth.) Brenan, 196, 267
        ss. *nilotica*, 108
        ss. *subalata* (Vatke) Brenan, 114, 121, 129
    A. nubica Benth., 121, 208, 210, 212
    A. pennivenia Balf.f., 255
    A. permixta Burtt Davy, 96
    A. polyacantha Willd., 63, 89, 128, 129, 172, 209
        ss. *campylacantha* (Hochst. ex A. Rich.) Brenan, 91, 95, 105, 174
    A. pseudofistula Harms, 116
    A. redacta J. H. Ross, 140
    A. reficiens Wawra, 120, 121
        ss. *misera* (Vatke) Brenan, 114, 120
        ss. *reficiens*, 145, 193, 194
    A. rehmanniana Schinz, 96
    A. robusta Burchell, 196, 201

Acacia (Leguminosae: Mimosoideae)—*contd*
    ss. *clavigera* (E. Meyer) Brenan, 91, 95, 129
    ss. *robusta*, 96
    ss. *usambarensis* (Taubert) Brenan, 117, 128, 188
  A. robynsiana Merxm. & A. Schreiber, 140
  A. senegal (L.) Willd., 105, 107, 120, 128, 182, 189, 201, 203, 206, 207, 208, 210, 211, 212, 213
    var. *kerensis* Schweinf., 121
  A. seyal Del., 62, 105, 107, 108, 115, 116, 120, 121, 128, 203, 209, 210, 211, 212, 214, 222
  A. sieberana DC., 95, 96, 105, 107, 128, 129, 174, 196, 201, 209, 210, 211
  A. stenocarpa Hochst. ex A. Rich.: see *A. seyal*
  A. tanganyikensis Brenan, 116
  A. tenuispina I. Verdoorn, 96
  A. thomasii Harms, 114
  A. tortilis (Forssk.) Hayne, 31, 91, 94, 95, 96, 113, 114, 116, 117, 120, 121, 127, 128, 145, 193, 206, 207, 208, 212, 213, 218, 219, 220, 221, 224
    ss. *raddiana* (Savi) Brenan, 219
  A. welwitschii Oliver, 90
  A. xanthophloea Benth., 30, 31, 91, 128, 129, 267
  A. zanzibarica (S. Moore) Taubert, 189
Acalypha chirindica S. Moore (Euphorbiaceae), 90, 98
Acanthosicyos horridus Welw. ex Hook.f. (Cucurbitaceae), 144, 145
Acanthospermum hispidum DC. (Compositae), 252
Acanthus (Acanthaceae), 158
  A. eminens C. B. Clarke, 122
  A. mollis L., 149
Acer campestre L. (Aceraceae), 147, 149, 156
  A. granatense Boiss., 156
  A. monspessulanum L., 149, 151, 156, 157
  A. obtusatum Waldst. & Kit. ex Willd., 156
Aceras (Orchidaceae), 158
Achyranthes aspera L. (Amaranthaceae), 90, 182
Acokanthera (Apocynaceae), 115
  A. schimperi (A.DC.) Oliver, 115
Acridocarpus excelsus Adr. Juss. (Malpighiaceae), 243
  A. smeathmannii (DC.) Guill. & Perr., 81
Acrocephalus sericeus Briq. (Labiatae), 100
Acroceras macrum Stapf (Gramineae), 100
Acrostichum aureum L. (Pteridaceae), 188, 262, 263
Actiniopteris (Actiniopteridaceae), 238
Adansonia (Bombacaceae), 241, 242
  A. digitata L., 31, 62, 90, 94, 95, 96, 114, 116, 128, 174, 189, 212, 213
  A. fony Baillon, 241, 242
  A. grandidieri Baillon, 241
  A. madagascariensis Baillon, 255
  A. rubrostipa Jumelle & Perrier, 241
  A. za Baillon, 241, 242
Adenia (Passifloraceae), 242
  A. globosa Engl., 114, 188
  A. pechuellii (Engl.) Harms, 142
  A. venenata Forssk., 114
Adenium multiflorum Klotzsch (Apocynaceae): see *A. obesum*
  A. obesum (Forssk.) Roem. & Schult., 94, 114
  A. socotranum Vierh., 115, 155, 255
Adenocarpus bacquei Battand. & Pitard (Leguminosae: Papilionoideae), 153
  A. foliolosus (Aiton) DC., 249
  A. viscosus Webb & Berth., 248
Adenolobus (Leguminosae: Caesalpinioideae), 137
  A. garipensis (E. Meyer) Torre & Hillcoat, 140

  A. pechuelii (Kuntze) Torre & Hillcoat, 140, 144
Adiantum capillus-veneris L. (Adiantaceae), 224
  A. vogelii Mett. ex Keys., 75
Adina microcephala (Del.) Hiern (Rubiaceae), 91, 105
Aeluropus lagopoides (L.) Trin. ex Thwaites (Gramineae), 230
  A. repens (Desf.) Parl.: see *A. lagopoides*
Aeonium (Crassulaceae), 246, 249
  A. arboreum (L.) Webb, 226, 228
Aerva javanica (Burm.f.) Juss. ex Schultes (Amaranthaceae), 115
  A. persica (Burm.f.) Merr., 221, 252
  A. tomentosa Forssk.: see *A. javanica*
Aeschynomene elaphroxylon (Guill. & Perr.) Taub. (Leguminosae: Papilionoideae), 266
  A. pfundii Taub., 266
  A. trigonocarpa Taub., 98
Aframomum angustifolium K. Schum. (Zingiberaceae), 236
  A. biauriculatum K. Schum., 96
Afrobrunnichia (Polygonaceae), 74
Afrocrania (Cornaceae), 162
  A. volkensii (Harms) Hutch., 167
Afrormosia angolensis (Baker) De Wild.: see *Pericopsis angolensis*
  A. elata Harms: see *P. elata*
  A. laxiflora Benth. ex Baker: see *P. laxiflora*
Afrosersalisia cerasifera (Welw.) Aubrév. (Sapotaceae), 187
Afrotrilepis pilosa (Boeckeler) J. Raynal (Cyperaceae), 81
Afzelia africana Smith (Leguminosae: Caesalpinioideae), 79, 80, 83, 85, 105, 107, 178
  A. bipindensis Harms, 74
  A. quanzensis Welw., 93, 95, 96, 97, 99, 117, 187, 188, 189, 200, 201
Agathophora (Chenopodiaceae), 218
Agathosma (Rutaceae), 132
Agauria (Ericaceae), 237
  A. buxifolia (Lam.) Cordem., 258
  A. salicifolia (Lam.) Hook.f. ex Oliver, 237, 238, 258
Agave (Agavaceae), 159
Agelaea (Connaraceae), 75, 253
Ageratum conyzoides L. (Compositae), 252
Agrostis (Gramineae), 169
  A. azorica (Hochst.) Tutin & E. Warb., 246
  A. elliotii Hackel, 239
Aichryson (Crassulaceae), 246
Aizoanthemum dinteri (Schinz) Friedrich (Aizoaceae), 142
Aizoon (Aizoaceae), 266
  A. canariense L., 229
  A. dinteri Schinz: see *Aizoanthemum dinteri*
  A. mossamedense Welw. ex Oliver, 144
  A. virgatum Welw. ex Oliver, 144
Ajuga (Labiatae), 236
  A. ophrydis Burchell ex Benth., 194
Alangium chinense (Lour.) Harms (Alangiaceae), 181, 187
Alberta (Rubiaceae), 237
  A. minor Baillon ex K. Schum., 237, 238
Albizia (Leguminosae: Mimosoideae), 98, 181, 241
  A. adianthifolia (Schumach.) W. F. Wight, 74, 83, 173, 178, 187, 199
  A. amara (Roxb.) Boivin, 95, 106, 116, 208, 209, 210, 212, 213, 214
  A. anthelmintica Brongn., 188, 191, 193, 194, 212
  A. antunesiana Harms, 96, 97
  A. aylmeri Hutch., 210
  A. brachycalyx Oliver: see *A. petersiana*
  A. chevalieri Harms, 105, 107
  A. falcata (L.) Back. ex Merr., 257

A. ferruginea (Guill. & Perr.) Benth., 81, 178
A. forbesii Benth., 200
A. glaberrima (Schumach. & Thonn.) Benth., 117
A. grandibracteata Taubert, 182
A. gummifera (J. F. Gmelin) C. A. Smith, 130, 166, 181
A. harveyi Fourn., 95, 116, 129
A. lebbeck (L.) Benth., 257
A. malacophylla (Steud. ex A. Rich.) Walp., 211
A. petersiana (Bolle) Oliver, 97, 188
A. sericocephala Benth.: see *A. amara*
A. tanganyicensis Baker f., 96
A. versicolor Welw. ex Oliver, 91, 95, 173, 174, 201
A. zimmermannii Harms, 117
A. zygia (DC.) J. F. Macbr., 83, 105, 172, 178, 211
Alchemilla (Rosaceae), 236
Alchornea cordifolia (Schumach. & Thonn.) Muell. Arg. (Euphorbiaceae), 83
  A. occidentalis (Muell. Arg.) Pax & K. Hoffm., 90
Allanblackia stuhlmannii (Engl.) Engl. (Guttiferae), 187
Allium (Alliaceae), 158
Allmaniopsis (Amaranthaceae), 111
Allophylus abyssinicus (Hochst.) Radlk. (Sapindaceae), 122
  A. africanus P. Beauv., 182
Alloteropsis cimicina (Retz.) Stapf (Gramineae), 213
  A. semialata (R. Br.) Hitchcock, 194, 202, 239
Alluaudia (Didiereaceae), 242
  A. ascendens Drake, 242
  A. procera Drake, 242
Alluaudiopsis (Didiereaceae), 242
Alnus glutinosa (L.) Gaertner (Betulaceae), 147, 149, 152, 156
Aloe (Liliaceae), 111, 114, 115, 116, 129, 137, 140, 174, 198, 201, 238, 242
  A. arborescens Miller, 96, 196
  A. asperifolia Berger, 142
  A. bainesii Dyer, 200
  A. ballyi Reynolds, 117
  A. breviscapa Reynolds & Bally, 116
  A. candelabrum Berger, 201
  A. capitata Baker
    var. *cipolinicola* H. Perrier, 243
  A. dichotoma Masson, 140
  A. eminens Reynolds & Bally, 115
  A. ferox Miller, 135, 195, 201
  A. kedongensis Reynolds, 115
  A. littoralis Baker, 145
  A. marlothii Berger, 96, 201
  A. perryi Baker, 115, 255
  A. pillansii L. Guthrie, 140
  A. plicatilis (L.) Miller, 134
  A. rigens Reynolds & Bally, 116
  A. scobinifolia Reynolds & Bally, 116
  A. speciosa Baker, 137, 201
  A. spectabilis Reynolds, 201
  A. volkensii Engl., 129
Alstonia boonei De Wild. (Apocynaceae), 81, 181
  A. congensis Engl., 82
Alyssum serpyllifolium Desf. (Cruciferae), 230
  A. spinosum L., 158
Amanoa bracteosa Planchon (Euphorbiaceae), 82
Amaranthus graecizans L. (Amaranthaceae), 212
Amblygonocarpus andongensis (Welw. ex Oliver) Exell (Leguminosae: Mimosoideae), 87, 96, 97, 105
Amelanchier ovalis Medicus (Rosaceae), 158
Ammannia gracilis Guill. & Perr. (Lythraceae), 204
Ammodaucus (Umbelliferae), 221

Ampelodesma mauretanicum (Poir.) Th. Durand & Schinz (Gramineae), 149, 154, 155, 157, 158, 159, 229
Amphimas (Leguminosae: Caesalpinioideae), 74
Anabasis aphylla L. (Chenopodiaceae), 218, 229, 266
  A. aretioides Moq. & Coss.: see *Fredolia aretioides*
  A. articulata (Forssk.) Moq., 221, 222, 224
  A. oropediorum Maire, 230
Anacampseros (Portulacaceae), 137, 140
  A. albissima Marloth, 142
Anacamptis (Orchidaceae), 158
Anacardium (Anacardiaceae), 189
  A. occidentale L., 174
Anadelphia afzeliana (Rendle) Stapf (Gramineae), 84
  A. leptocoma (Trin.) Pilger, 84
  A. trispiculata Stapf, 84
Anagyris (Leguminosae: Papilionoideae), 147
  A. foetida L., 158
Anastatica (Cruciferae), 218
  A. hierochuntica L., 221
Anastrabe (Scrophulariaceae), 199
  A. integerrima E. Meyer ex Benth., 199, 200
Ancistrophyllum (Palmaceae), 82
Androcymbium (Liliaceae), 221
Andropogon (Gramineae), 169
  A. amplectens Nees, 196
  A. appendiculatus Nees, 194
  A. brazzae Franchet, 100
  A. canaliculatus Schumach., 178
  A. curvifolius W. D. Clayton, 84
  A. distachyos L., 208, 211
  A. eucomus Nees, 239
  A. gayanus Kunth, 85, 97, 108, 206, 209, 210
  A. greenwayi Napper, 126, 127
  A. kelleri Hackel, 116
  A. perligulatus Stapf, 84
  A. schirensis Hochst. ex A. Rich., 85, 100, 101, 173, 194, 196
  A. tectorum Schumach. & Thonn., 83, 84, 85
  A. trichozygus Baker, 239
Androstachys (Euphorbiaceae), 87
  A. johnsonii Prain, 96
Aneilema johnstonii K. Schum. (Commelinaceae), 90
Aneulophus (Linaceae), 74
Angkalanthus (Acanthaceae), 111
Angraecum (Orchidaceae), 238
Aningeria adolfi-friedericii (Engl.) Robyns & G. Gilbert (Sapotaceae), 85, 164, 187
  A. altissima (A. Chev.) Aubrév. & Pellegr., 74, 79, 90, 181
  A. pseudoracemosa J. H. Hemsley, 186
  A. robusta (A. Chev.) Aubrév. & Pellegr., 79
Anisophyllea boehmii Engl. (Rhizophoraceae), 97
  A. cabole Henriq., 253
  A. gossweileri Engl. & v. Brehm., 173
  A. pomifera Engl. & v. Brehm., 93
  A. quangensis Engl., 173
Annona senegalensis Pers. (Annonaceae), 83, 85, 105, 107, 174, 189
Anogeissus (Combretaceae), 62, 106
  A. leiocarpus (DC.) Guill. & Perr., 55, 80, 83, 105, 106, 107, 203, 208, 209, 210, 211, 214
Anonidium usambarense R. E. Fries (Annonaceae), 187
Anopyxis (Rhizophoraceae), 74
Ansellia gigantea Reichb.f. (Orchidaceae), 93, 201
  A. nilotica (Baker) N. E. Br.: see *A. gigantea*
Anthephora argentea Goossens (Gramineae), 191, 193

Anthephora (Gramineae)—*contd*
  A. lynesii Stapf & C. E. Hubbard, 209, 210
  A. pubescens Nees, 193, 194
  A. schinzii Hackel, 143
Anthocleista (Loganiaceae), 80
  A. nobilis G. Don, 83
  A. schweinfurthii Gilg, 91, 181
Anthonotha (Leguminosae: Caesalpinioideae), 74
  A. obanensis (Baker f.) J. Léonard, 81
  A. pynaertii (De Wild.) J. Léonard, 181
Anthospermum rigidum Eckl. & Zeyh. (Rubiaceae), 194
Anthostema (Euphorbiaceae), 235
  A. aubryanum Baillon, 253
Anthoxanthum madagascariense Stapf (Gramineae), 239
Anthyllis cytisoides L. (Leguminosae: Papilionoideae), 154
Antiaris africana Engl. (Moraceae): see *A. toxicaria*
  A. toxicaria (Rumph. ex Pers.) Leschen., 82, 83, 105, 172, 178, 181, 186, 187
Antidesma venosum E. Meyer ex Tul. (Euphorbiaceae), 105, 189
Antrocaryon (Anacardiaceae), 74
  A. micraster A. Chev. & Guillaumin, 81
Aphania senegalensis (Poir.) Radlk. (Sapindaceae), 129
Aphanocalyx (Leguminosae: Caesalpinioideae), 74
Aphloia (Flacourtiaceae), 237
  A. theiformis (Vahl) Benn., 257, 258
Aphyllanthes (Liliaceae), 147
Apodocephala (Compositae), 232, 237
Apodytes dimidiata E. Meyer ex Arn. (Icacinaceae), 122, 165, 166, 181, 187, 196, 201, 255, 259
Apollonias barbujana (Cav.) Bornm. (Lauraceae), 246, 247, 249
Aporrhiza nitida Gilg ex Engl. (Sapindaceae), 91
Aptosimum (Scrophulariaceae), 193
  A. depressum Burchell ex Benth.: see *A. procumbens*
  A. procumbens (Lehm.) Steud., 195
Aquilegia vulgaris L. (Ranunculaceae), 156
Arachis (Leguminosae, Papilionoideae), 178
Arbutus canariensis Veill. (Ericaceae), 246
  A. pavarii Pampan., 159, 226
  A. unedo L., 147, 149, 152, 159
Archidium capense Hornschuch (Archidiaceae), 182
Arctotis (Compositae), 141
Ardisiandra (Primulaceae), 162
Arenaria dyris Humbert (Caryophyllaceae), 158
  A. pungens Clemente ex Lagasca, 158
Argania (Sapotaceae), 226, 228, 229
  A. spinosa (L.) Skeels, 54, 64, 149, 150, 154, 157, 225, 226, 227, 228, 229
Argemone mexicana L. (Papaveraceae), 254
Argyranthemum (Compositae), 246
Argyroderma (Aizoaceae), 140
Aristida (Gramineae), 140, 144, 193, 194, 210, 213, 221, 238, 239, 243
  A. acutiflora Trin. & Rupr.: see *Stipagrostis acutiflora*
  A. adoensis Hochst., 208
  A. adscensionis L., 114, 116, 120, 208, 209, 210, 212, 243, 252, 254, 267
  A. barbicollis Trin. & Rupr., 114
  A. cardosoi Cout., 252
  A. 'coerulescens', 222
  A. congesta Roem. & Schult., 194, 195, 211, 242, 243
  A. diffusa Trin., 141, 195
  A. funiculata Trin. & Rupr., 212, 213, 252
  A. graciliflora Pilger: see *A. stipitata*
  A. hordeacea Kunth, 145
  A. junciformis Trin. & Rupr., 194, 202

  A. mutabilis Trin. & Rupr., 113, 116, 120, 211, 212
  A. pallida Steud.: see *Aristida sieberana*
  A. pungens Desf.: see *Stipagrostis pungens*
  A. rhiniochloa Hochst., 145, 208, 209, 210
  A. rufescens Steud., 239, 242, 243
  A. sieberana Trin., 176, 204, 207, 211, 212, 213
  A. similis Steud., 239
  A. stipitata Hackel, 101
  A. stipoides Lam., 206, 207
  A. vanderystii De Wild., 173
Artemisia (Compositae), 129
  A. afra Willd., 129
  A. campestris L., 229, 230
    ss. *glutinosa* (J. Gay) Battand., 222
  A. gorgonum Webb, 252
  A. herba-alba Asso, 153, 222, 229, 230, 266
  A. inculta Del.: see *A. herba-alba*
  A. tilhoana Quézel, 222
Arthraerua (Amaranthaceae), 137
  A. leubnitziae (Kuntze) Schinz, 142, 144
Arthraxon lancifolius Hochst. (Gramineae), 255
Arthrocarpum (Leguminosae: Papilionoideae), 111
Arthrocnemum (Chenopodiaceae), 267
  A. dunense Moss ex Adamson, 144
  A. glaucum (Del.) Ungern-Sternb., 224, 230
  A. indicum (Willd.) Moq., 144, 223, 263
Arthropteris orientalis (J. F. Gmelin) Posthumus (Oleandraceae), 99
Arundinaria (Gramineae), 237
  A. alpina K. Schum., 55, 130, 167
  A. marojejyensis A. Camus, 237
  A. tesselata (Nees) Munro, 55, 167
Arundo donax L. (Gramineae), 159
Ascarina (Chloranthaceae), 237
Ascarinopsis (Chloranthaceae), 232
  A. coursii Humbert & Capuron, 237
Asclepias multicaulis Schltr. (Asclepiadaceae), 194
Ascolepis anthemiflora (Welw.) Welw. (Cyperaceae), 100
  A. elata Welw., 100
Aspalathus (Leguminosae: Papilionoideae), 132, 134
Asparagus (Liliaceae), 98, 156, 201
  A. acutifolius L., 151, 152
  A. albus L., 153, 158
  A. pastorianus Webb & Berth., 229
  A. stipularis Forssk., 158, 227, 266
  A. warneckei (Engl.) Hutch., 83
Asphodelus (Liliaceae), 158, 221, 228
  A. aestivus Brot., 152
  A. fistulosus L., 266
  A. microcarpus Salzm. & Viv., 152, 158, 160
  A. tenuifolius Cav., 229
Aspidium aculeatum Swartz (Aspidiaceae), 157
Aspilia mossambicensis (Oliver) Wild (Compositae), 115, 122
Asplenium (Aspleniaceae), 75
  A. adiantum-nigrum L., 157
  A. dregeanum Kunze, 81
  A. nidus L., 235
Aster (Compositae), 140
Asteriscus graveolens (Forssk.) DC. (Compositae), 221
Asteropeia densiflora Baker (Asteropeiaceae), 237
Asthenatherum forskalii (Vahl) Nevski (Gramineae), 144, 220
  A. glaucum (Nees) Nevski, 191, 193
  A. mossamedense (Rendle) Conert, 144
Astragalus (Leguminosae: Papilionoideae), 147
Asystasia gangetica (L.) T. Anderson (Acanthaceae), 182

Ataenidia (Marantaceae), 75
Atalaya (Sapindaceae), 199
    A. natalensis R. A. Dyer, 199
Athyrium filix-femina (L.) Roth (Athyriaceae), 151, 157
Atriplex (Chenopodiaceae), 224, 267
    A. halimus L., 144, 223, 228, 229, 230, 266
    A. mollis L., 230
    A. vestita (Thunb.) Aellen, 267
Aubrevillea (Leguminosae: Mimosoideae), 74
    A. kerstingii (Harms) Pellegr., 79
Aucoumea (Burseraceae), 74
    A. klaineana Pierre, 77
Augea (Zygophyllaceae), 137
Auxopus (Orchidaceae), 75
Avena bromoides (Gouan) Trabut (Gramineae), 230
Avicennia (Avicenniaceae), 254, 261, 262, 263, 264
    A. africana P. Beauv.: see *A. germinans*
    A. germinans (L.) L., 261, 262
    A. marina (Forssk.) Vierh., 253, 259, 261, 263, 264
    A. nitida Jacq.: see *A. germinans*
Azanza garckeana (F. Hoffm.) Exell & Hillcoat (Malvaceae), 95, 210
Azima tetracantha Lam. (Salvadoraceae), 182, 201
Azolla africana Desv. (Azollaceae), 265

Babiana (Iridaceae), 137, 141
Bachmannia (Capparidaceae), 199
Bafodeya benna (Scott Elliot) Prance (Chrysobalanaceae), 176
Baikiaea (Leguminosae: Caesalpinioideae), 90, 98
    B. eminii Taubert: see *B. insignis*
    B. insignis Benth., 181
    B. plurijuga Harms, 90, 97, 98, 101
Baillonella (Sapotaceae), 74
Baissea wulfhorstii Schinz (Apocynaceae), 90
Balanites aegyptiaca (L.) Del. (Balanitaceae), 62, 87, 105, 107, 108, 121, 203, 206, 207, 209, 210, 212, 213, 214, 219, 222, 224
    B. angolensis (Welw.) Welw. ex Exell, 91, 95
    B. maughamii Sprague, 90, 96, 200
    B. orbicularis Sprague, 114, 121
    B. wilsoniana Dawe & Sprague, 74, 187, 188
Ballochia (Acanthaceae), 111
Ballota (Labiatae), 228
    B. hispanica (L.) Munby, 227
Balthasaria (Theaceae), 162
    B. mannii (Oliver) Verdc., 246, 253
    B. schliebenii (Melchior) Verdc., 246
Bambusa vulgaris Schrad. (Gramineae), 55
Baphia burttii Baker f. (Leguminosae: Papilionoideae), 97
    B. massaiensis Taubert, 90, 97
    B. obovata Schinz: see *B. massaiensis*
Barbeya (Barbeyaceae), 162
    B. oleoides Schweinf., 115
Barleria (Acanthaceae), 140
    B. hochstetteri Nees, 204
    B. macrostegia Nees, 194
    B. solitaria P. G. Meyer, 142
Barringtonia racemosa (L.) Sprengel (Lecythidaceae), 188, 261, 264
Barteria fistulosa Masters (Passifloraceae), 31
Bassia muricata (L.) Asch. (Chenopodiaceae), 230
Bathiaea (Leguminosae: Caesalpinioideae), 241
Bauhinia macrantha Oliver (Leguminosae: Caesalpinioideae): see *B. petersiana*
    B. natalensis Oliver, 201
    B. petersiana C. Bolle, 90, 99
    B. rufescens Lam., 105, 206
    B. taitensis Taubert, 114
    B. tomentosa L., 91
Beckeropsis uniseta (Nees) K. Schum. (Gramineae): see *Pennisetum unisetum*
Begonia (Begoniaceae), 165, 253
Beilschmiedia natalensis J. H. Ross (Lauraceae), 199
Bellevalia (Liliaceae), 158
Bequaertiodendron natalense (Sond.) Heine & J. H. Hemsley (Sapotaceae), 200
Berberis hispanica Boiss. & Reuter (Berberidaceae), 158
Berchemia discolor (Klotzsch) Hemsley (Rhamnaceae), 90, 95, 200, 201
    B. zeyheri (Sond.) Grubov, 96
Berkheya (Compositae), 140, 194
    B. onopordifolia (DC.) O. Hoffm. ex Burtt Davy, 194
    B. rigida (Thunb.) Bolus & Wolley Dod ex Adamson & T. M. Salter, 194
Berkheyopsis angolensis O. Hoffm. (Compositae), 144
Berlinia auriculata Benth. (Leguminosae: Caesalpinioideae), 83
    B. giorgii De Wild., 90, 173
    B. grandiflora (Vahl) Hutch. & Dalz., 176
    B. occidentalis Keay, 77
Berzelia lanuginosa Brongn. (Bruniaceae), 134
Betula alba auct.: see *B. pendula*, Betulaceae
    B. pendula Roth, 147, 149, 156
Bidens pilosa L. (Compositae), 252
Biscutella (Cruciferae), 228
Bivinia (Flacourtiaceae), 186
    B. jalbertii Tul., 188
Blaeria (Ericaceae), 132, 168
    B. mannii (Engl.) Engl., 74
    B. spicata Hochst. ex A. Rich., 211
Blechnum spicant (L.) Roth (Blechnaceae), 157
Blepharis (Acanthaceae), 140
    B. acanthoides sensu D. B. Burtt, 116
    B. ciliaris (L.) B. L. Burtt, 108, 204
    B. edulis (Forssk.) Pers.: see *B. ciliaris*
    B. linariifolia Pers., 120, 212
    B. maderaspatensis (L.) Roth, 90
Blighia unijugata Baker (Sapindaceae), 199
Boerhavia coccinea Miller (Nyctaginaceae), 206
    B. repens L., 252
Bolbitis (Lomariopsidaceae), 75
Bolusanthus (Leguminosae: Papilionoideae), 87
    B. speciosus (Bolus) Harms, 96
Bombax costatum Pellegr. & Vuillet (Bombacaceae), 105, 109
Bonamia poranoides Hallier f. (Convolvulaceae), 182
Borassus aethiopum Martius (Palmaceae), 83, 85, 95, 105, 107, 108, 189, 201, 209
    B. madagascariensis Bojer, 243
Boscia (Capparidaceae), 111
    B. albitrunca (Burchell) Gilg & C. Benedict, 90, 137, 140, 191, 193, 194
    B. angustifolia A. Rich., 87, 207
    B. coriacea Pax, 114, 119
    B. foetida Schinz, 140
    B. microphylla Oliv., 95
    B. rehmanniana Pest, 95
    B. salicifolia Oliver, 87, 105, 109, 210, 222
    B. senegalensis (Pers.) Lam. ex Poir., 206, 207, 210, 212, 213, 216
Bosqueia angolensis Ficalho: see *Trilepisium madagascariense*
    B. phoberos: see *T. madagascariense*

Boswellia (Burseraceae), 111
    B. ameero Balf.f., 115
    B. dalzielii Hutch., 105, 107
    B. elongata Balf.f., 115
    B. hildenbrandtii Engl.: see *B. neglecta*
    B. neglecta S. Moore, 114, 120
    B. papyrifera (Del.) Hochst., 107, 203, 209, 210, 211
    B. socotrana Balf.f., 115
Bothriochloa insculpta (Hochst.) A. Camus: see *Dichanthium insculptum*
Bottegoa (Sapindaceae), 111
Bowringia mildbraedii Harms (Leguminosae: Papilionoideae), 81
Brabeium stellatifolium L. (Proteaceae), 135
Brachiaria (Gramineae), 169
    B. brizantha (Hochst. ex A. Rich.) Stapf, 85, 101, 173, 174, 210
    B. eruciformis (Smith) Griseb., 114
    B. falcifera (Trin.) Stapf, 178
    B. fulva Stapf: see *B. jubata*
    B. jubata (Fig. & De Not.) Stapf, 107
    B. lata (Schumach.) C. E. Hubbard, 209
    B. leersioides (Hochst.) Stapf, 114
    B. mutica (Forssk.) Stapf, 266
    B. nana Stapf, 243
    B. nigropedata (Munro ex Ficalho & Hiern) Stapf, 193
    B. ramosa L., 243
    B. serrata (Thunb.) Stapf, 194, 196
Brachylaena (Compositae), 236
    B. discolor DC., 166
    B. huillensis O. Hoffm., 117, 188
    B. hutchinsii Hutch.: see *B. huillensis*
    B. ilicifolia (Lam.) E. P. Phillips & Schweick., 201
    B. microphylla Humbert, 237
    B. uniflora Harv., 199
Brachypodium perrieri A. Camus (Gramineae), 239
    B. ramosum (L.) Roem. & Schult., 154
Brachystegia (Leguminosae: Caesalpinioideae), 87, 92, 96, 97, 106
    B. allenii Burtt Davy & Hutch, 92, 93
    B. angustistipulata De Wild., 92
    B. bakerana Burtt Davy & Hutch., 63, 89, 92, 98
    B. boehmii Taubert, 30, 54, 92, 93, 99
    B. bussei Harms, 92, 93
    B. cynometroides Harms, 77
    B. floribunda Benth., 93, 99
    B. glaberrima R. E. Fries, 92, 93
    B. glaucescens Burtt Davy & Hutch.: see *B. microphylla*
    B. laurentii (De Wild.) Louis ex Hoyle, 77, 78
    B. leonensis Burtt Davy & Hutch., 77
    B. longifolia Benth., 91, 92, 93, 97
    B. manga De Wild., 92, 93
    B. microphylla Harms, 92, 93, 99
    B. mildbraedii Harms, 77
    B. puberula Burtt Davy & Hutch., 92, 97
    B. russelliae I. M. Johnston, 92
    B. spiciformis Benth., 54, 91, 92, 93, 97, 98, 99, 173, 188
    B. stipulata De Wild., 92, 99
    B. tamarindoides Welw. ex Benth., 92, 93
    B. taxifolia Harms, 91, 92, 93, 99
    B. torrei Hoyle, 92
    B. utilis Burtt Davy & Hutch., 92, 93
    B. wangermeeana De Wild., 92, 93, 97, 173
Brackenridgea arenaria (De Wild. & Th. Durand) N. Robson (Ochnaceae), 173

Brenania (Rubiaceae), 74
Breonadia microcephala (Del.) Ridsdale: see *Adina microcephala*
Breonia sp. (Rubiaceae), 255
Bridelia ferruginea Benth. (Euphorbiaceae), 85, 105, 174
    B. taitensis Pax, 114
Bromus erectus Hudson (Gramineae), 157
    B. madritensis L., 228
    B. rubens L., 228
    B. speciosus Nees, 169
Broussonetia greveana (Baillon) C. C. Berg (Moraceae), 241
Brucea antidysenterica Miller (Simaroubaceae), 122
Bruguiera (Rhizophoraceae), 261
    B. gymnorrhiza (L.) Lam., 259, 261, 263, 264
Brunia (Bruniaceae), 134
Bryonia dioica Jacq. (Cucurbitaceae), 227
Bryum argenteum Hedwig (Bryaceae), 182
Buchholzia (Capparidaceae), 74
Buchnerodendron speciosum Gürke (Flacourtiaceae), 81
Buddleja corrugata (Benth.) E. P. Phillips (Loganiaceae), 195
    B. saligna Willd., 193, 194, 195
    B. salviifolia Lam., 195
Bulbine (Liliaceae), 141
Bulbophyllum (Orchidaceae), 236, 237
    B. leptostachysum Schltr., 238
Bulbostylis abortiva (Steud.) C. B. Clarke (Cyperaceae), 84
    B. basalis Fosberg, 259
    B. cinnamomea C. B. Clarke, 100
    B. firingalavensis Chermezon, 243
    B. laniceps C. B. Clarke ex Th. Durand & Schinz, 84
    B. xerophila Chermezon, 243
Bupleurum spinosum Gouan (Umbelliferae), 156, 158
Burchellia (Rubiaceae), 199
Burkea africana Hook. (Leguminosae: Caesalpinioideae), 85, 87, 95, 96, 97, 99, 105, 106, 107, 173, 196
Burmannia (Burmanniaceae), 75, 84
Burttdavya nyasica Hoyle (Rubiaceae), 97, 186
Burttia prunoides Baker f. & Exell (Connaraceae), 97
Bussea massaiensis (Taubert) Harms (Leguminosae: Caesalpinioideae), 97
Butyrospermum (Sapotaceae), 103
    B. paradoxum (Gaertner f.) Hepper, 55, 83, 85, 105, 107, 108
    B. parkii (G. Don) Kotschy: see *B. paradoxum*
Buxus balearica Lam. (Buxaceae), 153
    B. hildebrandtii Baillon, 115
    B. sempervirens L., 156
Byrsocarpus orientalis (Baillon) Baker (Connaraceae), 90, 98

Cadaba (Capparidaceae), 111, 242
    C. aphylla (Thunb.) Wild, 96, 201
    C. farinosa Forssk., 114, 206
    C. glandulosa Forssk., 113, 204, 212, 213
    C. heterotricha Hook., 114
Cadia purpurea (Picciv.) Aiton (Leguminosae: Papilionoideae), 115
Caesalpinia trothae Harms (Leguminosae: Caesalpinioideae), 114
Cajanus cajan (Leguminosae: Papilionoideae), 252
Calamus (Palmaceae), 83
Calendula algeriensis Boiss. & Reuter (Compositae), 158, 227
    C. murbeckii Lanza, 229
Calicotome intermedia C. Presl. (Leguminosae: Papilionoideae): see *C. villosa*
    C. villosa (Poir.) Link, 152, 158, 226
Calligonum (Polygonaceae), 220

C. comosum L'Hér., 204, 207, 219, 224
Calluna vulgaris (L.) Hull (Ericaceae), 147, 152, 247
Calodendrum capense (L.f.) Thunb. (Rutaceae), 115, 130, 166, 196, 199
Caloncoba glauca (P. Beauv.) Gilg (Flacourtiaceae), 81
  C. welwitschii (Oliver) Gilg, 80
Calophyllum eputamen P. F. Stevens (Guttiferae), 258
  C. inophyllum L., 257
  C. tacamahaca Willd., 258
Calotropis procera (Aiton) Aiton f. (Asclepiadaceae), 121, 212, 219, 251, 252
Calpocalyx (Leguminosae: Mimosoideae), 74
Calvaria galeata A.W. Hill: see *Sideroxylon galeatum*
  C. major Gaertner f.: see *S. majus*
Calyptrotheca (Portulacaceae), 111
  C. somalensis Gilg, 114
  C. taitensis (Pax & Vatke) Brenan, 114
Campnosperma seychellarum Marchand (Anacardiaceae), 257
Campylanthus salsoloides (L.f.) Roth (Scrophulariaceae), 252
Canarina abyssinica Engl. (Campanulaceae), 246
  C. canariensis (L.) Vatke, 246
  C. eminii Asch. ex Schweinf., 246
Canarium (Burseraceae), 235, 236
  C. mauritianum Blume: see *C. paniculatum*
  C. paniculatum (Lam.) Benth. ex Engl., 258
  C. schweinfurthii Engl., 78, 79, 80, 81, 82, 91, 172, 173, 181
Canavalia rosea (Swartz) DC. (Leguminosae: Papilionoideae), 253
Canthium (Rubiaceae), 83, 98
  C. bibracteatum (Baker) Hiern, 259
  C. burttii Bullock, 98, 99
  C. frangula S. Moore, 90
  C. keniense Bullock, 115
  C. lactescens Hiern, 99
  C. martinii Dunkley, 90
  C. schimperanum A. Rich., 182
  C. vulgare (K. Schum.) Bullock, 182
Capitanya (Labiatae), 111
Capparis (Capparidaceae), 189, 219
  C. decidua (Forssk.) Edgew., 219, 224
  C. elaeagnoides Gilg: see *C. fascicularis*
  C. erythrocarpos Isert, 91, 129, 178
  C. fascicularis DC., 115, 176, 182
  C. sepiaria L., 201
  C. tomentosa Lam., 182, 183
Caperonia palustris (L.) A. St. Hil. (Euphorbiaceae), 108
Caralluma (Asclepiadaceae), 114, 116, 139
  C. edithae N. E. Br., 116
  C. penicillata (Defl.) N. E. Br., 116
Carapa grandiflora Sprague (Meliaceae), 85, 181
  C. procera DC., 82, 83
Cardamine (Cruciferae), 236
Carex capillaris L. (Cyperaceae), 147, 158
  C. distachya Desf., 152
  C. distans L., 156
  C. leporina L., 155
Carica papaya L. (Caricaceae), 252
Carissa (Apocynaceae), 188, 189
  C. bispinosa (L.) Desf. ex Brenan, 96, 135, 201
  C. edulis Vahl, 98, 109, 115, 121, 129, 182
  C. haematocarpa (Eckl.) A.DC., 137, 201
  C. xylopicron Thouars, 258
Carphalea glaucescens (Klotzsch) Verdc. (Rubiaceae), 114
Carpodiptera africana Masters (Tiliaceae), 189
Carthamnus fruticosus Maire (Compositae), 153

Carum verticillatum (L.) Koch (Umbelliferae), 155
Casearia barteri Masters (Flacourtiaceae), 211
  C. battiscombei R. E. Fries, 122
  C. gladiiformis Masters, 200
Cassia (Leguminosae: Caesalpinioideae), 98, 241
  C. abbreviata Oliver, 95, 96, 99
    ss. *kassneri* (Baker) Brenan, 114
  C. aschrek Forssk.: see *C. italica*
  C. italica (Miller) Lam. ex F.W. Andrews, 219
  C. mimosoides L., 178
  C. sieberana DC., 105
  C. singueana Del., 189
  C. tora L., 212
Cassine (Celastraceae), 98
  C. aethiopica Thunb., 91, 182, 201, 243, 259
  C. buchananii Loes., 115, 129
  C. parvifolia Sond., 134
  C. peragua L., 135
Cassinopsis ilicifolia (Hochst.) Kuntze (Icacinaceae), 195
Cassipourea (Rhizophoraceae), 187
  C. annobonensis Mildbr., 253
  C. congoensis R. Br. ex DC., 83, 122, 166
  C. euryoides Alston, 188
  C. gerrardii (Schinz) Alston, 200
  C. gossweileri Exell, 98
  C. gummiflua Tul., 253
  C. malosana (Baker) Alston: see *C. congoensis*
Casuarina equisetifolia L. (Casuarinaceae), 259
Catophractes alexandri D. Don (Bignoniaceae), 95
Caucanthus albidus (Niedenzu) Niedenzu (Malphigiaceae), 114
Cavacoa quintasii (Pax & Hoffm.) J. Léonard (Euphorbiaceae), 253
Caylusea canescens (Murray) Webb (Resedaceae): see *C. hexagyna*
  C. hexagyna (Forssk.) M. L. Green, 219, 220
Cedrus atlantica (Endl.) Carrière (Pinaceae), 60, 146, 147, 149, 150, 151, 152, 154, 155, 156, 157
  C. brevifolia (Hook.f.) A. Henry, 155
  C. deodara Loudon, 155
  C. libani A. Rich., 155
Ceiba pentandra (L.) Gaertner (Bombacaceae), 81, 83, 105, 178, 253
Celsia insularis Murb.: see *Verbascum capitis-viridis*
Celtis (Ulmaceae), 181
  C. africana Burm.f., 80, 191, 194, 195, 196, 200
  C. australis L., 149
  C. brownii Rendle, 80, 83
  C. durandii Engl.: see *C. gomphophylla*
  C. gomphophylla Baker, 90, 199, 253
  C. integrifolia Lam., 105, 209
  C. mildbraedii Engl., 79, 178, 199, 253
  C. philippensis Blanco: see *C. brownii*
  C. prantlii Priemer ex Engl., 253
  C. wightii Planchon, 187
  C. zenkeri Engl., 79, 80, 172
Cenchrus biflorus Roxb. (Gramineae), 206, 207, 210, 212, 216
  C. ciliaris L., 114, 116
  C. prieurii (Kunth) Maire, 210
Centaurea (Compositae), 147
Centauropsis (Compositae), 232, 237
Cephaelis peduncularis Salisb.: see *Psychotria peduncularis*
Cephalocroton socotranus Balf.f. (Euphorbiaceae), 115
Cephalopentandra (Cucurbitaceae), 111
Cephalosphaera (Myristicaceae), 186
  C. usambarensis (Warb.) Warb., 187

Ceraria (Portulacaceae), 137
    C. longepedunculata Merxm. & Podl., 140
    C. namaquensis (Sond.) H. Pearson, 140
Cerastium (Caryophyllaceae), 236
Ceratonia (Leguminosae: Caesalpinioideae), 147, 226
    C. siliqua L., 149, 152, 158, 159, 226, 228, 252
Ceratophyllum demersum L. (Ceratophyllaceae), 265
Ceriops tagal (Pers.) C. B. Robinson (Rhizophoraceae), 259, 261, 263, 264
Ceropegia (Asclepiadaceae), 111
    C. dimorpha Humbert, 238
Chaetacme aristata Planchon (Ulmaceae), 129, 166, 199, 253
Chaetocarpus africanus Pax (Euphorbiaceae), 80
Chamaemeles (Rosaceae), 247
Chamaerops (Palmaceae), 147
    C. humilis L., 149, 152, 157, 158, 160, 228
Chascanum marrubifolium Fenzl ex Walp. (Verbenaceae), 204
Cheilanthes (Sinopteridaceae), 239
Chenolea tomentosa (Lowe) Maire (Chenopodiaceae), 229
Chenopodium (Chenopodiaceae), 122
    C. ambrosioides L., 145
Chidlowia (Leguminosae: Caesalpinioideae), 74
Chionanthus foveolatus (E. Meyer) Stearn (Oleaceae), 135, 165
Chionothrix (Amaranthaceae), 111
Chloris gayana Kunth (Gramineae), 126, 208
    C. prieurii Kunth, 176
    C. roxburghiana Schultes, 114, 116
    C. virgata Swartz, 194, 212, 213, 243
Chlorophora (Moraceae), 187
    C. excelsa (Welw.) Benth., 74, 79, 80, 81, 172, 181, 186, 187, 188, 189, 253
    C. greveana (Baillon) Leandri: see *Broussonetia greveana*
    C. regia A. Chev., 178
Chondropetalum mucronatum (Masters) Pillans (Restionaceae), 134
Chrozophora brocchiana Vis. (Euphorbiaceae), 204, 219
Chrysalidocarpus (Palmaceae), 232, 237
    C. acuminum Jumelle, 238
    C. decipiens Becc., 237
Chrysanthemoides monilifera (L.) Norlindh (Compositae), 135
Chrysanthemum (Compositae), 158
Chrysithrix (Cyperaceae), 132
Chrysocoma (Compositae), 135, 137, 139, 140
    C. tenuifolia Bergius, 193, 194, 195, 196
Chrysophyllum albidum G. Don (Sapotaceae), 181, 253
    C. boivinianum (Pierre) J. H. Hemsley, 255
    C. gorungosanum Engl., 164, 181
    C. perpulchrum Mildbr. ex Hutch. & Dalz., 74, 79, 82, 187
    C. viridifolium Wood & Franks, 167, 199
Chrysopogon aucheri (Boiss.) Stapf (Gramineae): see *C. plumulosus*
    C. plumulosus Hochst., 116, 121, 222
Cicca disticha L. (Leguminosae: Papilionoideae), 252
Cincinnobotrys (Melastomataceae), 162
Cinnamomum zeylanicum Nees (Lauraceae), 257
Cissus (Vitaceae), 242
    C. cactiformis Gilg, 128
    C. petiolata Hook.f., 182
    C. quadrangularis L., 81, 114, 117, 128, 182, 201
    C. rotundifolia (Forssk.) Vahl, 114, 182
Cistanche phelipaea (L.) Cout. (Orobanchaceae), 229
Cistus (Cistaceae), 147, 226, 228
    C. clusii Dunal, 154
    C. crispus L., 152
    C. laurifolius L., 157

    C. parviflorus Lam., 159
    C. populifolius L., 152
    C. salviifolius L., 152
    C. symphytifolius Lam., 249
    C. villosus L., 154
Citropsis daweana Swingle & M. Kellerman (Rutaceae), 90
Citrullus colocynthis (L.) Schrader (Cucurbitaceae), 219
    C. ecirrhosus Cogn., 144
Citrus (Rutaceae), 159
Cladium mariscus (L.) Pohl (Cyperaceae), 265
Cladonia (Cladoniaceae), 237
    C. medusina (Bory) Nylander, 189
    C. pycnoclada (Persoon), Nylander, 239
Cladostigma (Convolvulaceae), 111
Clausena anisata (Willd.) Hook.f. ex Benth (Rutaceae), 81, 122, 173
Cleistanthus polystachyus Hook. ex Planchon (Euphorbiaceae), 187
    C. schlechteri (Pax) Hutch., 200
Cleistochlamys (Annonaceae), 87
Clematis cirrhosa L. (Ranunculaceae), 151, 152, 153, 154, 227
    C. flammula L., 158
Cleome (Capparidaceae), 144
    C. scaposa DC., 204
    C. viscosa L., 252
Clerodendrum (Verbenaceae), 237
    C. glabrum E. Meyer, 200
Clethra arborea Aiton (Clethraceae), 246, 247
Cliffortia (Rosaceae), 132, 134
    C. arborea Marloth, 134
    C. grandifolia Eckl. & Zeyh., 134
Cocos (Palmaceae), 189
    C. nucifera L., 257, 259
Coelocaryon (Myristicaceae), 74
    C. botryoides Vermoesen, 83
Coffea arabica L. (Rubiaceae), 252
Cola clavata Masters (Sterculiaceae), 187
    C. cordifolia (Cav.) R. Br., 178
    C. digitata Masters, 253
    C. gigantea A. Chev., 79
    C. greenwayi Brenan, 164
    C. laurifolia Masters, 176
    C. natalensis Oliver, 199
Colchicum (Liliaceae), 158
Colea seychellarum Seem., (Bignoniaceae), 257
Coleochloa setifera (Ridley) Gilly (Cyperaceae), 238
Colocasia antiquorum Schott (Araceae): see *C. esculenta*
    C. esculenta (L.) Schott, 252
Colocynthis vulgaris Schrader: see *Citrullus colocynthis*
Colophospermum (Leguminosae: Caesalpinioideae), 87
    C. mopane (Kirk ex Benth.) J. Léonard, 31, 54, 61, 62, 89, 94, 140, 141, 143, 144, 191
Combretodendron africanum (Welw. ex Benth.) Exell: see *Petersianthus macrocarpum*
    C. macrocarpum P. Beauv.: see *P. macrocarpum*
Combretum (Combretaceae), 75, 98, 110, 129, 173, 241
    C. aculeatum Vent., 113, 114, 141, 209, 212
    C. apiculatum Sond., 95, 96, 193, 201
    C. camporum Engl., 90, 172
    C. celastroides Welw. ex Lawson, 90, 97, 98
        ss. *laxiflorum* (Welw. ex Lawson) Exell, 173
        ss. *orientale* Exell, 97
    C. collinum Fresen., 85, 87, 90, 95, 96, 97, 98, 105, 107, 109, 189, 193, 201, 209, 210
    C. cordofanum Engl. & Diels: see *C. glutinosum*

C. elaeagnoides Klotzsch, 90
C. erythrophyllum (Sond.) Burchell, 141
C. fragrans F. Hoffm., 95, 105
C. ghasalense Engl. & Diels: see *C. fragrans*
C. glutinosum DC., 105, 106, 107, 108, 203, 209, 210, 212, 213
C. hartmannianum Schweinf., 62, 107
C. hereroense Schinz, 96
C. imberbe Wawra, 91, 95, 96, 141, 201
C. kraussii Hochst., 165, 199
C. mechowianum O. Hoffm.: see *C. collinum*
C. micranthum G. Don, 109
C. molle R. Br. ex G. Don, 87, 95, 96, 105, 121, 129, 196, 201, 211
C. mossambicense (Klotzsch) Engl., 90
C. mucronatum Schumach. & Thonn., 81
C. nigricans Lepr. ex Guill. & Perr., 105, 107
C. oxystachyum Welw. ex Lawson, 95
C. paniculatum Vent., 81, 209
C. psidioides Welw., 97, 173
C. racemosum P. Beauv., 81
C. schumannii Engl., 187, 188
C. trothae Engl. & Diels: see *C. celastroides* ss. *orientale*
C. zeyheri Sond., 96, 97, 201
Commelina benghalensis L. (Commelinaceae), 191
Commicarpus (Nyctaginaceae), 111
C. verticillatus (Poir.) Standley, 252
Commidendrum (Compositae), 254
C. robustum DC., 254
C. rugosum (Aiton) DC., 254
C. spurium DC., 254
Commiphora (Burseraceae), 31, 63, 98, 110, 111, 113, 114, 115, 116, 117, 120, 121, 125, 128, 140, 144, 145, 181, 188, 189, 191
C. africana (A. Rich.) Engl., 87, 105, 107, 114, 193, 203, 206, 207, 208, 209, 210, 211, 212, 216
C. anacardiifolia Dinter & Engl., 95
C. angolensis Engl., 90, 95, 193
C. baluensis Engl., 116
C. boiviniana Engl., 114
C. campestris Engl., 114, 116
C. capensis (Sond.) Engl., 140
C. dalzielii Hutch., 176
C. dulcis Engl.: see *C. saxicola*
C. engleri Guillaumin, 116
C. erythraea (Ehrenb.) Engl., 114
C. gracilifrondosa Dinter ex Van der Walt, 140
C. harveyi (Engl.) Engl., 196, 199, 200
C. madagascariensis Jacq., 128
C. merkeri Engl., 116, 128
C. mollis (Oliver) Engl., 96, 114
C. monstruosa (H. Perrier) Capuron, 242
C. mossambicensis (Oliver) Engl., 99
C. namaensis Schinz, 140
C. oblanceolata Schinz, 140
C. pedunculata (Kotschy & Peyr.) Engl., 105
C. pyracanthoides Engl., 95, 96
C. riparia Engl.: see *C. mollis*
C. saxicola Engl., 140
C. schimperi (Berger) Engl., 114, 116, 128
C. trothae Engl.: see *C. schimperi*
Conocarpus erectus L. (Combretaceae), 253, 261, 262
Convolvulus gharbensis Battand. & Pitard (Convolvulaceae), 158
C. trabutianus Schweinf. & Muschler, 229
C. tricolor L., 158

Conyza (Compositae), 237
C. pinnata (L.f.) Kuntze, 194
Corchorus (Tiliaceae), 252
C. tridens L., 206
Cordeauxia (Leguminosae: Caesalpinioideae), 111
Cordia abyssinica R. Br. (Boraginaceae), 209, 210, 211
C. caffra Sond., 199, 200
C. gharaf (Forssk.) Ehrenb. ex Asch.: see *C. sinensis*
C. millenii Baker, 181
C. ovalis R. Br. ex DC., 114, 128, 129, 182
C. rothii Roem. & Schult.: see *C. sinensis*
C. sinensis Lam., 114, 128, 144, 206, 212, 222
C. subcordata Lam., 257
Cordyla africana Lour. (Leguminosae: Caesalpinioideae), 96, 186, 187
C. madagascariensis R. Viguier, 241
Coriandrum (Umbelliferae), 147
Cornulaca monacantha Del. (Chenopodiaceae), 204, 216, 220, 221, 224
Coronilla glauca L. (Leguminosae: Papilionoideae): see *C. valentina*
C. valentina L., 151
Cosmos (Compositae), 243
Costus (Zingiberaceae), 75
Cotoneaster fontanesii Spach (Rosaceae), 151
Cotula (Compositae), 141
C. coronopifolia L., 145
Cotyledon (Crassulaceae), 137, 140
C. decussata Sims, 195
C. orbiculata L., 142
C. paniculata L., 140
Coula (Olacaceae), 74
C. edulis Baillon, 77
Crabbea acaulis N. E. Br. (Acanthaceae), 194
Craibia brevicaudata (Vatke) Dunn (Leguminosae: Papilionoideae)
ss. *burtii* (Baker f.) J. B. Gillett, 97
C. zimmermannii (Harms) Harms ex Dunn, 200
Craspedorhachis africana Benth. (Gramineae), 239
Crassula (Crassulaceae), 115, 132, 137, 140, 201
C. arborescens (Miller) Willd., 140
C. portulacea Lam., 137, 201
Crataegus (Rosaceae), 156
C. azarolus L., 149
C. laciniata Ucria, 156
C. monogyna Jacq., 149, 151, 157, 158
Craterispermum laurinum (Poir.) Benth. s.l. (Rubiaceae), 82, 91
C. montanum Hiern, 253
Crocus (Iridaceae), 158
C. boulosii Greuter, 226
Crossandra nilotica Oliver (Acanthaceae), 176
Crossopteryx febrifuga (Afzel. ex G. Don) Benth. (Rubiaceae), 83, 85, 105, 189
Crotalaria (Leguminosae: Papilionoideae), 111, 140, 243
C. agatiflora Schweinf.
ss. *imperialis* (Taub.) Polhill, 130
C. microphylla Vahl, 204
C. podocarpa DC., 191
C. retusa L., 252
Croton (Euphorbiaceae), 236
C. dichogamus Pax, 115, 128, 129, 182
C. dybowskii Hutch., 172
C. gratissimus Burchell, 90, 193, 200
C. macrostachyus Hochst. ex Del., 130
C. megalobotrys Muell. Arg., 91, 95

Croton (Euphorbiaceae)—contd
  C. megalocarpus Hutch., 167, 181
  C. mubango Muell. Arg., 81
  C. pseudopulchellus Pax, 90, 188
  C. scheffleri Pax, 90
  C. socotranus Balf.f., 255
  C. stelluliferus Hutch., 253
  C. sylvaticus Hochst., 74, 199
Crotonogyne (Euphorbiaceae), 74
Crudia gabonensis Pierre ex Harms (Leguminosae: Caesalpinioideae), 76
Cryptocarya angustifolia E. Meyer ex Meissner (Lauraceae), 134
  C. latifolia Sond., 165
  C. woodii Engl., 165
Cryptosepalum (Leguminosae: Caesalpinioideae), 63
  C. pseudotaxus Baker f., 57, 90, 97
  C. staudtii Harms, 77
  C. tetraphyllum (Hook.f.) Benth., 82
Ctenium concinnum Nees (Gramineae), 194, 239
  C. elegans Kunth, 209
  C. newtonii Hack., 85, 173, 210
  C. somalense (Chiov.) Chiov., 210
Cunonia capensis L. (Cunoniaceae), 135
Cuphocarpus (Araliaceae), 232, 236
Cupressus atlantica Gaussen (Cupressaceae), 146, 149, 150, 152, 153
  C. dupreziana A. Camus, 47, 218, 222
  C. sempervirens L., 146, 149, 150, 152, 153, 159, 226
Curtisia (Cornaceae), 162
  C. dentata (Burm.f.) C. A. Smith, 165
  C. faginea Aiton: see *C. dentata*
Cussonia (Araliaceae), 201, 237
  C. arborea Hochst. ex A. Rich., 83, 85, 105
  C. barteri Seemann: see *C. arborea*
  C. holstii Engl., 115
  C. kirkii Seemann: see *C. arborea*
  C. paniculata Eckl. & Zeyh., 194, 195
  C. sessilis Lebrun, 173
  C. spicata Thunb., 135, 194
  C. zimmermannii Harms, 117, 188, 189
Cyanotis nodiflora Kunth (Commelinaceae), 238
Cyathea (Cyatheaceae), 165
  C. manniana Hook., 82
Cyclamen (Primulaceae), 147
  C. rohlfsianum Asch., 226
Cyclopia (Leguminosae: Papilionoideae), 132
Cydonia oblonga Miller (Rosaceae), 252
Cylicodiscus (Leguminosae: Mimosoideae), 74
Cylicomorpha parviflora Urban (Caricaceae), 164, 187
Cymbopogon (Gramineae), 50, 193, 210
  C. commutatus (Steudel) Stapf, 208
  C. excavatus (Hochst.) Stapf ex Burtt Davy, 202, 210
  C. giganteus (Hochst.) Chiov., 210
  C. nervatus Chiov., 108, 212, 213
  C. plicatus Stapf, 239
  C. plurinodis (Stapf) Stapf ex Burtt Davy, 137, 194
  C. pospischilii (K. Schum.) C. E. Hubbard: see *C. plurinodis*
  C. proximus (Hochst. ex A. Rich.) Stapf: see *C. schoenanthus*
  C. schoenanthus (L.) Sprengel, 208, 209, 211, 212, 221
  C. validus Stapf ex Burtt Davy, 201
Cynodon dactylon (L.) Pers. (Gramineae), 50, 116, 126, 127, 128, 194, 210, 246, 253
  C. hirsutus Stent, 195
  C. incompletus Nees, 194

Cynometra (Leguminosae: Caesalpinioideae), 235
  C. alexandri C. H. Wright, 78, 85, 172, 181, 182
  C. ananta Hutch. & Dalz., 77
  C. hankei Harms, 77
  C. leonensis Hutch. & Dalz., 77
  C. mannii Oliver, 253
  C. megalophylla Harms, 178
  C. vogelii Hook.f., 176
  C. webberi Baker f., 188
Cynomorium coccineum L. (Cynomoriaceae), 229
Cynorkis (Orchidaceae), 238
Cynosurus echinatus L. (Gramineae), 157
Cyperus esculentus L. (Cyperaceae), 100
  C. haspan L., 265, 266
  C. laevigatus L., 144, 145, 223, 267
  C. margaritaceus Vahl, 100
  C. obtusiflorus Vahl, 194
  C. papyrus L., 55, 264, 265, 266
  C. platycaulis Baker, 100
Cyphostemma currorii (Hook.f.) Descoings (Vitaceae), 140
Cyrtosperma senegalense (Schott) Engl. (Araceae), 83, 266
Cytisus albidus DC. (Leguminosae: Papilionoideae), 227
  C. arboreus (Desf.) DC., 151
  C. balansae Ball, 158
  C. battandieri Maire, 151, 156, 157
  C. linifolius Lam., 151
  C. maurus Humbert & Maire, 152
  C. monspessulanus L., 152
  C. proliferus L.f., 248
  C. stenopetalus (Webb & Berth.) Christ, 246, 251, 252
  C. triflorus L'Hérit.: see *C. villosus*
  C. villosus Pourret, 152, 157

Daboecia azorica Tutin & E. Warb. (Ericaceae), 247
Dacryodes edulis (G. Don) H. J. Lam (Burseraceae), 91, 172, 253
Dactylis glomerata L. (Gramineae), 152
Dactyloctenium (Gramineae), 267
  D. aegyptium (L.) Willd., 210
  D. geminatum Hack., 263
  D. giganteum Fisher & Schweick., 101
  D. pilosum Stapf, 259
  D. robecchii Chiov., 116
  D. sp., 126
Dalbergia (Leguminosae: Papilionoideae), 235, 236, 241
  D. armata E. Meyer, 96, 200, 201
  D. boehmii Taubert, 95
  D. ecastaphyllum (L.) Taubert, 253
  D. hostilis Benth., 109
  D. martinii F. White, 90
  D. melanoxylon Guill. & Perr., 96, 105, 107, 116, 189, 208, 209, 210, 212, 213
  D. obovata E. Meyer, 196, 200
Daniellia alsteeniana Duvign. (Leguminosae: Caesalpinioideae), 90, 173
  D. ogea (Harms) Rolfe ex Holland, 178
  D. oliveri (Rolfe) Hutch. & Dalz., 83, 85, 105, 106, 107
Danthonia (Gramineae), 134
  D. forskalii (Vahl) R. Br.: see *Asthenatherum forskalii*
  D. macowanii Stapf: see *Merxmuellera macowanii*
  D. mossamedensis Rendle: see *A. mossamedensis*
Danthoniopsis dinteri (Pilger) C. E. Hubbard (Gramineae), 145
Daphne gnidium L. (Thymelaeaceae), 151, 157, 158, 249
  D. laureola L., 151, 156
Dasysphaera (Amaranthaceae), 111
  D. prostrata (Gilg) Cavaco, 120

Daucus (Umbelliferae), 221
Decaryia madagascariensis Choux (Didiereaceae), 242
Deckenia (Palmaceae), 257
    D. nobilis (Moore) H. A. Wendl. ex Balf.f., 257
Decorsella (Violaceae), 74
Delonix adansonioides (R. Viguier) Capuron (Leguminosae: Caesalpinioideae), 242
    D. elata (L.) Gamble, 114
    D. regia (Bojer) Rafin., 241
Dendrosicyos (Cucurbitaceae), 111
    D. socotranus Balf.f., 255
Desbordesia (Irvingiaceae), 74
Deschampsia (Gramineae), 169
Desmanthus virgatus Willd. (Leguminosae: Mimosoideae), 252
Desmodium tortuosum (Swartz) DC. (Leguminosae: Papilionoideae), 252
Detarium microcarpum Guill. & Perr. (Leguminosae: Caesalpinioideae), 31, 105, 107
    D. senegalense J. F. Gmelin, 85, 178
Dialium engleranum Henriques (Leguminosae: Caesalpinioideae), 85, 97, 173
    D. guineense Willd., 105, 178, 253
    D. schlechteri Harms, 200
Dichanthium insculptum (A. Rich.) W. D. Clayton (Gramineae), 121, 128
Dichapetalum (Dichapetalaceae), 241
Dichrostachys (Leguminosae: Mimosoideae), 242
    D. cinerea (L.) Wight & Arn., 84, 85, 96, 105, 107, 189, 193, 207, 208, 209, 210, 212, 214, 251
Dicksonia arborescens L'Hérit. (Dicksoniaceae), 254
Dicoma carbonaria Humbert (Compositae), 242
    D. foliosa O. Hoffm., 145
    D. incana (Baker) O. Hoffm., 237, 242, 243
    D. macrocephala DC., 194
    D. oleifolia Humbert, 243
Dicoryphe (Hamamelidaceae), 232
    D. viticoides Baker, 237
Dicraeopetalum (Leguminosae: Papilionoideae), 111
Dicranopteris linearis (Burm.) Underw. (Gleicheniaceae), 257
Dictyosperma album (Bory) H. A. Wendl. (Palmaceae), 257, 258
    D. aureum Balf.f.: see *D. album*
Didelotia (Leguminosae: Caesalpinioideae), 74
    D. brevipaniculata J. Léonard, 77
    D. idae Oldeman, de Wit & J. Léonard, 76
    D. unifoliolata J. Léonard, 77
Didelta (Compositae), 137, 140
Didierea madagascariensis Baillon (Didiereaceae), 241, 242
    D. trollii Capuron & Rauh, 242
Digitalis purpurea L. (Scrophulariaceae), 147, 155, 157
Digitaria (Gramineae), 169, 202
    D. adscendens (Kunth) Henrard, 254
    D. ankaratrensis A. Camus, 239
    D. argyrograpta (Nees) Stapf, 194
    D. biformis Willd., 243
    D. brazzae (Franchet) Stapf, 173
    D. diagonalis (Nees) Stapf, 173, 174, 194
    D. humbertii A. Camus, 239
    D. macroblephara (Hackel) Stapf, 126, 128
    D. milanjiana (Rendle) Stapf, 101
    D. monodactyla (Nees) Stapf, 194
    D. pentzii Stent, 193
    D. tricholaenoides Stapf, 194
    D. uniglumis (Hochst. ex A. Rich.) Stapf: see *D. diagonalis*
Diheteropogon (Gramineae), 129
    D. amplectens (Nees) W. D. Clayton, 202

D. emarginatus (De Wild.) Robyns: see *D. grandiflorus*
D. grandiflorus (Hackel) Stapf, 173
Dillenia ferruginea (Baillon) Gilg (Dilleniaceae), 257
Dilobeia (Proteaceae), 232, 235, 236
Dimorphotheca (Compositae), 141
Diosma (Rutaceae), 132
Diospyros (Ebenaceae), 25, 74, 187, 235, 236, 241
    D. abyssinica (Hiern) F. White, 109, 122, 129, 164, 167, 178, 187
    D. acocksii (de Winter) de Winter, 141
    D. austro-africana de Winter, 194, 195
        ss. *rugosa* (E. Meyer ex A.DC.) de Winter, 135
    D. batocana Hiern, 93, 97, 173
    D. chevalieri De Wild., 77
    D. comorensis Hiern, 255
    D. consolatae Chiov., 188, 189
    D. cornii Chiov., 188, 189
    D. dichrophylla (Gandoger) de Winter, 135, 200, 201
    D. diversifolia Hiern, 258
    D. elliotii (Hiern) F. White, 176
    D. feliciana Letouzey & F. White, 176
    D. ferrea (Willd.) Bakh., 77, 109
    D. gabunensis Gürke, 77, 181
    D. galpinii Hiern, 201
    D. glabra (L.) de Winter, 134
    D. grex F. White, 172
    D. heterotricha (B. L. Burtt) F. White, 172
    D. hoyleana F. White, 77
    D. inhacaenis F. White, 199
    D. latispathulata H. Perrier, 242
    D. longiflora Letouzey & F. White, 83
    D. lycioides Desf., 98, 137, 141, 191, 193, 194, 195, 202
    D. melanida Poir., 258
    D. mespiliformis Hochst. ex A.DC., 55, 81, 83, 90, 91, 95, 96, 99, 105, 107, 108, 117, 178, 186, 187, 206, 211
    D. monbuttensis Gürke, 81
    D. natalensis (Harv.) Brenan, 200
    D. perrieri Jumelle, 241
    D. pseudomespilus Mildbr., 74
    D. quiloensis (Hiern) F. White, 90
    D. ramulosa (E. Meyer ex A.DC.) de Winter, 140
    D. scabrida (Harv. ex Hiern) de Winter, 201
    D. seychellarum (Hiern) Kostermans, 257
    D. simii (Kuntze) de Winter, 201
    D. squarrosa Klotzsch, 189
    D. tesselaria Poir., 257
    D. villosa (L.) de Winter, 96
    D. wagemansii F. White, 172
    D. whyteana (Hiern) F. White, 195, 196, 201
Dipcadi (Liliaceae), 158
    D. serotinum (L.) Medik., 152
    D. thollonianum Hua, 100
Diplachne fusca (L.) P. Beauv. ex Stapf (Gramineae), 262
    D. paucinervis (Nees) Stapf: see *Odyssea paucinervis*
Diplorhynchus (Apocynaceae), 87
    D. condylocarpon (Muell. Arg.) Pichon, 64, 96, 97, 99, 173
Diplotaxis (Cruciferae), 158
    D. tenuisiliqua Del., 227
Dirachma socotrana Schweinf. (Dirachmaceae), 111
Dirichletia glaucescens Hiern: see *Carphalea glaucescens*
Discoclaoxylon occidentale (Muell. Arg.) Pax & Hoffm. (Euphorbiaceae), 253
Discoglypremna (Euphorbiaceae), 74
    D. caloneura (Pax) Prain, 81, 253
Dissotis canescens (Graham) Hook.f. (Melastomataceae), 266

Dissotis (Melastomataceae)—*contd*
    D. incana (E. Meyer ex Hochst.) Triana: see *D. canescens*
    D. rotundifolia (Smith) Triana, 265
Distemonanthus (Leguminosae: Caesalpinioideae), 74
Dobera glabra (Forssk.) Poir. (Salvadoraceae), 113, 114, 117, 188, 189
    D. loranthifolia (Warb.) Harms, 114
Dodonaea madagascariensis Radlk. (Sapindaceae), 237, 238
    D. viscosa Jacq., 109, 115, 137, 140, 251, 257
Dolichos lablab L. (Leguminosae: Papilionoideae), 252
Dombeya (Sterculiaceae), 236, 237
    D. burgessiae Gerrard, 115
    D. cymosa Harv., 200
    D. goetzenii K. Schum., 122, 167
    D. kirkii Masters, 182
    D. mukole Sprague: see *D. kirkii*
    D. quinqueseta (Del.) Exell, 210
    D. rotundifolia (Hochst.) Planchon, 95, 96, 193, 196, 201
    D. shupangae K. Schum., 174
Dorstenia foetida Schweinf. (Moraceae), 113
    D. gigas Schweinf. ex Balf.f., 255
    D. gypsophila Lavranos, 116
Dovea mucronata Masters: see *Chondropetalum mucronatum*
Dovyalis abyssinica (A. Rich.) Warb. (Flacourtiaceae), 121
Dracaena (Agavaceae), 235, 241
    D. arborea (Willd.) Link, 82
    D. camerooniana Baker, 91
    D. cinnabari Balf.f., 115, 246, 255
    D. draco (L.) L., 246, 247, 251
    D. ellenbeckiana Engl., 115
    D. hookerana K. Koch, 200
    D. ombet Kotschy & Peyr., 224, 246
    D. refexa Lam., 238
    D. schizantha Baker, 115
Drakebrockmannia (Gramineae), 111
    D. somalensis Stapf, 120
Drosanthemum luederitzii (Engl.) Schwantes (Aizoaceae), 142
    D. paxianum (Schltr. & Diels) Schwantes: see *D. luederitzii*
Drosera (Droseraceae), 84, 108
Dryopteris filix-mas (L.) Schott (Aspidiaceae), 156
    D. parasitica (L.) Kuntze, 252
Drypetes floribunda (Muell. Arg.) Hutch. (Euphorbiaceae), 178
    D. gerrardii Hutch., 129, 164, 167, 199
    D. glabra (Pax) Hutch., 253
    D. leonensis Pax, 82
    D. parvifolia (Muell. Arg.) Pax & K. Hoffm., 178
    D. principum (Muell. Arg.) Hutch., 253
Duboscia (Tiliaceae), 74
Dumoria africana (Pierre) Dubard: see *Tieghemella africana*
    D. heckelii A. Chev.: see *T. heckelii*
Duosperma eremophilum (Milne-Redh.) Napper (Acanthaceae), 119, 120
Duvalia (Asclepiadaceae), 113
Dypsis (Palmaceae), 232, 235

Ebenus pinnata L. (Leguminosae: Papilionoideae), 154
Eberlanzia spinosa Schwantes (Aizoaceae), 195
Ecbolium amplexicaule S. Moore (Acanthaceae), 114
    E. revolutum C. B. Clarke, 114
Echidnopsis (Asclepiadaceae), 114, 116
Echinocarpus (Elaeocarpaceae), 235
Echinocloa colora (L.) Link (Gramineae), 108, 191
    E. pyramidalis (Lam.) Hitchcock & Chase, 100, 108, 265, 266
    E. scabra (Lam.) Roem & Schultes, 100, 265
    E. stagnina auct. non (Retz.) P. Beauv.: see *E. scabra*

Echium (Boraginaceae), 158, 246, 251
    E. boissieri Steud., 158
    E. hypertropicum Webb, 251
    E. pomponium Boiss.: see *E. boissieri*
    E. stenosiphon Webb, 252
    E. vulcanorum A. Chev., 252
Ectadium virgatum E. Meyer (Asclepiadaceae), 142
Ectropothecium (Hypnaceae), 81
Edithcolea (Asclepiadaceae), 114
Ehretia rigida (Thunb.) Druce (Boraginaceae), 140, 193, 194, 195, 201
    E. teitensis Gürke, 114
Ehrharta (Gramineae), 134
    E. erecta Lam., 115
Eichhornia crassipes (Martius) Solms-Laub. (Pontederiaceae), 265
    E. natans (P. Beauv.) Solms-Laub., 265
Ekebergia capensis Sparrman (Meliaceae), 74, 105, 129, 200, 201
    E. pterophylla (C.DC.) Hofmeyr, 96, 196
    E. senegalensis Adr. Juss.: see *E. capensis*
Elaeis guineensis Jacq. (Palmaceae), 80, 83, 174, 188
Elaeodendron buchananii (Loes.) Loes.: see *Cassine buchananii*
    E. orientale Jacq., 257, 258
Elaeophorbia (Euphorbiaceae), 54
    E. drupifera (Thonn.) Stapf, 178
Eleocharis (Cyperaceae), 262
    E. acutangula (Roxb.) Schultes, 265
Eleusine jaegeri Pilger (Gramineae), 130
Elionurus argenteus Nees (Gramineae), 101, 169, 173, 194, 196
    E. hirtifolius Hackel, 210
    E. royleanus Nees ex A. Rich., 252
    E. tristis Hackel, 239
Elvira biflora (L.) DC. (Compositae), 252
Elymandra androphila (Stapf) (Gramineae), 84, 174
Elytropappus (Compositae), 132
    E. rhinocerotis Less., 32, 132, 139
Enantia kummeriae Engl. & Diels (Annonaceae), 187
Encephalartos (Zamiaceae), 199
    E. altensteinii Lehm., 200
    E. ferox Bertol.f., 200
    E. hildebrandtii A. Braun & Bouché, 188
    E. villosus Lemaire, 200
Endodesmia (Guttiferae), 74
Englerodendron (Leguminosae: Caesalpinioideae), 186
    E. usambarense Harms, 187
Enneapogon (Gramineae), 221
    E. brachystachyus (Jaub. & Spach) Stapf: see *E. desvauxii*
    E. cenchroides C. E. Hubbard., 145, 254
    E. desvauxii P. Beauv., 195, 221
    E. scaber Lehm., 221
Entada abyssinica Steud. ex A. Rich (Leguminosae: Mimosoideae), 85, 174
    E. africana Guill. & Perr., 107
    E. mannii (Oliver) Tisserant, 81
    E. pursaetha DC., 81
    E. spicata (E. Meyer) Druce, 200, 201
Entandrophragma angolense (Welw.) C.DC. (Meliaceae), 77, 172, 181
    E. candollei Harms, 77
    E. caudatum (Sprague) Sprague, 90, 99, 201
    E. cylindricum (Sprague) Sprague, 77, 181
    E. delevoyi De Wild., 90, 96
    E. excelsum (Dawe & Sprague) Sprague, 85, 164
    E. palustre Staner, 83
    E. utile (Dawe & Sprague) Sprague, 75, 77, 82, 181

Enterospermum (Rubiaceae), 237
Entolasia imbricata Stapf (Gramineae), 100
Ephedra alata DC. (Ephedraceae), 220
    E. altissima Desf., 152, 227
    E. fragilis Desf., 153, 154
    E. tilhoana Maire, 222
Ephippiandra (Monimiaceae), 232, 236
Ephippiocarpa (Apocynaceae), 199
Equisetum ramosissimum Desf. (Equisetaceae), 222, 252
Eragrostis (Gramineae), 116, 144, 193
    E. atherstonei Stapf, 194
    E. biflora Hackel, 193
    E. capensis (Thunb.) Trin., 194
    E. chalcantha Trin.: see *E. racemosa*
    E. chloromelas Steud., 194
    E. ciliaris (L.) R. Br., 101, 193
    E. curvula (Schrader) Nees, 195
    E. cyperoides (Thunb.) P. Beauv., 142
    E. decumbens Renvoize, 259
    E. gummiflua Nees, 194
    E. haraensis Chiov., 115
    E. kohorica Quézel, 222
    E. lateritica Bosser, 239, 243
    E. lehmanniana Nees, 141, 193, 194, 195
    E. micrantha Hackel, 194
    E. nindensis Ficalho & Hiern, 142
    E. obtusa Munro ex Ficalho & Hiern, 194, 195
    E. pallens Hackel, 193
    E. papposa (Dufour) Steud., 222
    E. plana Nees, 194
    E. porosa Nees, 144
    E. racemosa (Thunb.) Steud., 194, 196
    E. sclerantha Nees, 194
    E. scotelliana Rendle, 84
    E. superba Peyr., 194
    E. tenuifolia (A. Rich.) Steud., 122, 126
    E. tremula (Lam.) Hochst. ex Steud., 209, 210, 211, 212, 213
Eremospatha (Palmaceae), 83
Erica (Ericaceae), 132, 134, 149, 168, 195
    E. arborea L., 129, 147, 152, 157, 159, 216, 222, 246, 248, 249
    E. caffra L., 134
    E. caterviflora Salisb., 134
    E. inconstans Zahlbr., 134
    E. multiflora L., 154
    E. scoparia L.
        ss. *azorica* (Hochst.) D. A. Webb, 246, 247
    E. umbellata L., 152
Erinacea anthyllis Link (Leguminosae: Papilionoideae), 156, 158
Eriobotrya japonica (Thunb.) Lindley (Rosaceae), 252
Eriobroma oblongum (Masters) Bodard: see *Sterculia oblonga*
Eriocephalus (Compositae), 135, 137, 139, 140, 193
    E. racemosus L., 135
    E. spinescens Burchell, 195
Eriochloa meyerana (Nees) Pilger (Gramineae), 267
Eriosema (Leguminosae: Papilionoideae), 243
Eriospermum abyssinicum Baker (Liliaceae), 100
Erismadelphus (Vochysiaceae), 74
Erodium glaucophyllum (Geraniaceae), 221
Eryngium ilicifolium Lam. (Umbelliferae), 229
    E. tricuspidatum L., 152
Erysimum caboverdeanum (A. Chev.) Sund. (Cruciferae), 252

Erythrina abyssinica DC. (Leguminosae: Papilionoideae), 107, 174
    E. baumii Harms, 173
    E. caffra Thunb., 199
    E. excelsa Baker, 181
    E. sacleuxii Hua, 187
    E. sigmoidea Hua, 211
Erythrochlamys (Labiatae), 111
    E. spectabilis Gürke, 114
Erythrococca bongensis Pax (Euphorbiaceae), 182
    E. menyharthii (Pax) Prain, 90
Erythrophleum africanum (Welw.) Harms (Leguminosae: Caesalpinioideae), 87, 91, 93, 96, 97, 105, 106, 173
    E. guineense G. Don: see *E. suaveolens*
    E. lasianthum Corbishley, 200
    E. suaveolens (Guill. & Perr.) Brenan, 74, 91, 178, 187, 188
Erythrophysa (Sapindaceae), 241
Erythrostictus (Liliaceae), 158
Erythroxylum acranthum Hemsley (Erythroxylaceae), 259
    E. emarginatum Thonn., 99, 109
    E. lanceum Bojer, 255
    E. platycladum Bojer, 243
Eucalyptus (Myrtaceae), 160, 231
Euclea (Ebenaceae), 98, 115, 201
    E. coriacea A.DC., 195
    E. crispa (Thunb.) Gürke, 137, 194, 195
        ss. *ovata* (Burchell) F. White, 193, 195
    E. divinorum Hiern, 115, 129, 167
    E. lancea Thunb., 135
    E. natalensis A.DC., 99, 188, 189, 200
        ss. *capensis* F. White, 135
    E. pseudebenus E. Meyer ex A.DC., 141, 144
    E. racemosa Murray, 135, 200
        ss. *schimperi* (A.DC.) F. White, 115, 121, 129, 182, 188, 224
    E. schimperi (A.DC.) Dandy: see *E. racemosa* ss. *schimperi*
    E. tomentosa E. Meyer ex A.DC., 135
    E. undulata Thunb., 135, 137, 140, 141, 193, 201
Eugenia (Myrtaceae), 236
    E. capensis (Eckl. & Zeyh.) Sond., 202
    E. leonensis Engl. & v. Brehm., 82
    E. sp., 258
Eulalia villosa (Thunb.) Nees (Gramineae), 202
Euonymus latifolius (L.) Miller (Celastraceae), 151
Euphorbia (Euphorbiaceae), 50, 54, 111, 115, 116, 117, 137, 139, 147, 149, 174, 198, 201, 228, 229, 238, 241, 242, 249
    E. arbuscula Balf.f., 255
    E. avasmontana Dinter, 140, 193
    E. azorica Seub., 246
    E. balsamifera Aiton, 207
    E. beaumierana Hook.f. & Cosson, 225, 226, 227, 228
    E. bellica Hiern, 144
    E. bilocularis N. E. Br., 97
    E. calycina N. E. Br.: see *E. candelabrum*
    E. candelabrum Trémaux ex Kotschy, 98, 99, 115, 128, 129, 182, 188, 207, 210
    E. clavarioides Boiss., 195
    E. columnaris Bally, 116
    E. conspicua N. E. Br., 90, 91
    E. cuneata Vahl, 116
    E. currorii N. E. Br., 140
    E. dawei N. E. Br., 182, 183
    E. desmondii Keay & Milne-Redh., 62, 109
    E. dinteri Berger: see *E. virosa*
    E. echinus Hook.f. & Coss., 224, 225, 226, 227, 228, 229

Euphorbia (Euphorbiaceae)—*contd*
    E. eduardoi Leach, 140
    E. enterophora Drake, 241
    E. evansii Pax, 201
    E. grandicornis Goebel, 114, 188
    E. grandidens Haw., 137, 200, 201
    E. grandis Lemaire, 115
    E. gregaria Marloth, 140
    E. guerichiana Pax, 140
    E. gummifera Boiss., 140, 142
    E. inaequilatera Sond., 194
    E. ingens E. Meyer ex Boiss., 96, 201
    E. kamerunica Pax, 109
    E. mauritanica L., 135, 140, 195
    E. mosaica Bally & S. Carter, 116
    E. multiclava Bally & S. Carter, 116
    E. nyikae Pax, 114, 128, 188
    E. origanoides L., 254
    E. phillipsae N. E. Br., 113
    E. poissonii Pax, 109
    E. pyrifolia Lam., 259
    E. quinquecostata Volkens, 114
    E. regis-jubae Webb & Berth., 224, 225, 226, 228, 229
    E. resinifera Berger, 225, 226, 227, 228
    E. robecchii Pax, 114
    E. scheffleri Pax, 114
    E. schimperi Presl, 120
    E. sepulta Bally & S. Carter, 116
    E. socotrana Balf.f., 115
    E. spiralis Balf.f., 255
    E. stenoclada Baillon, 242
    E. striata Thunb., 194
    E. subsalsa Hiern, 145
    E. sudanica A. Chev., 109
    E. tetragona Haw., 200, 201
    E. tirucalli L., 128, 201
    E. triangularis Desf., 200, 201
    E. tuckeyana Steud., 246, 251, 252
    E. virosa Willd., 144
    E. wakefieldii N. E. Br., 188
Euryops (Compositae), 135, 140
Eurypetalum (Leguminosae: Caesalpinioideae), 74
Eustachys paspaloides (Vahl) Lanza & Mattei (Gramineae), 126, 128, 137, 194
Excoecaria bussei (Pax) Pax (Euphorbiaceae), 90
    E. venenifera Pax, 188
Exotheca abyssinica Anderson (Gramineae), 169

Fagara capensis Thunb.: see *Zanthoxylum capense*
    F. chalybea (Engl.) Engl.: see *Z. chalybeum*
    F. davyi I. Verdoorn: see *Z. davyi*
    F. macrophylla (Oliver) Engl.: see *Z. gilletii*
    F. trijuga Dunkley: see *Z. trijugum*
    F. xanthoxyloides Lam.: see *Z. xanthoxyloides*
Fagaropsis angolensis (Engl.) Dale (Rutaceae), 167
Fagonia (Zygophyllaceae), 230
    F. cretica L., 229
    F. flamandii Battand., 222
    F. glutinosa Del., 220
    F. latifolia Del., 221
    F. microphylla Pomel, 221
    F. mollis Del., 221
Farsetia (Cruciferae), 111
    F. aegyptiaca Turra, 221
    F. longisiliqua Decne., 115

    F. stenoptera Hochst., 204
Faurea (Proteaceae), 237
    F. forficuliflora Baker, 237, 238
    F. saligna Harv., 93, 96, 105, 167
    F. speciosa Welw., 99
Fedia (Valerianaceae), 158
Fegimanra (Anacardiaceae), 74
Felicia (Compositae), 141
    F. filifolia (Vent.) Burtt Davy, 194, 196
    F. muricata (Thunb.) Nees, 194
Feretia aeruginescens Stapf (Rubiaceae), 99
Fernandoa madagascariensis (Baker) A. Gentry (Bignoniaceae), 243
    F. magnifica Seemann, 117, 187
Ferula (Umbelliferae), 228, 230
    F. communis L., 158, 160
Festuca (Gramineae), 156, 169, 195
    F. abyssinica Hochst. ex A. Rich., 211
    F. camusiana St Yves, 239
    F. caprina Nees, 196
    F. costata Nees, 169
    F. hystrix Boiss., 156
    F. triflora Desf., 157
Ficalhoa (Theaceae), 162
    F. laurifolia Hiern, 164
Ficinia (Cyperaceae), 132
Ficus (Moraceae), 75, 98, 129, 209, 210, 259
    F. annobonensis Mildbr. & Hutch., 253
    F. capensis Thunb., 91, 105, 187, 196, 199, 251
    F. carica L., 252
    F. congensis Engl., 91, 181, 266
    F. cordata Thunb., 140, 193
    F. exasperata Vahl, 83
    F. fischeri Warb. ex Mildbr. & Burret, 90, 98
    F. glumosa Del., 105, 109
    F. guerichiana Engl., 140, 193
    F. ingens (Miq.) Miq., 96, 99, 117, 196, 222
    F. lecardii Warb., 109
    F. marmorata Bojer, 242
    F. natalensis Hochst., 199, 200
    F. populifolia Vahl, 207
    F. pseudosycomorus Decne., 224
    F. sagittifolia Warb. ex Mildbr. & Burret, 75
    F. salicifolia Vahl, 207, 222
    F. socotrana Balf.f., 115, 255
    F. soldanella Warb., 196
    F. sonderi Miq., 99, 196
    F. sycomorus L., 91, 95, 96, 105, 117, 141, 144, 201, 211, 222, 251, 252
    F. teloukat Battand., 222
    F. trichopoda Baker, 200
    F. vallis-choudae Del., 187
    F. verruculosa Warb., 266
    F. vogelii (Miq.) Miq., 83
Filicium decipiens (Wight & Arn.) Thw. (Sapindaceae), 255
Fimbristylis pilosa Vahl (Cyperaceae), 178
Fingerhuthia africana Lehm. (Gramineae), 137, 195
Fissidens sciophyllus Mitten (Fissidentaceae), 182
Flacourtia flavescens Willd. (Flacourtiaceae): see *F. indica*
    F. indica (Burm.f.) Merr., 178, 189, 241
Fleurydora felicis A. Chev. (Ochnaceae), 176
Foeniculum vulgare Miller (Umbelliferae), 158
Foetidia mauritiania Lam. (Foetidiaceae), 257
    F. rodriguesiana Friedmann, 258
Foleyola (Cruciferae), 218

Fomes annosus (Fries) Cooke (Polyporaceae), 78
Forgesia borbonica Pers. (Escalloniaceae), 258
Forsskålea tenacissima L. (Urticaceae), 204, 221
Frangula alnus Miller: see *Rhamnus frangula*
Frankenia (Frankeniaceae), 223, 266
    F. corymbosa Desf., 229
    F. laevis L., 230
    F. portulacifolia Spreng., 254
Fraxinus angustifolia Vahl (Oleaceae), 149, 150
    F. xanthoxyloides Wall., 149, 153, 156
Fredolia aretioides Moq. ex Coss. (Chenopodiaceae), 221
Freylinia oppositifolia Spin (Scrophulariaceae), 135
Friesodielsia obovata (Benth.) Verdc.: see *Popowia obovata*
Fuirena pubescens (Poir.) Kunth (Cyperaceae), 100
    F. umbellata Rottb., 84, 266
Funtumia africana (Benth.) Stapf (Apocynaceae), 74, 81, 187, 253
Furcraea foetida (L.) Haw. (Agavaceae), 257
    F. gigantea Vent.: see *F. foetida*

Gaertnera (Rubiaceae), 82
Gagea (Liliaceae), 158
Galenia (Aizoaceae), 140, 193
Galpinia (Lythraceae), 199
    G. transvaalica N. E. Br., 200
Garcinia chromocarpa Engl. (Guttiferae), 83
    G. echirensis Pellegr.: see *G. chromocarpa*
    G. livingstonei T. Andersson, 91, 117, 129, 201
    G. polyantha Oliver: see *G. smeathmannii*
    G. punctata Oliver, 74, 83
    G. smeathmannii (Planchon & Triana) Oliver, 82, 91
Gardenia (Rubiaceae), 242
    G. imperialis K. Schum., 91
    G. jovis-tonantis (Welw.) Hiern: see *G. ternifolia*
    G. lutea Fresen.: see *G. ternifolia*
    G. sokotensis Hutch., 107, 109
    G. ternifolia Schumach. & Thonn., 85, 107, 108, 174, 210, 211
Garuleum (Compositae), 140
Gasteria (Liliaceae), 137, 140
Geigeria (Compositae), 194
    G. alata (DC.) Benth. & Hook.f. ex Oliver & Hiern, 204
    G. aspera Harv., 194
    G. spinosa O. Hoffm., 144, 145
Genista (Leguminosae: Papilionoideae), 147, 229
    G. ferox Poir., 227
    G. myriantha Ball, 153
    G. saharae Coss. & Durieu, 220
    G. retamoides Spach, 154
    G. tricuspidata Desf., 157
Geopanax (Araliaceae), 257
Geophila (Rubiaceae), 75
Gerrardanthus lobatus (Cogn.) C. Jeffrey (Cucurbitaceae), 114
Geum sylvaticum Pourret (Rosaceae), 157
Gilbertiodendron (Leguminosae: Caesalpinioideae), 74
    G. bilineatum (Hutch. & Dalz.) J. Léonard, 77
    G. brachystegioides (Harms) J. Léonard, 77
    G. dewevrei (De Wild.) J. Léonard, 78, 79
    G. ogoouense (Pellegr.) J. Léonard, 79
    G. preussii (Harms) J. Léonard, 77
    G. splendidum (A. Chev. ex Hutch. & Dalz.) J. Léonard, 77
Gilletiodendron glandulosum (Portères) J. Léonard (Leguminosae: Caesalpinioideae), 103
Givotia gosai Radcl.-Smith (Euphorbiaceae), 114
    G. madagascariensis Baillon, 241

Gladiolus (Iridaceae), 158
    G. byzantinus Miller, 158
Globularia alypum L. (Globulariaceae), 153, 155, 222, 226, 229
    G. amygdalifolia Webb, 252
Glossonema boveanum (Decne.) Decne. (Asclepiadaceae), 204
Glumea ivorensis Aubrév. & Pellegr. (Sapotaceae), 76, 77
Gnidia (Thymelaeaceae), 194
    G. glauca (Fresen.) Gilg, 130
    G. kraussiana Meissner, 173, 194
    G. polycephala (C. A. Meyer) Gilg, 195
    G. subcordata Meissner, 115
Gossweilerodendron (Leguminosae: Caesalpinioideae), 74
Gossypium hirsutum L. (Malvaceae), 252
    G. somalense (Gürke) J. B. Hutch., 204
Grandidiera (Flacourtiaceae), 186
    G. boivinii Jaub., 188
Grangeria borbonica Lam. (Chrysobalanaceae), 258
Greenwayodendron suaveolens (Engl. & Diels) Verdc. (Annonaceae), 74
    ss. *usambaricum* Verdc., 187
Grewia (Tiliaceae), 98, 113, 236, 242
    G. avellana Hiern, 90
    G. bicolor Juss., 182
    G. burttii Exell, 97
    G. carpinifolia Juss., 91, 178
    G. fallax K. Schum., 114, 128
    G. flava DC., 96, 140, 193
    G. flavescens Juss., 90, 210, 211, 213
    G. megalocarpa Juss., 176
    G. mollis Juss., 174, 182, 211
    G. occidentalis L., 194, 195, 201
    G. plagiophylla K. Schum., 189
    G. robusta Burch., 201
    G. similis K. Schum., 115, 121, 129, 182
    G. tembensis Fresen., 114, 115
    G. tenax (Forssk.) Fiori, 114, 213, 222
    G. trichocarpa A. Rich., 129
    G. truncata Masters, 189
    G. villosa Willd., 95, 114, 176, 211, 251
Grielum (Neuradaceae), 137, 141
Griffonia simplicifolia (Vahl ex DC.) Baillon (Leguminosae: Caesalpinioideae), 178
Grimmia campestris Burchell ex Hooker (Grimmiaceae), 53
    G. ovalis (Hedwig) Lindberg, 53
    G. ovata Weber & Mohr: see *G. ovalis*
Grossera (Euphorbiaceae), 74
Grubbia (Grubbiaceae), 132
Guarea cedrata (A. Chev.) Pellegr. (Meliaceae), 77, 82
    G. thompsonii Sprague & Hutch., 77
Guibourtia copallifera Bennett (Leguminosae: Caesalpinioideae), 103, 109
    G. demeusei (Harms) J. Léonard, 83
Guiera senegalensis J. F. Gmel. (Combretaceae), 203, 210, 212
Gymnorinorea: see *Decorsella*
Gymnosiphon (Burmanniaceae), 75
Gyrocarpus americanus Jacq. (Hernandiaceae), 109, 242
Gyroptera (Chenopodiaceae), 111

Haematostaphis (Anacardiaceae), 103
    H. barteri Hook.f., 105
Hagenia (Rosaceae), 162
    H. abyssinica (Bruce) J. F. Gmelin, 47, 130, 161, 165, 166, 167
Hakea acicularis (Vent.) Knight (Proteaceae), 135

Halimium (Cistaceae), 147
    H. atlanticum Humbert & Maire, 157
    H. halimiifolium (L.) Willk., 151
    H. lasiocalycinum (Boiss. & Reuter) Maire, 152
    H. libanotis Lange, 151
Halleria lucida L., (Scrophulariaceae), 165, 195, 196, 201
Halocnemum strobilaceum (Pallas) M. Bieb. (Chenopodiaceae), 233, 224, 230
Haloxylon scoparium Pomel (Chenopodiaceae), 221, 228, 229, 230
Haplocarpha (Compositae), 194
    H. scaposa Harv., 194
Haplocoelum foliolosum (Hiern) Bullock (Sapindaceae), 98, 99, 129, 188
    H. inoploeum Radlk., 188, 189
Harmsia (Sterculiaceae), 111
Harpachne (Gramineae), 111
Harpagophytum (Pedaliaceae), 241
Harpechloa falx (L.f.) Kuntze (Gramineae), 194
Harpephyllum (Anacardiaceae), 199
    H. caffrum Bernh., 199, 200, 201
Harrisonia abyssinica Oliver (Simaroubaceae), 189
Hartogia capensis L.f. (Celastraceae), 135
Harungana madagascariensis Lam. ex Poir. (Guttiferae), 80, 81, 83, 105, 173, 236
Haworthia (Liliaceae), 137, 140
    H. tesselata Haw., 195
Haya Balf.f. (Caryophyllaceae), 111
Heberdenia bahamensis (Gaertner) Sprague (Myrsinaceae): see *H. excelsa*
    H. excelsa (Aiton) Banks ex DC., 246, 247
Heckeldora (Meliaceae), 74
Hedera helix L. (Araliaceae), 151, 156
Hedycaryopsis (Monimiaceae), 232, 236
Hedychium coronarium Koenig (Zingiberaceae), 236
Hedyotis adscensionis DC. (Rubiaceae), 254
    H. arborea Roxb., 254
Heeria argentea (Thunb.) Meissner (Anacardiaceae), 134
    H. concolor Presl ex Sond., 140
    H. crassinervia (Engl.) Engl., 140, 193
    H. reticulata (Baker f.) Engl., 96, 97, 129, 189
Heisteria parvifolia Smith (Olacaceae), 253
Helianthemum (Cistaceae), 147
    H. canariense Pers., 153, 227, 228
    H. gorgoneum Webb, 252
    H. kahiricum Del., 221
    H. lavandulifolium Miller, 154
    H. pergamaceum Pomel, 230
Helichrysum (Compositae), 132, 140, 168, 194, 237, 238, 239
    H. dregeanum Sond. & Harv., 194
    H. glumaceum DC., 116
    H. latifolium (Thunb.) Less., 194
    H. oreophilum Klatt, 194
    H. rugulosum Less., 194
    H. yuccifolium Lam., 258
Heliophila (Cruciferae), 141
Heliotropium curassavicum L. (Boraginaceae), 145
    H. rariflorum Stocks, 204
Heritiera littoralis Dryander (Sterculiaceae), 261, 262, 263, 264
    H. utilis (Sprague) Sprague: see *Tarrietia utilis*
Hermannia (Sterculiaceae), 135, 137, 140, 141, 143
    H. betonicifolia Eckl. & Zeyh., 194
    H. candidissima Spreng.f., 195
    H. coccocarpa Kuntze, 194, 195
    H. depressa N. E. Br., 194

Hernandia ovigera L. (Hernandiaceae), 257
    H. voyroni Jumelle, 241
Heteromorpha (Umbelliferae), 237
    H. arborescens (Sprengel) Cham. & Schlechtd., 194
Heteropogon (Gramineae), 238
    H. contortus (L.) P. Beauv. ex Roem. & Schult., 193, 194, 195, 196, 202, 208, 239, 243, 252
Hexalobus monopetalus (A. Rich.) Engl. & Diels (Annonaceae), 109
Heywoodia lucens Sim (Euphorbiaceae), 199
Hibiscus (Malvaceae), 140, 220
    H. asper Hook.f., 108
    H. diversifolius Jacq., 265
    H. marlothianus K. Schum., 195
    H. micranthus L.f., 145
    H. tiliaceus L., 188, 257, 264
Hildebrandtia (Convolvulaceae), 111
Hildegardia barteri (Masters) Kosterm. (Sterculiaceae), 79, 83
Hippobromus (Sapindaceae), 199
    H. pauciflorus (L.f.) Radlk., 200
Hippocratea indica Willd. (Celastraceae), 99
    H. parviflora N. E. Br., 90
Hippocrepis (Leguminosae: Papilionoideae), 228
Hirtella (Chrysobalanaceae), 186
Holarrhena floribunda (G. Don) Dur. & Schinz (Apocynaceae), 81, 83
Holoptelea grandis (Hutch.) Mildbr. (Ulmaceae), 79, 81, 181
Homalium (Flacourtiaceae), 83
    H. dentatum (Harv.) Warb., 199
Homeria (Iridaceae), 141
Hoodia (Asclepiadaceae), 139
    H. currori (Hook.) Decne., 142, 145
Hornea mauritania Baker (Sapindaceae), 257
Huernia (Asclepiadaceae), 139
Humbertochloa bambusiuscula A. Camus & Stapf (Gramineae), 242
Hyaenanche globosa (Gaertn.) Lambert (Euphorbiaceae), 134
Hydrilla verticillata Caspary (Hydrocharitaceae), 265
Hydrocotyle (Umbelliferae), 236
Hydrodea bossiana Dinter: see *Mesembryanthemum cryptanthum*
    H. cryptantha (Hook.f.) N. E. Br.: see *M. cryptanthum*
Hygrophila auriculata (Schumach.) Heine (Acanthaceae), 108
Hylodendron (Leguminosae: Caesalpinioideae), 74
Hymenaea (Leguminosae: Caesalpinioideae), 186
    H. verrucosa Gaertn., 188
Hymenocardia acida Tul. (Euphorbiaceae), 85, 97, 99, 105, 173, 174
    H. ulmoides Oliver, 83, 200
Hymenocoleus (Rubiaceae), 75
Hymenodictyon floribundum (Steud. & Hochst.) B. L. Robinson (Rubiaceae), 81, 82, 83
    H. parvifolium Oliver, 114
Hymenostegia (Leguminosae: Caesalpinioideae), 74
    H. afzelii (Oliver) Harms, 77, 80
    H. laxiflora (Benth.) Harms, 172
Hyophorbe (Palmaceae), 257
    H. verschaffeltii H. A. Wendl., 258
Hyoscyamus muticus L. (Solanaceae), 219
Hyparrhenia (Gramineae), 50, 97, 129, 169, 189, 194, 209, 210, 238
    H. anthistirioides (Hochst.) Andersson ex Asch. & Schweinf., 108, 210, 212
    H. bracteata (Willd.) Stapf, 100
    H. confinis (A. Rich.) Stapf, 173, 209
    H. cyanescens (Stapf) Stapf, 107

H. cymbaria (L.) Stapf, 128, 243
H. dichroa (Steud.) Stapf, 101
H. diplandra (Hackel) Stapf, 85, 100, 173
H. familiaris (Steud.) Stapf, 85, 173
H. filipendula (Hochst.) Stapf, 128, 174, 202, 210
H. hirta (L.) Stapf, 128, 135, 137, 193, 195, 196, 208, 211, 252, 255
H. lecomtei (Franchet) Stapf: see *H. newtonii*
H. multiplex (Hochst. ex A. Rich) Andersson ex Stapf, 211
H. mutica W. D. Clayton, 84
H. newtonii (Hackel) Stapf, 100, 101, 174, 239
H. nyassae (Rendle) Stapf, 85, 239
H. pachystachya Stapf: see *H. diplandra*
H. papillipes (Hochst.) Andersson ex Asch. & Schweinf., 208
H. petiolata Stapf, 212
H. pseudocymbaria (Steud.) Stapf: see *H. anthistirioides*
H. rufa (Nees) Stapf, 85, 108, 174, 209, 239, 243
H. ruprechtii Fourn.: see *Hyperthelia dissoluta*
H. schimperi (Hochst. ex A. Rich.) Andersson, 243
H. subplumosa Stapf, 32, 85
Hypericum lalandii Choisy (Guttiferae), 266
H. lanceolatum Lam.: see *H. revolutum*
H. revolutum Vahl, 166, 258
H. roeperanum Schimp. ex A. Rich., 74
Hyperthelia dissoluta (Nees ex Steud.) W. D. Clayton (Gramineae), 101, 107, 128, 174, 239, 243
Hyphaene (Palmaceae), 216, 219, 220
H. benguellensis Welw.: see *H. ventricosa*
H. compressa H. A. Wendl., 188, 189
H. coriacea Gaertner, 123
H. natalensis Kuntze, 201
H. petersiana Klotzsch: see *H. ventricosa*
H. shatan Bojer, 243
H. thebaica (L.) Martius, 55, 107, 108, 219, 220, 223
H. ventricosa Kirk, 95
Hypodaphnis (Lauraceae), 74
Hypoestes verticillaris (L.f.) R. Br. (Acanthaceae), 90
Hypolytrum (Cyperaceae), 75
Hypoxis angustifolia Lam. (Hypoxidaceae), 100
H. rigidula Baker, 194
H. rooperi S. Moore, 194

Icomum lineare Burkill (Labiatae), 100
Ifloga spicata (Forssk.) Schultes Bip. (Compositae), 220
Ilex aquifolium L. (Aquifoliaceae), 149, 151, 156, 157
I. canariensis Poir., 246
I. mitis (L.) Radlk., 74, 91, 122, 135, 165, 167, 181, 195, 196, 237, 238
I. perado Aiton, 246
ss. *azorica* (Loes.) Tutin, 247
ss. *platyphylla* (Webb & Berth.) Tutin, 246
Imbricaria seychellarum Oliver (Sapotaceae), 257
Impatiens (Balsaminaceae), 165, 167, 236, 238
I. irvingii Hook.f. ex Oliver, 265
Imperata cylindrica (L.) P. Beauv. (Gramineae), 50, 83, 85, 174, 224, 236, 239, 243
Indigofera (Leguminosae: Papilionoideae), 111, 140, 243
I. alternans DC., 194
I. cordifolia Heyne ex Roth, 204
I. cunenensis Torre, 142
I. daleoides Benth., 144
I. disjuncta J. B. Gillett, 204
I. rhynchocarpa Welw. ex Baker, 98
I. rostrata Bolus, 194

I. senegalensis Lam., 204
I. sokotrana Vierh., 115
I. spinosa Forssk., 116, 119
I. subcorymbosa Baker, 98
I. teixeirae Torre, 145
Indokingia (Araliaceae), 257
Inhambanella henriquesii (Engl. & Warb.) Dubard (Sapotaceae), 187, 200
Intsia bijuga (Colebr.) Kuntze (Leguminosae: Caesalpinioideae), 257
Iphiona (Compositae), 242
Ipomoea (Convolvulaceae), 111, 114, 265
I. crassipes Hook., 194
I. pes-caprae (L.) R. Br., 253, 254, 257, 258
I. sultani Chiov., 116
I. verbascoidea Choisy, 211
Iris (Iridaceae), 158
Irvingia gabonensis (Aubry-Lecomte ex O'Rorke) Baillon (Irvingiaceae), 253
I. smithii Hook.f., 83
Isalus (Gramineae), 238
Ischaemum (Gramineae), 189
Isoberlinia (Leguminosae: Caesalpinioideae), 54, 58, 59, 61, 92, 96, 102, 103, 106, 107, 176
I. angolensis (Welw. ex Benth.) Hoyle & Brenan, 87, 91, 92, 105, 106
I. doka Craib & Stapf, 105, 106, 108
I. scheffleri (Harms) Greenway, 187
I. tomentosa (Harms) Craib. & Stapf: see *I. angolensis*
Isolona heinsenii Engl. & Diels (Annonaceae), 187

Jardinea congoensis (Hackel) Franchet (Gramineae), 84
J. gabonensis Steudel, 84
Jasminum fluminense Vell. (Oleaceae), 182
J. fruticans L., 153, 154, 158
J. mauritianum Bojer ex DC.: see *J. fluminense*
Jatropha (Euphorbiaceae), 111, 242
J. curcas L., 252, 255
J. glandulosa Vahl: see *J. pelargoniifolia*
J. gossypiifolia L., 252
J. pelargoniifolia Courb., 115
J. unicostata Balf.f., 255
J. villosa (Forssk.) Muell. Arg.: see *J. pelargoniifolia*
Jubaeopsis (Palmaceae), 199
Julbernardia (Leguminosae: Caesalpinioideae), 92, 96, 97, 106
J. globiflora (Benth.) Troupin, 92, 93, 99
J. magnistipulata (Harms) Troupin, 187, 188
J. paniculata (Benth.) Troupin, 91, 92, 97
J. pellegriniana Troupin, 77
J. seretii (De Wild.) Troupin, 77, 78, 79, 85
Juncus acutus L. (Juncaceae), 160, 219, 246, 266
J. arabicus (Aschers. & Buchenau) Adamson, 224
J. bufonius L., 222
J. effusus L., 191
J. maritimus Lam., 222, 223, 266, 267
Juniperus (Cupressaceae), 115
J. brevifolia (Seub.) Antoine, 246, 247
J. cedrus Webb & Berth., 246
J. communis L., 153, 158
J. oxycedrus L., 149, 151, 152, 154, 156, 158, 226
J. phoenicea L., 146, 149, 150, 151, 152, 153, 154, 155, 157, 159, 226, 229, 248, 249
J. procera Hochst. ex Endl., 51, 115, 122, 130, 161, 164, 165, 166, 167
J. thurifera L., 146, 149, 150, 152, 153, 155, 156

Justicia flava (Forssk.) Vahl (Acanthaceae), 182

Kaempferia rosea Schweinf. ex Benth. & Hook.f. (Zingiberaceae), 90
Kalanchoe (Crassulaceae), 114, 115, 129, 201, 236, 238, 241, 242
    K. robusta Balf.f., 255
Kanahia (Asclepiadaceae), 111
Kaokochloa (Gramineae), 137
    K. nigrirostis de Winter, 142
Kaoue stapfiana (A. Chev.) Pellegr. (Leguminosae: Caesalpinioideae), 77
Kedrostis gijef (J. F. Gmelin) C. Jeffrey (Cucurbitaceae), 114
Kelleronia (Zygophyllaceae), 111
    K. quadricornuta Chiov., 116
Khaya anthotheca (Welw.) C.DC. (Meliaceae), 81, 172, 181
    K. comorensis Legris nom. nud., 255
    K. grandifoliola C.DC., 79, 82, 181
    K. nyasica Stapf ex Baker f., 91, 117, 186
    K. senegalensis (Desr.) Adr. Juss, 105, 178, 210, 211, 214
Kigelia africana (Lam.) Benth. (Bignoniaceae), 91, 95, 105, 117, 209
Kigelianthe madagascariensis (Baker) Sprague: see *Fernandoa madagascariensis*
Kiggelaria (Flacourtiaceae), 162
    K. africana L., 135, 165, 195, 199
Kirkia acuminata Oliver (Simaroubaceae), 95, 96, 99
    K. wilmsii Engl., 96
Kissenia (Loasaceae), 113
    K. capensis Endl., 113
Klainedoxa gabonensis Pierre ex Engl. (Irvingiaceae), 80, 172, 181
Kleinia (Compositae), 116, 139
    K. cliffordiana (Hutch.) C. D. Adams, 109
    K. kleinioides (Schultz-Bip.) M. R. F. Taylor, 120
    K. scottii (Balf.f.) Chiov., 255
Koeleria (Gramineae), 169
    K. pubescens (Lam.) P. Beauv., 230
    K. vallesiana (Honck.) Bertol., 230
Kohautia amatymbica Eckl. & Zeyh. (Rubiaceae), 194
    K. aspera (Roth) Bremek., 113
Kotschya africana Endl. (Leguminosae: Papilionoideae), 266
Kyllinga (Cyperaceae), 126
    K. alba Nees, 116
    K. erecta Schumach., 100

Lablab niger Medic.: see *Dolichos lablab*
    L. purpureus (L.) Sweet: see *D. lablab*
Laburnum platycarpum Maire (Leguminosae: Papilionoideae), 227
Lachanodes (Compositae), 254
    L. arborea (Roxb.) R. B. Nordenstam, 254
Lachenalia (Liliaceae), 141
Lachnocapsa (Cruciferae), 111
Lagarosiphon (Hydrocharitaceae), 265
Lagenantha nogalensis Chiov. (Chenopodiaceae), 120
Laguncularia racemosa Gaertner (Combretaceae), 261, 262
Lamarckia aurea (L.) Moench (Gramineae), 227
Landolphia (Apocynaceae), 241
    L. camptoloba (K. Schum.) Pichon, 173
    L. parvifolia K. Schum., 99
Lannea alata (Engl.) Engl. (Anacardiaceae), 114
    L. amaniensis Engl. & K. Krause: see *L. welwitschii*
    L. antiscorbutica (Hiern) Engl., 173
    L. discolor (Sond.) Engl., 96, 97, 98, 99, 196
    L. fruticosa (Hochst. ex A. Rich.) Engl., 209

    L. humilis (Oliver) Engl., 105, 116, 208, 209, 212
    L. microcarpa Engl. & K. Krause, 107
    L. schimperi (Hochst. ex A. Rich.) Engl., 105, 107
    L. stuhlmannii (Engl.) Engl., 95, 129, 189, 201
    L. triphylla (Hochst. ex A. Rich.) Engl., 114
    L. welwitschii (Hiern) Engl., 187, 253
Lantana (Verbenaceae), 189
    L. camara L., 236, 252
Lapeirousia (Iridaceae), 141
Lasiochloa (Gramineae), 134
    L. echinata (Thunb.) Adamson, 135
Lasiocorys argyrophylla Vatke (Labiatae), 116
Lasiurus hirsutus (Forssk.) Boiss. (Gramineae), 221, 224
Latania commersonii J.F. Gmelin (Palmaceae): see *L. lontaroides*
    L. lontaroides (Gaertner) H. E. Moore, 257
    L. verschaffeltii Lemaire, 258
Lathyrus (Leguminosae: Papilionoideae), 157
Launaea arborescens (Battand.) Maire (Compositae), 224, 229
    L. chevalieri O. Hoffm. & Muschler, 204
Laurophyllus capensis Thunb. (Anacardiaceae), 134
Laurus azorica (Seub.) Franco (Lauraceae), 246, 247, 249
    L. nobilis L., 147, 149, 150, 226
Lavandula (Labiatae), 147, 221, 228, 229
    L. coronopifolia Poir.: see *L. stricta*
    L. dentata L., 152, 154, 227, 252
    L. maroccana Murbeck, 228
    L. multifida L., 153, 154, 227
    L. pubescens Decne., 211, 222
    L. rotundifolia Benth., 252
    L. stoechas L., 151, 152
    L. stricta Del., 219
Lebeckia (Leguminosae: Papilionoideae), 140
    L. macrantha Harv., 193
Lebrunia bushaie Staner (Guttiferae), 79, 85
Lecaniodiscus fraxinifolius Baker (Sapindaceae), 91, 117, 129
Leersia hexandra Swartz (Gramineae), 100, 266
Lemna (Lemnaceae), 265
    L. perpusilla Torrey, 265
Leonotis mollissima Gürke (Labiatae), 122
Lepidopilum callochlorum C. Mueller ex Broth. (Daltoniaceae), 81
Lepidotrichilia volkensii (Gürke) Leroy (Meliaceae), 122, 167
Lepisanthes senegalensis (Poir.) Leenhouts: see *Aphania senegalensis*
Leptadenia pyrotechnica (Forssk.) Decne. (Asclepiadaceae), 204, 206, 207, 212, 213, 219, 220
    L. reticulata Wight, 243
Leptaspis (Gramineae), 75
Leptochloa uniflora Hochst. ex A. Rich. (Gramineae), 90
Leptolaena bojerana (Baillon) Cavaco (Sarcolaenaceae), 237
    L. pauciflora Baker, 237
Leptothrium senegalense (Kunth) W. D. Clayton (Gramineae), 119, 121
Lepturus (Gramineae), 266
Leucadendron (Proteaceae), 132, 134
    L. argenteum (L.) R. Br., 134
    L. concinnum R. Br.: see *L. procerum*
    L. eucalyptifolium E. Mey. ex Meissner, 134
    L. nobile J. M. Williams, 134
    L. procerum (Salisb. ex Knight) J.M. Williams, 134
    L. sabulosum Salter, 134
    L. salicifolium J. M. Williams, 134, 135
    L. salignum R. Br., 135
Leucaena glauca auct. (Leguminosae: Mimosoideae): see *L. leucocephala*

L. leucocephala (Lam.) de Wit, 257
Leucojum (Amaryllidaceae), 158
Leucosidea (Rosaceae), 162
    L. sericea Eckl. & Zeyh., 195, 196
Leucospermum (Proteaceae), 132, 134
    L. conocarpodendron (L.) Buek, 134
Leucosphaera (Amaranthaceae), 137
    L. bainesii (Hook.f.) Gilg, 191, 193, 267
Leuzea conifera (L.) DC. (Compositae), 155, 226
Librevillea (Leguminosae: Caesalpinioideae), 74
Lightfootia (Campanulaceae), 135, 140
Ligustrum robustum Blume (Oleaceae), 257
    L. vulgare L., 151
Limnophyton obtusifolium (L.) Miq. (Alismataceae), 265
Limoniastrum feei (de Gir.) Battand. (Plumbaginaceae), 221
    L. guyonianum Durieu, 223
    L. ifniense (Caball.) Font Quer, 223
    L. monopetalum (L.) Boiss., 230
Limonium (Plumbaginaceae), 246
    L. cymuliferum (Boiss.) Sauvage & Vindt, 230
    L. fallax (Wangerin) Maire, 229
    L. pruinosum (L.) Kuntze, 223
Linaria (Scrophulariaceae), 147
    L. sagittata Steud., 229
Lindackeria dentata (Oliver) Gilg (Flacourtiaceae), 81
Linociera foveolata (E. Meyer) Knobl.: see *Chionanthus foveolatus*
Linum villarianum Pau (Linaceae), 157
Lippia ukambensis Vatke (Verbenaceae), 122
Lithops (Aizoaceae), 140, 142
Lobelia (Campanulaceae), 53
    L. bambuseti R. E. Fries, 167
    L. barnsii Exell, 253
Lobostemon (Boraginaceae), 132
Lochia (Caryophyllaceae), 111
Lodoicea (Palmaceae), 257
    L. maldivica (J. F. Gmelin) Pers., 257
Loesenera (Leguminosae: Caesalpinioideae), 74
Loewia (Turneraceae), 111
Lonchocarpus bussei Harms (Leguminosae: Papilionoideae), 189
    L. capassa Rolfe, 95, 96, 201
    L. laxiflorus Guill. & Perr., 105, 107, 211
    L. nelsii (Schinz) Schinz ex Heering & Grimme, 90
Lonicera arborea Boiss. (Caprifoliaceae), 156
    L. etrusca G. Santi, 151, 157
    L. pyrenaica L., 158
Lophiocarpus polystachyus Turcz. (Chenopodiaceae), 145
Lophira alata Banks ex Gaertner f. (Ochnaceae), 75, 77, 78
    L. lanceolata Van Tiegh. ex Keay, 85, 105
Loranthus (Loranthaceae), 95
Lotononis tenuis Baker (Leguminosae: Papilionoideae), 145
Lotus (Leguminosae: Papilionoideae), 246
    L. arabicus L., 144, 204
    L. glinoides Delarbre, 252
    L. mossamedensis Welw. ex Baker: see *L. arabicus*
Loudetia (Gramineae), 129, 238
    L. arundinacea (Hochst. ex A. Rich.) Steud., 85, 173, 174
    L. demeusii (De Wild.) C. E. Hubbard, 173
    L. filifolia Schweick.
        ss. *humbertiana* A. Camus, 243
    L. kagerensis (K. Schum.) C. E. Hubbard ex Hutch., 84
    L. phragmitoides (Peter) C. E. Hubbard, 85, 265, 266
    L. simplex (Nees) C. E. Hubbard, 40, 50, 52, 84, 85, 100, 101, 169, 173, 202, 209, 210

    ss. *stipoides* (Hackel) Bosser, 239, 243
    L. togoensis (Pilger) C. E. Hubbard, 208, 209, 210
Loudetiopsis ambiens (K. Schum.) Conert (Gramineae), 84
    L. glabrata (K. Schum.) Conert, 84
Lovoa swynnertonii Baker f. (Meliaceae), 186, 187
    L. trichilioides Harms, 77, 172
Loxostylis (Anacardiaceae), 199
Ludia (Flacourtiaceae), 186
    L. mauritiana Gmelin, 189
    L. sessiliflora Lam.: see *L. mauritiana*
Ludwigia (Onagraceae), 265
    L. erecta (L.) Hara, 265
    L. leptocarpa (Nutt.) Hara, 265
    L. octovalvis (Jacq.) Raven, 265
    L. stolonifera (Guill. & Perr.) Raven, 265
Lumnitzera racemosa Willd. (Combretaceae), 261, 263, 264
Lupinus pilosus L. (Leguminosae: Papilionoideae): see *L. varius*
    L. varius L., 147
Luzula fosteri (Smith) DC. (Juncaceae), 155
    L. multiflora (Retz.) Lej., 156
    L. sylvatica (Hudson) Gaudin, 156
Lycium (Solanaceae), 140, 143, 223
    L. austrinum Miers, 201
    L. decumbens Welw. ex Hiern, 145
    L. europaeum L., 116
    L. intricatum Boiss., 152, 223, 224, 228, 229, 266
    L. tetrandrum L.f., 144
Lycopodium affine Bory (Lycopodiaceae), 84, 108
    L. carolinianum L., 84
    L. cernuum L., 84
    L. mildbraedii Hert., 82
Lygeum (Gramineae), 230
    L. spartum L., 149, 226, 229, 230
Lytanthus amygdalifolius (Webb) Wettst.: see *Globularia amygdalifolia*

Macaranga (Euphorbiaceae), 236
    M. capensis (Baillon) T.R. Sim, 187
    M. kilimandscharica Pax, 181
    M. monandra Muell. Arg., 81, 181
    M. pynaertii De Wild., 181
    M. schweinfurthii Pax, 181
    M. spinosa Muell. Arg., 81
Maerua (Capparidaceae), 111, 219
    M. angolensis DC., 87, 105, 145
    M. crassifolia Forssk., 119, 204, 206, 207, 212, 219
    M. denhardtiorum Gilg, 114
    M. filiformis Drake, 242
    M. mildbraedii Gilg & C. Benedict: see *M. triphylla*
    M. subcordata (Gilg) De Wolf, 114
    M. triphylla A. Rich., 182
Maesa lanceolata Forssk. (Myrsinaceae), 211, 253
Maesopsis eminii Engl. (Rhamnaceae), 81, 181
Magnistipula butayei De Wild. (Chrysobalanaceae), 74
    ss. *greenwayi* (Brenan) F. White, 187
Malacantha alnifolia (Baker) Pierre (Sapotaceae), 83, 187
Malcolmia aegyptiaca Sprengel (Cruciferae), 220
Malus domestica Borkh. (Rosaceae), 252
    M. sylvestris Mill.: see *M. domestica*
Mammea (Guttiferae), 235
    M. africana Sabine, 253
Mangifera (Anacardiaceae), 189
    M. indica L., 174, 252
Manilkara (Sapotaceae), 187
    M. concolor (Harv. ex C. H. Wright) Gerstner, 96, 199

Manilkara (Sapotaceae)—contd
  M. discolor (Sond.) J. H. Hemsley, 200
  M. mochisia (Baker) Dubard, 91, 188, 189
  M. obovata (Sabine & G. Don) J. H. Hemsley, 81, 83, 167, 178
  M. sansibarensis (Engl.) Dubard, 187, 188, 189
  M. sulcata (Engl.) Dubard, 116, 188
Mansonia altissima (A. Chev.) A. Chev. (Sterculiaceae), 79
Mapania (Cyperaceae), 75, 77
Maprounea africana Muell. Arg. (Euphorbiaceae), 97, 105, 173, 174
Maranthes glabra (Oliver) Prance (Chrysobalanaceae), 77, 85
  M. goetzeniana (Engl.) Prance, 186, 187
  M. polyandra (Benth.) Prance, 83, 85, 105
Marantochloa (Marantaceae), 75
Marattia (Marattiaceae), 82
Margaritaria discoidea (Baillon) Webster: see *Phyllanthus discoideus*
Mariscus deciduus C. B. Clarke (Cyperaceae), 100
Markhamia acuminata (Klotzsch) K. Schum. (Bignoniaceae), 90
  M. hildebrandtii (Baker) Sprague, 167
  M. obtusifolia (Baker) Sprague, 90, 95
Marquesia (Dipterocarpaceae), 173
  M. acuminata (Gilg) R. E. Fries, 90, 173
  M. macroura Gilg, 90, 91, 93, 173
Mascarena verschaffeltii (H. A. Wendl.) L. H. Bailey: see *Hyophorbe verschaffeltii*
Mathurina penduliflora Balf.f. (Turneraceae), 258
Matthiola kralikii Pomel (Cruciferae), 229
Maytenus (Celastraceae), 201
  M. acuminata (L.f.) Loes., 135, 195
  M. heterophylla (Eckl. & Zeyh.) N. Robson, 115, 135, 193, 195
  M. linearis (L.f.) Marais, 201, 243
  M. oleoides (Lam.) Loes., 134, 135
  M. polyacantha (Sond.) Marais, 195
  M. senegalensis (Lam.) Exell, 85, 91, 174, 189, 207, 222, 228, 259
  M. undata (Thunb.) Blakelock, 195
Medemia argun (Martius) Württemb. ex H. A. Wendl. (Palmaceae), 218
  M. nobilis Gallerand, 243
Medinilla (Melastomataceae), 236
Medusagyne oppositifolia Baker (Medusagynaceae), 257
Megalochlamys (Acanthaceae), 113
Megaloprotachne albescens C. E. Hubbard (Gramineae), 193
Megistostegium (Malvaceae), 232, 242
Melanodendron (Compositae), 254
  M. integrifolium (Roxb.) DC., 254
Melanthera scandens (Schumach. & Thonn.) Roberty (Compositae), 265
Melastomastrum segregatum (Benth.) A. & R. Fernandes (Melastomataceae), 265
Melhania melanoxylon Aiton: see *Trochetia melanoxylon*
Melia volkensii Gürke (Meliaceae), 114
Melinis minutiflora P. Beauv. (Gramineae), 252, 254
Mellissia (Solanaceae), 254
  M. begoniifolia (Roxb.) Hook.f., 254
Memecylon (Melastomataceae), 187
  M. eleagni Blume, 257
  M. sansibaricum Taubert, 188
  M. sapinii De Wild., 173
Merremia multisecta Hallier f. (Convolvulaceae), 142
Merxmuellera (Gramineae), 134
  M. disticha (Nees) Conert, 139

  M. macowanii (Stapf) Conert, 239
  M. stricta (Schrader) Conert, 139
Mesanthemum radicans (Benth.) Koern. (Eriocaulaceae), 84
Mesembryanthemum (Aizoaceae), 266
  M. cryptanthum Hook.f., 142, 254
Mesogyne henriquesii Engl. (Moraceae), 253
Metalasia (Compositae), 132, 134
  M. muricata (L.) Less., 134, 135
Metrosideros angustifolia (L.) Smith (Myrtaceae), 135
Michelsonia microphylla (Troupin) Hauman (Leguminosae: Caesalpinioideae), 78, 79
Microberlinia bisulcata A. Chev. (Leguminosae: Caesalpinioideae), 77
Microchloa caffra Nees (Gramineae), 194
  M. indica (L.f.) P. Beauv., 116, 213
  M. kunthii Desv., 126, 128, 209
Micromeria (Labiatae), 249
  M. forbesii Benth., 249
Mikania cordata (Burm.f.) B. L. Robinson (Compositae), 265
Mildbraediodendron excelsum Harms (Leguminosae: Caesalpinioideae), 181
Milium vernale M. Bieb. (Gramineae), 157
Millettia grandis (E. Mey.) Skeels (Leguminosae: Papilionoideae), 199, 200
  M. sutherlandii Harv., 199
  M. thonningii (Schumach. & Thonn.) Baker, 178
  M. usaramensis Taubert, 189
Mimetes (Proteaceae), 134
  M. fimbrifolius Salisb. ex Knight, 134
Mimosa pigra L. (Leguminosae: Mimosoideae), 266
Mimusops aedificatoria Mildbr. (Sapotaceae), 187
  M. caffra E. Mey. ex A.DC., 200
  M. maxima (Lam.) Vaughan, 258
  M. obovata Sond., 200
  M. petiolaris (DC.) Dubard, 258
  M. zeyheri Sond., 91, 99, 196
Miscanthus (Gramineae), 55
  M. teretifolius (Stapf) Stapf, 100
  M. violaceus (K. Schum.) Pilger, 265, 266
Mitolepis (Asclepiadaceae), 111
  M. intricata Balf.f., 115
Mitragyna ciliata Aubrév. & Pellegr. (Rubiaceae), 83
  M. inermis (Willd.) Kuntze, 105, 107, 108
  M. rubrostipulata (K. Schum.) Havil., 85, 164
  M. stipulosa (DC.) Kuntze, 83, 91, 181
Molinaea sp. (Sapindaceae), 258
Monadenium invenustum N. E. Br. (Euphorbiaceae), 114
Monanthes (Crassulaceae), 246
Monanthotaxis fornicata (Baillon) Verdc. (Annonaceae), 189
Monechma (Acanthaceae), 140
  M. genistifolium C. B. Clarke, 267
  M. tonsum P. G. Meyer, 267
Monelytrum (Gramineae), 137
Monocyclanthus (Annonaceae), 74
Monocymbium ceresiiforme (Nees) Stapf (Gramineae), 84, 85, 100, 101, 169, 173, 194, 196
Monodiella (Gentianaceae), 218
Monodora myristica (Gaertner) Dunal (Annonaceae), 181, 253
Monopetalanthus (Leguminosae: Caesalpinioideae), 74
  M. compactus Hutch. & Dalz., 77
  M. hedinii (A. Chev.) Pellegr., 77
  M. richardsiae J. Léonard, 91
  M. trapnellii J. Léonard, 91
Monotes (Dipterocarpaceae), 87, 99, 106, 176
  M. caloneurus Gilg, 174

M. dasyanthus Gilg, 173
M. kerstingii Gilg, 105
M. mutetetwa Duvign., 174
Monotheca buxifolia (Falcolner) A.DC.: see *Sideroxylon buxifolium*
Monsonia ignorata Merxm. & A. Schreiber (Geraniaceae), 142
M. nivea Webb, 220
M. senegalensis Guill. & Perr., 145
Montinia caryophyllacea Thunb. (Montiniaceae), 137, 193, 201
Moraea natalensis Baker (Iridaceae), 100
Moricandia arvensis (L.) DC. (Cruciferae), 221
Morinda asteroscepa K. Schum. (Rubiaceae), 187
Moringa (Moringaceae), 111, 242
M. ovalifolia Dinter & A. Berger, 140
M. peregrina (Forssk.) Fiori, 224
Morus lactea (Sim) Mildbr. (Moraceae): see *M. mesozygia*
M. mesozygia Stapf ex A. Chev., 79, 81, 82, 105, 178, 181, 199
Mucuna sloanei Fawcett & Rendle (Leguminosae: Papilionoideae), 253
Mundulea phylloxylon R. Viguier (Leguminosae: Papilionoideae), 238
M. sericea (Willd.) A. Chev., 210
Muraltia (Polygalaceae), 132, 134
Musanga cecropioides R. Br. (Moraceae), 25, 80, 81, 83, 181, 253
M. leo-errerae Hauman & J. Léonard, 85
Muscari (Liliaceae), 158
Myrianthus arboreus P. Beauv. (Moraceae), 81
M. holstii Engl., 164, 187
Myrica (Myricaceae), 237
M. faya Aiton, 246, 247, 248
Myrmecosicyos (Cucurbitaceae), 111
Myrothamnus (Myrothamnaceae), 238
M. flabellifolius (Sond.) Welw., 23, 99, 238
M. moschatus Baillon, 238
Myrsine africana L. (Myrsinaceae), 135, 195, 246, 247
Myrtus communis L. (Myrtaceae), 147, 158
M. nivellei Battand., 218, 222
Mystroxylum aethiopicum (Thunb.) Loes.: see *Cassine aethiopica*

Najas (Najadaceae), 265
Narcissus (Amaryllidaceae), 158
Nardurus cynosuroides (Desf.) Trabut (Gramineae), 230
Nardus stricta L. (Gramineae), 155
Nastus borbonicus J. F. Gmelin (Gramineae), 258
Nauclea diderrichii (De Wild. & Th. Durand) Merr. (Rubiaceae), 77
N. latifolia Smith, 83, 85, 105, 107, 108
N. pobeguinii (Pobéguin ex Pellegr.) Petit, 83, 91
Neoboutonia macrocalyx Pax (Euphorbiaceae), 181
N. mannii Benth., 253
Neocentema (Amaranthaceae), 111
Neodypsis (Palmaceae), 232, 237
Neophloga (Palmaceae), 232, 235
Nepenthes pervillei Blume (Nepenthaceae), 257
Nephrosperma (Palmaceae), 257
Neptunia oleracea Lour. (Leguminosae: Mimosoideae), 206
Nerium oleander L. (Apocynaceae), 147, 219, 222
Nesiota (Rhamnaceae), 254
N. elliptica (Roxb.) Hook.f., 254
Nesogordonia papaverifera (A. Chev.) Capuron (Steruliaceae), 79, 178
N. parvifolia (M. B. Moss) Capuron, 188
Nestlera (Compositae), 140
Neuracanthus (Acanthaceae), 111

Neurada (Neuradaceae), 218
N. procumbens L., 220
Neurotheca congolana De Wild. & Th. Durand (Gentianaceae), 84
Newbouldia laevis (P. Beauv.) Seemann ex Bureau (Bignoniaceae), 81
Newtonia aubrevillei (Pellegr.) Keay (Leguminosae: Mimosoideae), 82
N. buchananii (Baker) Gilbert & Boutique, 80, 85, 91, 167, 181, 186
N. erlangeri (Harms) Brenan, 188
N. hildebrandtii (Vatke) Torre, 90, 91, 117, 200
N. paucijuga (Harms) Brenan, 187, 188
Nicotiana glauca Graham (Solanaceae), 252
Nirarathamnos (Umbelliferae), 111
Nitella (Characeae), 265
Nitraria retusa (Forssk.) Asch. (Zygophyllaceae), 223, 224, 230
Northea seychellana Hook.f. (Sapotaceae), 257
Nostoc commune Vaucher (Nostocaceae), 53
Notelaea azorica Tutin: see *Picconia azorica*
N. excelsa (Aiton) Webb & Berth.: see *P. excelsa*
Notholaena (Sinopteridaceae), 238
Notonia (Compositae), 242
Nucularia (Chenopodiaceae), 218
N. perrinnii Battand., 223
Nuxia (Loganiaceae), 237
N. congesta R. Br. L ex Fresen., 74, 122, 165, 167, 176, 196, 199, 253
N. floribunda Benth., 165
N. pseudodentata Gilg, 255
N. verticillata Lam., 258
Nymania (Meliaceae), 137
N. capensis (Thunb.) Lindb., 140, 191
Nymphaea (Nymphaeaceae), 265
N. caerulea Savigny, 265
N. lotus L., 265
Nymphoides ezannoi Berhaut (Menyanthaceae), 204
N. indica (L.) Kuntze, 265

Ochlandra (Gramineae), 237
O. capitata Camus, 235, 236
Ochna (Ochnaceae), 83
O. afzelii R. Br. ex Oliver, 105
O. ciliata Lam., 259
O. holstii Engl., 164
O. leptoclada Oliver, 174
O. manikensis De Wild., 173
O. membranacea Oliver, 82
O. ovata F. Hoffm., 176
O. pulchra Hook., 96, 97, 193, 196
O. schweinfurthiana F. Hoffm., 99, 105, 109, 174
O. thomasiana Engl. & Gilg, 188
Ochradenus (Resedaceae), 218
O. baccatus Del., 116
Ochrocarpos: see *Mammea*
Ochthocosmus lemaireanus De Wild. & Th. Durand (Ixonanthaceae), 97
Ocimum (Labiatae), 122
O. suave Willd., 122
Ocotea (Lauraceae), 235, 237
O. borbonica auct.: see *O. obtusata*
O. bullata (Burchell) Baillon, 165, 246
O. comoriensis Kosterm., 255
O. foetens (Ait.) Benth. & Hook.f., 246, 247, 249
O. gabonensis R. Fouilloy, 83, 246

Ocotea (Lauraceae)—*contd*
    O. kenyensis (Chiov.) Robyns & R. Wilczek, 122, 165, 246
    O. michelsonii Robyns & R. Wilczek, 85
    O. obtusata (Nees) Kostermans, 258
    O. usambarensis Engl., 85, 164, 187
Odyssea jaegeri (Pilger) Robyns & Tournay: see *Psilolemma jaegeri*
    O. paucinervis (Nees) Stapf, 144, 267
Oldenburgia arbuscula DC. (Compositae), 134
Oldfieldia africana Benth. & Hook.f. (Euphorbiaceae), 77
    O. dactylophylla (Welw. ex Oliver) J. Léonard, 97, 173
    O. somalensis (Chiov.) Milne-Redh., 188
Olea (Oleaceae), 130, 226, 227, 228, 255
    O. africana Miller, 115, 121, 122, 129, 135, 167, 182, 193, 194, 195, 201, 224
    O. capensis L., 62, 105, 109, 122, 134, 164, 253
        ss. *macrocarpa* (C.H. Wright) I. Verdoorn, 200
    O. europaea L., 149, 152, 154, 157, 158, 159, 226, 227, 228, 229, 251
    O. foveolata E. Meyer: see *Chionanthus foveolatus*
    O. hochstetteri Baker: see *O. capensis*
    O. laperrinei Battand. & Trabut, 54, 204, 207, 208, 211, 222
    O. woodiana Knobl., 200
Oleandra articulata Presl (Oleandraceae), 235
Olinia (Oliniaceae), 135, 196
    O. emarginata Burtt Davy, 195
Oncinotis inhandensis J. M. Wood & Evans (Apocynaceae), 200
Oncostemum (Myrsinaceae), 232, 237
Onobrychis argentea Boiss. (Leguminosae: Papilionoideae), 230
Ononis (Leguminosae: Papilionoideae), 147
    O. atlantica Ball, 158
    O. polysperma Barr. & Murbeck, 227
Ophiobotrys (Flacourtiaceae), 74
Ophrys (Orchidaceae), 158
Opilia celtidifolia (Guill. & Perr.) Endl. ex Walp. (Opiliaceae), 109
Oplismenus hirtellus (L.) P. Beauv. (Gramineae), 90
Opuntia (Cactaceae), 159, 231, 254
Orbea (Asclepiadaceae), 113
Orchis (Orchidaceae), 158
Oreobambos buchwaldii K. Schum. (Gramineae), 55
Oricia bachmannii (Engl.) I. Verdoorn (Rutaceae), 199
Ormenis multicaulis Braun-Blanquet ex Maire (Compositae), 152
Ornithogalum (Liliaceae), 158
Orothamnus zeyheri Pappe (Proteaceae), 135
Oryza longistaminata Chev. & Roehr. (Gramineae), 100, 108, 265
    O. perennis auct.: see *O. longistaminata*
Oryzopsis caerulescens (Desf.) Hackel (Gramineae), 222
Osmunda regalis L. (Osmundaceae), 156
Osteospermum (Compositae), 141, 194
    O. scariosum DC., 194
Ostryoderris stuhlmannii (Taubert) Harms: see *Xeroderris stuhlmannii*
Osyris sp. (Santalaceae), 129, 135, 153, 154, 194, 195
Othonna protecta Dinter (Compositae), 142
Otoptera (Leguminosae: Papilionoideae), 243
Otostegia (Labiatae), 111
Ottelia ulvifolia (Planchon) Walp. (Hydrocharitaceae), 265
Oubanguia africana Baillon (Scytopetalaceae), 83
Ouratea (Ochnaceae), 74, 83
Oxalis (Oxalidaceae), 141, 194
    O. depressa Eckl. & Zeyh., 194
Oxystigma (Leguminosae: Caesalpinioideae), 74
    O. mannii (Baillon) Harms, 83

    O. oxyphyllum (Harms) J. Léonard, 77
Oxytenanthera abyssinica (A. Rich.) Munro (Gramineae), 55
Ozoroa crassinervia (Engl.) R. & A. Fernandes: see *Heeria crassinervia*
    O. reticulata (Baker f.) R. & A. Fernandes: see *H. reticulata*

Pachycarpus lineolatus (Decne.) Bullock (Asclepiadaceae), 100
Pachyelasma (Leguminosae: Caesalpinioideae), 74
Pachypodium (Apocynaceae), 238
    P. geayi Costantin & Bois, 242
    P. lamerei Drake, 242
    P. lealii Welw., 140
    P. namaquanum (Wyley ex Harv.) Welw., 140
    P. succulentum (L.f.) A.DC., 195
Pachystela brevipes (Baker) Baillon ex Engl. (Sapotaceae), 105, 109, 172, 188
    P. msolo (Engl.) Engl., 187
Paeonia atlantica Kralik ex Trabut (Paeoniaceae), 157
Pandanus (Pandanaceae), 188, 236, 243, 257
    P. alpestris Martius, 238
    P. candelabrum P. Beauv., 83, 262
    P. goetzei Warb., 189
    P. heterocarpus Balf.f., 258
    P. hornei Balf.f., 257
Pandiaka carsonii (Baker) C. B. Clarke (Amaranthaceae), 100
Panicum aldabrense Renvoize (Gramineae), 259
    P. baumannii K. Schum., 174
    P. coloratum L., 116, 126, 194
    P. deustum Thunb., 182
    P. dregeanum Nees, 239
    P. fulgens Stapf: see *P. baumannii*
    P. griffonii Franchet, 84
    P. heterostachyum Hackel, 90, 97
    P. kalaharense Mez, 193
    P. laetum Kunth, 204
    P. lanipes Mez, 193
    P. lindleyanum Nees ex Steud., 84
    P. luridum Hackel, 239
    P. maximum Jacq., 85, 101, 236, 243, 252
    P. natalense Hochst., 193
    P. parvifolium Lam., 84, 266
    P. phragmitoides Stapf, 85, 174
    P. pilgeri Mez, 84
    P. pusillum Hook.f., 211
    P. repens L., 55, 100
    P. subalbidum Kunth, 209, 266
    P. turgidum Forssk., 115, 204, 207, 212, 213, 216, 218, 219, 220, 221, 224
Pappea capensis Eckl. & Zeyh. (Sapindaceae), 98, 117, 121, 129, 137, 140, 141, 189, 201
Paramacrolobium coeruleum (Taubert) J. Léonard (Leguminosae: Caesalpinioideae), 74, 188
Parinari capensis Harv. (Chrysobalanaceae), 101, 173, 201
    P. congensis F. Didr., 83, 176
    P. congolana Th. & H. Durand, 83
    P. curatellifolia Planchon ex Benth., 85, 87, 93, 96, 97, 98, 99, 105, 129, 173, 174, 189
    P. excelsa Sabine, 71, 74, 81, 82, 85, 90, 92, 164, 176, 178, 181
    P. glabra Oliver: see *Maranthes glabra*
    P. goetzeniana Engl.: see *M. goetzeniana*
    P. polyandra Benth.: see *M. polyandra*
Parkia bicolor A. Chev. (Leguminosae: Mimosoideae), 55, 77, 82
    P. biglobosa (Jacq.) Benth., 83, 85, 105

P. clappertoniana Keay: see *P. biglobosa*
P. filicoidea Welw. ex Oliver, 74, 91, 171, 172, 181, 186, 187, 188
Parkinsonia aculeata L. (Leguminosae: Caesalpinioideae), 252
 P. africana Sond., 140, 143, 144, 191
Parmelia (Parmeliaceae), 142
 P. vagans Nylander, 53
Parnassia palustris L. (Parnassiaceae), 155
Paspalidium geminatum (Forssk.) Stapf (Gramineae), 55, 265
Paspalum commersonii Lam. (Gramineae): see *P. scrobiculatum*
 P. orbiculare Forster: see *P. scrobiculatum*
 P. scrobiculatum L., 100, 202, 266
 P. vaginatum Swartz, 262, 263
Passerina (Thymelaeaceae), 134, 167
 P. filiformis L., 134
 P. montana Thoday, 195
Paullinia pinnata L. (Sapindaceae), 83
Pavonia urens Cav. (Malvaceae), 122
Peddiea fischeri Engl. (Thymelaeaceae), 74
 P. thomensis Engl. & Gilg, 253
Peganum harmala L. (Zygophyllaceae), 227, 266
Pegolettia retrofracta (Thunb.) Kies (Compositae), 195
Pelargonium (Geraniaceae), 139
 P. cotyledonis (L.) L'Hérit., 254
 P. cristophoranum Verdc., 116
 P. otaviense Knuth, 142
 P. rössingense Dinter: see *P. otaviense*
Pellaea (Sinopteridaceae), 99, 238
Peltophorum africanum Sond. (Leguminosae: Caesalpinioideae), 95, 96, 98, 201
Pemphis acidula Forst. (Lythraceae), 259
Pennisetum (Gramineae), 169, 211
 P. mezianum Leeke, 126, 127, 128
 P. pedicellatum Trin., 209
 P. polystachion (L.) Schultes, 84, 107, 252
 P. purpureum Schumach., 50, 85
 P. ramosum (Hochst.) Schweinf., 209
 P. schimperi Steud., 130
 P. stramineum A. Peter, 126, 127
 P. unisetum (Nees) Benth., 107, 174
Pentaclethra macrophylla Benth. (Leguminosae: Mimosoideae), 81, 253
Pentadesma lebrunii Staner (Guttiferae), 79, 85
Pentanopsis (Rubiaceae), 111
Pentaschistis (Gramineae), 134, 169
 P. humbertii A. Camus, 239
 P. patula (Nees) Stapf, 135
 P. perrieri A. Camus, 239
 P. pictigluma (Steud.) Pilger, 211
 P. tysonii Stapf, 169
Pentzia (Compositae), 137, 139, 193
 P. globosa Less., 195
 P. incana (Thunb.) Kuntze, 193
 P. monodiana Maire, 222
 P. sphaerocephala DC., 195
Peperomia (Piperaceae), 165, 236, 237, 253
 P. fernandopoiana C.DC., 82
 P. staudtii Engl.: see *P. fernandopoiana*
Pergularia daemia (Forssk.) Chiov. (Asclepiadaceae), 114
Pericopsis angolensis (Baker) van Meeuwen (Leguminosae: Papilionoideae), 93, 95, 96, 98, 174
 P. elata (Harms) van Meeuwen, 77
 P. laxiflora (Benth. ex Baker) van Meeuwen, 85, 105, 107
Periploca laevigata Aiton (Asclepiadaceae), 153, 227, 228, 229, 251

Perotis patens Gand. (Gramineae), 243
Persea azorica Seub. (Lauraceae): see *Laurus azorica*
 P. indica (L.) Spreng., 246, 247, 249
Petalidium (Acanthaceae), 140
 P. angustifolium P. G. Meyer, 142
 P. engleranum C. B. Clarke, 267
 P. giessii P. G. Meyer, 142
Petersianthus macrocarpus (P. Beauv.) Liben (Lecythidaceae), 77, 81, 172
Petrobium (Compositae), 254
 P. arboreum R. Br., 254
Phaeoptilum (Nyctaginaceae), 137
 P. spinosum Radlk., 191, 193
Pharnaceum acidum Hook.f. (Aizoaceae), 254
Phaseolus lunatus L. (Leguminosae: Papilionoideae), 252
 P. vulgaris L., 252
Philippia (Ericaceae), 132, 167, 168, 189, 236, 237, 238, 258
 P. abietina (Willd.) Klotzsch, 258
 P. benguelensis (Welw. ex Engl.) Britten, 99
 P. chamissonis Klotzsch, 134
 P. comorensis Engl., 255
 P. mafiensis Engl., 189
 P. montana (Willd.) Klotzsch, 258
 P. simii S. Moore, 189
 P. thomensis Henriq., 253
Phillyrea angustifolia L. (Oleaceae), 149, 152, 153, 155, 158, 159, 226, 228
 P. latifolia L.: see *P. angustifolia*
 P. media L.: see *P. angustifolia*
Philoxerus vermicularis (L.) P. Beauv. (Amaranthaceae), 262
Phoenicophorium (Palmaceae), 257
Phoenix atlantica A. Chev. (Palmaceae), 246, 251
 P. canariensis Chabaud, 246
 P. dactylifera L., 216, 219, 226
 P. reclinata Jacq., 83, 122, 181, 188, 200, 211
Phormium tenax J. R. & G. Forster (Agavaceae), 254
Phragmites (Gramineae), 55
 P. australis (Cav.) Trin. ex Steud., 218, 219, 222, 223, 224, 265
 P. mauritianus Kunth, 265
Phyla nodiflora (L.) Greene (Verbenaceae), 145
Phylica (Rhamnaceae), 132, 134
 P. buxifolia L., 134
 P. leucocephala Cordem.: see *P. nitida*
 P. mauritiana Bojer ex Baker: see *P. nitida*
 P. nitida Lam., 258
 P. oleifolia Vent., 134
 P. paniculata Willd., 134
 P. ramosissima DC.
 P. villosa Thunb., 134
Phyllanthus comorensis Engl. (Euphorbiaceae), 255
 P. discoideus (Baillon) Muell. Arg., 98, 167, 199
 P. maderaspatensis L., 191
 P. muelleranus (Kuntze) Exell, 173
 P. verrucosus Thunb., 201
Phymaspermum (Compositae), 137
Phymatodes scolopendria (Burm.f.) Ching (Polypodiaceae), 189
Picconia (Oleaceae), 246
 P. azorica (Tutin) Knobl., 246, 247
 P. excelsa (Aiton) DC., 246, 247
Picralima nitida (Stapf) Th. & H. Durand (Apocynaceae), 83
Piliostigma reticulatum (DC.) Hochst. (Leguminosae: Caesalpinioideae), 105, 107, 108, 206
 P. thonningii (Schumach.) Milne-Redh., 83, 85, 87, 95, 96, 105, 107, 173, 174, 189, 210, 214

Pilostyles aethiopica Welw. (Rafflesiaceae), 93
Pilotrichella (Metioriaceae), 81
Pimpinella villosa Schousboe (Umbelliferae), 157
Pinguicula vulgaris L. (Lentibulariaceae), 155
Pinus canariensis Chr. Smith ex DC. (Pinaceae), 246, 248, 249
    P. halepensis Miller, 51, 146, 149, 150, 151, 152, 153, 154, 155, 157, 160, 226, 229, 231
    P. nigra Arnold, 152
    P. pinaster Aiton, 135, 146, 149, 150, 152, 155
Piper capense L.f. (Piperaceae), 74
Piptadeniastrum africanum (Hook.f.) Brenan (Leguminosae: Mimosoideae), 75, 78, 79, 82, 172, 181
Piptatherum coerulescens (Desf.) Hackel: see *Oryzopsis coerulescens*
Pisonia aculeata L. (Nyctaginaceae), 200
Pistacia (Anacardiaceae), 227
    P. atlantica Desf., 147, 149, 150, 157, 158, 222, 226, 227
    P. lentiscus L., 115, 121, 149, 151, 152, 153, 154, 157, 158, 159, 226, 227, 228
    P. terebinthus L., 149
Pistia stratiotes L. (Araceae), 265
Pittosporum (Pittosporaceae), 238
    P. coriaceum Dryander ex Aiton, 246, 247
    P. lanceolatum Cordem.: see *P. senacia*
    P. senacia Putterl., 258
    P. viridiflorum Sims, 62, 74, 195, 196
Pituranthos battandieri Maire (Umbelliferae), 221
Placopoda (Rubiaceae), 111
Pladaroxylon (Compositae), 254
    P. leucadendron (Forster f.) Hook.f., 254
Plagiochila (Plagiochilaceae), 81
Plagiochloa (Gramineae), 134
Plantago (Plantaginaceae), 236, 246
    P. ciliata Desf., 220
    P. coronopus L., 266
    P. robusta Roxb., 254
Platycelyphium (Leguminosae: Papilionoideae), 111
    P. voense (Engl.) Wild, 114
Platycerium (Polypodiaceae), 235
    P. elephantotis Schweinf., 182
Platypterocarpus (Celastraceae), 162
Plectranthus (Labiatae), 241
    P. ignarius (Schweinf.) Agnew, 120
Pleiomeris (Myrsinaceae), 246
    P. canariensis (Willd.) A.DC., 246
Pleuropterantha (Amaranthaceae), 111
Pleurostylia africana Loes. (Celastraceae), 188
Plinthus (Aizoaceae), 140, 193
Plumbago auriculata Lam. (Plumbaginaceae), 201
    P. capensis Thunb.: see *P. auriculata*
    P. zeylanica L., 90
Poa (Gramineae), 169
    P. ankaratrensis A. Camus, 239
    P. madecassa A. Camus, 239
Podalyria (Leguminosae: Papilionoideae), 132
Podocarpus (Podocarpaceae), 165, 237
    P. elongatus (Aiton) L'Hérit. ex Pers., 135
    P. ensiculus Melville: see *P. henkelii*
    P. falcatus (Thunb.) R. Br. ex Mirbel, 165, 181, 199, 200
    P. gracilior Pilger: see *P. falcatus*
    P. henkelii Stapf, 165
    P. latifolius (Thunb.) R. Br. ex Mirbel, 164, 165, 167, 181, 195, 196, 199
    P. madagascariensis Baker, 236
    P. mannii Hook.f., 253

    P. milanjianus Rendle: see *P. latifolius*
    P. rostratus Laurent, 238
    P. usambarensis Pilger var. *dawei* (Stapf) Melville: see *P. falcatus*
Poga (Rhizophoraceae), 74
Pogonarthria squarrosa (Licht. ex Roem. & Schult.) Pilger (Gramineae), 191, 193, 194, 243
Polycarpaea fragilis Del. (Caryophyllaceae): see *P. repens*
    P. repens (Forssk.) Asch. & Schweinf., 220
Polyceratocarpus scheffleri Engl. & Diels (Annonaceae), 187
Polygala (Polygalaceae), 132
    P. arenaria Willd., 178
    P. balansae Cosson, 154
    P. myrtifolia L., 134
Polygonum (Polygonaceae), 266
    P. acuminatum Kunth, 266
    P. pulchrum Blume, 265
    P. salicifolium Brouss. ex Willd., 265
    P. strigosum R. Br., 265
Polypogon monspeliensis (L.) Desf. (Gramineae), 246
Polyscias fulva (Hiern) Harms (Araliaceae), 74, 211
    P. quintasii Exell, 253
Polysphaeria multiflora Hiern (Rubiaceae), 259
Popowia (Annonaceae), 98
    P. obovata (Benth.) Engl. & Diels, 90, 98
Populus alba L. (Salicaceae), 150
    P. euphratica Oliver, 219
    P. ilicifolia (Engl.) Rouleau, 117
    P. tremula L., 149, 156
Portulaca oleracea L. (Portulacaceae), 254
Portulacaria afra Jacq. (Portulacaceae), 137, 140, 141, 201
Poskea (Globulariaceae), 111, 147
Potamogeton richardii Solms-Laub. (Potamogetonaceae), 265
    P. schweinfurthii A. Bennett, 265
Poterium spinosum L. (Rosaceae), 159
Premna hildebrandtii Gürke (Verbenaceae), 114
    P. quadrifolia Schumach. & Thonn., 178
    P. resinosa (Hochst.) Schauer, 114
Primula vulgaris Hudson (Primulaceae), 155
Prosopis africana (Guill. & Perr.) Taubert (Leguminosae: Mimosoideae), 105, 107, 209
Protarum (Araceae), 257
Protea (Proteaceae), 93, 99, 132, 134
    P. arborea Houtt., 134
    P. caffra Meissner, 96
    P. glabra Thunb., 134
    P. laurifolia Thunb., 134
    P. longiflora Lam., 134
    P. lorifolia (Salisb. ex Knight) Fourc., 134
    P. madiensis Oliver, 105
    P. nereifolia R. Br., 134
    P. obtusifolia Buek, 134
    P. petiolaris Welw. ex Engl., 173
    P. repens (L.) L., 134, 135
    P. susannae E. P. Phillips, 134
Protorhus (Anacardiaceae), 199, 236
    P. buxifolia H. Perrier, 237
    P. deflexa H. Perrier, 241
    P. humbertii H. Perrier, 241
    P. longifolia (Bernh.) Engl., 199, 201
    P. perrieri Courchet, 241
Prunus africana (Hook.f.) Kalkman (Rosaceae), 80, 122, 164, 165, 167, 181, 253, 255
    P. avium L., 149, 152, 157
    P. lusitanica L., 149

P. padus L., 147, 149, 156
P. persica (L.) Batsch, 252
P. prostrata Labill., 156, 158
Pseudagrostistachys africana (Muell. Arg.) Pax & Hoffm. (Euphorbiaceae), 253
Pseudocedrela (Meliaceae), 103
P. kotschyi (Schweinf.) Harms, 83, 85, 105, 107, 108, 209
Pseudolachnostylis (Euphorbiaceae), 87
P. maprouneifolia Pax, 31, 95, 96, 97
Pseudoprosopis fischeri (Taubert) Harms (Leguminosae: Mimosoideae), 97
Pseudosalacia (Celastraceae), 199
Pseudospondias microcarpa (A. Rich.) Engl. (Anacardiaceae), 181, 253
Psiadia (Compositae), 237, 238, 258
P. altissima (DC.) Benth. & Hook.f., 236
P. arabica Jaub. & Spach: see *P. punctulata*
P. punctulata (DC.) Vatke, 115
P. schweinfurthii Balf.f., 255
Psidium cattleianum Sabine (Myrtaceae), 236, 257
P. guajava L., 83, 236, 252, 254
Psilocaulon salicornioides (Pax) Schwantes (Aizoaceae), 142
Psilolemma jaegeri (Pilger) S. M. Phillips (Gramineae), 267
Psilonema (Cruciferae), 111
Psilotrichum (Amaranthaceae), 111
Psoralea obtusifolia DC. (Leguminosae: Papilionoideae), 144
P. pinnata L., 134
P. plicata Del., 219
Psorospermum febrifugum Spach (Guttiferae), 85, 174
Psychotria capensis (Eckl.) Vatke (Rubiaceae), 200
P. peduncularis (Salisb.) Steyerm., 91
Ptaeroxylon obliquum (Thunb.) Radlk. (Ptaeroxylaceae), 96, 165, 200
Pteleopsis anisoptera (Welw. ex Lawson) Engl. & Diels (Combretaceae), 98
P. diptera (Welw.) Engl. & Diels, 80, 172
P. myrtifolia (Lawson) Engl. & Diels, 200
P. suberosa Engl. & Diels, 107
Pteridium aquilinum (L.) Kuhn (Pteridiaceae), 96, 152, 157, 236
Pteris (Pteridiaceae), 75
P. vittata L., 252
Pterocarpus (Leguminosae: Papilionoideae), 98
P. angolensis DC., 93, 95, 96, 97, 98, 173, 174
P. antunesii (Taubert) Harms, 90
P. erinaceus Poir., 83, 85, 105
P. lucens Guill. & Perr., 209, 211
P. mildbraedii Harms, 74
ss. *usambarensis* (Verdc.) Polhill, 187
P. rotundifolius (Sond.) Druce, 95, 96, 98, 99, 201
P. santalinoides L'Hérit. ex DC., 176
Pterocelastrus (Celastraceae), 96, 196
P. tricuspidatus Sond., 135
Pterodiscus (Pedaliaceae), 140
Pterolobium stellatum (Forssk.) Brenan (Leguminosae: Caesalpinioideae), 115
Pteronia (Compositae), 135, 137
P. glauca Thunb., 140
Pterygota macrocarpa K. Schum. (Sterculiaceae), 79, 81
Ptilotrichum spinosum (L.) Boiss.: see *Alyssum spinosum*
Puccionia (Cruciferae), 111
Punica granatum L. (Punicaceae), 252
Pupalia lappacea (L.) Juss. (Amaranthaceae), 90
Putterlickia pyracantha (L.) Szyszyl. (Celastraceae), 135
Pycnanthus angolensis (Welw.) Warb. (Myristicaceae), 80, 81, 85, 172, 181

Pycnocoma littoralis Pax (Euphorbiaceae), 189
Pycreus aethiops (Welw. ex Ridley) C. B. Clarke (Cyperaceae), 100
Pygeum africanum Hook.f.: see *Prunus africana*
Pyrenacantha malvifolia Engl. (Icacinaceae), 114
Pyrus cossonii Rehder (Rosaceae), 149
P. gharbiana Trabut, 149
P. longipes Coss. & Durieu: see *P. cossonii*
P. mamorensis Trabut, 149, 151

Quercus afares Pomel (Fagaceae), 146, 149, 151, 156, 157
Q. calliprinos Webb: see *Q. coccifera*
Q. coccifera L., 146, 147, 149, 150, 152, 154, 158, 159, 226
Q. faginea Lam., 146, 149, 150, 151, 152, 155, 156, 157
Q. ilex L., 146, 147, 149, 150, 151, 152, 153, 154, 156, 157, 226, 229, 231
Q. lusitanica nom. ambig.: see *Q. faginea*
Q. pyrenaica Willd., 146, 149, 150, 156, 157
Q. suber L., 48, 146, 147, 149, 150, 151, 152, 154, 155, 157, 237
Q. toza Bast.: see *Q. pyrenaica*
Quivisia oppositifolia Cav.: see *Turraea oppositifolia*

Racopilum speluncae P. Beauv. (Racopilaceae), 182
Ramalina (Ramalinaceae), 224
Randonia africana Coss. (Resedaceae), 221
Ranunculus (Ranunculaceae), 236
Rapanea melanophloeos (L.) Mez (Myrsinaceae), 122, 135, 165, 167, 196, 200
Raphia (Palmaceae), 83, 91, 188
R. australis Oberm. & Strey, 200
R. farinifera (Gaertner) Hylander, 181
Rauvolfia caffra Sond. (Apocynaceae), 187, 200
R. nana E. A. Bruce, 173
R. vomitoria Afzel., 80, 253
Ravenala (Strelitziaceae), 232
R. madagascariensis Adans., 235, 236, 257
Ravensara (Lauraceae), 232, 235, 241
Reaumuria hirtella Jaub. & Spach (Tamaricaceae), 221
R. muricata Jaub. & Spach: see *R. vermiculata*
R. vermiculata L., 230
Redfieldia hitchcockii A. Camus (Gramineae), 238
Rendlia altera (Rendle) Chiov. (Gramineae), 196
Reseda battandieri Pitard (Resedaceae), 227
R. villosa Coss., 221
Restio (Restionaceae), 132
Retama bovei Spach (Leguminosae: Papilionoideae): see *R. monosperma*
R. monosperma (L.) Boiss., 152, 231
R. retam Webb, 147, 220, 224
Rhamnus alaternus L. (Rhamnaceae), 153
R. alpinus L., 158
R. catharticus L., 151
R. frangula L., 156
R. latifolia L'Hérit., 247
R. oleoides L., 153, 158, 227
R. prinoides L'Hérit., 195
R. staddo A. Rich., 121
Rhanterium (Compositae), 218
Rhigozum angolense Bamps (Bignoniaceae), 145
R. brevispinosum Kuntze, 95, 193
R. madagascariensis Drake, 242
R. obovatum Burchell, 96, 140, 193, 195
R. trichotomum Burchell, 139, 140, 191, 193
R. virgatum Merxm. & A. Schreiber, 95, 140

Rhipsalis (Cactaceae), 165, 174, 236
Rhizophora (Rhizophoraceae), 55, 261, 262, 263, 264
    R. harrisonii Leechman, 253, 261, 262
    R. mangle L., 261, 262
    R. mucronata Lam., 259, 261, 263, 264
    R. racemosa G. F. W. Meyer, 261, 262
Rhodognaphalon schumannianum A. Robyns (Bombacaceae), 188
Rhoicissus digitata (L.f.) Gilg & Brandt (Vitaceae), 201
    R. tomentosa (Lam.) Wild & R. B. Drummond, 200
    R. tridentata (L.f.) Wild & R. B. Drummond, 98, 201
    Rhus (Anacardiaceae), 96, 201
    R. albida Schousboe, 251
    R. chirindensis Baker f., 96
    R. ciliata Licht. ex Schultes, 193, 194, 195
    R. crenata Thunb., 135
    R. dregeana Sond., 193
    R. erosa Thunb., 193, 194, 195
    R. glauca Thunb., 135
    R. incana Mill., 222
    R. laevigata L., 135
    R. lancea L.f., 141, 144, 193, 194, 195
    R. leptodictya Diels, 96
    R. longipes Engl., 109
    R. lucida L., 135
    R. marlothii Engl., 193
    R. mucronata Thunb., 135
    R. natalensis Bernh. ex Krause, 109, 115, 129, 182
    R. oxyacantha Shousb., 227
    R. pentaphylla (Jacq.) Desf., 149, 158, 227, 228
    R. pyroides Burchell, 193, 195
    R. quartiniana A. Rich., 91
    R. somalensis Engl., 115
    R. taratana (Baker) H. Perrier, 237
    R. thyrsiflora Balf.f., 255
    R. tomentosa L., 135
    R. tripartita (Ucria) Grande, 222
    R. undulata Jacq., 140, 141, 193, 195
    R. vulgaris Meikle, 121, 207
Rhynchelytrum amethysteum (Franchet) Chiov. (Gramineae), 101, 173
    R. repens (Willd.) C. E. Hubbard, 101, 145, 252
    R. villosum (Parl. ex Hook.f.) Chiov.: see *R. repens*
Rhynchocalyx (Lythraceae), 199
Rhynchosia (Leguminosae: Papilionoideae), 220
    R. candida (Welw. ex Hiern) Torre, 144
    R. memnonia (Del.) Boiss., 222
    R. totta Thunb., 194
Rhynchospora candida (Nees) Boeck (Cyperaceae), 84
    R. corymbosa (L.) Britten, 31, 84
    R. holoschoenoides (L. C. Rich.) Herter, 84
    R. rubra (Lour.) Makino, 84
    R. rugosa (Vahl) Gale, 84
Rhytachne rottboellioides Desv. (Gramineae), 32, 84, 108
Ribes alpinum L. (Grossulariaceae), 158
    R. uva-crispa L., 158
Ricinodendron heudelotii (Baillon) Pierre ex Pax (Euphorbiaceae), 74, 78, 79, 81, 172, 186, 187, 188
    R. rautanenii Schinz, 90, 97
Ricinus communis L. (Euphorbiaceae), 252
Rinorea (Violaceae), 74, 235
    R. malembaensis Taton, 172
Riseleya griffithii Hemsley (Euphorbiaceae), 257
Romulea (Iridaceae), 158
Rosa (Rosaceae), 151, 157

Roscheria (Palmaceae), 257
    R. melanochaetes (H. A. Wendl.) H. A. Wendl. ex Balf.f., 257
Rosmarinus eriocalix Jordan & Fourr. (Labiatae), 155, 226, 229
    R. officinalis L., 154, 226
    R. tournefourtii De Noé ex Turrill: see *R. eriocalix*
Rotala pterocalyx A. Raynal (Lythraceae), 204
Rothia (Leguminosae: Papilionoideae), 243
Rubus (Rosaceae), 151, 167, 236, 254
    R. pinnatus Willd., 253
    R. ulmifolius Schott, 151, 157
Ruellia (Acanthaceae), 140
    R. insignis Balf.f., 115, 255
Ruschia (Aizoaceae), 139, 140
    R. unidens Schwantes, 195
Ruscus aculeatus L. (Ruscaceae), 151, 157
Ruta (Rutaceae), 228
Rytigynia umbellulata (Hiern) Robyns (Rubiaceae), 98

Sacciolepis africana C. E. Hubbard & Snowden (Gramineae), 100
Sacoglottis gabonensis (Baillon) Urban (Houmiriaceae), 76, 174
Salacia (Celastraceae), 75, 241
    S. kraussii (Harv.) Harv., 202
Salicornia (Chenopodiaceae), 267
    S. arabica L., 223, 230
    S. fruticosa L.: see *S. arabica*
Salix alba L. (Salicaceae), 150
    S. cinerea L., 151, 156
    S. purpurea L., 156
Salsola (Chenopodiaceae), 144, 193, 223, 224
    S. aphylla L.f., 139, 142
    S. baryosma (Schultes) Dandy, 223
    S. foetida Del. ex Sprengel: see *S. baryosma*
    S. longifolia Forssk., 230, 266
    S. nollothensis Aellen, 142
    S. oppositifolia Desf.: see *S. longifolia*
    S. sieberi Presl, 223, 229
    S. tetragona Del., 223, 229, 230
    S. tuberculata (Moq.) Schinz, 139, 267
    S. vermiculata L., 228, 229, 230, 266
    S. zeyheri (Moq.) Schinz, 144
Salvadora persica L. (Salvadoraceae), 114, 119, 121, 128, 144, 207, 219, 220, 224, 267
Salvia (Labiatae), 221
    S. aegyptiaca L., 221
Salvinia auriculata Aublet (Salviniaceae), 265
    S. molesta Mitchell, 265, 266
Sambucus africana Standley (Caprifoliaceae), 167
Samolus valerandi L. (Primulaceae), 145
Sanguisorba (Rosaceae), 246
Sanicula (Umbelliferae), 236
    S. europaea L., 156
Sansevieria (Agavaceae), 98, 114, 115, 116, 129, 188, 201
    S. arborescens Gérome & Labroy, 114
    S. cylindrica Bojer, 145, 174
    S. ehrenbergii Schweinf. ex Baker, 117, 128
    S. liberica Gérome & Labroy, 178
Santiria trimera (Oliver) Aubrév. (Burseraceae), 82, 83
Sapium bussei Pax: see *Excoecaria bussei*
    S. ellipticum (Hochst.) Pax, 74, 199, 253
Sarcocaulon (Geraniaceae), 137, 139
    S. marlothii Engl.: see *S. mossamedense*
    S. mossamedense (Welw. ex Oliver) Hiern, 142, 144, 145
    S. spinosum (Burm.f.) Kuntze, 142

Sarcolaena oblongifolia Gérard (Sarcolaenaceae), 237
Sarcophrynium (Marantaceae), 75
Sarcopoterium spinosum (L.) Spach: see *Poterium spinosum*
Sarcostemma (Asclepiadaceae), 116
  S. *daltonii* Decne., 251
  S. *viminale* (L.) Aiton f., 114, 115, 117, 191, 201
Saxymolobium holubii (Scott Elliot) Bullock (Asclepiadaceae), 100
Scabiosa columbaria L. (Dipsaceae), 194
Scaevola sp. (Goodeniaceae), 257, 258
Schefflera (Araliaceae), 236, 237
  S. *barteri* (Seemann) Harms, 82, 83
  S. *bojeri* R. Viguier, 237, 238
  S. *mannii* (Hook.f.) Harms, 253
  S. *umbellifera* (Sond.) Baillon, 165
Schefflerodendron usambarense Harms (Leguminosae: Papilionoideae), 74, 187
Schismus barbatus Juel (Gramineae), 50
Schizachyrium brevifolium (Swartz) Nees ex Büse (Gramineae), 174
  S. *exile* (Hochst.) Pilger, 210
  S. *platyphyllum* Stapf, 50
  S. *sanguineum* (Retz.) Alston, 85, 101
  S. *semiberbe* Nees: see *S. sanguineum*
  S. *thollonii* Stapf, 173
Schmidtia kalahariensis Stent (Gramineae), 145, 191, 193
  S. *pappophoroides* J. A. Schmidt, 114, 137, 145, 193, 206, 252
Schoenefeldia gracilis Kunth (Gramineae), 108, 176, 206, 207, 208, 212, 213
Schotia (Leguminosae: Caesalpinioideae), 201
  S. *afra* (L.) Thunb., 137, 140, 201
  S. *africana* (Baillon) Keay, 77
  S. *brachypetala* Sond., 96, 200
  S. *latifolia* Jacq., 137, 200, 201
Schouwia (Cruciferae), 218
Schrebera alata (Hochst.) Welw. (Oleaceae), 115, 129, 167
  S. *arborea* A. Chev., 176
  S. *trichoclada* Welw., 174
Schumanniophyton problematicum (A. Chev.) Aubrév. (Rubiaceae), 75
Scilla (Liliaceae), 158
  S. *nervosa* (Burchell) Jessop, 194
Scirpus cubensis Poeppig & Kunth (Cyperaceae), 266
  S. *holoschoenus* L., 218, 219, 222, 266
  S. *inclinatus* (Del.) Asch. & Graebner, 265
  S. *littoralis* Schrader, 144
  S. *microcephalus* (Steud.) Dandy, 100
Scleria aterrima (Ridley) Napper (Cyperaceae), 84
  S. *bulbifera* Hochst. ex A. Rich., 100
  S. *nutans* Kunth, 266
  S. *nyasensis* C. B. Clarke, 266
Sclerocarya (Anacardiaceae), 242
  S. *birrea* (A. Rich.) Hochst., 105, 107, 116, 129, 209, 210, 212, 213
  S. *caffra* Sond., 95, 96, 99, 174, 189, 201, 243
Sclerocephalus arabicus Boiss. (Caryophyllaceae), 252
Sclerodactylon macrostachyum (Benth.) A. Camus (Gramineae), 259
Sclerosciadium nodiflorum Ball (Umbelliferae), 229
Scolopia mundii (Eckl. & Zeyh.) Warb. (Flacourtiaceae), 165, 195, 196, 199
Scorodophloeus fischeri (Taubert) J. Léonard (Leguminosae: Caesalpinioideae), 117, 188
  S. *zenkeri* Harms, 77

Scutia myrtina (Burm.f.) Kurz (Rhamnaceae), 115, 121, 182, 201
Scytopetalum pierreanum (De Wild.) Van Tiegh. (Scytopetalaceae), 83
Securidaca longepedunculata Fres. (Polygalaceae), 85, 174, 189, 210, 211
Securinega virosa (Roxb. ex Willd.) Baillon (Euphorbiaceae), 98, 178, 222
  S. *seyrigii* Leandri, 241
Seddera latifolia Hochst. & Steud. (Convolvulaceae), 113
Sedum madagascariense H. Perrier (Crassulaceae), 239
Seetzenia africana R. Br. (Zygophyllaceae), 219
  S. *orientalis* Decne.: see *S. africana*
Sehima ischaemoides Forssk. (Gramineae), 108, 213
Selaginella (Selaginellaceae), 238, 242
  S. *echinata* Baker, 238
  S. *scandens* (P. Beauv.) Spring, 84
Selago (Scrophulariaceae), 135, 140
Senecio (Compositae), 53, 139, 141, 237, 238, 242
  subgen. *Dendrosenecio*, 169
  S. *anteuphorbium* (L.) Hook.f., 224, 227, 228, 229
  S. *bojeri* (DC.) Robyns, 182
  S. *coronatus* (Thunb.) Harv., 194
  S. *erubescens* Aiton, 194
  S. *leucadendron* (Forster f.) Hemsley: see *Pladaroxylon leucadendron*
  S. *longiflorus* (DC.) Schultz-Bip., 142
  S. *petitianus* A. Rich., 115
  S. *prenanthiflorus* (DC.) Hemsley: see *Lachanodes arborea*
  S. *redivivus* Mabberley: see *Lachanodes arborea*
  S. *stuhlmannii* Klatt, 182
Sericocomopsis (Amaranthaceae), 111
  S. *hildebrandtii* Schinz, 114, 120
  S. *pallida* (C. B. Clarke) Schinz, 114
Serruria (Proteaceae), 132
Sesamothamnus (Pedaliaceae), 113
  S. *benguellensis* Welw., 140
  S. *guerichii* (Engl.) E. A. Bruce, 140
  S. *lugardii* N. E. Br., 96
  S. *rivae* Engl., 114
Sesbania sesban (L.) Merrill (Leguminosae: Papilionoideae), 266, 267
Sesuvium (Aizoaceae), 144
  S. *digynum* Welw. ex Oliver: see *S. sesuvioides*
  S. *portulacastrum* L., 145, 262, 263
  S. *sesuvioides* (Fenzl) Verdc., 142
Setaria (Gramineae), 169, 189
  S. *anceps* Stapf ex Massey: see *S. sphacelata*
  S. *chevalieri* Stapf, 83, 129
  S. *flabellata* Stapf, 194
  S. *holstii* Herrm.: see *S. incrassata*
  S. *homonyma* (Steud.) Chiov., 90
  S. *incrassata* (Hochst.) Hackel, 108, 116
  S. *lynesii* Stapf & C. E. Hubbard, 209
  S. *nigrirostris* (Nees) Th. Durand & Schinz, 194
  S. *pallide-fusca* (Schumach.) Stapf & C. E. Hubbard, 209, 210
  S. *sphacelata* (Schumach.) Stapf & C. E. Hubbard ex M. B. Moss, 84, 100, 107, 122, 194, 239
  S. *verticillata* (L.) P. Beauv., 254
Sideroxylon (Sapotaceae), 251, 257
  S. *bojeranum* DC.: see *S. cinereum*
  S. *buxifolium* Hutch., 115
  S. *cinereum* Lam., 257
  S. *collinum* Lecomte, 241
  S. *galeatum* (A. W. Hill) Baehni, 258

Sideroxylon (Sapotaceae)—*contd*
    S. inerme L., 135, 188, 189, 200, 201, 259
    S. majus (Gaertner f.) Baehni, 258
    S. marmulano Banks ex Lowe, 246, 247, 251
Sieglingia decumbens (L.) Bernh. (Gramineae), 155
Silene (Caryophyllaceae), 147
Simocheilus (Ericaceae), 132
Sinapidendron (Cruciferae), 246
Sindoropsis (Leguminosae: Caesalpinioideae), 74
Sisyndite (Zygophyllaceae), 137
Sium helenianum Hook.f. (Umbelliferae), 254
Smilax aspera L. (Smilacaceae), 151, 153
    S. kraussiana Meissner, 96
Smithia elliotii Baker f. (Leguminosae: Papilionoideae), 266
Socotora (Asclepiadaceae), 111
    S. visciformis (Vatke) Bullock, 113
Socotranthus (Asclepiadaceae), 111
Solanum (Solanaceae), 254
    S. albicaule Kotschy ex Dunal, 204
    S. auriculatum Aiton, 236
    S. dubium Fresen., 212
    S. incanum L., 122
    S. indicum L.
      ss. *grandifrons* Bitter, 122
Solidago sempervirens L. (Compositae), 246
    S. virgaurea L., 155
Sonchus (Compositae), 246
    S. chevalieri (O. Hoffm. & Muschler) Dandy: see *Launea chevalieri*
    S. daltonii Webb, 252
    S. nanus Sond. ex Harv., 194
    S. pinnatifidus Cav., 226, 228
Sonneratia alba Smith (Sonneratiaceae), 259, 261, 263
Sorbus aria (L.) Crantz (Rosaceae), 149, 151, 156, 158
    S. domestica L., 149, 156, 157
    S. torminalis (L.) Crantz, 149, 151, 156
Sorghum (Gramineae), 211
    S. arundinaceum (Desv.) Stapf, 108
    S. purpureo-sericeum (Hochst. ex A. Rich.) Asch. & Schweinf., 108
Sorindeia (Anacardiaceae), 74
Soulamea terminalioides Baker (Simaroubaceae), 257
Soyauxia grandifolia Gilg & Stapf (Medusandraceae), 77
Spartium (Leguminosae: Papilionoideae), 147
    S. junceum L., 226
Spartocytisus nubigenus Webb & Berth. (Leguminosae: Papilionoideae), 248
Spathionema (Leguminosae: Papilionoideae), 111
Spathodea campanulata P. Beauv. (Bignoniaceae), 182
Spergularia maritima (Hill) Druce (Caryophyllaceae), 266
Sphagnum (Sphagnaceae), 45, 84, 237, 266
Sphenopus (Gramineae), 266
    S. divaricatus (Gouan) Reichenb., 228, 230
    S. gouanii Trin.: see *S. divaricatus*
Spirostachys africana Sond. (Euphorbiaceae), 95, 96, 200, 201
Spondianthus preussii Engl. (Euphorbiaceae), 83, 181
Spondias mombin L. (Anacardiaceae), 109
Sporobolus (Gramineae), 116, 126
    S. barbigerus Franchet: see *S. subtilis*
    S. centrifugus Nees, 239
    S. discosporus Nees, 194
    S. durus Brongn., 254
    S. festivus Hochst. ex A. Rich., 128, 209, 210, 243
    S. fimbriatus Nees, 194
    S. humifusus (Kunth) Kunth, 212, 213

    S. infirmus Mez, 84
    S. ioclados (Trin.) Nees, 126
    S. kentrophyllus (K. Schum.) W. D. Clayton, 126
    S. nitens Stent, 267
    S. pyramidalis P. Beauv., 101, 191
    S. robustus Kunth, 144, 223, 267
    S. sanguineus Rendle, 84
    S. spicatus (Vahl) Kunth, 116, 252, 267
    S. subtilis Kunth, 101
    S. subulatus Hackel ex Scott Elliot, 239
    S. tenellus (Sprengel) Kunth, 267
    S. testudinum Renvoize, 259
    S. virginicus (L.) Kunth, 253, 259, 263, 267
Stachys spathulata Burchell ex Benth. (Labiatae), 194
Stadmannia oppositifolia Poir. (Sapindaceae), 257
Stangeria (Stangeriaceae), 199
    S. eriopus (Kunze) Baillon, 200
Stapelia (Asclepiadaceae), 139
Stapfiella (Turneraceae), 162
Stathmostelma pauciflorum (Klotzsch) K. Schum. (Asclepiadaceae), 100
    S. welwitschii Britten & Rendle, 100
Statice (Plumbaginaceae), 266
    S. cyrtostachya Boiss. & Reut.: see *Limonium cymuliferum*
Staudtia stipitata Warb. (Myristicaceae), 79, 85, 172
Stauracanthus boivinii (Webb) Samp.: see *Ulex boivinii*
Steganotaenia araliacea Hochst (Umbelliferae), 96, 98, 99, 105
Stemonocoleus (Leguminosae: Caesalpinioideae), 74
Stenochlaena tenuifolia (Desv.) Moore (Blechnaceae), 200
Stenocline (Compositae), 238
Sterculia (Sterculiaceae), 187
    S. africana (Lour.) Fiori, 114, 141
    S. appendiculata K. Schum., 186, 188, 189
    S. oblonga Masters, 78, 79
    S. quinqueloba (Garcke) K. Schum., 173, 174
    S. rhinopetala K. Schum., 79
    S. rhynchocarpa K. Schum., 114, 189
    S. rogersii N. E. Br., 96
    S. setigera Del., 62, 90, 105, 107, 145, 211
    S. stenocarpa H. Winkler, 114, 116
    S. tragacantha Lindley, 74, 82, 178, 253
Stereospermum acuminatissimum K. Schum. (Bignoniaceae), 82
    S. euphorioides DC., 241, 243
    S. kunthianum Cham., 62, 83, 85, 105, 107, 173, 174, 189, 210, 211, 212
    S. variabile H. Perrier, 243
Sticherus flagellaris (Bory) St John (Gleicheniaceae), 236
Stipa (Gramineae), 230
    S. capensis Thunb., 222, 227, 228
    S. parviflora Desf., 222
    S. retorta Cav.: see *S. capensis*
    S. tenacissima L., 40, 51, 52, 149, 153, 155, 216, 226, 229, 230
    S. tortilis Desf.: see *S. capensis*
Stipagrostis (Gramineae), 137, 140, 142, 144, 193
    S. acutiflora (Trin. & Rupr.) de Winter, 212
    S. amabilis (Schweick.) de Winter, 191, 193
    S. brevifolia (Nees) de Winter, 141, 142
    S. ciliata (Desf.) de Winter, 141, 142, 193, 204, 221
    S. gonatostachys (Pilger) de Winter, 142
    S. hermannii (Mez) de Winter, 142
    S. hirtigluma (Steud. ex Trin. & Rupr.) de Winter, 113, 143, 144, 204
    S. hochstetterana (Beck. ex Hackel) de Winter, 143, 145
    S. namaquensis (Nees) de Winter, 141
    S. namibensis de Winter, 142

S. obtusa (Del.) Nees, 141, 142, 193, 221, 222
S. plumosa (L.) Munro ex T. Anderson, 221
S. pungens (Desf.) de Winter, 147, 204, 207, 216, 219, 220, 224, 230
S. ramulosa de Winter, 142
S. sabulicola (Pilger) de Winter, 142
S. subacaulis (Nees) de Winter, 143, 144
S. uniplumis (Licht. ex Roem. & Schult.) de Winter, 113, 141, 145, 193, 204
S. zitellii (Asch.) de Winter, 224
Stoebe (Compositae), 132, 167, 238
  S. passerinoides Willd., 258
Streptocarpus (Gesneriaceae), 165
Striga hermonthica (Del.) Benth. (Scrophulariaceae), 212
Strombosia grandifolia Hook.f. ex Benth. (Olacaceae), 85
  S. scheffleri Engl., 164, 181, 187
  S. sp., 253
Strychnos (Loganiaceae), 75
  S. cocculoides Baker, 173
  S. decussata (Pappe) Gilg, 200
  S. henningsii Gilg, 90, 129, 174, 200
  S. innocua Del.: see *S. madagascariensis*
  S. madagascariensis Poir., 85, 90, 99, 105, 189, 200, 211
  S. mellodora S. Moore: see *S. mitis*
  S. mitis S. Moore, 122, 187
  S. potatorum L.f., 90, 91, 98, 99, 182
  S. pungens Solered., 85, 96, 97, 173, 196
  S. spinosa Lam., 85, 107, 189, 210
  S. stuhlmannii Gilg: see *S. potatorum*
  S. usambarensis Gilg, 167
Stuhlmannia (Leguminosae: Caesalpinioideae), 186
Suaeda (Chenopodiaceae), 223, 224, 267
  S. articulata Aellen, 267
  S. fruticosa Forssk. ex J. F. Gmelin, 144, 223, 228, 230, 266
  S. ifniensis Caball., 223, 229
  S. mollis (Desf.) Del., 223, 229
  S. monodiana Maire, 223
  S. monoica Forssk. ex J. F. Gmelin, 31, 120, 224, 263, 267
  S. plumosa Aellen, 144
  S. vermiculata Forssk. ex J. F. Gmelin, 223
Suregada africana (Sond.) Kuntze (Euphorbiaceae), 200
  S. procera (Prain) Croizat, 129, 167, 181
  S. zanzibarensis Baillon, 188, 189
Sutera (Scrophulariaceae), 140
Swartzia madagascariensis Desv. (Leguminosae: Caesalpinioideae), 87, 97, 105, 173
Symphonia globulifera L.f. (Guttiferae), 83, 85, 181, 235, 236, 237, 253
Syzygium cordatum Hochst. ex Krauss (Myrtaceae), 91, 168, 181, 189, 196, 200, 202, 266
  S. guineense (Willd.) DC., 85, 181, 189, 211
    ss. *afromontanum* F. White, 90, 92, 164
    ss. *bamendae* F. White, 253
    ss. *barotsense* F. White, 91
    ss. *gerrardii* (Harv. ex Hook.f.) F. White, 200
    ss. *guineense*, 97, 99, 105
    ss. *occidentale* F. White, 82
  S. owariense (P. Beauv.) Benth., 74, 91
  S. sclerophyllum Brenan, 187

Tabernaemontana elegans Stapf (Apocynaceae), 189
  T. johnstonii (Stapf) Pichon, 164, 167
  T. stenosiphon Stapf, 253
Tagetes patula L. (Compositae), 252
Talbotiella (Leguminosae: Caesalpinioideae), 74

T. gentii Hutch. & Greenway, 176
Tamarindus indica L. (Leguminosae: Caesalpinioideae), 95, 105, 107, 117, 129, 182, 188, 189, 209, 241, 243, 251, 252, 255
Tamarix (Tamaricaceae), 47, 207, 216, 219 (and see footnote), 223, 224, 266, 267
  T. aphylla (L.) Karst., 113
  T. 'articulata', 219, 220
  T. canariensis Willd., 251
  T. 'gallica', 219
    ss. *'nilotica'*, 222
  T. 'mannifera', 224
  T. nilotica (Ehrenb.) Bunge, 113, 219
  T. usneoides E. Mey. ex Bunge, 141, 144, 191
  T. sp., 223
Tambourissa (Monimiaceae), 232, 235, 236, 258
  T. gracilis Baker, 238
Tamus communis L. (Dioscoreaceae), 151
Tapiphyllum floribundum Bullock (Rubiaceae), 97
Tarchonanthus camphoratus L. (Compositae), 115, 135, 191, 193, 195, 201
  T. galpinii Hutch. & E. P. Phillips, 96
  T. minor Less., 193
Tarenna graveolens (S. Moore) Bremek. (Rubiaceae), 129, 182
  T. luteola (Stapf) Bremek., 90
  T. neurophylla (S. Moore) Bremek., 98, 99
Tarietia utilis (Sprague) Sprague (Sterculiaceae), 76
Taxus baccata L. (Taxaceae), 149, 151, 152, 156, 247
Teclea (Rutaceae), 115, 167
  T. gerrardii I. Verdoorn, 200
  T. nobilis Del., 122, 129, 182, 211
  T. simplicifolia (Engl.) I. Verdoorn, 115, 121, 122, 129
  T. trichocarpa (Engl.) Engl., 129
Teline linifolia (L.) Webb & Berth. (Leguminosae: Papilionoideae): see *Genista linifolia*
  T. monspessulana (L.) K. Koch: see *Cytisus monspessulanus*
  T. stenopetala (Webb & Berth.) Webb & Berth.: see *Cytisus stenopetalus*
Teloschistes capensis (L.f.) Malme (Teloschistaceae), 142
Tephrosia (Leguminosae: Papilionoideae), 140, 220
  T. gracilipes Guill. & Perr., 204
  T. nubica (Boiss.) Baker, 204
  T. obcordata (Lam. ex Poir.) Baker, 204
  T. quartiniana Cuf., 204
  T. uniflora Pers., 206
Terminalia (Combretaceae), 110, 111
  T. avicennioides Guill. & Perr., 105, 107, 108
  T. bentzoe Pers., 257, 258
  T. benzoin L.f.: see *T. bentzoe*
  T. boivinii Tul., 259
  T. brachystemma Welw. ex Hiern, 99
  T. brownii Fresen., 107, 208, 209, 210, 211, 212, 213, 214
  T. glaucescens Planch. ex Benth., 83, 85, 105, 107
  T. laxiflora Engl., 85, 105, 107, 108, 203, 209, 210, 212
  T. macroptera Guill. & Perr., 105, 107, 108
  T. mollis Lawson, 95, 129, 173, 174
  T. orbicularis Engl. & Diels, 114
  T. parvula Pampan., 114
  T. prunioides Lawson, 95, 96, 143, 144, 191
  T. sambesiaca Engl. & Diels, 117, 186, 188
  T. sericea Burchell ex DC., 95, 96, 97, 191, 193, 194, 201
  T. seyrigii (H. Perrier) Capuron, 243
  T. spinosa Engl., 114, 188, 189
  T. stuhlmannii Engl., 116
  T. subserrata H. Perrier, 242
  T. superba Engl. & Diels, 57, 75, 78, 79, 81

Tessmannia (Leguminosae: Caesalpinioideae), 74
  T. burttii Harms, 91
Tetraberlinia (Leguminosae: Caesalpinioideae), 74
  T. bifoliolata (Harms) Hauman, 77
  T. polyphylla (Harms) J. Léonard, 77
  T. tubmaniana J. Léonard, 77
Tetracera (Dilleniaceae), 241
Tetraclinis (Cupressaceae), 147, 227
  T. articulata (Vahl) Masters, 48, 146, 149, 150, 151, 152, 153, 154, 155, 157, 227, 228, 229
Tetragonia (Aizoaceae), 137, 193
  T. reduplicata Welw. ex Oliver, 145
Tetrapleura tetraptera (Schumach. & Thonn.) Taubert (Leguminosae: Mimosoideae), 74, 82, 181, 253
Tetrapogon cenchriformis (A. Rich.) W. D. Clayton (Gramineae), 209, 210, 212
  T. tenellus (Roxb.) Chiov., 145
Tetrapterocarpon (Leguminosae: Caesalpinioideae), 232
  T. geayi Humbert, 242
Tetraria (Cyperaceae), 132
Tetrorchidium didymostemon (Baillon) Pax & Hoffm. (Euphorbiaceae), 80
Teucrium (Labiatae), 147
  T. fruticans L., 154, 158
  T. polium L., 152, 154
Thalia welwitschii Ridl. (Marantaceae), 108
Thamnochortus erectus (Thunb.) Masters (Restionaceae), 135
  T. spicigerus (Thunb.) R. Br., 135
Thelypteris confluens (Thunb.) Morton (Thelypteridaceae), 266
  T. striata (Schumach.) Schelpe, 265
Themeda quadrivalvis (L.) Kuntze (Gramineae), 243, 255
  T. triandra Forssk., 32, 100, 116, 121, 126, 127, 128, 129, 137, 168, 169, 191, 193, 194, 195, 196, 202, 208, 211
Thesium (Santalaceae), 238
Thespesia danis Oliver (Malvaceae), 188, 189
Thonningia (Balanophoraceae), 75
  T. sanguinea Vahl, 93
Thunbergia crispa Burkill (Acanthaceae), 99
  T. guerkeana Lindau, 114
Thylachium africanum (Capparidaceae), 188
  T. thomasii Gilg, 114
Thymelaea lythroides Barratte & Murb. (Thymelaeaceae), 152
  T. microphylla Coss. & Durieu, 230
  T. nitida Desf., 230
Thymus caespititius Brot. (Labiatae), 247
Tibestina (Compositae), 218
Tieghemella (Sapotaceae), 74
  T. africana Pierre, 77
  T. heckelii Pierre ex A. Chev., 77
Timonius seychellensis Summerhayes (Rubiaceae), 257
Tina (Sapindaceae), 232
  T. isoneura Radlk., 237
Tinnea aethiopica Kotschy & Peyr. (Labiatae), 115
Titanopsis (Aizoaceae), 140
Tournefortia argentea L.f. (Boraginaceae), 257
Trachylobium verrucosum (Gaertner) Oliver: see *Hymenaea verrucosa*
Trachypogon spicatus (L.f.) Kuntze (Gramineae), 100, 194, 196, 239
  T. thollonii Stapf, 101
Traganopsis glomerata (Maire) Wilczek (Chenopodiaceae), 229
Traganum nudatum Del. (Chenopodiaceae), 223
Tragus berteronianus Schult. (Gramineae), 194, 243
  T. koelerioides Asch., 194, 195
  T. racemosus (L.) All., 194, 204, 206, 207

Trapa natans L. (Trapaceae), 265
Treculia africana Decne. (Moraceae), 91, 172, 187, 253
Trema guineensis (Schumach. & Thonn.) Ficalho (Ulmaceae): see *T. orientalis*
  T. orientalis (L.) Bl., 25, 80, 83, 211
Trianthema hereroensis Schinz (Aizoaceae), 142
Tribulocarpus dimorphanthus (Pax) S. Moore (Aizoaceae), 113
Tribulus (Zygophyllaceae), 144
  T. terrestris L., 194, 206, 212
  T. zeyheri Sond., 141, 143
Tricalysia allenii (Stapf) Brenan (Rubiaceae), 90
Trichilia dregeana Sond. (Meliaceae), 187, 200
  T. emetica Vahl, 91, 95, 96, 105, 117, 188, 200, 201
  T. grandifolia Oliver, 253
  T. prieuriana Adr. Juss., 74, 90
Trichocalyx (Acanthaceae), 111
Trichocaulon (Asclepiadaceae), 139
  T. clavatum (Willd.) H. Huber, 142
  T. dinteri Berger: see *T. clavatum*
  T. pedicellatum Schinz, 142
Trichocladus (Hamamelidaceae), 162
  T. ellipticus Eckl. & Zeyh., 167, 181
Tricholaena monachne (Trin.) Stapf & C. E. Hubbard (Gramineae), 145
Trichomanes mannii Hook. (Hymenophyllaceae), 82
Trichoneura grandiglumis (Nees) Ekman (Gramineae), 194
Trichoscypha (Anacardiaceae), 74
Trifolium (Leguminosae: Papilionoideae), 147, 157, 166
Triglochin palustris L. (Juncaginaceae), 155
Trilepisium madagascariense DC. (Moraceae), 79, 172, 187, 188, 253
Trimeris (Campanulaceae), 254
  T. scaeviolifolia (Roxb.) Mabberley, 254
Triplocephalum holstii O. Hoffm., (Compositae), 267
Triplochiton scleroxylon K. Schum. (Sterculiaceae), 57, 79, 81
Tripogon leptophyllus (A. Rich.) Cuf. (Gramineae), 211
  T. minimus (A. Rich.) Hochst. ex Steud., 210
Triraphis andropogonoides (Steud.) E. P. Phillips (Gramineae), 194
Tristachya eylesii Stent & Rattray (Gramineae): see *T. nodiglumis*
  T. hispida (L.f.) K. Schum.: see *T. leucothrix*
  T. leucothrix Nees, 194, 196, 202
  T. nodiglumis K. Schum., 173
Tristemma incompletum R. Br. (Melastomataceae), 266
Triumfetta annua L. (Tiliaceae), 90
Trochetia erythroxylon (G. Forster) Benth. (Sterculiaceae), 254
  T. melanoxylon (Aiton) Benth. & Hook.f., 254
Turraea floribunda Hochst. (Meliaceae), 200
  T. ghanensis J. B. Hall, 176
  T. glomeruliflora Harms, 253
  T. holstii Gürke, 181
  T. mombassana Hiern ex C.DC., 115, 121
  T. nilotica Kotschy & Peyr., 182
  T. obtusifolia Hochst., 200
  T. oppositifolia (Cav.) Harms, 258
Turraeanthus (Meliaceae), 74
Typha (Typhaceae), 55, 219
  T. australis Schumach. & Thonn., 191, 222, 265
  T. latifolia L., 218, 265
Typhonodorum lindleyanum Schott (Araceae), 188, 236

Uapaca (Euphorbiaceae), 99, 106
  U. bojeri Baillon, 237
  U. chevalieri Beille, 82

U. guineensis Muell.-Arg., 83, 85, 91, 181
U. heudelotii Baillon, 83
U. kirkiana Muell.-Arg., 99
U. nitida Muell.-Arg., 173, 189
U. pilosa Hutch., 99
U. sansibarica Pax, 173, 189
U. togoensis Pax, 85, 105
Ulex boivinii Webb (Leguminosae: Papilionoideae), 151, 152
U. europaeus L., 254
Ulmus campestris L. (Ulmaceae), 149, 150
Umtiza (Leguminosae: Caesalpinioideae), 199
U. listerana Sim, 200
Uncarina (Pedaliaceae), 242
Urelytrum giganteum Pilger (Gramineae), 174
U. squarrosum Hackel, 239
Urena lobata L. (Malvaceae), 206
Urginea (Liliaceae), 158
U. maritima (L.) Baker, 152, 160
Ursinia (Compositae), 141
Usnea (Parmeliaceae), 99, 142, 181, 237
Utricularia (Lentibulariaceae), 108, 265
U. foliosa L., 265
U. gibba L., 266
Uvaria chamae P. Beauv. (Annonaceae), 81, 178
U. leptocladon Oliver, 189
Uvariodendron anisatum Verdc. (Annonaceae), 167

Vaccinium (Ericaceae), 168, 236, 237, 238
V. cylindraceum Smith, 247
Vahlia geminiflora (Del.) Bridson (Vahliaceae), 204
Vallisneria aethiopica Fenzl (Hydrocharitaceae), 265
V. spiralis L., 265
Vangueria venosa Hochst. ex Del. (Rubiaceae), 207
Vangueriopsis lanciflora (Hiern) Robyns (Rubiaceae), 97, 99
Vanilla roscheri Reichenb.f. (Orchidaceae), 114
Vateria (Dipterocarpaceae), 257
V. seychellarum Dyer, 257
Vella mairei Humbert (Cruciferae), 158
Venidium (Compositae), 142
Vepris heterophylla (Engl.) Letouzey (Rutaceae), 178
V. undulata (Thunb.) I. Verdoorn & C. A. Smith, 200
Verbascum capitis-viridis Huber-Mor. (Scrophulariaceae), 252
Vernonia (Compositae), 236, 237, 238
V. amygdalina Del., 253
V. auriculifera Hiern, 130
V. brachycalyx O. Hoffm., 182
V. conferta Benth., 80
V. oligocephala (DC.) Schultz.-Bip. ex Walp., 194
Verschaffeltia (Palmaceae), 257
V. splendida H. A. Wendl., 257
Vetiveria fulvibarbis (Trin.) Stapf (Gramineae), 178
V. nigritana (Benth.) Stapf, 108
Viburnum lantana L. (Caprifoliaceae), 151
V. tinus L., 151, 157, 247
Vicia (Leguminosae: Papilionoideae), 147, 157
Vigna luteola (Jacq.) Benth. (Leguminosae: Papilionoideae), 265
V. unguiculata (L.) Walp., 252
Viola (Violaceae), 236
V. arborescens L., 154
V. palustris L., 155
Viridivia (Passifloraceae), 87
Visnea (Theaceae), 246
V. mocanera L.f., 246, 247
Vitellariopsis marginata (N. E. Br.) Aubrév. (Sapotaceae), 200
Vitex (Verbenaceae), 237
V. agnus-castus L., 147, 219
V. doniana Sweet, 83, 85, 105, 107
V. humbertii Moldenke, 238
V. madiensis Oliver, 85
V. mombassae Vatke, 189
Vitis (Vitaceae), 159
Vittaria elongata Swartz (Vittariaceae), 235
Voacanga thouarsii Roem. & Schult. (Apocynaceae), 83, 181, 188, 200
Volkensinia (Amaranthaceae), 111
Vossia cuspidata Griff. (Gramineae), 55, 100, 265, 266
Vulpia bromoides (L.) S. F. Gray (Gramineae), 211

Wahlenbergia (Campanulaceae), 140, 254
W. angustifolia (Roxb.) A.DC., 254
W. linifolia (Roxb.) A.DC., 254
Walafrida (Scrophulariaceae), 135, 140, 194
W. densiflora Rolfe, 194
W. saxatilis Rolfe, 194, 195
Waltheria indica L. (Solanaceae), 206
Warburgia salutaris (Bertol.f.) Chiov. (Canellaceae), 167, 181
W. ugandensis Sprague: see *W. salutaris*
Warionia (Compositae), 218
Weinmannia (Cunoniaceae), 236, 237, 238, 255
Wellstedia (Boraginaceae), 113
Welwitschia bainesii (Hook.f.) Carrière (Welwitschiaceae), 53, 94, 136, 137, 143, 144, 145, 191
W. mirabilis Hook.f.: see *W. bainesii*
Wiborgia sericea Thunb. (Leguminosae: Papilionoideae), 134
Widdringtonia (Cupressaceae), 135, 167
W. cedarbergensis J. A. Marsh, 134
W. cupressoides (L.) Endl., 51, 134, 161, 165, 166
W. nodiflora (L.) Powrie: see *W. cupressoides*
W. schwarzii (Marloth) Masters, 134
W. whytei Rendle: see *W. cupressoides*
Willdenowia striata Thunb. (Restionaceae), 135
Wissmannia (Palmaceae), 111
W. carinensis (Chiov.) Burret, 113
Withania frutescens (L.) Pauquy (Solanaceae), 152, 227, 228
Wolffia arrhiza (L.) Horkel ex Wimm. (Lemnaceae), 265

Xanthocercis (Leguminosae: Papilionoideae), 87
X. zambesiaca (Baker) Dumaz-le Grand, 95
Xerocladia (Leguminosae: Mimosoideae), 137
Xeroderris stuhlmannii (Taubert) Mendonça & E. P. Sousa (Leguminosae: Papilionoideae), 95, 106
Xeromphis nilotica (Stapf) Keay (Rubiaceae), 188
X. rudis (E. Meyer ex Harv.) Codd, 201
Xerophyta (Velloziaceae), 238, 239, 242
X. dasylirioides Baker, 238
X. humilis (Baker) T. Durand & Schinz, 113
Xerosicyos (Cucurbitaceae), 232, 242
Ximenia (Olacaceae), 98
X. americana L., 91, 95, 107, 253
X. caffra Sond., 95
Xylia hildebrandtii Baillon (Leguminosae: Mimosoideae), 241
Xylocalyx (Scrophulariaceae), 111
Xylocarpus (Meliaceae), 261
X. granatum Koen., 259, 261, 263, 264
X. moluccensis (Lam.) M.J. Roem., 259, 261
Xylopia aethiopica (Dunal) A. Rich. (Annonaceae), 74, 81, 91, 253
X. holtzii Engl.: see *X. parviflora*
X. odoratissima Welw. ex Oliver, 97
X. parviflora (A. Rich.) Benth., 188

Xylopia aethiopica (Dunal) A. Rich. (Annonaceae)—*contd*
    X. rubescens Oliv., 91
Xymalos (Monimiaceae), 162
    X. monospora (Harv.) Baillon, 122, 164, 165, 181, 187, 199
Xyris (Xyridaceae), 84, 108

Zanha africana (Radlk.) Exell (Sapindaceae), 97
    Z. golungensis Hiern, 109
Zanthoxylum capense (Thunb.) Harv. (Rutaceae), 200, 201
    Z. chalybeum Engl., 189
    Z. davyi (I. Verdoorn) Waterman, 199
    Z. gilletii (de Wild.) Waterman, 81, 253
    Z. trijugum (Dunkley) Waterman, 90
    Z. xanthoxyloides (Lam.) ?, 109, 178
Zea mays L. (Gramineae), 252
Zenkerella capparidacea (Taubert) J. Léonard (Leguminosae: Caesalpinioideae), 187
Zilla (Cruciferae), 218
    Z. spinosa (L.) Prantl, 221, 222, 224
Ziziphus (Rhamnaceae), 98, 208, 210
    Z. abyssinica Hochst. ex A. Rich., 95, 106, 210
    Z. lotus (L.) Desf., 147, 149, 158, 160, 204, 219, 225, 226, 227, 228, 231
    Z. mauritiana Lam., 105, 211, 219, 252
    Z. mucronata Willd., 95, 96, 105, 106, 107, 129, 137, 141, 189, 193, 194, 195, 200, 201
    Z. pubescens Oliver, 129
    Z. spina-christi (L.) Desf., 209, 210, 211, 212, 214
    Z. zeyherana Sond., 194
Zornia glochidiata Reichb. ex DC. (Leguminosae: Papilionoideae), 212
Zygophyllum (Zygophyllaceae), 137, 140, 223
    Z. album L., 223, 224, 230
    Z. cornutum Coss., 223
    Z. gaetulum Emberger & Maire, 223, 229
    Z. hildebrandtii Engl., 116
    Z. morgsana L., 135
    Z. orbiculatum Welw. ex Oliver, 144, 145
    Z. retrofractum Thunb., 142
    Z. simplex L., 143, 144, 204, 252
    Z. stapfii Schinz, 142, 144
    Z. waterlotii Maire, 223